P9-AGH-079

CONSTRUCTION SURVEYING AND LAYOUT

A STEP-BY-STEP FIELD ENGINEERING METHODS MANUAL

WESLEY G. CRAWFORD

Creative Construction Publishing, Inc.
2720 South River Road
West Lafayette, IN 47906-4347

Creative Construction Publishing, Inc.
2720 South River Road
West Lafayette, IN 47906-4347

10 9 8 7 6 5 4 3 2

ISBN 0-9647421-0-1 (previously ISBN 0-9624124-3-0)

Printed and bound in the United States of America by Thomson/Shore Inc., Dexter, MI

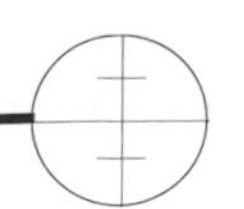

DEDICATION

This text is dedicated to the Field Engineers of Hensel Phelps Construction Company with whom I have worked as a faculty intern for many years. Without their assistance, insight and critical review this book would not have been possible. Also many thanks to the President and officers of Hensel Phelps for their support and encouragement to develop my knowledge about the many activities of field engineering. Thank you.

SPECIAL THANKS

A special note of gratitude to Merry and Paul Crawford (Mom and Dad) who taught me the great feeling of accomplishment that only comes from working hard and staying with a task until it is done. And to my wife, Bonnie, and my son, Matthew, who persevered through my absences from family activities and events while writing this book.

ACKNOWLEDGMENTS

EDITOR: BONNIE CRAWFORD

COVER DESIGN, INTERIOR DESIGN, ILLUSTRATIONS:

JONATHAN DAVID HUMPHERYS

ILLUSTRATORS: DALE JACKSON

ANDREW MIKESELL

JAMIE MOHLER

REVIEWERS :

Mark Beck, Lee Bender, Earl Burkholder, Scott Conrad, Chad Hobson, Cliff Lipman, Al McConahay, John Meyer, Matt Pietryka, Sharyn Switzer, Jerry Thomas, and other friends. And, of course, the many students in my classes.

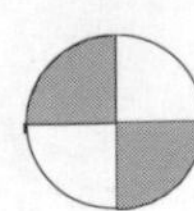

TABLE OF CONTENTS

PART ONE - FIELD ENGINEERING BASICS

CHAPTER 1

SCOPE AND RESPONSIBILITIES

CHAPTER 2

COMMUNICATIONS AND FIELD ENGINEERING

CHAPTER 3

GETTING STARTED AND ORGANIZED

CHAPTER 4

FIELDWORK BASICS

CHAPTER 5

OFFICEWORK BASICS

PART TWO - MEASUREMENT BASICS

CHAPTER 6

DISTANCE MEASUREMENT - CHAINING

CHAPTER 7

DISTANCE MEASUREMENT - EDM

CHAPTER 8

ANGLE MEASUREMENT

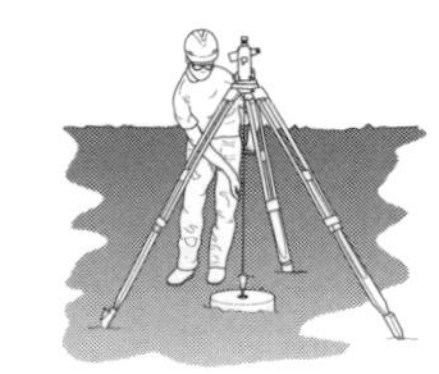

CHAPTER 9

LEVELING

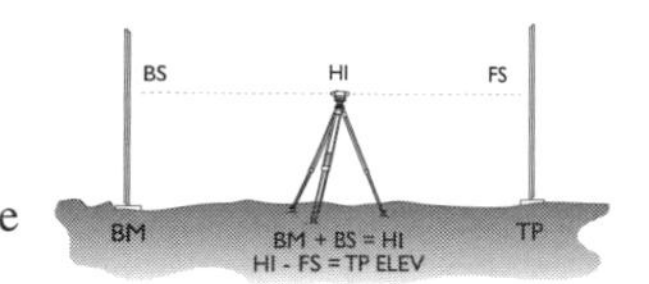

CHAPTER 10

TOTAL STATION

CHAPTER 11

CONSTRUCTION LASERS FOR LINE AND GRADE

CHAPTER 12

EQUIPMENT CALIBRATION

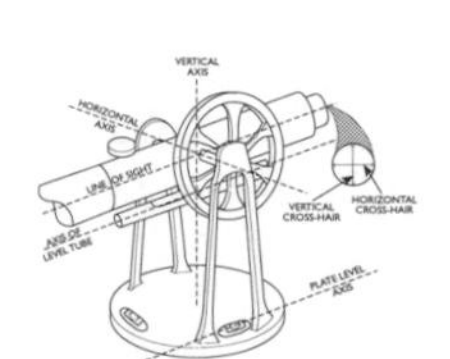

PART THREE - CONSTRUCTION SURVEYING CALCULATIONS

CHAPTER 13

MATH REVIEW AND CONVERSIONS

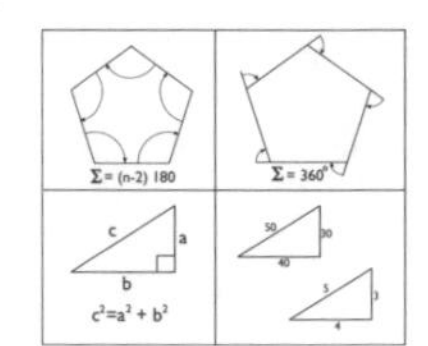

CHAPTER 14

DISTANCE CORRECTIONS

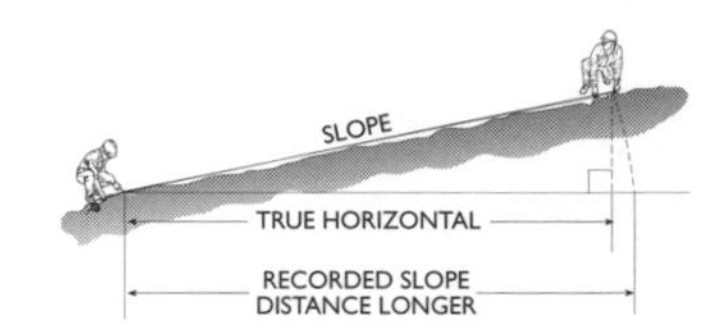

CHAPTER 15

TRAVERSE COMPUTATIONS

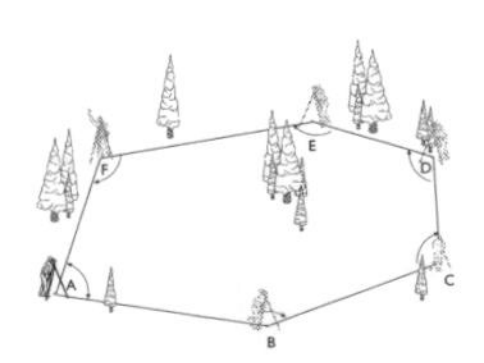

CHAPTER 16

COORDINATES IN CONSTRUCTION

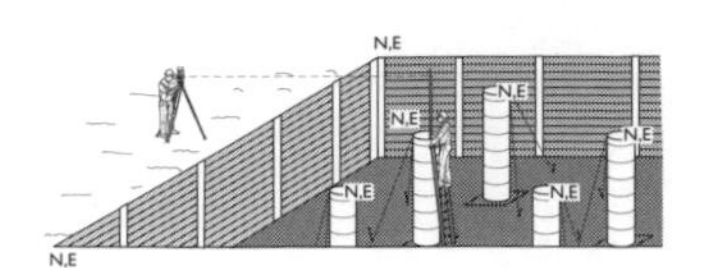

CHAPTER 17

HORIZONTAL CURVES

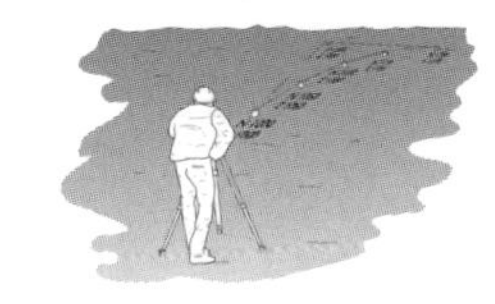

CHAPTER 18

VERTICAL CURVES

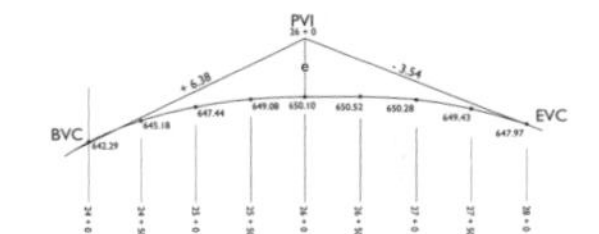

CHAPTER 19

EXCAVATIONS

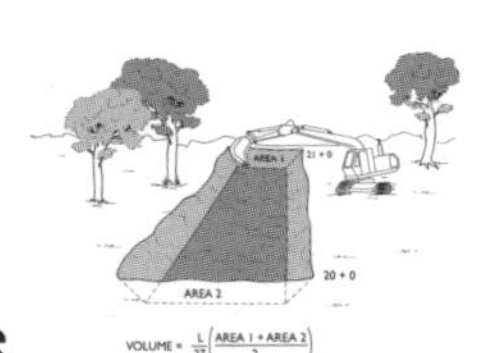

PART FOUR - CONSTRUCTION LAYOUT APPLICATIONS

CHAPTER 20

DISTANCE APPLICATIONS

CHAPTER 21

ANGLES & DIRECTION

CHAPTER 22

LEVELING APPLICATIONS

CHAPTER 23

CONSTRUCTION LAYOUT TECHNIQUES

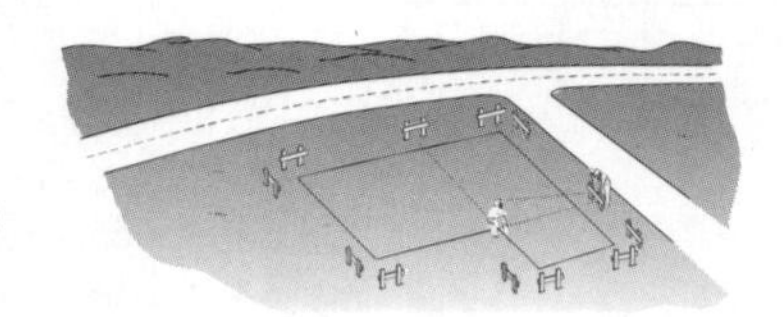

CHAPTER 24

BUILDING CONTROL & LAYOUT

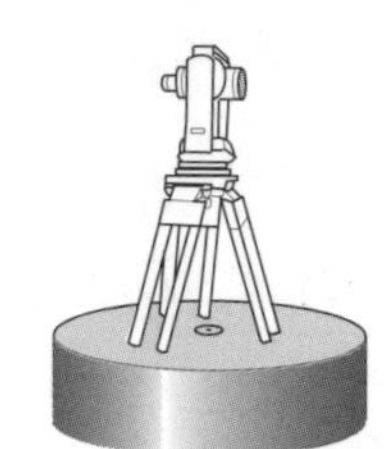

CHAPTER 25

ONE-PERSON SURVEYING TECHNIQUES

CHAPTER 26

THE PUNCH LIST

APPENDICES

PREFACE

This book was written to present a practical guide to construction surveying and layout that will actually show the reader "how-to" perform many of the measurement activities and calculations that are required in the field. It is intended to prepare individuals for the exciting and dynamic position of a field engineer.

This book is not theoretical. In most instances, only one of the many methods of performing a task or a calculation has been discussed. If a more theoretical or mathematical approach to surveying is required or other methods are needed, please refer to other fine books on surveying. Included in Appendix D is a listing of some other textbooks on surveying that are available.

As a teacher of surveying, I have always attempted to use simple explanations and lots of illustrations to make learning easy and interesting. This book has been written and designed so that it would follow that principle. Over 1000 illustrations showing the essence of field engineering activities have been included to enhance understanding. Icons have been used in the margins to identify special areas of interest to the reader. The Reference icon shows the reader at a glance where to turn in the book for additional information. The Tolerance icon shows the standard tolerance for the layout activity. The Clipboards are included to briefly state an important fact. The Thumb icon points out a basic rule of thumb that should always be followed.

The book is organized into four major parts that attempt to emulate the learning progression that a field engineer will experience. The first part, Field Engineering Basics, prepares the field engineer for performing general field and office practices. In the second part, Measurement Basics, the field engineer will learn how to use the equipment of surveying. Part three, Construction Surveying Calculations introduces some of the calculations that are used in field engineering. The final part, Construction Layout Applications presents the field engineer real applications in construction. There are also appendices covering sample problems, suggested field activities, Global Positioning Systems and Construction, a list of other surveying texts, and a glossary.

When writing this book I have assumed that the reader is interested in construction surveying and layout. It is also assumed that the reader has a background in basic algebra, basic geometry, and the fundamentals of trigonometry and that the reader has a basic knowledge of the terminology of construction. To learn from this book it is expected that the reader will "read and do." That is, the reader will read and study the material and will perform the described methods to really understand the process that has been discussed. Remember the saying. "He who hears, forgets; He who sees, remembers; He who does, understands."

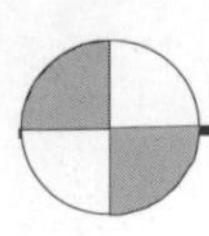

For many years I have been trying to get started on writing an illustrated surveying textbook. I had many starts and failures because I couldn't get the illustrations that I needed. Fortunately, last year I was able to make contact with some very talented illustrators who were able to turn my very rough sketches into finished illustrations. This book would not have been possible without them. I am forever indebted to Dale Jackson, Andy Mikesell, Jamie Mohler, and especially to Jon Humphreys who brought this all together. The illustrations are the key to my success in producing this book.

Wes Crawford
August 1994

PREFACE TO THE SECOND EDITION

Bringing out a new edition of this book so soon after the original publication wasn't anticipated. However, several events have caused this second edition to become necessary. First, I have changed publishers requiring among other things, a new ISBN number for the book. Second, it was necessary to add some important concepts I originally overlooked. Third, several mistakes in the text, calculations, and problems eluded my reviewers and myself and were unacceptable to my standards. With these important changes, I found it necessary to come out with the second edition at this time.

Many thanks to the many students in my classes for their assistance in pointing out mistakes that were overlooked in the first edition. If you as a reader have suggestions for additional construction layout material related to field engineering or would like to point out where the author missed an important concept, please contact: Creative Construction Publishing, Inc., 2720 South River Road, West Lafayette, IN 47906-4347 Phone 317-743-9704

Wes Crawford
June 1995

ABOUT THE AUTHOR

Wesley G. Crawford, is a professor at Purdue University where he has taught field engineering and construction surveying in the Department of Building Construction and Contracting since 1980. He is a registered surveyor in the State of Pennsylvania. In the Spring of 1995, Mr. Crawford was awarded the first-ever Outstanding Educator Award from the Associated Schools of Construction at their 31st Annual Conference of national and international construction educators in recognition of his Excellence in Teaching, Contribution to Construction Education, and Dedication to the Construction Profession.. Professor Crawford was selected for this award for his innovative and realistic teaching methods in the classroom, resulting in students who rapidly develop competency and success in field engineering applications.

He received an Associate of Science in Surveying Technology from Pennsylvania State University, a Bachelor of Science in Land Surveying Engineering and a Master of Science in Surveying and Mapping from Purdue University.

SCOPE AND RESPONSIBILITIES

1

PART 1 - FIELD ENGINEERING BASICS

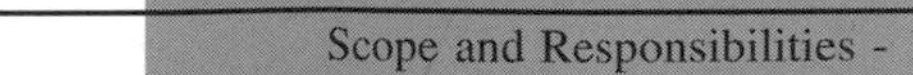

CHAPTER OBJECTIVES:

Describe what a field engineer is and list the duties and responsibilities of a person in that position.
Identify and discuss the characteristics necessary to be a successful field engineer.
Identify the various types of layout activities that a field engineer often performs.
Identify and discuss the characteristics necessary to be a successful office engineer.
Describe the primary job tasks of the office engineer.
List the joint duties of a field and office engineer.
Discuss why field engineers should not perform property surveys.
Describe the step-by-step process required to become a registered surveyor.

INDEX

RESPONSIBILITIES OF A FIELD ENGINEER

SCOPE

The position of field engineer introduces you to a variety of jobsite activities that will develop your understanding of the construction process and of company management systems. It is extremely beneficial that you work in this position before moving on to project engineering, area superintendent, or estimating. This position is flexible and is designed to fit the capabilities and background of the individual as well as the specific needs of each project.

Many companies place all of their newly-hired graduates into this position because there are many fundamentals about the construction process that can only be learned on the jobsite. As a field engineer, you'll be exposed to control and layout, excavation, concrete forming, placing and finishing, steel erection, mechanical and electrical installation, shop drawings, and other important aspects of project documentation. Performing field engineering duties well and understanding the construction process are very important to your success and to the success of your projects.

FIELD ENGINEER SUCCESS CHARACTERISTICS

To be a successful field engineer, you must possess the following characteristics:

BE COMMITTED TO CONSTRUCTION

The success of a construction company depends on individuals who enjoy the industry and are willing to dedicate themselves to being the best constructors possible.

BE WILLING TO WORK

You'll only be successful at field engineering if you dedicate yourself to working hard. A hard worker does whatever it takes to get the job done.

BE AN INITIATOR

Good constructors are initiators. Continually be looking for ways to improve safety, quality, and efficiency.

BE RESPONSIBLE

Your success on your projects depends on people who are willing to take personal responsibility for seeing that all aspects of the project are being successfully carried out. The successful field engineer doesn't assume that things are being done because they should be. A good field engineer checks it out.

BE A QUESTIONER

Take the time to learn every detail about your project. Ask questions. The more questions you ask, the more rapidly you will learn the construction business and the more quickly you will advance. You will be expected to know the project details.

BE A LEARNER

Listen to everyone on the jobsite. Your project engineer, superintendent, and project manager are interested in your development and in your understanding of the construction process. They will teach you at every opportunity. Pay special attention to the craftspeople and foremen. They are your best sources to learn the details of construction methods.

BE A COMMUNICATOR

Good communication skills are very important to your success in construction. If you can effectively communicate with others, you will be able to ask and answer questions quickly, accurately, and clearly. Being an effective communicator may be the single most important quality of a successful field engineer.

BE A LEADER

Set an example for other workers on the jobsite. If others see you going that extra step to achieve quality and efficiency, they will follow.

These qualities and characteristics will prove invaluable to you in successfully performing your primary duties and responsibilities as a field engineer and will establish habits which will carry you through your entire construction career.

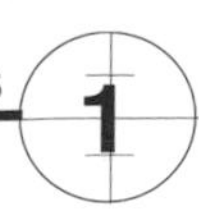

BEING A FIELD ENGINEER

The field engineer works directly for the superintendent. This position, together with the office engineer's position, is used as a part of the training leading to further advancement as a project engineer, area superintendent, or estimator.

ASSIST THE SUPERINTENDENT

As a field engineer, your most important function is to help the superintendent in any way possible. By working together as a team, projects will be successful and you will learn the construction process first hand from someone who was once in your shoes.

ENGINEERING LAYOUT

One of your primary duties as a field engineer will be engineering layout. This will include:

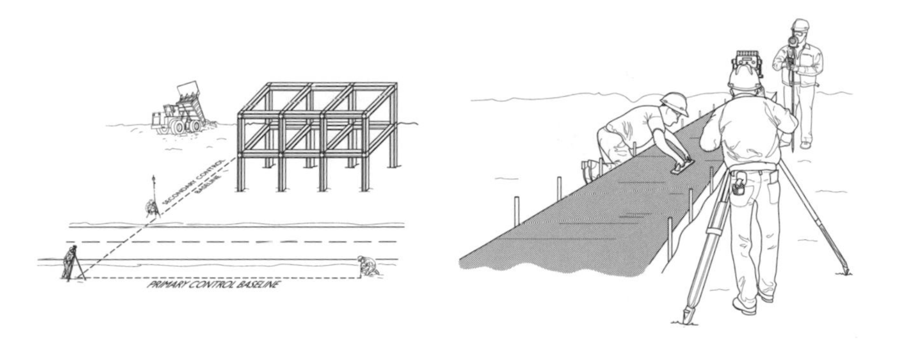

Primary, secondary, and working control points

Site work

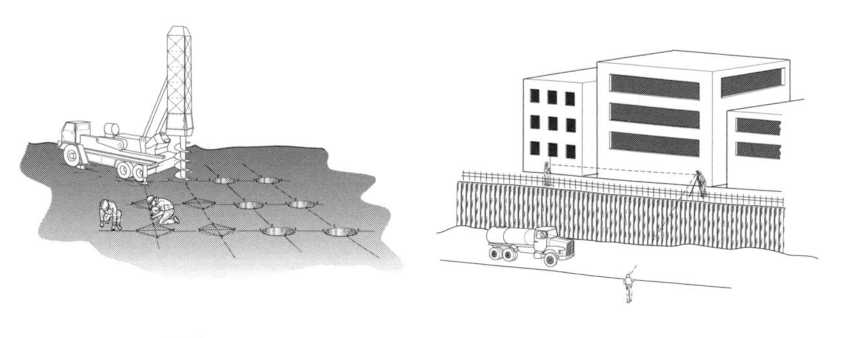

Caissons

Retaining systems

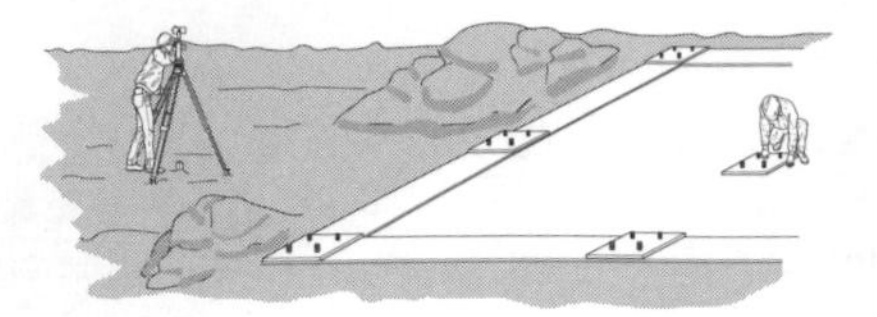

Footings and foundations

Anchor bolts

Structural steel

Structural concrete and metal decks

Layout of embedded items, sleeves, and blockouts

Concrete column layout

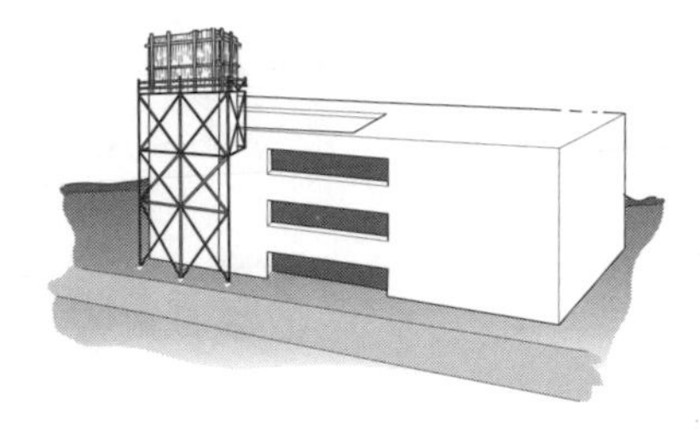

Control of elevator and stair cores, and sidewalks, curbs, and gutters

Utilities

As you can see, the layout duties of a field engineer are extensive. Guidelines for performing these activities can be found in this Manual.

QUALITY CONTROL

As a field engineer, it will be necessary for you to check the accuracy and adequacy of activities and work-in-place as the project progresses. You must also check the concrete forms for proper construction, and all embedded items for correct location before the concrete is placed. Continuous quality control will help to ensure that costly mistakes do not occur. Quality control activities include gathering production information which will be needed for the weekly and monthly cost control reports.

SUPERVISION

As you gain experience in the field, your superintendent may instruct you to direct a small work force, such as a labor crew or subcontractor crew. The skills you develop will be necessary for advancement within the company.

RESPONSIBILITIES OF AN OFFICE ENGINEER

SCOPE

Office engineering is complementary to field engineering. That is, the position of office engineer is intended to introduce a variety of jobsite office activities that will develop your understanding of the paperwork aspect of the construction process. Like field engineering, it is necessary that you work in this position before moving on to project engineering, area superintendent, or estimating. This position is flexible and is designed to fit the capabilities and background of the individual, as well as the specific needs of each project.

As an office engineer, you will be working with shop drawings, submittals, schedules, trend curves, and other important aspects of project documentation. Performing office engineering duties well and understanding the construction process are very important to your success and to the success of construction projects.

BEING AN OFFICE ENGINEER

The office engineer acts as the assistant to the project engineer and performs many of the same duties. The position of office engineer is part of the training leading to further advancement as a project engineer, area superintendent, estimator, or project manager.

PROJECT DOCUMENTATION

As an office engineer, it will be your function to assist in the recording, updating, conveying, interpolating, and maintaining of the project documentation. The accurate and timely processing of project documentation is very important to the success of the company and to your understanding of the project administration process. This constant review and follow through is a good method of tracking the materials that are needed on the jobsite. It ensures that materials are on the jobsite at the time that they are needed by the crafts.

PROCUREMENT OF MATERIALS

One of your activities as an office engineer is the procurement and tracking of materials. This will involve processing the submittals, shop drawings, and change orders in a timely manner to avoid project delays. These activities involve thorough review for completeness, accurate conformance to the plans and specifications, and coordination with subcontractors and suppliers. It will be your responsibility to ensure that these are up to date and available for review at a moment's notice.

INFORMATION REQUESTS

As the project progresses, assist the office supervision team (office engineer, foremen, and superintendent) in the preparation and follow through of information requests by conveying questions to and from the architect and owner. Your ability to satisfy the owner depends on understanding and effective communication of the owner's requirements.

FIELD/OFFICE ENGINEER JOINT DUTIES

SCOPE

Although there are duties specific to the field engineer and duties specific to the office engineer, there are also several related activities that are performed by persons in both positions working together. From the moment a field engineer and an office engineer are assigned to a project, they will work closely to support each other's work and will complete many tasks together.

GENERAL

PLANNING FOR SAFETY

Every project involves many people working toward a common goal. As a field engineer or office engineer, you are responsible for the coordination of their activities and for their safety. This is where careful planning is of prime importance. It is the policy of most companies to perform work in the safest possible manner, consistent with good construction practices. To meet this policy, an organized safety program must be administered. One of the first activities of the field or office engineer is to become involved in the safety program.

MATERIAL HANDLING

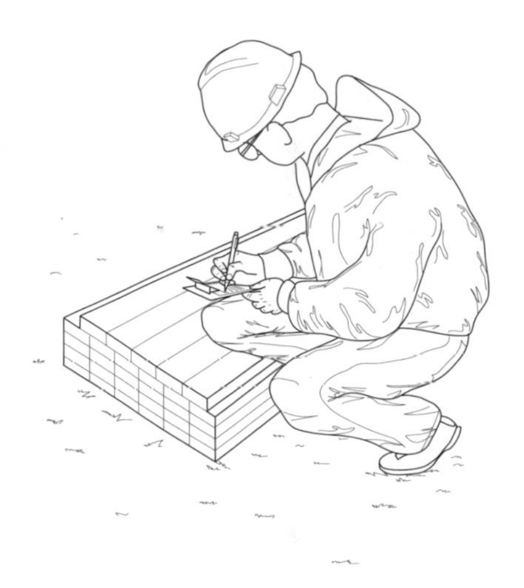

Just as there will be many people on the jobsite, there will be many materials. This is where planning becomes necessary. It will be up to you to check the material requirements in advance, and then to coordinate the deliveries of the materials. As the materials are delivered to the jobsite, you must be prepared to check the delivery for proper quantity, quality, and condition of the materials. Keep a record of these deliveries and of productivity and manpower. Because these records are the basis for issuing checks for work performed, it is essential that they be accurate. The records are also used in estimating similar projects. If you keep good records, that means your future estimates will be accurate. While paperwork is certainly not the most exciting part of your job, it is very necessary and valuable. Set aside time daily to fill out production records so that your reports are as precise as possible.

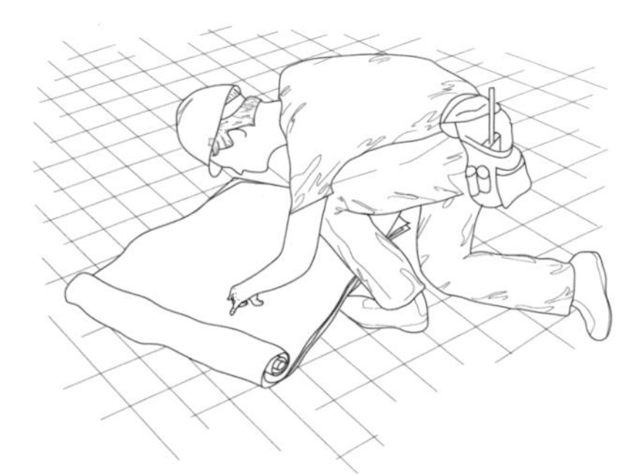

INTERPRETATION OF PLANS

In addition to your other responsibilities, you must be prepared to quickly and accurately assist the craftspersons or subcontractor with the interpretation of the plans and specifications. In order to do this, you must know your project inside and out.

MAINTAINING THE SCHEDULE

Attention to detail is very important in the area of scheduling. As a field or office engineer, you'll be responsible for developing detailed short-interval schedules, both daily and weekly. These schedules will be the tools that will be used to complete projects on time and within the budget. Their accuracy will depend on your input and communication with the project superintendent.

PROBLEM PREVENTION

In performing all of your duties as a field or office engineer, it is most important that you work to prevent problems. It takes a full grasp of the processes involved in a project to prevent problems. This full understanding will be gained through experience and by asking questions whenever necessary. The most successful people in any industry are those who can anticipate problems. By recognizing potential problems, costly delays can be avoided.

TEAMWORK REQUIREMENT

As you are doing your job, always remember that you are part of a team; you are working with your project manager, area superintendent, superintendent, and project engineer. Work closely to avoid duplication of effort. Continually evaluate and correct your efforts as necessary.

Remember also that at this point you are in a learning situation. You are expanding your knowledge so that you can move up in the company. Remember the resources you have in your project manager, superintendent, project engineer, craftspersons, subcontractors, and this Field Engineer Methods Manual. By remembering these things, you can look forward to a full conceptualization of the construction process, greater knowledge, and ultimately, personal satisfaction.

SUMMARY

As you can see, the positions of field engineer and office engineer are challenging and exciting. They are your entry into the rewarding world of construction. Work hard at applying the personal qualities that have been identified and you will build the proper foundation for a successful and rewarding career in construction.

FIELD ENGINEERS ARE NOT REGISTERED LAND SURVEYORS

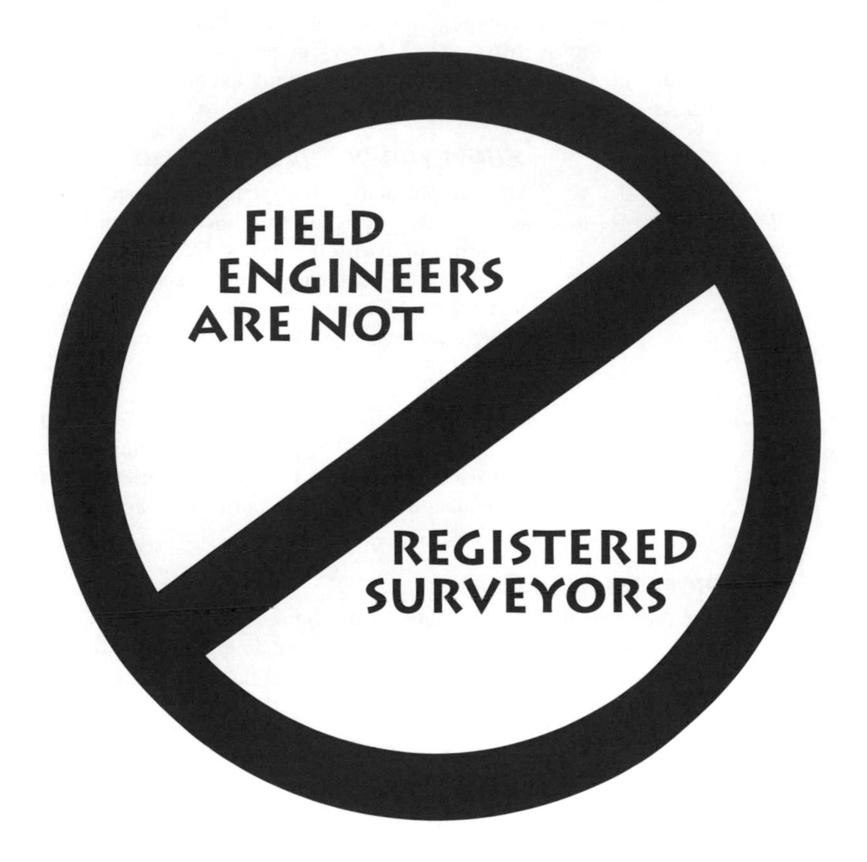

SCOPE

A major mistake by some field engineers and others in construction is to think they are capable of determining the legal boundaries of a property. They feel since they are knowledgeable about measuring, they can establish corners to build from. They think they can save time and money by assuming the fence at the edge of the property is the property line, and they can use it as a reference to locate the structure.

They are dead wrong!

BECOMING A REGISTERED SURVEYOR

In most states, property surveying must be performed by registered surveyors. To become a registered surveyor requires a combination of education and experience under the direction of a registered surveyor. All states are different. Consult the state board of registration for exact requirements. However, some typical requirements are listed here in the order in which they are attained.

EDUCATION

A four-year degree in an accredited surveying program is becoming a requirement of most states. This four-year degree includes mathematics, physics, communications, etc., in addition to specific courses on Land Survey Systems, Surveying Computations, Legal Aspects of Surveying, Instrumentation, Subdivision Design, Description Writing, Courthouse Research, etc. Surveying programs are designed to shorten the amount of time an individual must spend learning the broad range of topics that are necessary to become registered.

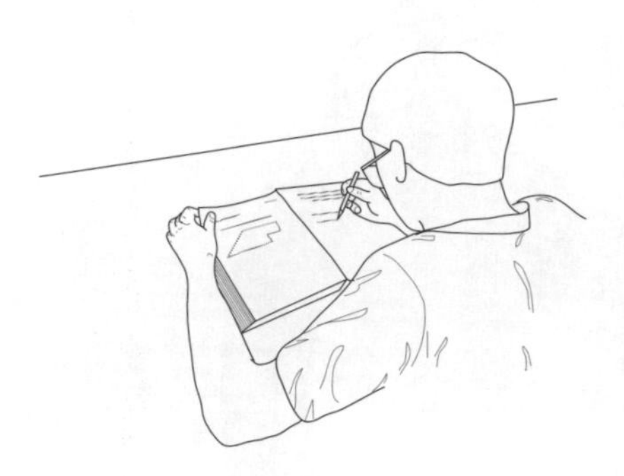

SURVEYOR-IN-TRAINING EXAM

Upon completion of a four-year degree in surveying, individuals are qualified to take an eight-hour Surveyor-in-Training (SIT) examination which covers the topics studied in the undergraduate degree program. If successful, an individual is then identified as a surveyor-in-training and must complete the experience requirements for registration.

EXPERIENCE

After passing the SIT, the typical state requires four years of land surveying work experience under the direct supervision of a registered land surveyor. This experience must expose the individual to all of the tasks that are expected of a practicing surveyor.

PROFESSIONAL EXAM

After spending the required time in the field, SIT becomes eligible to take the professional licensing exam. This is typically an eight-hour exam specific to the duties of the surveyor and specific to the laws of the state. If successful, the person becomes a registered surveyor!

AN ALTERNATIVE METHOD

Some states still allow individuals to become registered without obtaining a four-year surveying degree. However, the number of years of experience required is greatly increased, typically to at least 12 years.

So, the most direct method requires four years of education, passing the SIT, four years of experience, and passing the professional exam. The minimum time required to become a registered surveyor is eight years and without the education, 12 years. The reason for strict registration requirements is to protect the public from persons who are not qualified to perform property surveys.

For a field engineer to feel fully qualified to establish property surveys after having had one or two surveying courses is very foolish. Don't think it! Don't do it!

LIABILITY

If you are directed to perform surveys that establish property lines, you are exposing yourself and the construction company for whom you work to tremendous liability. If you locate a project from property lines that you establish, you are responsible if the building ends up on the wrong property, or at the very least, violates setback regulations established by the local government.

CONCLUSION

Hire a professional land surveyor for any measurement that relates to property lines.

Communications and Field Engineering

2

Part 1 - Field Engineering Basics

CHAPTER OBJECTIVES:

List some barriers to communications that the field engineer may experience on the jobsite.
Discuss the basics of communication.
List some factors that improve the listening ability of field engineers.
Identify the good habits that a field engineer can follow to more effectively communicate on the phone or radio.
Outline the steps involved in making meetings more effective and productive.
Describe the factors which affect how to deal with people.
Demonstrate the hand signals for general surveying communications.
Describe the basic principles of communicating via construction staking.
Identify the required tolerance for the various types of construction stakes.
Identify the "right" and "wrong" way of placing and communicating with construction marks.

INDEX

THE MOST IMPORTANT PART OF EVERY ACTION

SCOPE

A field engineer has to be competent in two distinct areas to be successful. One is the primary reason for this textbook: being technically competent and able to perform construction surveying and layout. The other distinct skill of the field engineer is to be able to communicate the results of those technical tasks that are performed. The field engineer will have to discuss layout with the crafts and be able to provide lines and grade for their needs. The field engineer will have to communicate with the superintendent the points that have been laid out and the plans for future layout. The field engineer may have to communicate with the owner giving an overall description of where parts of the project are located and how the phases of construction will occur. In short, the field engineer will constantly be communicating. The field engineer must be just as competent in communication skills as in technical skills.

This chapter provides reminders of some of the rules of thumb regarding the basics of written and oral communication. Also presented are several communications methods the field engineer uses including hand signals, surveying stakes information, marks and lines, abbreviations, symbols, and crane signals.

GENERAL

Communicating appears to be simple, doesn't it? But, as a field engineer, how do you do it well? How do you ensure that the other person understands you? And how do you know that you say what you mean? How do you cause others to complete the task at hand? Every day, all day long, your contacts as a field engineer require that you have skills in dealing with your bosses, fellow workers, subordinates, subcontractors, and customers. Your contacts require that you know how to deal effectively in delegating, discussing, explaining, listening, managing, motivating, negotiating, reporting, and writing. You must be able to interact courteously, criticize effectively, gather information, make decisions, deal with conflicts, hold successful meetings, and solve problems.

This section is offered as a brief collection of recognized skills and ideas to remember in your business dealings.

BASICS

First and foremost, the field engineer needs to understand some basic principles of effective communication within the workplace.

BE HONEST

Don't ever compromise integrity or honesty. Once you have been classified as dishonest, it is almost impossible to overcome it. Be willing to admit your mistake and move on to improvement. Attempting to cover up a blunder is only temporary because it will be discovered.

REMEMBER GOOD MANNERS

Answer correspondence and telephone messages promptly; use good judgment; maintain a good rapport with co-workers; do not postpone thank you's; offer praise, congratulations, sympathy, and assistance; and use good table manners. Leave an area cleaner than found; promptly return borrowed items; and develop structured work habits.

THINK, TALK, AND ACT LIKE A BUSINESS PERSON

Present yourself in a way that makes your priorities clear to everyone. Be perceived as someone who knows what the construction business needs and deliver it. Anticipate problems and take care of them quickly. Extend courtesy to all you meet as you never know when you might need their assistance. Be accessible. Dress for success and think for success. A proper professional manner will gain support and respect from others and will help to build social bridges. Others will work better because they are being treated with dignity and respect, and you will feel better and act better.

ALWAYS DOUBLE CHECK WHEN IN DOUBT

Throughout this book it will be stressed that the field engineer must always double check measurements. But, not only must the technical aspects of construction surveying and layout be double checked for accuracy, communications must also be double checked for accuracy. Is this what the architect intended? Is it what I heard at the meetings? Is it what the superintendent has said? Is it what I have communicated to others? Check, check, and double-check.

ORAL COMMUNICATIONS

The key to oral communication is listening. Are you listening? Did you hear what I said, I know you think you heard what I said, but I did not say what you think I said.

These statements are expressed wherever there are people working together. Effectively communicating orally is a skill which must be cultivated. You may think you are discussing and listening, but are you?

Most of us do not listen well because we are thinking about what we are going to say next. Others have a hard time remembering names of the people to whom they are introduced because they are waiting to hear their own names during the introduction process. Effective verbal communication is a key to the success of a field engineer.

LISTENING

Are you listening well? Truly effective people have mastered the art of listening well. It is a crucial skill in business and on the jobsite. Listening well requires personal strength and self-control. An especially difficult atmosphere in which to listen well occurs when there are strong differences of opinion present. However, by using the following suggestions, even the poorest listener can learn to listen effectively.

Effective Listening Techniques	
1	Put the other person at ease.
2	Show that you are actively listening.
3	Make a conscious effort to concentrate.
4	Remove distractions.
5	Occasionally repeat or paraphrase what the speaker has said.
6	Be understanding and patient.
7	Try not to focus on the appearance or abilities of the speaker; remember that thought-speed and speech-speed are different.
8	Watch your temper and the temptation to argue or criticize.

DISCUSSING

Discussion must be handled in a cooperative manner. It must also be clear (What are they to do with the information, and what are you to do with the information?). The field engineer must be well prepared to provide appropriate background information and descriptions. In addition, the field engineer must be prepared to listen well to the other person's point of view, ask good questions, encourage feedback, and be flexible and adaptable. The field engineer must know the audience, handle negative feedback with care, and effectively follow through on the agreed upon action.

SPEAKING ON THE TELEPHONE

Telephone communication is critical communication. Precise messages must be given and heard in the same manner. Good telephone habits must be cultivated. The suggestions below will help the field engineer to deliver clear telephone messages and conduct clear discussions.

Good Telephone Habits
Be prepared -- Know your material.
Call the person who can give the right answer.
Know what a satisfactory answer will be.
Exchange information, questions and confirm.
Write down the date, time, caller, and take notes.
Follow up with a written memo to confirm the details.
Make short, uncomplicated calls.
Monitor your feelings, voice, and language.
Use words and examples which are familiar to the listener.
Speak slowly, enunciating each word, especially when speaking with one who speaks another language.
Use your own phone when possible.
Limit personal calls on company time to emergency calls.
Make occasional friendly calls to fellow business people to maintain rapport.

SPEAKING ON A PORTABLE RADIOPHONE

Radio communication is also critical communication. Precise messages must be given and heard in the same manner. Good radiophone habits must be cultivated. The suggestions below will help the field engineer to deliver clear radiophone messages and conduct clear discussions.

Using a Two-Way Radio
Hold the radio microphone four to six inches away from your chin.
Refrain from yelling into the microphone.
Speak slowly, enunciating each word, especially where more than one language is spoken.
Be prepared; give organized messages in a short, uncomplicated, and concise manner.
Exchange information, questions, and confirm.
Use words and examples which are familiar to the listener.
Use the word "out" to indicate end of discussion.

WRITTEN COMMUNICATIONS

When you create memos, letters, reports, or letters of transmittal, you are presenting certain facts to your reader. In order for your document to be understood, and in order to present a proper impression, you must present those facts in the most correct manner. A document that is poorly structured and full of unclear paragraphs, incorrect words, and misspellings, can easily give the reader the impression that you are not a reliable person in business. This wrong impression can easily be avoided by proofreading your document. In addition, errors in your work may even cause future work to be in error. Although the study of the English language and its writing principles cannot be covered in this manual, you will find presented here a discussion of a few common areas which might be of assistance in producing more clear and correct messages.

WRITING AN EFFECTIVE LETTER

An effective letter or memo accomplishes what the writer intends—to carry out a purpose or to persuade someone to agree or to take action. Effective letters and memos outline procedures, answer questions, help make decisions, provide information, promote confidence, and maintain good will.

Effective writing is built upon coherence, logical thinking, and order of presentation. A document should provide reasons, be concise, be easy to note important items, make sense, contain all necessary information, and allow no misunderstanding. The tone should be direct, tactful, clear, and persuasive, with attention to tone of voice, point of view, word choice, crowding, punctuation, repetition, and correct use of graphics and other additional information. Remember to use "his or her" when referring to gender, or turn sentences around so that gender is not mentioned.

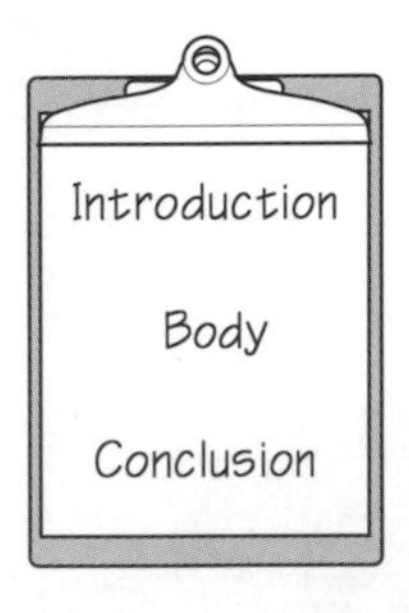

STRUCTURING WRITTEN DOCUMENTS

Business people expect letters and memos to be structured in a three-part manner. If you don't follow the expected structure, your audience may glance right over the important facts. Every letter or memo should contain an introduction, a body, and a conclusion. In addition, every individual paragraph within the document should contain an introduction, a body, and a conclusion. In other words, state what you are going to state, state it, and state what you stated.

DOUBLE CHECKING FOR ERRORS

Before finishing a document, remember always to run a spell check, read over for word usage (through, threw; it's, its; then, than; affect, effect; etc.), and, probably most important, ask someone to look over the document. Another person will see errors you might not and will read for understanding.

MEETINGS

At times you will feel as though your life revolves around meetings- whether between two people or among many. They are a vital part of the operations of a firm. Before meeting with foremen, rod persons, supervisors, and others, prepare yourself.

PLAN

Prior to the meeting, think about the goal of the meeting, identify the reasons for the meeting and think about what will be expected of you at the meeting. Identify what information or actions are necessary before and during the meeting, and identify a personal purpose for attending the meeting.

ARRIVE WELL PREPARED

Bring to the meeting up-to-date information on recent accomplishment, what is in place, what contacts have been made, what follow-ups have been made, and what problems there are. Verify all facts and prepare them in a well organized, easy to follow fashion.

ADOPT GOOD MEETING HABITS

Speak honestly and courteously. Speak less then one minute when not making a report. Avoid interrupting others and carrying on side conversations. Listen carefully and observe facts, feeling, opinions, silences, mistakes and ideas. Learn from the disagreements, and always be patient.

ACTIVELY PARTICIPATE

Study the needs and desires of the audience; collect feedback and make appropriate and positive comments. Ask for updates from subcontractors and others and deliver the message in an interesting, but brief and simple manner, with a clear use of directives.

PERSONAL CONSIDERATIONS

Be aware of your personal appearance. Maintain good eye contact, monitor your personal voice quality, and use appropriate words.

DEALING WITH PEOPLE

Everyone is different—thankfully. Therefore, the method of dealing with everyone is different. Here are a few things to consider when dealing with people.

DETERMINE TYPES OF PERSONALITIES

A first step in gaining successful communication with another is to discern how he or she communicates, and how he or she receives your messages. As a field engineer, you will work with several types of personalities, some of whom you will find to be a challenge. There will be the aggressive, take-charge type of person. There will be the one who shifts responsibility to others, the one who remains detached and independent, or the one who is only interested in following rules. You will find others who are great work companions who understand the importance of maintaining a good morale, good work ethics, a sense of humor, and a proper pace. It will be necessary to learn to deal with each of these personalities throughout the course of your career.

WATCH FOR NONVERBAL SIGNALS

You have probably heard that the way in which something is said is often more significant than what is said. Not only should you watch for nonverbal signals from others to help determine the meaning to their messages, but you should remember your nonverbal signals as well.

Non Verbal Signals
Movements of body and face.
Tone of voice, inflections, silences.
Eye contact (expected in American culture).
Touching (person in charge generally initiates in American culture).
Demeanor, emotions, culture and background.
How close we stand or sit.

LEARN ABOUT YOUR AUDIENCE

Another step in effective communication is finding out what matters to your audience. When discussing a task to be completed on a jobsite, you must know what may be affecting those to whom you are speaking. People will differ in how they understand, in their work experience, in how they draw their own concepts and meanings, and in how they fit new information into what they already know.

DECISION MAKING

Part of learning to be a leader in construction is the ability to make decisions. Field engineers must constantly make decisions that affect the progress of the project. Here are some helpful tables of some ideas related to decision making.

Principles of Good Decision Making	
1	Refuse to be pressured.
2	Consider one decision at a time.
3	Accept the risk of deciding.
4	Always include an alternative.
5	Match the decision with action.
6	If the decision is wrong, reverse it.
7	Review it for next time.

Step-by-Step Decision Making Process	
Step 1	Identify the decision to be made.
Step 2	Gather information.
Step 3	Identify the alternatives.
Step 4	Weigh the evidence.
Step 5	Choose among alternatives.
Step 6	Take action.
Step 7	Review the decision.

GIVING AND RECEIVING CRITICISM

Criticism can be positive as well as negative—as much can be learned from a complaint as from a compliment. Establish your viewpoint, understand the other person's viewpoint, and remember both of your viewpoints during the talk.

WHEN YOU CRITICIZE

When you criticize, direct your criticism at the behavior, not at the person, and try not to criticize when angry. Instead of screaming "You shouldn't have done that!" perhaps saying "This did not seem to work out, we need a different approach." would result in less damaged feelings and in better future efforts. If you first attempt to determine what the other person is trying to achieve, and then imagine how that person arrived at the particular action, you can then better temper your comments and criticize more correctly and effectively. Make a good, specific criticism that is understood without lecturing. Set a pleasant tone. Show understanding. Offer support. Show confidence that the person will better his or her performance.

WHEN YOU RECEIVE CRITICISM

When you receive criticism, try to reevaluate your views or your methods. Try to gain insight, try to learn from others, and try not to become self-defensive. When one allows self-defensive feelings to creep in, judgment and reasoning capabilities become clouded, defeating the positive possibilities which can grow from criticism.

RECOGNIZE COMMUNICATION BARRIERS

There are many reasons why a person will not listen to you or will not understand you. It is wise to attempt to overcome the following barriers to communicating effectively.

DIFFERENT BACKGROUNDS AND LANGUAGES

People of different cultures will apply different meanings to their communication. If you are to work in another country, or on any jobsite with people of another culture, it is extremely important to learn the meanings they will place on your words.

COMPETING NOISES

If possible, walk away from noisy equipment. Get behind a wall to break the sound. Use hand signals if possible.

EMPLOYEE STATUS LEVELS

Show respect for those around you. Some will have greater responsibilities than you and some will have fewer. All personnel on the jobsite have a part in completing the project. Recognize this importance.

LACK OF BASIC KNOWLEDGE

Recognize that everyone will not have the technical knowledge of construction surveying as does a field engineer. Be prepared to explain and teach the basics of measurement and applications.

DIFFERING WORD DEFINITIONS

The same word may mean different thing to different people. Ensure that everyone understands the same definition for the terms and measurements you are using.

INSUFFICIENT INTEREST

Overwhelming someone about the details of an item will lose their interest. Explain things one step at a time.

LACK OF PRAISE

People need to be told they are doing a good job. Not remembering to thank or praise for previous effort may result in silence.

POOR LISTENING ABILITY

Some people just have a short attention span and may have lesser hearing capability.

POOR HABITS

Refrain from continuously moving or making noises such as tapping, chewing, cracking, scratching, swinging, or moving closer and closer. Refrain from conducting personal hygiene in public places, misusing property, arriving late, and avoiding your fair share of responsibility.

PROCRASTINATION

Everyone, from your co-workers to the customer, tires of having to call repeatedly while phone calls and messages remain unanswered or unreturned. They also tire of finding it necessary to repeat information because of inadequate notetaking. Poor response time to requests is also offensive, especially when one fails to ensure that the other person remain informed of the progress.

COMMUNICATING WITH HAND SIGNALS

SCOPE

From the moment the first construction project was established at the beginning of time, people have been communicating with hand signals and will continue to do so on construction sites until the end of time. It is often easier to use hand signals than it is to speak. It is often necessary to use hand signals because of equipment noise or distance between the individuals. Hand signals are often the only method of communication between persons who speak different languages. Without hand signals, construction would be more difficult and more time consuming. It is the responsibility of the field engineer to quickly learn hand signals that are commonly used on the jobsite. Some examples are shown on the following pages.

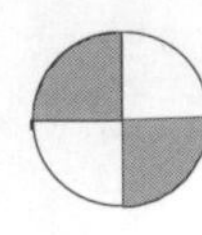

HAND SIGNALS FOR NUMBERS

There are several methods of communicating numbers by hand signals. The method shown below is one which has been in use for several years. It is a logical method. Also, the recipient of the signals doesn't have to think if the right arm or the left arm is being used to create the signal because it doesn't matter. Therefore, many of the illustrations below show two persons, one showing the signal with the right arm, and the other showing the signal with the left arm. Each is a correct way to show the same signal. In this method, also, it can be noted that the signals for numbers one through five are combined to create the signals for numbers six through nine. It is a very easy system to learn and use.

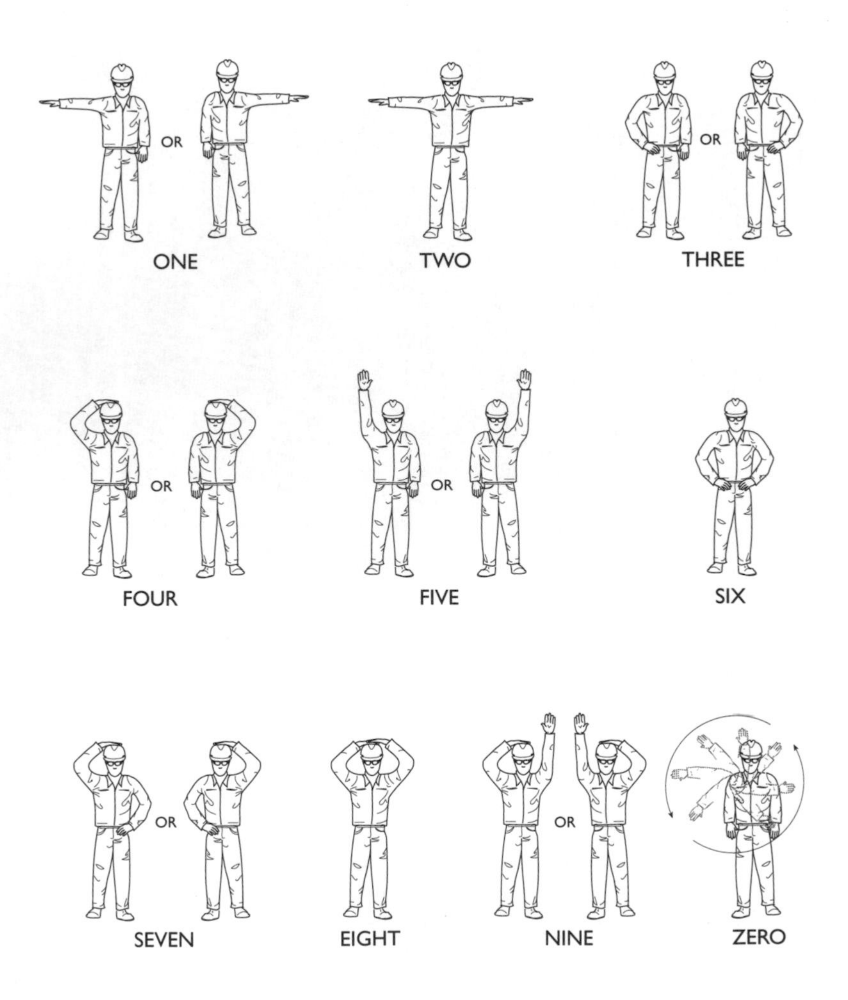

HAND SIGNALS FOR FIELD OPERATIONS

The following hand signals have evolved over years of use by field engineers all over the country. They are rather generic and fairly self explanatory.

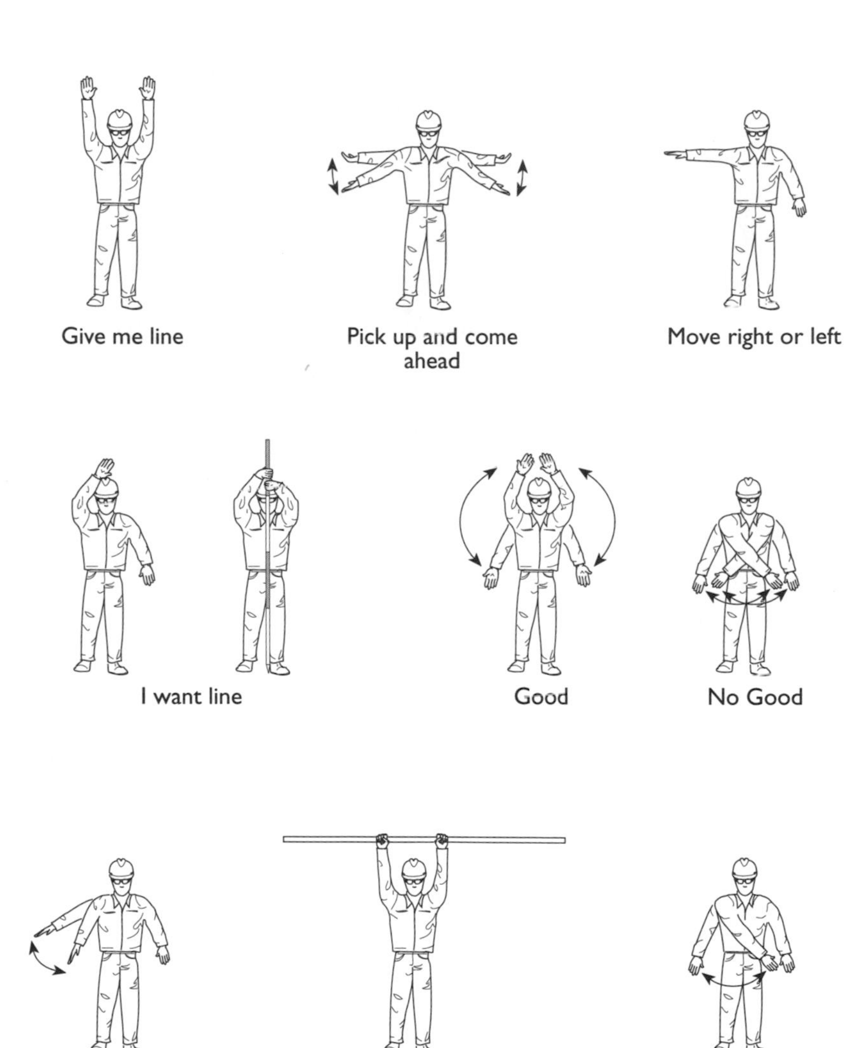

Close to grade

This is a turning point or I need a turning point

Plumb rod or prism pole (move arm slowly)

Cut

Fill

Foot

Point

COMMUNICATING WITH A CRANE OPERATOR

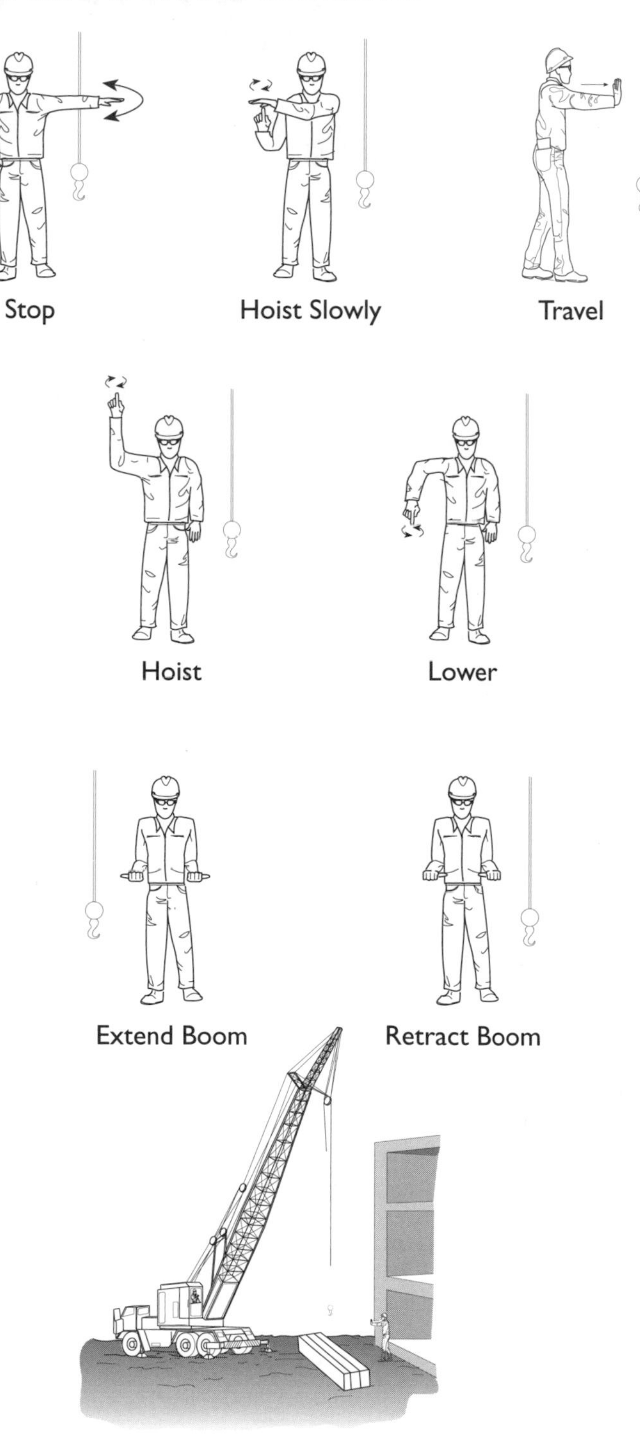

COMMUNICATING ON STAKES AND LATH

SCOPE

Construction stakes are the first on-site evidence that a project is about to begin. Nothing is built until a project is staked. Nothing is built accurately until the stakes are located correctly and the information on the stakes is conveyed clearly. Communicating to the crafts on construction stakes is one of the most important and most difficult aspects of field engineering.

Learning to communicate clearly on construction stakes is attained by experience as a field engineer. But like the "chicken and the egg" and which happened first, how can a field engineer communicate well on stakes without experience?

This section will cover some principles on dealing with what should be written on stakes and some guidelines to be used with typical stakes generally required on the construction site.

PRINCIPLES OF CONSTRUCTION STAKING

Anytime stakes are used by a field engineer, the persons who will be using them will have certain expectations about the type of stakes that are used and how they are placed.

FACE STAKES IN THE CORRECT DIRECTION

Face stakes so they can be read from the direction of use. For highways, center line stakes are placed so they can be read from the beginning of the project. Place offset stakes to be read from a person standing at center line. For buildings, place corner stakes so they all can be read from the same direction, such as when a person is reading from the front of the building. Offset stakes are set so they are read from inside the building.

Every project is unique, so the field engineer should think about stake placement in the planning phase and develop and communicate a system that is to be used by all.

USE THE PROPER SIZE STAKE FOR THE JOB

The size of the stake often indicates its use. There are three sizes that are typically used by field engineers.

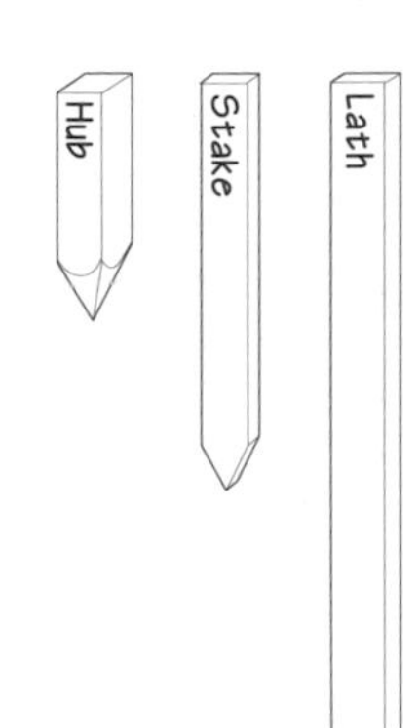

Hub

Typically a 1-1/2 inch by 1-1/2 inch piece of wood that can be obtained in lengths of 6", 12", 18", or 24". A hub is generally used to mark a specific point such as center line points, control points, radius points, curb and gutter lines, blue tops, etc.

Stake

Typically a 3/4 inch by 1-1/2 inch piece of wood that varies in length from 12 to 48 inches. Stakes are used for center lines, offset lines, slope stakes, and as information stakes next to hubs, blue tops, etc.

Lath

1/2 inch by 1-3/4 inches by 48 inches. Laths are used predominately as guards for hubs and stakes. Laths may also be used to indicate the limits of a clearing or as a rough location of easements.

OFFSET STAKES FOR PROTECTION

The best method of protecting stakes from being destroyed by construction is to offset them. Offsetting is defined as the setting of stakes away or "off" of the point of need. The amount of offset depends on the type of construction and will be discussed through the book.

SET STAKES WITHIN TOLERANCE

Stakes must be set to the accuracy (tolerance) required for the work. Some stakes can be set within a tenth while others must be within a hundredth. See *Chapter 4, Fieldwork Basics* for a discussion of tolerances.

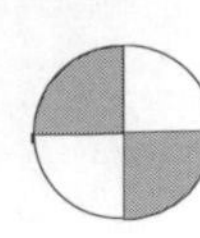

PLACE STAKES SOLIDLY INTO THE GROUND

A stake that falls down in a typical wind wasn't placed properly. Hammer dirt or rocks around a stake placed in soft ground. In hard or rocky ground, use a rock chisel to start a pilot hole.

PLACE STAKES AND LATH TO BE PLUMB

Care should be taken so that all stakes and laths are placed so that they are close to plumb. Here may be a reference mark for grade information at the bottom of the stake which may become confusing if the stake is leaning. Taking the extra time to position stakes to be plumb communicates to superiors that the field engineer cares about quality.

CENTER HUBS AND STAKES

When a stake or hub is driven to represent a point, time should be taken to drive it so the exact point is close to the center of the hub and on line and distance. If remeasurement shows the point is at the very edge of the hub, take the extra time needed to position it correctly.

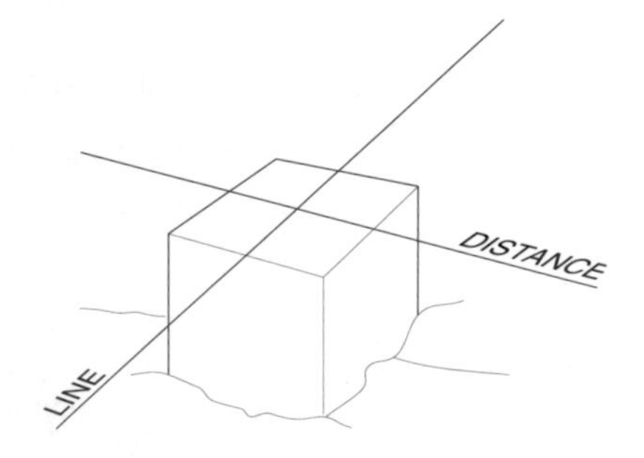

COLOR CODE THE FLAGGING

Plastic flagging can very effectively be used to communicate the type of stake and type of point that is being located. One can use three colors of flagging, such as red, white, and blue to communicate different types of points (see example below). Recognizing there are dozens of different colors of flagging available, the combinations available to the field engineer are almost limitless.

Color Code Flagging

Survey Point	Flagging Color(s)
Primary Control	Red/White/Blue
Secondary Control	Red/White
Building Control	Red
Benchmark	White/Blue
Temporary Benchmark	White
Slope Stake	Red/Blue
Finished Grade	Blue

PRINCIPLES OF COMMUNICATING ON STAKES

No one will be able to understand what you place on stakes if you are not consistent in how you mark them. The following principles should be followed carefully.

Write from top to bottom

PRINT NEATLY

Follow the lettering style that is used in the field book (all uppercase lettering, consistently formed, slightly slanted, evenly spaced, etc.).

WRITE SO IT IS EASY TO READ THE INFORMATION

Don't crowd the words and numbers. Write from the top to the bottom of the stake every time.

USE ABBREVIATIONS THAT EVERYONE UNDERSTANDS

If possible, write the entire words, not abbreviations (i.e., B.C. could be read as Beginning of Curb or as Back of Curb). A field engineer cannot afford to have a contractor misunderstand the verbiage on a stake or lath. Some word abbreviations are familiar throughout the industry and will not be misunderstood. However, there are many abbreviations that have more than one meaning, and the only way to remedy the situation is to record as many letters as possible on the layout markers.

If abbreviations are to be used, discuss with others on the jobsite to determine an acceptable abbreviation for common terms that are being used. A few common abbreviations include:

Stake Abbreviations

Abbreviation	What it means
Toe	Toe of Slope
Top	Top of Slope
P.I.	Point of Intersection
P.C.	Point of Curvature
E.O.P	Edge of Pavement
B.C.	Back of Curb
F.B.D.	Flat Bottom Ditch
T.B.M.	Temporary Benchmark
B.M.	Benchmark

There are as many abbreviations as there are words. Field engineers can be creative in using abbreviations as long as they are communicated to everyone on the jobsite. See the section on abbreviations later in this chapter.

2-30

USE ALL SIDES OF THE STAKE

When necessary a field engineer should use all sides of a stake, but the primary identifying information should be facing the direction of use.

STAKE INFORMATION

Stake Communication
Station number
General information description
Alignment information
Centerline
Offset
Super elevation data
Cut or fill data
Slope
Reference information
Specific information
Grade information

SETTING STAKES ON THE SITE

There are many types of stakes used on the construction site - many more than can be mentioned in a few pages of this manual. The following are a few examples of some stakes, the tolerance, and a description of their use.

SITE-CLEARING LIMITS

Tolerance:

Horizontal +\- 1.0'

Set stakes within a tolerance that includes the vegetation that cannot be cut by specie or per specification. On some projects, simply cutting one small plant can be detrimental. There really is no set spacing between the stakes, but one must consider fluctuating the clearing boundaries. When a boundary moves in and out, there must be stakes at each turn of the clearing line. The stakes that are placed must be visible before, during, and after cutting.

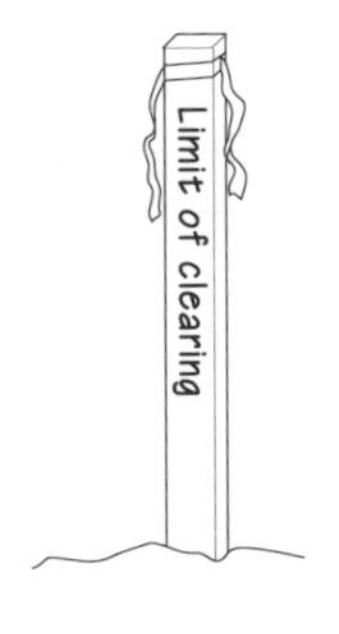

If the boundary permits, stakes may be set every 100' or more. In site clearing, offset stakes are not necessary because the stakes should not be removed from the ground. Offsets become necessary only when the marker stakes will be removed or are in danger of being removed in the course of normal construction.

ROUGH GRADE

Tolerance:

Horizontal	+\- 0.1'
Vertical	+\- 0.2'

Stakes will be set on offsets from the centerline at locations determined by the contractor or plans. 100' intervals will be the maximum the stakes should be placed.

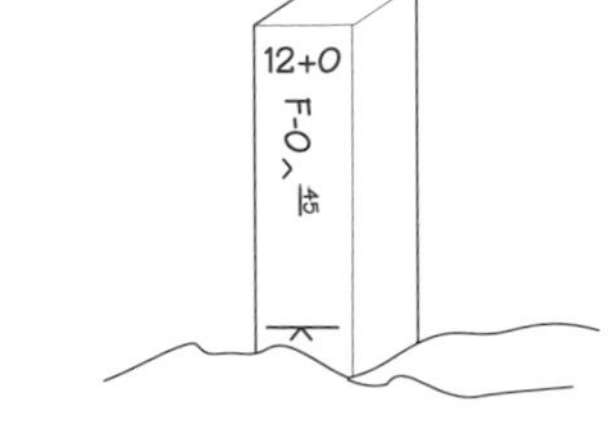

FINE GRADE

Tolerance:

Horizontal	+\- 0.1'
Vertical	+\- 0.01'

These stakes may be set as offsets or may be set as the centerline where elevation measurements will be taken off of the top of them. Cut or fill will be noted on each stake. The intervals between these stakes should be 50' or less. Noted on the stake will be station number, elevation of top of stake, alignment designation, and cut or fill.

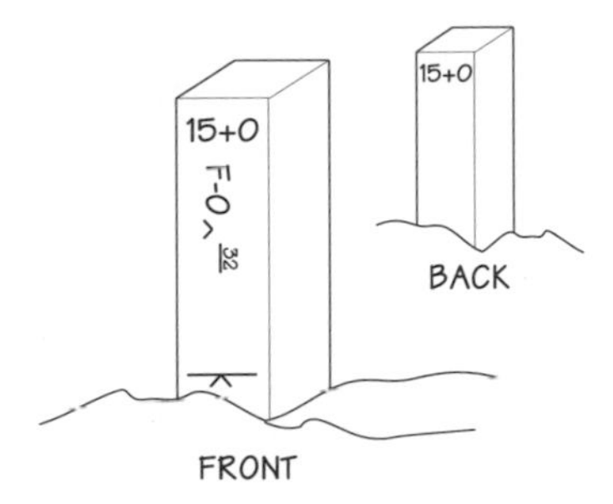

SLOPE

Tolerance:

Horizontal	+\- 0.1'
Vertical	+\- 0.1'

Since these stakes may be in danger of being removed from the digging, they should be at a 10' minimum offset. Never scale stakes from the plans, rather use mathematical calculations to determine field location and information to be written on them. Slope stakes should have the following labels attached to them: offset distance, total fill or cut from existing to proposed, total horizontal slope distance, slope ratio, total distance to and from the offset stake, offset elevation difference from existing centerline elevation (+ or -) and, station number.

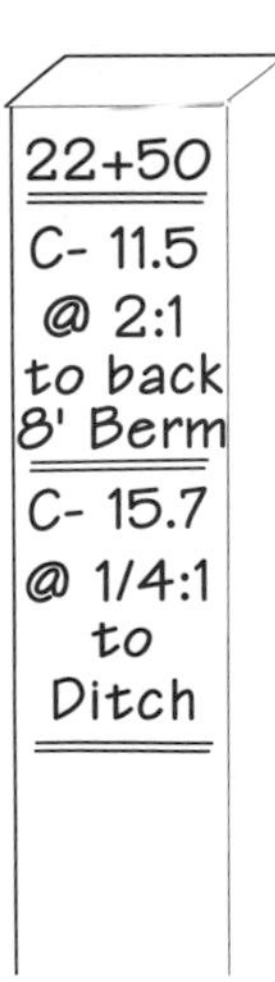

CURB AND GUTTER

Tolerance:

Horizontal	+\- 0.01'
Vertical	+\- 0.01'

This offset should be relatively close to the centerline, somewhere less than 6'. Grade elevations should be written as top of curb, and offsets should be written as back of curb. Hubs with a tack should be used on the offset line for referencing vertical and horizontal control. Stakes should have the following labels: offset to back of curb, cut or fill to top of curb, gutter slope, super elevation of pavement, and station number.

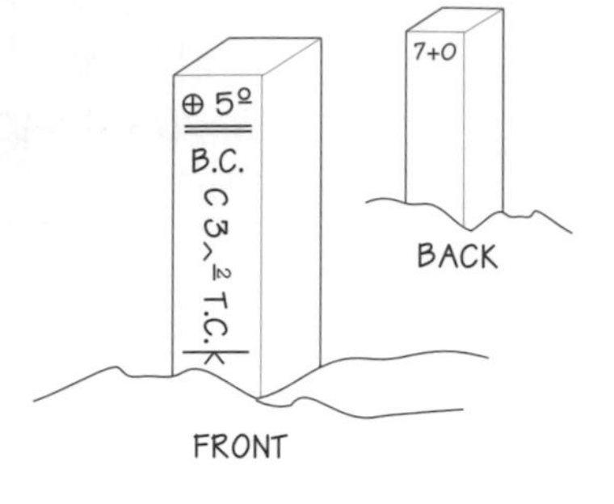

PIPE

Tolerance:

Horizontal	+\- 0.01'
Vertical	+\- 0.01'

There are two kinds of pipe stakes: hubs and laths. Hubs are used for alignment and grade while laths are used in writing information such as percent grade, offset distance, station, type and size of pipe (dia.), and cut or fill from hub to the invert. All runs of pipe should require a minimum of two reference lines: one at the beginning of the run and one at the end. The reference lines should be set perpendicular to the pipe run so as not to interfere with digging and dumping operations. The offset hub should be at least 15 feet from the reference hub at either end of the run. Pipe runs greater than 100' should have intermediate centerline staking. Pipe runs that enter drainage structures will be staked to a point where the total perpendicular circumference of the pipe is inside the structure. The lath at each hub should have the following information: pipe station, percent grade ("+" at outlet end and "-" at the inlet), cut or fill from hub, type and size of pipe, length of pipe run, and any necessary structures (i.e., catch basin, inlet, tee, flare, elbow, etc.).

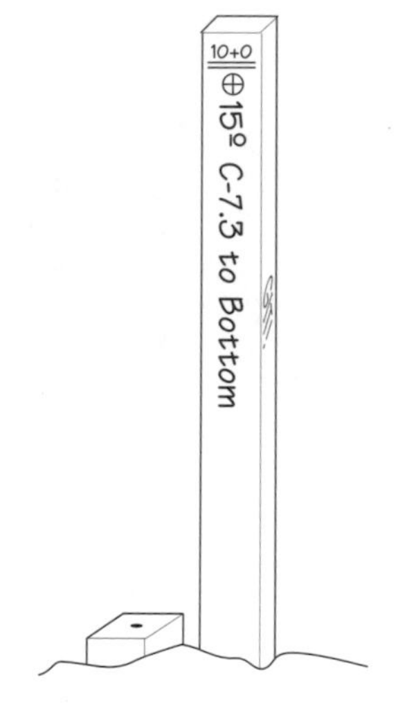

MINOR STRUCTURES

Tolerance:

Horizontal +\- 0.01'
Vertical +\- 0.01

Headwalls, inlets, culverts, Catch basins, and junction boxes should require a reference line with hubs on either side. Reference lines should refer to the center of a junction box, the center of an inlet, the inside back wall of a catch basin, and the front face of a headwall. Grades should come from the plans or be calculated to the inverts of the structures themselves, not the inverts of the pipes running into them. Lath should contain the following information: station, offset distance to the structure's exact location, elevation of hub, and cut or fill distance to the structure's invert or even to the bottom of the structure. Sometimes the latter is not possible because the sumps of the catch basins may differ from catch basin to catch basin.

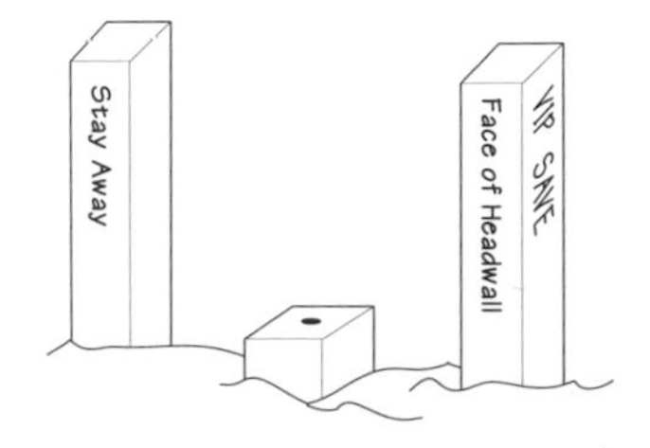

CONTROL

Tolerance:

Horizontal ±0.01
Vertical ±0.01

Hubs are sometimes used as control points. However, concrete will be used for critical control points on the construction site. If hubs are used, they should be as long as possible to resist movement in the normal freeze/thaw cycle that occurs in many parts of the country and should be protected by laths marking them.

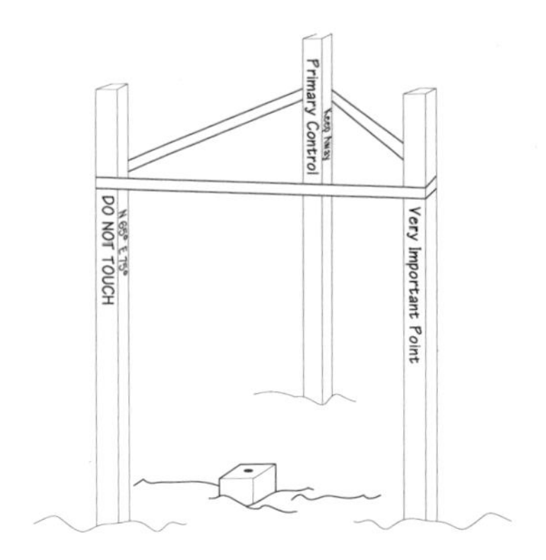

COMMUNICATING LINE AND GRADE WITH MARKS

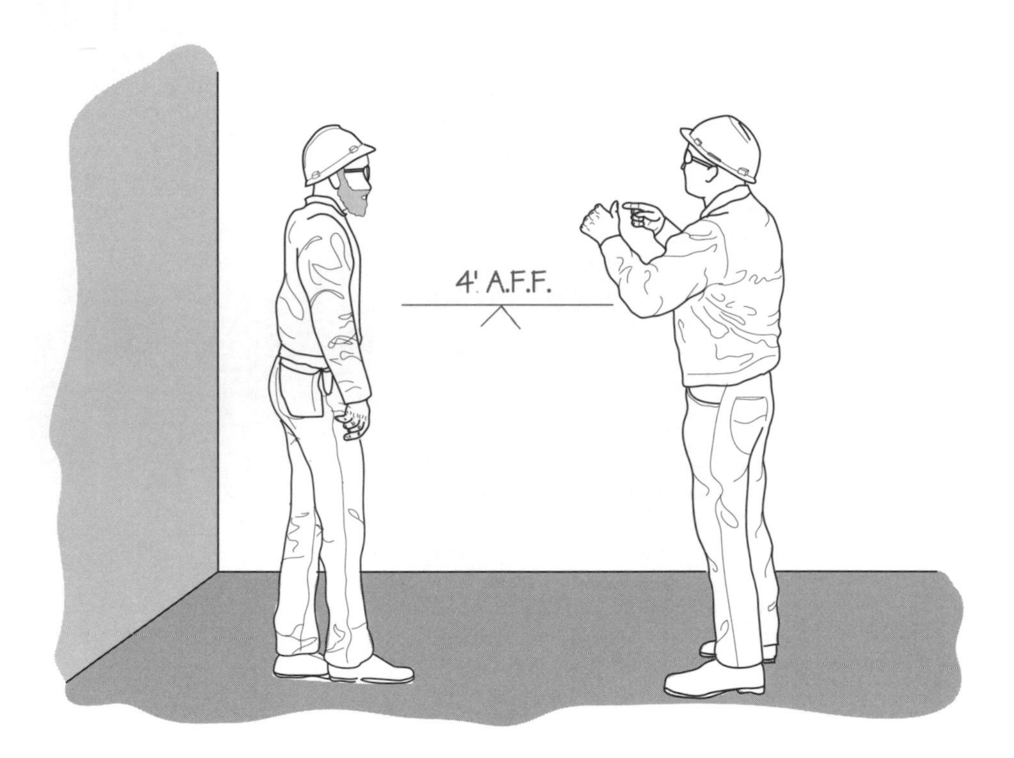

SCOPE

Visit a construction site and everywhere you look will be evidence of the effort of the field engineer to communicate to the craftsperson. In addition to stakes there will be lines and marks that indicate the working control line and grade needed for the crafts to build the structure. Pencils, lumber crayons, spray paint, chalk lines, nails, paint sticks, ink markers— just about anything that can be used to mark a line is used to make the marks. Some field engineers are very meticulous and precise in making marks that last and are easily understood, and other field engineers sloppily strike a line that no one understands. This section will list some principles of how to make marks and identify them, and will present a few examples of good and poor marks that have been observed on projects.

BASIC PRINCIPLES OF MARKING

Straight, level, and plumb! These terms are used to describe how a building is to be constructed. These terms also apply to how marks should be made for the craftsperson to use. See the illustrations below for examples of what to do and what not to do.

STRAIGHT

When a line is being placed on the floor to be used as a reference in the construction of interior walls, it must be "on line." In other words, it should not be skewed one way or the other and it should not be half on line and half off line. It should be exactly on line. When the person at the instrument looks at the line, the crosshair should split the line for its entire length.

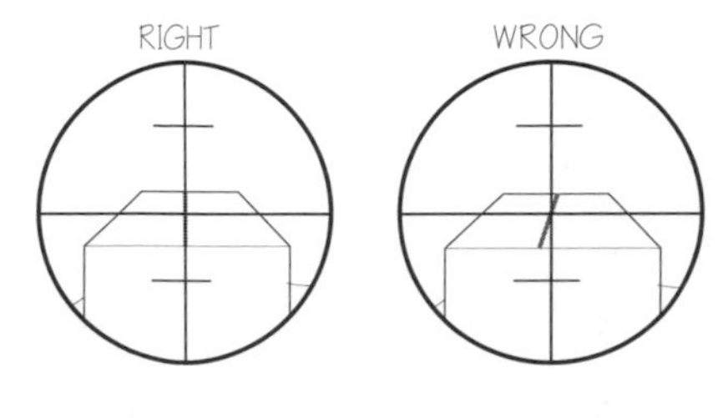

LEVEL

Lines placed on a wall to be used as a reference for elevation should be placed exactly horizontally. Lines should not slope at all. If they do, at which point are they to be used as a reference? In the middle? At the right end? At the left end? Mark them horizontaly and these questions won't arise. See the examples for marking an elevation on a wall.

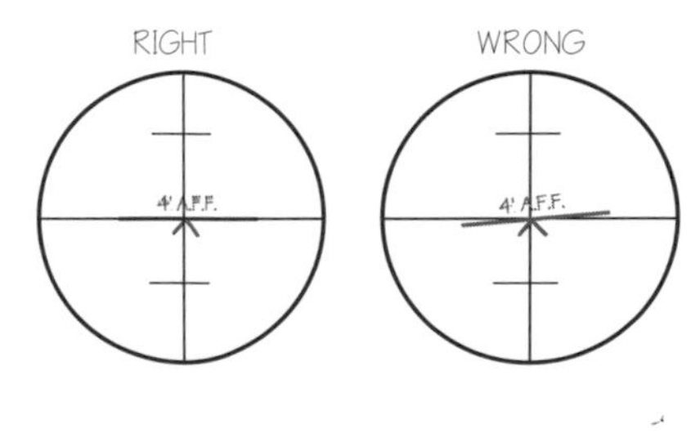

PLUMB

If a line is placed on a wall as a future target, the line should be plumb. It should not slant one way or the other so a person using an instrument in the future has to guess which part of the line to sight on. See the examples for proper and improper targets on a wall.

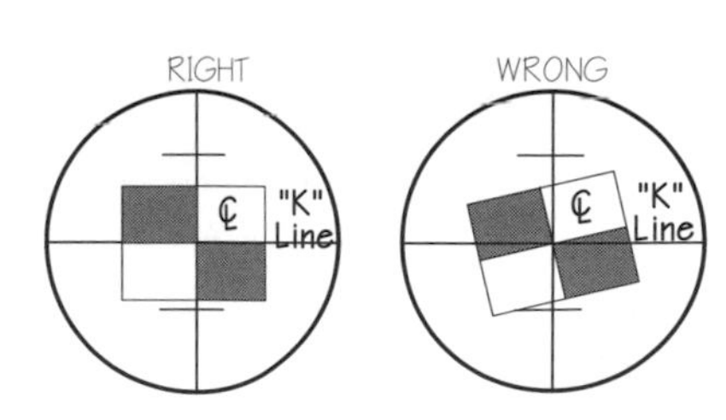

COMMUNICATING THE MEANING OF A POINT OR LINE

When communicating the meaning of a mark to others on the construction site, clarity is the best policy. This statement refers to the information included next to a mark that is made for alignment, elevation, or distance. Too much information may be confusing, and not enough information may stop the work. An experienced field engineer will know what needs to be written next to a mark to communicate to the crafts. An inexperienced field engineer should go to the foremen or craftspersons to find out exactly what is needed and ensure that all necessary information is included next to the mark.

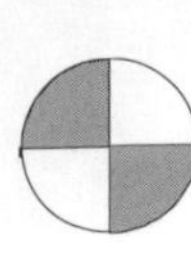

Illustrated below is an example of a well-written stake communication, and next to it, an example of a poorly written one. Note the missing words and the confusing information in the poorly written communication.

RIGHT	WRONG
2" Offset to South of "H" Line	2" Off of "H" Line (NORTH, SOUTH, EAST, WEST?)
4' Above Finish Floor	4' Above Floor (ROUGH, FINISH?)
Cut 10.4 at a 1:1 to back of 6' Flat Bottom Ditch	Cut 10.4 to back of Ditch (TYPE OR DITCH?)
12" Off Dowel Center to Face of Form	Measure 12.00" off the average center of the dowel rods to the flush face of the face of the typical column form...and then...

EXAMPLES OF COMMUNICATING LINE

In addition to the monuments, hubs, stakes, and other control points around the project site, the field engineer marks line for the crafts wherever there is something solid and on line. Examples of solid and on line marks generally include walls, footings, foundations, floors, and other places that are already poured or are in place. The following illustrations are examples of the correct way field engineers mark to communicate line.

Targets placed on a wall as backsights.

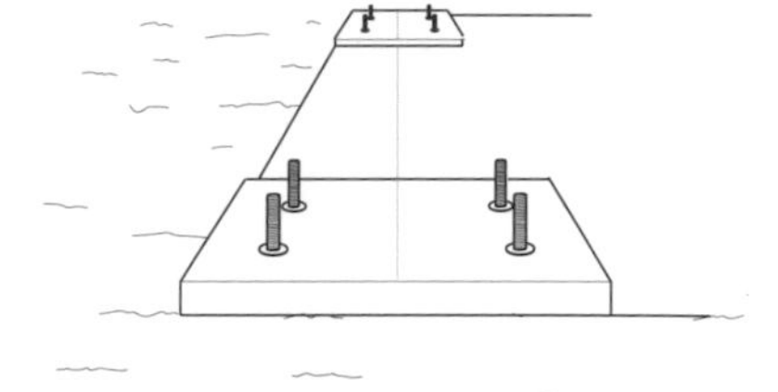

Chalkline marked on concrete between anchor bolt centers.

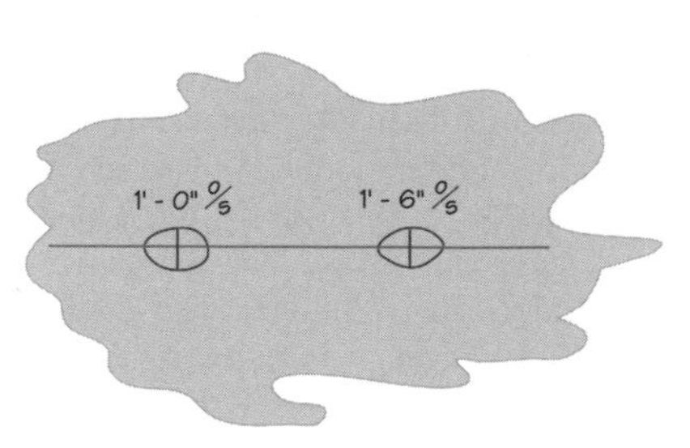

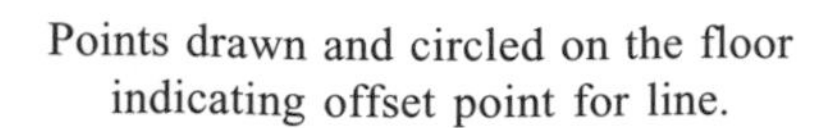

Points drawn and circled on the floor indicating offset point for line.

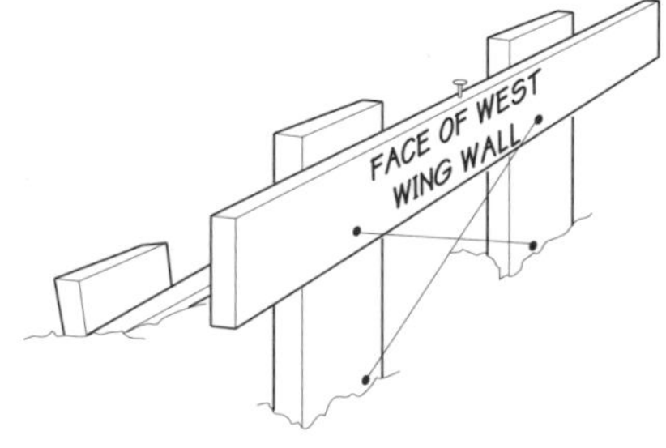

Nail on a batter board indicating face of wall.

EXAMPLES OF COMMUNICATING GRADE

In addition to monuments, railroad spikes, and other benchmarks around the project site, the field engineer marks elevations for the crafts wherever there is something solid. This generally means on walls, footings, foundations, floors, columns, and other solid places that are already poured or are in place. The following illustrations are examples of the correct way field engineers mark to communicate grade.

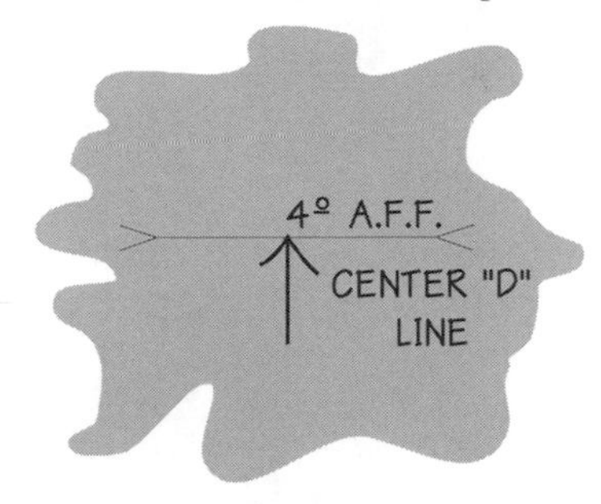

Grade for finish floor and control line information.

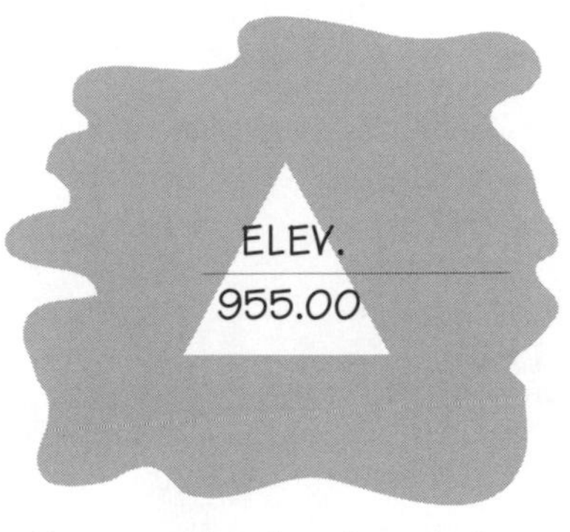

Temporary benchmark marked on a wall.

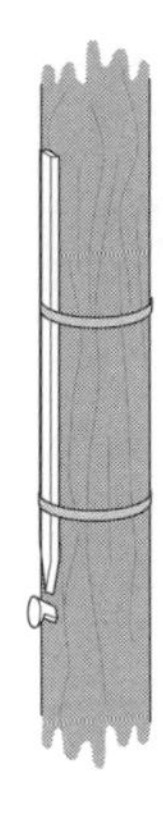

Spike driven into a pole and prominently marked with a lath.

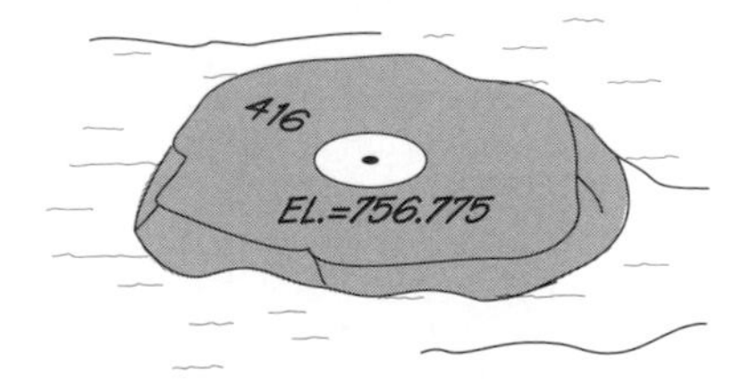

Control point number and elevation on a concrete monument.

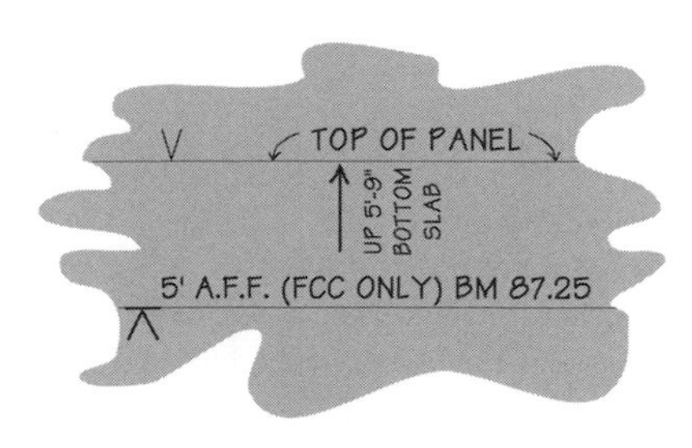

Marking a horizontal line as a reference.

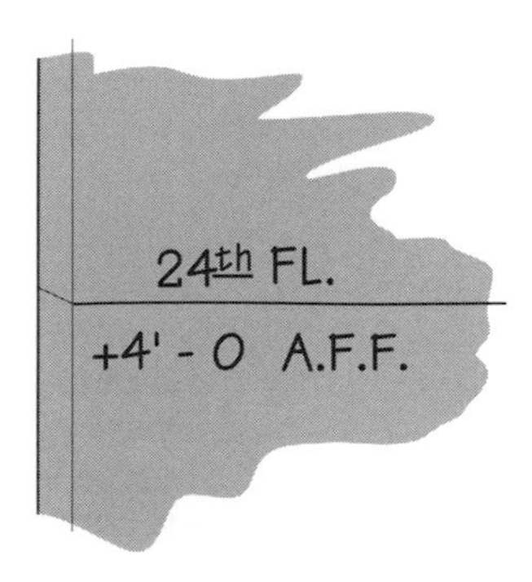

Grade for finishes on a high-rise structure.

CONSTRUCTION ABBREVIATIONS AND SYMBOLS

SCOPE

In the process of communicating, it is often the practice of the field engineer and those in construction to use abbreviations or symbols. The use of these become a part of every memo and every stake that is set. The list of abbreviations that follows are some of the more common ones that will be encountered. Symbols are used on stakes, lift drawings and in the field book. Often times the best method of representing objects is to use a symbol.

ABBREVIATIONS

"A"

Above Mean Sea Level	ABMSL
Abutment	abt.
Adjusted	adj.
Ahead	ahd.
And	&
Alignment	align.
Approximate	approx.
At	@
Avenue	Ave.
Average	avg.
Azimuth	az.

"B"

Back	bk.
Back of Sidewalk	BSW
Back of Walk	BW
Backsight	BS
Barbed Wire Fence	BWFe.
Bearing	brg.
Begin Curb Return	BCR
Begin Horizontal Curve	BC
Begin Vertical Curve	BVC
Bench Mark	BM
Between	betw.
Block	blk.
Boat Nail	BN
Boat Spike	bt. spk.
Book	bk.
Bottom	bot.
Boulevard	Blvd.
Boundary	bndry.
Brass	bra.
Bridge	br.
Building	bldg.
Bureau of Land Mgmt.	BLM
By Intersection	by int.

"C"

Calculated	calc.
Casing	csg.
Cast Iron Pipe	CIP
Catch Basin	CB
Catch Point	CP
Cement Treated Base	CTB
Center of Curve	CC
Channel	chnl.
Chord	chd.
Chiseled Cross	"X"
Circle	cir.
Clean Out	CO
Concrete Block Wall	CBW
Conduit (specify type)	cond.(tel.)
Construction	const.
Control Point	CP
Coordinate	coord.
Corner	cor.
Corrected	corr.
Corrugated Steel Pipe	CSP
County	Co.
Court	Ct.
Creek	cr.
Crossing	xng.
Cross Sections	X sec.
Culvert	culv.
Curb	cb.
Curb and Gutter	C&G
Cut	C

"D"

Deflection	def.
Degree	deg.
Description	desc.
Destroyed	dest.
Detour	det.
Diameter	dia. or D
Direct	D
Distance	dist.
Distance	D
Distance, horizontal	Dh
District	Dist.
Ditch	dit.
Double	dbl.
Down	dn.
Drill Hole	DH
Drive	Dr.
Driveway	drwy.
Drop Inlet	DI

"E"

East	E
Easterly	Ely.
Edge of Gutter	EG
Edge of Pavement	EP
Edge of Shoulder	ES
Electronic Distance Measurement	EDM
Elevaton	el.
End of Horizontal Curve	EC
End of Vertical Curve	EVC

End Wall	EW
Equation	eqn.
Existing	exist.
Expressway	Exwy.
External	E

"F"

Fahrenheit	F
Fence	fe.
Fence Post	FP
Feet	ft.
Field Book	FB
fill	F
Finish Grade	FG
Fire Hydrant	FH
Flow Line	FL
Foot	ft.
Footing	ftg.
Foresight	FS
Found	fd.
Foundation	fdn.
Freeway	Fwy.
Frontage Road	FR

"G"

Galvanized	galv.
Galvanized Steel Pipe	GSP
Gas Line	GL
Gas Valve	GV
Geodetic	geod.
Grid	grd.
Ground	grnd.
Gutter	gtr.

"H"

Headwall	hdwl.
Height	ht.
Height of Instrument	HI
Highway	Hwy.
Hinge Point	HP
Horizontal	hor.
Hub & Track	H&T

"I"

Inch	in.
Inside Diameter	ID
Instrument	inst.
Interchange	intch.
Intersection	int
Invert	inv.
Iron Pipe	IP
Irrigation Pipe	irr.P

"J"

Junction	jct.

"K"

Kilometre	km

"L"

Land Surveyor	LS
Lane	ln.
Left	lt.
Length of Curve or Length	L
Long Chord	LC

"M"

Manhole	MH
Marker	mkr.
Maximum	max.
Measured	meas.
Median	med.
Metre	m
Mid-Ordinate of Curve	M
Mid-Point of curve	MPC
Mile	mi.
Millimeter	mm
Minimum	min.
Minute	min.
Monument	mon.

"N"

Nail	N
National Geodetic Survey	NGS
National Oceanic and Atmospheric Administration	NOAA
National Ocean Survey	NOS
North	N
Northerly	Nly
Number	# or no.

"O"

Offset	O/S
Original Ground	OG
Outside Diameter	OD
Overhead	OH

"P"

Page	p.
Pages	pp.
Party Chief	PC
Pavement	pvmt.
Perforated Metal Pipe	PMP
Pipe	P
Place	pl.

Plastic	plas.
Point	pt.
Point of Compound Curve	PCC
Point of Compound Vertical Curve	PCVC
Point of Intersection	PI
Point of Reverse Curve	PRC
Point of Reverse Vertical Curve	PRVC
Point of Tangency	PT
Point on Horizontal Curve	POC
Point on Semi-Tangent	POST
Point on Tangent	POT
Point of Vertical Curve	POVC
Portland Cement Concrete	PCC
Power Pole	PP
Pressure	press.
Private	pvt.
Profile Grade	PG
Project Control Map	PCM
Project Control Survey	PCS
Property Line	PL
Punch Mark	PM

"R"

Radial	rdl.
Radius	R
Radius Point	rad.pt./RP
Railroad	RR
Railroad Spike	RRspk.
Record	rec.
Read Head Nail	RH
Reference	ref.
Reference Monument	RM
Reference Point	RP
Refraction	r
Reinforced Concrete Pipe	RCP
Retaining Wall	ret.W
Reverse	R
Right	rt.
Right of Way	R/W
River	Riv.
Road	rd.
Roadway	rdwy.
Rock	rk.
Route	Rte.

"S"

Sea Level Datum	SLD
Seconds	sec.
Section	S
Semi-Tangent	T
Sewer Line (Sanitary)	SS
Shoulder	shldr.
Sidewalk	SW
Slope Stake	SS
South	S
Southerly	Sly.
Spike	spk.
Stake	stk.
Standard	std.
Stand Pipe	SP
Station	sta.
Steel	stl.
Storm Drain	SDr.
Street	St.
Structure	str.
Subdivision	subd.
Subgrade	SG
Superelevation	SE
Surfacing	surf.

"T"

Tack	tk.
Tangent	Tan.
Telephone Cable	tel.C.
Telephone Pole	tel.P.
Temperature	temp.
Temporary Bench Mark	TBM
Top Back of Curb	TBC
Top of Bank	TB
Top of Curb	TC
Township	T
Tract	tr.
Transmission Tower	TT
Traverse	trav.
Triangulation	tria.
Turning Point	TP

"V"

Vertical	vert.
Vertical Angle	V
Vertical Control Monumentation	VCM
Vertical Curve	VC
Vitrified Clay Pipe	VCP

"W"

Water Line	WL
Water Valve	WV
Westerly	Wly.

Wing Wall	WW
With	w/
Witness Corner	WC

Z"

Zenith Angle	z

SYMBOLS

Symbols			
Utility Pole		Hedge	
Fence & Gate		Brush	
RR Tracks		Trees	
Culvert & Headwalls		Flow Line	
Bridge		Lake or Pond	
Trail (labeled)	Trail	Stream or River	
Unpaved Road (labeled)	Dirt	Water Line	W
Paved Road		Gas Line	G
Curb & Drive Way		Sewer Line	S
Foundation	CONC	Electrical Line	E
Building Lines		Irrigation Line	M
Sight		Center Line	℄
Deadman		City Line; County Line	Bedford Co. Crawford Co.

Reflector		Offset	⊕	Right Angle	
Instrument		Property Corner	⬡	Project control surv. pt.	△
Angle		Survey mon. in a well	◎	Center Line	℄
Chainman		Subdivision Corner		Note Keeper	
T-bar & Cap	Ⓣ	Round Survey mon.	○		
Rodman		Rectangular Surv. mon.	□		

GETTING STARTED AND ORGANIZED

3

PART 1 - FIELD ENGINEERING BASICS

CHAPTER OBJECTIVES:

Identify the most important activity of the field engineer.
List the common personal surveying tools used by most field engineers.
List considerations that a field engineer must make in selecting surveying software and computers.
State where a field engineer can find "consumer reports" guides to surveying equipment.
Describe the purpose of preparing a daily or weekly schedule of field engineering activities.
Prepare an daily schedule for the activities of a field engineer using any construction plans.

INDEX

GETTING STARTED AND ORGANIZED AS A FIELD ENGINEER

SCOPE

As a new field engineer, you will typically start to work a few weeks after graduation from college. Usually, this means that your first assignment will be to go to a job that is already under way. Someone else will have gone through the start-up phase of the project, selected the personal and project equipment for the work, and organized the work assignments with daily and two-week schedules. Although these activities will be completed when you arrive, close attention should be paid to what it takes to equip a crew because your next assignment will more than likely be on a job that is just starting up. Starting a job and organizing the daily work load is extremely important to your success as a field engineer. If you don't have the right equipment, it will be difficult to perform the layout work required for the project.

PLANNING

It is stated throughout this manual that planning is the most important of all construction activities. Without planning, conflicts will occur ten times, or probably one hundred times, more than they already do. Your success as a field engineer is directly related to your ability to plan your daily and weekly work activities. Your plan will be dependent on the plans of others. You will have to plan to plan. That seems repetitious, but it is true. In addition to planning your work activities, you will also have to plan your equipment, both personal and for the job. Checklists are included in this section to assist in your planning.

PERSONAL EQUIPMENT

"Be Prepared!" That may be the Boy Scout motto, but it couldn't be stated any better for field engineering. Nothing is worse for a field engineer performing a layout in front of a crew of carpenters than to suddenly discover that he doesn't have the tool needed to get the job done. Don't be embarrassed, be prepared! The personal equipment required for field engineering varies with the type of work, the phase of the work, and the preferences of the individual. Standard equipment used by most field engineers includes:

Personal Equipment			
	Field Book (preferably bound)		Calculator for quick checks of pre-calculated layout data
	4H pencil or lead holder with 4H lead		Straight edge
	Small protractor (C-thru Ruler Model 3751 or equivalent)		Sighting target
	Plumb Bob (24oz. minimum)		Marking pen
	Gammon reel		Several roles of plastic colored flagging
	Hand level		Chalk
	Carbide tipped etching tool		Marking crayons
	25-foot tape (preferably graduated in inches and tenths)		Hammer for driving nails
	6-foot engineer's rule		Orange marking paint
	Tack-ball for cupped hub tacks		Clear Coat (varnish) to preserve layout lines
	Red-eyes (nails stuck through a 1" square of plastic flagging to mark chaining points)		And, of course, a belt and some type of sheath or leather carpenter's apron

SURVEYING EQUIPMENT CHECKLIST

A field engineer should have a thorough understanding of the capabilities of the various surveying equipment available so that proper equipment is selected to perform the work. Information on the capabilities of various types of equipment can be obtained from the manufacturers. They will provide detailed specification sheets that can be used to analyze the capabilities of the equipment.

However, to get an unbiased evaluation of all of the equipment available, the field engineer should consult the various trade magazines available to the surveying profession. These magazines publish extensive comparisons of specific categories of equipment. Annually, comparisons of EDMs, total stations, theodolites, levels, data collectors, GPS units, etc., are published in these magazines. Contact a local surveyor to obtain the current names and addresses of magazines that prepare these comparisons.

EDM Checklist

Manufacturer/Distributor and Model	Company A	Company B
Range:Meters (Norm. Cond) To single prism		
Accuracy - MSE		
Mounting (Tilt Angle)		
Telescope (± Tilt)		
Pointing Aids		
Telescope (Power)		
Meter		
Measuring Times (Seconds)		
Single Reading Mode		
Display Type/Height of Numbers (mm)		
Ambient Temp Range		
Corrections Atmospheric - Steps (ppm)		
Slope Reduction Auto or Manual (accuracy)		
Accessories Battery/Type/Avg. # of Meas.		
Charger		
Hours to Recharge/Overcharge Protection		
Cables EDM to Battery (Spec.)		
Mounting Hardware		
PPM Correction Chart		
Carrying Case		
Other Features/Stake-Out Mode		
Weight (kg)/Instrument		
Size - H x W x D (mm)		
Warranty/Months		

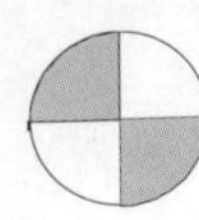

SURVEYING SOFTWARE AND COMPUTERS

More and more jobs are of the complexity that computers are required for the computation of layout data for the work. A short discussion of software and computer equipment is part of this section.

Many people buy computers and then think about software. That is just the opposite of what should be done. Software should be selected and then the computer which is needed to run it efficiently should be selected. Software is selected to do your tasks efficiently and easily. The computer is just the dumb electronic box that waits for instructions.

SURVEYING SOFTWARE

There is a plethora of surveying software on the market today. It will run on all types of computers, from the hand held, to the PC, to the mainframe. Your job should dictate the type of software needed. It would be nice if a specific software was available for construction, but there isn't one. Most surveying software is very broad in scope and tries to cover every calculation that could ever be encountered. When software is like this, it is usually expensive and is a memory hog on the computer.

Fortunately, some of the powerful software comes in modules that will allow you to install only what you need at the time, with the option of upgrading to more modules at a future date. This is the type of program that should be considered. Again, P.O.B. reviews surveying software annually. Their latest comparison of surveying software listed programs from more than 100 companies.

COMPUTERS, PRINTERS, AND PLOTTERS

With the ever evolving technological advances in computers, it is not possible to recommend specific computer hardware. What is written today is obsolete tomorrow. Just a few general comments about selection philosophy are possible.

Buy a computer that has the capability to run the software that is planned for use throughout the project. If a specific surveying program is being considered, call up the author and ask if upgrades to the software are going to require more powerful hardware. If it still isn't possible to project what will be needed, with the prices of computers today, it makes the most sense to buy the most powerful computer that the budget can afford. Clones are often just as dependable as big name brands, sometimes even more so. Check with others in the company for acceptable products.

Select a printer based on the software and its expected use. In surveying, a graphics printer is necessary. Today, most printers can emulate the more popular printers by doing graphics. Some printers can understand plotter language and can plot in small scale the images that the software sends to plotters.

In companies where funds for plotters are scarce, a printer which can plot may be a good compromise.

DAILY SCHEDULING

One of the most difficult activities of the field engineer will be time management. Everyone on the jobsite will want field engineering activities—NOW! A field engineer who is continually putting out fires by trying to respond in an instant to someone's perceived needs will be run ragged. It is better to take control of your time and plan. Daily schedules are essential to a field engineer's organization of time. Using a daily schedule to communicate your plan to others on the jobsite will make your time as a field engineer much easier and more enjoyable. The field engineer must plan work to allow for unforcseen problems which usually seem to occur. Time must be budgeted daily for activities such as layout, record keeping, subcontractor meetings, staff meetings, and inspections. The field engineer must plan continuously to stay ahead of the work requirements of others on the jobsite..

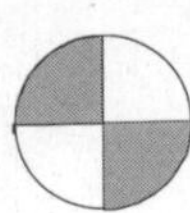

SAMPLE DAILY SCHEDULE

The bar chart schedule below shows the daily activities of a field engineer for three days. The planned sequence of activities and their duration is indicated by the length of the bar. These same activities are repeated every fourth day.

PROJECT DALLAS COUNTY COURT
JOB NO 78777 AREA DALLAS DIV
DATE 08-03-87 PREPARED BY HH

DAY 1
DAY 2
DAY 3

Exterior Grade
Control Lines
Level Core to Grade
Set Lasers
Locate Int. Control
CK Embeds Blockouts
Keyway Inserts Etc.
Preliminary Wall Loc.
Chain & String

Set Lasers
Loc/Alighn Prev. Level
Loc. embeds Next Lvl
Set Int Wall
Final Check Keyway
Embeds & Inserts

Chck Steel Clrnce.
Chck Untyp. Details
Set Lasers
Lk End Dimensions
Final Int Wall Loc
Quick Embed Lk
Check Int Beams
Check Ext. Control
String Ext Walls
Final Check

FIELDWORK BASICS

4

PART 1 - FIELD ENGINEERING BASICS

CHAPTER OBJECTIVES:

Describe the most important factor of being safe on the jobsite.
State the basic rule that should be followed when using cutting tools.
List the basic personal protective equipment of field engineers.
Discuss how to avoid natural hazards that field engineers encounter .
Distinguish between errors and mistakes.
Develop a list of 10 mistakes or blunders that could occur in construction measurement.
Discuss which errors can be eliminated and which errors can only be reduced in size by refined techniques.
Differentiate between accuracy and precision related to construction layout.
State the basic rule of thumb regarding tolerances in construction measurement.
List and discuss the cardinal rules of field book use.
State why erasures are absolutely not allowed in a fieldbook.
Identify three methods of notekeeping.
State the simple rules regarding transporting an instrument in a vehicle.
List and discuss the general rules for instrument care and use.
Discuss special instrument storage precautions should be used when the climate is very cold or very hot?
Describe how to care for and maintain surveying tools.
Describe Parallax.
State the left thumb rule for leveling an instrument.
Describe the process that is used to check for proper adjustment of the level bubble.
Describe the process used to set up an instrument over a point with a plumb bob.
Describe the process used to "Quick Setup" an instrument over a point with an optical plummet.

INDEX

SAFETY FOR FIELD ENGINEERS

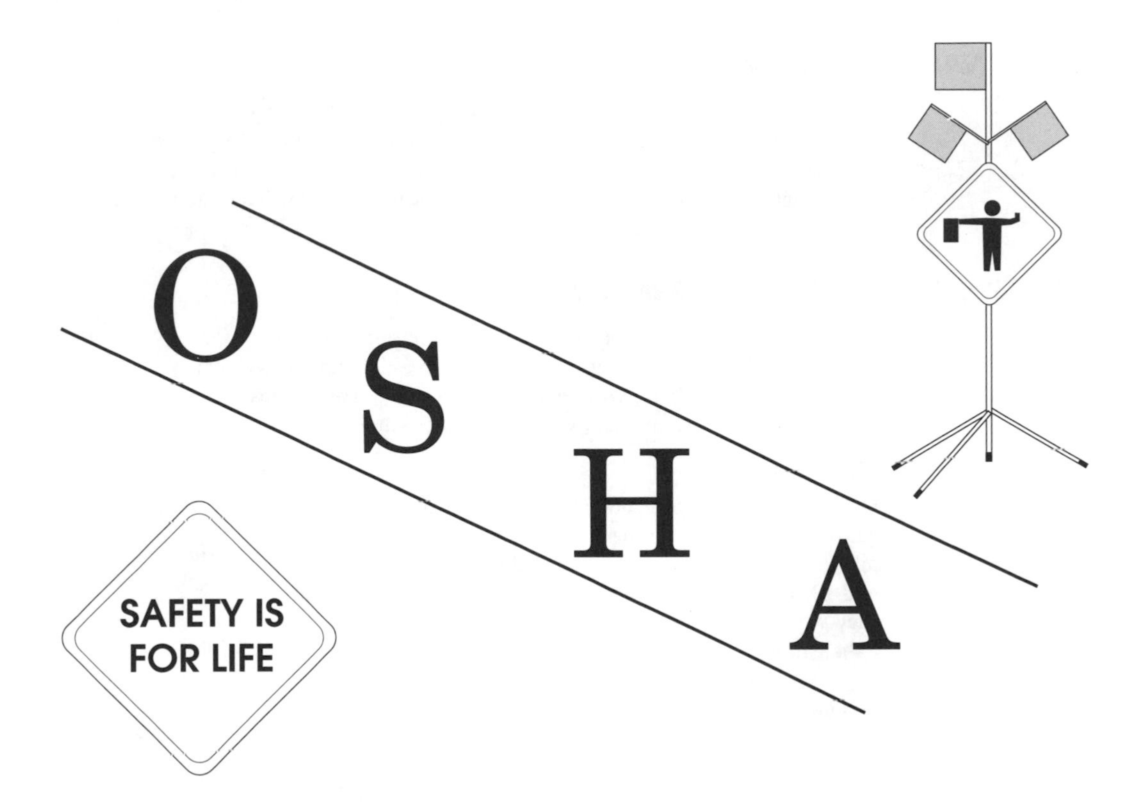

SCOPE

Safety is a common-sense topic which is often taken for granted by people who work in construction. Work-related injuries, including fatalities, occur 54% more often in construction than in any other industry. People die from accidents. If that doesn't make safety in this hazardous industry a serious matter, then consider that construction accidents cost the industry billions of dollars annually. This needless waste of manpower and money is a serious concern of management. This section cannot cover all aspects of safety that relate to construction surveying activities. However, what will be presented are those topics which will commonly be encountered and involve the greatest risk to the field engineer responsible for construction surveying and layout.

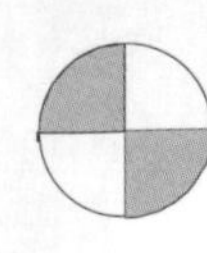

A RESPONSIBLE ATTITUDE

As a field engineer, it is your responsibility to observe standard safety practices and to cover these with your crew members and the craftspersons on the jobsite. Attitude is everything! Your attitude is very important when discussing safety details. If you have a good, positive attitude about safety practices, it is more likely that you will help others to become more concerned. Allowing an unsafe attitude to exist on company grounds will result in accidents. Numerous studies have proven this time and again.

PERSONAL PROTECTIVE EQUIPMENT

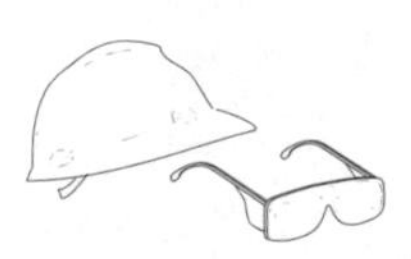

Safety equipment includes: hard hats, safety goggles, leather gloves, safety belts, lifelines, fluorescent safety vests, and ear protection. Some necessary traffic control devices are: traffic cones, flashers, signs, and barricades. The proper common equipment such as tool belt, hammer, tape, and walkie-talkie will also allow common tasks to be performed more safely.

DRESSING FOR THE ENVIRONMENT

As a field engineer, you must be concerned with more than just safety. When you work in the field, you will want to be comfortable, and the weather will play an important role in helping you decide what to wear. The greatest weather concerns are summer heat and sun, and winter cold and wind.

SUMMER

Comfort and safety go hand in hand when it comes to clothes. The proper attire for summer heat is light colored, light weight, and loose fitting clothing. You should wear a shirt, long pants, and a hat. The proper clothes will not only provide comfort; they will also protect you from sunburn, sun stroke, and heat exhaustion.

In addition to wearing the proper clothing in hot weather, it is important that you drink plenty of water. You should not, however, drink ice water in large quantities. Ice shocks your system and is not thirst quenching. You will receive the greatest refreshment from a cool glass of water rather than an iced one. Always remember to drink in moderation.

In most locations, hot weather lasts only half the year; therefore, as a field engineer you will also be concerned with dressing properly for winter weather.

WINTER

The proper clothing for winter is clothing which is warm and cuts the wind. It is best to dress in layers that fit loosely so your circulation is not restricted.

One of the major dangers of surveying in cold weather is the danger of frostbite. Frostbite is common in high wind and affects the extremities such as your nose, cheeks, ears, fingers, and toes. Proper clothing will protect you from frostbite and other winter hazards such as hypothermia. The wind, again, is a major hazard in cold weather.

HAND TOOL HAZARDS

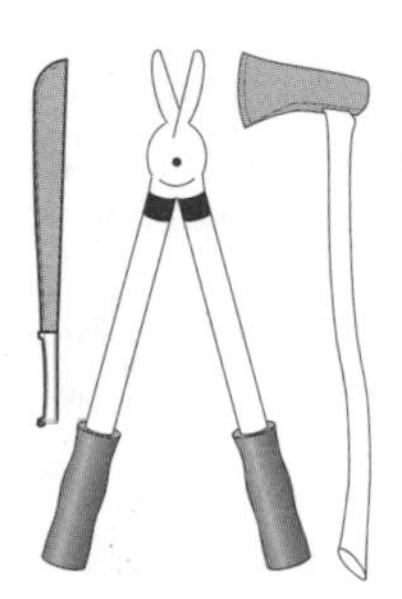

It will be your employer's responsibility to see that the surveying crew is provided with the proper safety equipment to carry out the assigned tasks.

The crew should be trained in the proper care, operation, and transportation of this safety equipment. With a trained crew and proper equipment, the field engineering activities that you will be performing will be both productive and safe.

At the start of a project, brush and tree clearing is the first activity where hazardous conditions may be encountered by the crew. Tools commonly used for clearing include a machete, ax, lopping pruner, or chain saw. Because of the nature of these tools and their ability to cause injury, care must be taken when using them.

The main rule to remember is that the safest cutting tool is a sharp one. The amount of effort required to use a sharp tool is far less than that required for a dull one. So, keep your tools sharp!

MACHETE

The machete is made for light brush cutting and should have a dull tip. When using a machete, you must always ensure that no one is closer to you than 10 feet. You must never use the tool while in a tree, and you must stay clear of all overhead obstructions which could cause the tool to be wrenched from your grasp.

CHAIN SAW

For heavier clearing jobs, a chain saw may be used. Some precautions that must be taken to avoid injury are the use of eye and ear protection, gloves, and snug-fitting clothes. Loose or baggy clothes present the possibility of becoming caught in the chain. When operating the chain saw, be certain you are on firm ground and that observers are well out of the way. Start the chain saw on the ground, and always use both hands when operating the saw.

You have now cleared a line so that surveying can begin. Notice that the area you have cleared has many stubs you might trip over, or worse yet, fall on. It is important that you walk carefully to avoid being injured.

NATURAL HAZARDS

In addition to weather and equipment hazards, surveying in the field will expose you to dangers presented by plants, animals, snakes, and insects.

PLANTS

Plants which you should identify and avoid in the field include poison ivy, poison oak, thorn bushes, and sticker patches. Again, the proper clothing will help protect you from the ill effects of accidental contact.

Clothes you should wear when plant hazards are present include heavy gloves, long sleeved shirts, and long pants which are bloused at the bottom below the top of your sock.

WILD ANIMALS

Not only are plants a concern in the field, but as a field engineer you may find yourself confronted with many different animals. In addition to common domestic dogs, it is not uncommon to encounter skunks, foxes, or bats.

When you are confronted by a dog, face it and stand still. Speak to it confidently, but do not try to pet it. Retreat slowly while facing the animal. Wild animals are often as frightened of you as you may be of them unless they are infected with rabies, in which case they are unpredictable and very dangerous.

If you are bitten by a wild animal try to kill it so it can be determined if rabies is present. A domestic animal should be captured so it can be observed for rabies.

SNAKES

The reptile world also presents hazards to the field engineer. Many species of poisonous and nonpoisonous snakes exist and may be encountered on the jobsite. It is the responsibility of the field engineer to be able to distinguish between the two types.

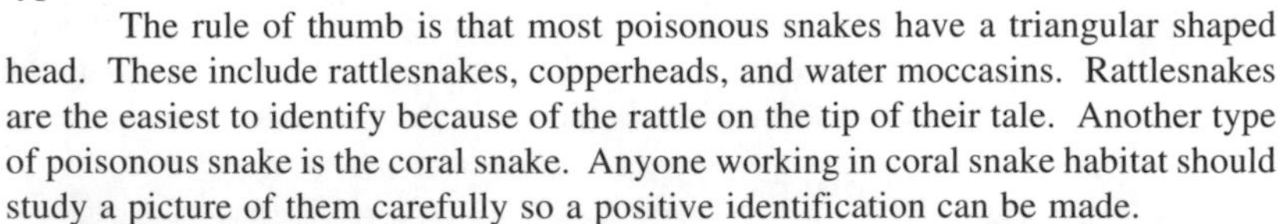

The rule of thumb is that most poisonous snakes have a triangular shaped head. These include rattlesnakes, copperheads, and water moccasins. Rattlesnakes are the easiest to identify because of the rattle on the tip of their tale. Another type of poisonous snake is the coral snake. Anyone working in coral snake habitat should study a picture of them carefully so a positive identification can be made.

It is wise when working in areas where snakes are known to exist to use all precautions to avoid their likely locations. They may be under rocks, in dense grass, around abandoned buildings, etc. If working near any of these, watch your step and hand placement. Wear knee high boots or leggings for protection. If driving stakes, carefully scan the area where you will be kneeling. Carry a stick and wave it around rocks, grass, etc., to see if anything moves or rattles.

If bitten by a snake, be calm and identify it as poisonous or non-poisonous. As calmly as possible, get to a hospital for anti-venom treatment. Excited movement accelerated the heart rate and spreads the venom more quickly throughout the body.

INSECTS

Another hazard in the field is insects. With no warning whatsoever, you may find yourself in a hornets' nest, so to speak. A few of the many insects to be aware of are chiggers, mosquitoes, spiders, ticks, wasps, and, of course, hornets.

More people die annually from bee stings than snake bites.

A good practice is to determine which insects are present in the area where you will be working so you can be prepared for them. Precautions you should take include proper clothing and insect repellent. The use of insect repellent is a good general practice when in the field. Insects such as chiggers cannot easily be seen and may become a nuisance only after they have embedded themselves in your pores. Mosquitoes can also be handled effectively with repellent, thus avoiding not only pain and discomfort, but also the diseases they may carry.

Be aware of those persons on the jobsite who may be allergic to bee stings. They must be injected with anti-sting medicine within minutes of the sting. Ask for anyone who reacts dangerously to bee stings to identify themselves. Make sure others on the job are aware and can administer the injection.

HIGHWAY HAZARDS

Now that many of the hazards of nature have been considered, you must be careful of man-made dangers such as those encountered when surveying on public roads. These include the danger of being run over, damaged equipment, or causing a motorist to have an accident. Precautions include signs, barriers, traffic cones, and orange vests.

It is your responsibility to see that men and equipment under your control are safe when working around traffic. Fluorescent vests will make crew members much more visible to the average driver.

The use of hand flags as well as warning signs will be very important if there is a need for a flagman. You should assign a flagman if work you are conducting interferes with the normal flow of traffic. It is the flagman's responsibility to direct traffic so both the crew and motorists are safe.

In addition to a flagman, signs may be needed if the surveying work will be conducted for extended periods of time or if the volume of traffic is high.

A sign is intended to forewarn approaching motorists. Therefore, the sign should be placed a safe distance from the point of work. Safe distances for such signs to be effective range from 150 to 600 feet. The safe distance a sign should be placed from the point of work is determined by the average stopping distance of a vehicle traveling at a given speed. For instance, a vehicle traveling at 20 miles per hour requires approximately 47 feet to stop and one traveling at 60 miles per hour requires 366 feet. It is, therefore, important to determine the speed limit of the road you are working on and review the stopping distances to properly locate your signs.

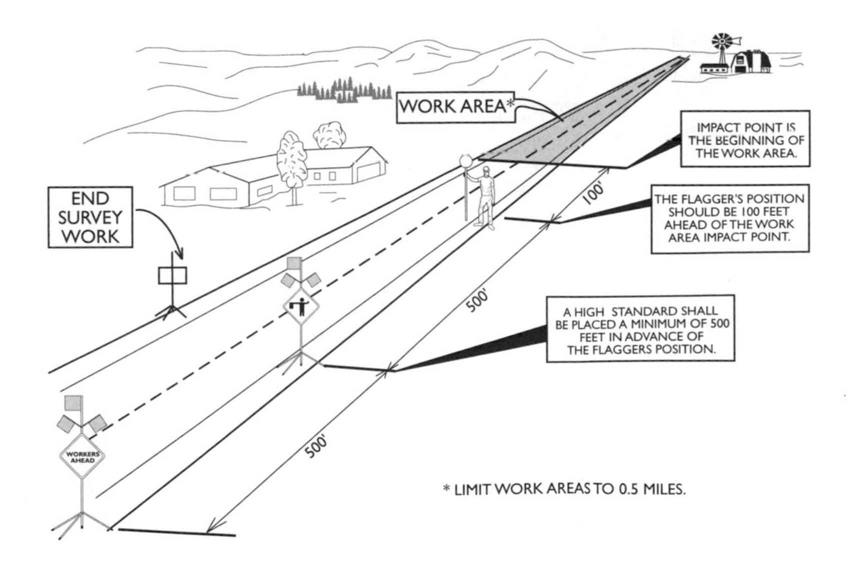

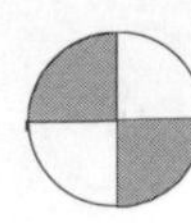

CONSTRUCTION SITE HAZARDS

As a field engineer, you will be working around more that just natural and highway hazards. There are also construction site hazards. A major part of your work will be on the construction site. You must remember that the construction site is continuously changing and is always hazardous. Some of these hazards include high voltage power lines, heavy equipment, holes, nails, stubs, materials, and trenches.

HIGH VOLTAGE

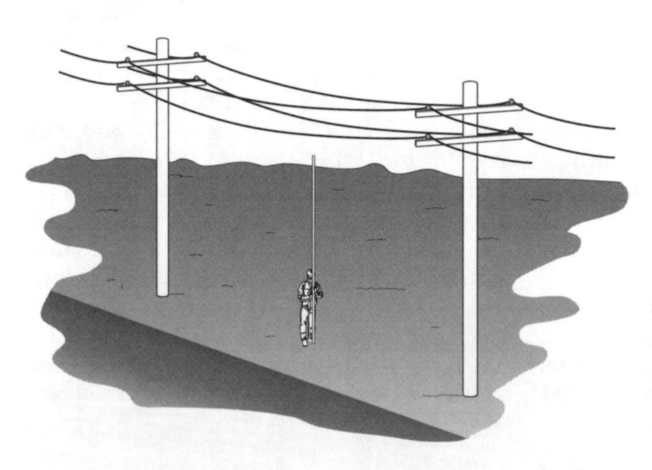

High voltage power lines are very dangerous. Electricity can arc to the ground if a rod is brought near the line. The use of rods, even wooden ones, is very dangerous, and you should only use one which is clean and dry. Wooden rods are the safest, but you must never bring the top of the rod nearer than 10 feet from any power line. Never touch any power line with any type of surveying rod. Unknowingly it may still conduct electricity.

If there is ever the need to measure the height of a power pole, you should use indirect methods using trigonometry. Measuring the height directly is very dangerous.

In addition to overhead power, you must beware of buried power lines when you are digging holes or driving pins. It is best to call your utility company to locate any suspected buried lines.

EQUIPMENT AND TRENCHING

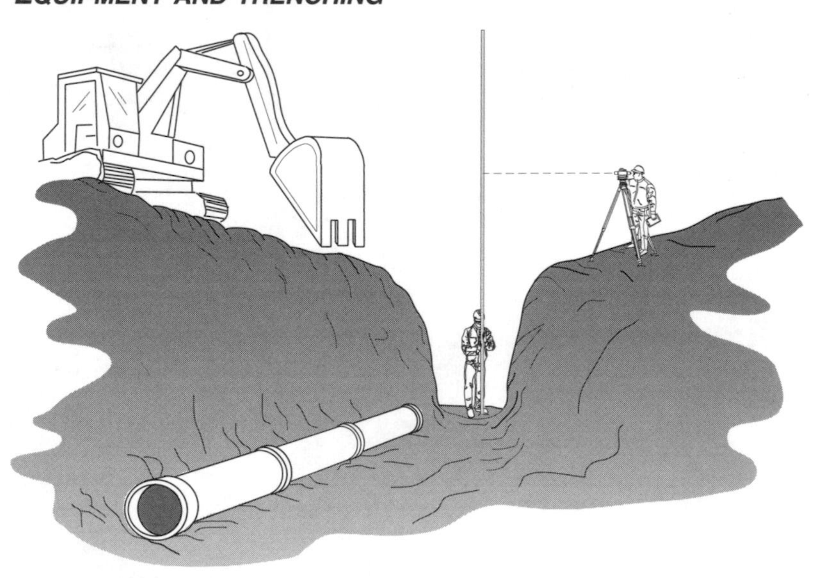

You must remember that heavy equipment presents one of the greatest dangers on the construction site. It is essential that equipment operators know when surveying work is taking place in their area. This is especially important when you are out of the direct view of the operator such as in a trench.

Working in and around trenches and holes is very dangerous. Not only are you generally out of an equipment operator's view, but the dangerous nature of trenches is also a major concern. Trenches must be properly shored or cut at the proper angle of repose. Holes or trenches in heavy work areas should be surrounded by a fence or guard rail. It is your option to refuse to enter an unsafe hole or trench. An unsafe situation, such as an unshored trench, should be reported to your supervisor immediately.

SAFETY STARTS WITH YOU!

Hazards that exist on a construction site are numerous and always present. We've looked at safety hazards brought about by natural causes, inclement weather, insects or animals, highways, and the ever-changing construction site itself.

As a field engineer, you will be at the first level of responsibility for safety. You must observe safe practices yourself and see that these same practices are followed by your peers and every craftsman and subcontractor under your control. Every construction operation can be conducted safely if the proper precautions are taken.

RESOURCES

Most companies are deeply committed to providing as safe a work environment as possible to their employees. If you have any questions about any aspect of safety, contact your safety director.

MEASUREMENT PRINCIPLES

CHECK, CHECK, CHECK - CHECK ALL DISTANCES TWICE - DIRECT AND REVERSE ANGLES - LONG BS AND SHORT FS - OBTAIN H.I. FROM 2 B.M.S

SCOPE

A field engineer must have the knowledge to perform standard surveying measurements on the jobsite. Standard surveying knowledge includes measuring distances, angles, and elevations, in addition to understanding mistakes and errors, preparing proper notes, and caring for equipment. However, everyone does not have the same basic surveying knowledge. This section is intended to point out the important aspects of the basics of construction surveying so everyone will have the same fundamental knowledge. It is important to remember that you won't know everything about surveying measurement. In the event that you don't know something, ask someone who does to obtain help!

The bottom line for any field engineer is eliminating mistakes and reducing the size of errors in measurement. This can only be attained thought a thorough understanding of measurement principles, and by being driven to check measurements an INFINITE number of times to ensure correctness.

TOTAL HONESTY REQUIRED!

Surveying measurement requires total honesty in the measurements that are made. Report what you measure, not what you have calculated that you should measure, or what you think someone wants to hear. There should be no cover-up of mistakes. Covering up a mistake almost always results in impacting some other aspect of the work. All erroneous measurements will be discovered eventually. Later, when the mistake is discovered, it will cost more to correct.

If your equipment is faulty, report it. Don't try to use equipment that is suspected of not being able to measure within tolerances. Repairing or calibrating an instrument is cheaper than taking out a column or footing. Reference *Chapter 12, Equipment Calibration.*

Just as in all other aspects of construction, one of the most important activities of the field engineer is to be a good communicator. Reference *Chapter 2, Commucication and Field Engineering.* This involves listening to the foremen and craftspersons to be able to perform the layout they want and communicating the work performed to the project engineer and superintendent. You should advise everyone of the meaning of information written on stakes or other survey marks so the correct work is performed.

MISTAKES

Mistakes and errors always exist in surveying measurement. A field engineer must be able to distinguish between the two. Errors are something we must put up with, but we can't live with mistakes.

Mistakes (also called blunders) are large. They occur because of carelessness, lack of understanding of the plans, lack of knowledge about measurement techniques, and lack of knowledge about measuring equipment. Examples are: dropping a hundred feet when measuring a large distance with a chain, forgetting that a foot was cut when using the chain, reading a rod one foot off, setting an instrument one degree off, etc. They are large, but if proper measurement techniques are used, they will be discovered before they can do any harm.

Mistakes are simply too large. Always remember to Check, Check, Check, and ReCheck your work. Measure all distances twice, turn all angles direct and reverse, check into two benchmarks, calculate all coordinates twice by different methods, etc.

Do whatever it takes to remember that mistakes/blunders are only discovered by checking the work performed! Mistakes can be avoided if proper procedures are used.

ERRORS

ERRORS DIFFER FROM MISTAKES

Errors will always exist in every measurement that is made. Always! They also differ from mistakes in size. Where mistakes are large, errors are typically very small. They are due primarily to human limitations or slight imperfections in the manufacture of surveying instruments. Human limitations are eyesight, sense of feel or touch, physical ability, dexterity, overall mechanical talent, etc. Typical imperfections in instruments are a chain length varying slightly from the marked length, the line of sight not being horizontal on a level, and perpendicular relationships not existing within an angle turning instrument, to name a few.

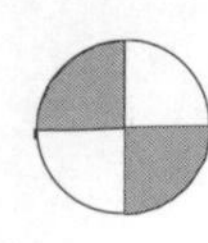

HUMAN OR PERSONAL ERRORS

Human errors are typically small in size and follow the laws of probability. That is, they are random and will be equally high or equally low. They cannot be eliminated. They always exist. If a distance is measured 100 times, half of the measurements will fall short of the true value and half will be more than the true value. Therefore, they tend to cancel each other out if measurements are repeated. We don't like to live with errors, but we must! We can minimize the effect of personal errors if proper measurement procedures are used. They can be reduced in size, but never eliminated. Reduction in the size of personal errors occurs because better equipment is available, and the measurement techniques are honed and refined to near perfection.

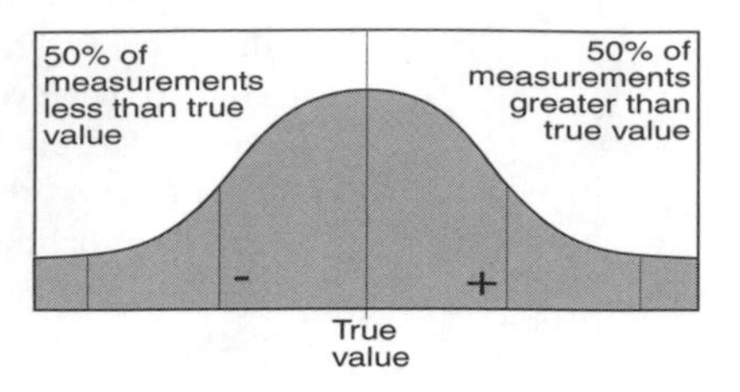

INSTRUMENTAL ERRORS

Instrumental errors are systematic in nature. That is, they follow the laws of physics, and *they can be eliminated by mathematical formula*, or by following instrumental procedures which cancel their effect.

The tape is an excellent example of how the laws of physics apply to correction calculations. The length of the tape is established under controlled environmental conditions in the factory. As soon as the tape leaves the factory and is no longer at 68 degrees Fahrenheit, the tape is no longer the length it was at the time of manufacture. Temperatures above 68 degrees cause it to expand in length, and temperatures below 68 degrees cause it to contract. Since the coefficient of expansion of the steel in tapes is known, and the change in temperature is recorded, a simple formula can be written to eliminate the effect varying temperature has on distance measurement. The same analysis holds true for other conditions that might affect the tape. A mathematical formula can be written to eliminate them.

A transit is a perfect example of how proper instrumental procedures will eliminate errors. Errors in the axis of the instrument will occur systematically in normal use. That is, they will be the same every time an angle is measured. If the procedure of measuring angles first with the scope in the direct position and then in the reverse position is followed, any error will occur one way the first time and the opposite way the second time. Therefore, by splitting the difference and eliminating the error all together, the correct angle is found. This topic will be discussed further in *Chapter 8, Angle Measurement,: Direct and Reverse Angles by Repetition* (p 8-18) and *Chapter 12, Equipment Calibration, Field Testing the Transit and Theodolite* (p 12-12).

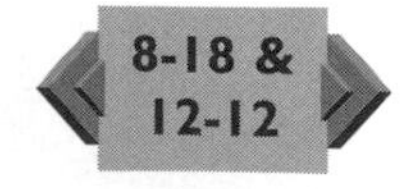

In summary, mistakes must be eliminated by doing the work twice. Human errors always exist, but they can be reduced in size by proper techniques. Instrumental errors can be eliminated by mathematical formulas or proper procedures. Field engineers must do all they can to eliminate or reduce in size all mistakes or errors that occur so control will be good, structures will be properly located, and the work on the project will be level, square, and plumb.

ACCURACY VS. PRECISION

Some people think that accuracy and precision are one and the same. That is far from true. In measurements you can have accuracy without precision, or, you can have precision without accuracy.

ACCURACY

Accuracy is best defined as being able to get the true value with the measurements taken. For instance, if a distance is measured ten times, the recorded values may not be very close to each other. However, out of the ten measurements, you would get very close to the true value for the measurement by averaging the ten measurements. Some will be less than the true value and some will be more, but the true value will be reasonably known.

PRECISION

Precision on the other hand, is the closeness of measurements to each other. That is, all of the measurements will almost be the same. There won't be a large spread. But even though there is a closeness to each other, that doesn't guarantee accuracy. For example, a tape that is 1 foot short can still obtain measurements close to each other, but the results will all be one foot off of the true value!

TARGET A

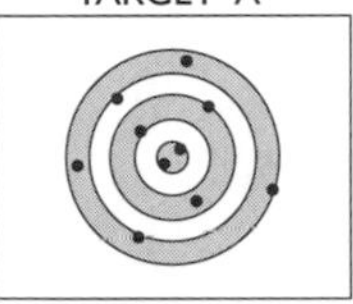

A common analogy of accuracy and precision is the archer shooting at a target. With an old bow and arrows, 10 shots are taken at a target. This old bow is shot using instinct, no sights are available. Some hit the center and the rest are all over the target. But the archer did hit the bull's-eye. So although not precise, there is some accuracy in the shots, because the true value (the bull's-eye) was hit. See target A.

The archer then shoots 10 arrows with a modern bow using special sighting equipment. The shots are all very close to each other. There is great **precision**, but no accuracy since the bull's-eye was not hit. See Target B.

TARGET B

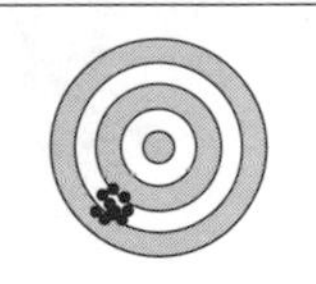

TARGET C

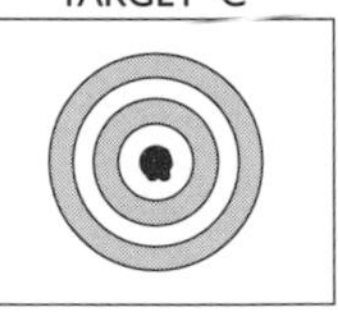

Of course, an adjustment could be made to the modern sighting equipment so that the bull's-eye would be hit practically every time **(accuracy)**, and the **precision** is maintained. See Target C.

In Construction layout, we want target C! We want to achieve both accuracy and precision in our measurements. Accuracy—we want to obtain the true value. Precision—we want our measurements to be close to each other to indicate that good techniques are being followed. In simple terms, we want to get close to the true value consistently. We want to hit the bull's-eye every time we measure!

TOLERANCES FOR CONSTRUCTION LAYOUT

±1' ±NONE ±1/4"

±0.00'

±0.01'

±0.005' ±1"

±

±0.1' ±10'

SCOPE

Tolerance in construction layout can be defined as how far off of the design location something can be built and still be accepted by the owner. If the specifications say that all anchor bolts must be located to within +0.01 feet of the design location, it means the anchor bolts can vary one hundredth either way from the exact design location. In another example, when setting rough grade on a parking area, the specifications might indicate that grade cannot vary more than one tenth from the design elevation. If the elevation at a specific spot is to be 545.2 and you check it to be 545.0, you know that it doesn't meet the tolerance required, and more dirt will have to be added.

GENERAL

Tolerances are necessary to control the quality of the work that is placed. If tolerances were very large, after awhile it would be difficult to fit the various components of the structure together. When tolerances are tight, it takes extra care in all phases of the construction process to make sure the specified tolerances are met and that the components fit easily.

For a field engineer, it is a good practice to always measure as if tolerances are going to be tight. That way, if slight deviations in distances, angles, or elevations occur, tolerances can still be met. However, always trying to measure as precisely as possible can be carried to an extreme. If the superintendent says to put in some stakes to locate a temporary parking lot, and the field engineer is measuring to the nearest hundredth when the nearest foot would do, then time is being wasted. Common sense must prevail, and the field engineer should use the method of measuring that quickly gets to the nearest correct measurement.

RULE OF THUMB

A rule of thumb in measuring is that you should always measure to one-half of what is coming behind you. That is, if the carpenters are going to locate forms to the nearest half-inch, then the field engineer should locate the layout lines for the forms to half of that, or to the nearest quarter. If an operator can excavate to the nearest tenth, then the grade stakes should be placed to the nearest half tenth.

COMMON TOLERANCES FOR CONSTRUCTION LAYOUT

Specified tolerances vary from job to job so the field engineer should read each specific specification to determine what is allowed. Common tolerances for layout activities are listed on the next page.

Notice that the tolerances vary from zero to a foot. If the allowable tolerance isn't known, it is the responsibility of the field engineer to ask the superintendent what it is.

These represent just a small sample of the various layout activities that will be encountered on the jobsite. They may not be the tolerances allowed on a specific project. Check specific project documentation for tolerances.

REFERENCES

American Institute of Steel Construction Green Manual

American Concrete Institute ACI-318

Project Specifications

Common Tolerances for Construction Layout	
Layout Activity	***Tolerance***
Rough grade	Tenth
Blue tops	Two Hundredths
Slope stakes	Tenth
Primary control	Zero
Secondary control	Hundredth
Anchor bolts	Hundredth
Locating the office trailer	Foot
Aligning a wall	Hundredth
Plumbing a wall	Hundredth in ten feet
Plumbing a steel structure	1/4" in ten feet
Radius points for sidewalk curb	Hundredth
Interior partitions	1/2"
Fine grade	1/4"
Locating rebar	Five Hundredths
Locating a pipe sleeve	Hundredth
Measuring a crane pick	Foot

FIELD NOTEKEEPING PRACTICES

SCOPE

You must leave evidence of your work on the ground from which others will build and on paper for your reference and defense. Many field engineers do an excellent job of laying out the points needed for the project, but they are negligent in keeping good notes. Most of the time, lack of good notes may not hurt them; if what they do is never questioned, a paper record isn't needed. But, it only takes one question about the correctness and integrity of the work performed to cause a field engineer to realize the fallacy of poor notekeeping. It only takes one question in a court of law to realize that notes should be kept and should never be altered in any way.

BASIC PRACTICES

Nearly all engineering surveys require the recording of observed data. Other than doing the actual work, keeping field notes is the most important activity of the field engineer. It can't be stressed enough that it is a necessary part of the permanent record of the project to record work and layout activities as they are completed.

RECORD DATA EXACTLY

No fudging in the field book. The field book must be honest. It must be admissible in a court of law. Any erasures or other questionable recording of the data measured or work performed may result in the entire book being thrown out of court.

KEEP FIELD BOOK SAFE

Imagine the value of the field book in work-hours if each page represents one day of work. The value would be in the thousands of dollars. Be sure to keep it safe when on the jobsite and lock it up in a fireproof safe at night.

LEAVE NO ROOM FOR INTERPRETATION

Perfect legibility must be preserved. Remember the words of the old professor who admonished his students: "Always put your notes down as if you expected to die before morning and wanted to leave them in such good condition that in ten year's time, a stranger, with none of the old party to help him, could take your book and proceed on the job without delay."

MAKE REFERENCES IN FIELD NOTES

When material has been copied from another source, refer to the book and page numbers of the original. All data should be recorded in the field book at the time of observation and never left to memory or recorded on a loose piece of paper. The information should be complete enough to answer any questions.

USE A 4H PENCIL

To ensure permanence, a hard pencil is necessary. A soft pencil line will smudge and will not be permanently impressed in the paper. A soft pencil will require more sharpening than a hard pencil. The field notebook must be completed with all information before the surveyor leaves the field. The items listed below are necessary to ensure an accurate and complete record of survey data, as well as to ensure efficient handling of the notebook by other people.

CARDINAL RULES OF FIELD BOOK USE

IDENTIFY WHO OWNS THE FIELD BOOK

Write the name, address, and phone number of the company, the field engineers' names, and any other identifying information on the outside or inside cover. Hopefully, this information will get the field book back to you in the event that it is lost, dropped, or otherwise misplaced.

ESTABLISH STANDARD NOTE FORMS FOR THE PROJECT

All crews should be advised to record data using similar methods. This will help to ensure that field books can be exchanged between crews and easily understood.

PROVIDE PAGES FOR A TABLE OF CONTENTS

Nothing is more frustrating than having to page through a book looking for specific information. It only takes a minute to write down descriptive information in a table of contents. The first three pages of the field notebook shall contain the title page and the table of contents The table of contents should follow chronological sequence, containing the number and title of each problem, and the page number indicating where each problem can be found. Number only the right-hand pages in the upper right-hand corner of the page.

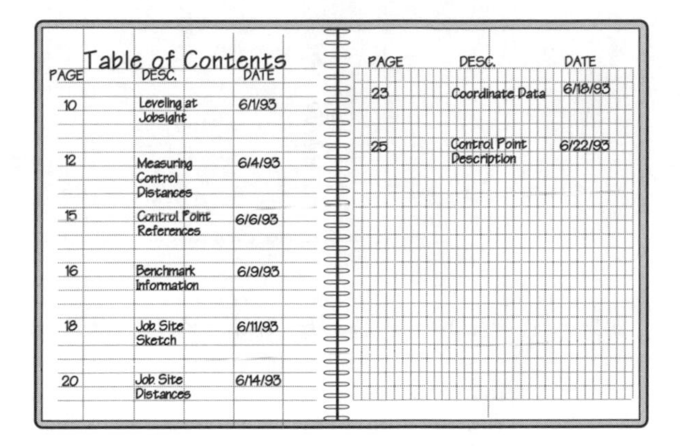

PROVIDE PAGES FOR A LEGEND OF SYMBOLS

With limited space on a field book page, it is advisable to use standard symbols or to develop your own symbols to identify objects on the sketch. Having a central location for the symbols facilitates their use and understanding.

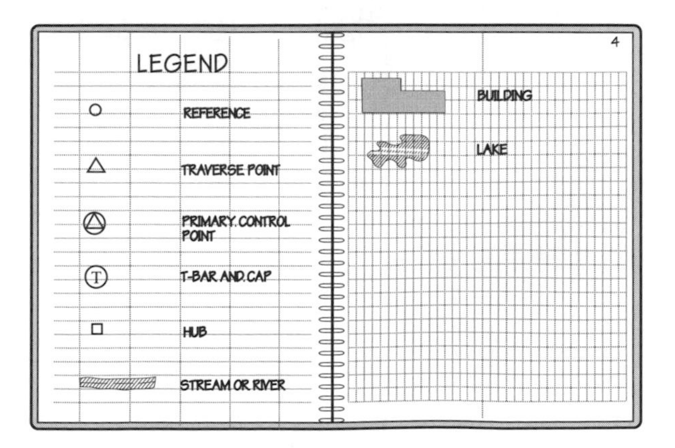

ALWAYS PROVIDE A NORTH ARROW

A North arrow should accompany all sketches for orientation to the worksite.

SKETCHES SHOULD BE USED FREELY

A separate sketch of small details, drawn to a larger scale, will sometimes facilitate the interpretation of the notes. It is not recommended that sketches be drawn to precise scale in the field book; it consumes too much time. Drawing to scale is better left to the drafting room. Make your sketches as large as the space will allow. A north arrow should accompany each sketch.

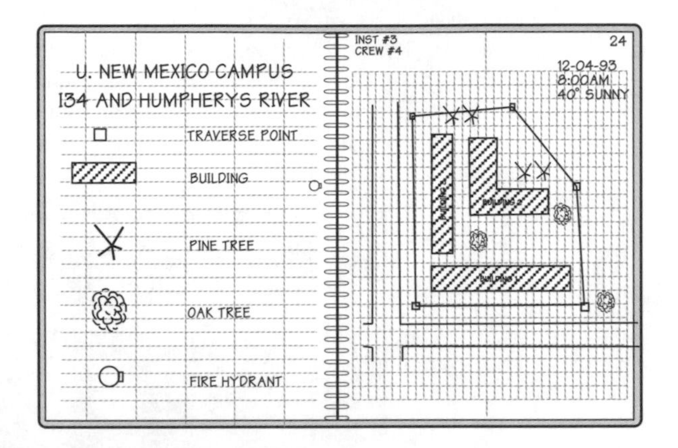

PROVIDE DATE, TIME, AND WEATHER

Recording this information should be first, before any data is recorded. The date and time is necessary to establish when the work occurred. If ever questioned about the work in a court of law, this information will be of utmost importance to back up your testimony. Recording the weather information helps to identify any effect the weather may have had on the measurements. This should be listed at the upper right-hand corner of the first right-hand page used for the work activity.

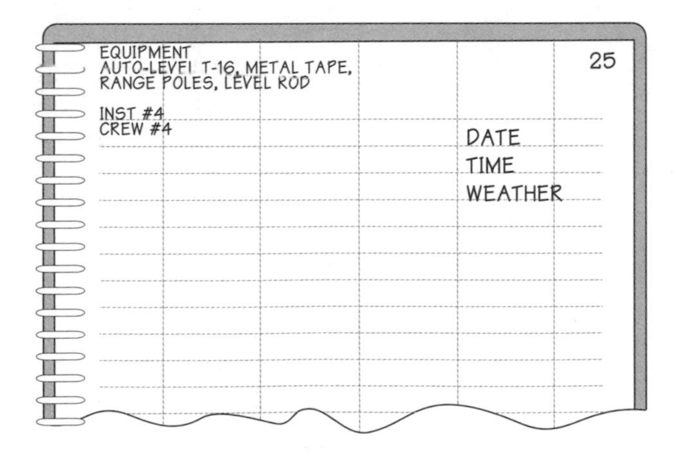

PROVIDE IDENTIFICATION INFORMATION

The names of persons performing the work are used to identify all who could help to clarify what work was done. Identifying the equipment used can be beneficial in isolating problems that have occurred in the measurements. If a particular instrument was found to be out of adjustment, all work done with it may need to be checked. Recording what instrument was used on a particular day is very important. The number and title of each problem should be indicated at the top of the first left-hand page listed for the problem.

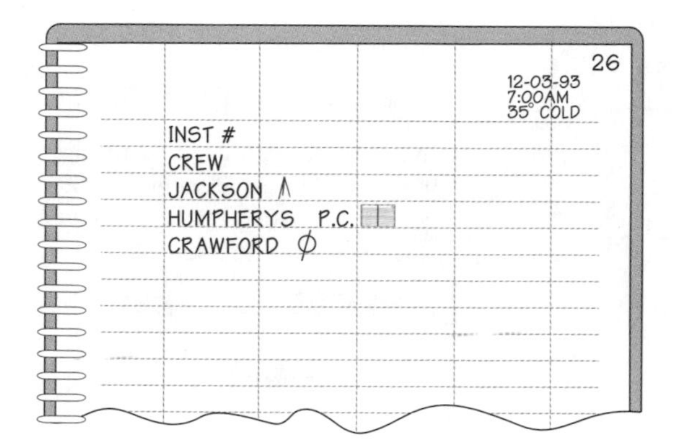

DO NOT ECONOMIZE ON PAPER BY CROWDING THE DATA

Yes, it is important to conserve and recycle paper, however, not at the expense of making it difficult to record all necessary information about the fieldwork. Use several pages if necessary for each project.

ERASING

Do Not Erase!! It cannot be stressed enough that field books that contain erasures are suspected of alterations, forgery, etc. They will be thrown aside if there is any question about the authenticity of the work. DO NOT ERASE!!! Draw a single line through a mis-entry and write the correct data above.

RECORD EVERYTHING

If in doubt, do it! If in doubt, record it! Leave nothing for possible misinterpretation. Record in your field book in such a way that anyone with the slightest knowledge of surveying could understand your work.

GUIDELINES TO LEGIBLE NOTES

USE STANDARD LETTERING TECHNIQUES

It is recommended that all capital letters be used for clarity and consistency.

Always PRINT 1/2 space high so that there is room to cross out mis-entries and write above them.

A hard pencil (4h) or harder should be used to avoid smudges and to ensure permanence.

Arrange columns so that all decimals are aligned.

In recording angles, it is customary to use two digits each in the minutes and seconds columns. For example, if the angle read was forty-two degrees, seven minutes, and thirty-one seconds, it would be recorded: 42° 07' 31".

CORRECT	INCORRECT
17 / ~~15~~	15
6.7	6.70
5.20	5.20
9.13	9.13
42 °07' 31"	42 °7' 31"

USE STANDARD DRAFTING TECHNIQUES

Use standard symbols for clarity and understanding. Use a straight edge on sketches for neatness. Carry a small protractor to use in drawing angles. Try to make sketches proportionally correct, if possible.

METHODS OF NOTEKEEPING

There are several methods of keeping notes. They include sketches, tabulations of data, descriptions, and a combination of those mentioned. A sample notebook containing each method is shown on the following pages.

SAMPLE SKETCH

A picture is worth a thousand words is very true when it comes to notekeeping. Some information simply cannot be listed or described adequately. Sketches are the only way to convey the information. Because of this, sketches must be clear and completely capable of standing on their own in most instances.

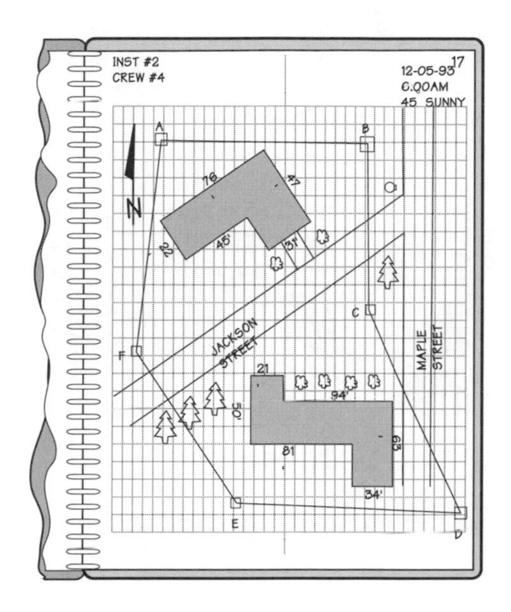

SAMPLE TABULATION

POINT	BS	HI	FS	SS	ELEVATION
		BM. A TO. TBM. 1			
BM. A					5280.00
INST. 1	6.77	5286.77			
TP. 1			4.23		5282.54
INST. 2	7.45	5289.99			
TP. 2			5.12		5284.87
INST. 3	7.07	5291.94			
TBM 1			3.48		5288.46

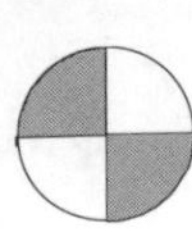

SAMPLE DESCRIPTION

INST #5
CREW #4

19

11-23-93
7:30AM
55° RAINING

DESCRIPTION

POINT MAPLE IS LOCATED 1.3 MILES DUE EAST OF THE 400 MILE MARKER ON JACKSON STREET. BEGINNING AT THE AIRPORT, TRAVEL 2 MILES ON 134 TO THE SCHOOL. IMMEDIATELY FOLLOWING THE SCHOOL IS WILLIAMSON BLVD. TURN RIGHT ONTO THE BLVD AND CONTINUE FOR 3.5 MILES. UPON REACHING A STOP SIGN AT 3.5 MILES THE INTERSECTION WILL BE THAT OF WILLIAMSON AND JACKSON STREET. TURN RIGHT CONTINUE FOR 2.6 MILES.

NOTES: LARGE DOG AT LOCATION

SAMPLE COMBINATION

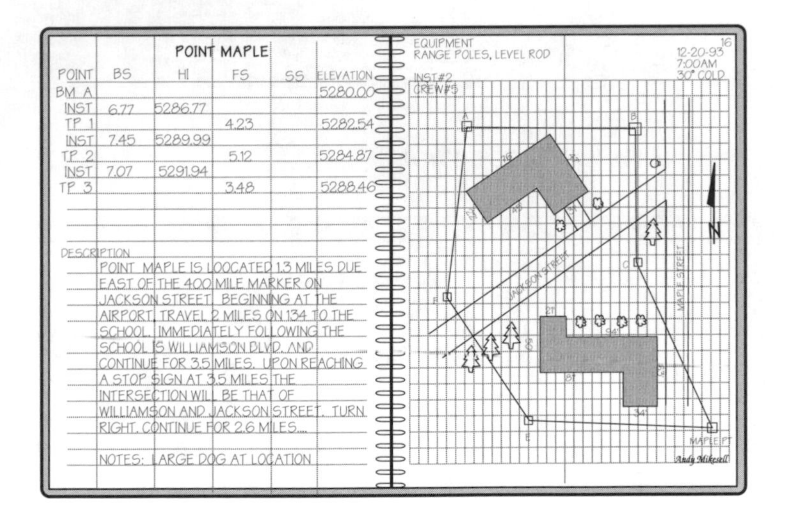

POINT MAPLE

POINT	BS	HI	FS	SS	ELEVATION
BM A					5280.00
INST	6.77	5286.77			
TP 1			4.23		5282.54
INST	7.45	5289.99			
TP 2			5.12		5284.87
INST	7.07	5291.94			
TP 3			3.48		5288.46

DESCRIPTION

POINT MAPLE IS LOOCATED 1.3 MILES DUE EAST OF THE 400 MILE MARKER ON JACKSON STREET. BEGINNING AT THE AIRPORT, TRAVEL 2 MILES ON 134 TO THE SCHOOL. IMMEDIATELY FOLLOWING THE SCHOOL IS WILLIAMSON BLVD. AND CONTINUE FOR 3.5 MILES. UPON REACHING A STOP SIGN AT 3.5 MILES THE INTERSECTION WILL BE THAT OF WILLIAMSON AND JACKSON STREET. TURN RIGHT. CONTINUE FOR 2.6 MILES...

NOTES: LARGE DOG AT LOCATION

NOTEKEEPING CHECKS OF MATH

Whenever possible, the measurements on the jobsite should be performed so that mathematical checks can be made on the work completed. Examples are: summing the interior angles of a traverse to meet the (n-2)180 geometric requirement; providing an arithmetic check on differential leveling notes; averaging distances; etc. Always perform all possible checks before leaving the field so that data can be re-measured at that time.

GENERAL INSTRUMENT CARE

SCOPE

The old adage "cleanliness is next to godliness" very much applies to field engineering. "Instruments should be used, but not abused" is another saying that applies to how field engineers should treat surveying equipment. "Good work can only come from good, clean, well-cared-for equipment" also applies.

How equipment is being kept is something that superintendents and managers look at in their evaluation of the performance of a field engineer. Dirty, poorly maintained equipment is an indicator of a field engineer who doesn't care or is lackadaisical. Having a person with that attitude in the field engineering position is not very comforting.

Although built to withstand much use, precision surveying equipment performs accurately only if it is well cared for. Field engineers should always take the time to clean and care for the equipment. It will pay off with good measurements and good performance. The following are tips on the use, care, and continuing everyday maintenance of surveying equipment.

SURVEYING INSTRUMENTS

Dumpy levels, automatic levels, transits, theodolites, EDM's, and total stations are surveying instruments that are used to measure angles, distances, or elevations. Although each instrument is different in the measurements it can obtain, the very basic parts of the instruments are the same. Each has a telescope that contains glass lenses; each has clamps and tangent screws; and each has leveling screws. Each of these parts is delicate and must be treated with great care while operating and handling.

A surveying instrument will perform properly only if it is in good condition, is properly calibrated, and is used according to procedures recommended by the manufacturer. Proper care and handling is easy to do, but it takes discipline on the part of the user.

Fundamentally, instruments are subjected to three situations which may expose them to hazards and require proper care. These are transportation, use, and storage.

TRANSPORTING SURVEYING INSTRUMENTS

Transporting typically involves carrying the surveying equipment from storage to the survey vehicle and then to the setup point. When carrying the equipment on foot, it is recommended that it be carried in its protective case. This is the safest method of insuring that it isn't damaged.

However, it is common practice to pick up the instrument and tripod and carry it over your shoulder. If any obstacles are close by, such as tree limbs, buildings, etc., this can be very dangerous to the instrument. For those times when it is necessary to carry an instrument while on the tripod, put one tripod leg in front and two to the back to keep the instrument upright and close to your body.

Although surveying equipment can generally be hand-carried from one location to another on a small jobsite, often it is necessary to transport it in a vehicle. This is one of the most common sources of damage to surveying instruments. To avoid problems, a few simple rules should be followed:

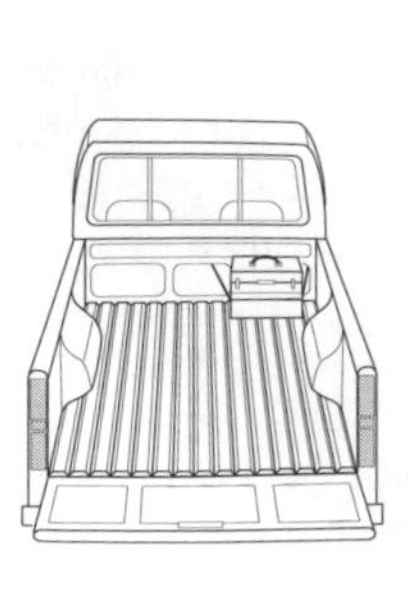

> Always place the instrument in its protective case and make sure it is securely closed before transporting it in a car or truck. Some manufacturers recommend tightening the clamps on the instrument, others recommend allowing them to be loose. Follow the manufacturer's exact instructions.

Place the case in a location in the vehicle that will prevent it from bouncing up and down or sliding back and forth. Wise field engineers build a storage box in the vehicle, line it with foam, and provide straps to secure the case in place.

Avoid holding the instrument on your lap. This could create a dangerous situation if an accident occurs.

Check the instrument manual for any special handling instructions.

CARRY THE INSTRUMENT IN ITS CARRYING CASE

Many maladjustments and damages are caused by transporting equipment which is mounted on tripods, particularly when transporting the equipment in vehicles and when walking for long distances.

KEEP THE CASE CLOSED WHEN NOT IN USE

Allowing an instrument case to be open while the instrument is in use on the tripod is a poor practice because the inside of the case will collect dust. Later, when the instrument is put in the case, it will be like a dust storm inside of the case while the instrument is being carried. Some of the dirt will eventually penetrate the bearing surfaces in the instrument, requiring an expensive cleaning and repair bill. Keep an empty case closed!

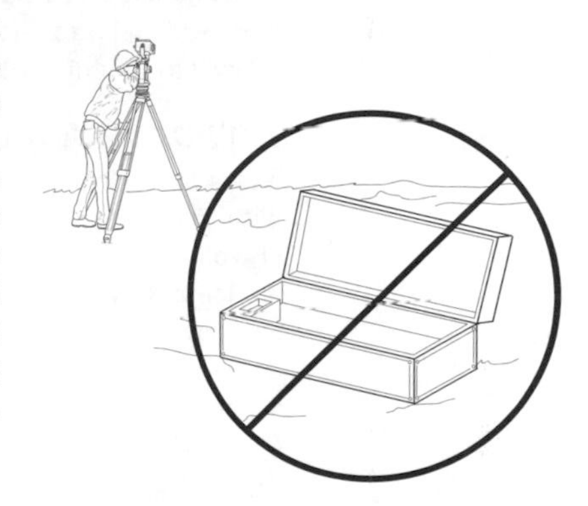

CARRYING THE INSTRUMENT

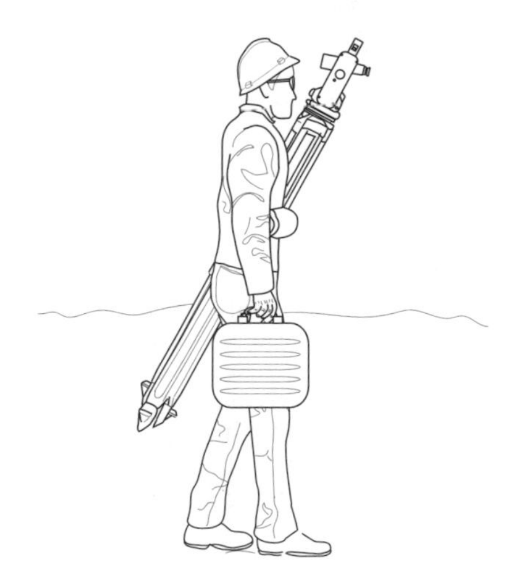

The instrument can be carried on the tripod for short distances, but carry the instrument in front of the body, under your arm with the legs of the tripod trailing behind in case there are tree limbs, brush, or other obstructions. If it is carried on the shoulders, care must be taken that the instrument does not strike any obstruction. The instrument should always be kept as close to vertical as possible (60 degrees or more) Clamp the motions lightly so the equipment does not rotate, but is loose enough to give.

GENERAL INSTRUMENT CARE AND USE

While in use, an instrument is exposed to the elements as well as to hazards encountered on the jobsite. The weather may range from the hot dry winds of the desert, to the humidity of a rain forest, or the frigid cold of winter. Jobsite hazards to instruments include exposure to workers and equipment, dirt and dust, and excessive vibrations. The user of the instrument must operate it properly, in addition to protecting it from weather conditions. Some common weather practices include:

ATTACH INSTRUMENT SNUGLY TO THE TRIPOD

Attach the instrument just enough so that it will not move, but not unnecessarily tight so that the tripod head or spring plate become warped.

GRASP THE INSTRUMENT FIRMLY

Grasp the instrument by the standards and tribrach when placing it on the tripod, not by the telescope or other parts. Secure it immediately to the tripod. Do not allow it to sit atop the tripod unclamped.

DO NOT TOUCH LENSES

Never touch the lenses of an instrument with your fingers or a common cloth. Use only lens cleaning tissues and fluids. All lenses have optical coatings on them and use of anything other than proper material to wipe them may result in scratches and damage to this soft optical coating.

Do not handle the equipment with muddy, excessively dirty, or greasy hands or gloves.

Accumulations of dirt, dust, and foreign material will eventually penetrate into the motions and cause sticking. When possible, use the plastic hood for protection against dust when working in such conditions.

TRY TO KEEP EQUIPMENT DRY

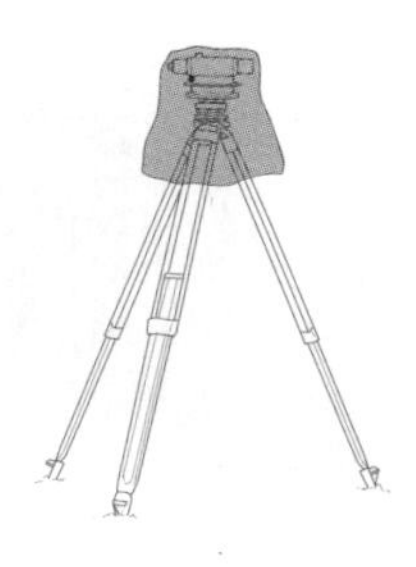

Given the weather conditions on many jobs, keeping the equipment dry is practically impossible. However, with careful attention to weather and jobsite conditions, the instrument can be protected when moisture is present. If caught in temporary showers where the instrument must be left setting up, cover it with the plastic hood provided. Wipe the tripod completely dry later, but allow the instrument to air dry by leaving the hood off of the case. Don't rub off the finish.

IF AN INSTRUMENT DOES GET WET

When an instrument is exposed to moisture or high humidity, care should be taken to insure that the instrument is allowed to dry properly. Do not store it in an airproof case overnight. The moisture will condense on the inside of the instrument. Leave the case open so air can circulate around the instrument.

NEVER LEAVE THE INSTRUMENT UNATTENDED

Do not leave the instrument unattended, either while set up during a survey or while it is in unlocked vehicles, etc.

ESTABLISH A WIDE FOUNDATION

Always spread the tripod legs at least 36 inches evenly apart, and plant tips firmly when parking the instrument, even temporarily. Never lean the tripod against vehicles, buildings, or trees, even when the instrument is unattached.

NEVER FORCE ANY MOTION ON AN INSTRUMENT

When using the motions of an instrument, the clamps should be just snug (not tight). Micrometer tight is appropriate. The idea is to clamp the motion just snug enough so that rotation about the axis will not occur during slight pressure.

NEVER FORCE THE INSTRUMENT

When rotating about the axis, be sure the clamp is loose. Rotate about the axis gently with fingertips, rather than forcefully with a grip. If screws operate too tightly, arrange for readjustment by qualified technicians. Never remove any plate or make any adjustment without understanding thoroughly what you are doing beforehand.

STORAGE

RETURN THE INSTRUMENT TO IT'S CASE

When placing the instrument back in its case, be sure illuminating mirrors are closed, place telescope with objective lens down, clamp motions lightly, and secure the base. When transporting the instrument in a vehicle for long trips, it is suggested that the instrument in its storage case be stored in its shipping case for added protection. Close the case when its not in use.

CLIMATIZE THE INSTRUMENT

When the instrument is to be in continual use outside in very hot or very cold weather, store the instrument nightly in its case outside so there is no significant instrument temperature change overnight. This will help to prevent condensation from occurring inside the instrument.

ELECTRONIC SURVEYING INSTRUMENTS

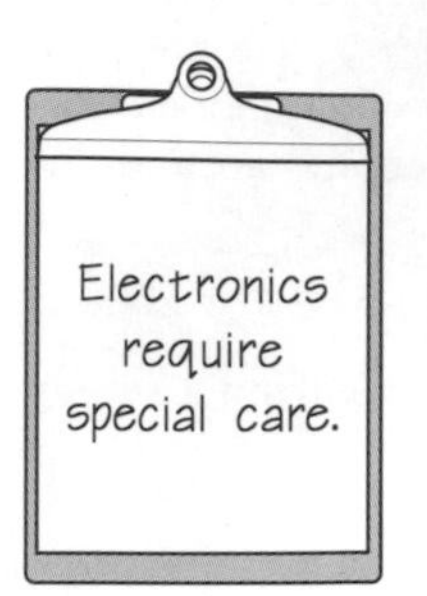

In addition to the mechanical parts and optics, electronic instruments contain transistors, wires, and computer chips. Because of the electronics involved, these devices require some special care to maintain their usefulness on the jobsite. Hazards include: humidity, rain, snow, cold, heat, vibrations, etc.

The comments made about the care and handling of optical instruments should be followed for electronic instruments, especially those about exposure to moisture. Electrical contacts are very susceptible to corrosion, especially from moisture that is encountered around sea water. Humidity along the sea contains salt elements and can penetrate an instrument. Electronic instruments being used around salt water areas should be taken to a local repair shop to be conditioned for the exposure that will be encountered.

TRIPODS

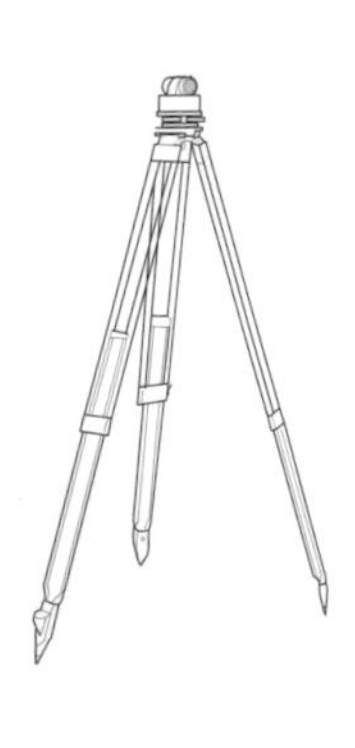

Just as a solid foundation is needed for houses and bridges, a tripod provides a solid foundation for instrument setups. Unfortunately, they are not generally well cared for. Like level rods, they are thrown into the backs of pickups, left lying on the ground, left out in the rain, seldom cleaned, etc. A foundation in poor shape will not support good work.

There are essentially two types of tripods: fixed leg and adjustable leg. The fixed leg is solid, and can be neither lengthened nor shortened to assist in the setup of the instrument. The adjustable leg tripod features legs that can be shortened or lengthened to make setups over a point easier. A clamp on the leg of the adjustable leg tripod is used by the operator to set the leg at the desired length. Regardless of the type, all tripods have similar parts. A tripod consists of a head for attaching the instrument, wooden or metal legs, and metal points with foot pads to help force the leg points into the ground.

The tripod is often overlooked in discussions of equipment. Unfortunately, when a discrepancy occurs in an observation, it is easy to assume that it is the fault of the operator, or the instrument, and never the tripod. However, its use and care may greatly affect the results of a survey. The following are rules for the care and handling of a tripod.

POSITION TRIPOD LEGS PROPERLY

Push the legs into the ground just far enough for a firm setup, but do not exercise unnecessary force. Always push directly longitudinal to the leg, and do not push the leg so that it flexes.

KEEP IT HORIZONTAL

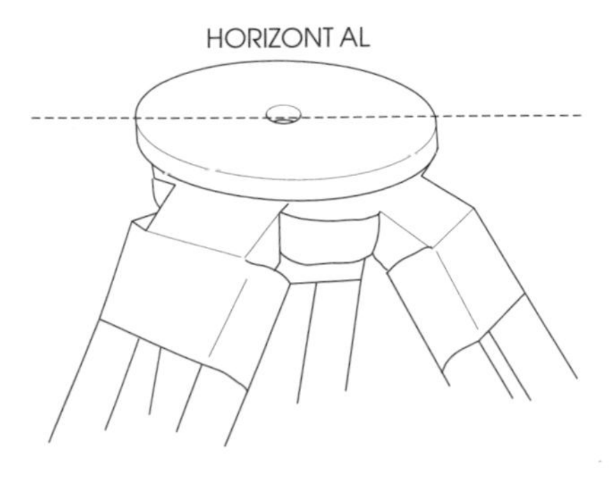

Place the head of the tripod in a horizontal position. Inexperienced users have a tendency to disregard the relative position of the tripod head when setting it up. That is, the head is not parallel to the horizontal reference plane. This causes the leveling screws in an instrument to be overextended, making it more difficult to level. If the instrument is not level, errors occur!

KEEP IT TIGHT

Firmly attach the tripod legs to the head. Regular use of the tripod may cause the connection of the legs to the head to become loose and wobbly. This will cause inconsistency in measurements. Taking a few minutes to tighten a wing nut or bolt will eliminate this potential error.

ENSURE A SOLID SETUP

NO

Firmly insert the tripod legs into the ground. This is the "root" of the solid foundation. Setting the tripod gently on the ground invites settlement. As the instrument settles, it can shift out of level, affecting the measurement. By firmly inserting the points of the tripod legs into the ground before measurements are taken, most settlement can be avoided. If the tripod is being set on a smooth floor, be sure to set it on something which will prevent the legs from slipping and possibly damaging the instrument.

SECURE ADJUSTABLE LEGS

YES

Be certain the clamp on the tripod adjustable legs holds them in position securely. Most tripods used in surveying have adjustable legs. It is often said, "A tripod with adjustable legs can be set over any point." For the most part, this is true. Adjustable legs have made difficult setups easier. One problem with adjustable legs is a tendency for them to slip if they are not securely clamped.

MAINTAIN PROPER TENSION

Keep tension on the legs so they will fall slowly or swing freely. If the legs are extended straight out from the tripod head, the hinge they pivot about should be tight enough to almost hold the leg in that position. The leg should fall or swing slowly to the ground. If the leg stays straight out from the tripod head, it is too tight and will wear excessively. It should be loosened and adjusted. Also, if the hinge where the leg is attached is loose and there is excessive movement, the hinge should be tightened. Some tripods require a special wrench to make this adjustment.

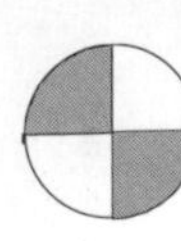

WHEN LUBRICATING TRIPOD, APPLY SILICONE LUBRICANT

Frequently lubricating joints will prevent excessive wear. Also, lubricate the legs of an adjustable leg tripod occasionally to keep them from sticking. A silicon-based lubricant is recommended because it will not attract dirt as do petroleum lubricants such as oil or grease. Clean the tripod after use Carry some cleaning cloths, a wire brush, etc., to get the dirt off of the tripod. Mud and dirt cling to the tripod leg points and should be wiped off between uses.

CHECK ALL SCREWS AND BOLTS

Loose screws are not good. The tripod will start to wobble when being used. Check them frequently to ensure rigidity. Even though most of the screws or bolts on the tripod are steel, the threads could be stripped if over tightened. If tightening is needed, use the proper tool.

DO NOT USE EQUIPMENT FOR UNAUTHORIZED PURPOSES

Do not use tripod for prying, vaulting, self support, hammering, prodding, digging, etc.

PROTECT THE HEAD OF THE TRIPOD

If the head is dented or scratched, the instrument may not fit the tripod head accurately or securely. Avoid leaning tripods against anything. Slight movements may cause the tripod to fall, damaging the tripod or other things.

DO NOT THROW THE TRIPOD INTO THE BACK OF A PICKUP

For the same reasons as stated in the discussion on level rods, try to protect tripods from heavy objects or other damage when carrying them in vehicles. Never load other material (stakes, monuments, etc.) on top of tripods, range poles, and other instruments. Objects such as stakes, monuments, etc., may damage a tripod if it is thrown into the back of a truck. Preferably, store a tripod in a separate, clean storage bin.

SUMMARY

The suggested care and handling listed are not intended to be inclusive. However, the list should give you some basic guidelines to follow in everyday use and maintenance of optical and electronic equipment and tripods.

USE AND CARE OF SURVEYING HAND TOOLS

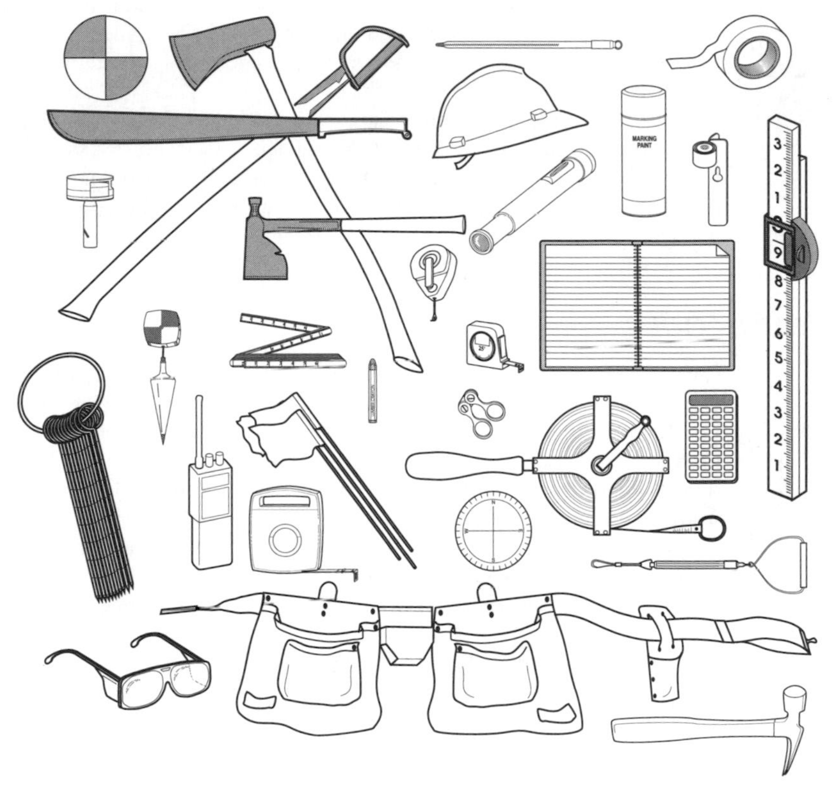

SCOPE

This section will quickly review the most common types of surveying equipment and tools. It will explain the basic purpose of the equipment and will summarize how you should use and care for surveying equipment. As a result of this information, you will be much better prepared to comprehend later chapters that describe the procedures for properly measuring distances and laying out structures.

If you have already studied or performed surveying, much of this chapter will be review. However, you should quickly read through the chapter to refresh your memory and to pick up tips on the safe use of equipment. If you have NOT had previous training in surveying, study this chapter carefully.

SURVEYING EQUIPMENT IMPORTANCE

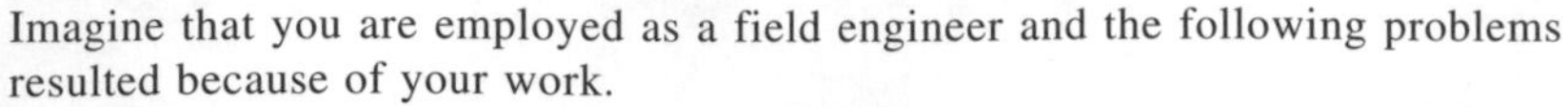

Imagine that you are employed as a field engineer and the following problems resulted because of your work.

Maintain your equipment

Problem: The building project was proceeding smoothly until the day the specially fabricated, steel I-beams were delivered. To your displeasure, they did not fit! The beams were too short for the distance between the established footings.

Problem: A survey of a subdivision had been completed months earlier. When recalculating bearings for each line, it was found that the angles had an error of two degrees. Several houses were partially constructed too close to the property line!

Problem: A sewer line was connected from a main sewer line to a new house, using your measurements. To the dismay of the new home owners, the waste water flowed from the sewer line to the new house. Expensive repairs to the house and sewer rework were required!

What happened in each of these cases? Were there mistakes in procedures, or was the equipment improperly maintained? Certainly, it could have been either. In this section, you will learn the proper care of equipment so that problems similar to these examples do not result from poor maintenance. In other chapters, you will study proper taping, leveling, and angle turning procedures which will also prevent problems like these.

KNOW YOUR TOOLS

The craftsperson who works with tools must know each tool's function, limitations, utilization, and above all, proper care. Equipment should never be subjected to misuse or lack of maintenance. This applies to any profession, be it woodworking, art sculpting, machining, or field engineering! The finished product is only as good as the operator and his or her equipment. If either is lacking, the finished product will not be top quality.

It is imperative that the field engineer know and use proper procedures when working with surveying equipment. If basic principles are adhered to, the steel will fit, the traverse will close, and the sewer will operate as planned.

GENERAL RULES OF USE FOR SURVEYING EQUIPMENT

Surveying equipment principles and guidelines are needed so your surveying work will be consistent and meet specifications.

ONLY USE EQUIPMENT IF YOU KNOW HOW TO OPERATE IT

Learn proper procedures by studying the equipment manual, studying your surveying text, and by asking for help about proper operation when needed.

DO NOT FORCE EQUIPMENT PARTS

If a part does not work easily, find out why, ask for help, or refer to the manual. Forcing may damage a thread or bearing, or may break something on the equipment.

KEEP THE EQUIPMENT CLEAN

Lightly brush dust and dirt off of instruments with a soft bristle brush. To prevent scratches on lenses, use camera lens cleaning supplies.

USE ANY PROTECTIVE CASE

If a protective case is provided, use it when the equipment is not being used. This helps keep it clean and serves to protect it from other hazards.

GENERAL SURVEYING EQUIPMENT

In this section, you will learn about the basic purpose, construction, care, and handling of some general surveying equipment. Remember, the responsibility of becoming familiar with the operation of any equipment rests with you, the operator.

General surveying equipment includes items that are used in several types of surveying activities. These tools include: the plumb bob and gammon reel, hand level, range pole, prism pole, chaining pins, brush clearing equipment, and hammers. Each will be explained.

PLUMB BOB

Plumb bobs, usually weighing 8 to 24 ounces, are made of brass and have a replaceable steel tip. A cord of at least six feet is attached to the plumb bob to make it useful for many situations. The plumb bob is used to create a vertical line or a point to be used as a reference. The plumb bob is the most fundamental instrument used by field engineers in taping, turning angles, laying out--the list is endless! From the plumb bob reference line, you can position yourself directly over a point, align a column, set a form, etc. Plumb bobs are manufactured in a variety of shapes and sizes. The most common is the 16 oz. brass with a replaceable steel tip. Remember these rules about the plumb bob:

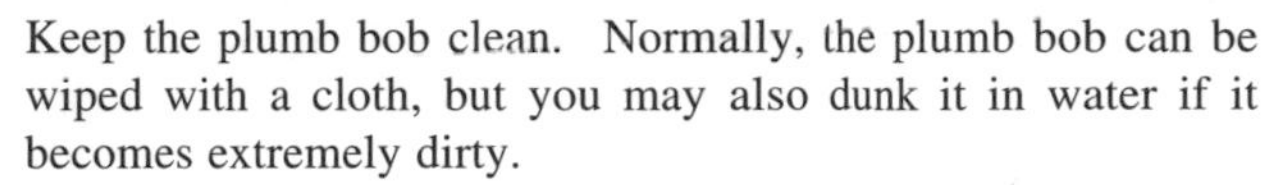

Keep the plumb bob clean. Normally, the plumb bob can be wiped with a cloth, but you may also dunk it in water if it becomes extremely dirty.

Never use a plumb bob as a hammer. Brass dents easily and the plumb bob will become unsightly with nail head marks all over it.

Do not use a plumb bob as a scribe. Damage to the point will result if the plumb bob is used to mark hard surfaces. Replace the point if it becomes damaged.

SIGHT LEVEL

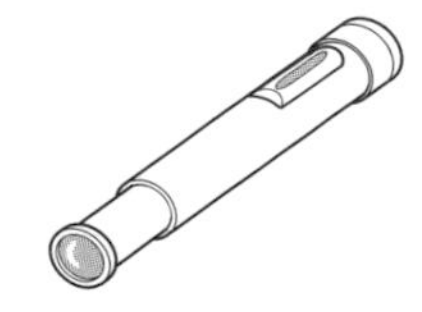

In conjunction with the plumb bob, the sight level is one of the most common instruments of surveying. The sight level is used in taping to keep the tape horizontal; and in leveling to keep from setting the leveling instrument above or below the level rod.

The sight level consists of a metal tube on which a level vial (similar to those found on a common carpenter's level) is attached. A mirror, or prism, inside the tube allows you to center the bubble and look through the tube at an object at the same time. A horizontal line is formed from the observer's eye to where the cross hair of the sight level hits the ground or any other object. A few rules for the use and care of the sight level include:

Wipe the lens with a clean cloth as needed.

Keep the sight level in a protective case when not in use.

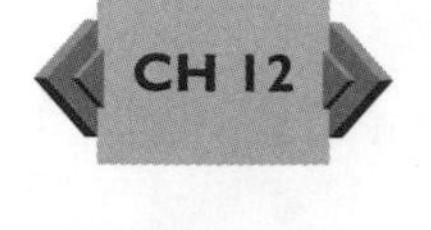

Frequently use the sight level test described in *Chapter 12, Equipment Calibraion*. This test on calibration is used to check if the line of sight is parallel with the axis of the level vial.

Be careful not to drop a sight level.

GAMMON REEL

The gammon reel is used to store the plumb bob string. This plastic device is spring loaded to hold up to 8 feet of plumb bob string. The string automatically retracts into the reel to prevent the string from dragging or tangling in your feet. It also eliminates the need for the string to be stored around your neck which could result in an abrasion burn when pulling the string to use the plumb bob. In addition to holding the string, the case on the gammon reel has a target which can be used for sighting. A stud on the back of the case allows the user to hook the string to it and suspend the plumb bob and gammon reel from under an instrument. The gammon reel is a convenient tool and has become a standard item. Some rules for the use and care of the gammon reel are:

Keep the string clean. Allowing dirt to get inside the reel will affect its ability to perform properly.

Check the string for wear. Replace it if string breakage appears imminent. Follow the instructions provided by the manufacturer.

CHAINING PINS

Chaining pins, also called field engineer's arrows, are used to mark intermediate points when taping. Heavy gauge wire is used in manufacturing them. After being bent into shape, the tip is sharpened so that they can easily be inserted into the ground. They are typically painted in alternate stripes to make them highly visible and to help guard against loss. Concerning the use and care of chaining pins, remember these rules:

Clean chaining pins regularly.

Straighten the shaft of a chaining pin if it becomes bent when being pushed into hard ground.

Repaint chaining pins when the paint becomes faded.

RANGE POLES

Range poles are used to make points in a surveying job more visible. This helps maintain alignment for taping and gives the instrument person a target to sight on when turning an angle.

Range poles are made of wood, fiberglass, or metal and are available in various lengths. They are painted in alternate one-foot bands of white and red or orange to make them easy to see. The most useful range poles are ones that are in sections which can be added together to create poles of unlimited length, but are easily transportable. Some rules for the use and care of range poles include:

Clean dirt and mud from the tips of range poles after each use.

Store a range pole in its protective case when it's not in use.

Repaint range poles as needed.

Maintain a sharp point. If replaceable, put a new point on when needed. If permanent, sharpen the point with a file.

PRISM POLES

Prism poles are used with electronic instruments to measure distances and to make points visible. They seem similar to range poles, however, they typically consist of a pointed, hollow tube that allows a graduated rod to telescope in and out. This allows the rodperson to establish the height of the prism quickly.

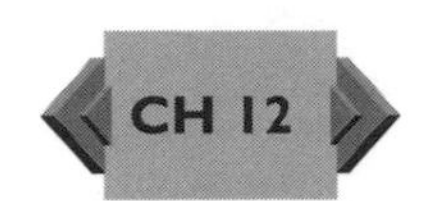

The standard prism pole also has a bull's-eye bubble permanently attached so the pole can be kept plumb while a measurement is being made. Calibration of this bubble will be discussed in *Chapter 12, Equipment Calibration*. A few rules for the use and maintenance of prism poles are:

Keep the telescoping rod clean. Any dirt on the rod will act as an abrasive while the rod is telescoped in and out, and will wear off the graduations.

Clean mud and dirt from the prism pole point after each use.

Store the prism pole in a protective case.

Remove the prism from the prism pole after use and store it in its protective case.

Do not oil the telescoping rod. Oil attracts dirt and only makes operation more difficult. Keeping the rod clean is the most important factor for continued trouble-free operation.

Don't touch graduations with soiled hands.

BRUSH CLEARING EQUIPMENT

Brush clearing equipment includes the axe, hatchet, brush hook, machete, and chain saw. Often in surveying, it is necessary to clear brush and trees from the line that is being surveyed. It is beyond the scope of this text to cover these items in detail, but below is a list of some important safety considerations. Remember these rules for brush clearing:

Wear safety goggles! You only have two eyes--protect them!

Be aware of activity in and around your work area.

Warn others who enter that danger exists. Any tool with a cutting edge is dangerous!

Do not use brush clearing equipment unless you have been instructed about proper use and safety.

Chain saws are very dangerous! Extreme care must be exercised when using chain saws.

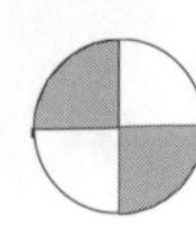

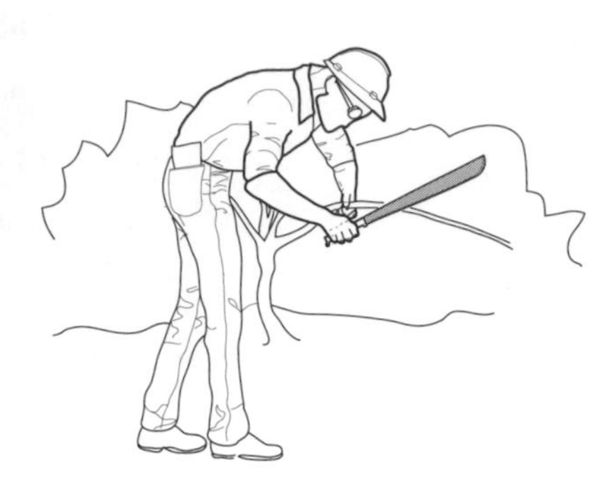

Keep all tools sharp! A sharp tool is much safer than a dull one because it requires less effort and is easier to control.

Keep all brush clearing tools clean! Wipe off dirt and apply a light coat of oil to protect them from rust.

Store tools in their protective cases. With sharp cutting edges present, this is very important.

Protect your ears! Wear ear plugs to reduce noise levels when using a chain saw.

Wear proper clothing! Proper clothing for brush clearing from head to foot includes: a hardhat, close fitting shirts and pants, and leather work shoes. The hard hat protects from falling objects and provides protection from the elements. Close fitting clothing resists snagging on briars and becoming caught by equipment. Steel-toed shoes protect the feet, provide lateral support, and give good traction.

HAMMERS

Field engineers often use hammers to drive tacks into stakes, stakes into the ground, and spikes into trees or utility poles for temporary benchmarks. They are used almost every day for numerous tasks. Hammer sizes and types vary with the type of activity. A hammer with a 12-ounce head will drive nails and tacks, while a 10 pound sledge may be needed to drive stakes into hard ground. A few rules for the care and safe use of hammers are:

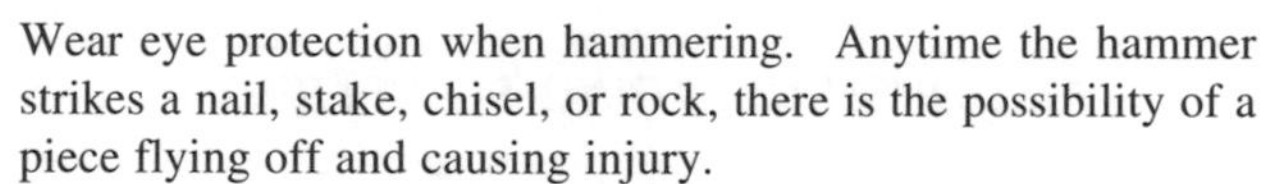

Wear eye protection when hammering. Anytime the hammer strikes a nail, stake, chisel, or rock, there is the possibility of a piece flying off and causing injury.

Check the hammer for cracks! If any sign of weakness or damage is observed, replace the hammer.

Replace or tighten splintered handles and loose hammer heads. A flying hammer head could become a dangerous missile. Tighten hammer heads by inserting a wedge between the metal and the wood.

ROCK CHISEL

The field engineer follows dimensions obtained from plans and must install stakes where the measurements indicate. More often than not, the point falls on hard ground or on a rock. While it is possible to offset the point, it is preferable to place it at the indicated location. The rock chisel can be used to start a hole in hard ground so you can drive a stake. Rock chisels are found in many shapes and sizes. The best rock chisels are those made from the drill bits of a jack hammer. They are of a hardened steel that is capable of being driven by a hammer through extremely hard surfaces. To use and care for rock chisels, follow these rules:

Wear eye protection! When two metal objects are struck together, the possibility of a piece breaking apart and flying off is great.

Inspect the chisel. After a period of use, a flange, or "mushroom" will begin to develop on the head of the chisel. As soon as this becomes evident, it should be removed with a grinder. The sharp flange can cut your hand or bits of metal can fly when striking the chisel.

STEEL CHAIN (TAPE)

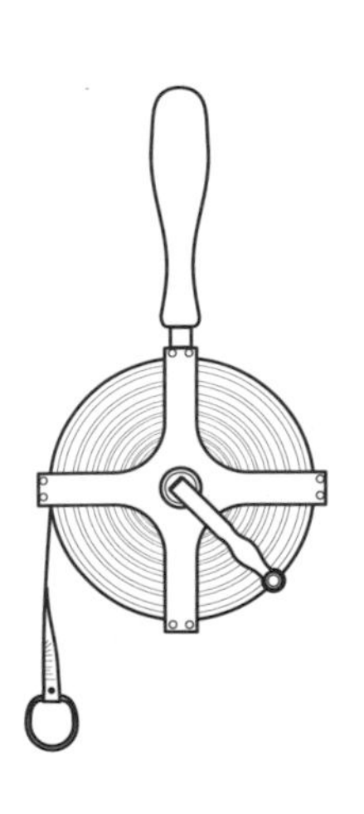

The steel chain is used for measuring long distances. The measurement may be to obtain the distance between two known points, or it may be to lay out a distance on the ground from a set of plans. Most are commonly 100 feet long, although they can be obtained in increments of 25, 50, and 200 feet or longer. Metric chains are also available.

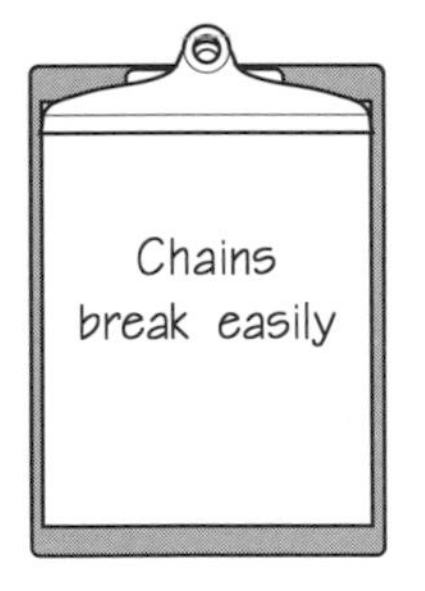

Chains are graduated in many different styles and are purchased to suit a variety of needs. The 100-foot, fully graduated chain is the most common one used by the field engineer. Whatever chain is used, it is the responsibility of the chain person to use and care for it properly.

The physical nature of the chain makes it vulnerable to various hazards. Being only a few thousandths of an inch thick it is very fragile. Therefore, it must be handled with great care. Some possible hazards to the steel chain are:

Getting the chain wet and allowing it to rust.

Allowing it to be run over by a vehicle and broken.

Pulling a loop until the chain becomes kinked or broken.

Inexperienced and poorly instructed persons using the chain.

Becoming permanently elongated from too much tension.

Of course, this list is not inclusive. There are many other hazards that can occur. You want to avoid as many of them as possible. In addition to the general fundamental guidelines previously discussed, the following guidelines for the steel chain are also important. These are a few rules for the care and handling of the steel chain:

Keep the chain on the reel or rolled up when not in use. Allowing it to lie loosely on the ground invites trouble.

Dry the chain when wet. Most are steel and will rust when constantly wet. So carry cloths or paper towels to dry the chain when finished. The chain is most easily dried by two people. Once dry, it is a good practice to wipe it with an oily rag for added protection against rust.

Watch out for cars. Many surveying jobs require that chaining be conducted close to roads or highways. As you can imagine, it is difficult to stop traffic to allow you to chain. Therefore, it is up to you to avoid the traffic. Some field engineers allow vehicles to run over their chain while they are holding them flat on a smooth surface. This may only cause minimal damage. However, this is a very poor policy because the chain can jump and catch on a part of a vehicle. This would cause the chain to jerk violently and cause severe injury to the field engineer. It is strongly recommended that you never allow cars to run over a chain.

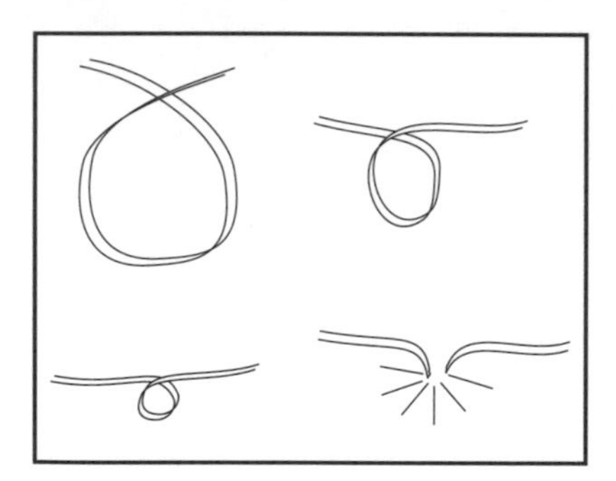

Remove all loops in a chain immediately. Very few things become more of a hazard as they get smaller, however, the loop happens to be one of those things. It starts out as a simple loop, but, as the chain is pulled, the loop becomes smaller and smaller until a kink develops and the chain becomes permanently deformed. If a kink in the chain is disregarded, it will break.

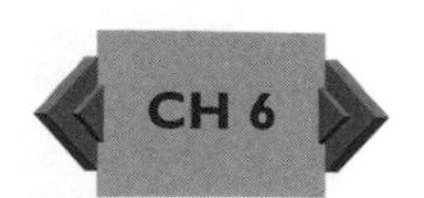

Pull hard enough to take up slack, but not excessively. In *Chapter 6, Distance Measurement - Chaining*, you will learn that a certain amount of tension must be applied to the chain to obtain good results. This tension is predetermined at the time the chain is manufactured. Pulling it with more than the required tension will eventually stretch, and permanently elongate it. There is a mathematical formula to correct the elongation, but why continually cause this unnecessary step?

Following the above guidelines will ensure that your equipment is in good condition. Once you learn to follow the procedures for chaining, you will experience great results.

CLOTH TAPE

The field engineer uses the cloth tape for locating points in areas where steel tapes and precise distances are not necessary. Usually, these tapes are 100 feet long. Some cloth tapes are woven with strong yarns which are covered with a plastic coating to prevent moisture damage. Other cloth tapes have fine metallic wires woven lengthwise. The addition of these metallic wires cuts down on the stretching that occurs with cloth materials. It is this natural characteristic of cloth that makes these tapes unsuitable for determining precise distances. For care and handling of the cloth tape, follow these rules:

Use a cloth tape only when measurement tolerances are only required to the nearest tenth of a foot, or longer.

Wind the cloth tape on the spool when it's not in use.

Wash soiled cloth tapes by immersing in a bucket of water.

Check the tape frequently for worn areas. Replace if necessary.

LEVEL RODS

After tripods, level rods are probably the most abused piece of surveying equipment. They are thrown into the backs of pickup trucks, laid down on the ground or in the mud, improperly held, etc. As expensive as they are, it seems as though they would receive better care.

Leveling is a process that involves setting up a level and reading a graduated rod to determine differences in elevation. The graduated rod that is used is called a level rod. It is of various shapes and sizes and is made out of a variety of materials. Although made to withstand the rigors of everyday use on the construction site, it is still a delicate part of the leveling process that must be handled and cared for properly. These are some things to remember when handling level rods.

Avoid touching the face. Avoid wrapping your hands around the face of the level rod. If you continually touch the face of the rod, you will eventually rub off the numbers and markings. See the illustration to the left.

Keep the level rod clean. Clean the face of the level rod, the joints where the rod slides together, and the bottom of the level rod, constantly throughout the day. Pay special attention to the bottom of the level rod so that buildup of dirt does not cause incorrect elevation. Neglecting to maintain a clean level rod will result in elevations unacceptable for construction purposes.

Keep it tight. Regularly check all screws, bolts, clamps, etc. that may become loose and affect the operation and use of the level rod.

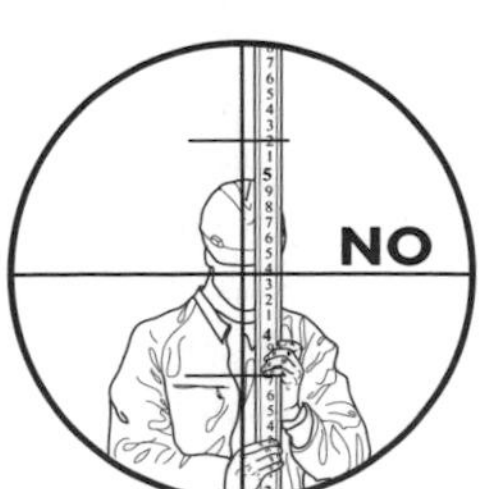

Never throw the rod into the back of a pickup or leave it protruding from a trunk. In either instance, it is exposed to abuse. A rod in the back of a pickup will soon have other items such as the tripod, hammer, or stakes thrown on top of it. It will soon be damaged by treating it this way.

Protect the rod with cloth covers or put it in a protective tube. Some rods have a cloth cover for protection. Most of them are immediately lost. The best way to protect a rod is to obtain a solid tube in which to place the rod. A popular method is to put together some 4" PVC pipe with some end covers. Place the rod in this at the end of the day. This tube can be permanently attached to the survey vehicle if desired.

Do not abuse. The level rod is not a stake driver or a pry bar and it is not used for pole vaulting.

If it is necessary to lubricate, use silicon. Periodically apply a lubricant on moving parts to ensure ease of operation. An oil-based lubricant should be avoided for it will attract dirt. A silicon-based lubricant is recommended. Definitely do not use oil or some other lubricant that attracts dirt or dust. Never lubricating at all is better than using something that will soon have abrasive dirt in it.

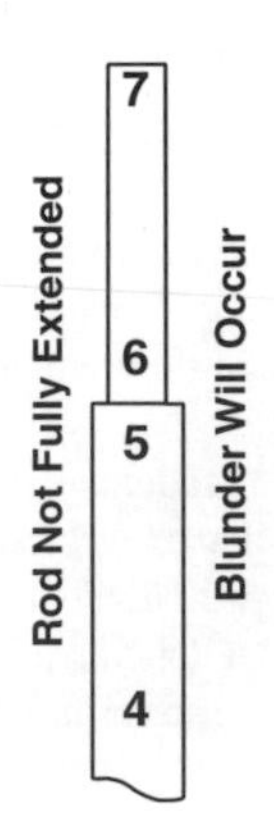

Use the level rod as intended! A general principle regarding the care of any piece of equipment is that it should be used only for the purpose for which it was designed and in a manner according to accepted manuals or other instruction.

Use full extension. When using level rods, care should be taken to make sure they are fully extended. That is, some level rods telescope to achieve their maximum length. Careless users may not extend them fully and this will result in major errors in the leveling activity.

HAND-HELD COMPUTERS / CALCULATORS

Surveying will require you to perform a wide variety of geometric, algebraic, and mathematical functions. An electronic hand-held computer will make a dramatic difference in your ability to perform these functions quickly, easily, and with remarkable accuracy.

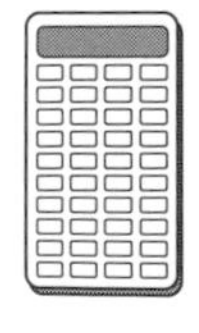

The hand-held computers are also called programmable calculators. They are available from a number of manufacturers in a wide range of models, styles, and prices. They can be expensive. However, the cost of field and office work will be decreased by increasing productivity and the accuracy of your measurements.

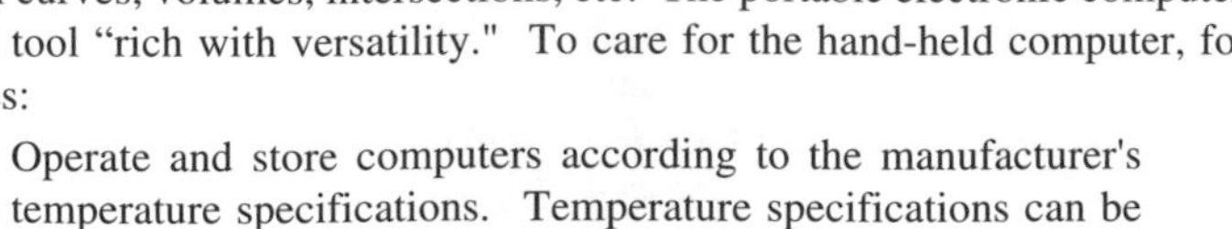

Many come with modules that can be inserted into the case and provide the operator with specific calculation needs. For the construction field engineer, several modules are available to assist in calculating horizontal control, vertical curves, horizontal curves, volumes, intersections, etc. The portable electronic computer has become a tool "rich with versatility." To care for the hand-held computer, follow these rules:

Operate and store computers according to the manufacturer's temperature specifications. Temperature specifications can be found in the owner's manual.

Always turn off the computer to remove or insert modules. To avoid damage, replacements must be made while the unit is off.

Do not insert fingers or foreign objects into an input/output port on the computer. You could damage the unit with some models, by interrupting the operation of the memory functions.

Pay careful attention to low power and battery indicators. Power interruptions of sufficient length can cause difficulty with some memory functions.

Turn off instruments when replacing batteries

Most computer batteries are not rechargeable. Refer to the manual for direct battery requirements. Do not attempt to recharge conventional batteries.

Turn off the computer before attempting to remove and replace batteries.

Always replace all batteries with new ones. Leaving one old battery inside the unit may cause damaging battery leakage and acid corrosion to the computer.

BATTERIES

Much of the equipment you will encounter and use as a field engineer will be dependent upon batteries for power. The batteries are an important accessory to all equipment. Without them, field operations will become impossible. For this reason, batteries should be respected and cared for according to the specifications set forth by the manufacturer of each piece of equipment.

Much of the equipment will have nickel cadmium batteries designed to be recharged. Nickel cadmium batteries are typically very sensitive. Using and recharging them properly is especially important because replacement can be quite costly. Of course, they will eventually have to be replaced, even after proper usage, but conscientious care will help to prolong battery life. Remember these rules for the care of equipment batteries:

Turn off equipment when lengthy delays in operation are expected, as this will prolong the charge on the batteries.

Generally, try to use the charge in the battery fully before recharging. This may even require you to leave equipment on at the end of the day to fully discharge the battery. Nickel cadmium batteries can develop a memory pattern if they are not fully discharged, which will eventually shorten the immediate time/ usage capability of the battery.

Turn off the equipment before removing or inserting batteries.

Charge batteries under the conditions specified by the manufacturer. This includes adhering to all time limits specified for recharging. Generally, it is best never to charge for more than 24 hours, but an ideal time will be documented. This information will be available in most equipment manuals.

SUMMARY

Regardless of how conscientiously operations are performed, the construction field engineer can only be as effective as his or her equipment. Therefore, the field engineer must always operate and maintain equipment with great respect for its value.

The Fundamental Guidelines for Handling All Tools of Field Engineering	
	Only use equipment if you know how to operate it.
	Do not force a part if it is difficult to move.
	Keep equipment clean.
	Use protective cases.
	When an instrument is wet, let it dry naturally.
	When using an instrument in cold areas, store it in a similar climate at night.
	Always attend the equipment.
Each individual piece of equipment requires that additional specific use and care guidelines be followed. This will require you to regularly refer to the standards provided in the manufacturer's operating manuals.	

LEVELING AN INSTRUMENT

SCOPE

Leveling an instrument looks like an easy process, and it is if a few rules are followed. However, observation indicates that many people either have forgotten how, or never knew how, to level. They simply take too long to level an instrument and get ready to use it. This section needs to be understood before moving on to the "Setting over a point with a plumb bob" and "Setting over a point with an optical plummet" sections. Being able to set up an instrument depends on the ability to level it.

GENERAL PRINCIPLES

The leveling system on instruments is simple and easily understood. There are instruments with one level vial and instruments with two level vials, and there are three or four leveling screws depending on the type of instrument. Simply use the leveling screws to raise or lower the ends of the level vial until the level bubble is centered.

It is simple, but some users make it difficult because they don't follow these basic principles about the leveling procedure.

THE HEAD OF THE TRIPOD SHOULD BE HORIZONTAL

Start the leveling process with the head of the tripod close to horizontal. A tripod head that is not horizontal requires more movement of the leveling screws taking more time to complete the process. Insert the tripod legs into the ground or adjust the length of the tripod legs up or down to get the head of the tripod horizontal and over the point.

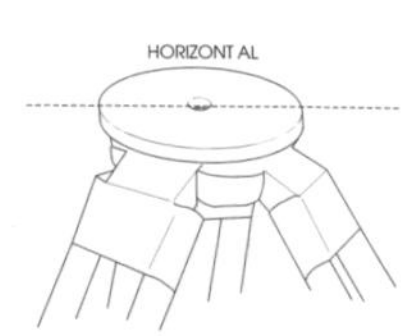

LEVEL VIAL POSITIONING

Position the level vial over the leveling screws properly so the manipulation of the leveling screws is effective. The proper position depends on the type of leveling head. See the step-by-step procedure for the three screw or four screw leveling heads described later.

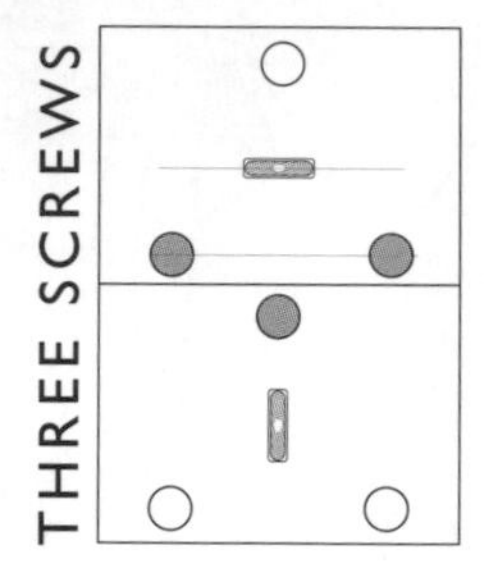

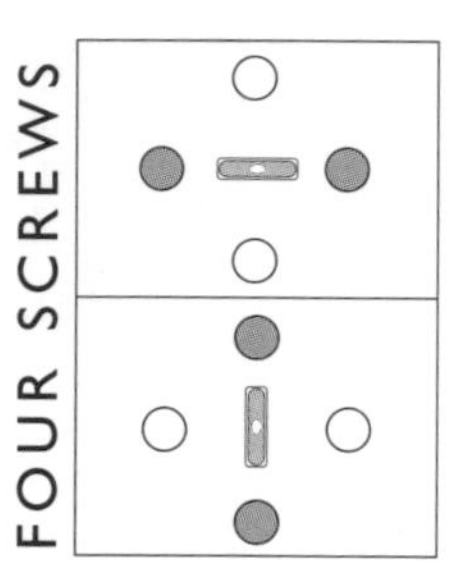

LEFT THUMB RULE

Both thumbs in, both thumbs out, the bubble follows the left thumb. This rule applies for all types of leveling systems because all manufacturers use the same type of threads in their leveling screws. This is so simple; but repeatedly, people tend to try to use a random method for leveling the instrument which takes much longer. A graphical illustration of the rule is shown.

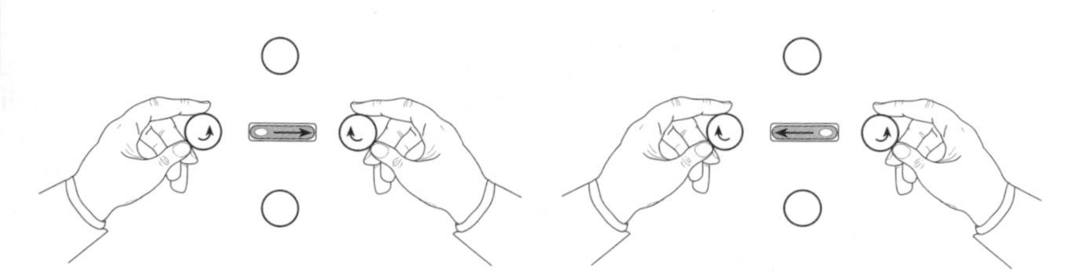

ALWAYS ROTATE 180 DEGREES AS A CHECK

Continually rotate the instrument 180° as a check of the proper adjustment of the level bubble to ensure that the bubble stays centered.

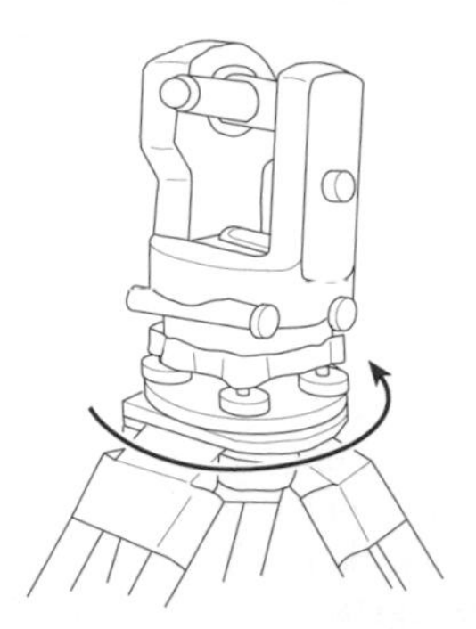

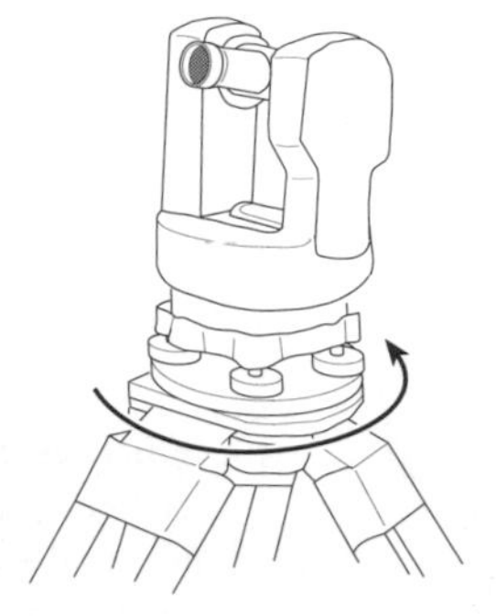

TURN LEVELING SCREWS EQUALLY

When using two leveling screws together, rotate them the same amount, in opposite directions, at the same rate. This is extremely important with the four-screw leveling system. If they are not operated exactly the same, they will become too tight or too loose.

When the leveling screws become too tight on a four screw leveling system, it is difficult to operate them. And if the operator continues to operate them incorrectly, they will become tighter and tighter. This will result in not being able to move them at all. In an extreme case, the structure of the leveling head itself could be damaged or the threads on the leveling screws could be stripped.

If improper operation causes the leveling screws to become too loose, the instrument will rock on the leveling head. This will make it difficult to keep the instrument level during the operation of turning angles. Again, rotate leveling screws the same amount at the same rate.

There are two types of leveling head systems which have been used on instruments—the original four-screw system, and the more recent three-screw system.

FOUR-SCREW LEVELING HEAD

The four-screw leveling head system allows the instrument to be re-leveled without changing the height of the instrument. The reason for this is that the four-screw system involves the pivoting of the instrument around a half ball at the bottom of the spindle. This is very useful when using the dumpy level because the instrument user can re-level between readings if necessary. See the following illustrations.

STEP-BY-STEP PROCEDURE

STEP 1

Position the plate level bubble so it is directly over two opposite leveling screws.

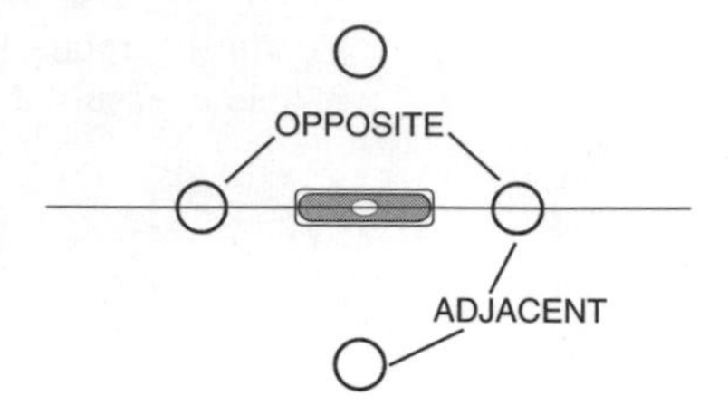

STEP 2

Observe the location (left or right of the center of the vial) of the level bubble and, using the left thumb rule, move both leveling screws equal amounts in opposite directions to move the level bubble towards the center of the level vial. Center it exactly.

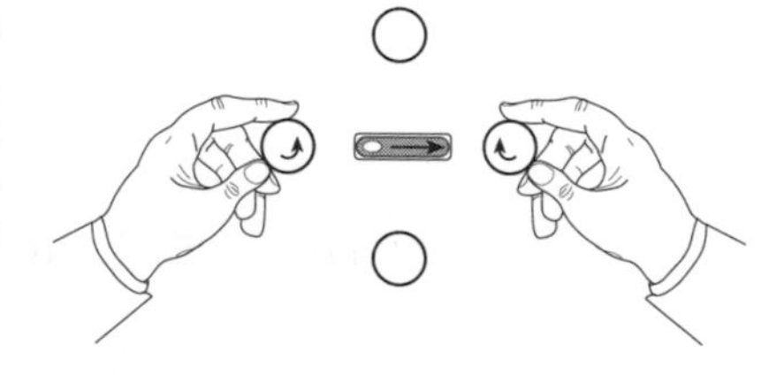

STEP 3

Rotate the instrument 90° so the bubble is directly over the other two leveling screws. Observe the location (left or right of the center of the vial) of the level bubble and, using the left-thumb rule, move both leveling screws equal amounts in opposite directions to move the level bubble towards the center of the level vial. Center it exactly.

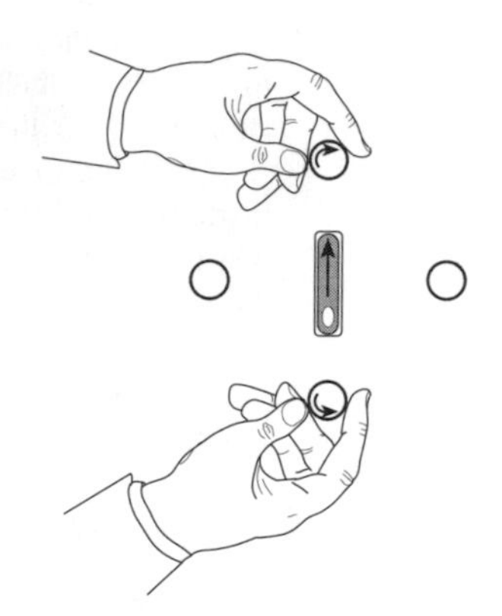

STEP 4

Rotate the instrument 90° again so the level bubble is over the original leveling screws again. Check the location of the bubble and make any necessary adjustments using those leveling screws as needed to center the bubble.

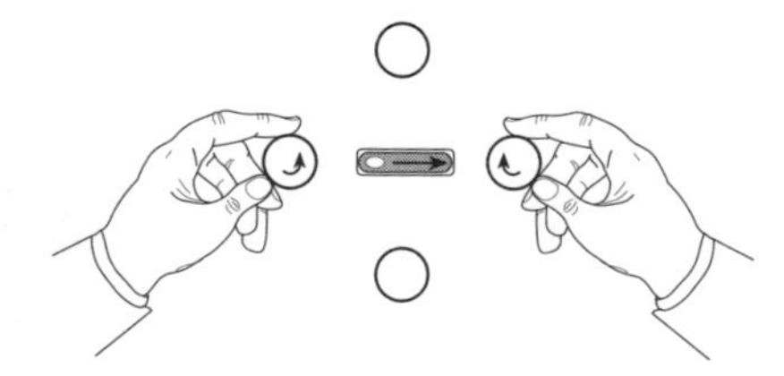

STEP 5

Rotate the instrument 90º again so the level bubble is over the second set of leveling screws again. Check the location of the bubble and make any necessary adjustments using those leveling screws as needed to center the bubble.

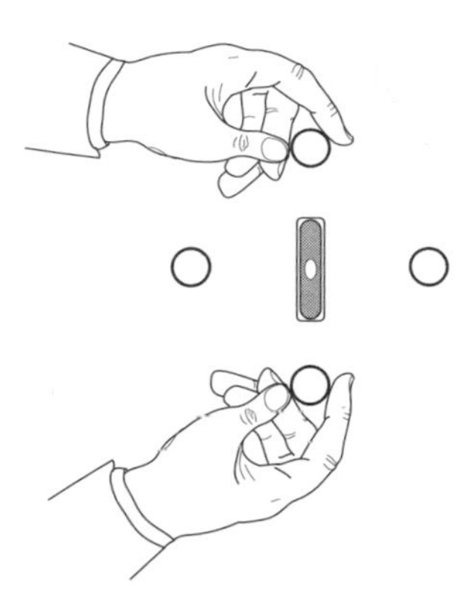

STEP 6

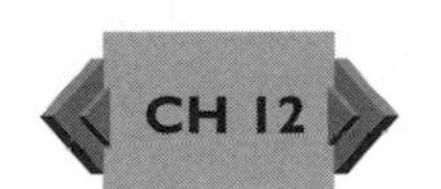

Repeat this process until the bubble stays centered as the instrument is rotated. If the bubble fails to stay centered when the instrument is rotated 180º, the bubble needs adjustment. See *Chapter 12, Equipment Calibration* for an explanation of this process.

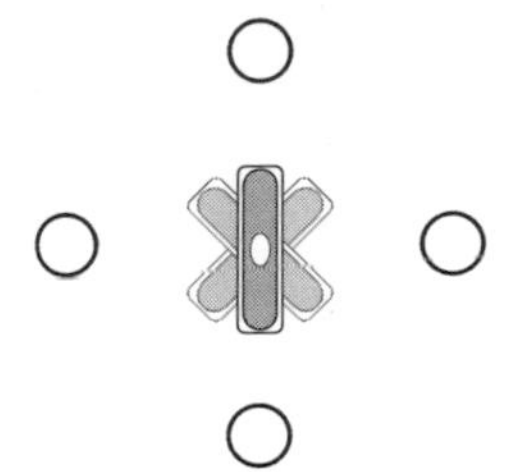

THREE-SCREW LEVELING HEAD

The advantage of the three-screw leveling head system is that an instrument can be leveled more quickly. It is indeed quicker if a systematic approach to leveling the instrument is followed. Persons have been observed going around in circles trying to level this type of system because they were using it improperly. It has taken 30 minutes or more to do what should have taken 2 or 3 minutes. The following procedure should be used.

STEP-BY-STEP PROCEDURE

STEP 1

Align the level vial so its axis is parallel to a line through two of the leveling screws.

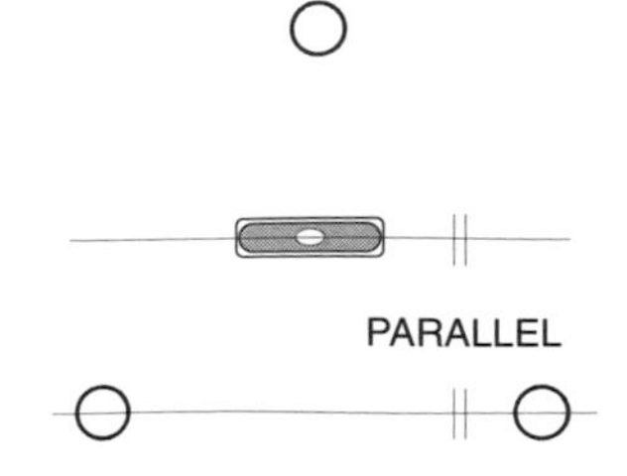

STEP 2

Observe the location (left or right of the center of the vial) of the level bubble and, using the left-thumb rule, move both leveling screws equal amounts in opposite directions to move the level bubble towards the center of the level vial. Center it exactly.

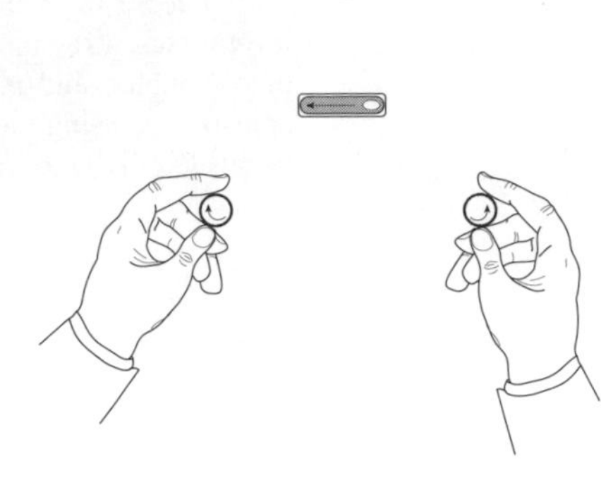

STEP 3

Turn the instrument 90° so the bubble is aligned over the top of the third remaining screw.

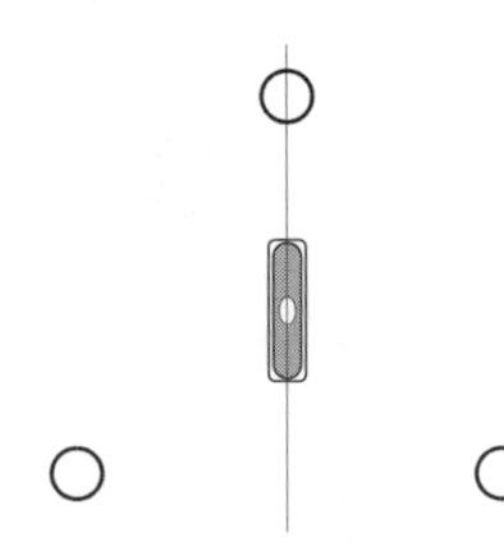

STEP 4

Observe the location (left or right of the center of the vial) of the level bubble and following the left-thumb rule <u>use that screw only</u> to center the bubble in this direction. Center it exactly.

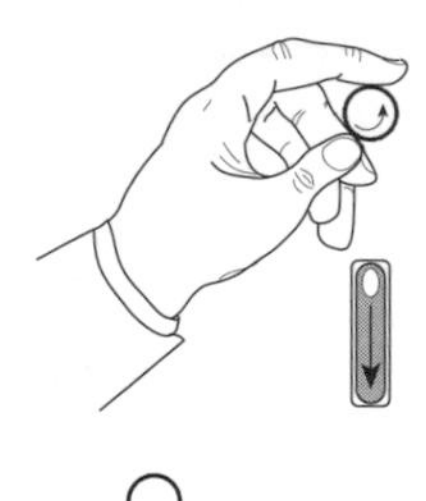

STEP 5

Turn the instrument 90° again to align the bubble so it is over the original two screws. Adjust these two screws and re-center the bubble if necessary.

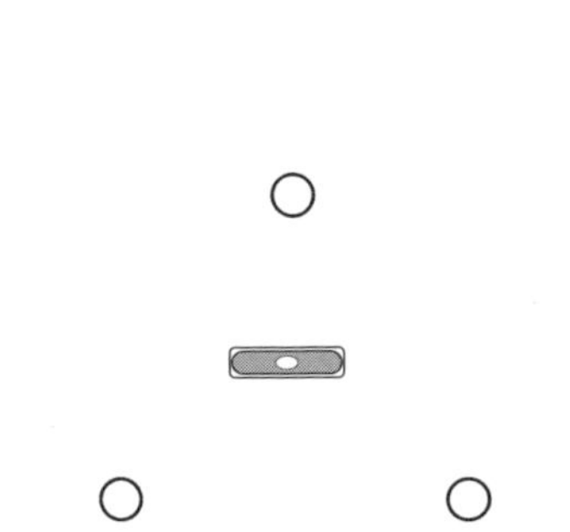

STEP 6

Turn the instrument 90° again to align the bubble so it is over the remaining screw. Adjust this screw and re-center the bubble if necessary.

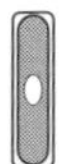

STEP 7

CH 12

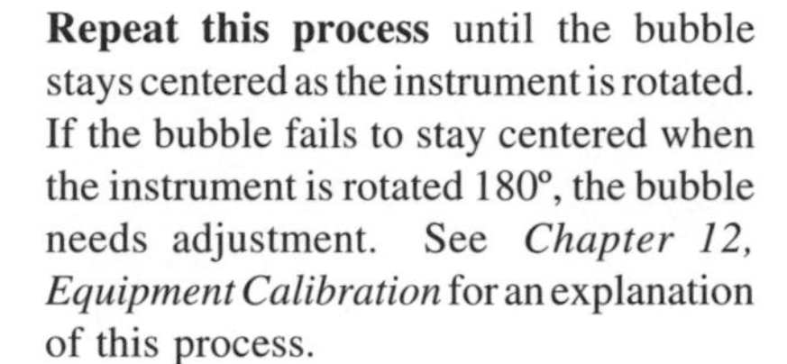

Repeat this process until the bubble stays centered as the instrument is rotated. If the bubble fails to stay centered when the instrument is rotated 180°, the bubble needs adjustment. See *Chapter 12, Equipment Calibration* for an explanation of this process.

SETTING INSTRUMENTS OVER A POINT WITH A PLUMB BOB

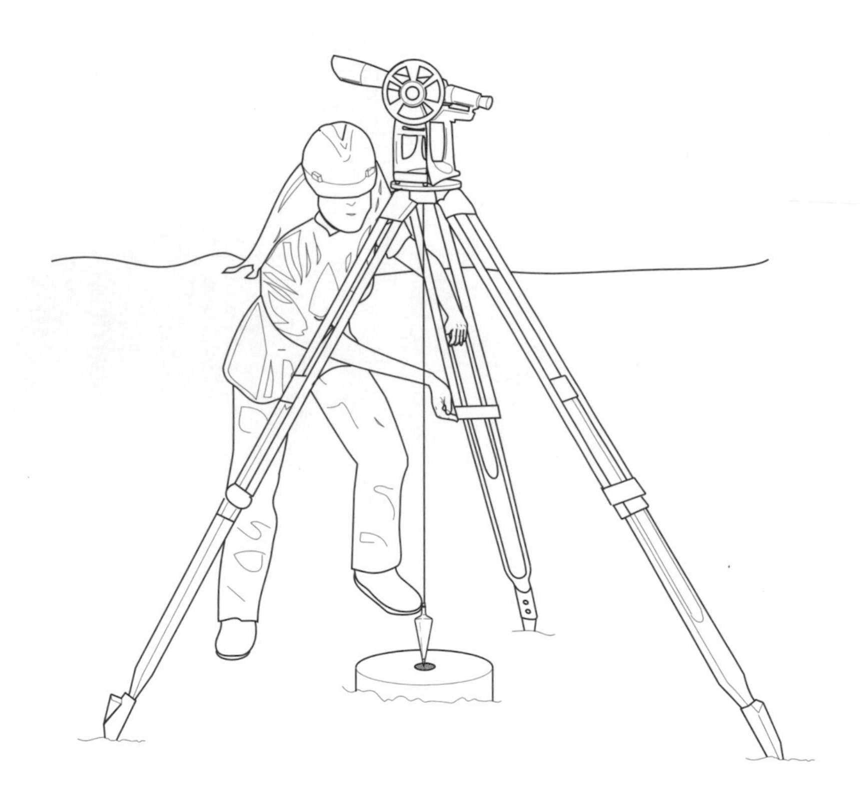

SCOPE

One of the fundamental tools of the field engineer, the plumb bob, has been used since the beginning of time to set instruments over a point. Sketches from ancient Egypt show plumb bobs hanging from wooden instruments. It has only been recently, with the advent of the optical plummet (see the next section), that the plumb bob has not been used exclusively in setting an instrument over a point. Although optical plummets are found on most instruments today, all field engineers should still be aware of the process used to set up an instrument over a point with a plumb bob.

STEP-BY-STEP PROCEDURE

STEP 1

Roughly set the tripod over the point. By observation, rough set the tripod over the point. Hold two legs, and place the third one past the point. Move the two you are holding until the head of the tripod is approximately over the point and horizontal. Stand on the tripod fcct to sink them firmly into the ground. Take the instrument out of its box and place it on the tripod head. Screw or clamp the instrument to the tripod as indicated by the type of instrument.

STEP 2

Attach the plumb bob to the chain or hook hanging down out of the bottom of the instrument. **Adjust the plumb bob** (using the gammon reel) so the tip of the plumb bob is as close lo touching the point as possible, but not touching.

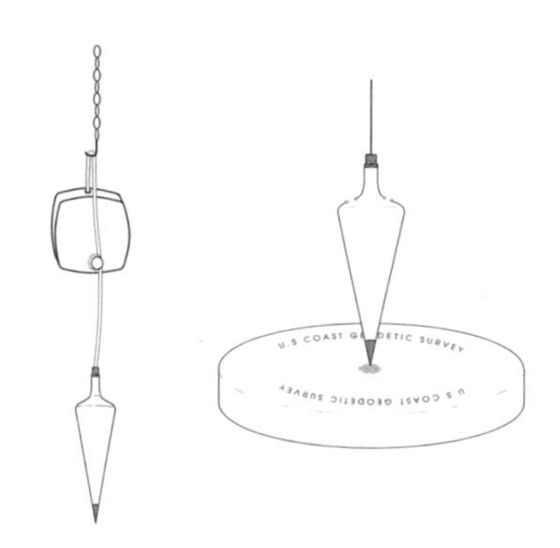

STEP 3

Observe the location of the plumb bob and the tripod legs in relation to the point. Determine which tripod legs need to be adjusted to move the plumb bob over the point. Sometimes this takes moving one tripod leg, and other times it takes moving two. Carefully and slowly adjust the tripod legs so the plumb bob moves to within a quarter of an inch of the point. It may be necessary to adjust the length of the gammon reel string at this time to keep the plumb bob just over the point.

STEP 4

Level the instrument with the leveling screws. Fine adjust over the point with the instrument. Check the location of the plumb bob to the point. If it is still within a quarter of an inch, move to the instrument head and make a final adjustment to get the plumb bob exactly over the point. With a four-screw leveling head instrument, slightly loosen two adjacent screws and carefully slide the instrument over the point. Then tighten the screws and relevel. With an instrument that clamps from the bottom, loosen the clamp and slide the instrument over the point. Tighten the clamp and re-level as needed. For either type of instrument, **Do not rotate the instrument** as doing so will require complete releveling instead of simply fine adjusting.

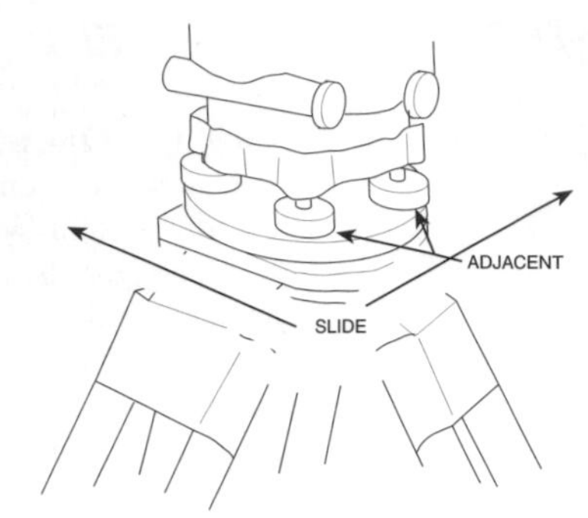

STEP 5

Check and repeat the leveling and shifting of the instrument until it is level and over the point.

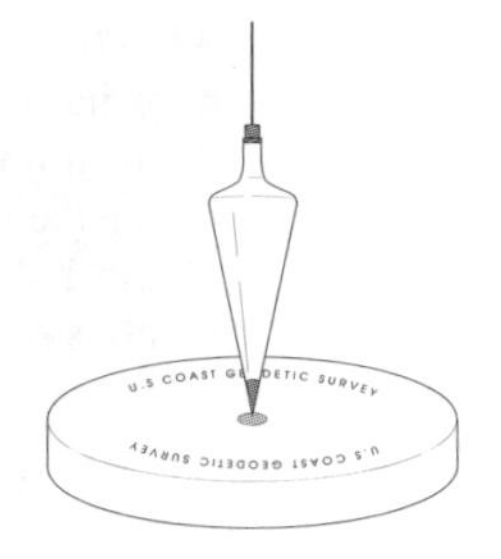

SUMMARY

This is the standard approach to setting an instrument over a point with a plumb bob. It works well and any field engineer who follows it should be able to have an instrument set up over a point and level within a few minutes.

QUICK SETUP WITH AN OPTICAL PLUMMET

SCOPE

The ability to set up quickly over a point is not considered an art. Anyone can set up quickly if exact procedures are followed. This is another of the basic procedures where some persons can set up an instrument in 2 minutes and it takes others 30 minutes. Needless to say, on the construction site with a crew of carpenters waiting for line, a field engineer must know how to make a quick setup! This section illustrates a method which many field engineers have found to work well.

STEP-BY-STEP PROCEDURE

Any instrument with an optical plummet can be set up quickly and efficiently by following these steps.

STEP 1

Rough set the tripod, by observation, over the point by moving the legs in or out on the ground. Hold two legs and place the third one past the point. Move the two you are holding until the head of the tripod is approximately over the point and horizontal.

STEP 2

Attach the instrument to the tripod. Look through the optical plummet and determine its location relative to the point (place your foot next to the point to assist in determining location). Adjust the location of the tripod, if necessary, to make the point visible in the optical plummet.

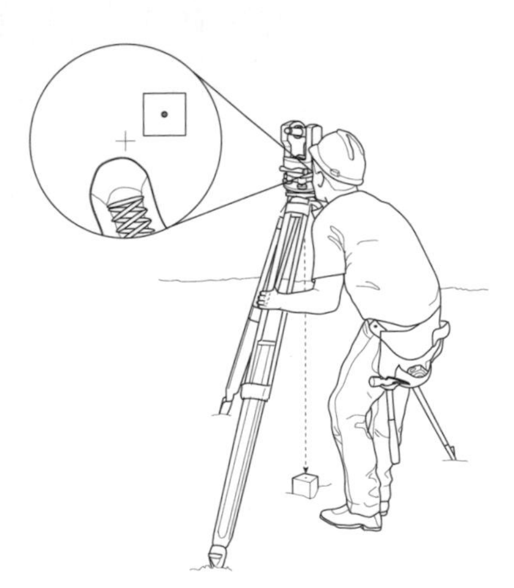

STEP 3

Center the optical plummet. When the point is visible in the optical plummet, use the instrument's leveling screws to center the optical plummet exactly on the point.

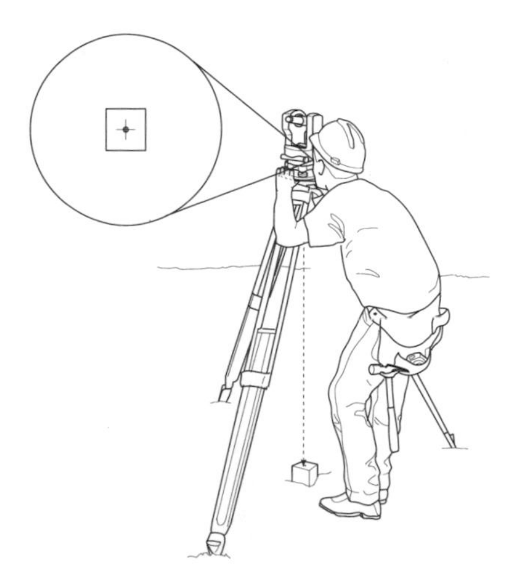

STEP 4

Adjust the tripod legs up or down to center the bull's-eye bubble on the instrument. Only work with one leg at a time!

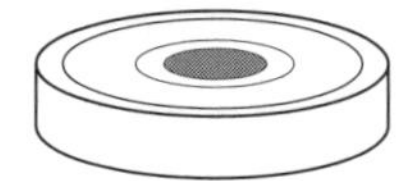

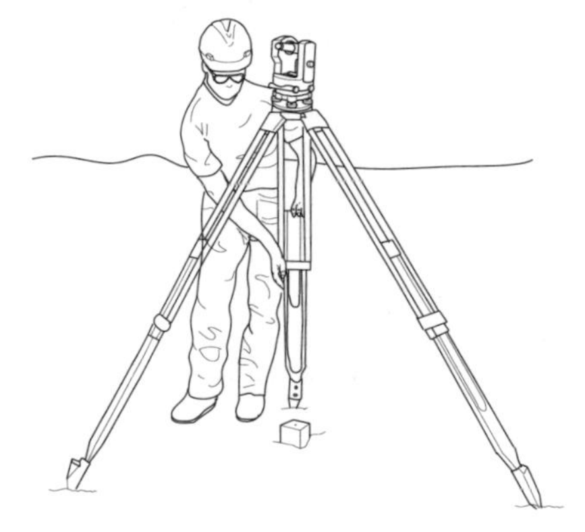

STEP 5

Confirm the position. Check to confirm that the optical plummet is still on or is very close to the point.

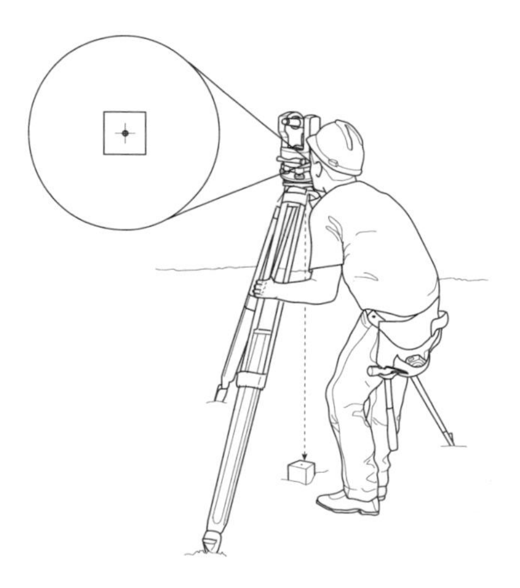

STEP 6

Center the bubble. Return to the leveling screws and use them to exactly center the plate bubble on the instrument. The plate bubble should remain centered when the instrument is rotated about its vertical axis.

STEP 7

Check the relation of the optical plummet to the point. It will probably be off the point slightly.

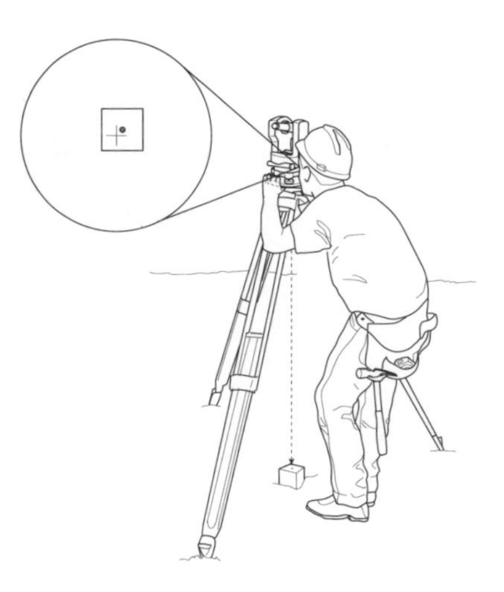

STEP 8

Loosen the instrument attachment clamp on the tripod slightly and while looking through the optical plummetslide the instrument until it is exactly on the point. Do not rotate the instrument!

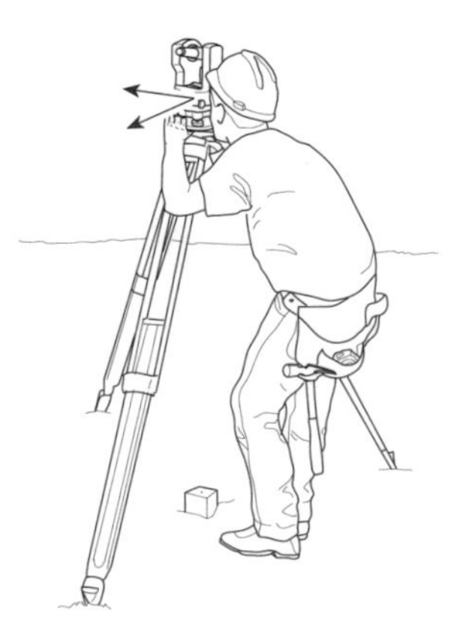

STEP 9

Tighten the attachment clamp and re-check the plate level.

STEP 10

Re-level and repeat, if necessary, until the instrument is exactly level and directly over the point. When this is achieved, check the instrument frequently to ensure that it remains level and over the point.

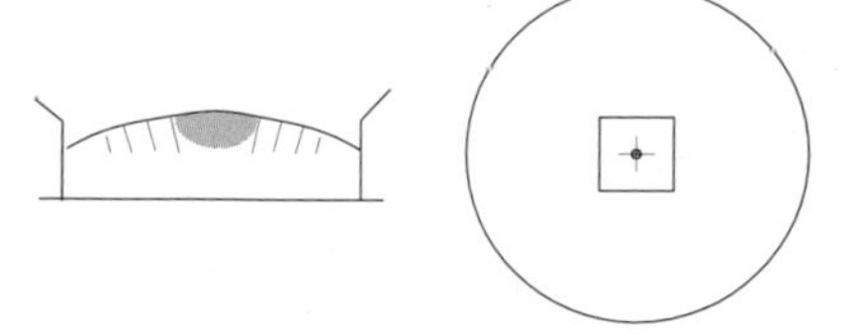

OFFICEWORK BASICS

5

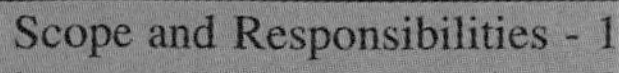

PART 1 - FIELD ENGINEERING BASICS

CHAPTER OBJECTIVES:

List the basic principles of drafting.
Illustrate different line weights and line types.
State the keys to good lettering skills.
List the basic information that should be shown on a site drawing.
State why a graphical scale should be used on all drawings.
Distinguish between existing and proposed contours.
Define contour line.
Describe the common characteristics of contours.
State the purpose of lift drawings.
Describe the importance of daily reports for the field engineer.
State the purpose of trend charts.
Describe how a field engineer could use a trend chart.
Describe the quality control responsibilities of a field engineer.

INDEX

FIELD ENGINEERING DRAFTING AND CAD FUNDAMENTALS

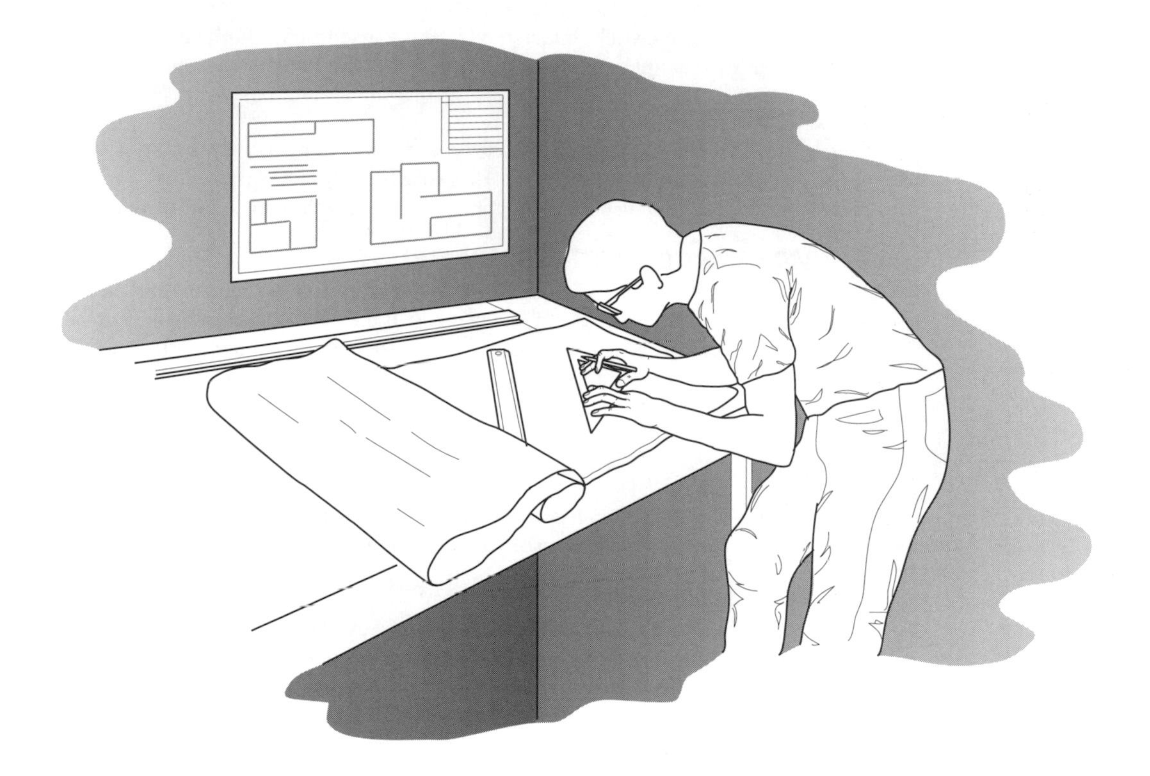

SCOPE

One of the tasks of the field engineer with many companies may be to prepare layout drawings, lift drawings, site plan drawings, or as-built drawings. Although it will not be expected that these drawings be the quality of those that are prepared by an architectural firm, it will be expected that they follow some basic principles of drafting or CAD. They may have to be presented to an owner for consideration of a proposed change order. This section briefly presents an overview of drafting principles. Many fine books on drafting or CAD are available as references.

BASIC PRINCIPLES OF DRAFTING

Producing a good drawing is more of an art than a science. Some individuals have a knack for drafting and can produce drawings that are well arranged, clear, and beautiful, while other persons' drawings look like they have been produced with crayons. Even though both individuals had the same drafting equipment, the artistic ability is different. For those field engineers who do not have artistic ability and for those artistic people who could brush up, some basic principles of drafting are presented here and should be followed in order to produce quality drawings.

FOLLOW INDUSTRY STANDARDS

Just about every drafting text available has standards for line weight, line type, lettering, dimensioning, sectioning, use of equipment, etc. Follow those standards, and drawings produced will be understood by those in industry. Don't be creative and develop your own unique methods; they will most likely be misinterpreted.

BE CONSISTENT

Consistency in lettering, line quality, line type, dimensioning, layout, and style is the key to producing an attractive drawing. Everyone is going to have his or her own style and ability; but, if there is consistency in technique, the drawings will be readable.

PROVIDE THE CORRECT INFORMATION

The most beautiful drawing will be thrown into the trash if the information on it is incorrect. If dimensions are wrong, in the trash. If the views are wrong, in the trash. If the coordinates are wrong, in the trash. Check, double check, and triple check all drawings for correct information. Have someone else check it before it goes to those who are going to build from it. It must be correct.

DRAFTING EQUIPMENT

The following is a suggested list of the minimum drafting equipment necessary for a field engineer to be able to prepare drawings on the jobsite.

DRAWING SURFACE

Some type of table dedicated to drafting should be available for the field engineer. Recommended table size is 30"x48" with a bonded vinyl drawing board cover. It should include a T-Square or a parallel bar.

DRAWING TOOLS

Drawing Pencils for drawing are available in many styles and shapes. Common wood/lead pencils, lead holders, or mechanical pencils are available. The type of pencil selected depends upon personal preference.

Whichever type of pencil is selected for use, it is almost certain that several types of lead weight will be needed. Common leads used are the 4H (Hard), 2H (Medium), H (Soft), and HB (Extra-soft). The type of lead necessary is dependent upon the drafting techniques and skill of the individual. Some persons do all of their work with just two lead weights while others require all that are available.

Scales: The engineer's and architect's scales are the standard measurement tools used for drafting. The engineer's scale is used for civil-type construction work which includes roadways, bridges, site work, etc. The architect's scale is used for architectural work including structural dimensions, overall building dimensions, finish dimensions, etc.

Triangles: They are available in all sizes, from those that will fit into a wallet to the ones which are as large as a drafting table. Size selection depends on the use. Typical triangles for a field engineer are: 8 to 10 inch - 45 degree, 8 to 10 inch - 30/60 degree, and adjustable.

Templates: They are convenient time savers. Some typical templates include a circle template, a French curve template, an architectural symbols template, etc.

Compass: Select a compass that has a layout radius of at least 6" in order to be able to plot many of the curves that will be encountered on site plans.

Protractor: For field engineering, use a 360-degree protractor. If points are going to be laid out radially in the field with an instrument that will be turning angles from 0 to 360°, this type of protractor can be used effectively to plot points on paper.

Eraser: Several different types should be available to be able to erase on a variety of drafting surfaces. Soft white, hard green, and hard red are common. An electric eraser is very nice if available. An erasing shield should also be available.

Sharpener: The type of sharpener is dependent upon the type of pencil selected. If a wooden pencil or lead holder is used, a pointer will work well.

Brush and Drafting Powder: Maintaining a clean drawing is one of the keys to a good drawing. Dust from sharpeners and graphite from drawing lines will soil a drawing. Brush often and use drafting powder all of the time.

DRAFTING TECHNIQUES

The objective of drafting is to produce a drawing that is easy to read and follow. One of the main factors in preparing a drawing that meets these criteria is having a good understanding of the types and weights of lines that are used on the drawing.

LINE WEIGHTS

There are three line weights produced by using the proper type of pencil lead and by using the correct width of line.

Heavy, Dark

This line is produced by using a soft pencil such as H or HB. It may be necessary to go over the line more than once to obtain the width needed. Be consistent. This type of line is used for outlines of objects, profiles, section cuts, lettering, etc. to bring the object to the attention of the reader.

Medium

This line is produced by using a medium pencil such as H or 2H. If sharpened properly, it should only be necessary to go over the line once to make the crisp line required. This type of line is used for elevations, sketches, phantom lines, some hidden lines, etc. This is the standard line used on most of the drawing.

Light
This line is produced by using a hard pencil such as 4H. Because it is hard, this pencil won't need to be sharpened as often as the softer ones. This line is very sharp and defined. It is used for grids, layout lines, dimensions, arrow lines, etc. This line is unobtrusive. It doesn't bring attention to itself.

LINE TYPES

Many types of lines, in various line weights, can be used on a drawing to illustrate particular items. A few of the most common line types are illustrated below in medium line weight.

Object line
Used for outline of plan view, section cuts, profiles, elevations

Hidden lines
Used for elements below a solid surface

Center lines
Used for center of objects, grid system, round openings

Dimension lines
Used for indicating distance from point to point

Phantom lines
Used to show an object's future location

Utility lines
Used to indicate the location and type of utility

s

BASIC PRINCIPLES OF LETTERING

The ability to letter neatly is a learned skill. It takes practice and perseverance to become good at lettering. The problem with many individuals is not that they don't know how to letter, it is that they don't practice. Computer use is not helping because we are practicing less. However, if some basic principles are followed, good lettering can be accomplished.

USE HORIZONTAL GUIDELINES

Lettering that goes up and down like a roller coaster is not very attractive. It is not easy to read and leaves the reader with a queasy feeling about the quality of the drawing. It only takes a few seconds to draw horizontal guidelines to maintain letters which are uniform in height.

USE VERTICAL OR SLANTED GUIDELINES

Often, the primary difference between good lettering and poor lettering is the failure to take the time to add vertical or slanted guidelines. Drawing vertical guidelines will force the letters to be formed so they are all parallel.

STICK WITH A STANDARD STYLE

A single-stroke style of lettering has proven, through the years, to be the optimum in legibility, speed, and ease in the forming of letters. A simple style is recommended for use by field engineers on drawings as well as in the field book. A sample of standard lettering is shown below.

ABCDEFGHIJKLMNOPQRSTUVWXYZ

abcdefghijklmnopqrstuvwxyz

1234567890

BASIC PRINCIPLES OF CAD

When working with a CAD system, there are several principles that are different from manual drafting. Although there are principles for manual drafting which can be applied to using a CAD system, there are also principles unique to CAD.

LAYERS

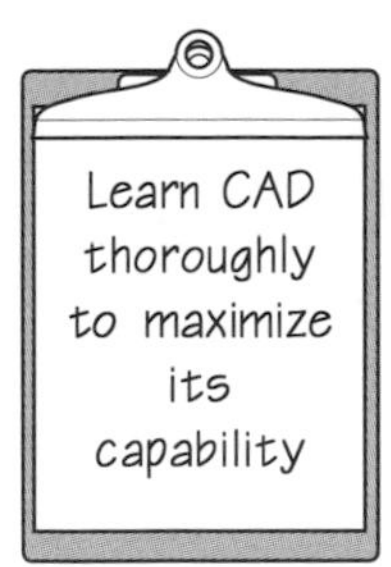

When using a computer aided drafting system, the designer can represent individual parts of an entire drawing separately by using layers. The computer can align these layers like a person could with clear drafting film. Any layer can be separately chosen or all layers can be viewed at once.

MEASUREMENT UNITS AND ACCURACY

A CAD system allows the designer to select the unit of measurement and the accuracy the drawing will have. It is not uncommon for a CAD drawing to be accurate to 1/1000th of an inch. However, the more accurate a drawing is, the smaller the available drawing space becomes.

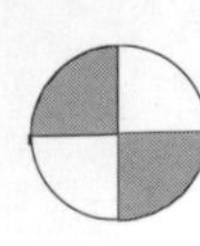

VIEW MANIPULATION

With most CAD systems today, view manipulation is very easy. The designer has choices such as zoom, rotate, pan, and scroll. These choices change the way the designer views the database he or she has created.

COLOR

For many complex drawings, different colors can be used to distinguish separate objects. In many cases, certain colors will represent certain materials and objects (i.e. blue = steel). The use of color has many of the same advantages of using layers.

LIBRARIES

When dealing with common manufacturing objects, it is likely that a commercial library of such objects is available. For larger projects, it is recommended that you create your own library of commonly used objects for easy accessibility.

UNDERSTANDING SITE DRAWINGS

SCOPE

Site plans are often utilized by field engineers to determine what exists on the site, to determine the general location of the structure on the site, and to visualize what the site will look like after it has been completed. Initial discussion of a project often starts with the site plan. How does the structure fit on the property, how is it oriented to the road, how is there access to the site, what is the impact to the existing trees, what will be happening with the drainage, how is the site going to be landscaped? These and many more questions are asked and answered by looking at the site plan for a project. A field engineer must have a good understanding of the lines shown on a site plan and what they represent.

INFORMATION ON SITE PLANS

Existing planimetric (manmade) and topographic (natural) features as well as planned project information is shown on a site plan. Planimetric features will include roads, sidewalks, utilities, buildings, etc. Topographic features will include trees, streams, springs, existing contours, etc. Project information will include the building outline, general utility information, proposed sidewalks, parking areas, roads, landscape information, proposed contours, and any other information that will convey what is to be constructed or changed on the site.

The purpose of a site plan is to convey information related to site activities.

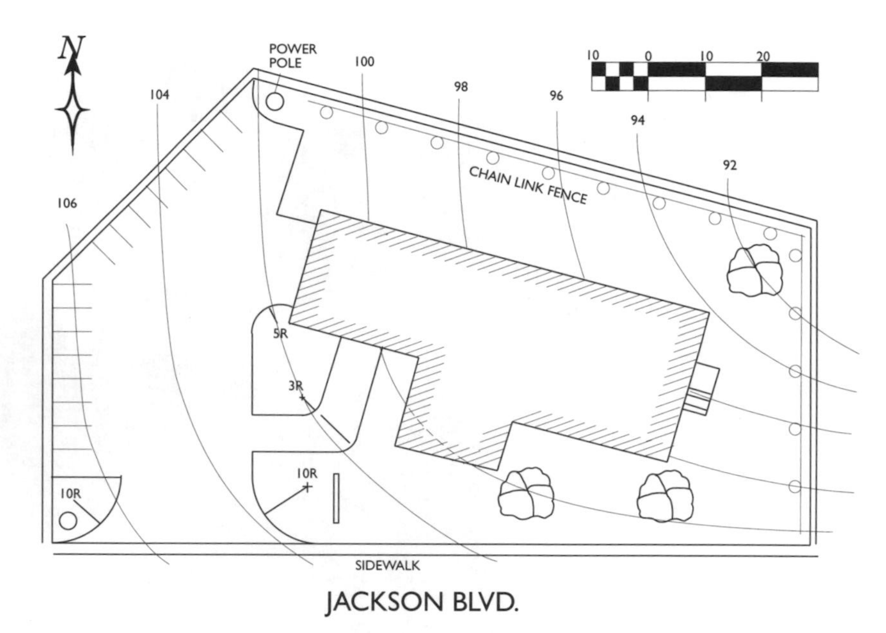

The illustration above shows the information typically found on a site plan. Information important to the field engineer includes:

DIRECTION OF NORTH

A North direction should always be shown on site plans for orientation purposes. Displaying it prominently is the standard practice.

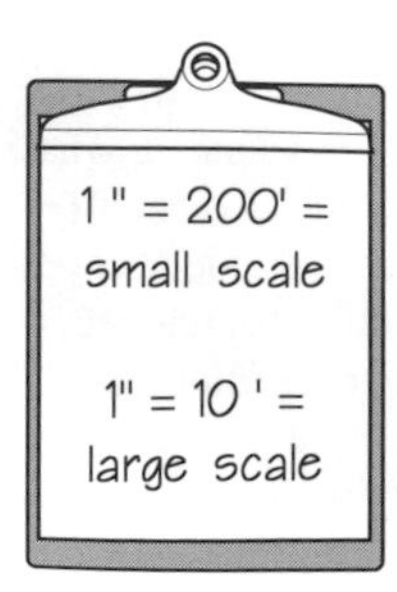

SCALE

It is impossible to make a site drawing to full scale because the drawing would have to be the same size as the project. So, in order to be usable, site drawings must be scaled down to fit on standard-size drawing media. The scale used on site drawings is typically an “engineer’s” scale. That is, distances are represented in feet and tenths of a foot, as opposed to the "architect's" scale where distances are represented in feet, inches, and fractions.

The scale used on a site plan will depend on the size of the project. A project covering a large area will have a small scale such as 1" = 100'. A project on a small site will have a large scale of 1" = 20' or similar. Field engineers must be familiar with scales used and be able to interpret data from the site plan using the given scale.

GRAPHICAL SCALE

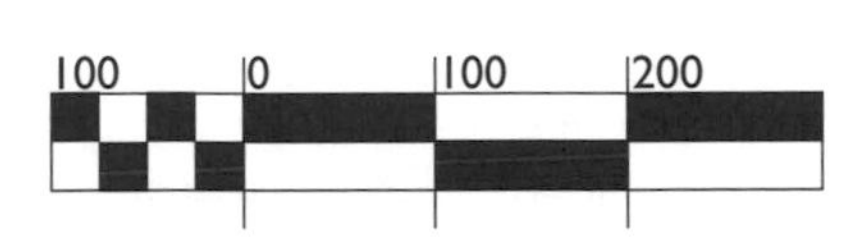

Scales shown on site plans are typically shown in graphical form on the drawing so the scale can still be used if the drawing is enlarged or reduced. For example, if only a numerical scale is given on the plan, reduction by a copy machine would render it useless. If a graphical scale is shown, it would still be there on the reduced copy and would be useable by the field engineer to check the rough location of items on the plan. A sample graphical scale is shown to the left.

LEGEND

It is difficult to show the amount of information required on site drawings if symbols are not used. A good site drawing will include a legend of symbols used, especially those that are non-standard. Utilize symbols to describe types of roads, bridges, buildings, survey data, utility lines, trees, etc. For a more complete listing of symbols reference *Chapter 2, Communications and Field Engieering.*

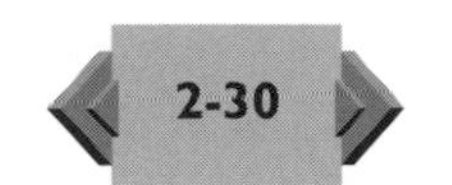

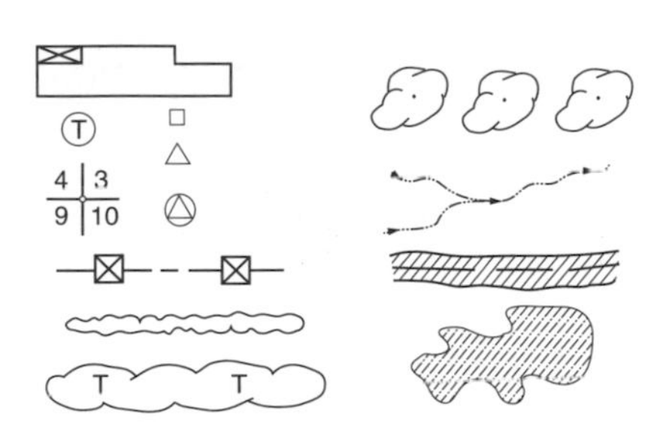

VICINITY MAP

It is advisable for a site plan to contain a large scale map of the overall area and to indicate where the project is located on the site.

TITLE BLOCK

All drawings should have title information about the project, owner, architect, and others involved in the work so they can be contacted if needed for information.

NAME		LIFT#
SCALE		REVISIONS
CHECKED		
PM		
SUPT		
O/A		
TO	KEY	

EXISTING CONTOURS WITH INDEX CONTOURS

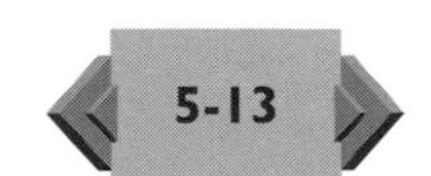

Contours are shown to facilitate an understanding of the general lay of the land. For a complete description of conturs, see the next section titled *Understanding and Using Contours.*

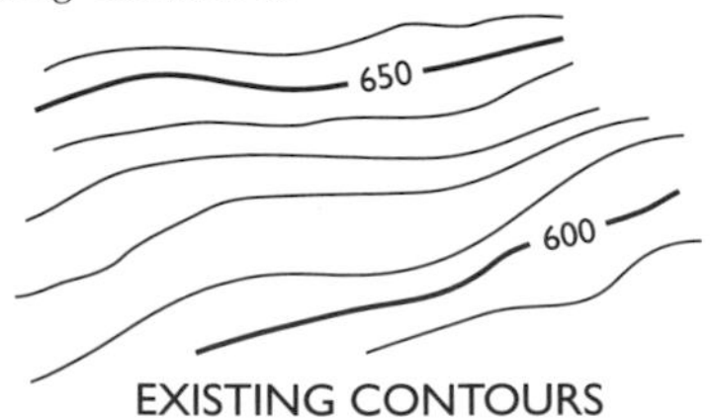

EXISTING CONTOURS

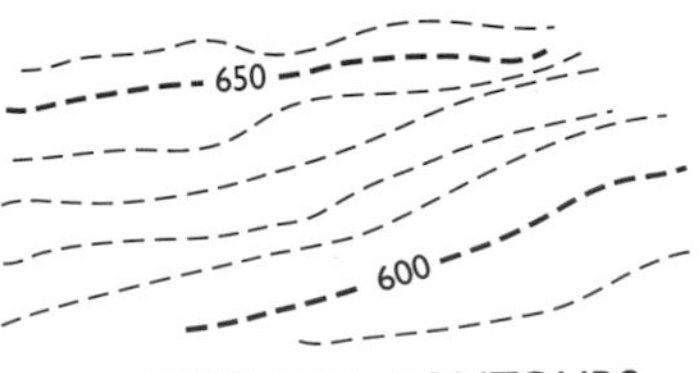

PROPOSED CONTOURS

COORDINATES

Coordinates of control points or property corners are shown or referenced for use by anyone performing layout on the site.

N 10050.00
E 9321.00

DIRECTION AND LENGTH OF LINES

The direction and length of property lines or control lines should be given. This information can be used by the field engineer to check the field measurements between points.

N 65°31'E
323.51ft

DESCRIPTION OF MONUMENTS

All control and property monuments should be fully described on the site plan or referenced to a field book so the points can be found when needed. The field engineer should walk the site and locate all property corners and control monuments (if provided).

ROAD NAMES

All roads shown on the site plan should be labeled as a reference in describing project entrances, etc.

UNDERSTANDING AND USING CONTOURS

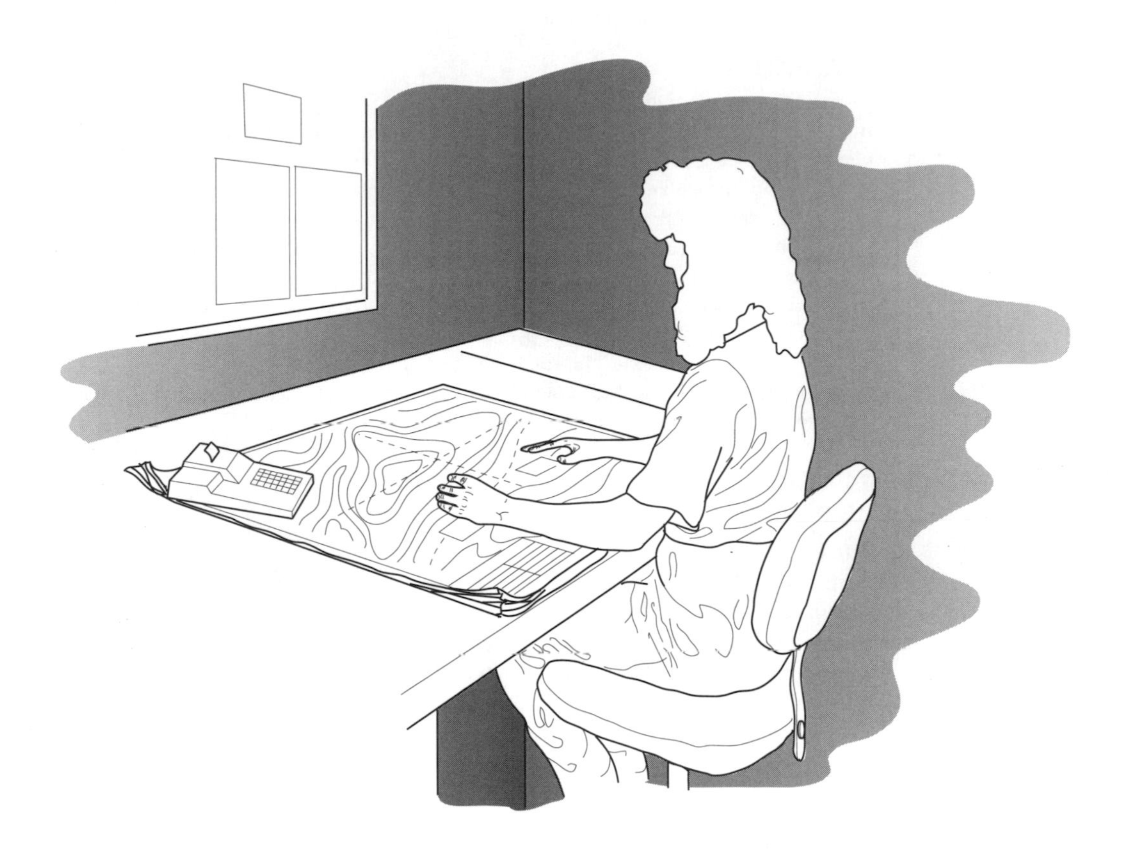

SCOPE

Contours are used by field engineers to determine the amount of cut or fill that will be needed on the project site. Close examination of the contours on a site reveal what the site looks like before the project starts and what it will look like after the project is completed. The knowledge of contours is extremely useful to anyone on the jobsite who will be dealing with the sitework and landscaping. The field engineer will be involved with contours throughout the project.

CONTOURS

DEFINED

The term "contour" is used to describe a line that is drawn on a map to represent a line of constant elevation on the ground. A contour is a line that would be formed if the surface of the ground was intersected by a level surface. For example, a contour that is visible to all is the edge of the water in a lake. When the water is still, a single line of constant elevation exists. But, as field engineers, we are not interested in just one contour along the edge of the lake. We are interested in the shape of the ground where we are going to be building.

INTERVALS

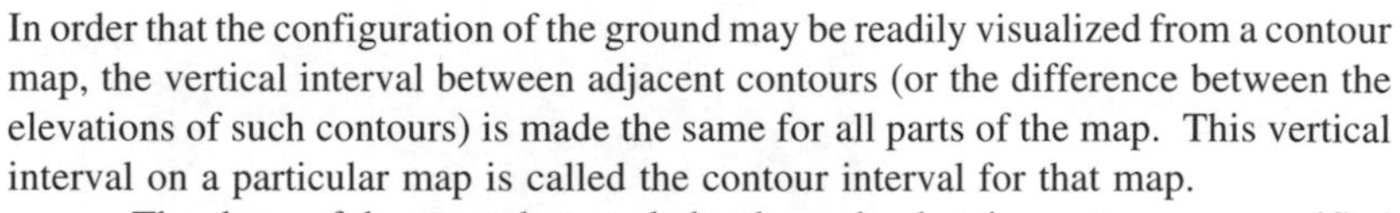

In order that the configuration of the ground may be readily visualized from a contour map, the vertical interval between adjacent contours (or the difference between the elevations of such contours) is made the same for all parts of the map. This vertical interval on a particular map is called the contour interval for that map.

The shape of the ground can only be shown by drawing contours at specified intervals over the area of interest. An interval might be 1\2-foot, 1-foot, 2-foot, 5-foot, 10 foot or even more. The amount of elevation detail desired is the determining factor in contour interval. If a great deal of detail is desired, the interval might be 1\2 foot. If the contours are just being used as a rough guide, then a 5-foot interval might be sufficient. The best contour interval depends upon the degree of accuracy desired, the slope of the ground, and the scale of the map. Properly located contours indicate elevations with a relatively high degree of accuracy.

SAMPLE FORMATIONS REPRESENTED BY CONTOURS

HILL

In the case of a hill or an island, each contour has the general form of a loop as shown here. Obviously, the higher contours are smaller than the lower contours and are entirely enclosed by them. The contour lines are drawn freehand and are made more or less wavy to conform better to the usual irregularities of the ground surface.

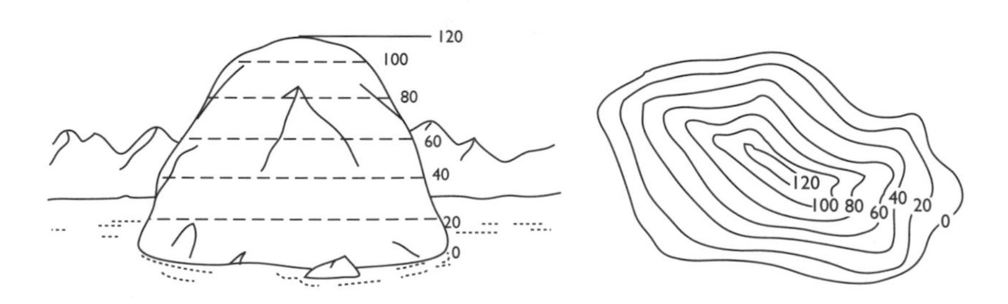

DEPRESSION

Contours of a depression or hollow in the ground have the same general appearance as contours at a hill. In the case of a depression, the lower contours are smaller than and enclosed within the higher ones.

If both a hill and a depression are shown on a contour map, it is customary to make an obvious distinction between the hill and the depression by drawing short lines perpendicular to the contours at the depression as shown here.

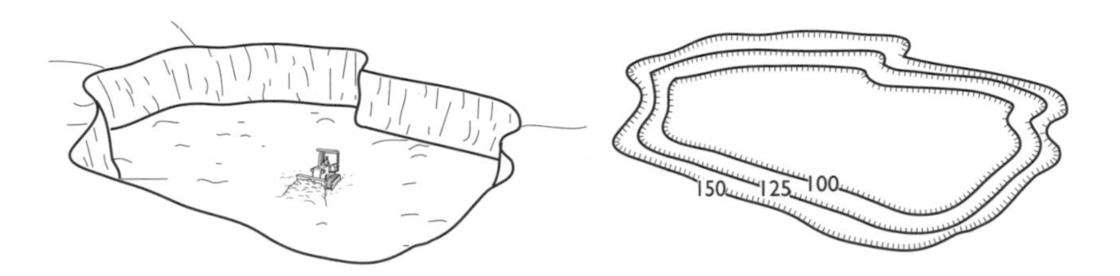

HIGHWAY CUT

In this illustration, a portion of a hill has been excavated to accommodate a highway. Notice where the ground is steep (as in the cut near the road), the contours are closer together. Where the ground is flatter, the contours are further apart.

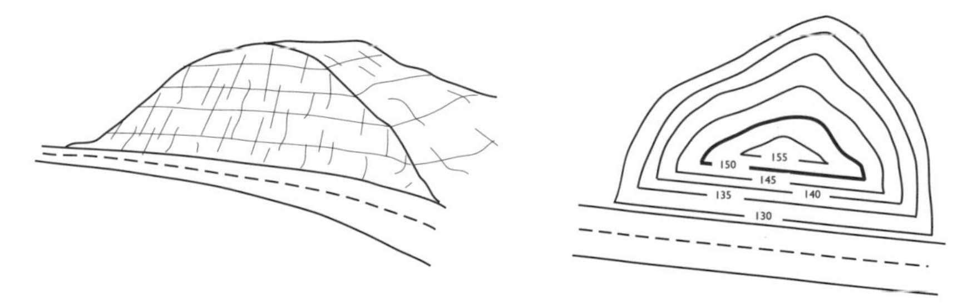

HIGHWAY FILL

In this illustration, a valley has been filled with dirt from a highway cut.

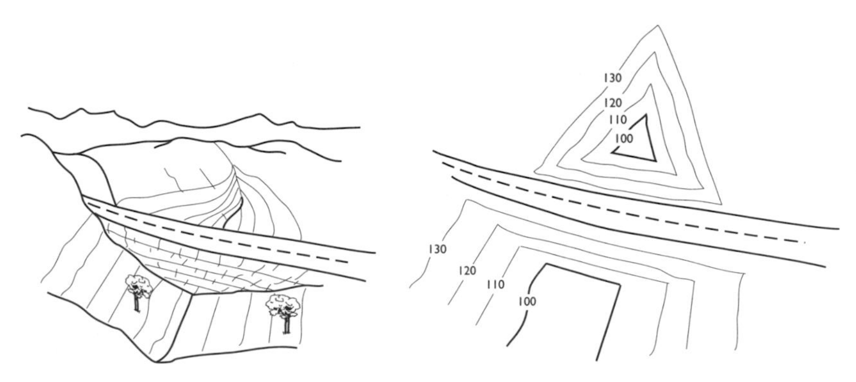

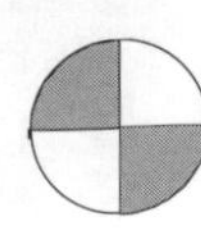

COMBINATION

The following illustrates various topographic features. Notice, for example, how a mountain saddle, a ridge, a stream, and a flat area are shown with contour lines.

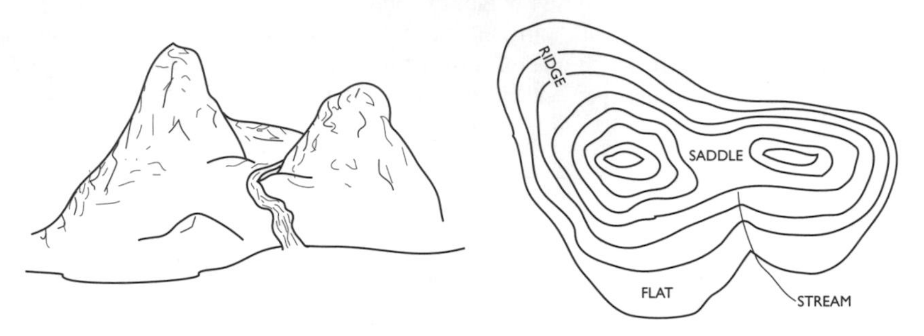

INDEXES

Indexing contours means to draw every fifth contour darker and more bold so the user of the site plan can easily follow the contour and refer to it when trying to determine the elevation of an adjacent contour.

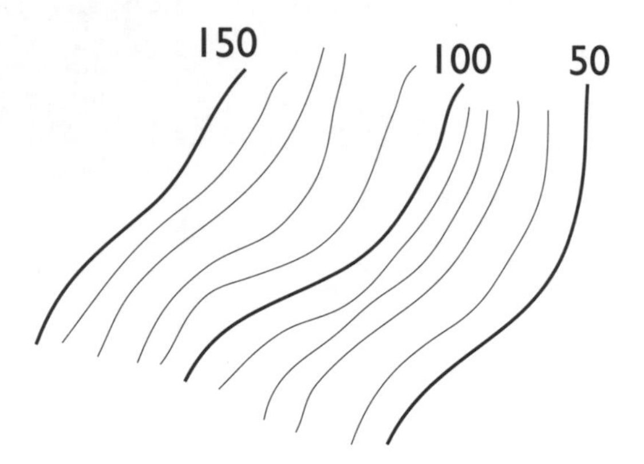

CHARACTERISTICS OF CONTOURS

ELEVATION

Contours represent a line of constant elevation. Everywhere on the contour line the elevation is the same. A good represenation of this is the edge of the water along a pond or lake. This line is essentially a contour that can actually be seen. It is the same elevation all around the lake.

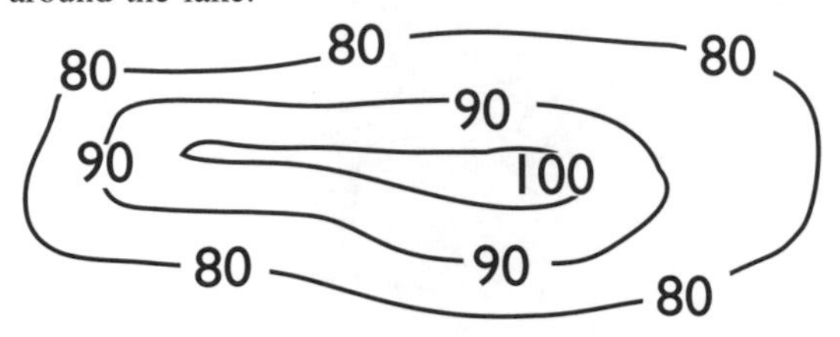

A CONTOUR ON THE GROUND CLOSES ON ITSELF

In other words, if a person starts at any point on a contour and follows the path of the contour, he will eventually return to the starting point. A contour may close on a site plan as indicated, or it may be discontinued at any two points at the borders of the plan, as shown in the illustration below. Such points mark the limits of the contour on the map, but the contour does not end at those points.

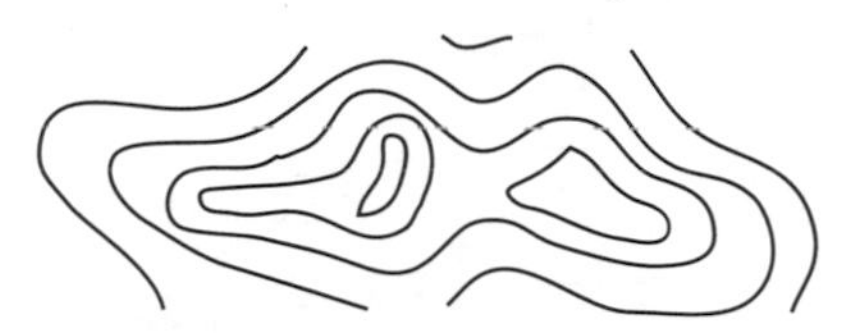

CONTOURS ARE AT RIGHT ANGLES TO THE SLOPE

The direction of a contour at any point is at right angles to the direction of the slope of the ground at that point. The direction of a contour at any point is at right angles to the slope of the ground at that point. That is, along a hillside, the contours run along the side of the hill and the slope is perpendicular to the contours.

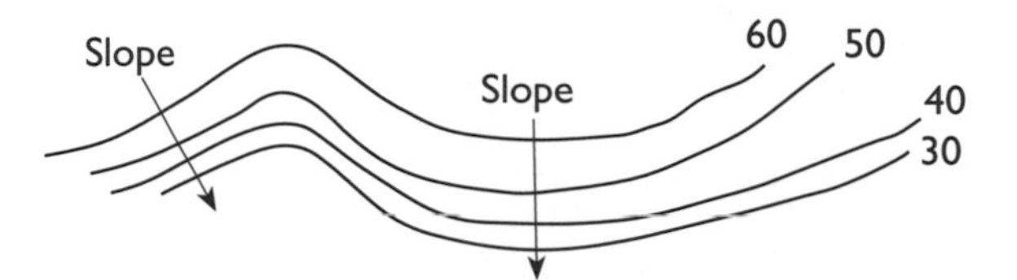

THE GROUND IS ALWAYS HIGHER ON THE SAME SIDE

As a contour line is followed on a topographic map, it will readily be seen that the ground higher than the contour is always higher on the same side of the contour. This is a very important characteristic that must be followed exactly when a field engineer is attmpting to draw contours from spot or grid elevations. If this is followed, there can only be one interpretation of where the contour is located.

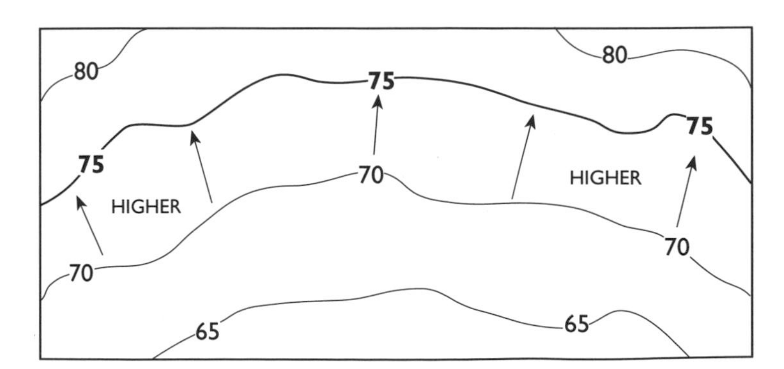

CONTOURS WON'T CROSS

Contours on the ground cannot cross one another, nor can contours having different elevations come together and continue as one line. However, where an overhanging cliff or a cave is represented on a map, contours on the map may cross. The lower contour or contours must then be shown as a dashed line. At a vertical ledge or wall, two or more contours may merge. A final check should be made of a newly completed contour map to avoid crossing or branching contours.

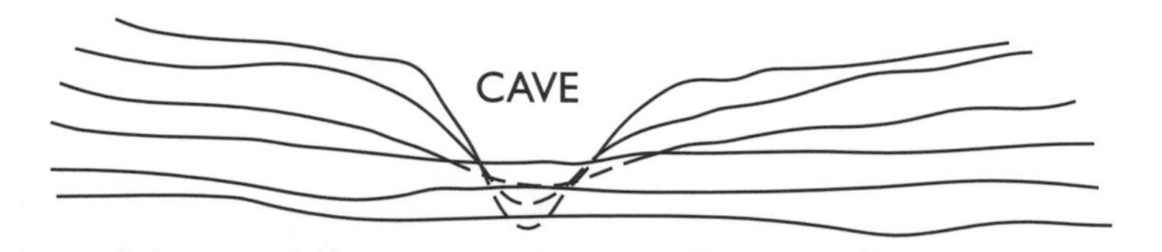

CONTOURS DISPLAY STEEPNESS OR FLATNESS

The horizontal distance between adjacent contours indicates the steepness of the slope of the ground between the contours. When the contours are relatively close together, the slope is comparatively steep, and when the contours are far apart, the slope is gentle.

Also, when the spaces between contours are equal, the slope is uniform; and when the spaces are unequal, the slope is not uniform. Review some of the illustrations shown to observe variable slopes.

CONTOURS POINT UPSTREAM

As a contour approaches a stream, the contour turns upstream until it intersects the shoreline. It then crosses the stream and turns back along the opposite bank of the stream. If the stream has an appreciable width on the map, the contour is not drawn across the stream but is discontinued at the shore and reappears on the opposite side of the stream.

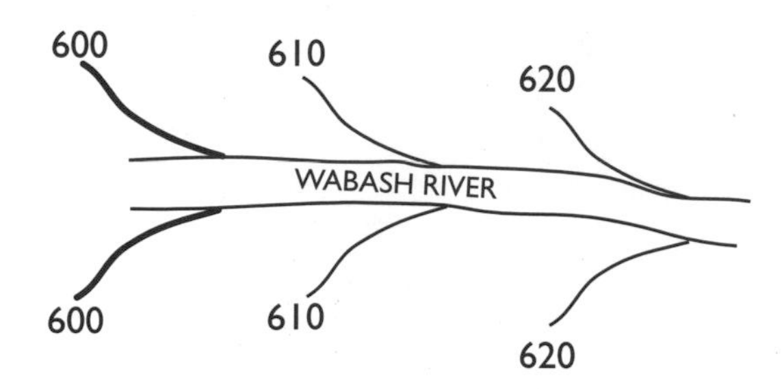

PREPARING CONCRETE PLACEMENT (LIFT) DRAWINGS

SCOPE

"Jobs that don't use lift drawings to save a few bucks in overhead are just kidding themselves!" This is a quote from one of the officers of a company. It has been shown that using lift drawings saves time and money in the field by clarifying the construction and, thereby, eliminating mistakes. Lift drawings are nothing more than a communications tool that puts everyone on the same page of the work—i.e., the lift drawing sheet. Care must be taken in the production of the lift drawings for a project. This section is intended to guide the field/office engineer in preparing lift drawings accurately.

GENERAL

PURPOSE

The purpose of a lift drawing is to accurately convey to the foremen what to construct, what to embed, and where to locate. Lift drawings are so named because information is "lifted" from numerous contract drawings and other sources to create a single set of drawings. This should be on one or two sheets and will serve to the foremen as a substitute for all other drawings pertaining to the lift with the exception of form and reinforcing drawings. The lift drawings also provide a bill of material for use by the warehouse, for pour checkout, for labor coding, for production reporting, and for recording of pay quantities. The use of lift drawings reduces the probability of errors or omissions.

PLANNING

The essential part of the drawing is the "plan" of the lift. The plan should be drawn to as large a scale as possible or practical. "Sections" should be taken at appropriate locations to illustrate the extent of all concrete outlines. NOTE: Sections should be located to show the maximum detail possible for each section. Offset sections should be avoided. Details should be made to further clarify information that appears on "plan" or "sections."

Before starting, determine the approximate area required for "sections" and "details" and choose the scale for the plan accordingly.

TYPICAL LAYOUT OF DRAWING

Plan in upper left of drawing.

Front elevation, if taken, below plan if space allows.

Sections to side of plan at right.

Details in lower right center. A separate drawing may be used for details, if required.

Key Plan or Section in the extreme upper right of drawing.

Notes along right hand margin.

Title in the lower right corner.

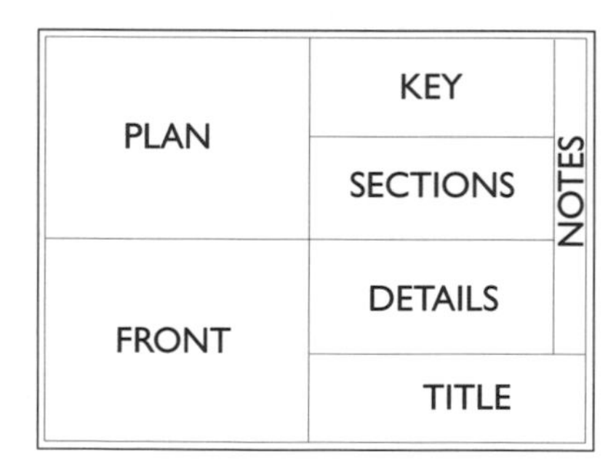

SOURCES

Information for lift drawings can come from a variety of different types of **drawings** including: Contract drawings, manufacturer's drawings (shop drawings), and sub-contractor drawings. Information can also be obtained from **suppliers'** catalogs and cut sheets. The **craft superintendents** can provide helpful information regarding installation dimensioning for such items as drainage piping, electrical conduit, etc.

INFORMATION SHOWN ON DRAWINGS

See examples at the end of this section.

BASICS

All concrete outline information is to be shown including openings, passages, and recesses, complete with individual and locating dimensions.

REFERENCE LINES

Every drawing should have at least one reference line which is used for actual field location, such as:

centerline of unit

an offset line

appropriate station

Drawing dimensions = field dimensions

DIMENSIONS AND ELEVATIONS

Dimensions used on drawings should be those which will actually be used in the field. Additional dimensions can be added to facilitate a check in the field or office. All dimensions should be closed.

Indicate elevations at top and bottom of "sections," and at other intermediate points as necessary. Dimension intervals between elevations. Dimensions are usually given to the nearest 0.01 and elevations to the nearest .01 of a foot.

EMBEDS

All embedded material information should be included but is not limited to:

piping

pipe sleeves

anchor bolts

inserts

frames

conduit

ground wire

welding plates

All of these items must be shown - complete with location dimensions. Identify embedded items on the drawing by stating their size, kind of material, supplier's mark number, and any other information that will aid field personnel in locating the item quickly.

Embedded pipe and electrical conduit should be detailed on the drawing. Pipe should be detailed with size, type, and elevation (centerline or flowline) at fitting locations and ends. Electrical conduit should be identified by size and run location. Although the embedded pipe and conduit may be relocated in the field, the lift drawing process will identify major conflicts that can be resolved prior to work in the field.

MATERIALS

Each drawing has a bill of materials showing all the items that will be embedded in the lift, plus the volume of concrete to be placed. The bill of materials should show the following:

BILL OF MATERIALS
BID ITEM
MARK NUMBER
ITEM DESCRIPTION
UNIT WEIGHTS, IF APPLICABLE
UNIT OF MEASURE
ESTIMATED QUANTITY

FINISHES

Indicate finished-surface outlines on "plan," as wood float, street trowel, or broomed as required. Indicate direction of concrete floor slopes and give appropriate elevations on "plan" at break points, hatch frames, and other openings and edges. Also indicate surface treatments and joint fillers as required.

PRODUCTION SUMMARY

Each drawing will have a production summary which is used for coding labor on time cards and for reporting production. The summary should include such things as the Labor Recap Code, Description, Unit of Measure, and Production Quantity.

REFERENCES

Reference drawings to adjacent lifts, identified by lift numbers. Shade portions of adjacent lifts as shown on the drawing. Put shading on the back side of the drawing if the drawing is being made into blueprints. Any color except yellow is satisfactory.

TITLE BLOCK INFORMATION

Every drawing should have a title block that includes basic informations such as the name of the structure, the work area that the drawing is for, and the number of lifts that are required to complete the work. If different drawings are used, the scale of each "plan," "section," and "detail" should be shown on each drawing. Title blocks should include a revision block identifying revisions made to the drawing after it is initially approved. Each revision should be identified on the drawing along with the revision date and a brief description noted in the revision block. Once the revised lift drawing is completed and checked by the approving parties (Project Engineer, Superintendent, Owner/Architect), the drawing will be stamped "Approved for Construction." The approval stamp should be dated and initialed. A distribution log showing the recipient's name, quantity of drawings, and the date given should be kept. Lastly, a key plan with the appropriate lift shaded in should be included.

NAME		LIFT#
SCALE		REVISIONS
CHECKED		
PM		
SUPT		
O/A		
TO	KEY	

DRAFTING AND LAYOUT TECHNIQUE

EASY READING

Use large, easy-to-read, views and lettering. Excessive sun and wet weather are very hard on lift drawings.

GOOD LINE WEIGHT

Line weight is important. Use extra heavy lines for concrete outlines. Use light lines for dimensions to provide good contrast. Maintain a good clearance between the concrete outline and dimension lines.

NOTES

All notes should be placed outside of dimension lines with arrow lines referring them to noted objects.

REFERENCE

Refer to the *Field Engineering Drafting and CAD Fundamentals* section in this chapter for more information on drafting techniques.

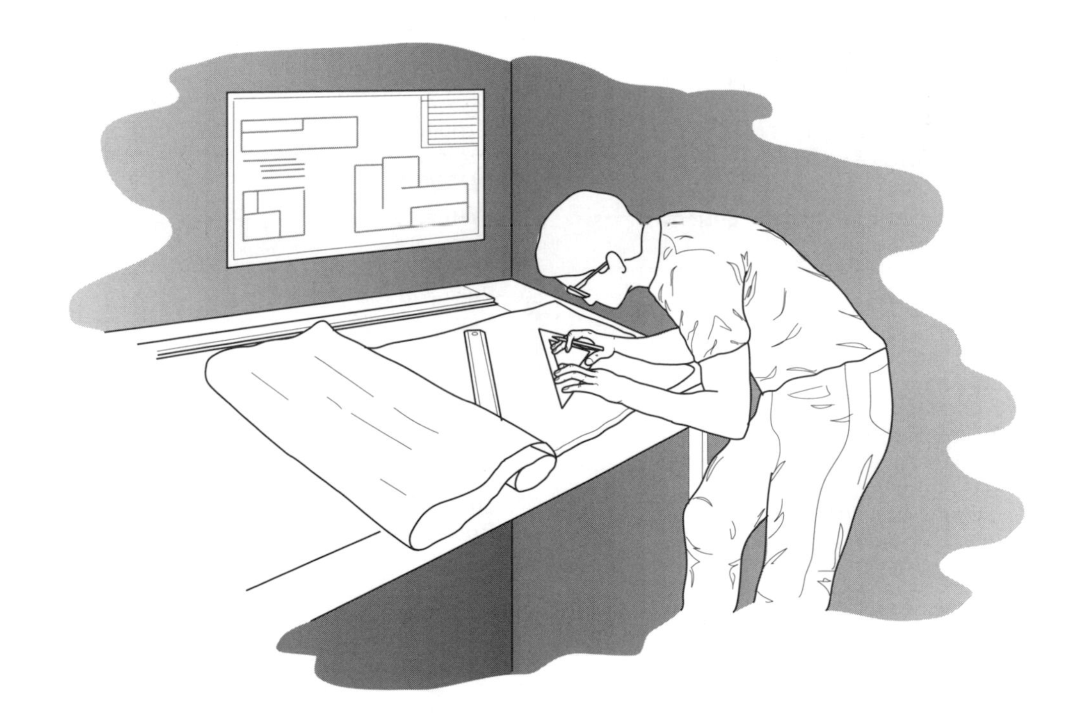

LIFT DRAWING CHECKLIST

Title Block
- ☐ Locator
- ☐ Scale
- ☐ Date
- ☐ Initials

Reference Drawings
- ☐ Structural
- ☐ Architectural
- ☐ Mechanical
- ☐ Electrical

Concrete Drawings
- ☐ Outside dimensions from column lines
- ☐ Blockout and bulkhead dimensions
- ☐ Location of blockouts from column lines
- ☐ Formed face dimensions
- ☐ Top and bottom elevations

Structural
- ☐ Horizontal waterstop and splices
- ☐ Vertical waterstop and splices
- ☐ Finishes
  1. Tops of slabs
  2. Tops of walls
  3. Sides of walls
- ☐ Keyway location and dimensions
- ☐ Adjacent pours

Architectural
- ☐ Door Frames
- ☐ Floor and room finishes
- ☐ Stairways
- ☐ Embedded frames
- ☐ Miscellaneous metals
- ☐ Mark numbers consistent with plans and shop drawings
- ☐ Dimensions from column lines

Pay Quantities
- ☐ Quantities match production
- ☐ Concrete to nearest tenth of a CY
- ☐ Cement and Pozzolan to nearest hundredth of a ton
- ☐ Round off all other quantities to the next largest integer

Mechanical
- ☐ Embedded Pipe
  1. Cast iron pipe and floor drains
  2. Ductile iron pipe and fabricated fittings with mark numbers
  3. Steel and others
  4. Check for form clearances
- ☐ Equipment Locations
  1. Centerline location of equipment
  2. Anchor bolt pattern
  3. Size of concrete base
- ☐ HVAC and fire protection
- ☐ Unistruct and embedded sleeves
- ☐ Exposed piping; check for additional embeds
- ☐ Plumbing

Electrical
- ☐ Conduit
- ☐ Fire protection
- ☐ Lighting
- ☐ Grounding

Production
- ☐ Waterstop
- ☐ Concrete
- ☐ Forming
- ☐ Bulkhead forming
- ☐ Keyway
- ☐ Piping and equipment
- ☐ Joint preparation
- ☐ Finishes (point and patch, floor hardener)
- ☐ Hanging forms
- ☐ Cure and protect
- ☐ Steel items
- ☐ Round quantities to integers

EXAMPLE LIFT DRAWINGS

Lift drawings for a vault are shown on the next few pages. Study them carefully for a summary of all of the information that must be included on lift drawings.

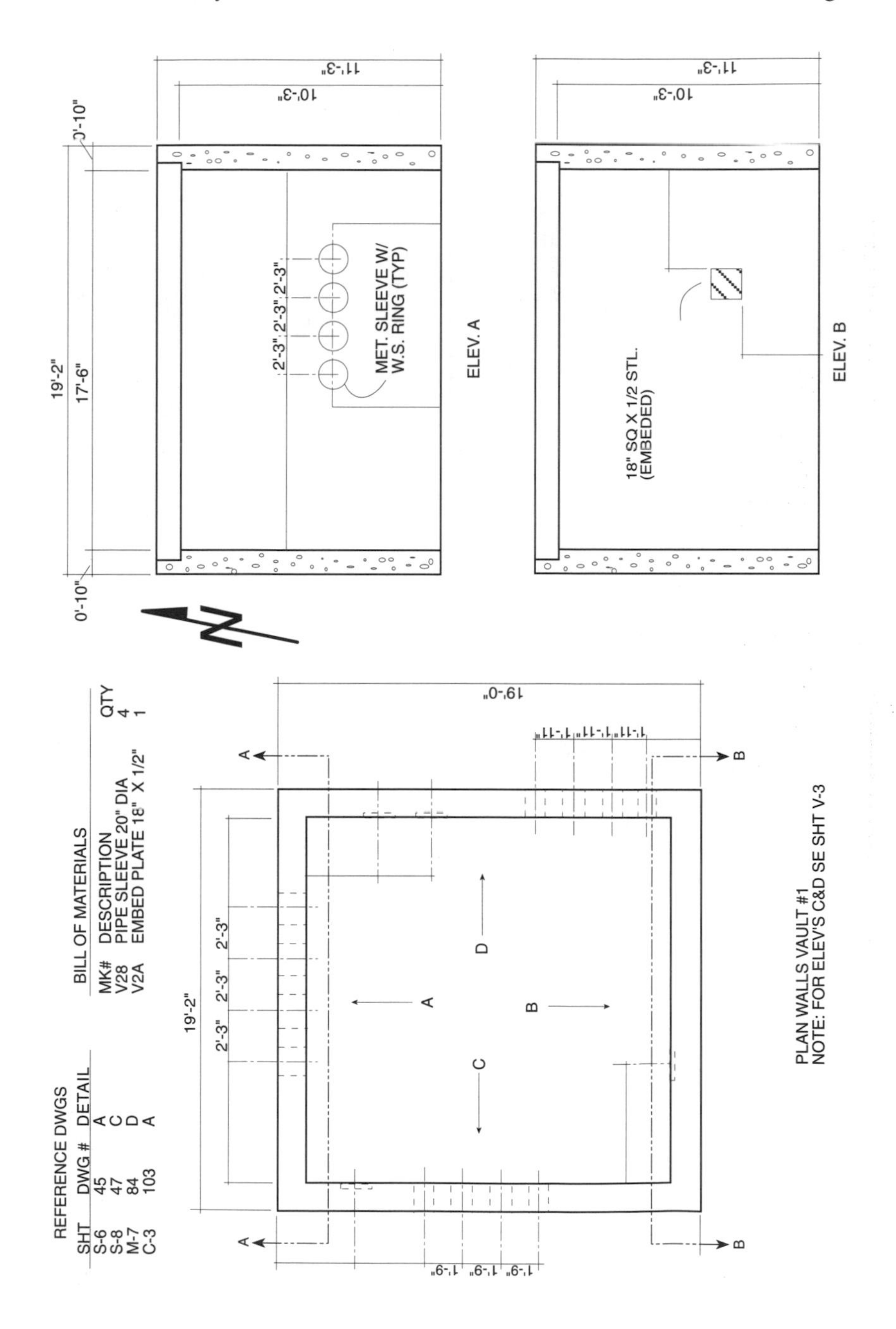

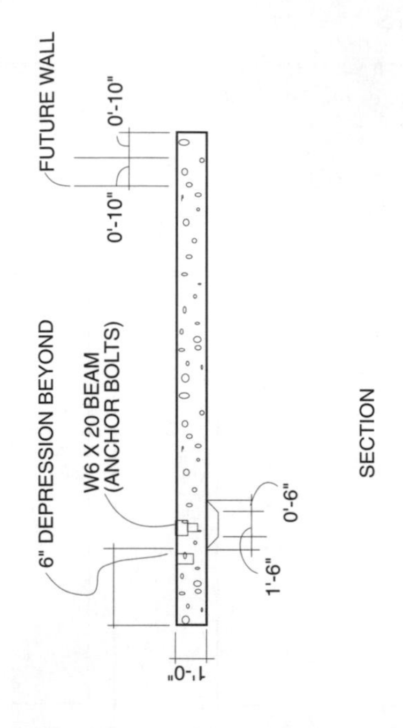

FUTURE WALL
0'-10"
0'-10"
6" DEPRESSION BEYOND
W6 X 20 BEAM
(ANCHOR BOLTS)
0'-6"
1'-6"
1'-0"
SECTION

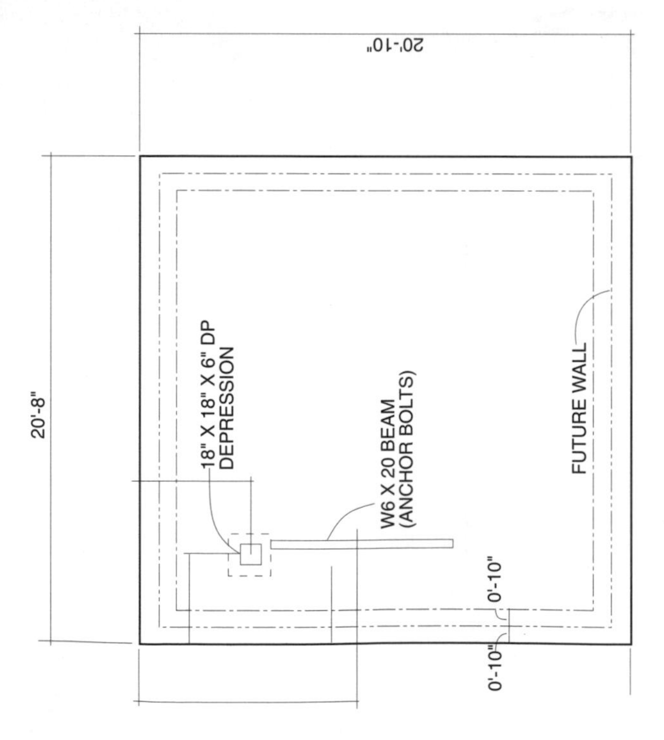

20'-10"
20'-8"
18" X 18" X 6" DP
DEPRESSION
W6 X 20 BEAM
(ANCHOR BOLTS)
FUTURE WALL
0'-10"
0'-10"
TOP PLAN VAULT NO. 1

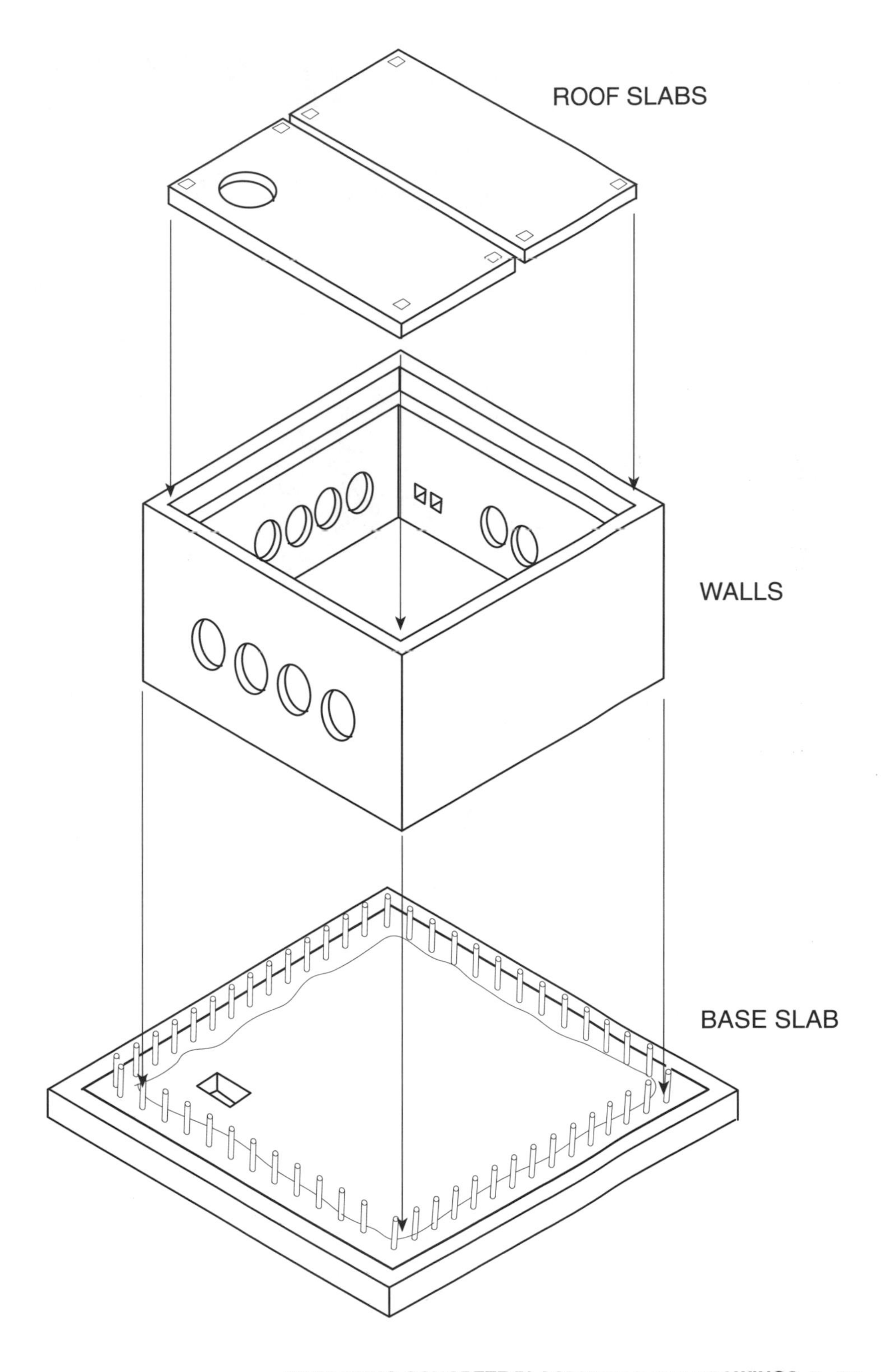
ROOF SLABS
WALLS
BASE SLAB

PREPARING REPORTS, LOGS, & DIARIES

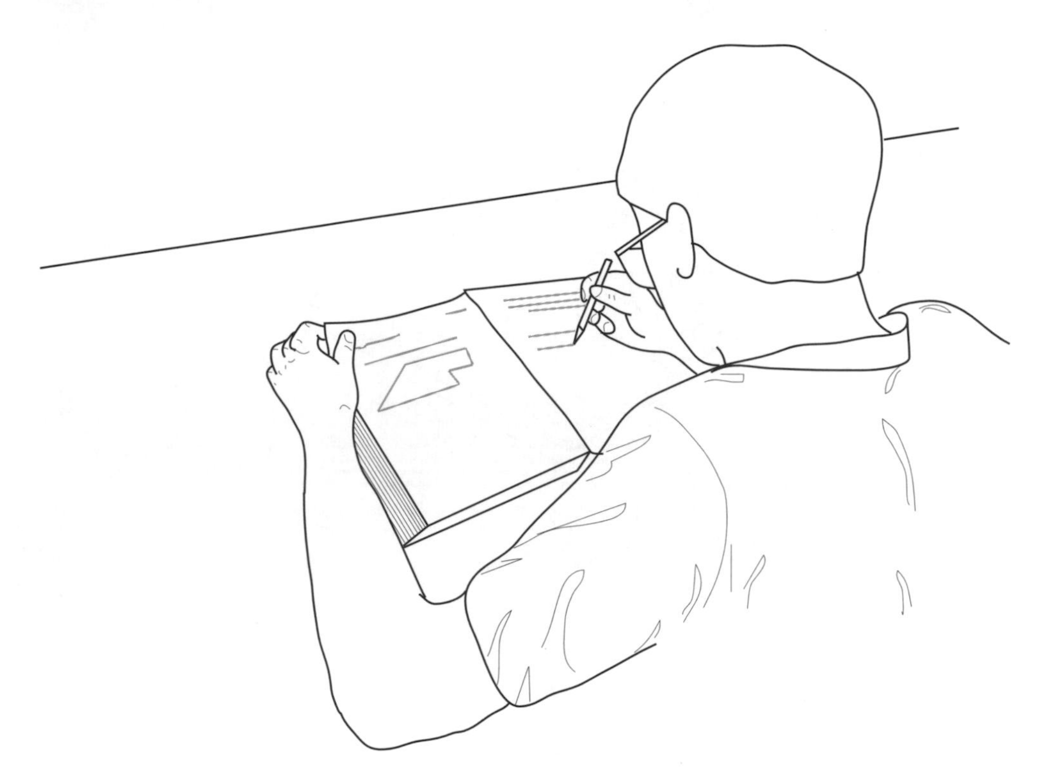

SCOPE

The duties of a field engineer go far beyond layout and control work. One of the most important responsibilities is that of documentation and recordkeeping. This will include the documentation of all engineering work performed in the field, as well as administrative records such as production reporting, project and subcontractor logs, caisson logs, concrete placing logs, and trend charts. These records are vital to the evaluation of project status and unit cost accounting.

The success of a project, and ultimately the success of the company, depend to a great extent on how well the cost monitoring system functions. The field engineers may be responsible for gathering the basic input information. This information will include calculating the labor time, the cost codes to which these times are charged, and quantities of work accomplished. The field engineer must be impressed with the importance of accurate reporting. Inaccurate cost records yield nothing of real value and often defeat the basic reasons for keeping records.

PLANNING

Paperwork may not be one of the more glamorous duties of a field engineer. However, it is one of the most important, and it is necessary that it be thoroughly completed on time. To accomplish this as needed will require careful planning of the daily schedule. Purposely set aside a time block each day for the sole task of completing the paperwork. A field engineer who is a true professional will treat recordkeeping as one of the most important duties of the day.

It is also the responsibility of the field engineer to maintain a well-stocked inventory of all necessary forms.

ENGINEER'S FIELD BOOK

The Engineer's Field Book should be considered the heart of all field engineering activities. It will be in this book that all pre-planning, control information, layout sketches, work activities, and project conflicts will be noted. The field book does not simply represent a means of notekeeping. Rather, it represents a legal document which may be subpoenaed to court should conflicts arise. It should be maintained in a neat and orderly fashion with consecutive pages dated and prepared in ink. All information should be described and stated clearly so it may be correctly interpreted by someone else. At the conclusion of the project, each Engineer's Field Book should be numbered and labeled before it is submitted with the project's permanent files.

OBSERVATION

While pouring concrete today at level 4, I noticed the rebar stiffups on level 5 were bent at 45 degrees rather than 30 degrees per specifications. Check with architectural representative to see if changes were made to material list.

Jan 20,1992

Also 6 workers in placing crew. Note that on production report.

Jon Humpherys

DAILY PROJECT LOGS

The Daily Project Log is the only complete report of day-to-day activities on the project. This report is designed to be as specific as possible while including all pertinent information. The project log should be completed by filling in all sections in detail. This will include weather conditions, concrete pour locations, employees hired and laid off, crew sizes, and descriptions of all subcontractor and project activities. These logs will then be reviewed by the superintendent, project manager, and area and district managers, if needed. When filling out logs, be sure to include all unusual or unique activities or circumstances which may have occurred that day. These are the items which may be of significant importance should this log ever be entered in court.

SUBCONTRACTOR DAILY REPORTS

Subcontractor Daily Reports are completed each day by the subcontractor's superintendent or jobsite foreman. These reports describe crew size, areas of activity, and any conflicts which need to be brought to the contractor's attention. These subcontractor reports will be maintained as part of the project's permanent files. The subcontractor may choose to make copies of this report and submit it to the main office as well.

Sample Daily Construction Report

Contract no.________________________ Date ________________________

Description and Location of Work____

Weather: ( )Clear, ( ) P. Cloudy, ( ) Cloudy. Temperature: Min._____ Max. _____

Rainfall: _____inches

Contractor / Subcontractors & Area of Responsibility w/Labor Count for Each

a. ____

b. ____

c. ____

d. ____

e. ____

Equipment Data: Indicate items of construction equipment (other than hand tools) at the jobsite, and whether or not used.

____

____

____

____

____

DAILY AND WEEKLY PRODUCTION REPORTS

At the end of each week, the weekly production report must be filled out with the production quantities completed during that week. Each cost code which has had time charged against it must have corresponding production reported. Setting up a check sheet of all cost codes will be helpful to ensure that all codes which have time charged against them have production recorded. The field engineer should request that all foremen submit their daily time cards for review at the end of each work day. This will enable the cost code check sheet to be current. Production should also be collected on a daily basis and tabulated on a form such as the one shown below. The collection of data on a daily basis will not only improve the accuracy of that data, but it will eliminate the need for late Friday afternoon quantity takeoffs.

SAMPLE PRODUCTION REPORT FORM

Daily production reporting forms take many shapes and may include very detailed questions. A simple form for reporting production is shown.

Sample Production Report

Date	Activity/Description	Quantity Reported	Quantity To Date	Balance

USING TREND CHARTS

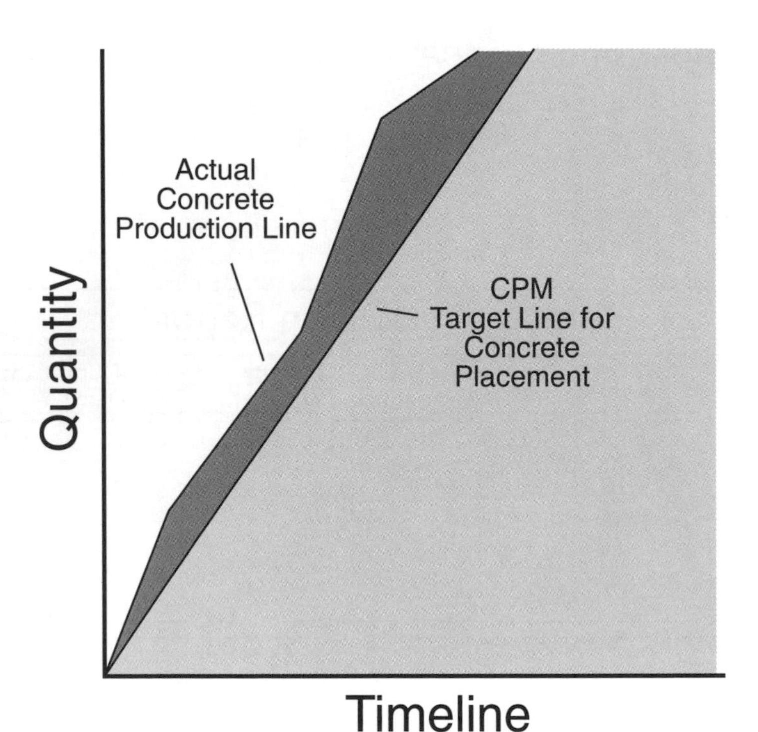

SCOPE

In order to complete a project on schedule, communicating the progress of the work in place is extremely important to everyone on the construction site. One of the tools that is used to communicate with everyone is a graphical representation of the progress called a trend chart.

Trend charts are simple but effective tools that show, in an instant, activities that are ahead of schedule, on schedule, or behind schedule. If activities are behind schedule, everyone knows it and action can be taken to accelerate the activity. Specifically, trend charts graphically depict the progress of construction activities and show schedule requirements as determined by the Critical Path Method (CPM) schedule.

BASIC SETUP OF TREND CHARTS

The principle lines on a trend chart are:

VERTICAL AXIS

The vertical axis is used for quantity of work in place. It can be cubic yards of concrete, tons of steel, pieces of steel, square feet of drywall, etc. Just about all of the activities on the project can be quantified by some unit.

HORIZONTAL AXIS

The horizontal axis represents the timeline in which the activity must be completed. Depending on the detail, it can be days, weeks, or months. During a critical time, it could even be hours.

CPM TARGET LINE

The CPM target line is determined from the CPM schedule to show the start and finish dates of the activity. Typically, Early Start and Early Finish Dates of the activity are shown.

ACTUAL PRODUCTION LINE

The actual production line is determined by daily or weekly monitoring of the work in place plotted from data shown on the production reports of the field engineer. The fundamental relationships of a trend chart are shown on the example below.

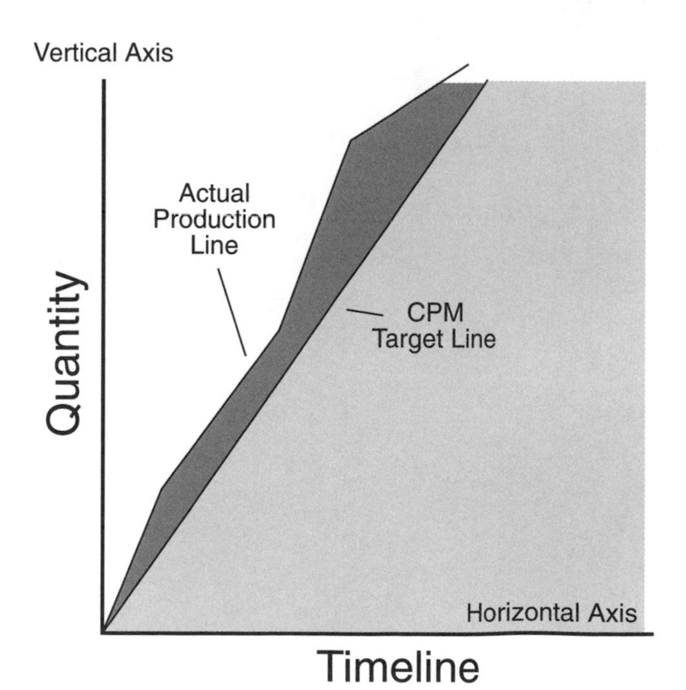

TREND CHART INDICATING SUCCESS

On a trend chart for concrete placement, it can be determined at a glance that the relationship between the actual production line and the CPM target line indicates that the production of concrete was good throughout the project, as it never dropped below the CPM Target Line. Undoubtedly, the project engineer and the superintendent paid close attention to this chart and were pleased throughout the project that it consistently showed positive production.

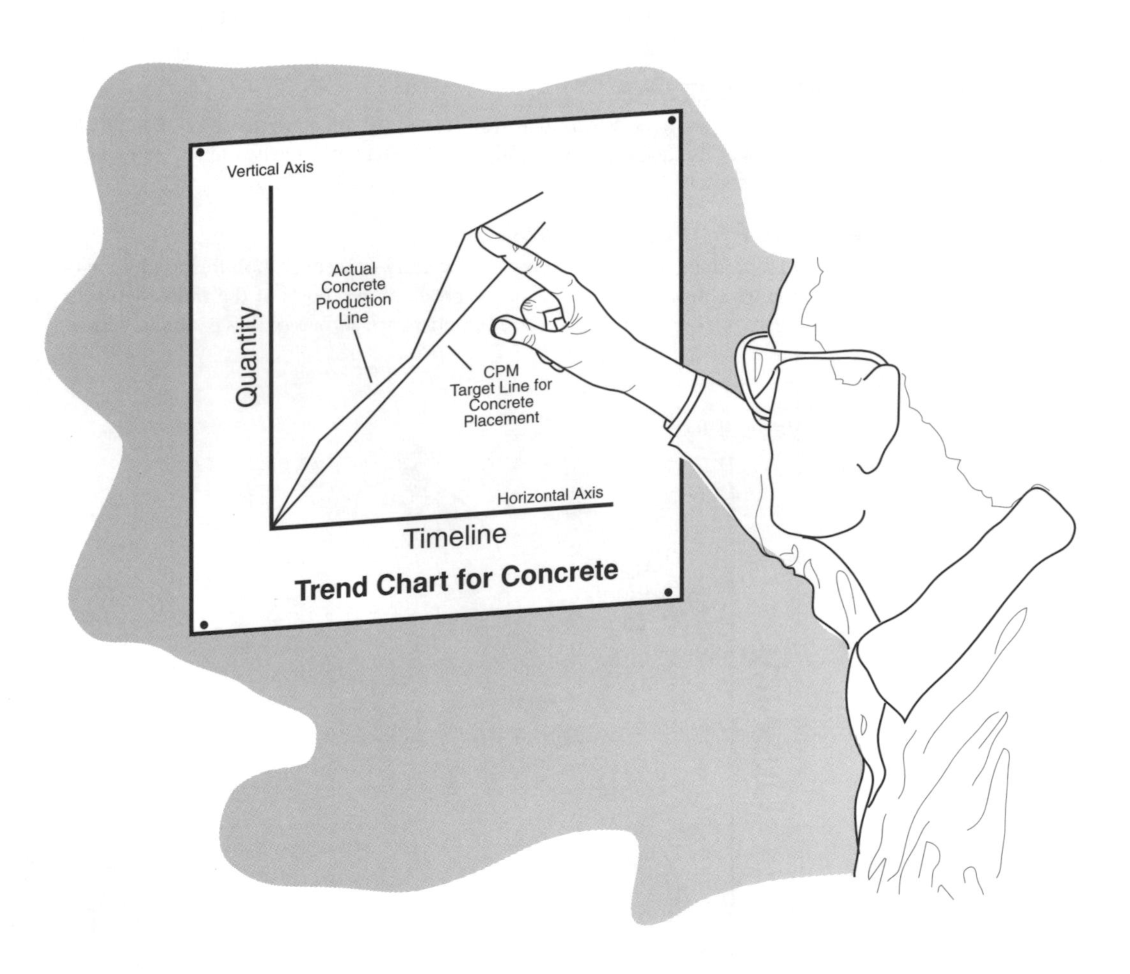

TREND CHART INDICATING A PROBLEM

An example that might cause some concern and action by the superintendent at the start and almost to the end of a project is shown below.

On a trend chart for a drywall subcontractor, it can be quickly seen that the production fell behind from the first day of the project. Action was taken, and the activity finally finished ahead of schedule. It is probable that the superintendent was working on this activity throughout the project and was relieved when it crossed over the CPM Line to be ahead of schedule.

It should be apparent by now that trend charts are indeed an excellent communication tool that can be used by all subcontractors and craftspeople on the job. Trend charts may be copied and passed around for everyone to take back to the office. A better method is to prepare them on a large sheet of paper and hang them in a place that is highly visible to everyone. They can be constantly monitored for the work progress every day.

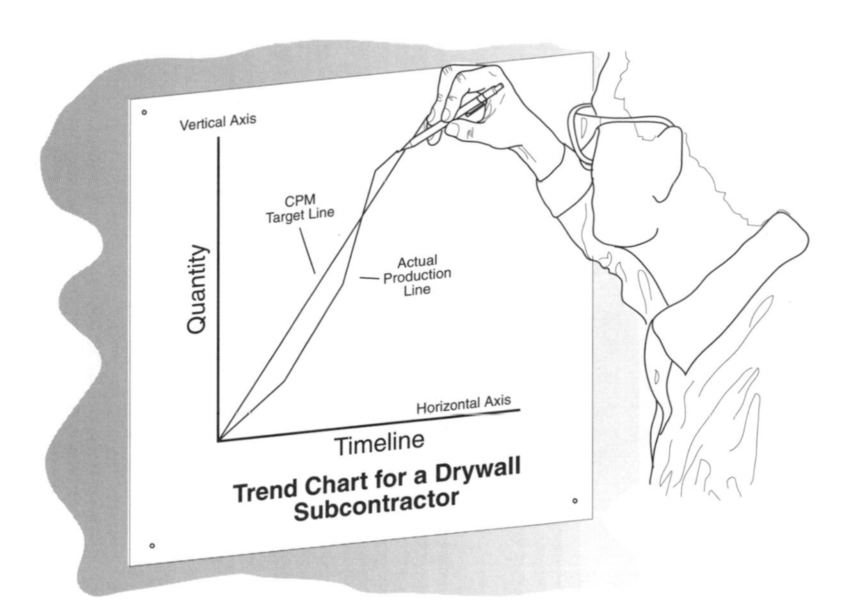

TREND CHARTS AS HISTORICAL DATA

The trend chart can also be used as historical data for use on future jobs. Any relevant information should be recorded on each trend chart such as the type and size of building, cost, location, weather conditions, subcontractors performing work, etc. Contact the Chief Estimator to see if the information is needed.

At the end of the job, it is ideal to prepare a manual with copies of each trend chart used on the job with all pertinent historical information so it can be filed with other job records.

QUALITY CONTROL RESPONSIBILITIES

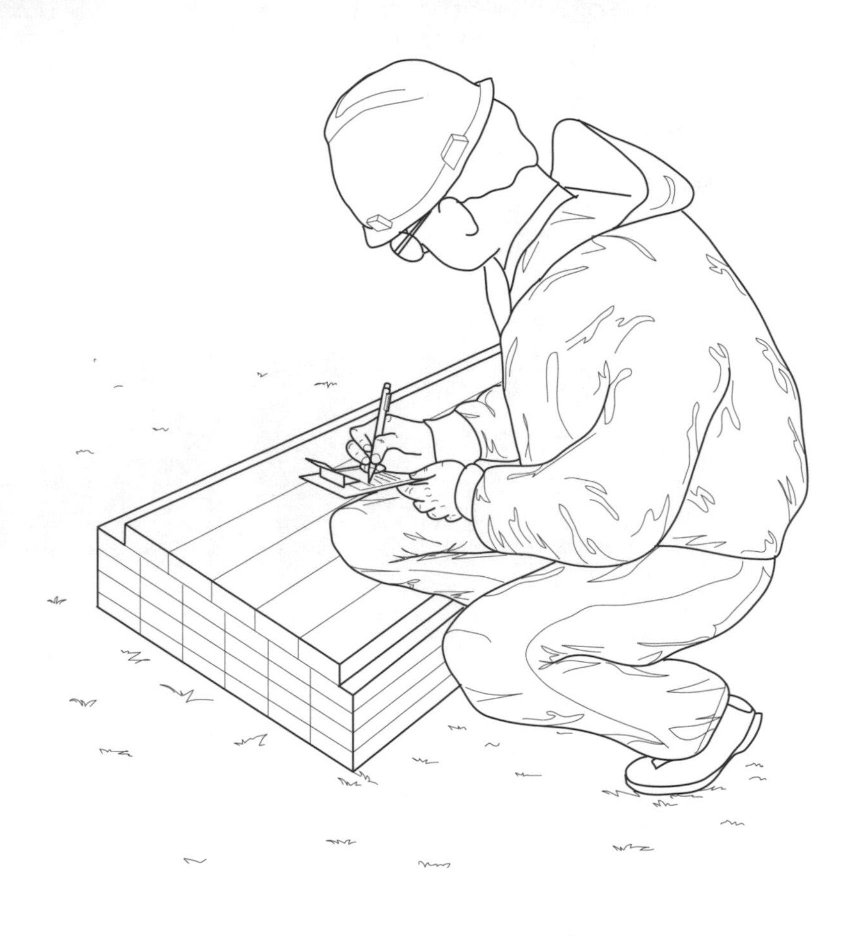

SCOPE

The duties of a field/office engineer extend into quality control inspection. Inspections consist of checking work for the proper dimensions, location, and materials. Quality control takes place in every phase of construction. Therefore, a field/office engineer is responsible for work performed by the general contractor and the subcontractors. However, a field/office engineer is not responsible for all inspections; part of the quality control duty is the coordination of inspections performed by others.

PLANNING

Planning quality control work is a most important step. Sufficient time should be allocated to quality control inspections each day. A field/office engineer must interpret schedules, know what work is occurring, and know what work to inspect. To accurately inspect work, the field/ office engineer must understand the plans and specifications for each work item. Therefore, it is essential that the field/ office engineer review the plans and specifications before the work begins. Because most work items form the base for future work items, these inspections must take place at the proper times. Planning these inspections is vital to preventing delays in construction.

LAYOUT

The quality of layout determines the quality of construction. A field/office engineer's primary duty is layout, and his or her first quality control inspection is to double check the layout. The rule of thumb is do everything twice. All layout work must be checked by measuring distances, turning angles, running level loops, and establishing control. Mistakes in layout must be detected to prevent rework. It is far less costly to double check layout than it is to demolish and rework an item.

INSPECTIONS BY OTHERS

Field/office engineers are not qualified to inspect all areas of construction. Mechanical, electrical, and specialty work should be inspected by qualified people. Part of a field/ office engineer's quality control work is the coordination of these inspections to occur at the proper times. Inspections must be timely to prevent delays in construction, and the field/office engineer must communicate with all subcontractors to plan these inspections. If proper communication does not take place, uninspected work might be covered up by other types of work.

PLACEMENT CHECKLISTS

A placement checklist can be developed for all types of work. This checklist enables the field/office engineer to systematically check work before it's started. It also ensures that similar work items are checked using the same criteria.

A prime example of work that needs a placement checklist is cast-in-place concrete. A checklist is needed because of the complexity and cost of this work. The checklist should be completed by a field/office engineer as well as a foreman. An example of a concrete placement checklist is shown on the next page.

CONCRETE PLACEMENT CHECKLIST

Some of the items on this checklist are not applicable to all jobs. On some jobs, items might need to be added to this checklist. Whatever the situation might be, a suitable checklist should be developed for each job.

Placement checklists should not be limited to concrete work. Checklists can also be developed for masonry, structural steel, drywall, and other areas of construction. Regardless type of work being inspected, a placement checklist will greatly assist the field/office engineer.

Concrete Placement Checklist

Location:____

Date:____________________ Time:____________________

Scheduled Pour Date: __________ Type Finish:________________

Mix Design:____________________ Slump:____________________

	Foreman	Engineer
Subgrade		
Elevation	☐	☐
Compaction	☐	☐
Temperature	☐	☐
Reinforcing		
Support	☐	☐
Clearance	☐	☐
Size	☐	☐
Spacing & Count	☐	☐
Clean/Grade	☐	☐
Forms		
Line & Grade	☐	☐
Support/ Bracing	☐	☐
Chamfer	☐	☐
Key way	☐	☐
Clean/Oil	☐	☐
Bulkhead	☐	☐
Architectural	☐	☐
Blockout		
Size	☐	☐
Location	☐	☐

	Foreman	Engineer
Anchor Bolts		
Line & Grade	☐	☐
Projection	☐	☐
Size / Type	☐	☐
Electrical		
Embeds	☐	☐
Blockout	☐	☐
Line & Grade	☐	☐
Mechanical		
Embeds	☐	☐
Blockout	☐	☐
Line & Grade	☐	☐
Water Stop	☐	☐
Specialties	☐	☐
Screeds	☐	☐
Other____________________		
Final Clean Up	☐	☐

Release to place concrete

Superintendent____________________ Time: __________ Date: ________

QC Inspector____________________ Time:__________ Date: ________

TESTING

Many times materials used in construction will be tested in the field to see if they meet the specifications. A field/office engineer should be familiar with the tests that will be performed. Again, concrete is a prime example of a material that will be tested on the jobsite. A field/office engineer should know how to perform a Slump Test ASTM C143, Temperature Test ASTM C1064, Yield Test ASTM C138, and an Air Content Test ASTM C231. The American Concrete Institute publishes books which will help a field/office engineer learn these tests.

AS-BUILT SURVEYS

As-built surveys are performed by a field/office engineer to determine if finished work meets the specified tolerances and has the proper size and location. Finished work should be surveyed immediately after the work is completed. In case there is a problem with the work, this would identify where corrective action is needed. Some areas that are normally checked are interior elevations, flatness of floors, and locations of embeds. Accurate notes of as-built surveys need to be kept in the field book. This ensures that the work has been checked and that the work was in accordance with all tolerances and specifications.

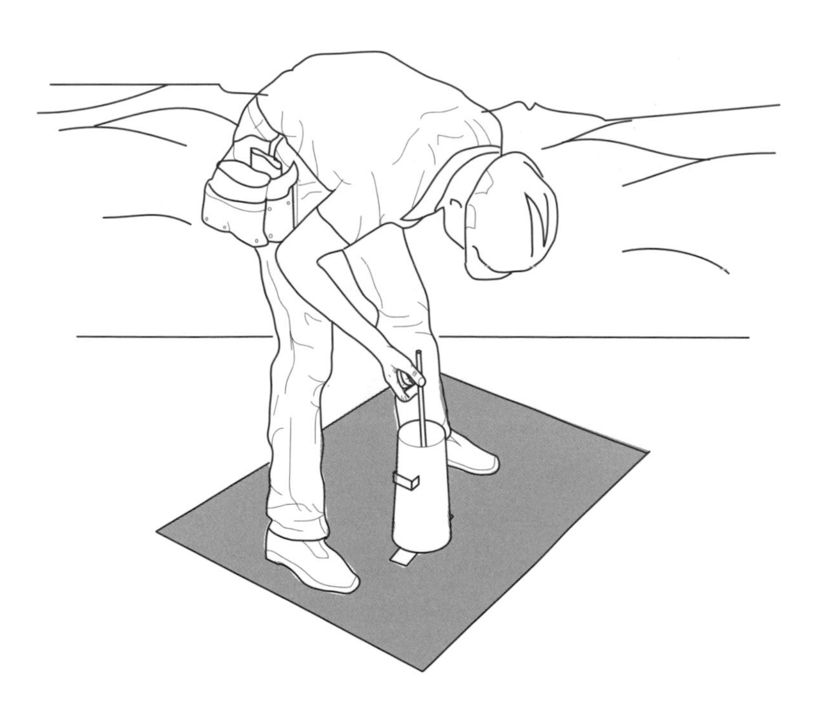

DISTANCE MEASUREMENT - CHAINING

6

CHAPTER OBJECTIVES:

List the tools a field engineer uses to measure a distance by chaining.
Describe the following chaining basics: good alignment, measure horizontally, proper tension, and measure twice.
State the key to good chaining.
Outline the step-by-step procedure for chaining a distance.
Explain the process used when lengths less than a full chain length are measure because of the slope.
Identify the relative precision that can be attained by precise chaining.
State when it is necessary to use a hand level while chaining.
List common mistakes that occur when chaining.
State and describe the three sources of error in chaining.
Distinguish between systematic error and random errors in chaining.
Explain the amount of tension to apply to a chain.
Describe how changes in temperature affect the chain length.
Prepare a list of the duties of a field engineer when chaining a distance.

INDEX

EQUIPMENT AND BASICS

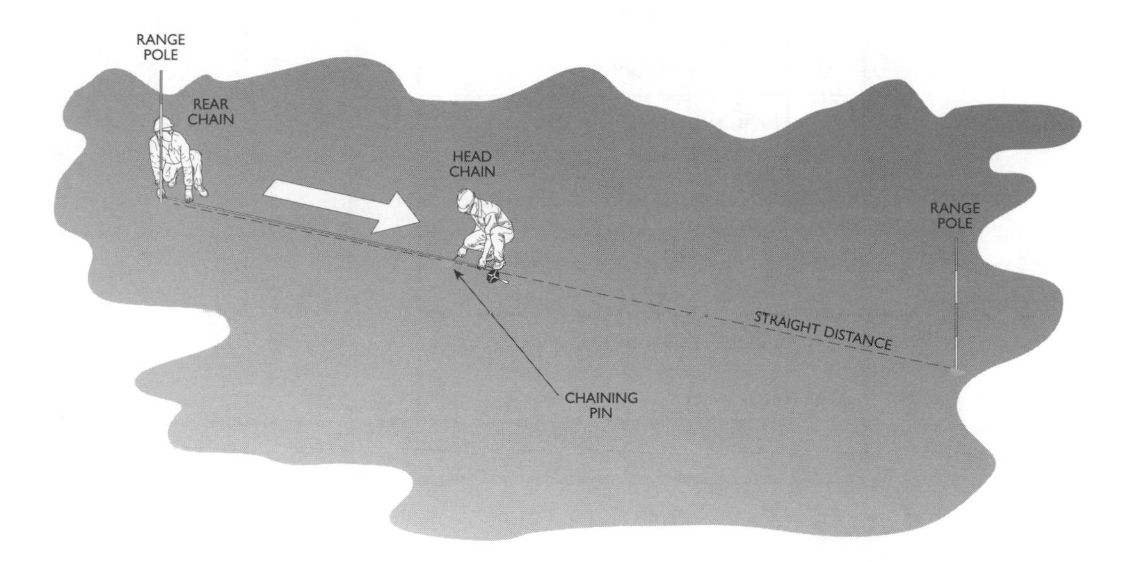

SCOPE

Measuring distances with a chain seems so simple. Hold one end of the chain on the starting point and hold the other end where you want to determine the distance. That's it! It does seem simple, doesn't it? Not so fast. Consider that the chain must be horizontal, straight between the points, properly tensioned, held exactly on zero, read correctly, etc. All of these activities must be considered and performed in the field in less than a minute. Yes, chaining is simple once you understand the procedure and get a feel for the physical skills required.

The field engineer, and everyone else associated with the project, is constantly measuring distances. Every control point, offset point, anchor bolt, column, etc. is located horizontally from the information given on the plans. Accurate locations of footings, columns, sewers, etc., depend upon the chaining ability of the field engineer. A field engineer, to be successful at measuring distances accurately and precisely, must be consistent in the chaining procedure.

It should be pointed out that the terms "tape" and "chain" are used interchangeably. This is especially true on the construction site. The author has chosen to use the term chain to distinquish it from the 25' tape that many carpenters use and call a tape.

CHAINING EQUIPMENT

Common Chaining Equipment

1	Field book
2	4H pencil
3	Two range poles
4	Two plumb bobs
5	Straight edge
6	Two hand levels
7	100-foot steel tape
8	Wooden stakes
9	Eleven chaining pins
10	Hammer

CHAINING BASICS

USE TWO PEOPLE

To chain a distance requires two persons working together and communicating, verbally and through body language, as they perform the various steps in the chaining process. Two persons who have chained together for a long time can actually measure distance after distance without speaking a word. They understand the process, and they understand each other's movements very well.

BECOME FAMILIAR WITH THE CHAIN BEING USED

Chains come in many sizes, shapes, and colors. Possibly the greatest source of error in chaining is that the users are not familiar with the chain. Where is zero? One would think there's a simple answer to that question--that zero is at the end of the chain. However, upon closer examination, it is found that zero is not always at the end of the chain. Therefore, always examine a chain closely before using it. A sample of chains is presented here. Note the positions of zero.

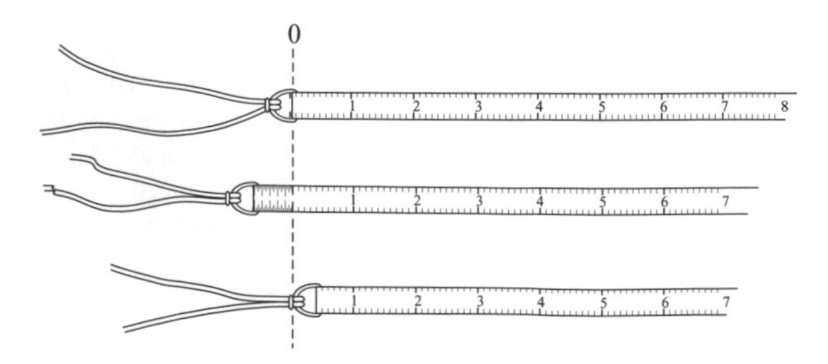

MAINTAIN GOOD ALIGNMENT

The old saying, "the shortest distance between two points is a straight line," is a fundamental principle by which distances must be measured. When measuring distances less than a chain length, this isn't a problem. However, when measuring beyond the length of a chain, intermediate chaining points are used. They should be directly on line between the points that are being measured. The procedure used to maintain this alignment is described in detail in the step-by-step procedure for chaining a distance.

MEASURE HORIZONTALLY

Distances shown on the plan view of plans are horizontal unless noted otherwise. Therefore, when chaining, the distances that are laid out for construction will be horizontal. Two devices are used when chaining to obtain horizontal measurement. They are the site level and the plumb bob. The site level is used to determine where horizontal is, and the plumb bob is used to plumb down to the point.

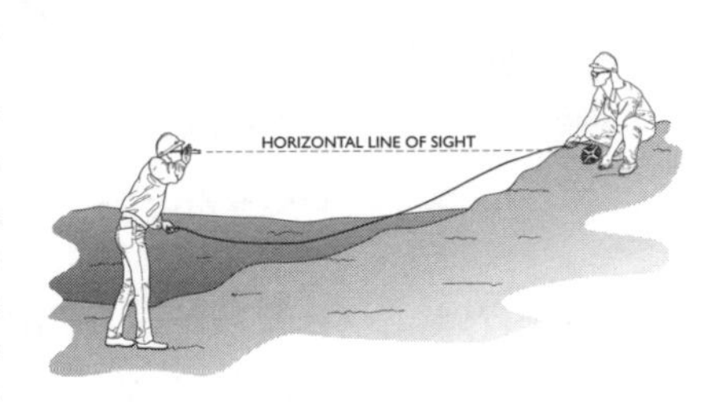

The sight level is a device consisting of a tube with a level bubble and a prism or mirror with a horizontal crosshair inside the tube. The basic principle of use is to look into the tube and center the crosshair on the bubble while looking through the tube at the object. Where the crosshair hits the object when the bubble is centered is horizontal from the user's line of sight. Thus, when chaining, a horizontal line can be determined between the persons at the ends of the chain.

The plumb bob is used in chaining to transfer the point or distance measured vertically to the horizontal measurement line. It is an ancient device used to assist in keeping buildings plumb. Whenever possible, while plumbing, do so from one end only to eliminate doubling errors that occur when a plumb bob is used.

APPLY PROPER TENSION

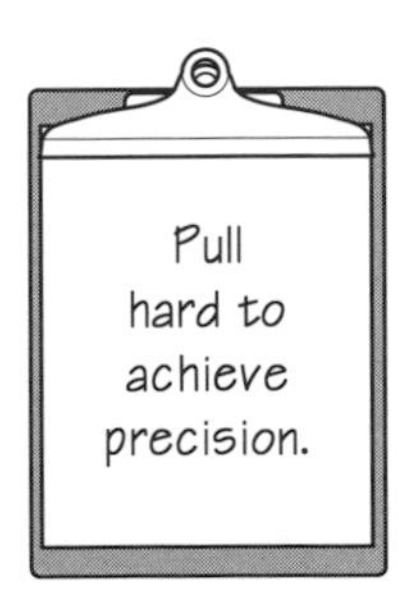

It isn't too tough to imagine that measuring a distance while the chain is sagging in the middle from its own weight is going to affect the distance. Chains must be pulled tight to achieve the precision expected for construction layout. The exact tension that must be applied varies with the weight of the chain being used. Some chains have a large cross-sectional area and, as a result, are very heavy per foot while other chains have a small cross-sectional area and weigh much less. The amount of tension depends upon the weight of the chain. A typical chain used in construction will require 20 to 25 pounds of tension to overcome the sag. When in doubt about the amount of tension to apply, PULL HARD! It is better to pull a little extra than not enough.

MEASURE ALL DISTANCES TWICE

Measuring just once in any situation risks making a mistake without discovering it. To avoid the mistakes that frequently occur in chaining, all distances must be measured twice. Preferably, measure from the beginning point forward to the end of the line, and then from the end point back to the beginning. Reversing the direction of measurement helps to eliminate the possibility of making the same mistakes twice, as often occurs, when measuring in the same direction.

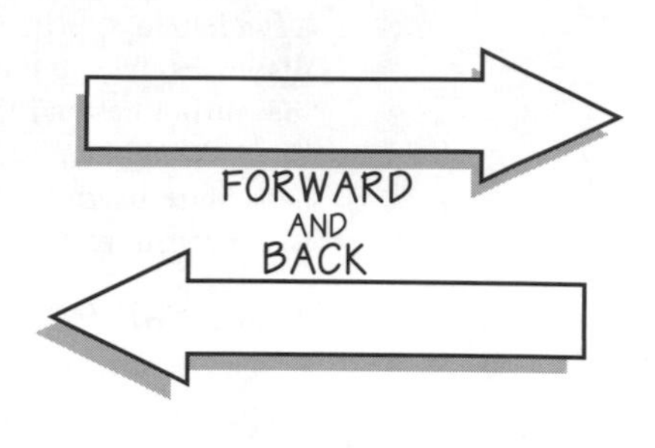

REMEMBER: FIELD ENGINEERS ALWAYS DO IT TWICE!

PROPERLY CARE FOR CHAINS AND CHAINING EQUIPMENT

Good results can only occur with well-cared-for equipment. This axiom is repeated throughout this book. Some basics of chain care include: keeping the chain on reel when not in use, removing loops from the chain, drying the chain when wet, cleaning off dirt and sand to prevent wearing off numbers, and wiping the chain with an oily rag to prevent rust

See *Chapter 4, Fieldwork Basics, (4-25)* for a detailed discussion on the care of chains and chaining equipment.

COMMUNICATION IS THE KEY TO GOOD CHAINING

For beginners to be successful at chaining, it is required that they communicate loudly and clearly to each other to acknowledge each stage in the process of measuring a distance. A step-by-step procedure of the chaining process for measuring the distance between two known points follows in the next section.

STEP-BY-STEP PROCEDURE FOR CHAINING A DISTANCE

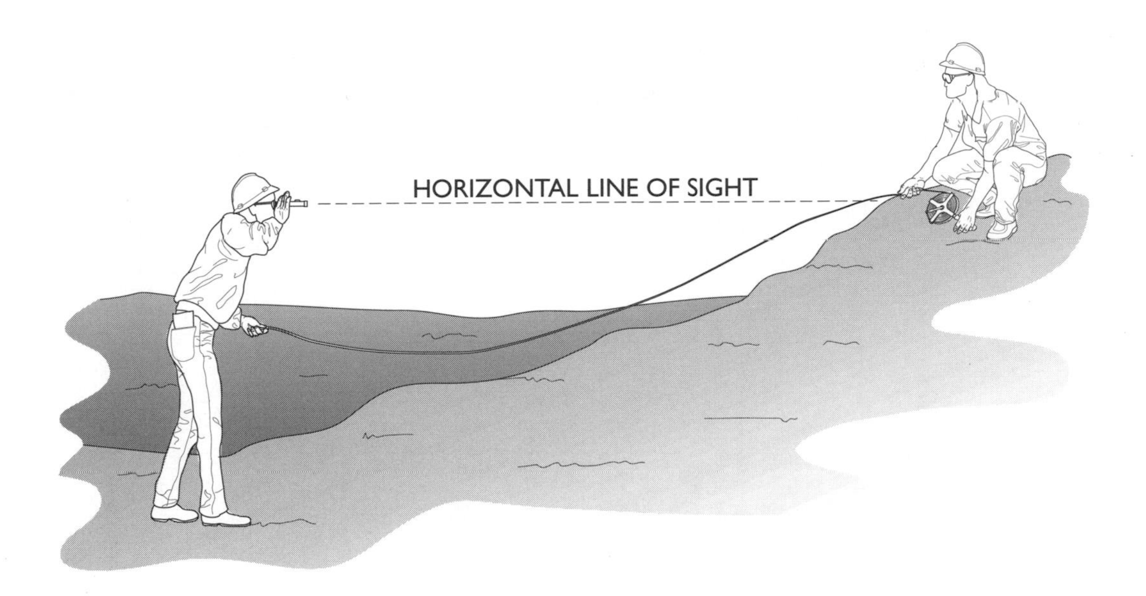

SCOPE

Even though it has been shown in the previous section that the chaining process involves simple steps, it is still involved and requires constant attention to the details of performing chaining properly. To be successful at measuring distances accurately and precisely requires consistency in the chaining procedure. This section will explain and demonstrate the step-by-step activities of the persons involved in the chaining process.

MEASURING A DISTANCE

	REAR CHAIN	***HEAD CHAIN***

STEP 1	Locates the beginning point and sets a range pole or lath with flagging to mark it.	Locates the end point and sets a range pole or lath with flagging to mark the point.

STEP 2	Holds the end of the chain and gives direction to the Head Chain verbally or by hand signals.	Walks from the beginning point toward the end point, clearing brush, etc., out of the chaining path. Allows the chain to come off the reel while walking.

	REAR CHAIN	*HEAD CHAIN*

STEP 3	Stands on the beginning point, sights on the range pole at the end, and puts the head chain on line.	Upon reaching the end of the chain, turns and looks to the rear chain for alignment directions. Gets on line as instructed.

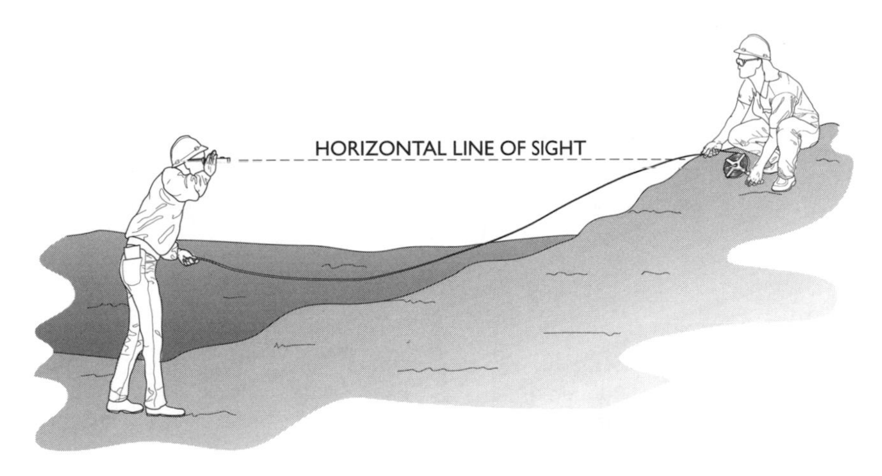

STEP 4

Whoever is downhill will take out a sight level and determine the proper height to plumb while measuring to ensure that the chain will be horizontal.

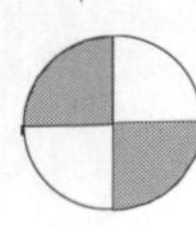

	REAR CHAIN	*HEAD CHAIN*

STEP 5

Rear chain: Resists the tension being applied and yells "good" or "mark" when the end of the chain is over the point.

Head chain: Applies the proper tension to the tape to eliminate sag. Identifies the spot on the ground that marks the end of the chain and releases tension when the point is marked with a chaining pin or nail.

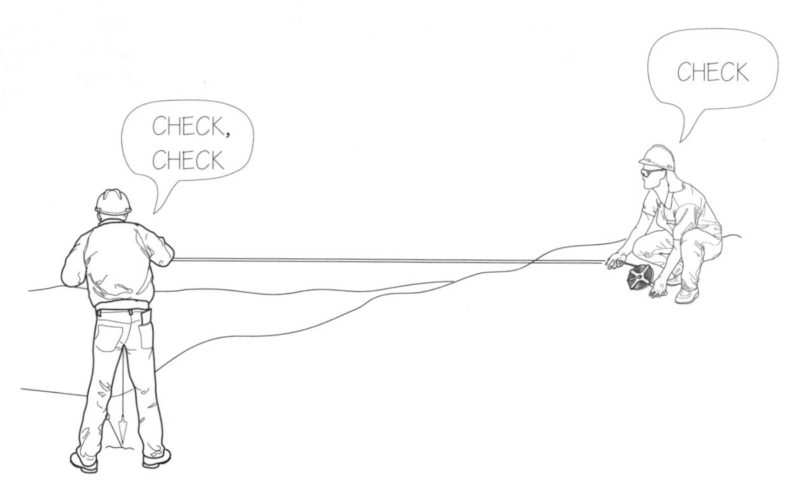

STEP 6

Rear chain: Holding at the prescribed height, plumbs if necessary, checks holding the ZERO over the point.

Head chain: Holding at the prescribed height—using a plumb bob if necessary—checks the 100 foot mark.

	REAR CHAIN	***HEAD CHAIN***
	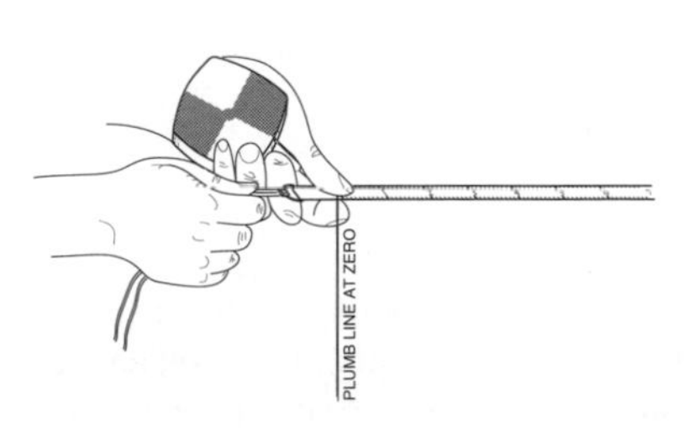	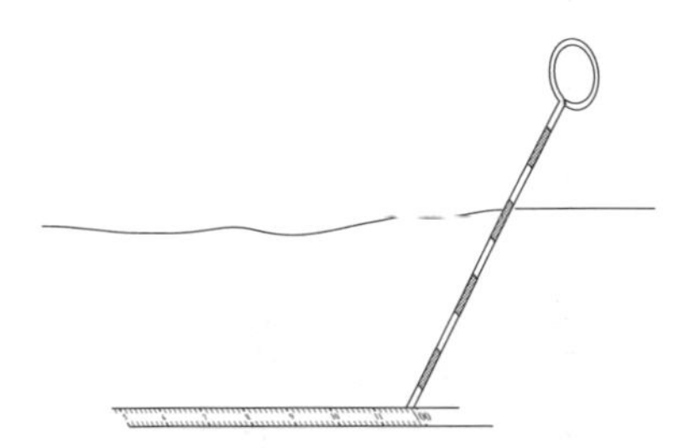
STEP 7	Rechecks alignment and makes sure ZERO was being held on the point.	Checks the chaining pin or nail in the ground at the new point. Checks to make sure exactly 100 was being held.

STEP 8

Steps 5 through 7 are repeated to check the measurement.

STEP 9	Gathers equipment and advances to the point just set. Watches the end of the chain and yells "Chain" when the end of the chain is near the point.	Moves ahead on line toward the end point dragging the tape behind. Stops when the Rear Chain yells "Chain."

REAR CHAIN | **HEAD CHAIN**

STEP 10

Steps 2 through 9 are repeated for each additional full chain length measured.

The following steps apply when the full length of the chain is not used.

STEP 11

When measuring the less-than-full chain length at the end of a line, don't forget to chain horizontally. Also, the amount of tension applied should be decreased slightly because of the smaller length of chain being used.

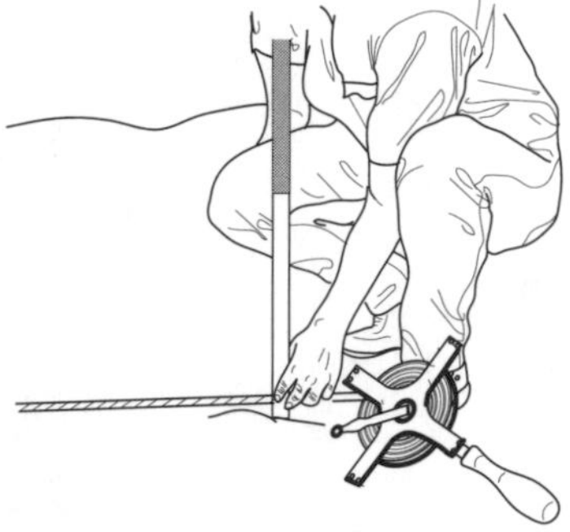

STEP 12

Rear chain: Gets in position at the last point marked.

Head chain: Walks to the end point of the line, reels up the extra chain, and prepares to read the chain.

STEP 13

Rear chain: Holding chain at the determined height horizontal at the last point marked, resists tension being applied and yells "good" when the end of the tape is over the point.

Head chain: Holding at the determined height to chain horizontal, applies tension and reads the chain when over the end point to the nearest 0.01 when the rear chain yells "good".

REAR CHAIN ***HEAD CHAIN***

STEP 14

Checks the last measurement. The head chain yells out the last reading for the note keeper to record. The note keeper adds the hundreds of feet to the final distance and repeats it back as a check as the head chain checks the reading.

STEP 15

The distance measured is the total of the number of chain lengths measured plus the reading the head chain made at the last measurement.

STEP 16

Repeat all of the above steps anytime a measurement between two known points is desired.

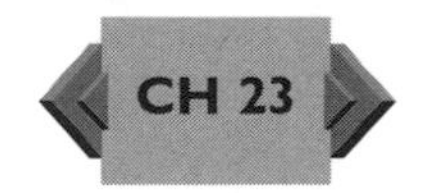

This is the basic process for chaining a distance between existing points. This detailed procedure should be studied and used every time a measurement is made until it becomes second nature . To lay out a point by chaining reference *Chapter 23, Construction Layout Techniques* in this manual.

CALCULATIONS

The forward and back distances are averaged by adding them together and dividing by 2 to get a mean distance. Some persons chaining also calculate a discrepancy ratio for each line measured. The calculated ratio is compared to an established standard to determine if the distance is acceptable or needs re-measured.

Forward - Back = Discrepancy

$$\frac{Discrepancy}{Mean} = \frac{1}{x}$$

For example, in the notes below for line 19-13.

$$\frac{168.29 - 168.25}{168.27} = \frac{1}{4200}$$

FIELD NOTES

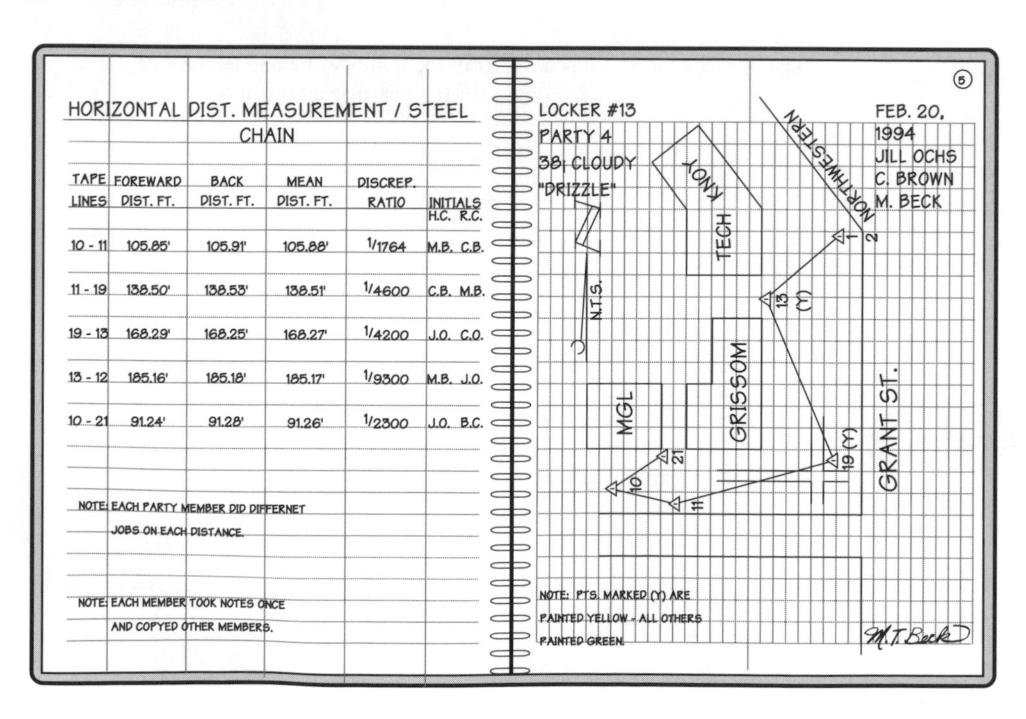

HORIZONTAL DIST. MEASUREMENT / STEEL CHAIN

TAPE LINES	FOREWARD DIST. FT.	BACK DIST. FT.	MEAN DIST. FT.	DISCREP. RATIO	INITIALS H.C. R.C.
10 - 11	105.85'	105.91'	105.88'	1/1764	M.B. C.B.
11 - 19	138.50'	138.53'	138.51'	1/4600	C.B. M.B.
19 - 13	168.29'	168.25'	168.27'	1/4200	J.O. C.O.
13 - 12	185.16'	185.18'	185.17'	1/9300	M.B. J.O.
10 - 21	91.24'	91.28'	91.26'	1/2300	J.O. B.C.

NOTE: EACH PARTY MEMBER DID DIFFERNET JOBS ON EACH DISTANCE.

NOTE: EACH MEMBER TOOK NOTES ONCE AND COPYED OTHER MEMBERS.

POTENTIAL MISTAKES AND ERRORS IN CHAINING

SCOPE

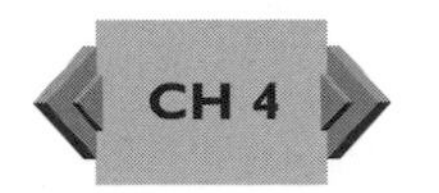

As stated earlier in *Chapter 4, Fieldwork Basics,* errors always exist in any measurement. This statement is especially true in chaining since there are two peopleinvolved and there are many steps that must be followed. Individuals who are measuring distances must follow exact procedures or mistakes will occur, and errors will multiply. Even though many of the mistakes and errors listed here have been discussed previously, it is prudent to "say it again."

COMMON MISTAKES

Recall, errors are small but mistakes are large and should be eliminated by following proper procedures. Mistakes that are allowed to exist in critical layout activities are inexcusable because proper checking of work will identify them, and they can be eliminated. Some examples of mistakes which you can correct are:

Chain not horizontal

Reading the tape incorrectly

Miscounting the number of full tape lengths

Transposing figures

Breaking the tape into two pieces

One person is "cutting a foot" and the other isn't

Check, check, and recheck your work to eliminate your chaining mistakes before they become costly! Remember, there are no excusable mistakes.

SOURCES OF ERROR IN CHAINING

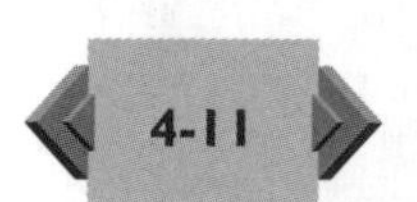

In *Chapter 4, Fieldwork Basics,* it was explained how errors are different than mistakes. They are small and may only become significant when they are allowed to accumulate. There are different sources of errors that occur when chaining.

INSTRUMENTAL

A tape may be different in length from its nominal length because of a defect in manufacture or repair or as a result of kinks.

NATURAL

The horizontal distance between end graduations of a tape varies because of the effects of temperature, wind, and the weight of the tape itself.

PERSONAL

Persons chaining may be careless in setting pins, reading the tape, and/or manipulating the equipment.

TYPES OF ERRORS

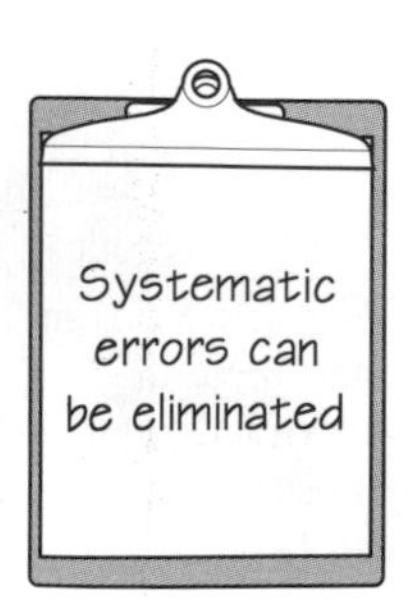

Errors can be classified into two different types, Systematic Errors and Random Errors.

SYSTEMATIC

Systematic errors are those errors which are predictable. They occur over and over. They follow the laws of physics and mathematics and can be removed by calculation. An example is:
A 100' tape is found to be 0.05' short, which makes the tape only 99.95'. Every time this tape is used, 0.05' must be added to obtain the correct distance.

RANDOM

Random errors are variable in nature. Every measurement made by human hands and observed by human eyes will contain random errors. If 100 measurements of a line are taken, half will be more than the average measurement and half will be less than the average measurement. If the 100 measurements are plotted relative to their frequency of occurrence, a bell-shaped curve will result. Thus, random errors can be evaluated through the statistical formulas of the laws of probability.

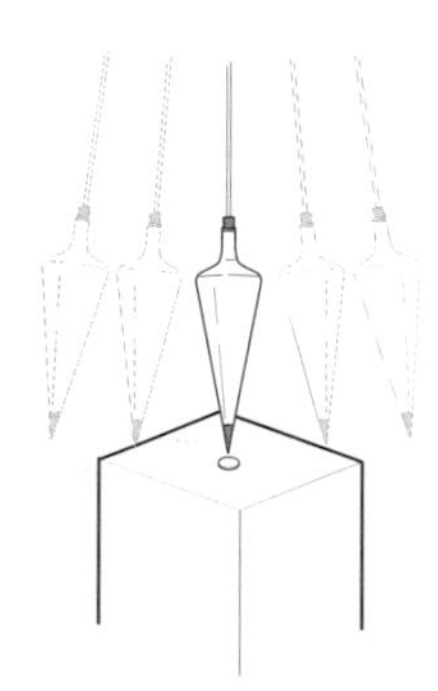

Random errors occur because of human limitations in the senses of sight and touch. An example of a random error is:

During chaining, "Good" is called out when the plumb bob has settled. However, since no one can hold the plumb bob perfectly still, it is really a judgement call. Therefore, most of the time, the plumb bob is slightly off of the point. Thus, a small random error exists.

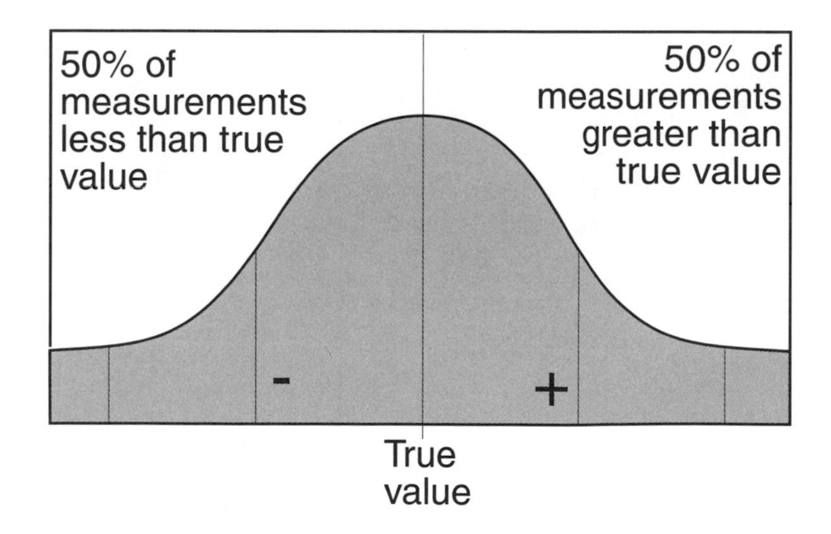

ANALYSIS OF COMMON CHAINING ERRORS

The following table lists and analyzes the errors that are encountered when chaining. It should be studied carefully to get a feel for the effect of each of the errors when chaining distances.

Analysis of Common Systematic and Random Chaining Errors

Error	Source	Type	Makes Chain Too...	Importance	Procedure to Eliminate
Chain Length	Instrumental	Systematic	Long or Short	Direct impact to record measurement. Always check	Calibrate chain and apply adjustment
Temperature	Natural	Systematic or Random	Long or Short	For a chain standardized at 68 degrees, 0.01 per 15 degrees per 100 foot chain	Observe temperature of chain, calculate and apply adjustment
Tension	Personal	Systematic or Random	Long or Short	Often not important if close to required pull	Apply the proper tension. When in doubt, PULL HARD!
Tape Not Level	Personal or Natural	Systematic	Short	Negligible on slopes less than 1%; must be calculated for greater than 1%	Breaking chain by using a hand level and plumb bob to determine horizontal; or correcting by formula for slope or elevation difference
Alignment	Personal	Systematic	Short	Minor if less than 1 foot off of line in 100 feet; major if 2 or more feet off of line	Stay on line; or determine amount off of line and calculate adjustment by formula
Sag	Personal or Natural	Systematic	Short	Large impact on recorded measurement. 0.01 if sag of .06 occurs at center of 100 foot chain	Apply proper tension; or calculate adjustment by formula
Plumbing	Personal	Random	Long or Short	Direct impact to recorded measurement	Avoid plumbing if possible; or plumb at one end of chain only
Interpolation	Personal	Random	Long or Short	Direct impact to recorded measurement	Check and recheck any measurement that requires estimating a reading
Improper Marking	Personal	Random	Long or Short	Direct impact to recorded measurement	Mark all points so they are distinct

SUMMARY OF FIELD DUTIES AND RESPONSIBILITIES

Field Engineers Chaining Shall	
Be safe	Exercise great care in ensuring that safety procedures are followed.
Maintain Equipment	Ensure that chains are properly cared for while in use; that loops will be removed and kinks avoided; that chains will be wound onto reels when not in use; and that chains will be regularly dried, cleaned, and oiled.
Understand the Chain	Become familiar with the type of chain used and the markings on the chain (feet, tenths, and hundredths, or feet and inches).
Chain Straight	Maintain good alignment in straight line measurement while chaining distances which are greater than a chain length.
Pull Hard	Utilize proper tension when pulling a chain (when in doubt about proper pulling tension -- WILL PULL HARD!).
Plumb	Plumb to ensure that all measurements are horizontal measurements. Plumb only at one end whenever possible in order to reduce errors.
Measure Twice	Measure all distances at least twice and in reverse order.
Set Solid Points	Drive stakes until they are solid and flush; and place tacks exactly on line and exactly at the measured distance.
Guard Stakes	Place guard lathe around stakes to warn others of the importance of the point; and mark lathe clearly with information that describes the use of the point.
Record	Record all measurements in the fieldbook as they are observed.
Check and Recheck	Record complete, concise, and properly arranged notes in the standard format; provide ample sketches ("a picture is worth a thousand words"); and check and correct all notekeeping to ensure accuracy before leaving the grounds.

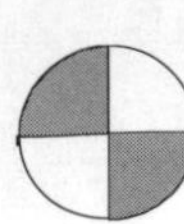

DISTANCE MEASUREMENT - EDM

7

PART 2 - MEASUREMENT BASICS

CHAPTER OBJECTIVES:

State what EDM is an acronym for.
List potential uses of an EDM in construction.
List and describe the various types of EDM's that are available.
Describe the effect of distance on the precision attainable with an EDM.
List and describe the measuring basics required for EDM measurement.
Describe the rule of thumb regarding EDM and slope or horizontal measurement.
Describe the sighting procedure with the EDM that can be used to eliminate complicated geometric calculations.
Outline the step-by-step procedure for measuring a distance with an EDM.
State the greatest source of error with EDM measurement.
Prepare a chart that lists the recommended uses for an EDM based on the length of the distance.
List five duties of the EDM operator and the prism pole holder in EDM measurement.

INDEX

EDM BASICS

SCOPE

Since the beginning of time, measurement has been a highly physical activity that first required the stretching of a rope, then, a chain, and more recently, a steel tape. Obtaining accurate measurement with any of these tools was time consuming and required a great deal of training and coordination between the individuals involved.

The ability to measure distances with the simple push of a button has been a boon to those in the surveying profession and to field engineers on the construction site. In construction, the Electronic Distance Measuring (EDM) instrument has radically changed the manner in which surveying is used on the jobsite. The EDM has enabled the field engineer to accurately and quickly measure distances for project control as well as to utilize the method of radial layout in locating the components of the project.

However, the introduction of this method of distance measurement has also introduced new sources of measurement error. The competent field engineer must be aware of these potential errors and must utilize techniques that expose them when they occur. Theses techniques should enable them to be eliminated or reduced to an acceptable magnitude.

EDM BASICS

An EDM is a device that sends out a signal (wavelength) and measures the time it takes for the signal to go out and come back. It displays distance, but the EDM really measures time. It contains a very accurate timing device which measures infinitesimal parts of a second. Knowing the wavelength of the signal sent out and the precise time, the basic formula of "Distance = Wavelength x Time" from physics is used within the instrument to calculate the distance which is then displayed to the operator.

Measurements that are displayed by the EDM are not just single measurements. In the few seconds it takes from the time the measure button is pushed until the distance is displayed, the EDM has measured the distance several hundred times and presents the average of the readings. Because of this averaging, most measurements with the EDM are typically very consistent and if the EDM is in proper calibration, very accurate.

TYPES OF EDM'S AVAILABLE

EDMs are like cars. There are many different models, styles, shapes, and sizes which are made by many companies. They range from basic one-button models that measure only slope distances to sophisticated models that contain computers for field calculation as well as electronic storage of data. Basic or advanced, the main function is the same - only the operation is different. It is the responsibility of the field engineer to read the operating manual and learn the step-by-step operation of the specific EDM being used so reliable measurements can be made and structures can be located with confidence.

The following is a brief summary of three types of EDMs available for use by the field engineer. The type or style is only briefly described because any equipment dealer can present a much more comprehensive hands-on demonstration of the various EDM's.

TOP-MOUNT

The top-mount EDM is common and is mounted on top of the standards of a theodolite. It measures the slope distance and must depend on a vertical angle measurement or difference in elevation for the horizontal distance to be calculated. Using the top-mount EDM is a rather time-consuming task because the scope on the theodolite must be sighted to determine the horizontal and vertical angles to the prism. Then the EDM must be sighted onto the prism. This may appear to be simple until compared to the operation of the total-station described later.

This EDM can also be mounted directly onto an azimuth base and placed on a tripod without a theodolite below it. In this case, only the elevation difference between the prism and the EDM can be used to calculate the horizontal distance since no vertical angle is available.

The top-mount EDM is often preferred by many in construction because it is placed on the theodolite only when distances are needed and is removed and safely stored when only angles are being measured. Thus, when the EDM is not in use, it is not exposed as much to the many hazards (such as material movement, heavy equipment, dust, rain, etc.) that exist daily on the construction site. It is also much less expensive than the total station especially if a theodolite is already available.

SEMI-TOTAL STATION

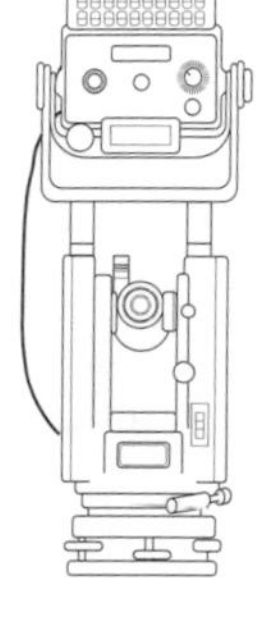

The semi-total station type of EDM is similar in appearance to the top-mount models, but it is placed on top of an Electronic Digital theodolite. Two separate sightings by the scopes of the instruments are still necessary. However, the two electronic instruments can be wired together, and the theodolite can obtain the slope distance from the EDM. This in turn allows the computer on the theodolite to calculate the horizontal distance. Therefore, this type of setup has some of the advantages of the top mount EDM and some of the advantages of the total station. Advantages include being able to remove the EDM from constant exposure to the hazards of construction. An additional advantage is eliminating operator error in calculating the horizontal distance. Many field engineers find this combination suits their layout needs very well.

ELECTRONIC TOTAL STATION

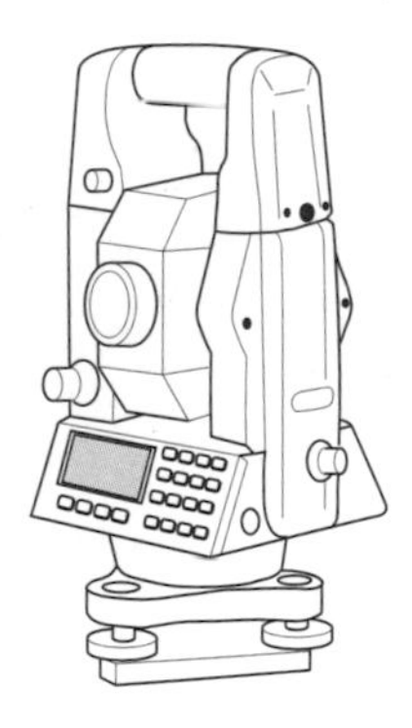

The electronic total station is the ultimate in the EDM/ theodolite combination. Field engineers who use this type of instrument never wish to return to the "top-mount" style of EDM. The EDM and the theodolite are totally integrated. They utilize the same telescope for operation, and time is saved because only one sighting is needed. The electronics of the instrument are integrated. Horizontal and vertical angles, slope, horizontal and vertical distances, and coordinates of points are available with the push of a button. Therefore, mistakes made in calculations are also eliminated. For the ultimate in elimination of field data mistakes, electronic fieldbooks (data collectors) can also be attached to the instrument. The total station is the instrument of today and the immediate future for field engineers. For more information on the Total Station, reference *Chapter 10, Total Station.*

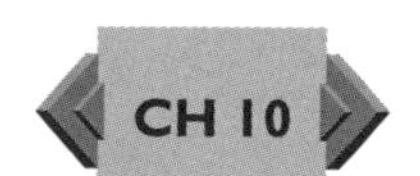

RELATIVE PRECISION OF EDMS

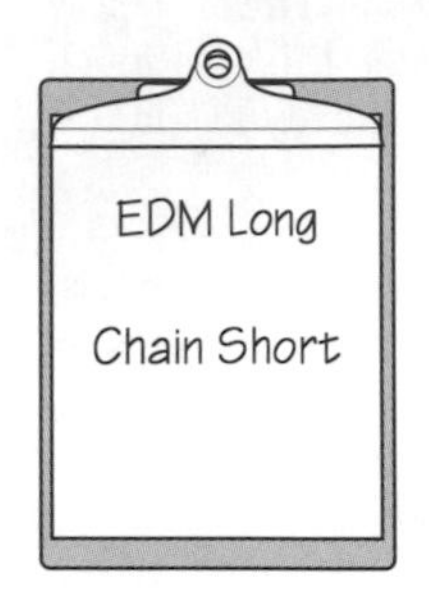

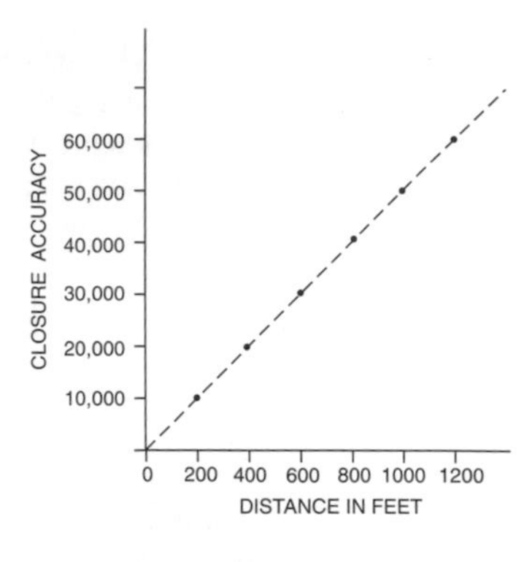

Most EDMs found on the construction site use infrared light as the source of their signal. These systems are reliable and durable as well as relatively inexpensive. The precision of these EDMs is usually found to be ± 0.02' for any distance measured. A distance of 10.00 feet displayed on the EDM will be 9.98 to 10.02. A distance of 1000 feet is 999.98 to 1000.02. Obviously, the EDM is better for measuring long distances because the constant ± 0.02 stays the same for all distances. See the graph for a visual relationship of the precision attainable to the distance measured.

Why does an EDM have this built-in error? It is a matter of economics. EDMs are available that measure to the thickness of a human hair, however they cost many tens of thousands of dollars more than the EDMs with the ± 0.02 error. Contractors can live with ± 0.02 accuracy of measurements. For accuracy, field engineers need to remember that EDMs are better for long distances, and chains are better for short distances.

Basic EDM Field Equipment	
Required	***Optional***
EDM	Two-way Radios
Tripod	Additional Tripods
Theodolite or Transit	Spare Batteries
Prism(s) with Pole or Tripod	Additional Prism Poles
Fieldbook, 4H Pencil, Straight edge	Hammer & wooden stakes
6-Foot Engineer's Rule	

EDM MEASURING BASICS

TWO PERSONS REQUIRED

To be accomplished efficiently, electronic distance measurement requires a minimum of two persons working together and communicating as the various steps involved in measuring distances electronically are performed. One person operates the EDM and one person holds the prism. If a great number of distances are to be measured, several persons can be assigned to hold prisms to speed the time between measurements.

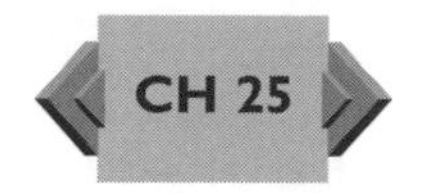

With proper planning and equipment, electronic distance measurement can also be accomplished by one person. See *Chapter 25, One-Person Surveying Techniques* for a detailed description of this process.

INTERVISIBILITY IS A MUST

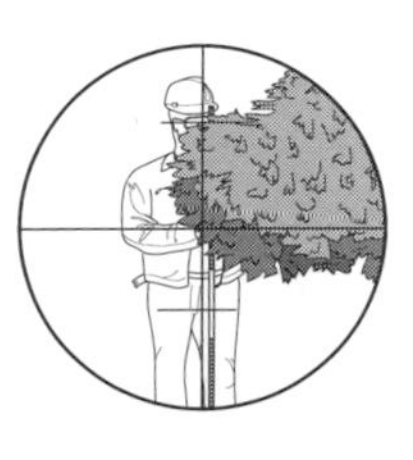

A clear line of sight is a basic requirement for EDM use. As stated earlier, a signal is sent from the EDM to the prism and reflected back to the EDM. For this signal to occur, there can be no constant obstructions between the EDM and the prism. There can be the occasional interruption of the signal by someone walking between the EDM and the prism or a leaf fluttering across the line of sight of the signal, but nothing that stays for more than a second or two can be in the way of the EDM or it will be unable to display the distances.

MEASURE ALL DISTANCES TWICE (FORWARD & BACK)

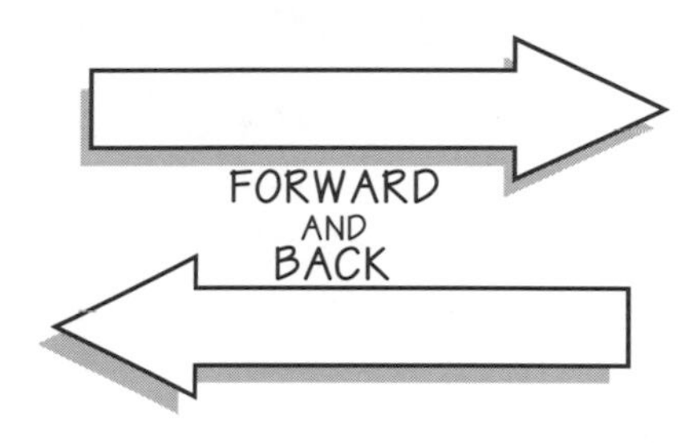

Only measuring once in any situation risks making a mistake and not discovering it. To avoid mistakes that occur in EDM operation, all distances should be measured twice. Measure from the beginning point forward to the end of the line and then from the end point back to the beginning. By reversing direction, mistakes that might occur when measuring a distance twice in the same direction might be discovered. **Remember: Field Engineers always do it twice!**

SLOPE DISTANCES

EDMs measure between the instrument and the prism. Unless the ground is level, the distance displayed will always be a slope distance. That is, if an EDM is at the bottom of a hill and the prism is at the top of a hill, the measurement parallels the slope of the hill. It should be noted the total station also measures slope distances; however, it has the capability to immediately calculate and display the horizontal distance.

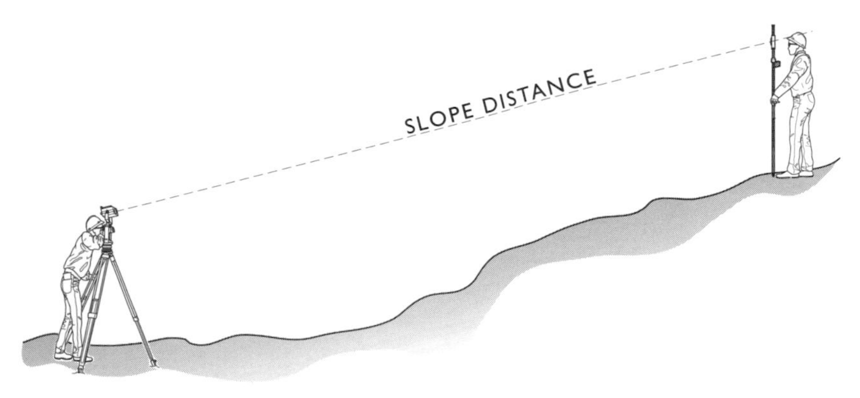

DISTANCES MUST BE CONVERTED

Dimensions shown on the plan view of drawings and thus used in construction layout are horizontal distances. Therefore, the slope distances measured by the EDM must be converted to horizontal distances to be usable. Converting slope distance to horizontal distance is dependent upon knowing either the measured slope angle or the difference in elevation between the instrument and the prism.

Given: EDM Slope Distance and Vertical angle off of the horizon.

$$\textit{Horizontal Distance} = \textit{Cosine}(\textit{Vertical Angle}) \times \textit{Slope Distance}$$

Given: EDM Slope Distance and Zenith Angle.

$$\textit{Horizontal Distance} = \textit{Sine}(\textit{Zenith Angle}) \times \textit{Slope Distance}$$

Given: EDM Slope Distance and Difference in elevation between instrument point and prism pole point.

$$\textit{Horizonal Dist.} = \sqrt{(\textit{Slope Dist.})^2 - (\textit{Difference in Elev})^2}$$

USING GEOMETRIC PRINCIPLES WHEN SIGHTING

When using a top-mount EDM, close observation of the geometry involved in the sighting reveals that messy calculations can be avoided if a very simple sighting procedure is followed. See the sketch below:

By sighting the horizontal crosshair of the theodolite at a distance below the prism (d_2) equal to the distance between the center of the EDM and the optical center of the theodolite (d1), parallel sighting lines are created. The vertical angle can be used directly with the slope distance in the formula to calculate the horizontal distance.

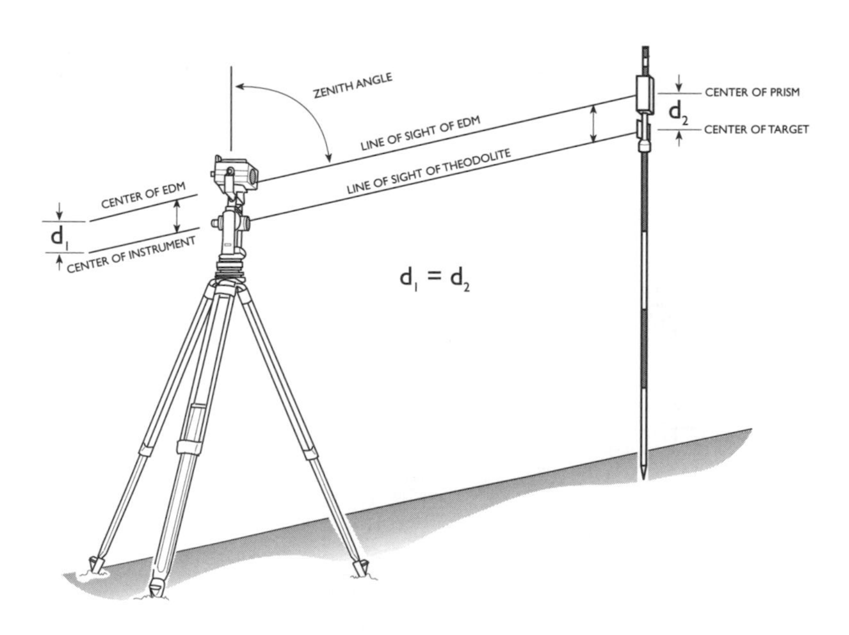

MEASURING A DISTANCE WITH AN EDM

SCOPE

Recall there are two types of distance measurement situations. The first type is measuring to determine or check a distance between two established points. This is used for checking between control points or making a final check of points laid out earlier. The second type is measuring to lay out new points. This procedure is used to lay out the initial control for excavation, the working control for establishing centerlines, and the location of footings, etc. Measuring between two established points will be described in this section.

STEP-BY-STEP PROCEDURE FOR MEASURING WITH AN EDM

Every point must be laid out, measured and checked. For a basic understanding of operating an EDM, the step-by-step procedure to measure a distance will be described here, and the step-by-step procedure for laying out a point with an EDM will be described in the next section.

EDM OPERATOR

PRISM POLE HOLDER

STEP 1

EDM Operator: Gathers all needed equipment and proceeds to the work area.

Prism Pole Holder: Gathers all needed equipment and proceeds to the work area.

STEP 2

EDM Operator: Locates point at one end of the line to be measured.

Prism Pole Holder: Locates the point at the other end of the line to be measured.

STEP 3

EDM Operator: Sets tripod over point and places the theodolite on the tripod.

Prism Pole Holder: Places the prism on the prism pole and waits for a signal to prepare for the measurement.

EDM OPERATOR | PRISM POLE HOLDER

STEP 4

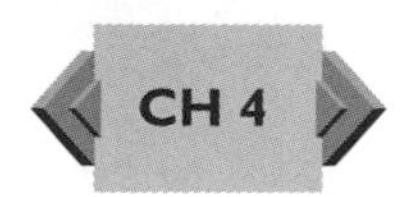

EDM Operator: Levels and centers theodolite over the point using the Optical Plummet Quick Set Up procedure described in *Chapter 4, Fieldwork Basics.*

Prism Pole Holder: Waits.

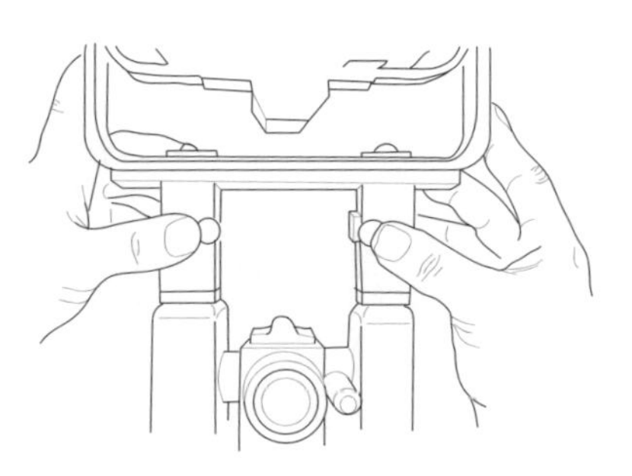

STEP 5

EDM Operator: Removes the EDM from its case and securely fastens it on top of the instrument.

Prism Pole Holder: Waits.

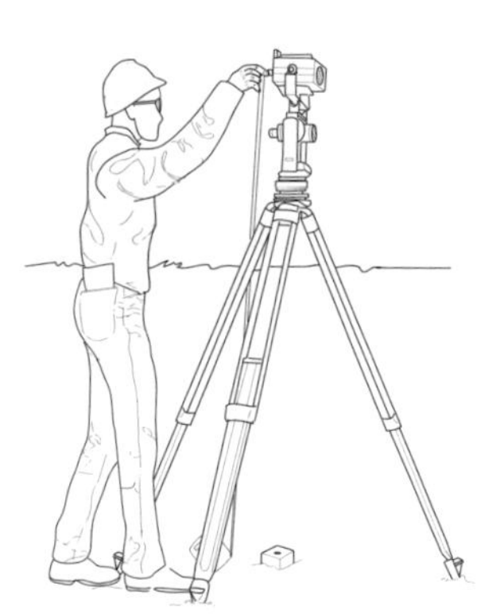

STEP 6

EDM Operator: Using a tape or rod, measures from the ground to the center of the EDM. Informs the prism holder of the height of the instrument.

Prism Pole Holder: Adjusts the prism pole until its height is the same as the EDM.

EDM OPERATOR | PRISM POLE HOLDER

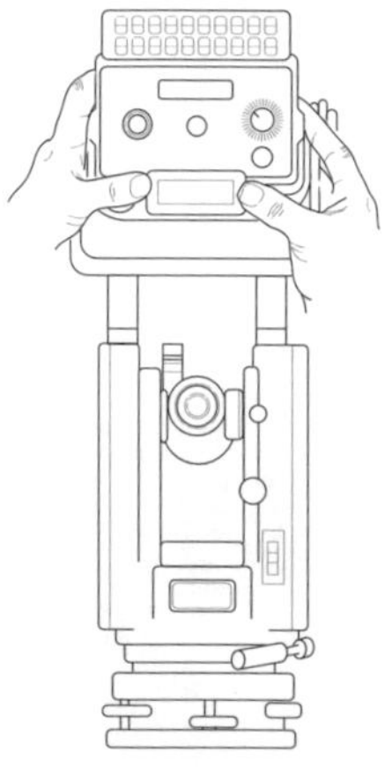

STEP 7

Inserts the battery into the EDM and checks its strength.

Waits.

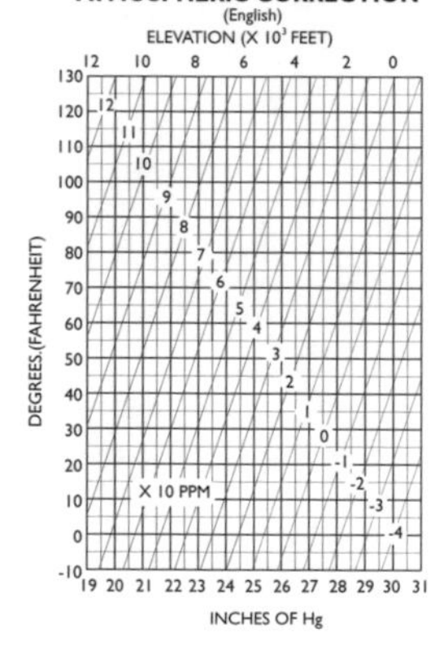

STEP 8

If required by the EDM, inputs a parts per million correction factor based on the temperature and barometric pressure or elevation. See the operating manual of the EDM for complete details.

Waits.

EDM OPERATOR | *PRISM POLE HOLDER*

STEP 9

EDM Operator: Communicates to the prism holder the readiness for measuring.

Prism Pole Holder: Facing the EDM, places the prism pole exactly on the point and holds the pole plumb by centering the bull's-eye bubble attached to the side of the pole.

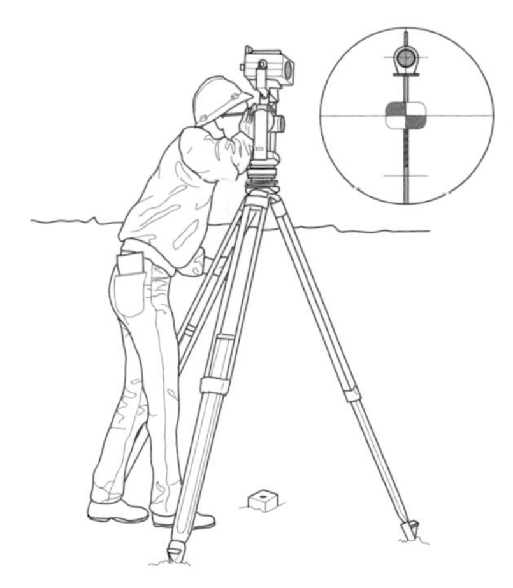

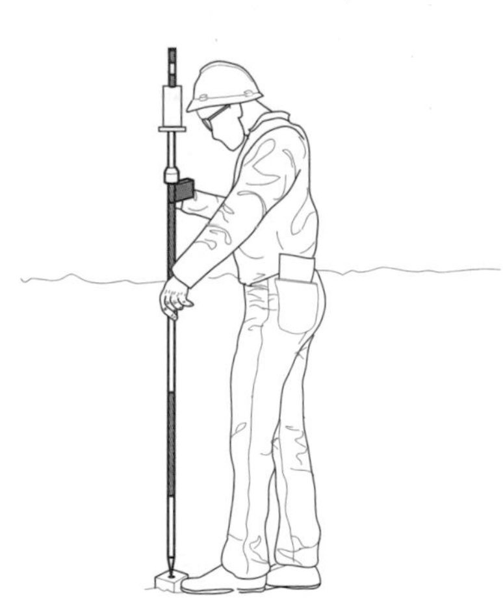

STEP 10

EDM Operator: Sights onto the prism pole target with the theodolite to determine the vertical angle. Reads and records the vertical angle for use in calculating the horizontal distance.

Prism Pole Holder: Holds the prism pole steady, waiting for a signal that the reading has been taken.

	EDM OPERATOR	PRISM POLE HOLDER
STEP 11	Sights onto the prism with the EDM scope, and using the vertical tangent screw on the EDM and the horizontal motion on the theodolite, maximizes the return signal as per manufacturer's instruction in the operations manual.	Continues to hold the prism pole plumb over the point.

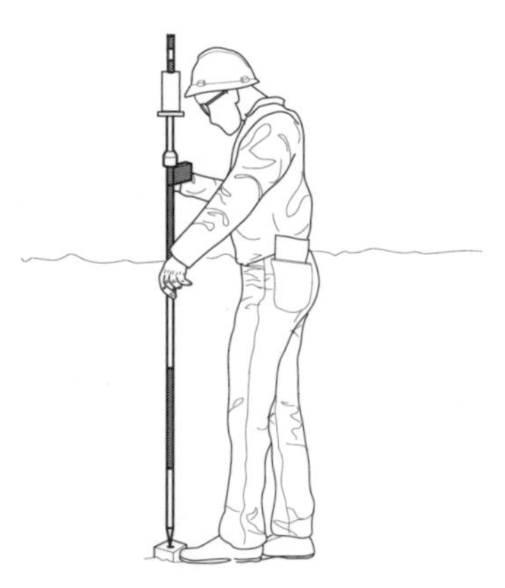

STEP 12	When the signal is maximized, pushes the measure button on the EDM and waits the few seconds necessary for the EDM to operate and display the distance. Reads and records the distance in a field book.	Continues to hold the prism pole plumb and over the point.

STEP 13	As soon as the distance is displayed, signals to the prism holder to relax.	Relaxes when signaled the measurement has been made.

EDM OPERATOR

PRISM POLE HOLDER

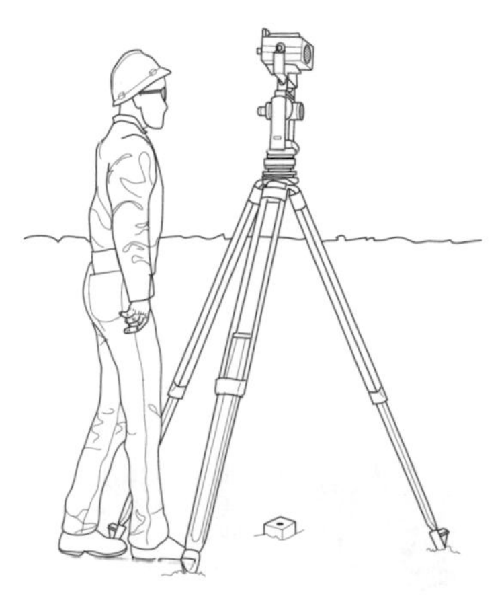

STEP 14

EDM Operator: Double checks to ensure that the instrument is level and over the point. Double checks the vertical sighting and angle measured. Double checks the sighting of the EDM. Signals to the prism pole holder that the distance is to be double checked.

Prism Pole Holder: Places the prism pole on the point, plumbs the pole, and waits for the EDM operator to re-measure the distance.

STEP 15

EDM Operator: Reads the distance and confirms it in the field book. Signals that the measurement is complete and prepares for the measurement to the next point. Moves the instrument if a traverse is being run.

Prism Pole Holder: Upon receiving a signal to move, picks up any hand tools and proceeds to the next point.

RADIAL LAYOUT WITH AN EDM

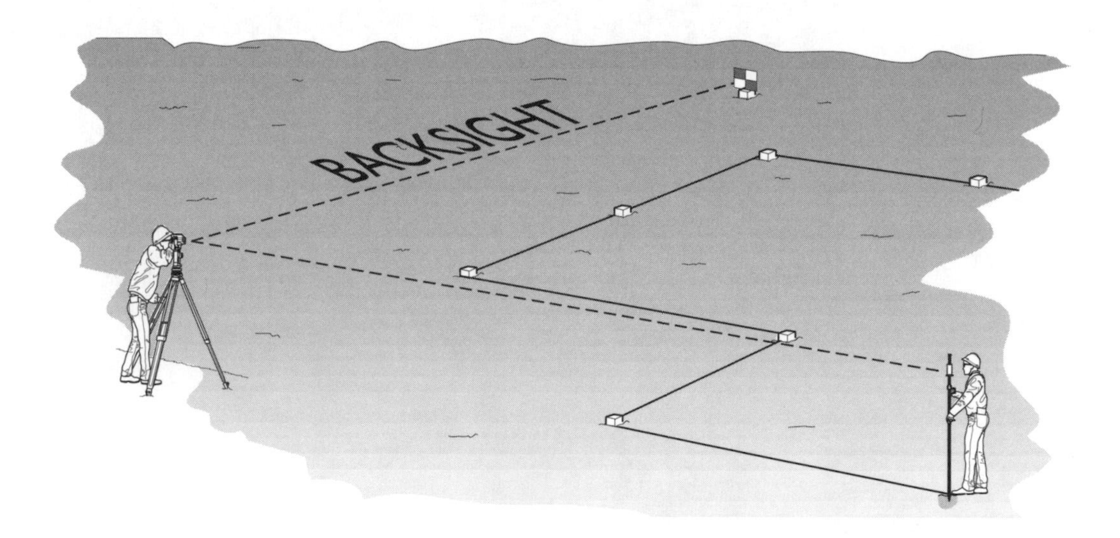

SCOPE

The most frequent use of the EDM by field engineers on a construction site is to lay out points. The step-by-step procedure for laying out a point with an EDM is described in this section.

MEASURING A DISTANCE WITH A TOP-MOUNT EDM

	EDM OPERATOR	PRISM POLE HOLDER
STEP 1	Gathers all needed equipment and proceeds to the work area.	Gathers all needed equipment and proceeds to the work area.

	EDM OPERATOR	PRISM POLE HOLDER
STEP 2	Locates the point that will be used to radially lay out the points. Indicates to the field engineer holding the prism pole where a backsight is needed.	Places the prism on the prism pole and walks to the backsight point as instructed.

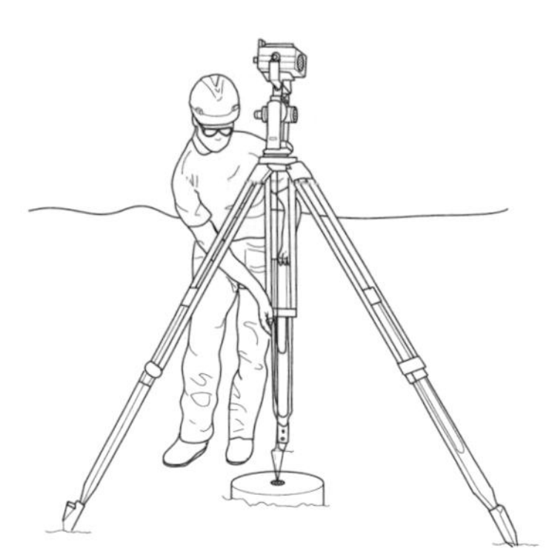

	EDM OPERATOR	PRISM POLE HOLDER
STEP 3	Sets tripod over the point and places the theodolite on the tripod. Levels and centers theodolite over the point. Removes the EDM from its case and securely fastens it on top of the instrument. Inserts the battery into the EDM and checks its strength. If required by the EDM, inputs a parts per million correction factor.	Waits at the backsight for a signal to give line.

EDM OPERATOR | PRISM POLE HOLDER

STEP 4

EDM Operator: Measures the height of the edm in case elevations are needed. Informs the prism holder of the height of the instrument.

Prism Pole Holder: Adjusts the prism pole until its height is the same as the EDMs.

STEP 5

EDM Operator: When the instrument is over the point and is level, sights to the backsight with the instrument set at zero. Refers to the construction layout data sheet. Turns on the theodolite the calculated angle for the point to be laid out.

Prism Pole Holder: Holds a target, plumb bob, or prism pole steady while the backsight is being taken. Advances, as directed by the EDM operator, to the area where the points will be laid out.

EDM OPERATOR

PRISM POLE HOLDER

STEP 6

EDM Operator: Communicates to the prism holder the direction of the line of sight and the approximate distance. Gets the prism on line by communicating verbally or through hand signals.

Prism Pole Holder: Looks back to the EDM operator for instructions. Estimates where the distance to the point will be and gets on line as directed. Facing the EDM, places the prism pole exactly on the point and holds the pole steady and plumb by centering the bull's-eye bubble attached to the side of the pole.

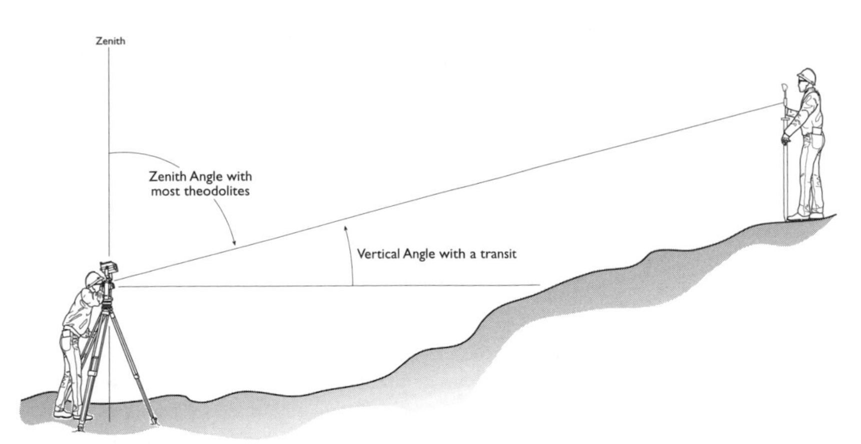

STEP 7

EDM Operator: Sights onto the prism pole with the theodolite to determine the vertical angle. Reads and records the vertical angle for use in calculating the horizontal distance.

Prism Pole Holder: Holds the prism pole steady, waiting for a signal that the reading has been taken.

EDM OPERATOR

PRISM POLE HOLDER

STEP 8

Sights onto the prism with the EDM scope. Using the vertical tangent screw on the EDM and the horizontal motion on the theodolite, maximizes the return signal as per manufacturer's instruction in the operation's manual. When the signal is maximized, pushes the measure button on the EDM and waits the few seconds necessary for the EDM to operate and display the distance.

Continues to hold the prism pole on line and plumb.

STEP 9

As soon as the distance is displayed, calculates the horizontal distance and compares the distance from where the prism presently is to where the point is to be set. Signals whether the point is closer or farther away from the EDM.

Upon hearing the distance and direction to the point, moves the prism pole to that location for another reading.

EDM OPERATOR | PRISM POLE HOLDER

STEP 10

EDM Operator: Gets the prism on line and repeats the measuring process to determine the distance to the new prism location. Determines how far the prism is from the point to be set now and communicates that to the prism holder.

Prism Pole Holder: Communicates when ready for a new measurment.

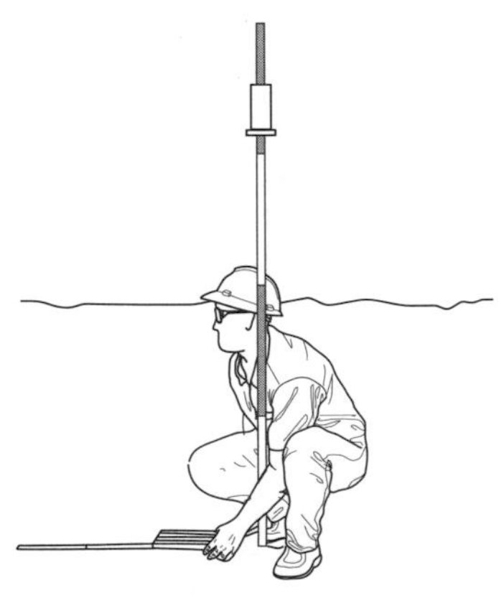

STEP 11

EDM Operator: Repeats the previous step until the point to be set is located and a stake is driven.

Prism Pole Holder: Repeats the previous step until the prism pole location is within a few feet of the desired point. Uses a 6-foot rule or tape to measure to the point and marks the point by driving in a stake or hub.

	EDM OPERATOR	PRISM POLE HOLDER
	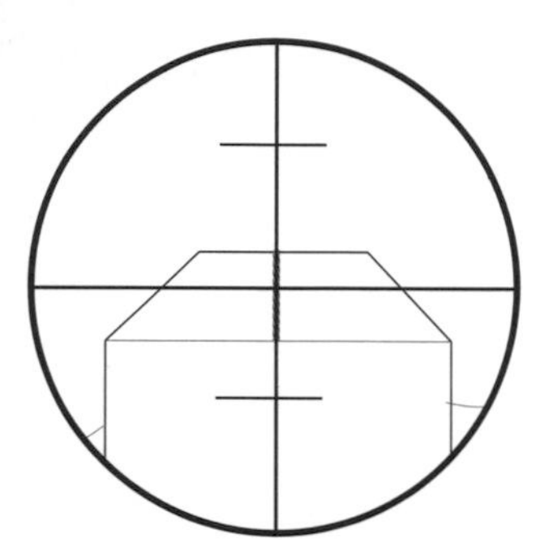	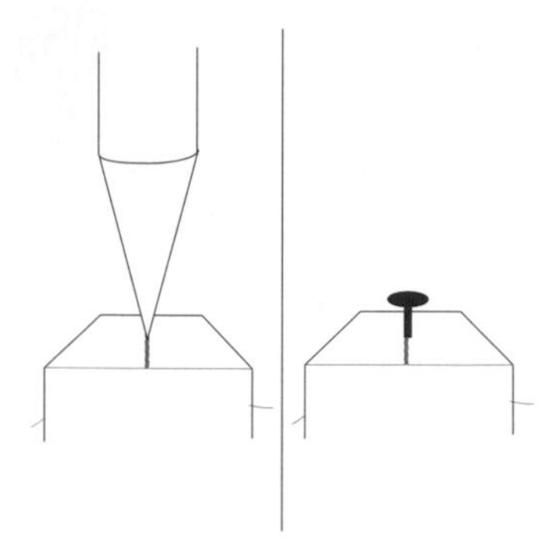
STEP 12	After the stake is driven, establishes line on it exactly. Remeasures the distance to the prism pole held on top of the stake. Confirms the distance on the data layout sheet and writes it in the field book.	Gets line exactly on the top of the stake. Holds the prism pole on top of the stake while the distance is measured. Measures with a rule and sets the exact point on top of the stake. Inserts a tack at the point.

STEP 13	Signals to the prism holder the measurement is complete and refers to the layout data sheet for the measurement to the next point. Repeats the entire process outlined above until all points are laid out.	Proceeds to the next point to be laid out. Repeats the process outlined above for all points to be laid out.

EDM OPERATOR

PRISM POLE HOLDER

STEP 14

EDM Operator: When the last point is established, turns the instrument to the backsight to confirm the instrument is still oriented properly.

Prism Pole Holder: After the last point is established, returns to the backsight to give line to the instrument operator.

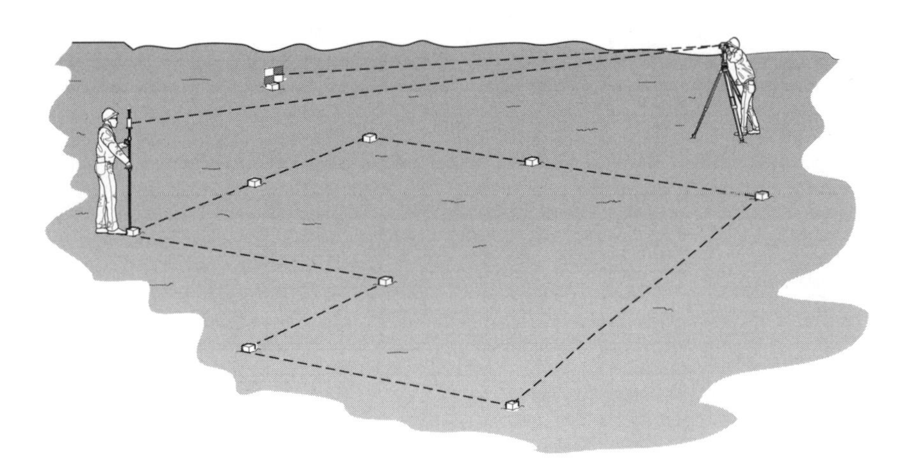

STEP 15

EDM Operator: As a check - Picks up the instrument and moves to another control point. Obtains a backsight from another control point and independently checks the location of some of the points established.

Prism Pole Holder: Goes, as directed, to a new backsight and to established points to check the correctness of their location.

FIELD NOTES

The following is a representative sample of field notes which can be used for radial layout

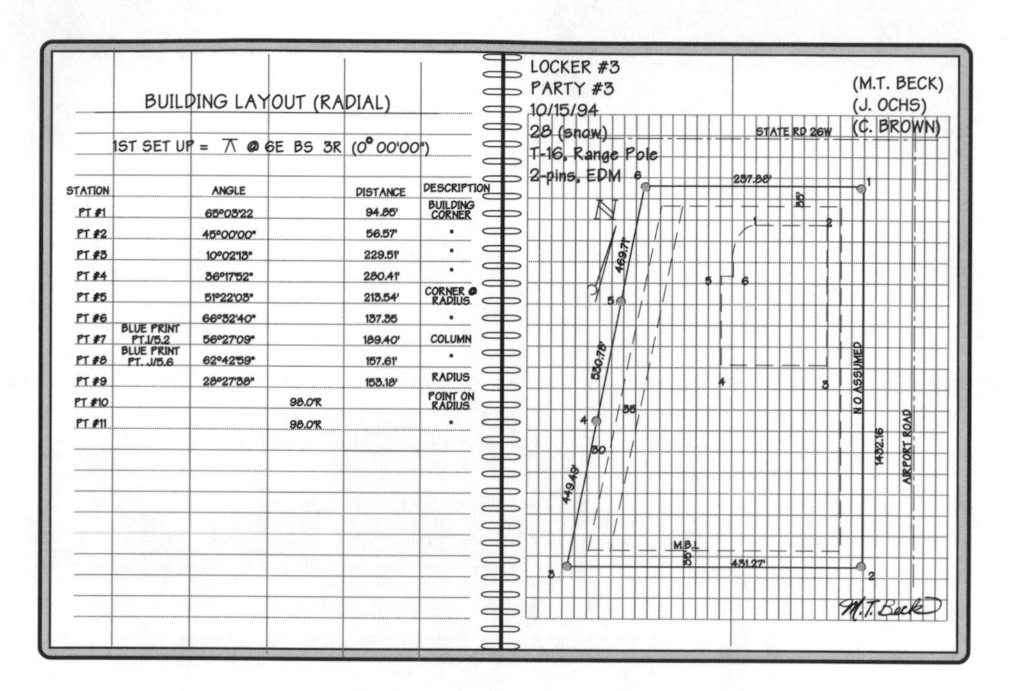

BUILDING LAYOUT (RADIAL)

1ST SET UP = ⊼ @ 6E BS 3R (0° 00'00")

STATION		ANGLE		DISTANCE	DESCRIPTION
PT #1		65°03'22		94.85'	BUILDING CORNER
PT #2		45°00'00"		56.57'	"
PT #3		10°02'13"		229.51'	"
PT #4		36°17'52"		280.41'	"
PT #5		51°22'03"		213.54'	CORNER @ RADIUS
PT #6		66°32'40"		137.35	"
PT #7	BLUE PRINT PT.I/5.2	56°27'09"		189.40'	COLUMN
PT #8	BLUE PRINT PT. J/5.6	62°42'59"		157.61'	"
PT #9		28°27'38"		153.18'	RADIUS
PT #10			98.0'R		POINT ON RADIUS
PT #11			98.0'R		"

LOCKER #3
PARTY #3
10/15/94
28 (snow)
T-16, Range Pole
2-pins, EDM

(M.T. BECK)
(J. OCHS)
(C. BROWN)

POSSIBLE MISTAKES WHEN USING AN EDM

SCOPE

Many users of the EDM treat it with a certain amount of reverence and think mistakes can't occur. In actuality, mistakes can and do occur in all measurement. The field engineer must be aware of these mistakes and do everything required to eliminate or reduce them in size.

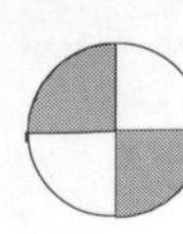

MISTAKES

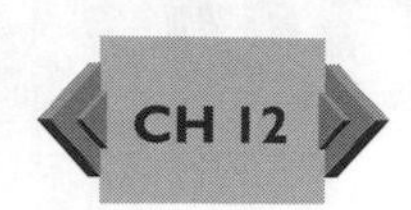

PRISM POLE HOLDER

In most cases the greatest source of error in EDM use is the person holding the prism pole. This person must hold the prism pole vertically and as motionless as possible to obtain a good reading from the EDM. The bull's-eye bubble on the pole should be checked periodically for proper calibration. See *Chapter 12, Equipment Calibration,* for the proper procedure.

CALIBRATION

The "black box" of the EDM creates a false sense of security. It becomes easy to depend on it as an infallible machine. To ensure that good measurements continue, frequently check the EDM against the calibration baseline.

INCORRECT REFLECTOR CONSTANT

An error that can occur when several different kinds of EDM instruments are available on the same jobsite is the reflector constant. Reflectors (prisms) are manufactured by several different companies, and there has been no standardization on the location of the prism in relation to the center of the prism pole. Other than a 0 mm offset, an offset of -30mm is often encountered. Using a 0 mm offset rather than -30mm would result in an automatic error of almost a tenth of a foot in every measurement. Needless to say, that is unacceptable for construction measurement and layout.

EXTRANEOUS REFLECTIONS

A field engineer using an EDM must be aware that the signal may be reflected by objects other than just the prism. EDM signals have been know to reflect off of the reflectors on highways, off of reflectors on the side of cars, off of mirrors, etc. Anything that can reflect light may reflect the EDM signal. If a distance doesn't look right, check to see if there is a reflector close to the line of sight that might be intercepting the signal before it gets to the prism pole.

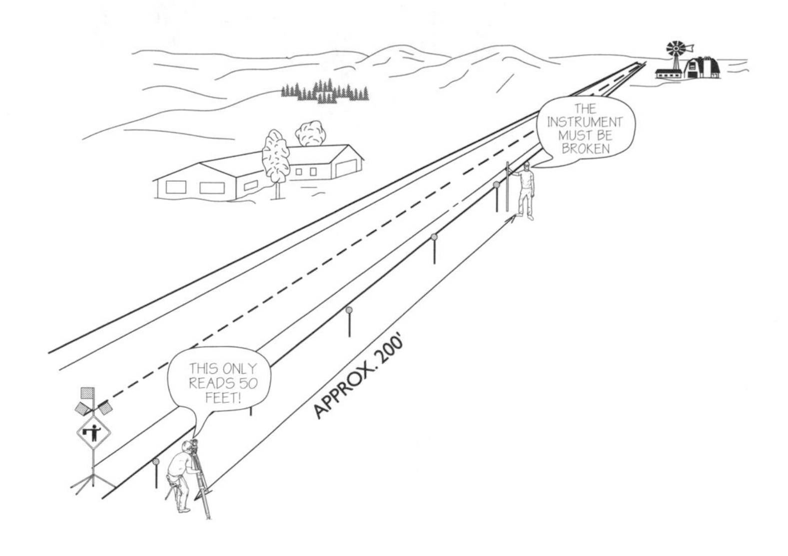

RECOGNIZE THE LIMITS OF THE EDM

Because of the built-in error range of plus or minus 0.02 in most EDMs, the rules of thumb for EDM measurement are the following:

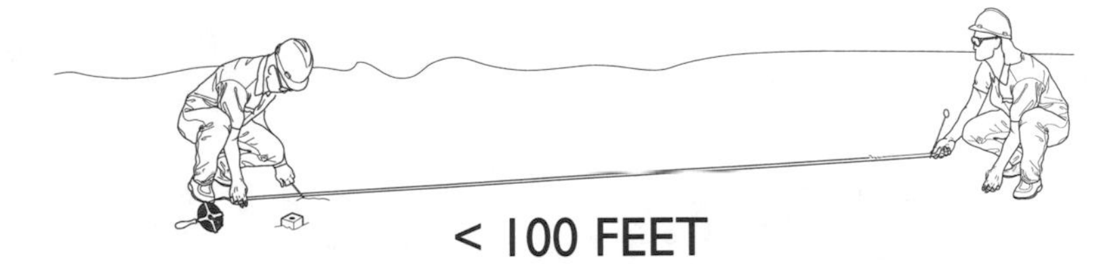

If the distance is less than 100 feet on level terrain, use the chain for more accuracy.

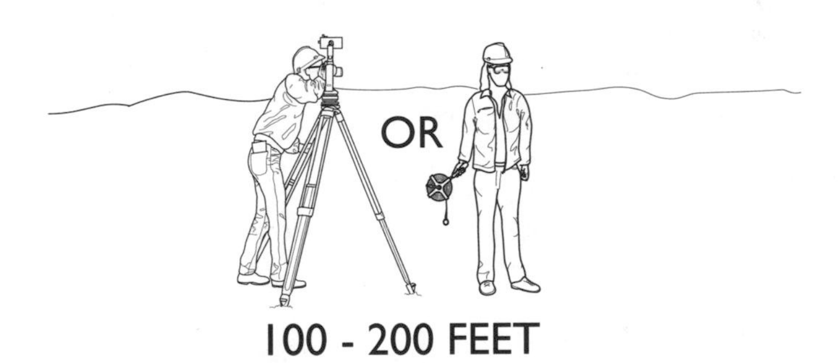

If the distance is between 100 and 200 feet on level terrain, use the chain or EDM.

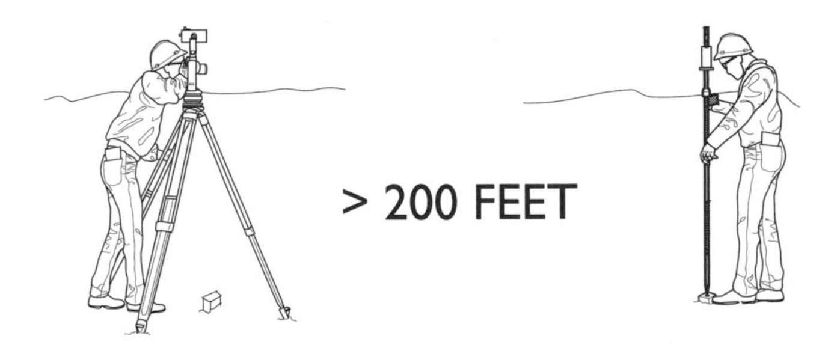

Any distances greater than 200 feet are measured more accurately with an EDM!

These rules of thumb apply for persons with average chaining ability. Those who are experts at chaining will use the chain for distances of up to 400 feet and will use the EDM if available for distances over 400 feet.

SUMMARY OF EDM FIELD DUTIES AND RESPONSIBILITIES

EDM Field Duties and Responsibilities	
A field engineer at the EDM shall:	***A field engineer holding the Prism shall:***
Communicate with the prism holder.	Communicate with the person operating the EDM.
Set up the instrument so it is exactly over the point and level.	Ensure that the prism pole remains plumb and steady so it does not drop off of the point while the reading is being taken.
Eliminate parallax.	Carefully place the point of the prism pole exactly on the point being measured.
Check the charge level of the batteries before leaving the office so no time is wasted due to dead or partially used batteries.	Check the calibration of the bull's-eye bubble on the prism pole for proper adjustment.
Provide proper care for EDM instrument.	Carefully store the prism pole(s) in a safe and secure location while transporting.
Read the operating manual for the EDM to fully understand its operating procedures.	Clean the prism pole by wiping mud, dirt, ice, and snow off of it immediately.
Be careful to follow all steps required to obtain accurate measurements.	Carefully store the prism pole(s) in a safe and secure location while they are not in use.
Utilize care when the EDM is set up in traffic.	Utilize care when readings are being taken in hazardous locations.
Check instrument calibration as recommended by the manufacturer.	Assist the person operating the EDM at all times.
Check instrument focus.	Carry marking pens and spray paint at all times and mark each point and its number on lath or stakes.

ANGLE MEASUREMENT

8

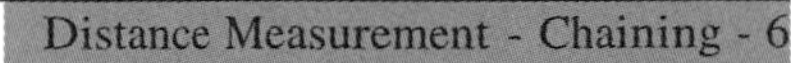

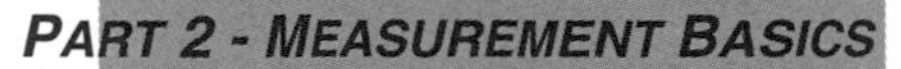

PART 2 - MEASUREMENT BASICS

CHAPTER OBJECTIVES:

Describe the fundamental features of the following instruments: transit, optical theodolite, digital theodolite, total station.
Identify the principle geometric lines of a transit/theodolite.
State the geometric relationships of any angle-turning instrument.
Identify the three basic parts of the transit.
Describe the function of the upper and lower clamp in a repetition angle measuring instrument.
State the basic order of the steps in turning a horizontal angle direct and reverse.
Identify the standard practices for the use of a transit or theodolite.
Describe the process of reading an angle on the horizontal circle and vernier of a transit.
Describe the process or reading angles on a theodolite.
Describe the function of tangent screws on any instrument.
Describe the process that eliminates systematic errors in a transit's horizontal axis.
Describe the step-by-step process of turning angles direct and reverse with a transit.
Distinguish between zenith angle and vertical angle.
Describe why zenith angle is used on modern theodolites.
Describe which of the geometric relationships within an instrument is most important for plumbing a structure.
Describe which of the geometric relationships within an instrument is most important when using a transit as a level for setting grade on forms, establishing elevation, etc.

INDEX

EQUIPMENT AND BASIC PRINCIPLES

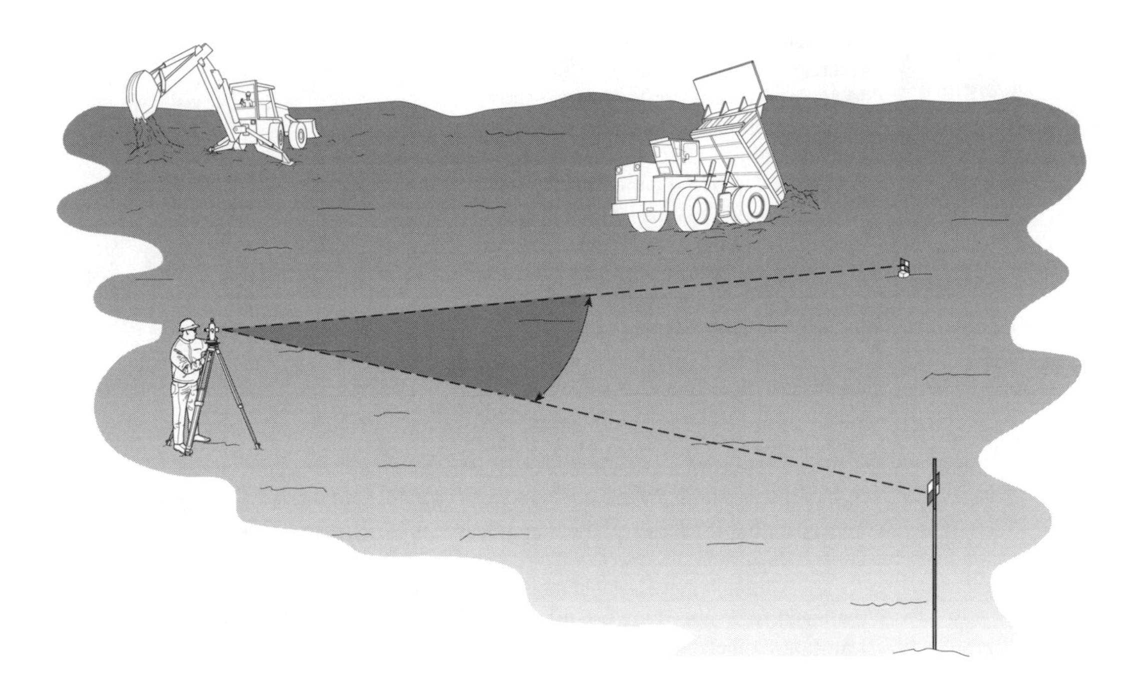

SCOPE

Plumb, square, and level are terms used to describe how we want buildings to be constructed. The transit or theodolite is one of the surveying instruments used to achieve those requirements. The field engineer must have an excellent understanding of the angle-turning equipment being used as well as the process of measuring the angles. If not understood and used properly, even the best instrument will result in angles which do not meet the exacting requirements of construction layout. Control will not close, structures will not fit together, accurate layout by coordinates will be impossible, etc. A good field engineer must understand the importance of becoming an excellent operator of angle-turning equipment.

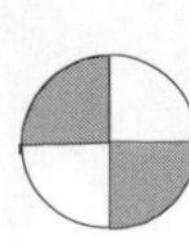

OVERVIEW OF ANGLE MEASURING INSTRUMENTS

There are many types of angle-measuring instruments in use on the construction site today. They include transits, theodolites, digital theodolites, and total stations. Even though the names are different, they look different, and the operation is slightly different, they all are used for the same purpose—angle measuring and layout.

TRANSIT

The transit was developed to its present form during the 1800's. It has been used on construction projects from railroads across the wild west to the skyscrapers of the modern city. A good, solid and reliable instrument, the transit is still used by many contractors today. However, optical theodolite technology, and now digital electronic technology, have passed it by. The major companies who manufactured transits have dropped them from their product line. They are slower to use than modern instruments, and they are not as easily adaptable to having an EDM attached to them.

Its major contribution to the field engineer today is that it is a good tool to learn the fundamentals of angle measurement. Its parts are exposed making it easier to see what is going on
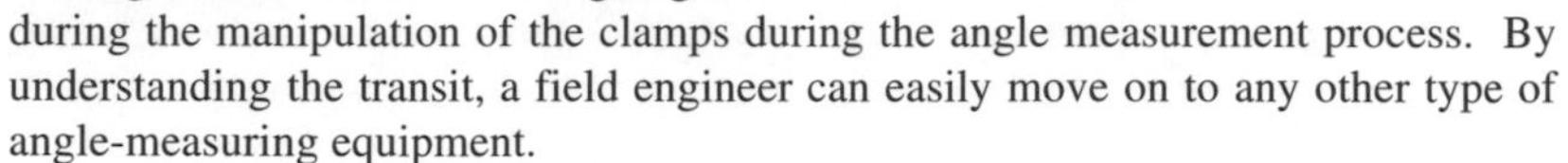
during the manipulation of the clamps during the angle measurement process. By understanding the transit, a field engineer can easily move on to any other type of angle-measuring equipment.

Although there are a number of differences between transits and theodolites, the major difference between them is in the method of reading angles. The typical transit has a metal horizontal circle which is marked from 0 to 360 and is graduated to the nearest 20 or 30 minutes of arc. A vernier is then used to obtain more precise readings to the nearest one minute or even 20 seconds of arc. The procedure for reading an angle off of the circle is straightforward, but it is cumbersome compared to the more modern systems.

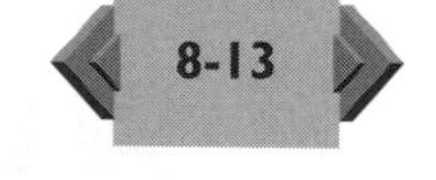

To read an angle on a vernier transit, refer to the section titled *Reading Angles (8-13).*

OPTICAL THEODOLITE

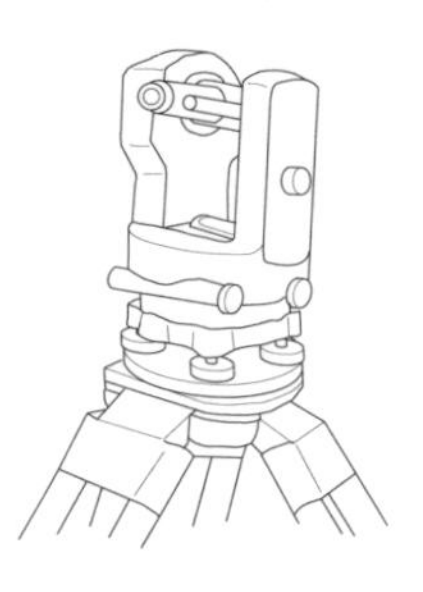

The optical theodolite is a term that was originally applied in Europe to instruments similar to the transit. However, as instrument technology progressed, theodolite became synonymous with a style of instrument that was enclosed, used a magnified optical system to read the angles, had a detachable tribrach with an optical plummet, used a three-screw leveling system, and was more precise than the transit. These features have made it much easier to use than the transit.

The better optical theodolites have been "delicate" workhorses since they were introduced. That is, if they are properly cared for, they seem as though they will last forever because of their excellent construction and quality materials. However, they must be handled gently and carefully. A typical optical theodolite may have as many as 20 prisms or lenses as part of the optical angle reading system. With a sharp bump, these can get out of alignment, and may render the instrument unusable. As with any surveying instrument, the theodolite cannot be exposed to inclement weather because of the optical system.

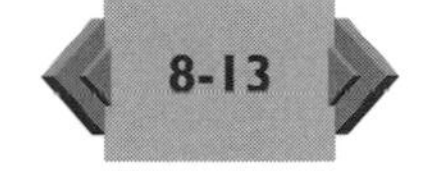

To read an angle on a theodolite, refer to the section titled *Reading Angles (8-13)*. Optical theodolites are excellent instruments; but just like the transit, they are also being surpassed by the technology of electronics.

DIGITAL THEODOLITE

Digital electronics has recently entered the arena of surveying instruments. Angles are no longer read on a circle or with a micrometer - they are displayed on a screen. The metal or optical circle of the past has been replaced with electronic sensors that determine the angle quickly and precisely. From a distance, the digital theodolite looks like the optical theodolite in size and overall shape. Actually, the manufacturers did build the original digital instruments on the same structure used for optical theodolites. The telescope, the clamping system, the tribrach, and the optical plummet are the same. Only the angle measuring and reading system is different.

Most digital theodolites are designed to be interfaced with top-mount EDMs. This essentially has the impact of turning them into what is commonly called a semi-total station which measures distances and angles and can be connected to a data collector for recording measurements. Sighting by both the instrument telescope and the EDM scope are accomplished separately with this setup. This is a slight disadvantage by making the process more time consuming. But this is also an advantage because the contractor can attach the EDM only when it is needed. This keeps the EDM from constantly being exposed to the hazards of the construction site.

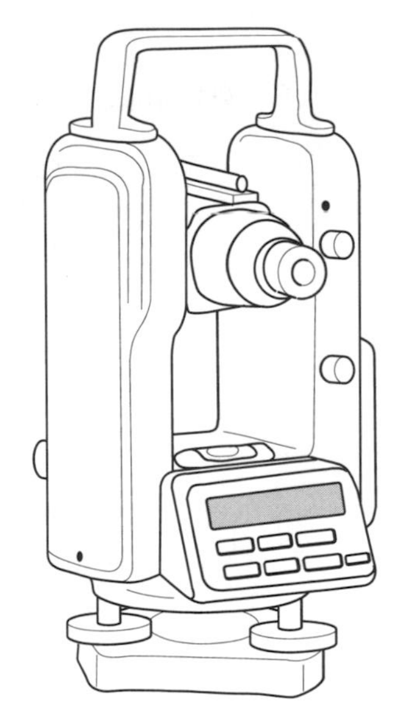

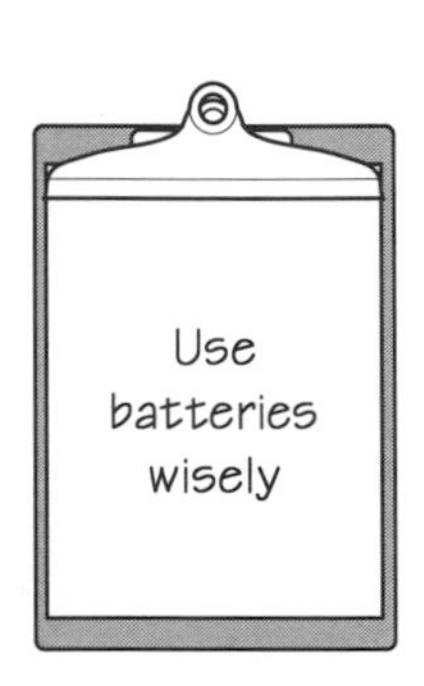

Because it is electronically based, the digital theodolite is like other electronic equipment--it either works or it doesn't work. If a circuit goes out, the instrument is useless. If the battery isn't charged, the instrument is useless. For this reason and more, many contractors are not purchasing digital theodolites as their primary angle measuring instrument. However, it is a great instrument when it is working. An advantage to the contractor is that it is typically less expensive than the optical theodolite.

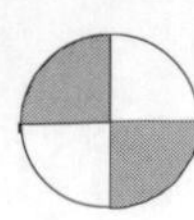

TOTAL STATIONS

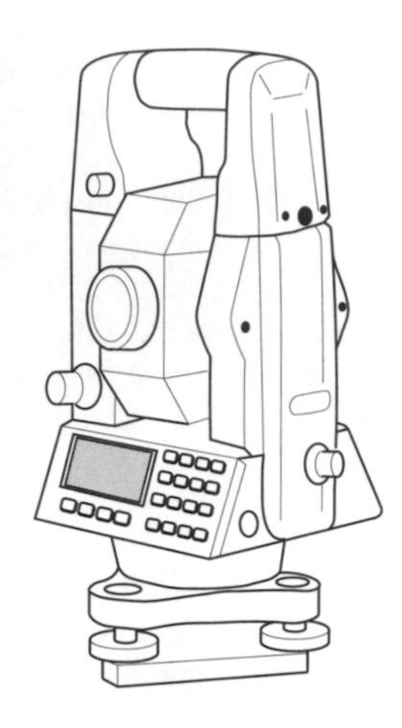

The total station is the ultimate surveying instrument.

The electronic total station is the ultimate in surveying measurement instruments. It is a combination digital theodolite and EDM which allows the user to measure distances and angles electronically, calculate coordinates of points, and attach an electronic field book to collect and record the data. The electronic total station does it all for the field engineer operating it; thus, it is called a total station.

Since the total station is a combination digital theodolite and EDM, everything that has been said about electronics applies to them. Since they are a combination of two instruments, their cost is quite high as compared to a single instrument. However, what they are capable of doing is simply phenomenal in comparison to the way field engineers had to measure just a few years ago. Complex projects can now be calculated on a computer in the office and the data uploaded to the data collector. The data collector is then taken to the field and connected to the total station where hundreds of points can be rapidly established. The total station is the present and future for field layout activities. For more information about the total station, reference *Chapter 10, Total Stations.*

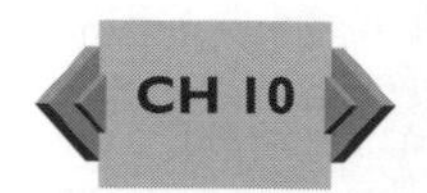

GEOMETRY OF AN ANGLE-MEASURING INSTRUMENT

To operate an instrument well, a field engineer needs to understand it's construction. All angle-turning instruments are the same! Yes, there are differences among a transit, an optical theodolite and the new electronic theodolites; but geometrically they are all the same. All instruments are designed around the same fundamental relationships or principle lines.

THE PRINCIPAL LINES ON A TRANSIT/THEODOLITE ARE:

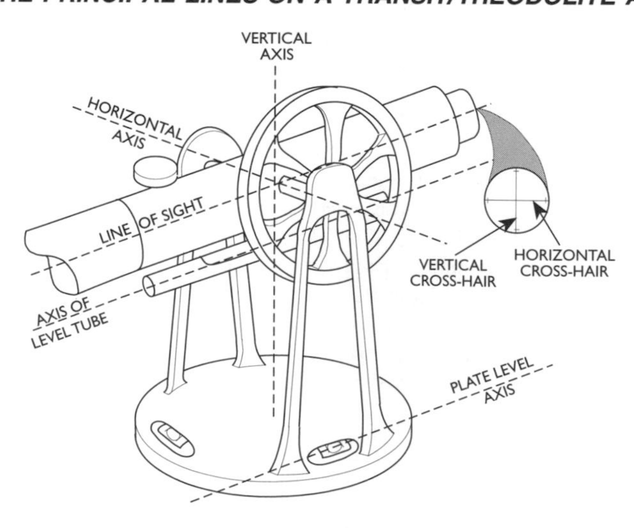

THE GEOMETRIC RELATIONSHIPS

The vertical axis should be perpendicular to the axis of the plate levels.

The vertical cross-hair should lie in a plane perpendicular to the horizontal axis.

The line of sight should be perpendicular to the horizontal axis.

The horizontal axis should be perpendicular to the vertical axis.

The axis of the telescope level should be parallel to the line of sight.

Again, all instruments—no matter how elaborate or how inexpensive—have these same principle lines and geometric relationships. A good instrument operator will be aware of these relationships and the effect they have on measuring angles or giving line. If the perpendicularities that should exist, do not exist, angles measured or laid out will be incorrect if proper procedures are not used. Buildings will not be plumb or square. An instrument where the geometric relationships do not exist is simply out of adjustment.

Can good work be done with an instrument that is out of adjustment? Can an instrument that is out of adjustment be used? The answer is a qualified "yes, it can be," if proper procedures are strictly adhered to. The fundamental principle in using any instrument is the "principle of reversion." That is, everything should be measured direct and reverse with the true value being the average of the two measurements. Instrument operators should always use the double-centering ability of the instrument to eliminate instrumental errors.

THREE COMPONENTS OF AN ANGLE MEASURING INSTRUMENT

Angle measuring instruments can be broken down into three major components. The upper plate assembly or the alidade, the lower plate assembly or the horizontal circle, and the leveling head.

THE ALIDADE ASSEMBLY

The alidade assembly consists of the telescope, the vertical circle, the vertical clamp and vertical tangent, the standards or structure that holds everything together, the verniers, plate bubbles, telescope bubble, and the upper tangent screw. A spindle at the bottom of the assembly fits down into a hollow spindle on the horizontal circle assembly.

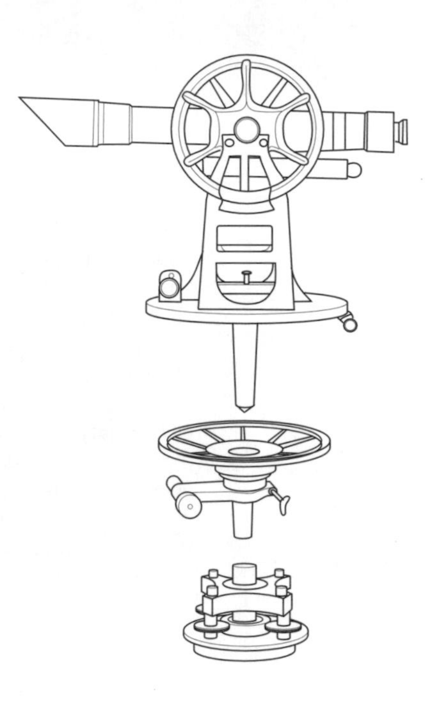

THE HORIZONTAL CIRCLE ASSEMBLY

The horizontal circle assembly is comprised of the horizontal circle, the upper clamp, and the hollow spindle that accepts the spindle from the alidade and fits into the leveling head.

THE LEVELING HEAD

The leveling head is the foundation which attaches the instrument assemblies to the tripod. It consists of leveling screws (4 on a transit, 3 on a theodolite), the lower clamp and tangent, and a threaded bracket for attaching to the tripod.

OPERATING THE UPPER CLAMP AND LOWER CLAMP

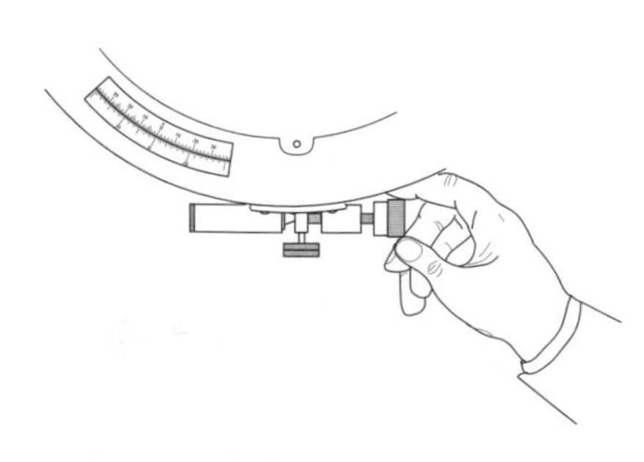

The clamping system is probably the most important feature on an instrument because it is designed so angles can be repeated and added on the instrument. By repeating angles, they can be averaged and greater accuracy and precision can be attained than from turning one angle. The clamping system really is the heart of the instrument operation.

Operating the clamps properly is simple if the process is properly understood. There used to be only one system of clamping the upper and lower assemblies found on the transit. With the optical reading instruments, came several new, simpler clamping systems and procedures. Now, the electronic theodolites present even easier clamping and angle turning systems.

Transits and some optical theodolites have two horizontal motion clamps. They are commonly called the upper motion and lower motion. Each has a specific purpose in the operation of the instrument.

THE LOWER CLAMP

The lower clamp is used to secure the line of sight onto the backsight (left side of angle).

THE UPPER CLAMP

The upper clamp sets the instrument to zero, and it secures the line of sight onto the foresight (right side of angle).

TIGHTENING THE CLAMPS

When using either clamp, which is nothing more than a brass screw with a knob on the end, be careful to operate it properly. They will not withstand excessive tightening. Finger tight is all that is necessary to avoid stripping the threads.

TANGENT SCREW FUNCTION AND OPERATION

Tangent screws are part of the lower, upper, and vertical clamping systems that allow the observer to "fine-tune" the instrument when sighting onto the backsight, setting zero, sighting the foresight, or measuring vertical angles.

When preparing to use an instrument, one activity that separates good instrument users from lackadaisical users is their operation of the tangent screws. A good user will look at the tangent screw to be sure the full range of the tangent screw will be available during the angle turning. By observation of the tangent screw assembly, one can see the cylinder and threads and determine if an equal amount of movement will be able to be made in both directions. On newer instruments, there is often a mark on the screw itself that is an indicator of the midpoint of available movement. Working from the center is important because if, while using the tangent screw, you suddenly run out of threads, you will have to stop and take time to correct the problem.

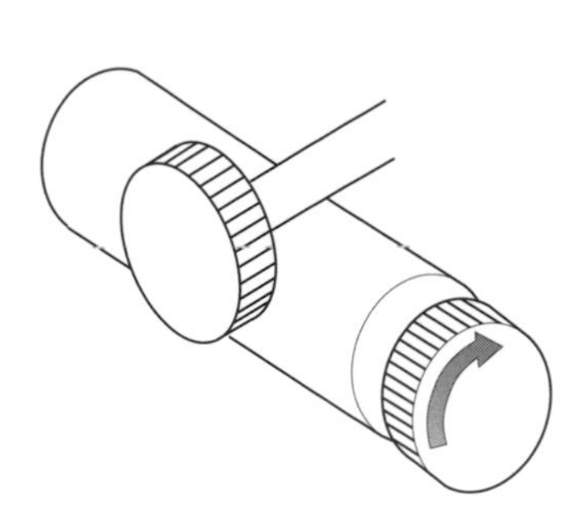

Another detail to observe when using the tangent screws is always be sure to make the last movement of the clamp clockwise. Making the last movement clockwise eliminates the possibility the spring-loaded cylinder isn't binding because of a piece of dirt or some other reason. If the last movement you make with the tangent screw is clockwise into the clamp, you will be putting pressure on the spring. This is a small detail that could be important to accurate angle measurement.

CLAMPING SYSTEMS ON THEODOLITES

Clamping systems on theodolites vary widely. Some theodolites operate exactly like a transit, while directional theodolites have only one clamp with a tangent screw that performs dual purposes. No attempt to explain those various systems will be made here. Just recognize several different ones exist, and they are based on the system devised for the transit. If the transit clamping system is thoroughly understood, its use can be readily adapted to any of the newer instruments.

TURNING AN ANGLE WITH AN UPPER AND LOWER CLAMP

Loosen both clamps.
Lock the upper clamp when the vernier is close to zero on the circle.

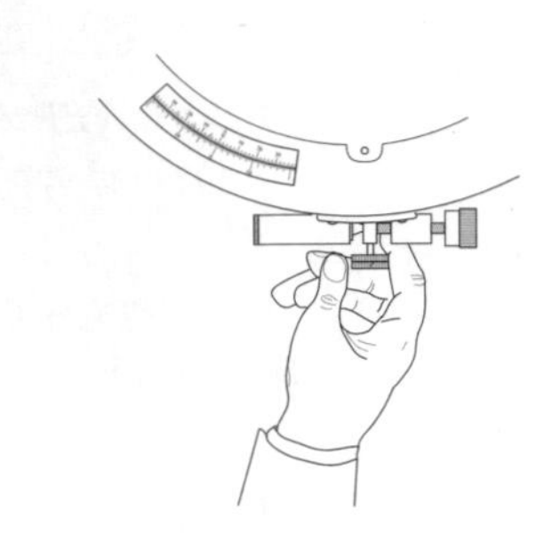

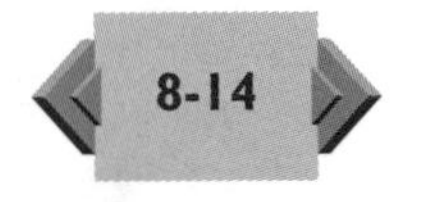

Use the upper tangent to set the vernier exactly to zero. This process is described in the next section.

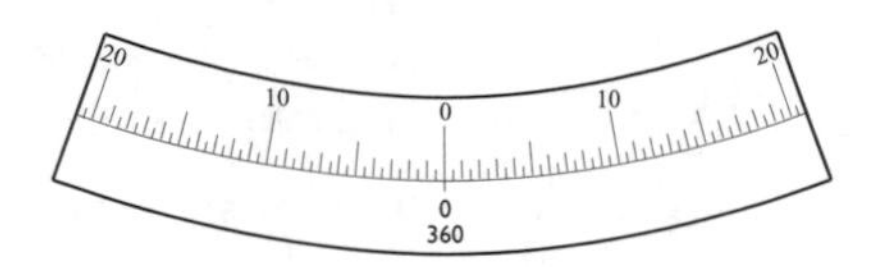

Sight close to the backsight and lock the lower clamp.

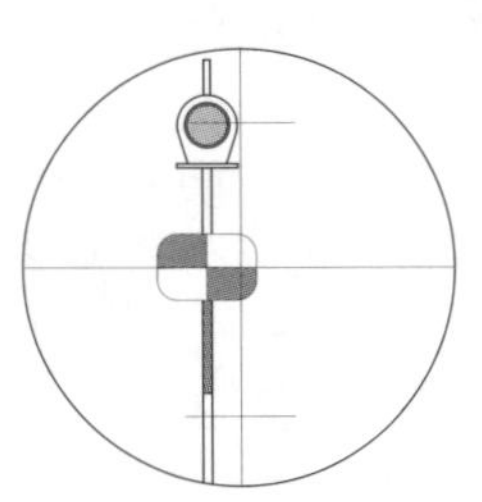

Use the lower tangent to sight the vertical cross-hair exactly on the backsight point.

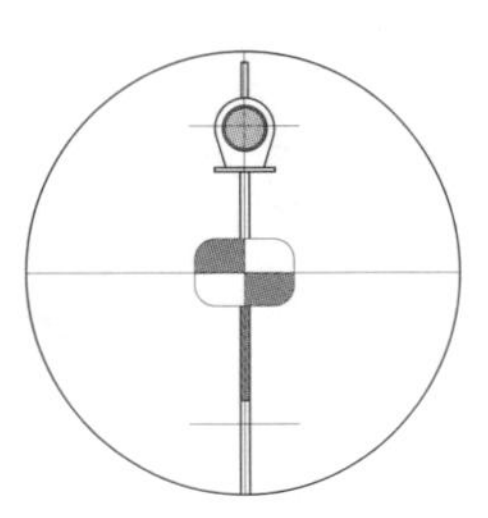

Loosen the upper clamp and sight the instrument close to the foresight point (right side of the angle).

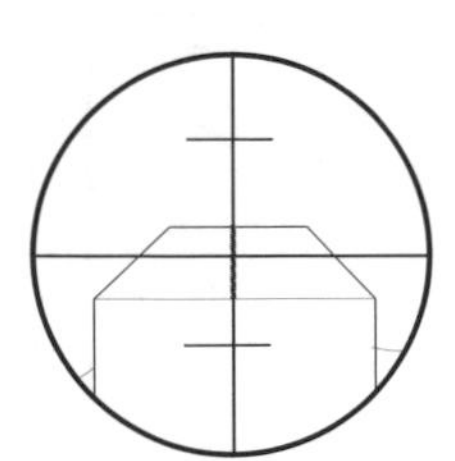

Tighten the upper clamp.
Use the upper tangent to sight the vertical cross-hair exactly on the point.

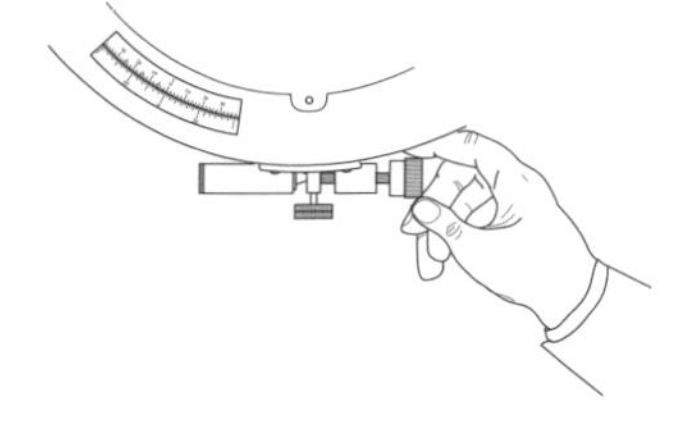

Read the angle on the vernier.

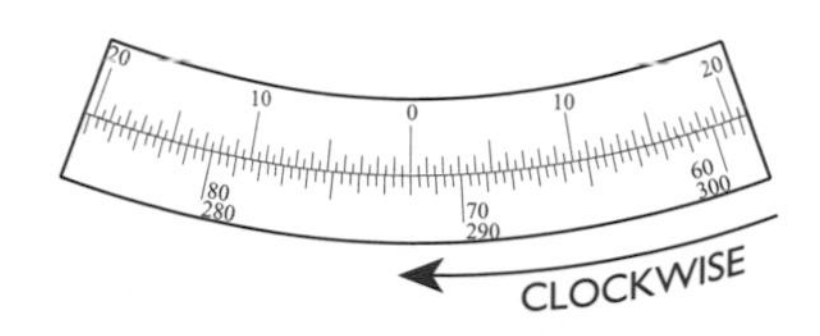

STANDARD TRANSIT OR THEODOLITE PRACTICES

SET OVER THE POINT

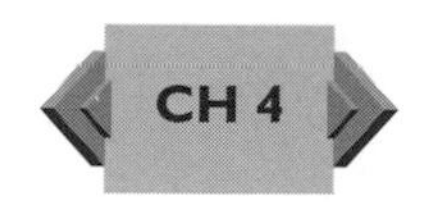

The transit or theodolite must be set over the point exactly, or as close as you can get it. Even a little off of the point will result in erroneous angles measured or laid out. See the sections in *Chapter 4, Fieldwork Basics, on Setting Over a Point with a Plumb Bob, and Optical Plummet Quick Set-Up* .

SOLID SETUPS

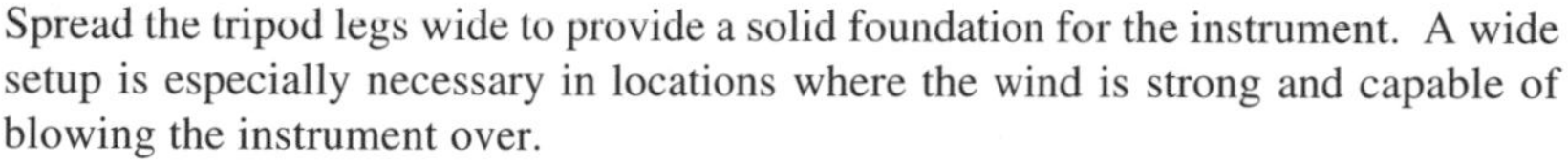

Spread the tripod legs wide to provide a solid foundation for the instrument. A wide setup is especially necessary in locations where the wind is strong and capable of blowing the instrument over.

Step on the foot pads and sink the points solidly into the ground. This is just one more little step to ensure that a good foundation is created with the tripod. If the setup is on soft ground, the weight of the instrument will cause constant settlement if the points aren't initially shoved deeply into the ground. If setting up on a hard surface, search for an indentation where the point of the tripod can be inserted to prevent slipping off the surface.

If necessary—when it is very windy and the setups are on hard surfaces—tie the tripod legs to concrete blocks. It is not necessary to tie them tightly—just have the concrete blocks serve as anchors in case the wind catches the instrument and it starts to fall over.

FOCUS

Good focusing is very important in obtaining good results. There are two things to focus on the typical instrument. The first is fairly obvious. You must be able to see the objective clearly. The second is you must be able to see the cross-hairs clearly. Focusing both of these will be an ongoing activity with each shot taken.

FOCUSING THE OBJECTIVE

Focusing onto the objective for each shot is necessary since the distance to each shot will more than likely be different. When focusing onto the objective, select a part that can be seen clearly and turn the focus knob until you can see it sharply. Look at several parts of the objective for fine details and adjust the focus knob as needed for the absolute best clarity.

FOCUSING THE CROSS-HAIRS

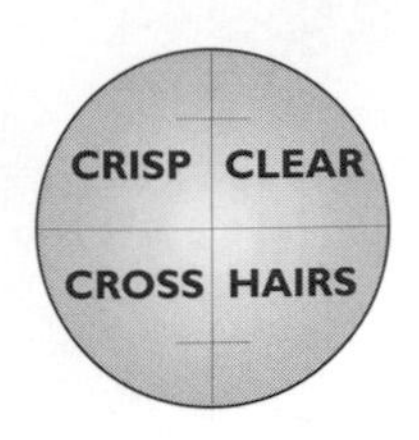

Focusing onto the cross-hairs often throughout the day will be necessary because the human eye tires as the day progresses, and its ability to focus changes. When focusing on the cross-hairs, look at a light colored object (a white house, a cloud in the sky, a piece of paper, etc.) Adjust the eyepiece on the instrument to make the cross-hairs as dark as possible. Look at the center of the cross-hair and progress towards the outer ends. Again, make them as dark and crisp as possible.

If the cross-hairs are not focused properly, a condition called parallax can occur. This is when the cross-hairs seem to move slightly on the object as your eye moves. Some people say the cross-hairs seem to jump. This can be critical when reading a level rod with the horizontal cross-hair or sighting onto a point using the vertical cross-hair. Parallax only occurs when the eye isn't focused properly to the cross-hairs. Be sure to pay close attention to this detail and eliminate parallax. For a more complete understanding of eliminating paralax, see *Chapter 9, Leveling*, on page 9-14.

HELPFUL HINTS FOR READING ANGLES ON A TRANSIT

READ BOTH WAYS

As a check when reading angles, read left and right horizontal circle, then add the angles to compare to 360 degrees. They should total exactly 360 degrees.

READ THE ANGLES IN THE CORRECT DIRECTION

Watch the direction the vernier moves when turning the angle and read the horizontal circle and vernier in that direction. Never read the horizontal circle one way and the vernier in the opposite direction.

USE A MAGNIFYING GLASS

When reading angles on a transit, always use a magnifying glass. Very few people have good enough eyesight to see the exact lines that are lining up between the circle and vernier.

CLOSE THE HORIZON

Close the horizon to confirm the angles turned add up to 360 degrees. If turning several angles, turn the final angle back to the original backsight as a check of 360 degrees.

USE DIRECT AND REVERSE

As stated earlier, errors in the calibration of any angle-turning instrument can only be eliminated by calibration or by using the direct and reverse procedure on all angles measured or laid out. See the section later in this chapter titled *Direct and Reverse Angle by Repetition* for a step-by-step explanation of this process

REFERENCE

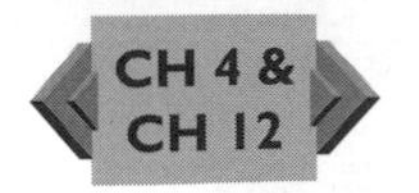

Check geometric relationships regularly. See *Chapter 12, Equipment Calibration* for a thorough explanation of the procedures to follow.

Reference *Chapter 4, Fieldwork Basics* for the step-by-step procedure for setup over a point with a plumb bob.

For the step-by-step procedure to setup with an optical plummet see *Chapter 4, Fieldwork Basics.*

ANGLE READING SYSTEMS

SCOPE

Although there are a number of differences in the construction and features among transits, optical theodolites, and digital theodolites, the major difference is in the method of reading angles. With the transit, the procedure for reading an angle off of the circle is straightforward, but it is cumbersome when compared to the scale-reading optical theodolite and difficult compared to the LCD display of the digital theodolite. The field engineer should become familiar with the instrument being used on the project.

THE TRANSIT VERNIER

The vernier is a device that is constructed so the graduations on it are slightly closer together than the graduations on the horizontal circle. For example, if a vernier has 30 graduations, for the same length on the horizontal circle there will be 29 graduations. Thus, the distance between graduations on the vernier are closer than those on the horizontal circle. By observing which line on the vernier coincides exactly with an opposite line on the horizontal circle, a more precise value of the angle can be determined. A description of how to read a horizontal circle and vernier is listed below.

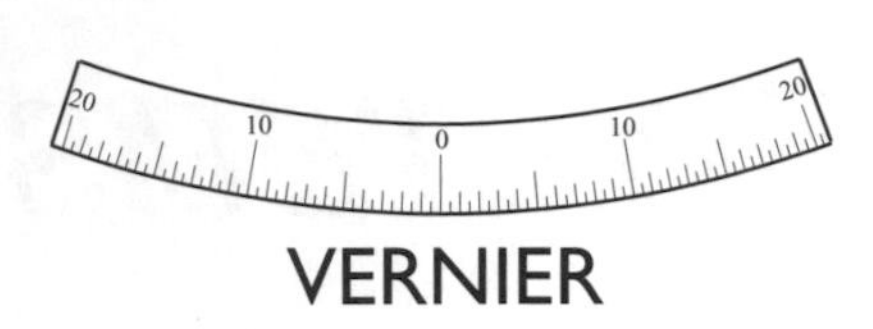

The typical transit has a metal horizontal circle marked from 0° to 360° and graduated to the nearest 20 or 30 minutes of arc. A vernier is then used to obtain more precise readings to the nearest one minute or even to 20 seconds of arc.

READING A CLOCKWISE ANGLE ON A VERNIER TRANSIT

READ THE ANGLE IN THE SAME DIRECTION IT WAS TURNED

Observe the direction the angle was turned. Everything is to be read in that same direction. If the angle was turned clockwise, the vernier and the circle must be read clockwise.

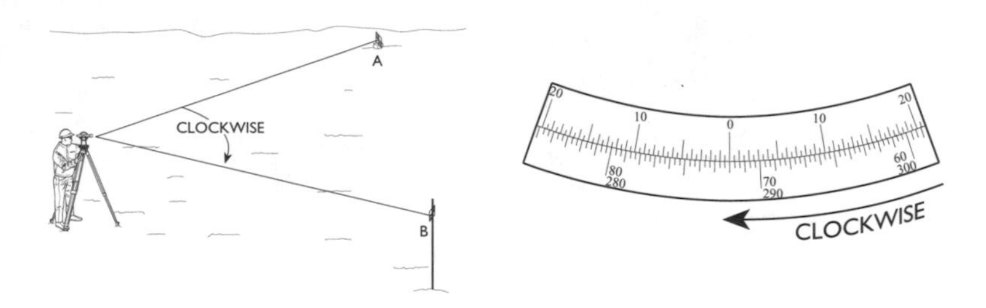

READ THE HORIZONTAL CIRCLE

Observe where the "0" on the vernier lines up on the horizontal circle. Looking at the horizontal circle, read from the nearest marked degree to determine the horizontal circle reading. Remember the reading from the horizontal circle or write it down on a piece of paper.

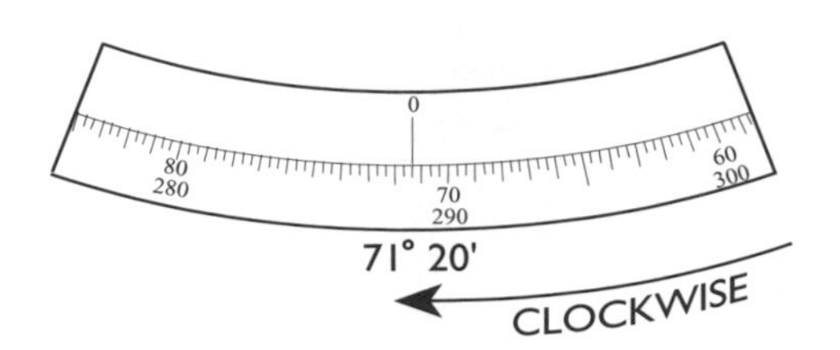

READ THE VERNIER

Go to the vernier and read in the same clockwise direction. Go across on the vernier until you observe where one of the lines on the vernier lines up with one of the lines on the horizontal circle. This is the reading that must be recorded from the vernier.

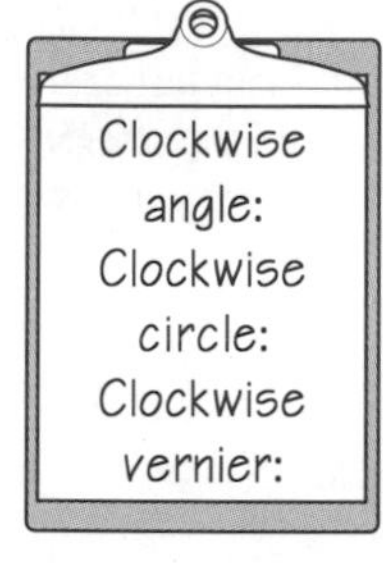

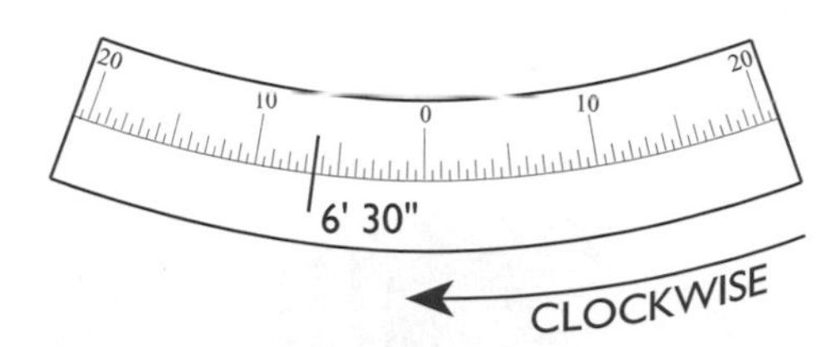

ADD THE READINGS

Add the vernier reading to the horizontal circle reading to obtain the total angle measured. 71°20' + 6'30" = 71°26'30"

OPTICAL THEODOLITES

A typical optical theodolite may have as many as 20 prisms or lenses as part of the optical angle reading system. Several different categories of optical theodolites are listed below.

THE SCALE READING OPTICAL THEODOLITE

The typical scale reading theodolite has a glass circle with a simple scale that is read. The scale is read where it is intersected by the degree readings from the circle. See the illustration for an example scale reading.

Simply read the degree that shows up in the window, and observe where the degree index mark intersects the scale. Both the horizontal circle and the vertical are generally observed at the same time.

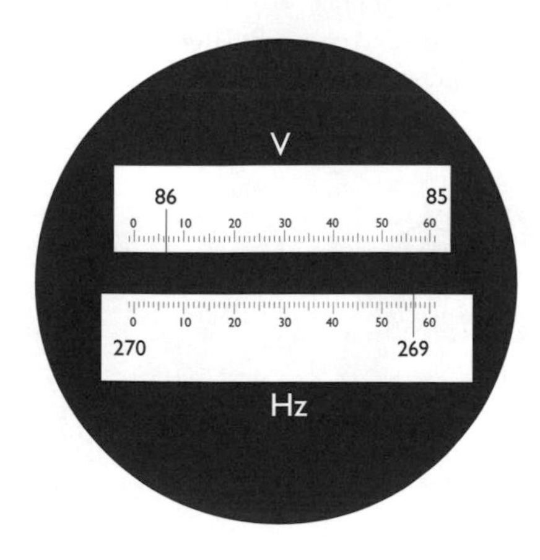

Vertical Angle = 86° 06' 30"

Horizontal Angle = 269° 56' 30"

THE REPEATING OPTICAL THEODOLITE

The repeating theodolite contains the same upper and lower clamp system as the transit; however, the reading of the angle is different because of the optical reading capability. This instrument also has a glass circle, but it does not have a scale. It has a micrometer used to precisely read the angle. The operator of the instrument turns the angle and then uses the micrometer to align the degree index marks. A micrometer is used to optically align the degree index marks, and the reading from the degree window and the micrometer window are added together to obtain the angle.

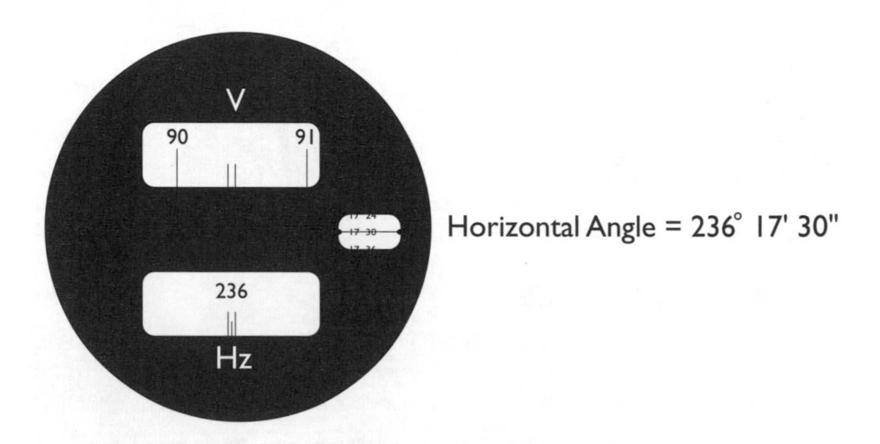

THE DIRECTIONAL OPTICAL THEODOLITE

The directional theodolite is different from other instruments as it does not have a lower motion clamp and tangent screw. Directions are observed and recorded, and then the directions are subtracted to obtain the angle between the directions. This type of system has generally been used only on the most precise instruments. Zero is usually not set on the instrument. The reading of the angle is similar to that of the repeating theodolite. A micrometer is used to optically align the degree index marks, and the reading from the degree window and the micrometer window are added together to obtain the angle.

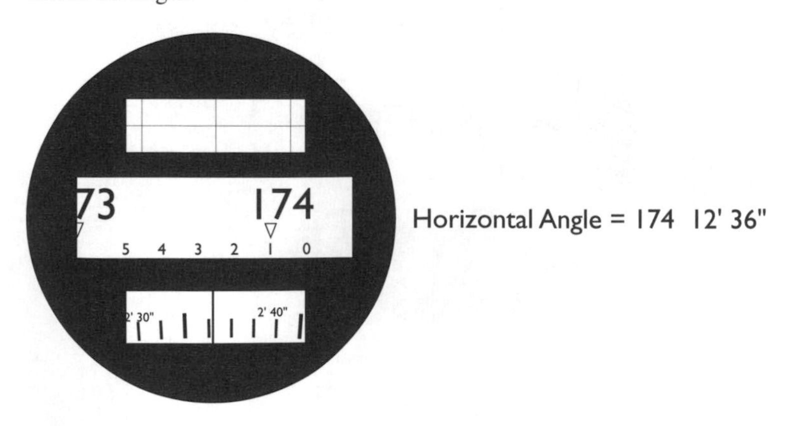

THE DIGITAL THEODOLITE

Digital theodolites are as easy to read as reading the display on a calculator. The angle reading is presented in degree, minute, and second format. The hardest part for the field engineer is to transfer the angle from the display to the field book correctly. There are no instructions for reading a digital display. A sample is shown in the illustration.

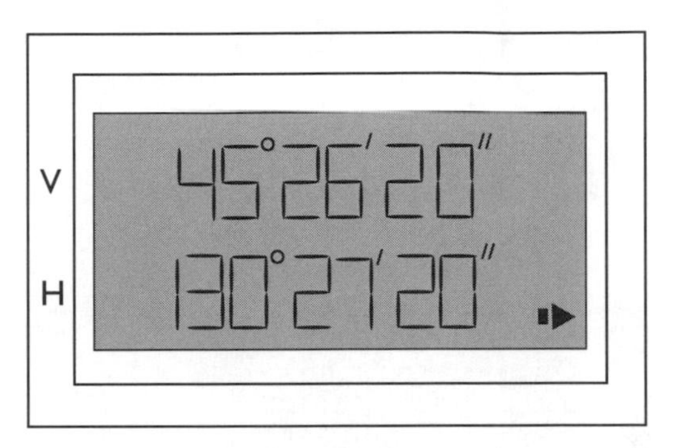

DIRECT AND REVERSE ANGLES BY REPETITION

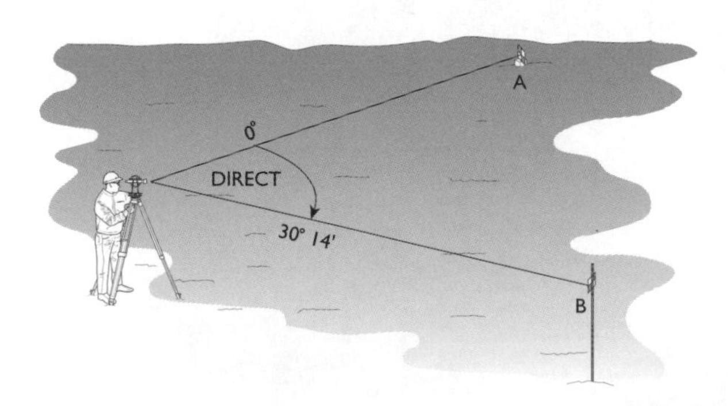

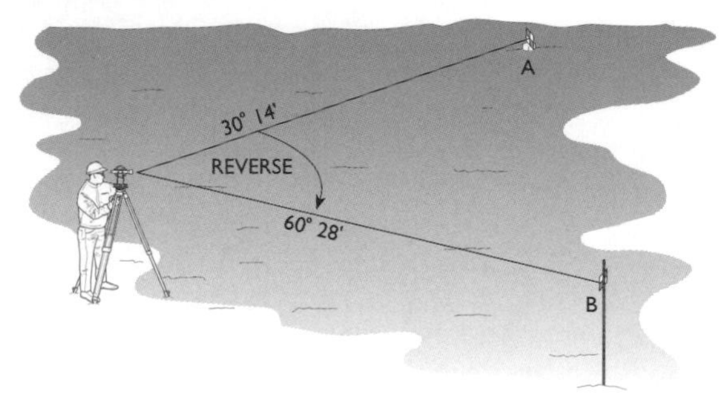

SCOPE

Measurement by repetition is a simple procedure used to increase the precision of angular measurements. An angle is repeated or mechanically added up to six, eight, or more times with the instrument. Half of these measurements are made with the instrument telescope in the direct (or normal) position and half in the reverse (inverted) position. This direct and reverse process has the effect of eliminating most instrumental errors that occur because the geometry of the instrument is out of adjustment. It also allows for the averaging of angles turned which increases the precision. After the desired number of direct and reverse measurements has been turned, the value of the angle is determined by dividing the final reading on the instrument by the number of times the angle was turned. (360° must be added to the final angle for each time 0° is surpassed on the circle.)

GENERAL

Measuring angles direct and reverse is essentially the same process with any instrument, although there are some differences. The step-by-step procedure outlined and illustrated below is for the old stand by, the engineer's transit. The procedure is pretty generic and can be applied to most instruments. However, some horizontal angles are measured with an instrument utilizing the upper and lower motions of the instrument. Field engineers may find it helpful to remember the following rules relative to the use of the upper and lower motion clamps and tangent screws. Recall from the previous section:

> The upper clamp and tangent screw are used for setting the instrument to zero, for setting the instrument to a given angle, and for foresighting.
>
> The lower clamp and tangent screw are used exclusively for backsighting.

Remember the objective of direct and reverse angles by repetition is to add the angles on the circle of the instrument.

EQUIPMENT

Angle Measuring Equipment	
Per crew	***Per person***
Engineer's transit	Plumb bob
Tripod	Field book
Range pole	Straight edge
	4H pencil

PROCEDURE FOR MEASURING DIRECT ANGLES

SETUP

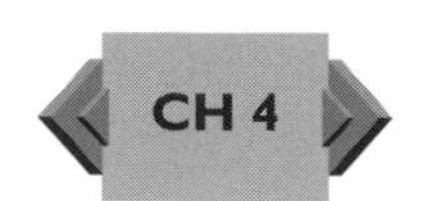

Set up the instrument over the point using the procedure outlined in *Chapter 4, Fieldwork Basics*. Test the telescope for parallax; and, if present re-focus the eyepiece lens to eliminate this condition before observing the angles.

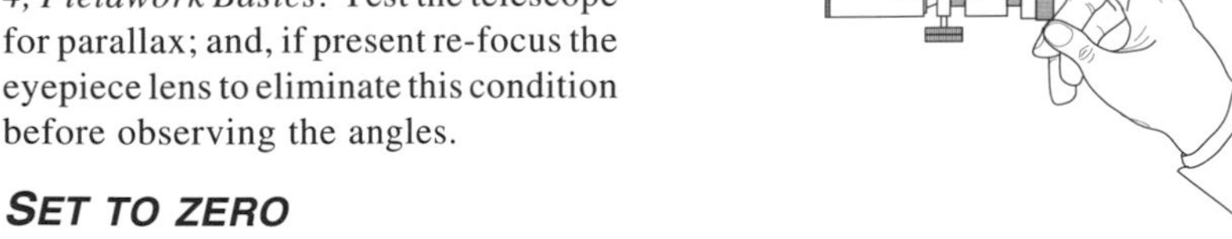

SET TO ZERO

Release both the upper motion clamp and the lower motion clamp. Rotate the instrument until zero on the horizontal and zero on the vernier are approximately on zero. Tighten the upper clamp, and using the upper tangent, set the horizontal circle and vernier precisely to zero.

POSITION THE TRANSIT TELESCOPE IN THE DIRECT POSITION

Looking at the transit, position the telescope so the vertical circle is to your left as you are looking through the eyepiece of the instrument.

SIGHT TO THE LEFT SIDE OF THE ANGLE FOR A BACKSIGHT

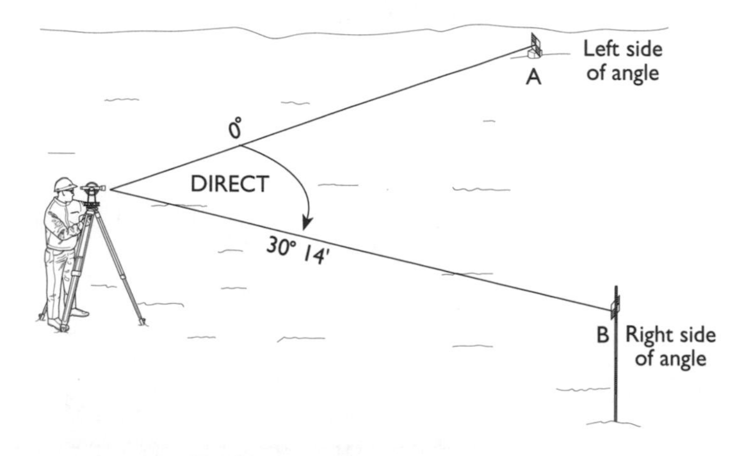

Aim the transit telescope so the left point of the angle to be measured is in the field of view. When the crosshairs are within a few tenths of the point, tighten the lower motion clamp. Set the vertical crosshair precisely on the rear station target using the lower tangent screw.

SIGHT TO THE RIGHT SIDE OF THE ANGLE FOR A FORESIGHT

Release the upper motion clamp and point the telescope so the right side of the angle is in the field of view. Retighten the upper motion clamp and set the vertical crosshair precisely on the forward station target using the upper motion tangent screw.

READ AND RECORD THE ANGLE

Record the reading from the vernier. This is the direct angle and should be recorded in the proper location in the field book.

PROCEDURE FOR MEASURING REVERSE ANGLES

RELEASE THE LOWER MOTION CLAMP

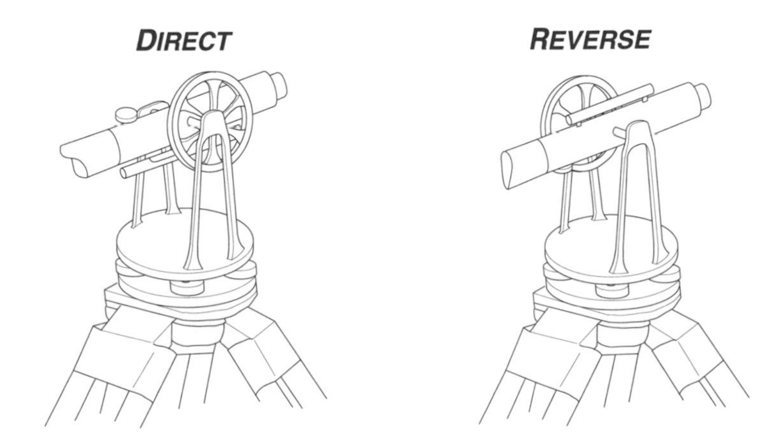

Remember the lower motion is used for backsighting only. Do not touch the upper motion clamp or tangent, or you will change the value of the direct angle just measured.

REVERSE (INVERT) THE TELESCOPE

Release the vertical motion clamp and pivot the telescope about the horizontal axis. The vertical circle will now be on the right side of the instrument as the scope is being sighted.

SIGHT TO THE LEFT SIDE OF THE ANGLE FOR A BACKSIGHT

Aim the transit telescope so the left point of the angle to be measured is in the field of view. When the crosshairs are within a few tenths of the point, tighten the lower motion clamp. Set the vertical crosshair precisely on the rear station target using the lower tangent screw.

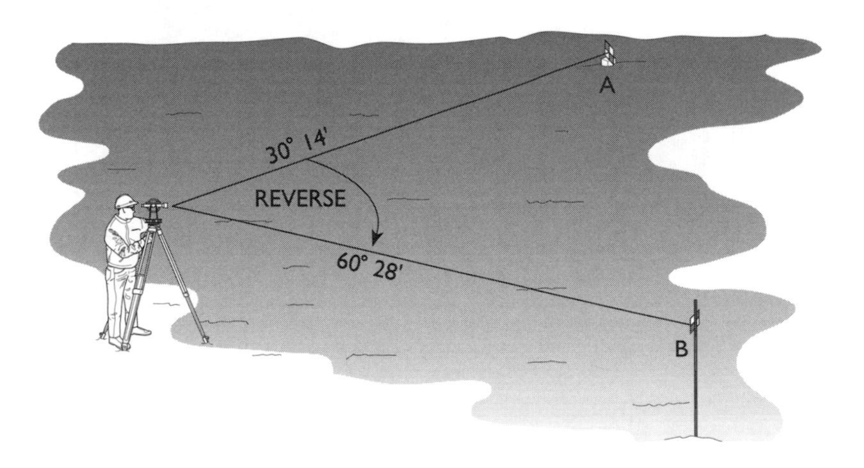

SIGHT TO THE RIGHT SIDE OF THE ANGLE

Release the upper motion clamp and point the telescope so the right side of the angle is in the field of view. Retighten the upper motion clamp and set the vertical crosshair precisely on the forward station target using the upper motion tangent screw.

READ AND RECORD

Record the reverse angle reading from the vernier. This is the reverse angle and should be recorded in the proper location in the field book. This angle should be double the first angle. That is, if the first angle was 30° 14', then this angle should be approximately 60° 28'.

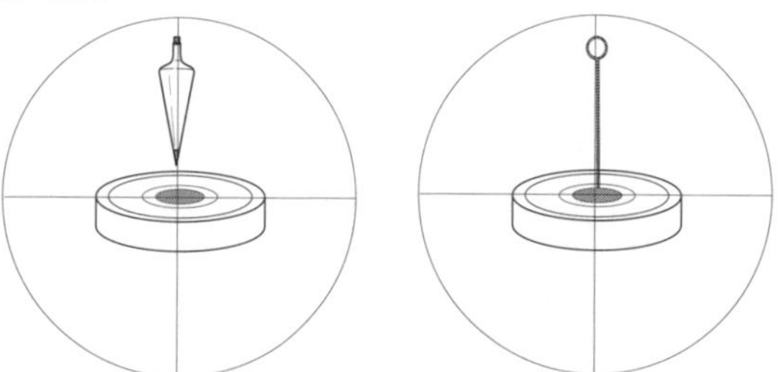

REPEAT THE DIRECT & REVERSE PROCESS

If a better estimate of the true value of the angle is required (i.e. greater precision), more sets of direct and reverse angles can be turned. For control work it is not uncommon to measure four direct and four reverse.

Read and record the final reverse angle measured . If the horizontal circle went past 360° one or more times, it may be necessary to add 360° or multiples of 360° to the final angle to obtain the total sum of the angles measured. Divide the final angle by the number of turns to obtain the average angle. This process can be repeated for the desired number of repetitions.

NOTE KEEPING

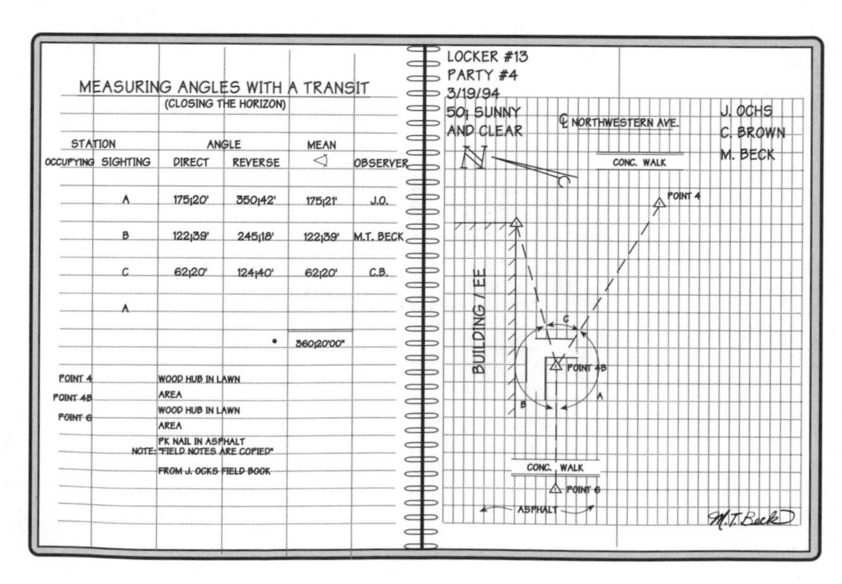

MEASURING ANGLES WITH A TRANSIT
(CLOSING THE HORIZON)

STATION		ANGLE		MEAN	
OCCUPYING	SIGHTING	DIRECT	REVERSE	◁	OBSERVER
	A	175°20'	350°42'	175°21'	J.O.
	B	122°39'	245°18'	122°39'	M.T. BECK
	C	62°20'	124°40'	62°20'	C.B.
	A				
			•	360°20'00"	

POINT 4 WOOD HUB IN LAWN AREA
POINT 4B WOOD HUB IN LAWN AREA
POINT 6 PK NAIL IN ASPHALT
NOTE: "FIELD NOTES ARE COPIED" FROM J. OCKS FIELD BOOK

VERTICAL ANGLES

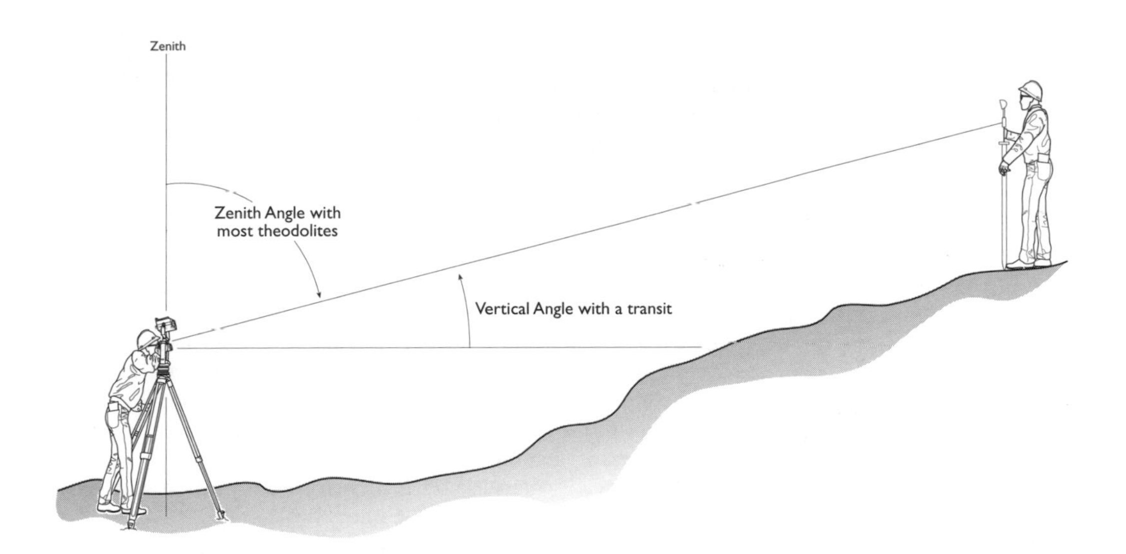

SCOPE

It may not be often, but occasionally the field engineer needs to measure vertical angles as part of the establishment of control or the layout of a distance, etc. Accurate vertical angles are necessary to precisely calculate horizontal distances when using top mount EDMs or when measuring distances on a slope with a chain. Again, all angle measuring instruments are different so the procedure outlined here will not apply in all cases, but the basic fundamentals will pertain.

GENERAL

TRANSIT

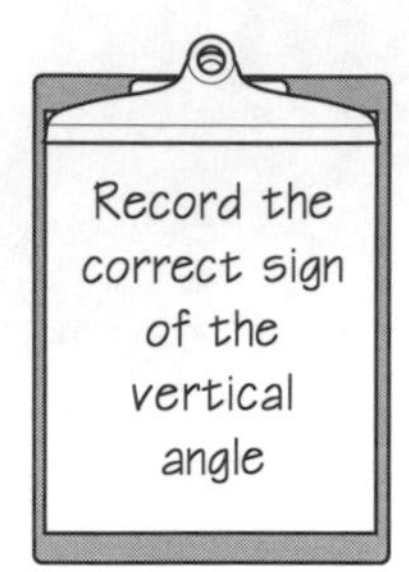

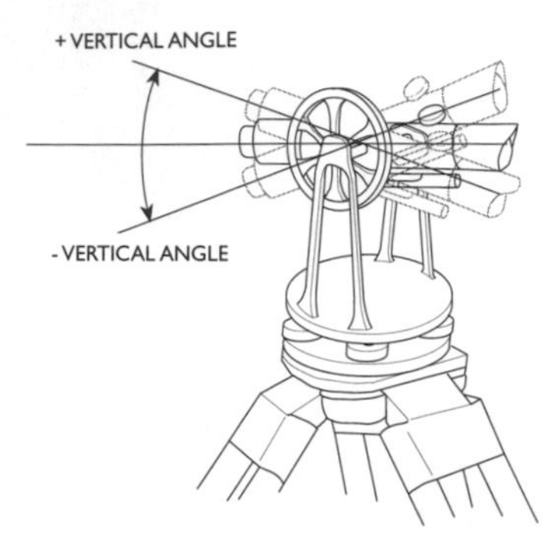

When vertical angles are to be measured with a transit and some theodolites, it is necessary to have the instrument precisely leveled to ensure the vertical axis of the instrument is truly vertical. The operator must be very careful in the leveling of the instrument to ensure this occurs. Some instruments have a telescope bubble that allows the operator to more precisely level the instrument to achieve a truly vertical axis.

Vertical angles are measured off of the horizon with most transits. The vertical angle is considered positive if it is an angle of elevation (above the horizon) and negative if it is an angle of depression (below the horizon). The instrument operator must be very careful to record the correct sign of the vertical angle when measuring with a transit.

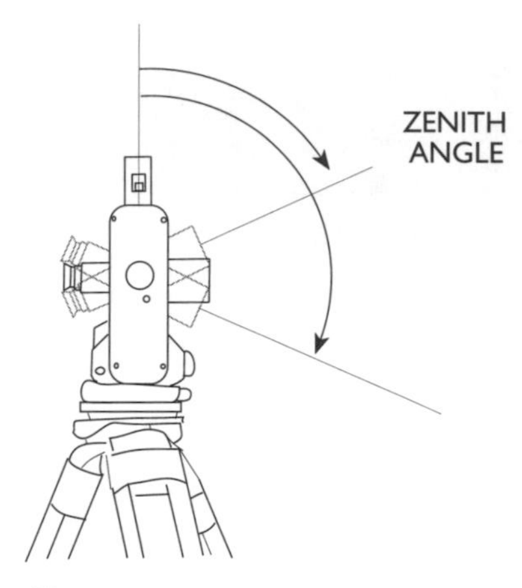

THEODOLITE

Better quality optical theodolites have automatic vertical collimators that mechanically index the vertical circle so it is always correctly oriented. More and more electronic theodolites and most total stations have automatic indexing of the vertical circle so vertical angles measured will be corrected if the instrument is out of level. However, it is good practice to follow the procedure outlined below even with instruments that provide automatic compensators so the vertical angle can be averaged to get a better estimate of the true value.

Vertical angles on theodolites are generally measured from the zenith (point straight above). This method eliminates the need to record a positive or negative sign with the vertical angle.

PROCEDURE

LEVEL THE INSTRUMENT

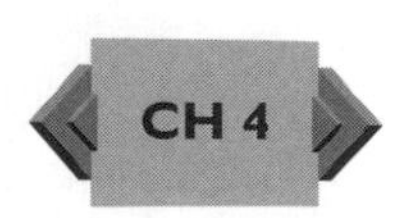

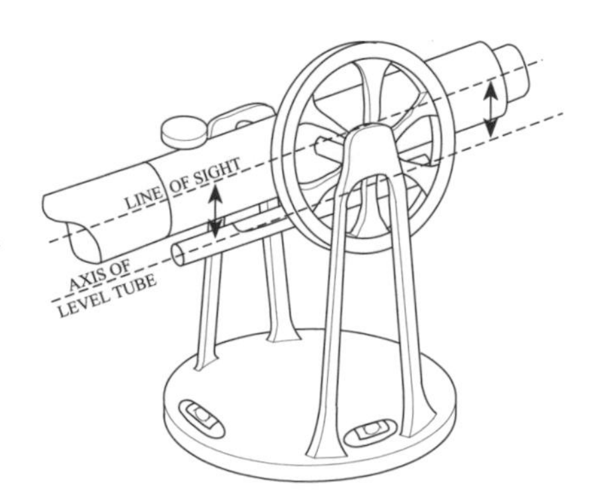

Using the plate level(s) in the usual manner, level the instrument following the procedures outlined in *Chapter 4, Fieldwork Basics*. If the instrument is equipped with a telescope bubble, align the telescope over an opposing pair of leveling screws, and center the telescope level bubble by using the vertical clamp and tangent screw. Rotate the telescope 180° about the vertical axis using the lower motion. If the telescope level bubble does not center, bring it halfway to the center by turning the vertical tangent screw, and then center the bubble by using the two leveling screws. Again, rotate the telescope 180° about the vertical axis. The bubble should remain centered. If it does not, the process should be repeated.

The same process should be repeated alternately over both pairs of opposing leveling screws until the telescope level bubble remains centered when the telescope is rotated 360° about its vertical axis. If the bubble still does not stay centered, split to each side of the center the amount it is off. It can be used this way, but it should be serviced. See *Chapter 12, Equipment Calibration.*

DIRECT SIGHT WITH THE HORIZONTAL CROSS-HAIR

With the telescope in the direct position, bring the line of sight approximately on the point and tighten the horizontal motions. Set the horizontal cross-hair exactly on the point observed, tighten the vertical clamp, and use the vertical tangent screw to get the horizontal cross-hair exactly on the point.

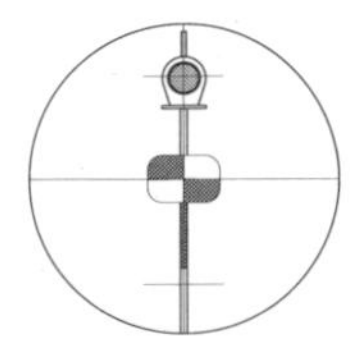

READ AND RECORD THE DIRECT VERTICAL ANGLE

Read the vertical circle and record the reading as the direct angular value in the field notes.

REVERSE SIGHT WITH THE HORIZONTAL CROSS-HAIR

Repeat the procedure with the telescope in the reversed position.

READ AND RECORD THE REVERSED VERTICAL ANGLE

Record the vertical circle reading as the reversed angular value.

AVERAGE THE ANGLES

The average of the direct and reversed readings is the correct value of the vertical angle. This double-sighting procedure eliminates the effects of instrumental errors due to non-parallelism between the line of sight and the axis of the telescope and displacement or lack of adjustment of the vertical vernier.

FIELDNOTES

The following is a representative sample of field notes for vertical angles. In this case, measured to determine the horizontal distance.

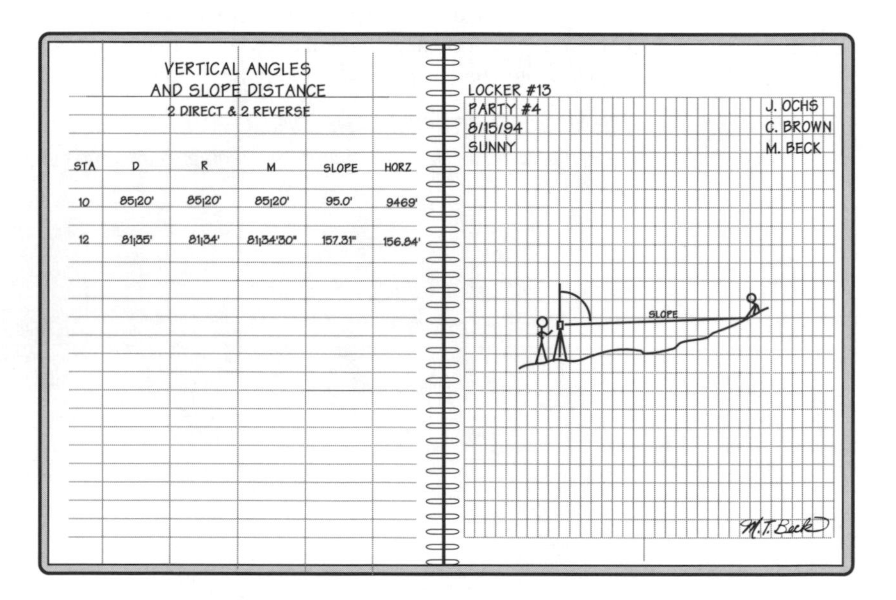

VERTICAL ANGLES
AND SLOPE DISTANCE
2 DIRECT & 2 REVERSE

STA	D	R	M	SLOPE	HORZ
10	85°20'	85°20'	85°20'	95.0'	9469'
12	81°35'	81°34'	81°34'30"	157.31"	156.84'

LOCKER #13
PARTY #4
8/15/94
SUNNY

J. OCHS
C. BROWN
M. BECK

MEASURING TRAVERSE ANGLES

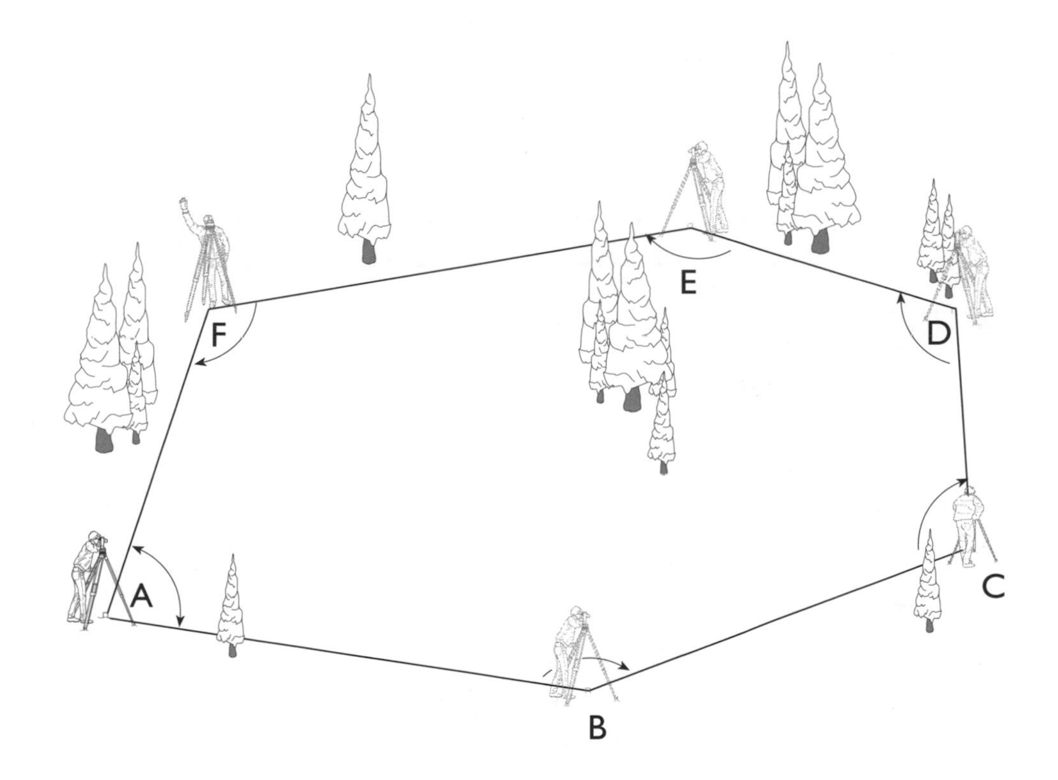

SCOPE

The traverse is a fundamental tool of field engineers. By knowing the angles and distances about a traverse, field engineers can calculate the precision of the work and the coordinates for each point, as well as the area to be enclosed. This information is used to control mapping work and to establish grid systems on large construction projects among other things. Because of this importance, angles and distances which comprise the traverse must be measured with a great amount of care.

To achieve the precision required for traversing, the field engineer must eliminate mistakes and reduce error by repetitiously turning angles with the telescope alternately in the direct and reverse positions. Mean angles are then added for a closed figure and the total is compared to specified requirements.

REFERENCE

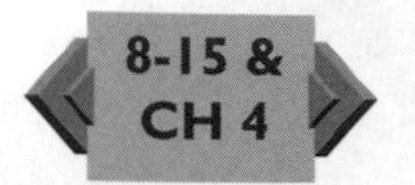

Study the section on *Direct and Reverse Angles by Repetition (8-15).* Review the *Chapter 4, Fieldwork Basics.*

EQUIPMENT

Angle Measuring Equipment	
Per crew	***Per person***
Engineer's transit	Plumb bob
Tripod	Field book
Range pole	Straight edge
	4H pencil

PROCEDURE

LOCATE THE TRAVERSE POINTS IN THE FIELD

Walk the traverse to determine the scope of work and to decide on a starting point.

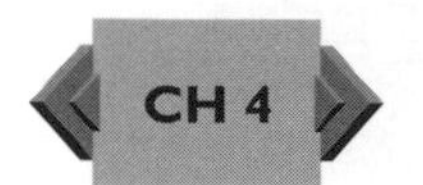

DETERMINE A STARTING POINT

Set up the instrument over the point using the procedure outlined in *Chapter 4, Fieldwork Basics.* Test the telescope for parallax and, if present, re-focus the eyepiece lens to eliminate this condition before observing the angles.

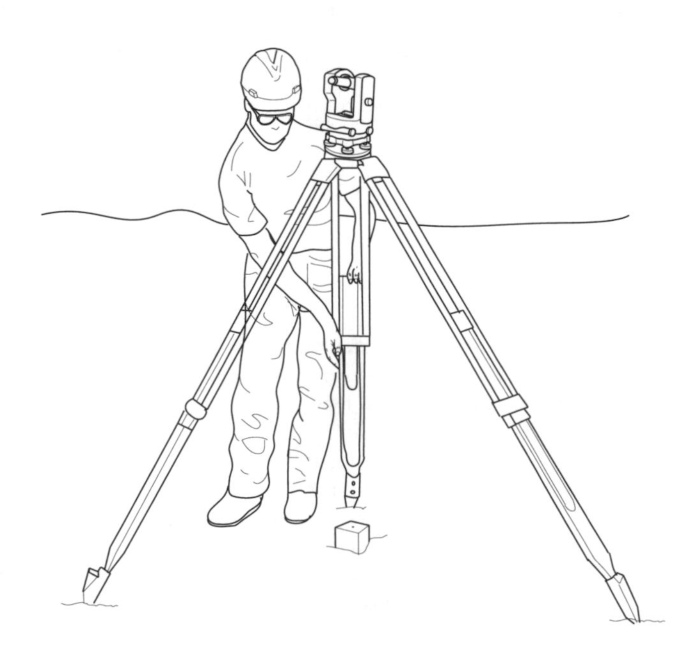

SET RANGE POLES OR TARGETS ON TWO ADJACENT POINTS

Place range poles or some other targets on or behind the points so that they can be seen from the instrument. If it is not possible to set a range pole or target, have other field engineers wait at the points to give line with a plumb bob.

TURN ANGLES DIRECT AND REVERSE

8-15

All traverse angles must be turned direct and reverse to average out any instrumental errors and to get a better value for the angle. Follow the procedure outlined in *Direct and Reverse Angles by Repetition (8-15).* When turning angles on a traverse, it is recommended that a minimum of two direct and two reverse angles be turned. If very high precision is required, it may be necessary to turn 3, 4, or more direct and reverse angles at each point. To turn more than one direct and one reverse angle, simply keep repeating the process. Always turn the same number of direct as reverse angles.

RECORD THE FIRST ANGLE AND THE LAST ANGLE

Only the first direct angle and the last reverse angle need to be recorded in the field book. The first angle is recorded as a reference, and the last angle is recorded.

AVERAGE THE ANGLE

If the horizontal circle went past 360° one or more times, it may be necessary to add 360° or multiples of 360° to the final angle to get the total sum of the angles measured. Divide the final angle by the number of turns to obtain the average angle. If two direct and two reverse angles were measured, the final reverse angle should be divided by 4.

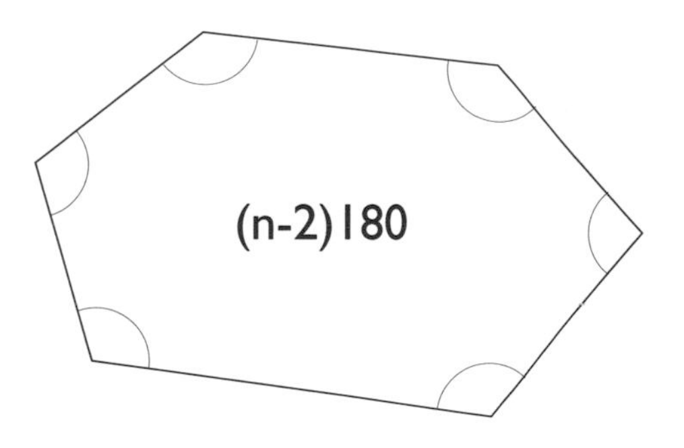

SUM ALL INTERIOR ANGLES AND COMPARE TO (N-2)180

After all angles have been measured, recorded, and averaged, the mean of the interior angles should be summed and compared to (n-2)180 (n is the number of sides).

RESULTS

Averaging and summing the angles allows the field engineer to review the data and determine if any angles do not meet specified standards of accuracy for the instrument used. See the instrument operating manual for expected results. If the angles turned do not meet expectations, re-measure the angle.

FIELD NOTES

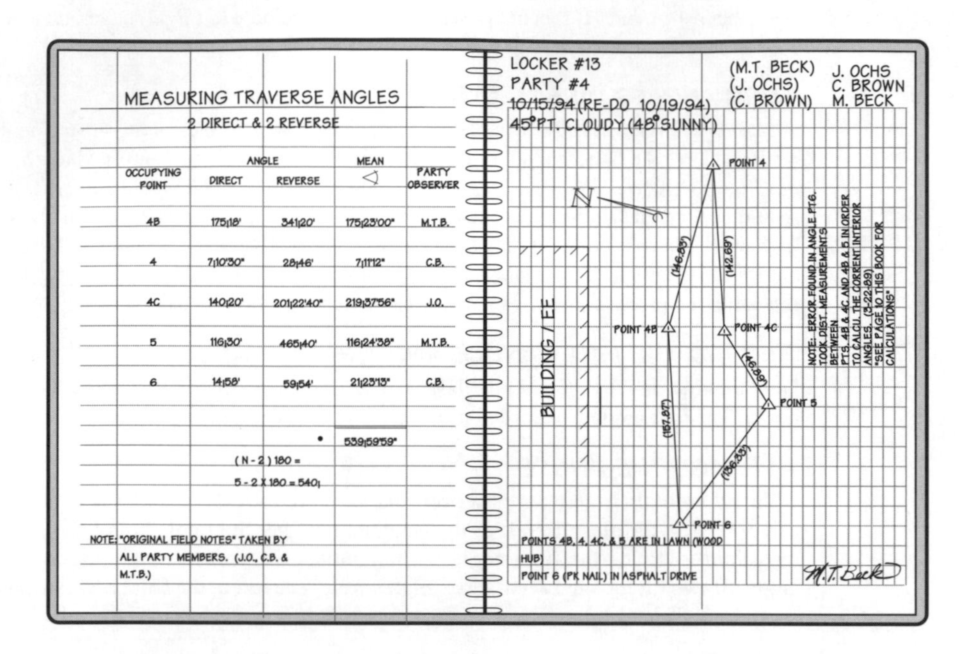

MEASURING TRAVERSE ANGLES

2 DIRECT & 2 REVERSE

OCCUPYING POINT	ANGLE DIRECT	ANGLE REVERSE	MEAN ∢	PARTY OBSERVER
4B	175°18'	341°20'	175°23'00"	M.T.B.
4	7°10'30"	28°46'	7°11'12"	C.B.
4C	140°20'	201°22'40"	219°37'56"	J.O.
5	116°30'	465°40'	116°24'38"	M.T.B.
6	14°58'	59°54'	21°23'13"	C.B.
			539°59'59"	

(N - 2) 180 =

5 - 2 X 180 = 540°

NOTE: "ORIGINAL FIELD NOTES" TAKEN BY ALL PARTY MEMBERS. (J.O., C.B. & M.T.B.)

LOCKER #13
PARTY #4
~~10/15/94~~ (RE-DO 10/19/94)
~~45° PT. CLOUDY~~ (48° SUNNY)

(M.T. BECK) J. OCHS
(J. OCHS) C. BROWN
(C. BROWN) M. BECK

NOTE: ERROR FOUND IN ANGLE PT.6. TOOK DIST. MEASUREMENTS BETWEEN PTS. 4B & 4C AND 4B & 5 IN ORDER TO CALCU. THE CORRENT INTERIOR ANGLES. (3-22-89) "SEE PAGE 10 THIS BOOK FOR CALCULATIONS"

POINTS 4B, 4, 4C, & 5 ARE IN LAWN (WOOD HUB)

POINT 6 (PK NAIL) IN ASPHALT DRIVE

M.T. Beck

SUMMARY OF ANGLE MEASUREMENT

Angle Measurement - Field Duties and Responsibilities	
A field engineer operating a Transit or Theodolite shall:	***Field engineer assisting in angle measurement shall:***
Communicate to the assistants any needs for backsights, plumb bob use, foresights, etc.	Communicate constantly to the person operating the instrument any irregularities observed in the angle measuring process.
Read the operations manual thouroughly. The understanding and skill of the operator makes a difference in the ultimate results that are obtained.	Upon departure from an office or vehicle, lay aside the necessary equipment for the day's operations and check that equipment is in working condition. Check the equipment again at the completion of the day's operations to see that all is complete and accounted for before leaving the jobsite.
Set up the instrument so it is exactly over the point and level.	Exercise great care in making sure the plumb bob or target is held over the EXACT point at which line is being taken.
Utilize protective covers when rainy conditions exist. Wet instruments should be dried lightly with a clean cloth being careful not to touch the lens. In cold weather or when instruments may be moist, they should be left out of their cases overnight to prevent moisture from penetrating the instrument.	Carefully store equipment in protective cases or boxes when not in use, maintain clean equipment during and after each day's work, and assist the field engineer in caring for instruments.
Set zero carefully and as exact as possible.	Utilize care when readings are being taken in hazardous locations.
Do not tighten clamps to tightly, thereby minimizing chances of stripping the threads. Keep the instrument in focus and remove parallax before use.	Carry marking pens and spray paint at all times and mark each point and its number on lathe or stakes.
Sight exactly onto the points where line is being taken or given.	Setup tripods, tribachs, prisms, and targets, taking care to be certain all level bubbles are centered and that the instrument is exactly over the point.
Check instrument calibration at the start of a project and anytime during the project when critical layout is occurring.	Place tacks or marks exactly on the line being given. Drive hubs straight and flush until they are solid.

LEVELING

9

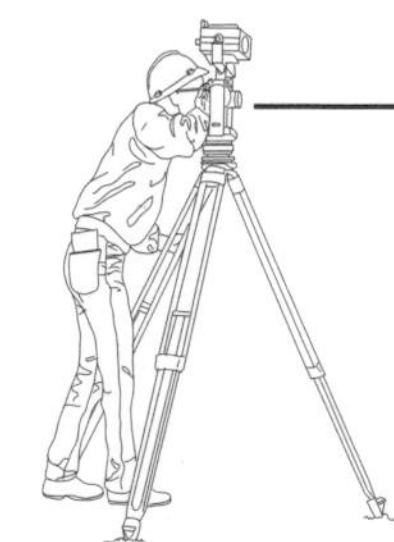

PART 2 - MEASUREMENT BASICS

IN THIS CHAPTER:

Illustrate the basic theory of differential leveling.
Define the following leveling terms: benchmark, backsight, height of instrument, foresight, turning point.
Develop the formulas used to calculate the HI and the elevation of any point.
Distinquish between the various types of instruments used for leveling.
Describe how the compensator works in an automatic level.
State how a rod is graduated and describe how to read a rod.
Explain how to use a rod level.
Analyze how to determine the correct height to set up a leveling instrument.
Describe how to eliminate parallax in an instrument.
Describe the step-by-step differential leveling procedure.
State why the distance from the instrument to the backsight and the instrument to the foresight must be the same. (Balanced)
Describe why a rod held out of plumb makes the rod reading larger.
Describe when the leveling instrument is moved around the position of the rod.

INDEX

LEVELING ON THE JOBSITE

SCOPE

Determining or establishing elevations is, at times, the most called for activity of the field engineer. Elevations are needed to set slope stakes, grade stakes, footings, anchor bolts, slabs, decks, sidewalks, curbs, etc. Just about everything located on the project requires elevation. Differential leveling is the process used to determine or establish those elevations.

DIFFERENTIAL LEVELING BASICS

Differential leveling is a very simple process based on the measurement of vertical distances from a level line. Elevations are transferred from one point to another through the process of using a leveling instrument to read a rod held vertically, on first, a point of known elevation, and then, on the point of unknown elevation. Simple addition and subtraction are used to calculate the unknown elevations. Perhaps this procedure can best be understood through a few illustrations.

THE THEORY OF DIFFERENTIAL LEVELING

The illustration below shows the typical leveling process that is repeated time and time again to transfer an elevation from one point to another. The known elevation is transferred up to the line of sight by reading the backsight. By definition, the line of sight is horizontal; therefore, the line of sight elevation can then be transferred down to the unknown elevation point by reading the foresight.

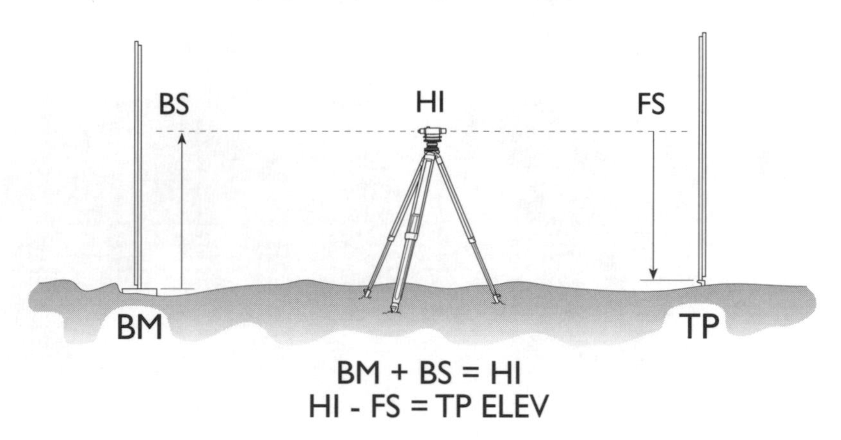

TRANSFERRING THE ELEVATION TO THE HI

In the illustration below, the elevation where the rod is held is known. That elevation is transferred vertically to the line of sight by reading the rod. By adding the known elevation and the backsight reading, the Height of Instrument can be determined. The formula for this is shown as:

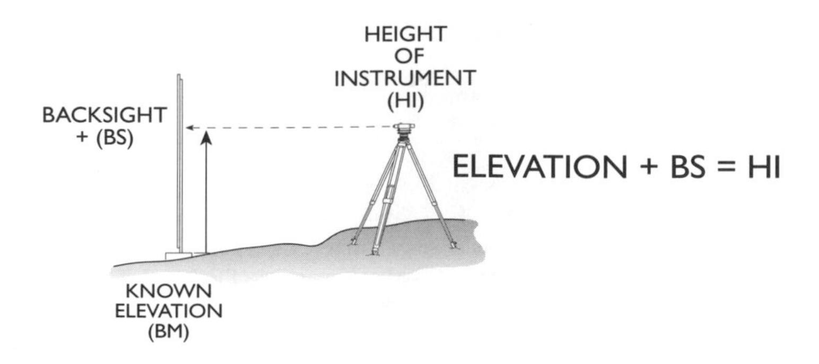

TRANSFERRING THE HI TO THE TURNING POINT

In the illustration below, the elevation of the HI is known. That elevation is transferred vertically down to the point on the ground by reading the rod. By subtracting the HI and the foresight reading, the elevation of the point on the ground can be determined. The formula for this is shown as:

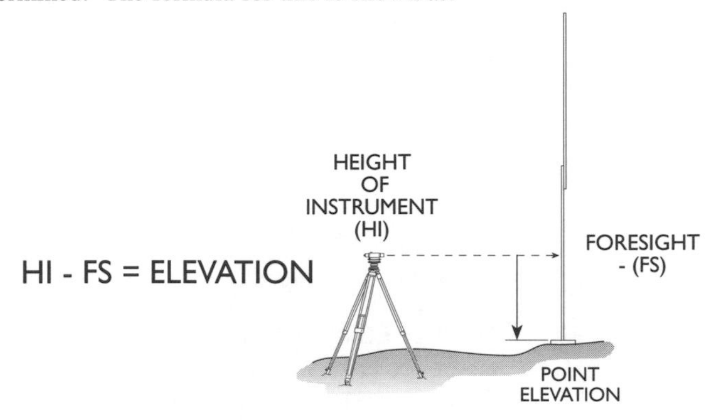

Now that the basic theory of differential leveling has been illustrated, it is time to define the terms that have been presented and look at the detailed procedure that should be followed to perform leveling accurately.

DEFINITIONS AND RELATIONSHIPS

ELEVATION

Elevation is the vertical distance above a zero datum. Datum in the case of leveling is generally based on the Ocean's Mean Sea Level (MSL). An elevation in the mile high city of Denver would be approximately 5280 feet. An elevation in the city of New York, which is close to the ocean level, might be 15 feet. A point in Yellowstone might be 8900 feet above MSL. There is a network of MSL monuments around the United States which have been established by the Government.

If you are at a construction site where there is no MSL monument with an elevation nearby, a datum point is established and an elevation is assigned. Generally, a number large enough to prevent any negative elevation numbers from occurring is picked. Usually 100.00, 500.00, or 1000.00 is used. If, at a later time, an MSL elevation is available, it is easy to convert any assumed elevation to MSL by determining the difference at the established datum point and adding that to each established elevation.

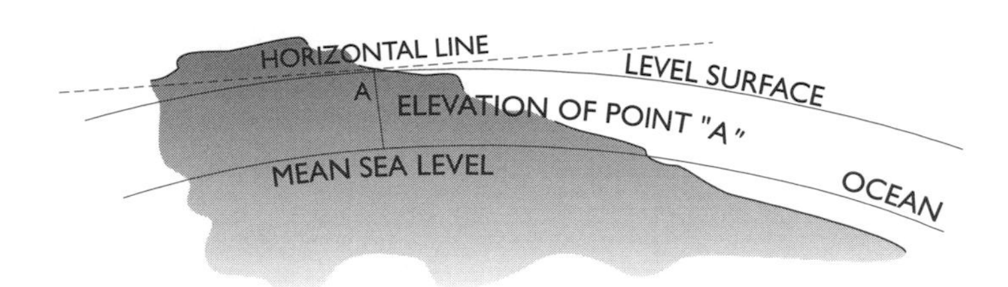

BENCHMARK (BM)

A benchmark is best described as a permanent, solid point of known elevation. Benchmarks can be concrete monuments with a brass disk in the middle, iron stakes driven into the ground, or railroad spikes driven into a tree, etc. Reference the section *Setting Benchmarks* in *Chapter 22, Leveling*

BACKSIGHT (BS)

A backsight is a reading on a rod held on a point whose elevation is known.

HEIGHT OF INSTRUMENT (HI)

The height of instrument is the elevation of the line of sight of the instrument. This is determined by adding the backsight to the known elevation.

FORESIGHT (FS)

A foresight is a reading on a rod held on a point whose elevation is unknown.

TURNING POINT (TP)

A turning point is a point used in the differential leveling process on which to temporarily transfer the elevation from the HI. The elevation of the TP is determined by subtracting the FS from the HI.

CLOSED LOOP

Closing a loop means to end the level loop at the benchmark where you began or at another benchmark. **"Always, always, always close your loop"** is a rule of thumb which must be strictly observed in leveling. If a loop isn't closed, there is no way of knowing if any mistakes were made in the leveling process.

QUICK DESCRIPTION OF THE LEVELING PROCESS

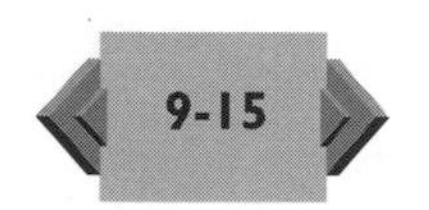

As can be seen, leveling is a very simple process of taking a backsight and a foresight from an instrument setup and repeating it over and over to transfer an elevation from one point to another. Indeed, it is simple, but in order to do it accurately and to achieve the precision required, much attention must be paid to the details. The following step-by-step description of the process of leveling attempts to cover the details that must be followed. Reference the section *Step-by-step Differential Leveling Procedure* later in this chapter.

Leveling Process at a Glance

1	Set up instrument and level it.	
2	Hold the rod on known elevation point and record to the nearest 0.01 ft. the backsight reading, with the rod person holding the level rod plumb over the point	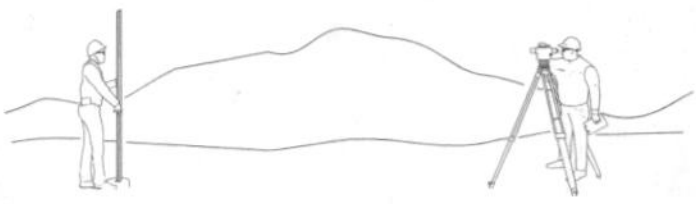
3	Advance the rod person to the turning point (an equal distance from the instrument as the backsight station).	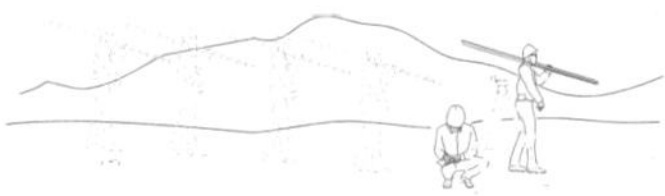
4	Hold the rod on the turning point and read and record the foresight reading.	
5	Repeat previous steps until the elevation of the designated point has been determined.	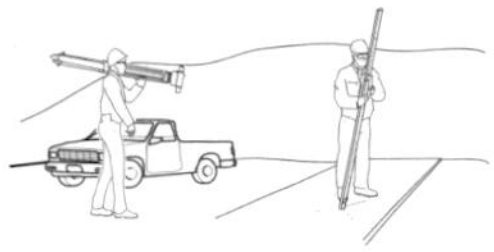
6	Close the loop by repeating the process back to the benchmark.	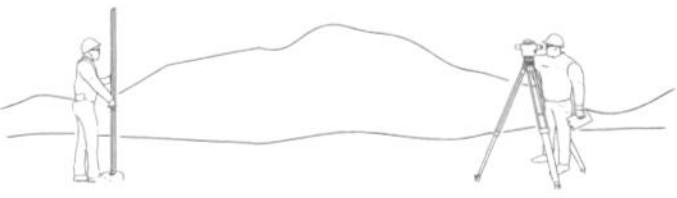

LEVELING EQUIPMENT AND STANDARD PRACTICES

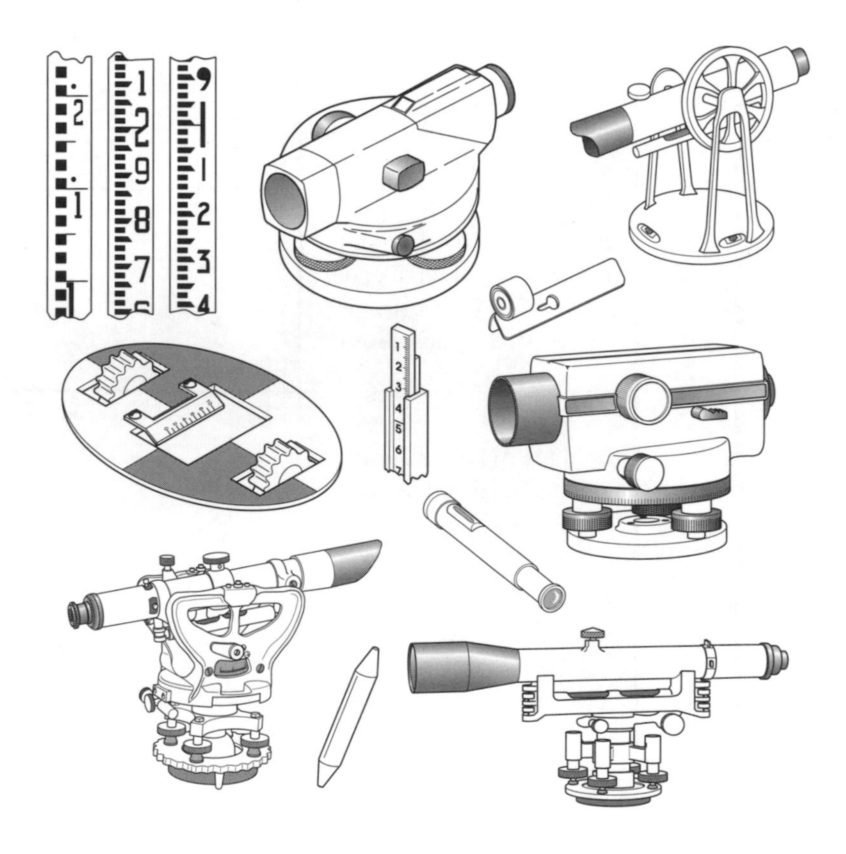

SCOPE

Many types of instruments have been used for leveling purposes. The instruments have ranged from water troughs, hoses, four-foot levels, and string line level bubbles, to instruments such as the dumpy level, automatic level, and laser. The field engineer should realize how each of these can be used to determine elevations for construction. This section specifically discusses the dumpy and automatic level and accompanying equipment that is used on a daily basis by the field engineer.

OVERVIEW OF LEVELING INSTRUMENTS

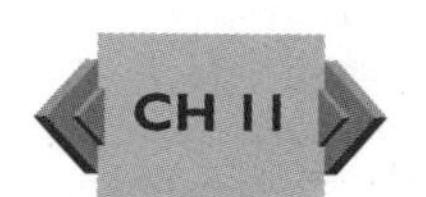

There are a number of kinds of leveling instruments, however; there are basically three types. One type uses a level bubble to obtain a horizontal line of sight; another uses an automatic compensator to establish a horizontal line of sight. An additional type is the laser level. See *Chapter 11, Construction Lasers for Line and Grade* for a discussion of lasers.

LEVEL BUBBLE INSTRUMENTS

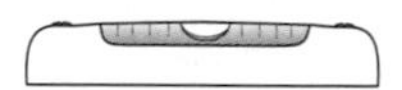

Level bubble instruments include the builder's transit level, the transit, and the dumpy level. Each of these instruments contains a level vial with a bubble that must be centered to be used for leveling. Each instrument consists of three main components: all contain a four-screw leveling head; all contain a level vial attached to the telescope; and all contain a telescope for magnification of the objective.

THE BUILDER'S LEVEL

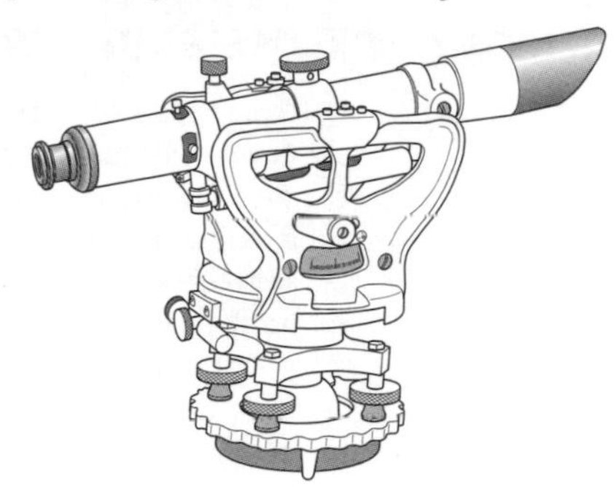

The builder's level is one of the most inexpensive instruments used by field engineers. It is a versatile instrument. In addition to being able to perform leveling operations, it can be used to turn angles, and the scope can be tilted for sighting. Many residential builders use this instrument because it serves their purpose of laying out a building very well.

THE TRANSIT

Although the primary functions of the transit are angle measurement and layout, it can also be used for leveling because there is a bubble on the telescope. In fact, many construction companies who don't have both an angle measuring instrument and a dumpy or automatic level, will do all of their leveling work with a transit. Some people prefer to use the transit for leveling because they are comfortable with it's operation.

The field engineer should be aware that the transit is versatile and is used on the jobsite in an infinite number of leveling and angle turning applications.

THE DUMPY LEVEL

The dumpy level has been the workhorse of leveling instruments for more than 150 years. It has been used extensively for many of the great railroad, canal, bridge, tunnel, building, and harbor projects for the last century and a half. It has been a mainstay of the industry in constructing projects that are level. Even with advancements in other leveling instruments such as the automatic level and the laser, the dumpy is still the instrument of choice by a number of persons in construction.

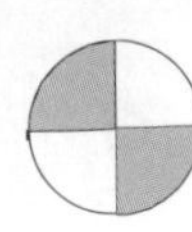

In some types of construction, such as high rise buildings, the dumpy level must be used. Vibrations in buildings occur as tower cranes and personnel hoists cause the compensator and line of sight of some automatic levels to be constantly moving making it practically impossible to take a reading on a rod. The line of sight of the dumpy level, however, basically holds steady in a similar situation. Thus, the dumpy level is the level of choice by many on high rise construction projects. On any type of project where there is going to be a great deal of vibration, such as pile driving, heavy equipment usage, etc., the dumpy may be the best choice for a leveling instrument.

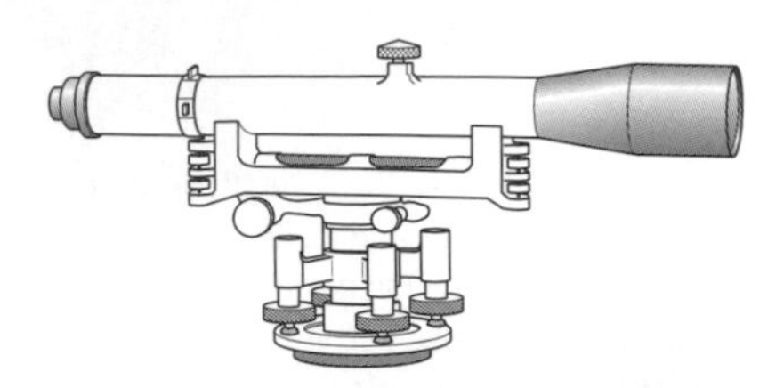

Many carpenters prefer to use a dumpy level rather than an automatic level because they can relate better to the bubble on the dumpy as opposed to the "magic" of the compensator on the automatic. In the dumpy, they can see the bubble and have confidence in the horizontal line of sight when it is centered.

AUTOMATIC COMPENSATOR INSTRUMENTS

Compensator instruments were developed about 50 years ago. Although each manufacturer may have developed a unique compensator, all compensators serve the same purpose-creating a horizontal line of sight. Compensator instruments are extremely fast to set up and level. An experienced person can easily have an automatic level ready for a backsight in less than ten seconds as compared to a minute or more with a bubble-based leveling system.

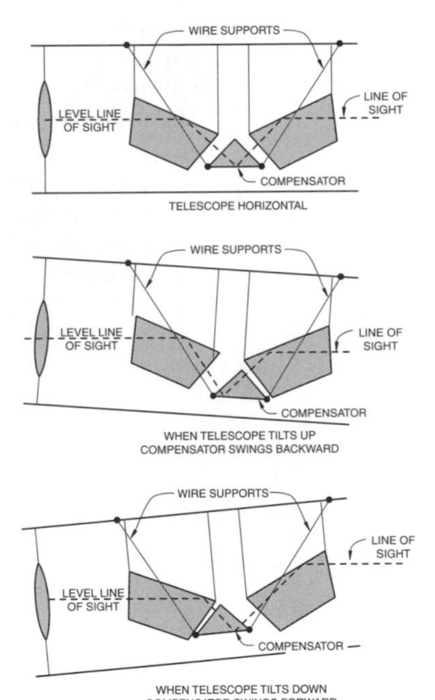

Compensators are available in several styles. Some are constructed by suspending a prism on wires. Some have a prism that is contained within a magnetic dampening system. Undoubtedly there are other methods of compensator construction. The illustration to the right shows how a wire suspension compensator allows the prism to swing freely and maintain a horizontal line of sight. Note that the instrument must be manually leveled to within the working range of the compensator by centering the bull's-eye bubble.

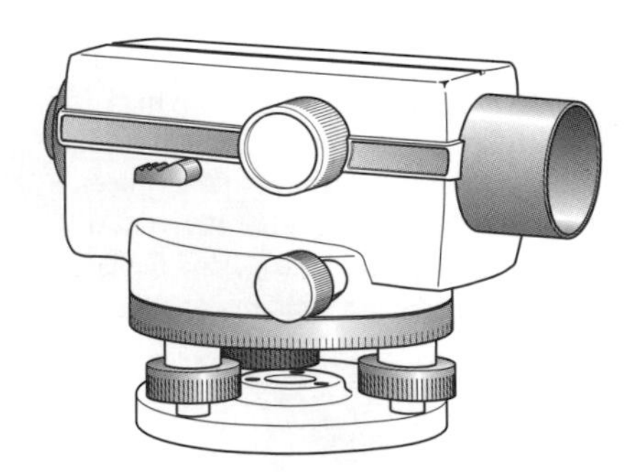

LEVEL RODS

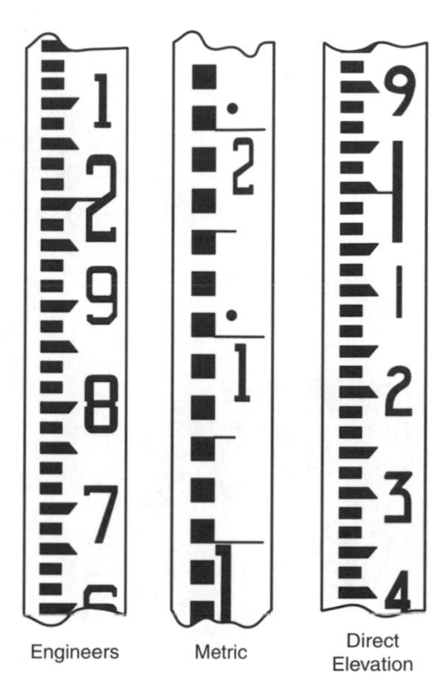

In addition to the leveling instrument, a level rod is required to be able to transfer elevations from one point to another. The level rod is nothing more than a graduated pole held vertically and read by the person at the instrument. The reading taken can be used to determine elevations.

Level rods are available in many sizes, shapes, and colors. They are made of wood, fiberglass, and metal, or a combination of these materials. There are one-piece rods, two-piece rods, three-piece rods, six-piece rods, etc. Some have a square cross-section and others are round or oval. Some are less than 10 feet long while others are up to 30 feet long. Practically whatever type of rod a field engineer needs is available. Level rods are named after the cities where they have been manufactured: they are called Philadelphia, Chicago, San Francisco, and probably other geographical names. The name has a connotation of a particular style of rod. The Philadelphia rod is a two-piece rod which can be extended to approximately 13 feet. The Chicago name is applied to rods with three or four sections which are placed together end to end. The San Francisco rod is similar to the "Philly" but it has two extensions. Rather than calling rods by a proper name, some people just call them "rods" or "storypoles."

A popular rod that doesn't seem to have a proper name is the telescoping rod. This rod is the rod of choice for field engineers working on projects where there is a great deal of change in elevation because telescoping rods are 25 or 30 feet long. Great rod lengths increase the elevation difference that can be transferred at one time. It has been argued that these types of rods wear rapidly; and, therefore, aren't as accurate as the more traditional rods. If telescoping rods are well cared for, they are excellent for construction use.

TYPICAL ROD GRADUATIONS

Level rods are graduated in feet, inches, and fractions; feet, tenths and hundredths; or meters and centimeters. The method of representing units of measurement onto the face of the rod also varies. The field engineer should, after studying the graduations on the face of the rod, be able to use any rod available. The illustration at the left shows the markings on a typical "engineer's rod" that is widely used on the construction site. Note that the rod is graduated to the nearest foot with large numbers which are usually painted red (thus the term "raise for red" when the instrument person cannot see the foot graduation). The feet are then graduated to the nearest tenth from 1 to 9. Each tenth is then graduated to the nearest one hundredth which is the width of the smallest mark on the rod.

To read the hundredths, the observer should keep in mind that the bottom of each graduation is an odd number and the top of each is an even number. Also, the longest graduation mark is for feet and tenths, which points upward. The next longest graduation is the half tenth or 0.05, 0.15, 0.25, etc., which points downward. The field engineer only has to learn to read the hundredths graduations of one tenth of the rod, because this pattern repeats itself for the length of the rod.

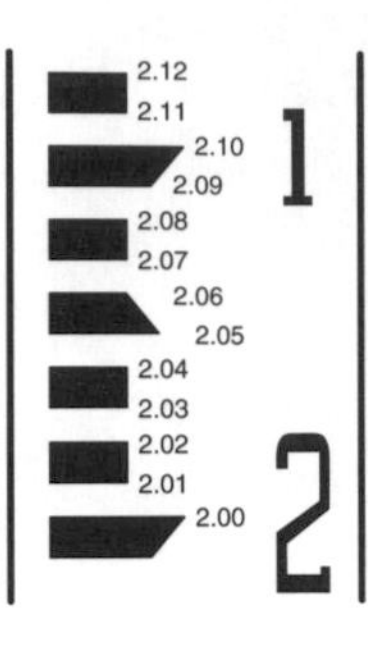

ROD TARGETS

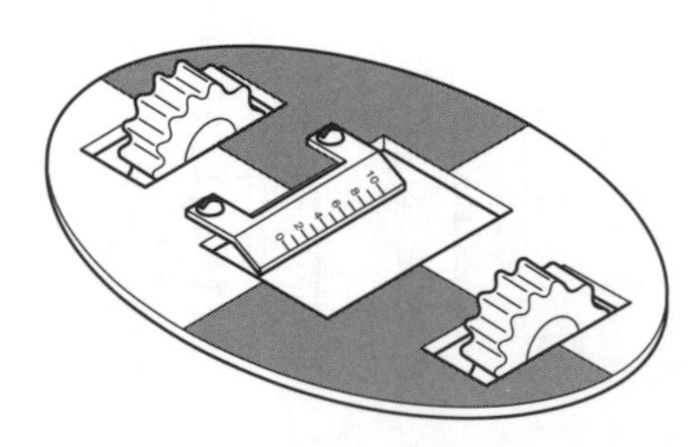

Rod targets are useful devices for several purposes, but few people use them. They forget them or don't have them when they are needed. The rod target can be used as a target by the person looking through the instrument to help locate the rod when visibility conditions are poor. If the rod target has a vernier, it can also be used to obtain a reading on the rod to the nearest thousandth of a foot. In this case, the instrument person communicates to the rodperson to move the target until it has centered on the horizontal crosshair. The rodperson can then read the rod and the vernier to obtain thousandths of a foot. This is sometimes required on very precise leveling work.

ROD LEVELS

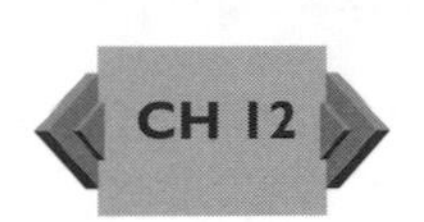

Rod levels are used to keep the rod plumb while the reading is being taken. Rod levels are simple devices made of metal or plastic and have a bull's-eye bubble attached. They are held along the edge of the level rod while the level rod is moved until the bubble is centered. If the rod level is in proper adjustment, the rod is plumb. See *Chapter 12, Equipment Calibration* for testing of a bull's-eye bubble.

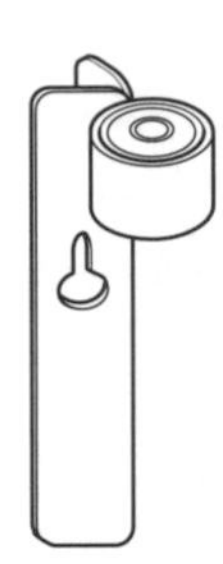

TURNING POINT PIN

If a solid natural point such as a sidewalk, large rock, etc., isn't available during the leveling process to be used as a turning point, then the field engineer will have to create one. This is accomplished by carrying something such as a railroad spike, piece of rebar, wooden stake, plumb bob, etc. that can be inserted solidly into the ground. The rod is placed on top of the solid point while the foresight and backsight readings are taken. These solid points are removed after each backsight to be used the next time a solid turning point is needed.

STANDARD LEVELING INSTRUMENT USE PRACTICES

Just like the transit/theodolite, the level instruments must be set up, leveled, and focused onto the objectives and crosshairs to be ready for operation.

SET UP AT THE CORRECT HEIGHT

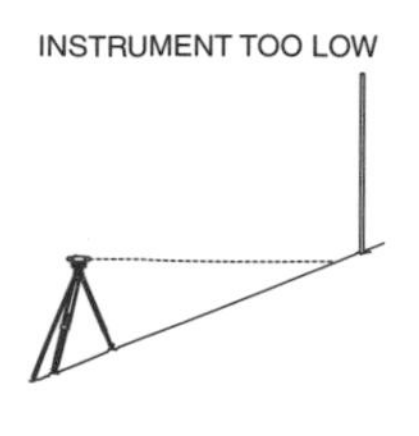

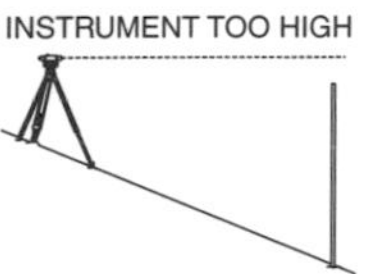

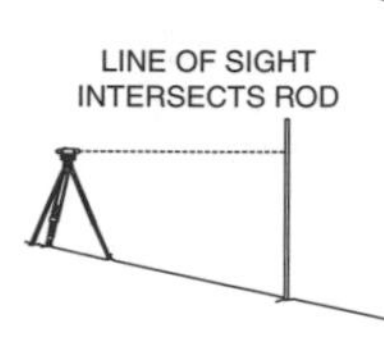

As opposed to angle turning instruments, levels DO NOT have to be set over a point. However, levels must be set up at the correct height to enable sighting on the level rod. That is, levels need to be set up so the horizontal line of sight of the leveling instrument intersects the level rod. If the level is too high, and is above the rod, the level must be moved downhill. If the level is too low, and the line of sight hits the ground in front of where the rod is being held, the level must be moved uphill. A hand level is a useful device used by the person at the instrument to quickly determine if the location is too high or too low.

SOLID, WIDE SETUPS

Spread the level's tripod legs wide to provide a solid foundation for the instrument. A wide setup is especially necessary in locations where the wind is strong and capable of toppling the instrument. Level setups must be solid. If the level is set on the ground, the field engineer must walk around and step on the foot pads to firmly sink the points of the tripod legs into the ground. If the ground is too soft for solid setups, stakes may be driven into the ground while the tripod legs set on top of the stakes. When the level must be set on a hard surface such as concrete, it may be necessary to cut out a tripod stand in which to set the legs to avoid slippage. If it is very windy, tie the tripod to concrete blocks or something that is solid or has considerable weight to keep the tripod from tipping over.

FOCUS ONTO THE LEVEL ROD

Good focusing onto the objective (in this case, the level rod) is very important in leveling work. If the level rod is not clear, a hundredth or two mistake in a reading will likely occur. This is unacceptable in leveling work. Focus onto the level rod by looking directly at the graduations on the rod. Each hundredth should be sharp and crisp. It will be necessary to focus each time the distance to the rod is changed.

BALANCE BACKSIGHTS AND FORESIGHTS

The distance from the instrument to the rod held at the backsight, and the distance from the instrument to the rod held at the foresight should always be equal. When the distances are equal, the errors that occur will also be equal and will be cancelled out when the basic equations for differential leveling are applied. Therefore, by balancing backsights and foresights, errors due to miscalibration of the instrument, curvature of the earth, and refraction of the line of sight can be eliminated.

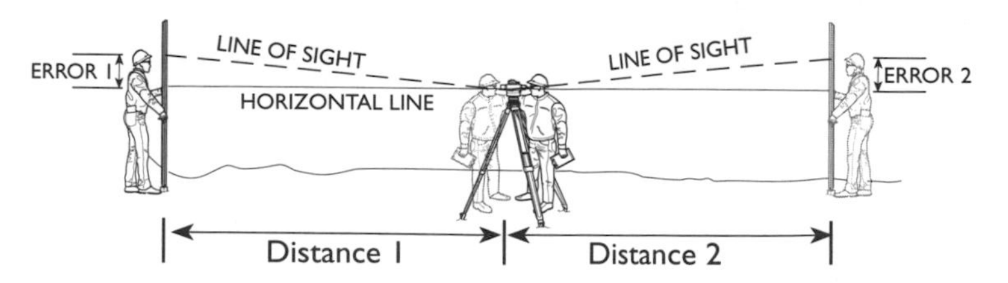

IF DISTANCE 1 = DISTANCE 2
THEN ERROR 1 = ERROR 2

IF DISTANCE 1 ≠ DISTANCE 2
THEN ERROR 1 ≠ ERROR 2

FOCUS ONTO THE CROSSHAIRS

It was pointed out earlier in Chapter 8 that failure to focus properly onto the crosshairs causes parallax. Recall, parallax occurs when the crosshairs seem to move slightly on the objective as the eye moves. If this occurs when reading a level rod, major errors can occur in readings causing the level loop to not close.

Eliminating Parallax		
1	Look through the instrument at a light-colored object: a white house, a piece of paper, someone's shirt, etc. The cross hairs may be fuzzy or almost not visible.	
2	Adjust the eyepiece until the crosshairs are as dark and crisp as they can possibly be. Look from end to end of the crosshair making it the darkest and crispest at the center.	
3	Look at the level rod and move your head slightly, looking at the crosshair. If it stays on the same spot on the rod, parallax has been eliminated for now.	
4	Throughout the day, as your eyes tire, the crosshairs will become fuzzy and parallax will return. Continually repeat the procedure described above to keep parallax in check.	

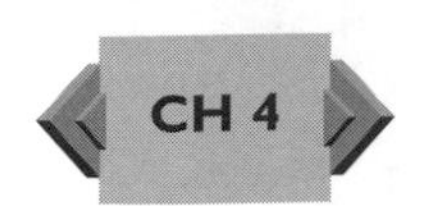

CENTERING THE BUBBLE

Refer to *Chapter 4, Fieldwork Basics* for detailed instructions on using a three-screw or four-screw leveling system to center the bubble.

STEP-BY-STEP DIFFERENTIAL LEVELING PROCEDURE

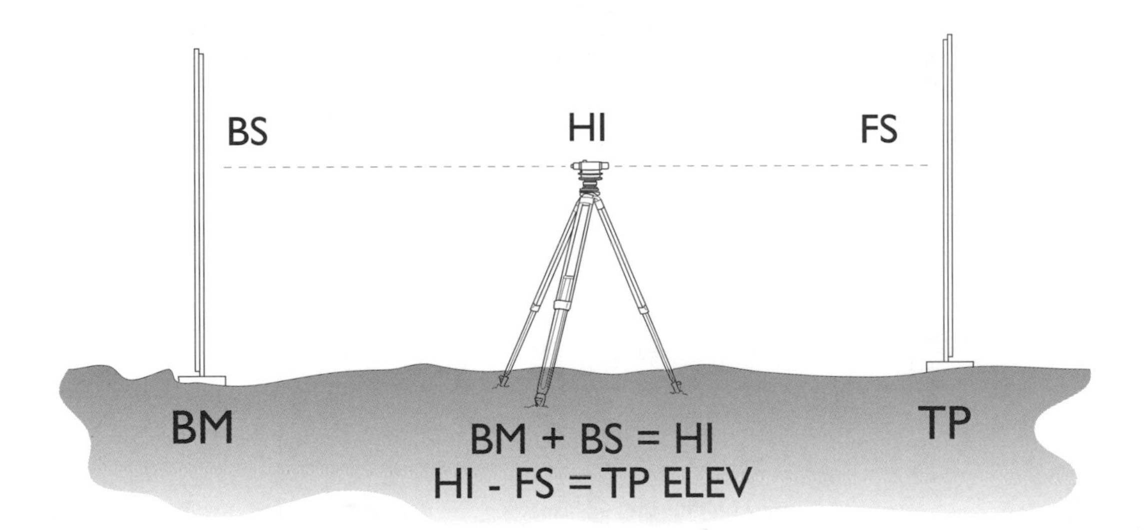

SCOPE

Beginning at a known benchmark, with a given elevation above mean sea level (AMSL), complete a level loop to a point where elevation is needed, closing again on the known benchmark. That is the objective of leveling.

Before you begin the leveling process, you must know where you are going to end before you know where to start. That is, knowing where an elevation is desired will allow you to select the closest benchmark to the work, thereby shortening the distance to level. The key to successful leveling is to know the step-by-step procedure forward and backward and to perform all of the steps with consistency.

STEP-BY-STEP PROCEDURE FOR DIFFERENTIAL LEVELING

The following step-by-step procedure for leveling has been illustrated to show the reader what each individual in the process of leveling is doing. Read the caption and carefully study the graphic to develop a detailed understanding of the process of leveling.

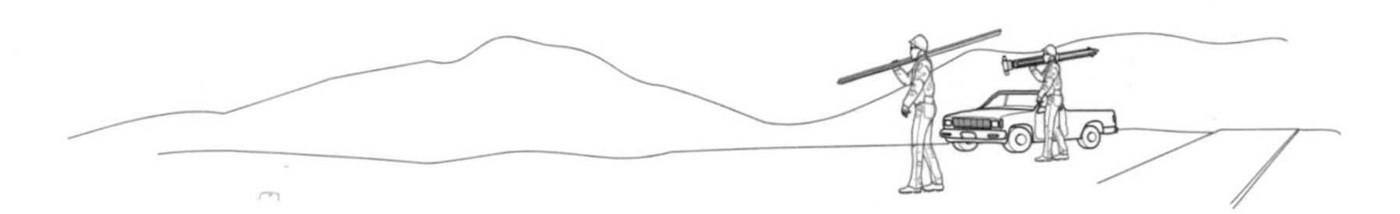

STEP 1

The person holding the rod removes the rod from its case, inspects it for dirt on the bottom, and inspects it for any broken or loose parts. Proceeds to the benchmark.

The person using the instrument sets up the tripod with the legs spread apart for a firm base. Inspects tripod for loose screws, etc. Takes level out of the case, inspects its condition, places it on the tripod, and clamps it securely. Proceeds to the vicinity of the benchmark.

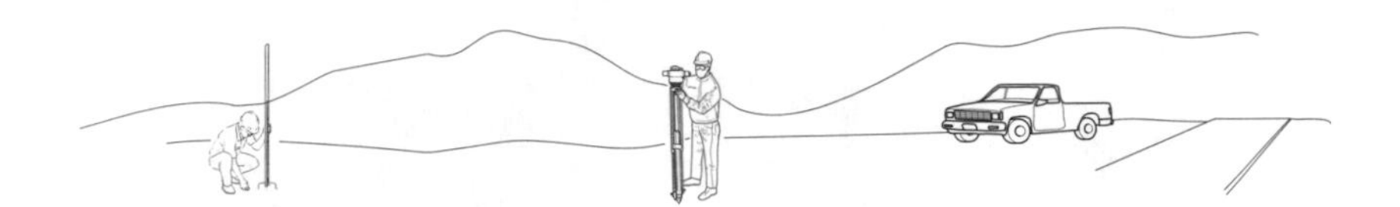

STEP 2

The person holding the rod inspects the benchmark once it has been located to see if it shows any signs of having been disturbed. Reports any irregularities to the field engineer in charge.

The person using the instrument observes the location of the benchmark. Notes the direction that must be taken to get to the point where the elevation is needed. Considers where the first turning point will be and determines the best location for setting up the level.

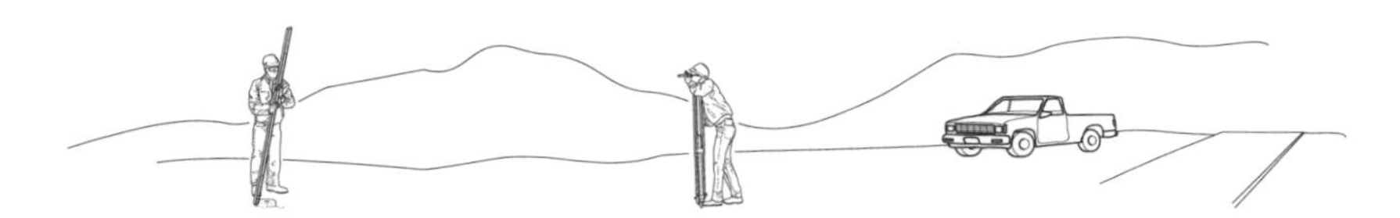

STEP 3

The person holding the rod stands at the benchmark, facing the instrument, waiting for the level to be set up. Extends the rod fully if downhill from the instrument.

The person using the instrument, at the selected set-up location, looks at the benchmark, and sights through a sight level to make sure the line of sight of the level will intersect the rod. If the line of sight is going to be above a fully extended rod or hits the ground in front of the rod at the benchmark, a new location must be selected.

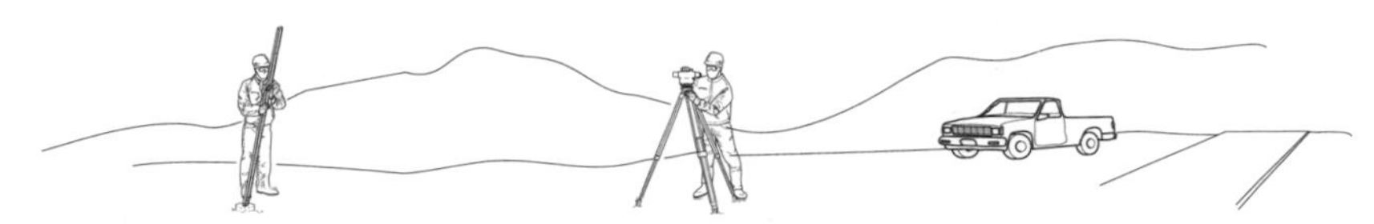

STEP 4

The person holding the rod waits.

The person using the instrument, positions the tripod with the legs wide apart, keeping the head of the tripod close to horizontal. When sure the line of sight will intersect the rod, walks around the tripod and stands on the tripod feet, sinking them firmly into the ground. Looks through the telescope at a light-colored object and focuses the eyepiece until the crosshairs are as dark and crisp as possible.

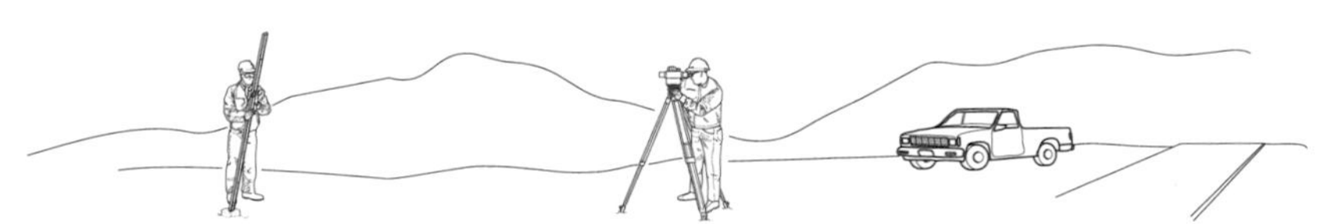

STEP 5

The person holding the rod waits.

The person using the instrument levels the instrument using the leveling screws, centering the bubble so the instrument is level in all directions turned.

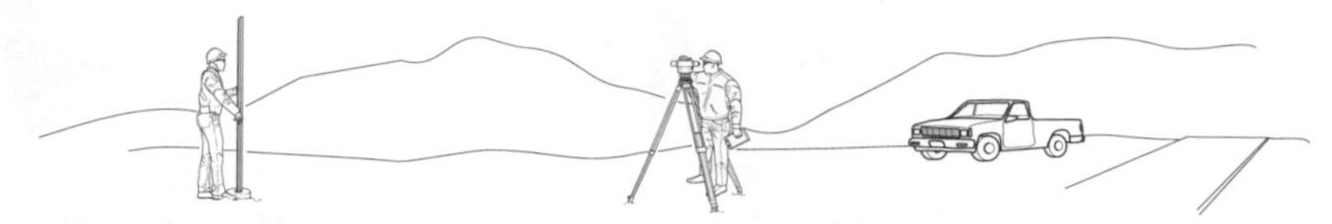

STEP 6

The person holding the rod holds it on the benchmark, keeps it plumb with a rod level, while the backsight reading is being taken, and waits for a signal to move.

The person using the instrument looks through the telescope and focuses on the rod held at the benchmark. Carefully reads the rod and records the **backsight** reading in the field book. Re-reads the rod and confirms what was recorded.

STEP 7

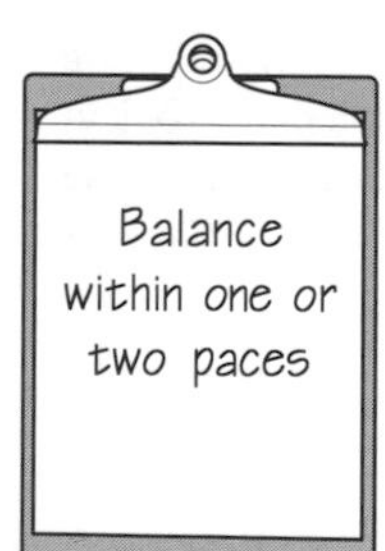

The person holding the rod, at the signal, begins to pace the distance to the instrument. Upon reaching the instrument, calls out the number of paces, and consults with the person using the instrument to determine where the loop is headed and begins pacing in that direction. When the number of paces in that direction is equal to the number of paces to the instrument (i.e., balanced) stops and looks for a suitable turning point, such as a solid rock, top of curb, bolt on fire hydrant, etc. If no solid point is available, hammers a metal pin from the supplies pack to create a solid point. If the turning point is on a sidewalk, carefully marks the exact location of the point with a crayon.

The person using the instrument signals to the person holding the rod that the reading has been taken. While the person holding the rod is moving to the turning point, the field book is calculated to determine the H.I. of the instrument. Turns the telescope in the direction the rod person is headed and waits for the signal (circular motion overhead with one hand) that a turning point has been located or established.

STEP 8

The person using the instrument looks through the telescope and focuses on the rod held at the turning point. Carefully reads the rod and records the foresight reading in the field book. Re-reads the rod and confirms what was recorded. Calculates the elevation of the turning point. Signals to the person holding the rod to relax while the instrument is being moved.

The person holding the rod signals readiness to the person using the instrument and holds the rod on the turning point. Keeps the rod plumb with the rod level until signaled that the reading has been taken. After being signaled, removes rod from the turning point.

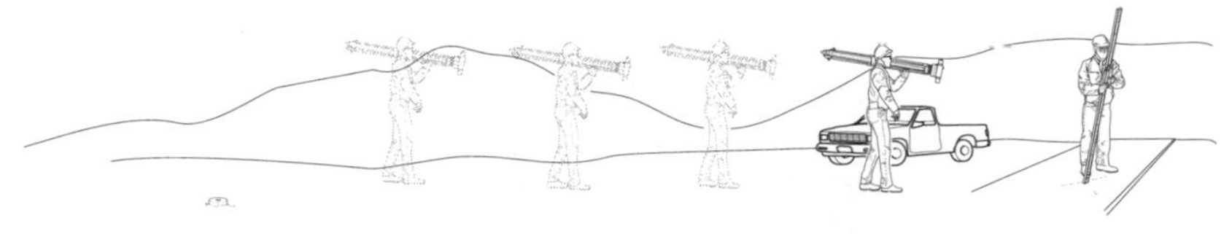

STEP 9

The person using the instrument walks, carrying the instrument past the position of the turning point, in the direction where the elevation is needed. Locates the best location for setting up the level.

The person holding the rod waits while the instrument is being advanced.

STEP 10

The person using the instrument repeats previous steps:
Sets up, Levels the instrument, Focuses on the rod, Reads the backsight, Records the backsight reading in the field book, Calculates the HI, Turns the instrument to the new turning point, Reads the foresight, AND Calculates the elevation of the turning point. Moves the instrument again.

The person holding the rod repeats previous steps:
Holds the rod plumb on the known elevation while the backsight reading is taken, Paces the distance to the instrument, paces to locate a turning point, selects a solid turning point, and holds the rod on the turning point while a foresight reading is taken.

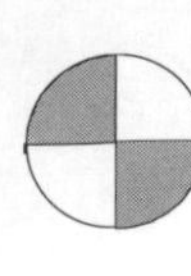

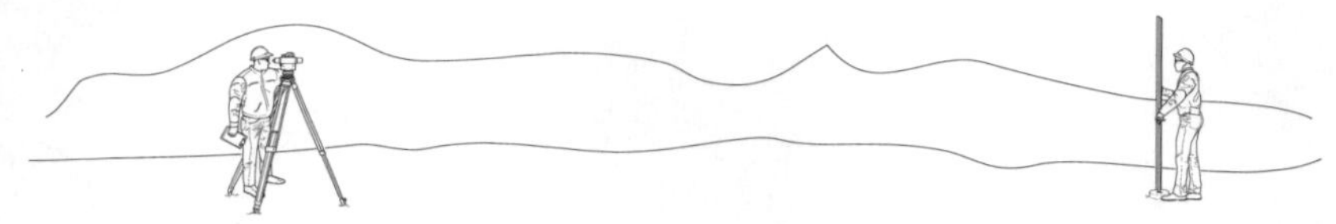

STEP 11

The person using the instrument repeats this entire process until the last foresight taken is to the point where elevation was needed. There may be a dozen or even a hundred or more setups required to transfer an elevation to where it is needed.

The person holding the rod repeats this entire process until the last foresight is taken to the point where elevation was needed. There may be many turning points that will be located or established. Using solid turning points throughout the leveling process is a major factor in closing the loop successfully.

STEP 12

The person holding the rod, after the needed elevation is determined, uses the same process described above to close the loop by repeating the differential leveling process back to the original starting benchmark.

The person using the instrument uses the same process described above to return to the original starting benchmark.

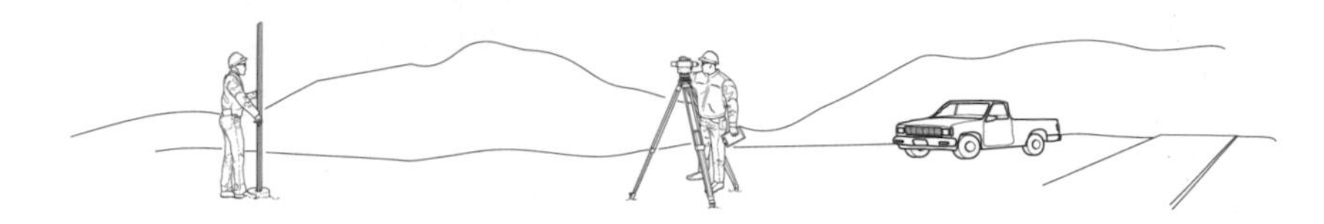

STEP 13

The person holding the rod, upon arriving at the original benchmark, waits to find out if the work meets acceptable closure. If it does, helps in putting away equipment to go to the next work activity. If it doesn't, repeats the level circuit just completed.

The person using the instrument, upon arriving at the original benchmark and having read the final foresight, compares the final calculated elevation with the starting elevation. If it meets the established closure standards, the process is complete. If it doesn't, the loop will have to be repeated.

FIELDNOTES FOR LEVELING

SCOPE

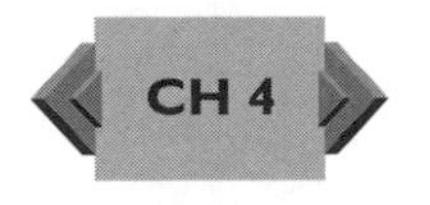

As stated earlier in *Chapter 4, Fieldwork Basics*, notekeeping can be in three forms—tabulation, description, or sketch. Because of the quantity of numbers that are read and recorded, tabulation is by far the best method of notekeeping for leveling work. The user should study it carefully and look at the sample set of notes to determine where numbers have been recorded and calculated.

RECORDING LEVELING DATA

Leveling notes follow a very distinct and methodical pattern. Data is recorded across, drop down to next line, left to right, drop down to next line, left to right, drop down to next line, etc. Once the pattern is recognized, leveling notes are very easy to keep and to understand. The note page below describes what is recorded in each column and shows the progression from point to point in each row. Remember the basic formulas for leveling: **Elevation + BS = HI and HI - FS = Elevation.**

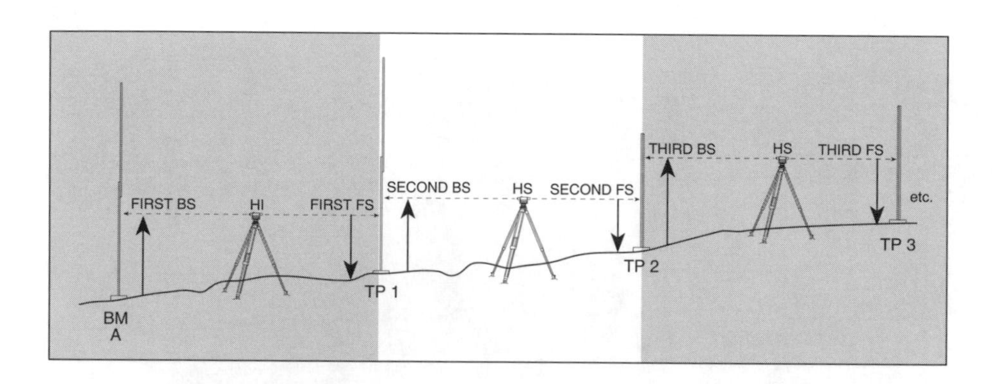

STATION	BS	HI	FS	SS	ELEVATION
Describe Points and Number	Record Backsight Readings	Add the Backsight to the Elevation to record the HI	Record Forsight Readings	Left blank for now. Will be used for profile leveling	Subtract Foresight from the HI and record Elevations
Starting Point					Starting Elevation
Instrument Setup #1	First Backsight	HI			
Turning Point #1			First Foresight		Turning Point #1 Elevation
Instrument Setup #2	Next Backsight	Calculated HI ELEV + BS = HI			
Turning Point #2			Next Foresight		Turning Point #2 Elevation
Instrument Setup #3	Next Backsight	Calculated HI ELEV + BS = HI			
Turning Point #3			Next Foresight		Turning Point #3 Elevation
etc	etc	etc			
etc			etc		etc

SAMPLE NOTES FOR A TYPICAL LEVEL LOOP

A SAMPLE SET OF NOTES FOR GOING FROM BM A TO TBM

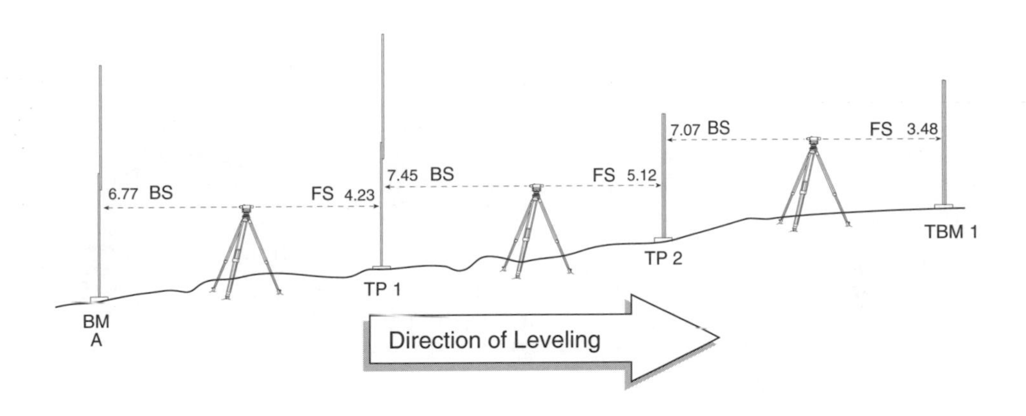

BM. A TO. TBM. 1

POINT	BS	HI	FS	SS	ELEVATION
BM. A					5280.00
INST. 1	6.77	5286.77			
TP. 1			4.23		5282.54
INST. 2	7.45	5289.99			
TP. 2			5.12		5284.87
INST. 3	7.07	5291.94			
TBM 1			3.48		5288.46

NOTES FOR CLOSING THE LOOP FROM TBM 1 BACK TO BM A

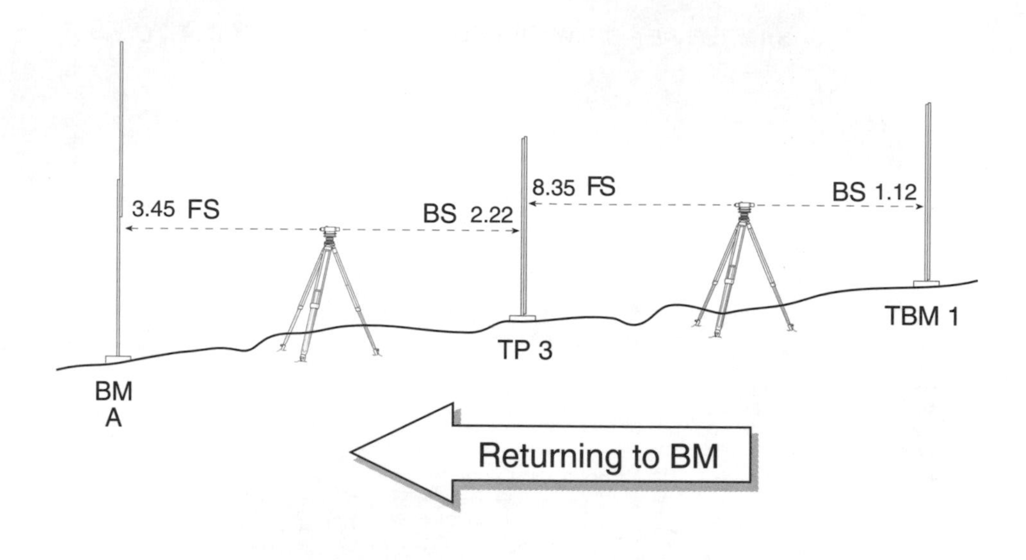

TBM 1 TO BM A

POINT	BS	HI	FS	SS	ELEVATION
TBM 1					5288.46
INST	1.12	5289.58			
TP. 3			8.35		5281.23
INST	2.22	5283.45			
BM. A			3.45		5280.00

ARITHMETIC CHECK

It is always a good practice to check level notes for arithmetic or calculator input errors. Simply sum the BS and FS columns and compare the difference between them with the starting and ending elevation for the level loop. An example is shown below.
The difference between the BS Sum and the FS Sum is 0.01.
The difference between the starting elevation of 5280.00 and the ending elevation of 5280.01 is 0.01.

Since the differences are the same, the arithmetic checks. An arithmetic error would exist if the differences had not been equal.

BM. A TO TBM. 1

POINT	BS	HI	FS	SS	ELEVATION
BM					5280.00
INST	6.77	5286.77			
TP. 1			4.23		5282.54
INST	7.45	5289.99			
TP. 2			5.12		5284.87
INST	7.07	5291.94			
TBM 1			3.48		5288.46

TBM. 1. TO. BM. A

POINT	BS	HI	FS	SS	ELEVATION
TBM 1					5288.46
INST	1.12	5289.58			
TP. 1			8.35		5281.23
INST	2.22	5283.45			
BM 1			3.44		5280.01

BS. SUM = 24.63 FS. SUM = 24.62

STARTING. BM. + BS. SUM. - FS. SUM. = END. ELEV

5280. + 24.63. - 24.62. = 5280.01

REMEMBER THE RULES RELATED TO FIELD BOOK USE

10 Field Book Rules	
1	Don't forget to add each activity to the table of contents.
2	Provide date, time, and weather.
3	Provide identification information.
4	Add to the legend any symbols used.
5	Use sketches freely.
6	Do not crowd the data.
7	Do not erase.
8	Record everything.
9	Use standard lettering techniques.
10	Use standard drafting techniques.

SUMMARY OF LEVELING

Leveling Field Duties and Responsibilities

The field engineer at the instrument shall:	*The field engineer holding the rod shall:*
Communicate with the rodperson.	Communicate with the instrument person.
Read and reread the rod to be sure the correct reading is made.	Ensure that the level rod is plumb when the reading is taken by using a rod level or by waving the rod through vertical.
Eliminate parallax.	Carefully select solid points for turns.
Assure the compensator is working in an automatic level.	Pace to balance backsights and foresights between turns.
Provide proper care for the leveling instrument.	Mark each turning point.
Level the instrument exactly.	Clean the bottom of the rod of mud, ice, and snow.
Assist in balancing the backsights and foresights.	Carefully select solid existing turning points or insert a spike solidly in the ground for a turning point.
Utilize care when the level is set up in traffic.	Check by tying into another point of known elevation. Remember to close the loop.
Check instrument calibration weekly.	Assist the instrument person at all times.
	Select turning points equal distances from the instrument. The distance will be determined by pacing.
	Carry marking pens and spray paint at all times and mark each turning point and its number.
	Ensure that the level rod is on the exact point of the benchmark and that it has not slipped off of the point during the reading.

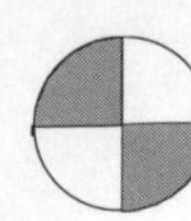

TOTAL STATION

10

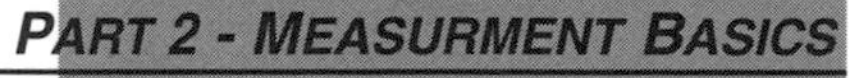

PART 2 - MEASURMENT BASICS

CHAPTER OBJECTIVES:

Discuss why the total station is the instrument of choice for the field engineer.
List and explain the advantages of the total station on the construction site.
List and explain the disadvantages of the total station.
Describe the discontinuing features of the total station.
List and review the measurement and computational capabilities of a total station.
Describe the advantages of the electronic field book when used with the total station on the construction site.
List and review the functional capabilities of the electronic field book.
Describe how to prepare a total station for use.
Explain the process involved in measuring a distance with a total station.
Explain the process involved in laying out a distance with a total station.

INDEX

INTRODUCTION TO THE TOTAL STATION

SCOPE

The electronic total station is the instrument of choice of the field engineer in surveying measurement today. It is the combination of a digital theodolite and an EDM that work together with a microprocessor to give the user the power to rapidly perform a variety of measurement tasks. Numerous measurement and computational functions are programmed into it so practically any distance, direction, or coordinate a field engineer needs can be determined. Most total stations have the capability to interface with electronic field books for storage of data or to provide layout information. The field engineer should be aware of the power and versatility of the total station and use its total capabilities.

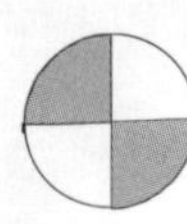

GENERAL

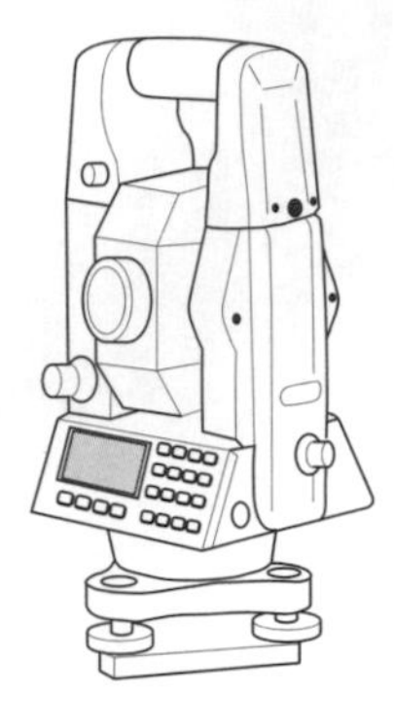

What makes a total station so powerful is the computer inside it. All total stations have microprocessor technology that serves as their "brain." This computer directs the functions of the EDM and the digital theodolite and combines their measurements. For instance the slope distance and zenith angle are measured at essentially the same time and the microprocessor performs calculations on the data. Because of this combination of data, the horizontal distance can be displayed instantly rather than having to perform hand calculations at a later time. Compared to the top-mount EDM system, the total station is faster to set up, faster to sight, faster to operate, faster at performing computations, and faster to output needed data. In the hands of an experienced user, the total station increases productivity of measurement and layout activity by 50 percent or more. If a project has numerous points to establish and a tight schedule, the total station is the instrument to use.

Advantages of the Total Station	
Single Sighting	Recall that with a top mount EDM, two scopes must be sighted: the theodolite scope and the EDM scope. With the total station, there is only one scope to sight. Obviously, this saves time and increases the number of points that can be set or located.
Instant Horizontal Distance	Since the vertical angle is constantly available, as soon as the slope distance is measured the microprocessor in the total station instantly computes the horizontal distance and displays it to the operator.
Quick Checking	Rapid operation allows for multiple checking of layout and measurement points which allows the field engineer to check and recheck layout work in approximately the same amount of time it takes to do the initial layout with a top-mount system.
Single Instrument	Carrying only one instrument in it's case to the field can be a great advantage if a two-person crew is working. There are so many tools, stakes, rods, etc., for the field engineer to carry having one less box is appreciated.
Easy to Learn	Most total stations are easy to use even though they may appear difficult at first because of all the buttons and functions that are available. Getting over the initial learning curve may prove to be the most difficult part of learning to operate the total station. Once the operator learns the basics of the system, most operations are common sense.

Advantages of the Total Station (continued)	
Computer in the Field	If calculations are needed to establish a point, many total stations have the calculating ability of a powerful hand-held scientific calculator. Programs designed to solve just about any measurement situation in the field are available.
Electronic Field Book	Total stations can be connected to an electronic field book to store data for thousands of points or to lay out previously calculated information. When fully implemented, this feature pays for itself many times over with increase productivity and the elimination of recording errors by a notekeeper.

Just as there are tremendous advantages to the total station, there are also a few potential disadvantages. Some of these may not be disadvantages to all users.

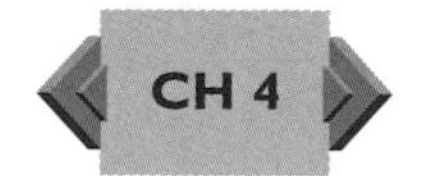

Disadvantages to the Total Station	
Expense	One disadvantage is that investment in total stations requires a huge initial outlay of capital. However, if the increase in productivity and error checking are taken into account, total stations are worth the investment.
Exposure	When used on a construction site, the total station is exposed to the hazards of material movement, heavy equipment, etc. With a top-mount system, the EDM can be removed when not in use, thus reducing the hazardous risk to part of the instrument. Just like any instrument, total stations must be protected at all times.
Batteries	Just as in all electronic usage, total stations are dependent on battery power. See Chapter 4, Equipment Care, for a discussion of maximizing the use of batteries.
Over Confidence	Unfortunately, some users of the total station quickly become dependent on it and do not follow the basic principles of measurement. They do not perform as many checks as they should in the extra time that is available.

DISTINGUISHING FEATURES

The total station appears similar to many digital theodolites because of its keyboard and digital display. Leveling screws, the optical plummet, and the clamping system are the same in total stations as they are in digital angle-turning instruments. However, the total station is slightly larger, has an EDM as an integral part of the telescope, and usually has more keys in the keyboard. The telescope of the total station serves as a combination optical scope for sighting purposes and a sending and receiving unit for the EDM. For sighting purposes, the telescope functions the same as any other. The EDM surrounds and is connected to the telescope with its electronics and necessary optics.

The keyboard on the total station usually has more keys on it to accommodate additional measurement functions. The entire alphabet is available for fully describing the points.

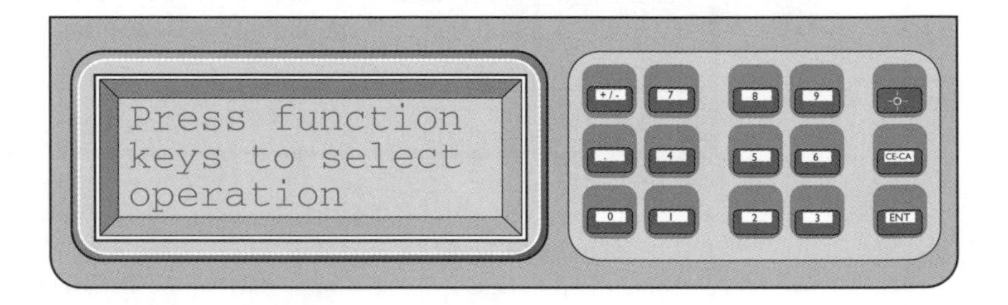

The display on the total station is typically a digital LCD. The type of angle turned, (horizontal or vertical) is displayed as well as the angle measured. The type of distance, (slope, vertical or horizontal) is also displayed with the numerical value. North and East coordinates as well as the elevation are also displayed at once. Many total stations display distinguishing information about angles, distances and coordinates.

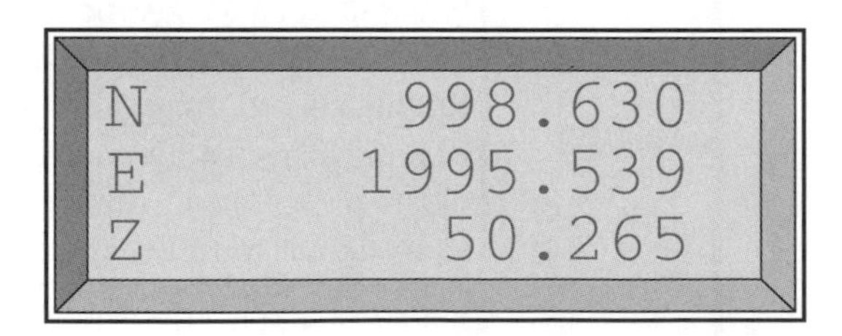

MEASUREMENT AND FUNCTIONAL CAPABILITIES

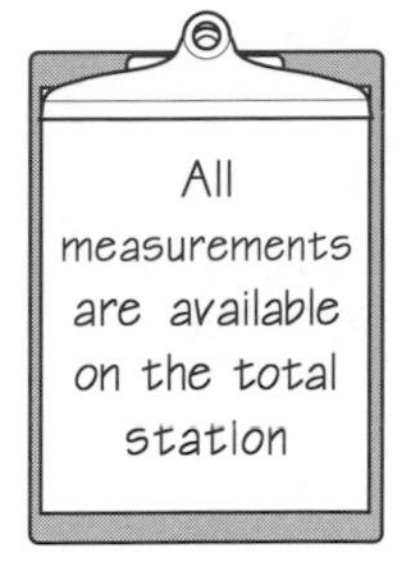

Just about any surveying measurement or calculation activity is available with a total station. Most total stations have the capability of measuring and displaying the following: horizontal and vertical angles, slope distances, vertical distances, and horizontal distances, inverses between two unoccupied points, elevations of remote objects, coordinates for any point occupied or sighted, traverse closure and adjustment, resection for the location of an occupied point, topography, data units in feet or meters, averaging of multiple sets of angles, corrections for prism constants, atmospheric pressure, and temperatures, curvature, and refraction corrections to elevations, and interface with an electronic field book

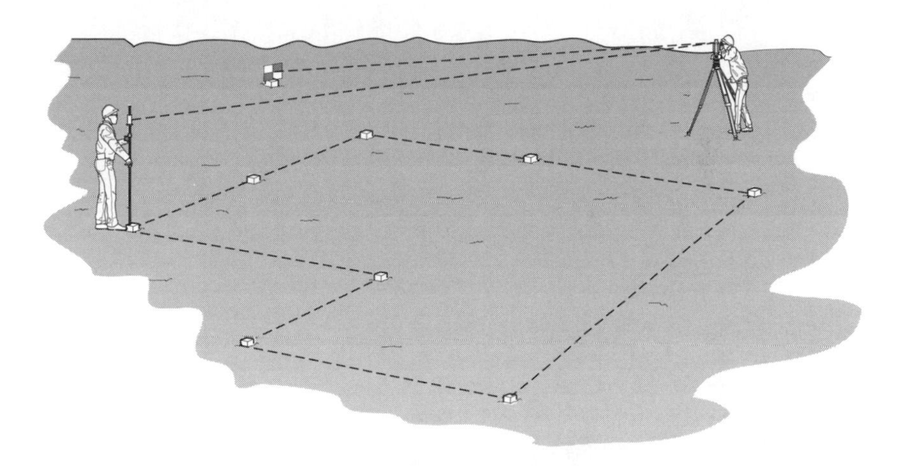

MAXIMIZE THE USE AND THE CAPABILITIES

Many operators of total stations do not tap into the vast number of functions that are available. Simply reading and studying the owner's manual and practicing with the total station will greatly expand the measurement opportunities. See the owner's manual for a complete list of functions for a particular total station.

OPERATING FUNDAMENTALS

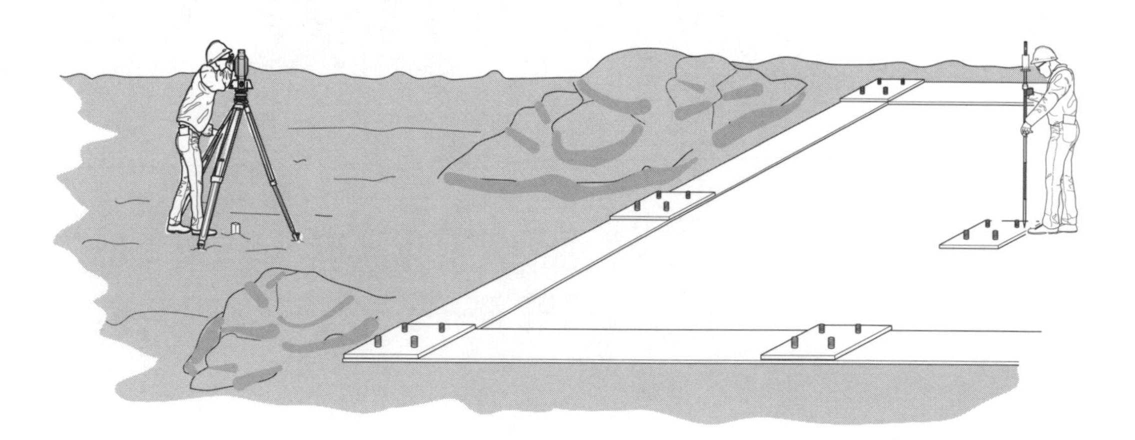

SCOPE

Operating a total station is a two-step process. Step one includes set up and preparation for measurement. Step two is the actual measurement process itself. To get accurate measurements, the total station must have a solid set up exactly over the point. The field engineer then must provide reference information to the total station so it can output desired results. This information will include the units of measurement, constants, a reference elevation, initial coordinates, and a reference direction.

PREPARING A TOTAL STATION FOR USE

Even though the total station instruments are capable of several functions, the manufacturers have generally designed them to be very simple to use. Many of the operations such as set up, focus, and clamp use are similar to other instruments.

SET UP

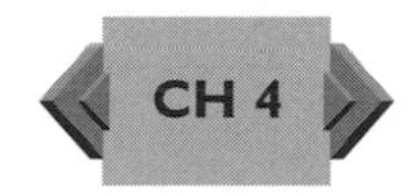

Set up, use of the optical plummet, and leveling a total station are similar to other instruments discussed. See *Chapter 4, Fieldwork Basics* for a discussion of setting up over a point and leveling. However, some total stations have automatic dual-axis compensators which sense any amount the instrument is out of level and adjusts the angles accordingly. This feature dramatically improves measurement accuracy. Always refer to the total station owner's manual.

FOCUS

Use of the telescope of the total station is identical to other instruments. Always remember to focus on the crosshairs to eliminate parallax and focus onto the objective as clearly as possible.

CLAMP USE

Total stations generally have clamping systems similar to other angle turning instruments. However, because of the push button zeroing capability and the ability to hold an angle with the push of a button, only one clamp may be needed for operation. See the owner's manual of the particular instrument for the most efficient method of manipulating the clamps in the angle turning process.

READING THE TOTAL STATION DISPLAY

The display of the total station is similar to the display of a calculator. Horizontal and vertical angles are given in degrees, minutes, or seconds. Distances are displayed in feet and decimal parts of a foot. Various icons are used to distinguish between horizontal, vertical, or slope distances. Coordinates are typically given as N and E. The display below show the zenith angle and the horizontal angle right (HAR).

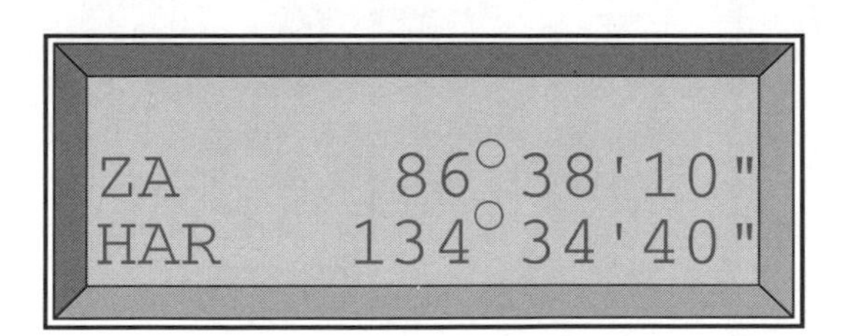

PREPARING FOR TOTAL STATION MEASUREMENT

The operation of many total stations is similar to simply using a combination of digital theodolite and EDM. See the owner's manual for an exact description of the process. However, input characteristics of the total station differ. When preparing for measurement with a total station, it is necessary to initialize the instrument with data for the point occupied and for the backsight. The following procedure can be used as a guide for most total stations.

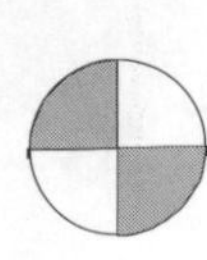

Typical Total Station Preparation for Measurement	
Turn the total station on.	Check the battery strength and replace if necessary.
Select the units	Degrees? Feet? Meters? Select on the keyboard.
Input correction factor	Atmospheric pressure and temperature are needed to determine the parts per million correction factor.
Check constants	Prism offset must be set accurately for the prism being used.
Input elevation	An elevation of the occupied point is generally inputted into the instrument or the electronic field book.
Input set up heights	Height of the total station and the height of the prism on the prism pole is needed to be able to accurately determine the elevation of the points where readings are taken.
Initiate point numbering	Decide if the numbering starts at 1, 10, 100, 1000 or some other number.
Enter point number and description of occupied point	Describe all occupied and back sighted points thoroughly. Many a survey has been lost because of inadequately described starting points.
Enter coordinates of occupied point	If the coordinates of the point are known from previous surveys, input them. If values for the starting coordinate are to be assumed, review the scope of the project and assume coordinates that are large enough to prevent negative coordinates from occurring.
Enter coordinates or direction to backsight.	When using points from a previous survey, use the coordinates of the backsight. If a new survey is being established, a direction is typically assumed.
Ready, set, Measure! or Layout.	At this point, the total station should be ready for measurement. Go to the next section for an illustrated explanation of this process.

MEASURING A DISTANCE AND AN ANGLE

SCOPE

The measurement of a distance and angle between point on a traverse with a total station is similar to the measurement process that takes place when measuring with a top-mount EDM. However, the single sighting and not having to record intermediate slope distances and vertical angles for calculation purposes makes the process very simple.

STEP-BY-STEP PROCEDURE FOR MEASURING A DISTANCE

STEP 1

Setup and prepare the total station for measurement by imputing constants, coordinates, elevation, etc.

STEP 2

Sight onto the backsight and input the direction to the backsight. After the total station has been initialized with all of the information for the occupied point and a backsight point, the instrument can be sighted onto the backsight and locked onto that point. Set the direction of the backsight or an assumed direction. (Often times zero).

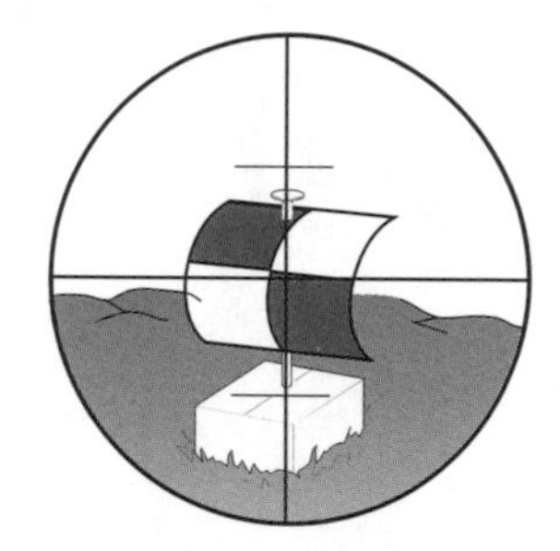

STEP 3

Loosen the clamp and turn to the point. The clamp can be loosened and the instrument turned and sighted onto the point where distance, direction, and coordinate data is desired.

STEP 4

Sight onto the prism. Sight the horizontal and vertical crosshair exactly onto the center of the prism so accurate vertical and horizontal angles can be measured.

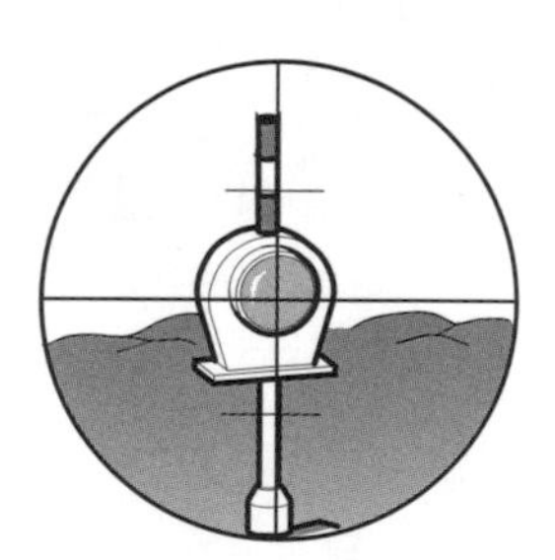

STEP 5

Press the Measure or Read button. A few seconds later a slope distance is displayed. With the push of a conversion button, the microprocessor calculates and displays the horizontal distance and the horizontal angle. Coordinates of the point measured to can also be determined with the simple push of a button on the keyboard.

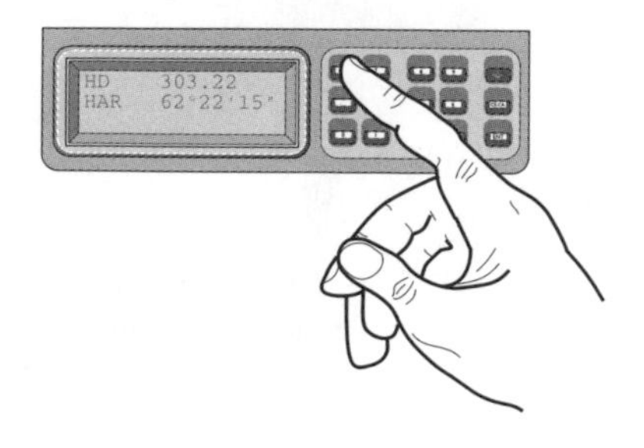

STEP 6

Read and record the data. Record in proper fieldbook form the distance, angle, and coordinates of the point.

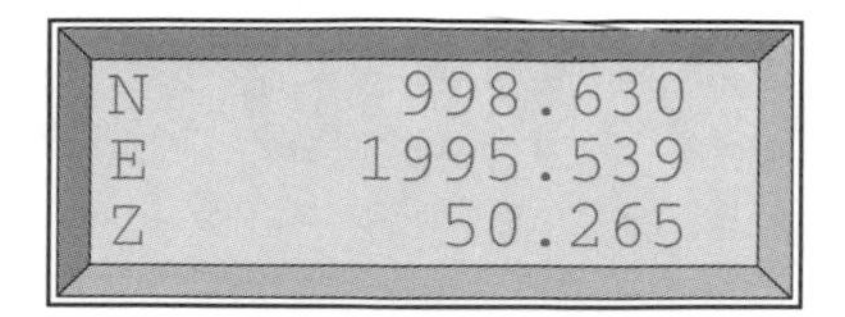

STEP 7

Double check. Double any angle turned and recheck the measurement. If setting up on control points sequentially around a traverse, always measure distances forward and back.

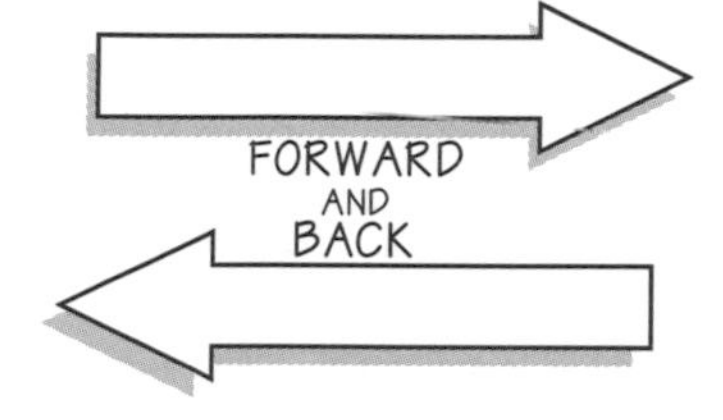

SUMMARY

Recall that when chaining a distance or measuring a distance with a top mount EDM, many steps were needed in order to complete the process. It can be seen that the total station performs the same measurement in fewer steps and, therefore, is much faster. The field engineer using a total station will be able to handle more layout work and will have the time to check and recheck to ensure accuracy.

FIELD NOTES

When using a total station the extent that field notes must be kept depends on weather or not a data collector is used. If no data collector is available, then complete field notes of all measurements must be recorded. If a data collector is used, field notes of random measurements should be recorded.

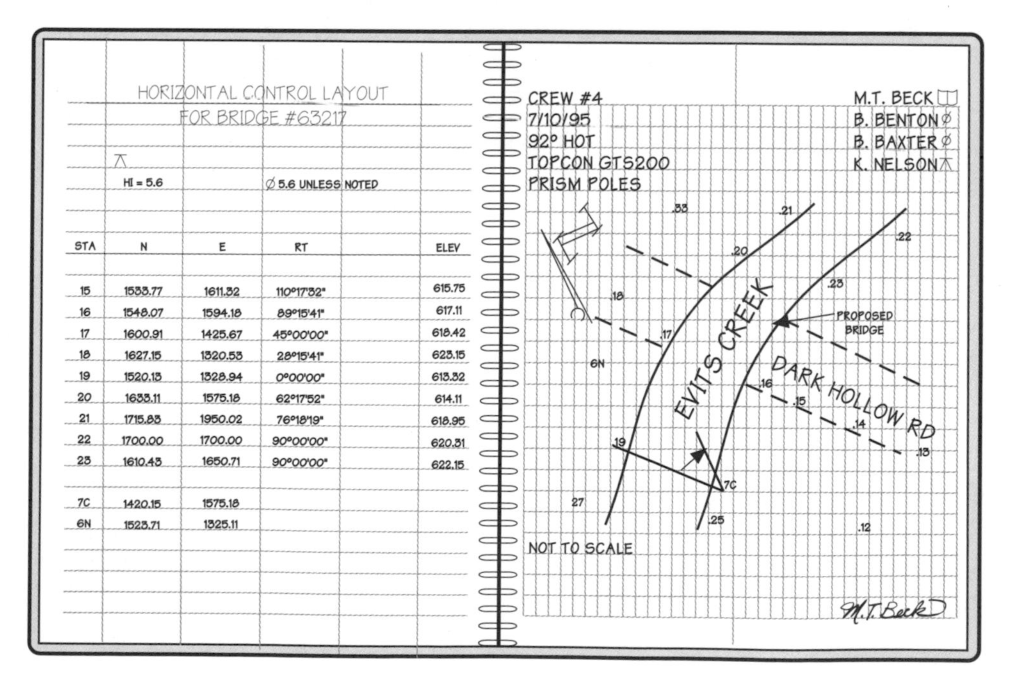

HORIZONTAL CONTROL LAYOUT
FOR BRIDGE #63217

HI = 5.6 ∅ 5.6 UNLESS NOTED

STA	N	E	RT	ELEV
15	1533.77	1611.32	110°17'32"	615.75
16	1548.07	1594.18	89°15'41"	617.11
17	1600.91	1425.67	45°00'00"	618.42
18	1627.15	1320.53	28°15'41"	623.15
19	1520.13	1328.94	0°00'00"	613.32
20	1633.11	1575.18	62°17'52"	614.11
21	1715.83	1950.02	76°18'19"	618.95
22	1700.00	1700.00	90°00'00"	620.31
23	1610.43	1650.71	90°00'00"	622.15
7C	1420.15	1575.18		
6N	1523.71	1325.11		

LAYOUT OF A POINT

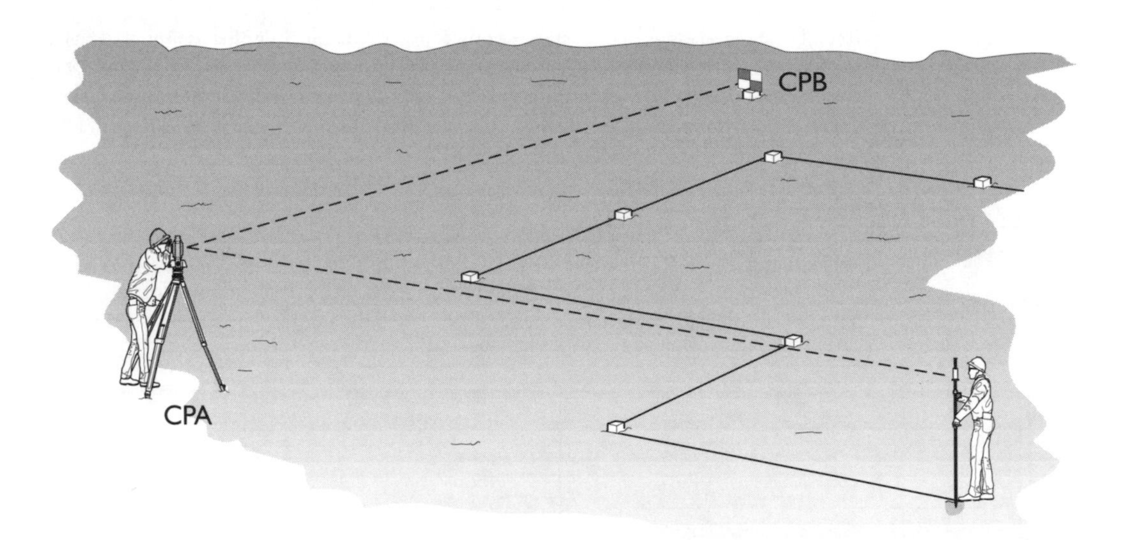

SCOPE

Again, the layout of a distance with a total station is similar to the layout process that takes place when using a top-mount EDM. However, the single sighting for horizontal and vertical angles speeds up the process tremendously. Communications with the prism pole holder is the key to rapid setting of points with the total station.

STEP-BY-STEP PROCEDURE FOR LAYING OUT A POINT

STEP 1

Sight onto the backsight. After the total station has been initialized with all of the information for the occupied point and a backsight point, the instrument can be sighted onto the backsight and locked onto that point.

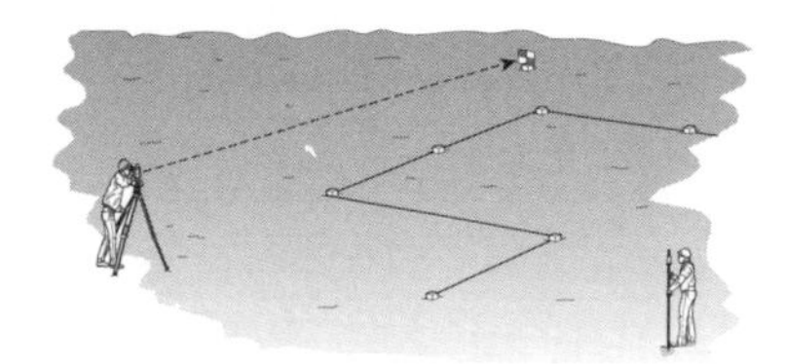

STEP 2

Input the direction to the backsight. Set the direction of the backsight or an assumed direction. (Typically zero).

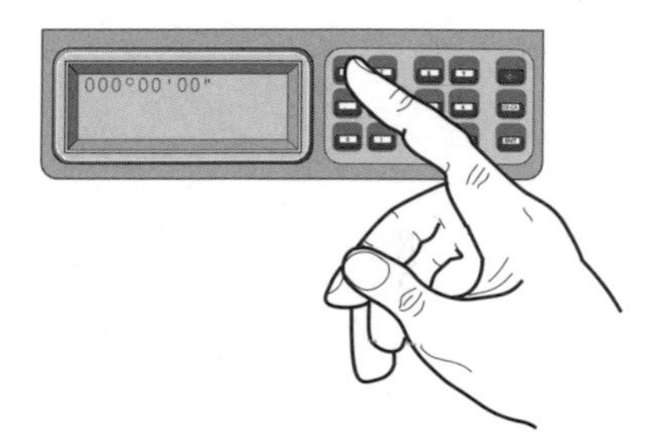

STEP 3

Loosen the clamp and turn to the angle needed. Refer to the Layout data sheet and turn the instrument until the angle desired is read on the display.

STEP 4

Direct the prism pole holder. Using hand signals or a radio, direct the holder of the prism pole to get on the line of the angle turned and at the estimated distance needed.

STEP 5

Press the measure or read button. Measure the distance and compare it to the desired layout distance.

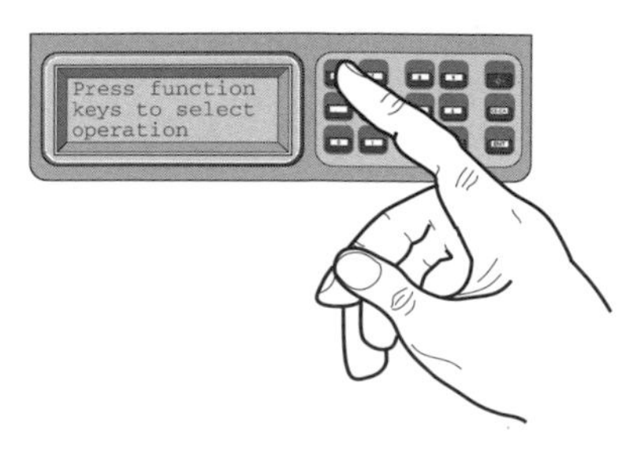

STEP 6

Direct the prism holder to move closer or farther away as needed.

STEP 7

Remeasure as needed. Continue to measure and move the prism holder until the distance desired is obtained. Set the point.

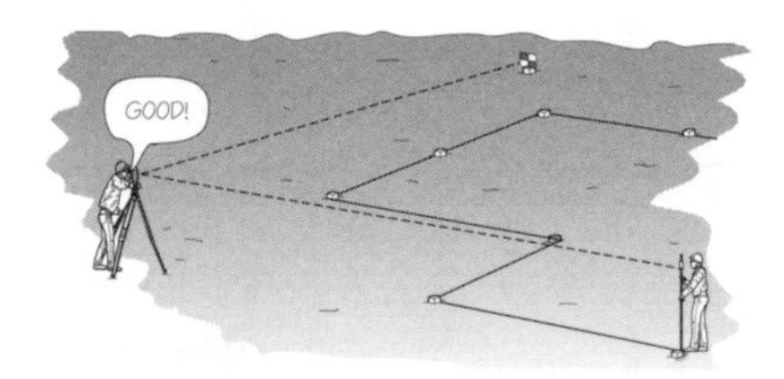

STEP 8

Double check. Remeasure any point set from another instrument set up to assure no blunders have been made.

FIELD NOTES

The following is a sample of field notes which can be used for layout of a point with a total station.

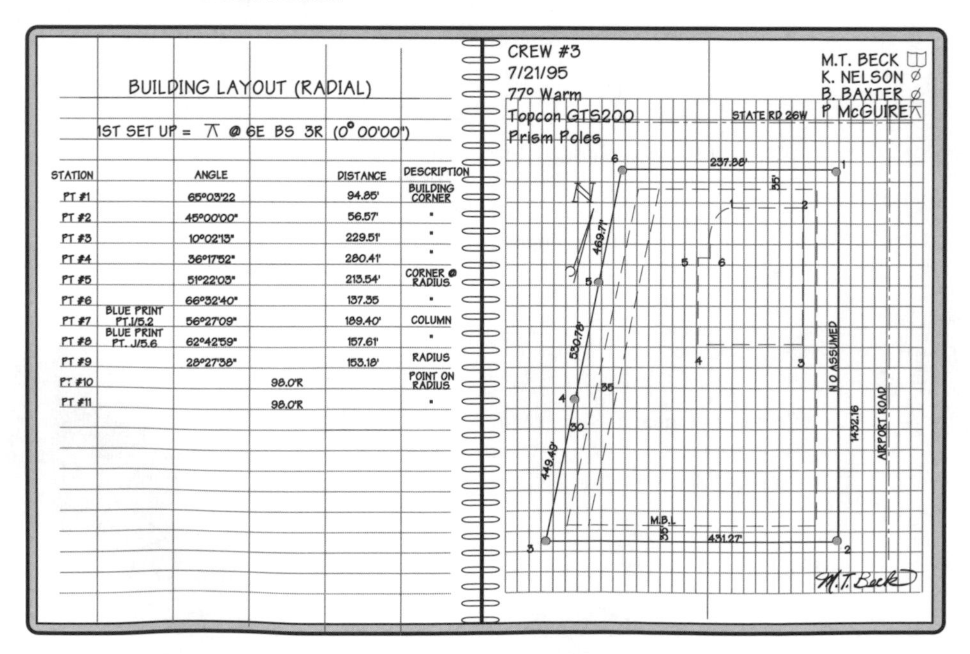

BUILDING LAYOUT (RADIAL)

1ST SET UP = ⊼ @ 6E BS 3R (0° 00'00")

STATION		ANGLE		DISTANCE	DESCRIPTION
PT #1		65°03'22		94.85'	BUILDING CORNER
PT #2		45°00'00"		56.57'	"
PT #3		10°02'13"		229.51'	"
PT #4		36°17'52"		280.41'	"
PT #5		51°22'03"		213.54'	CORNER @ RADIUS
PT #6		66°32'40"		137.35	"
PT #7	BLUE PRINT PT.I/5.2	56°27'09"		189.40'	COLUMN
PT #8	BLUE PRINT PT. J/5.6	62°42'59"		157.61'	"
PT #9		28°27'38"		153.18'	RADIUS
PT #10			98.0'R		POINT ON RADIUS
PT #11			98.0'R		"

THE ELECTRONIC FIELD BOOK

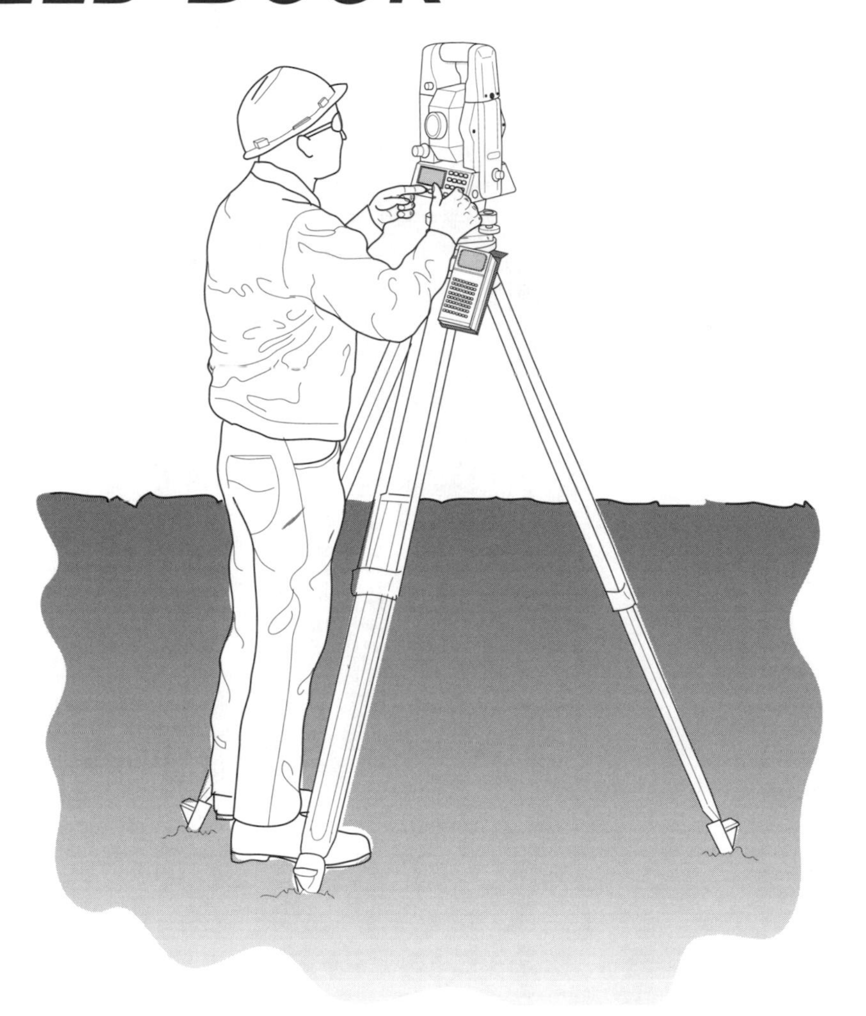

SCOPE

The electronic field book is an extremely powerful feature that compliments the total station. Used properly, an electronic fieldbook eliminates human error and saves time and money. An electronic fieldbook is essentially a computer that fits in your hand and adds to the features of the total station by allowing for extensive data collection, the availability of layout data, and additional computational functions in the field.

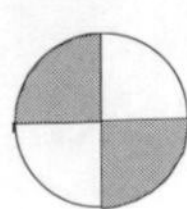

GENERAL

With a total station, electronic field book and associated construction surveying software, the field engineer will be able to work at a faster rate. Layout data can be calculated on a computer in the office and the data uploaded to the electronic field book. The electronic field book is taken to the field and connected to the total station when hundreds of points can be rapidly established. Or, the electronic field book can be used in the field as a data collector capable of storing distance, direction, elevation, coordinates, and descriptions for thousands of points. It can then be taken back to the office computer to download the information for calculations and plotting.

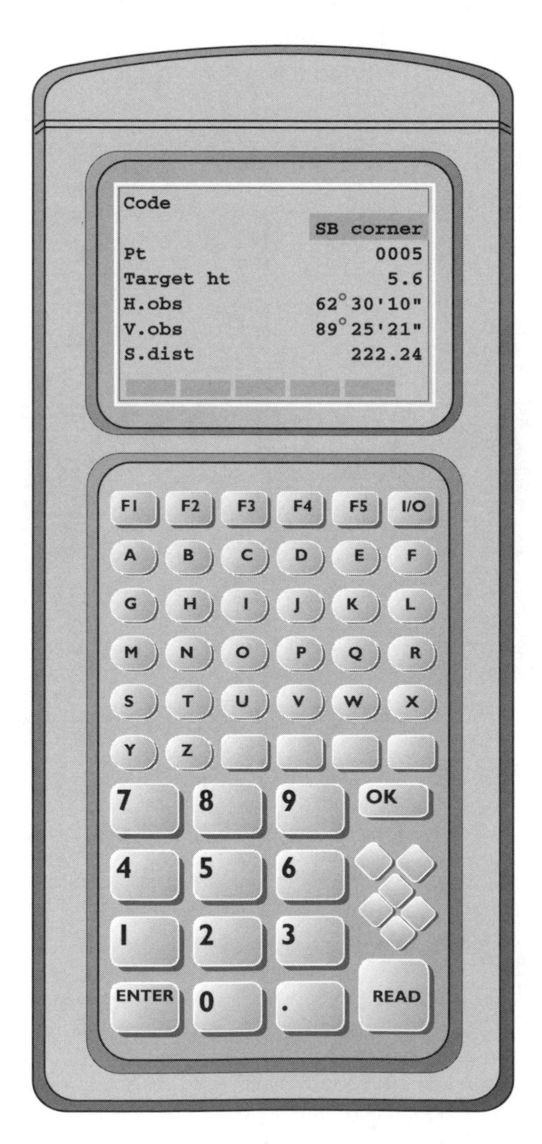

Advantages of Using an Electronic Field Book	
Eliminates human errors	Errors such as recording an angle or a distance in a fieldbook can be eliminated because the electronic field book records the distance and angle without human involvement other than to push the read-and-record button on the electronic field book. Numbers will not be transposed or forgotten.
Saves time	One manufacturer reports that the use of the electronic field book can reduce field time by 50 percent over recording notes in a fieldbook. This, combined with the time advantages already discussed regarding the total station, indicates that combining and using the two instruments together offers a tremendous opportunity for additional increased productivity to the field engineer.
Makes work easier	A full-featured electronic fieldbook enables the field engineer to work more easily by entering user-defined codes for objects that are being located.
Works efficiently	Most electronic field books are menu driven and enable the user to quickly obtain the programmed feature that is needed.
Processes data	The programmed features of the electronic field books enable data that is collected to be processed while in the field. After measuring the data for a traverse, immediately available will be the closure, precision, and area. In addition, other common fieldwork computations such as curve computations, intersections, inversing, etc., are available on the electronic field book.

Disadvantages of Using an Electronic Field Book	
Battery	Loss of battery power could mean loss of data. However, most electronic field books have a dual set of batteries to prevent memory loss. Check all batteries before use.
Expense	Electronic field books add thousands of dollars to an already expensive total station. However, it should be recognized that the increased productivity more than compensates for this expense. The cost can be recovered quickly by a savings in labor cost.

FUNCTIONAL CAPABILITIES

Electronic Field Books are all different in how they operate, but they are getting very close to standardization in their functions and capabilities. Some typical features include:

1) Easy selection of "programs" by use of menu structure.

2) Memorization of codes so only the first character or two of a code is needed to display the code.

3) Storage of data is available for instant recall. Once entered it cannot be inadvertently deleted or changed.

4) Lines and points can be described completely.

5) Coordinate Geometry (COGO) capabilities are available such as: setting out, resection, inverse, intersections, station and offset, and much more.

6) Topographic data can be collected for development of site plans.

7) Traversing can be performed that allows the electronic field book to keep track of the distance, direction and coordinate information for points occupied and points sighted. Traverse computations can be performed before leaving the field.

8) Least squares adjustment of the data may be available.

9) Coordinates can be rotated if needed.

10) Horizontal curve computations can be performed that calculate deflections, long chord lengths from the PC, offset information, coordinates of the curve stations, radial layout of the curve from any control point, etc.

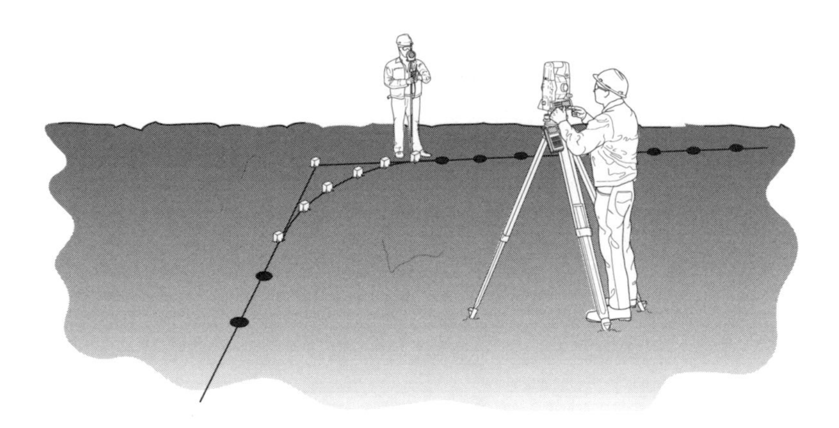

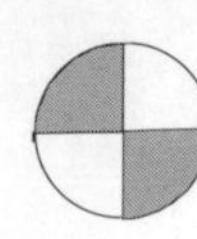

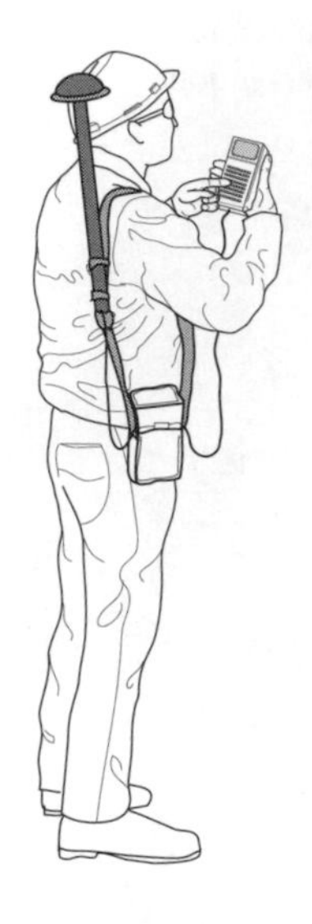

11) Vertical curve elevations can be calculated.

12) Cut and fill for any point within a defined road alignment can be determined.

13) Slope staking can be performed.

14) Exportation of data to surveying software and to popular drafting software is provided.

15) Interfacing to GPS (Global Positioning Systems) is possible.

16) Built-in calculator functions.

17) Communication capability to printers, computers and plotters is available.

18) In addition to those listed above, many other features may be available. See the electronic field book owner's manual for a complete list of functions for a particular total station.

ELECTRONIC FIELD BOOK VERSATILITY

Obviously, attaching an electronic field book is the ultimate in maximizing the electronic capabilities of the total station. Many electronic field books available will work with several types of total stations. Therefore, an electronic field book should be chosen for its versatility and overall power and features.

Construction Lasers for Line and Grade

11

Part 2 - Measurement Basics

CHAPTER OBJECTIVES:

Explain what LASER is an acronym for.
List five advantages of Lasers on the construction site.
List disadvantages of lasers as a construction tool.
Distinguish between the two types of lasers available.
Distinguish between fixed, rotating, and utility lasers.
Describe the primary applications of lasers in construction.
Describe how a laser can be used for vertical alignment.
Describe how it is possible to establish grade (elevations) with one person, using a lenker rod and laser.

INDEX

LASERS FOR CONSTRUCTION

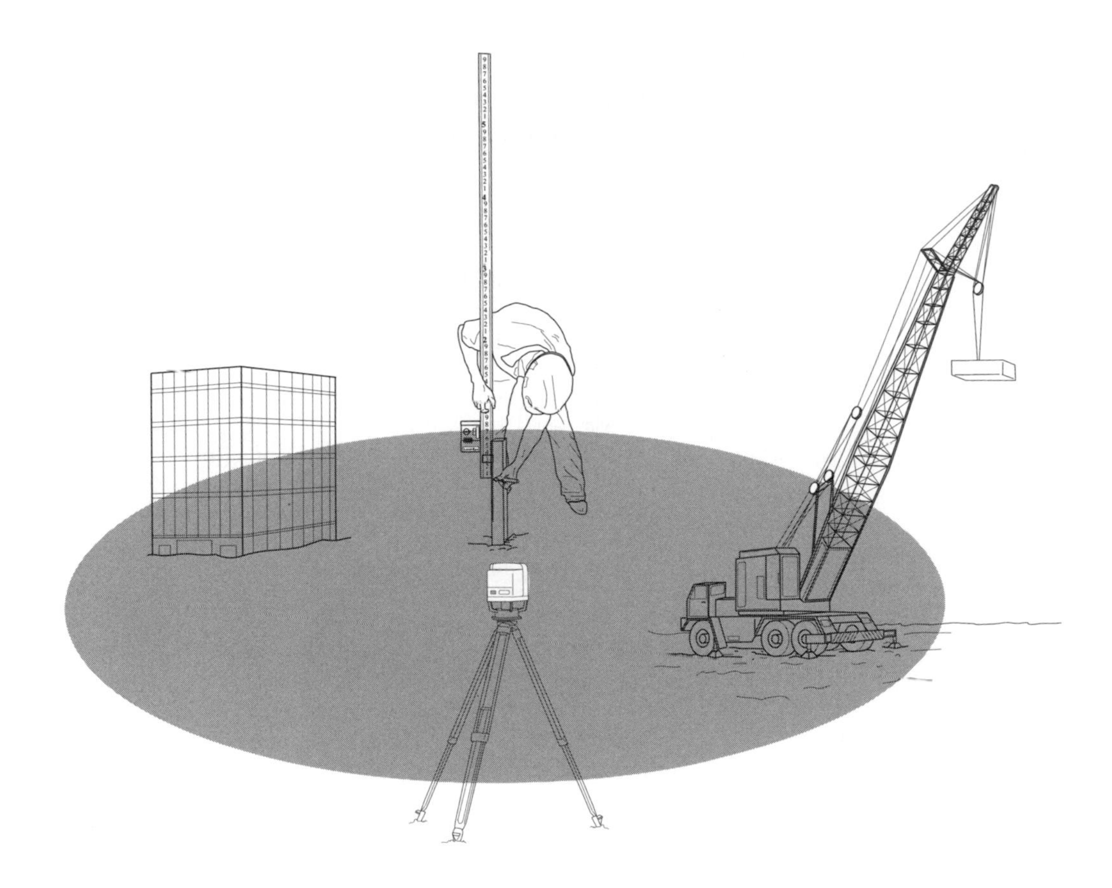

SCOPE

In today's world of construction, repetitive work which requires constant grade or alignment is best accomplished with a laser. By using a laser, the contractor reduces costs and increases productivity. The availability of lasers has truly revolutionized the methods of controlling the elevations on a jobsite. Although very simple to use, mistakes can be made with construction lasers. The field engineer must be aware of the proper operation of the laser, its limits, and its extensive opportunities for application in construction.

THE EVOLUTION OF LASERS IN CONSTRUCTION

DEFINITION

A laser is a light beam. More specifically, LASER is an acronym for "Light Amplification by Stimulated Emission of Radiation." It is a beam of light that contains only one color of the spectrum. Therefore, it is monochromatic. The beam emitted is coherent in that it does not scatter. The rays of light in the beam tend to stay almost parallel and do not diverge like ordinary light. In surveying, the term laser is also generically used to describe electronic levels that are not technically lasers.

EARLY APPLICATIONS OF LASERS

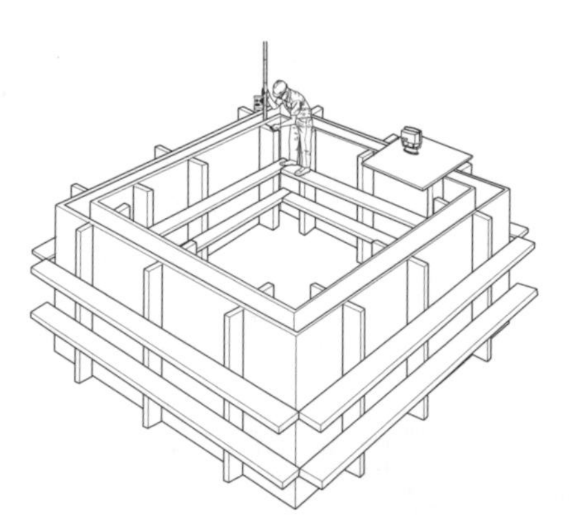

Lasers were first developed in the late 1950's and were more a curiosity than a tool. However, it wasn't long until American ingenuity began to develop a host of applications such as communications, weapons, medical equipment, and of course, construction.

In the mid 1960's, lasers began to be applied to surveying. Electronic Distance Measurement instruments such as the Laser Ranger was the first application. Later, it was recognized this durable tool could be used on the construction site for sewer alignment and grade work; sewer lasers were developed. Soon it was found that a rotating prism could be attached to the laser and a horizontal plane could be created. This breakthrough in applying the laser to construction made it a valuable tool for the building contractor. Since then, other construction applications have evolved, and the laser is becoming the line and grade tool of choice on the construction site.

THE FUTURE OF LASERS

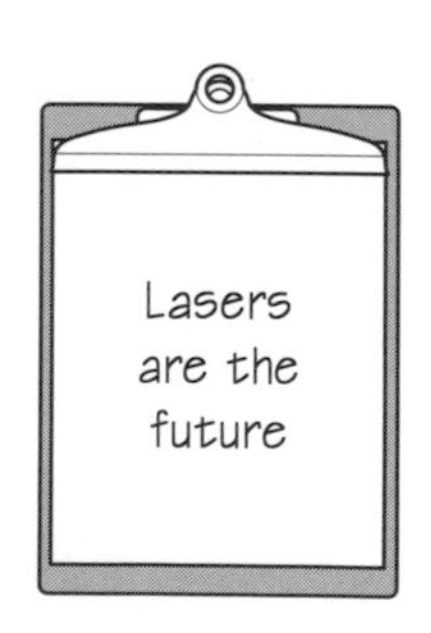

The future is here: Present-day lasers are so versatile and inexpensive that any contractor requiring constant line and grade will be a step behind if they do not have a laser available for everyday use. In the future, lasers will undoubtedly become even more prevalent on the jobsite as technology further advances the use of lasers for construction applications. There will probably be small pocket-size lasers every carpenter can carry on a tool belt just like a hammer. This pocket laser will be able to create a horizontal or vertical beam or plane in an instant. The 4-foot level would become obsolete as this pocket laser will more quickly and accurately establish level or plumb lines. The sensor for this laser will be the size and shape of a pencil so the carpenter will be able to instantly mark a point. It will become an electronic plumb bob able to establish a vertical line quickly and accurately. It will be durable and able to work for months on a small battery. And, it will be very inexpensive.

Although the future is here with present day lasers, there is an even brighter future for construction line and grade from the lasers and sensors of the future. Field Engineers must keep up to date with laser technology to maximize the effectiveness of lasers on the jobsite.

BENEFITS OF USING A LASER OR AN ELECTRONIC LEVEL

EASE OF USE

Simple! There isn't a better word to describe the lasers of today. They have been designed to be easy to use for anyone with basic knowledge of surveying instruments and the leveling process. The operator only has to roughly level the laser using standard leveling screws. The self-leveling system takes over and levels it exactly. Keeping the batteries charged is probably the most difficult operation in using a laser.

PRODUCTIVITY

Using a laser cuts labor costs on every reading. Because no one has to stay at the instrument, laser leveling can be a one-person activity. One person can perform those same tasks which require two persons using a conventional level and rod. Therefore, laser leveling frees a field engineer to perform other layout tasks.

FEWER HUMAN ERRORS

Errors caused by poor communication between the instrument person and rodperson are eliminated since laser leveling is a one-person operation. Errors from misreading numbers are reduced.

ACCURACY

Lasers provide a consistent, reliable reference plane. Accuracy of the signal can be set to ±1/16" of an inch at a distance of 200 feet.

OUT-OF-LEVEL WARNING CUTOFF SYSTEM

Most lasers automatically shut down if bumped out of level. If the laser is bumped or disturbed while being used, it will automatically turn off and indicate with a blinking light that it is out of level. This eliminates bad readings that might be taken.

DEPENDABILITY

Lasers are rugged. They are built to withstand the harsh construction environment. Most lasers are designed to operate from -20°F to 130°F. They are water resistant and can operate in a light rain shower. Lasers continue to work all day as long as there is power.

TYPES OF LASERS

VISIBLE LIGHT LASER

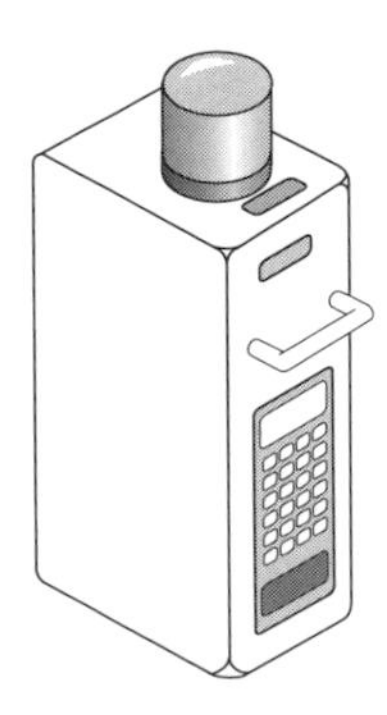

Helium-Neon is a gaseous type (helium & neon) of laser. These gases are trapped in a cylindrical tube and become active when electric current is applied which causes the neon to glow very brightly. The cylindrical tube is constructed with concave mirrors in the ends. These mirrors help focus and concentrate the intense light which has been created into a continuous beam. This beam emits from one end of the cylinder becoming a visible beam laser. This type of laser can be dangerous to the eye if stared at directly for long periods of time.

ELECTRONIC LEVEL

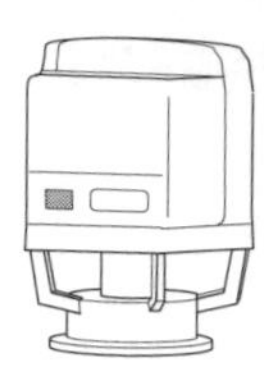

This relatively new laser uses an infrared energy source and is not visible to the observer. An electronic sensor must be used to locate the signal emitted. The infrared energy is rotated continuously in a complete circle around the instrument.

CLASSES OF LASERS AND THEIR USES

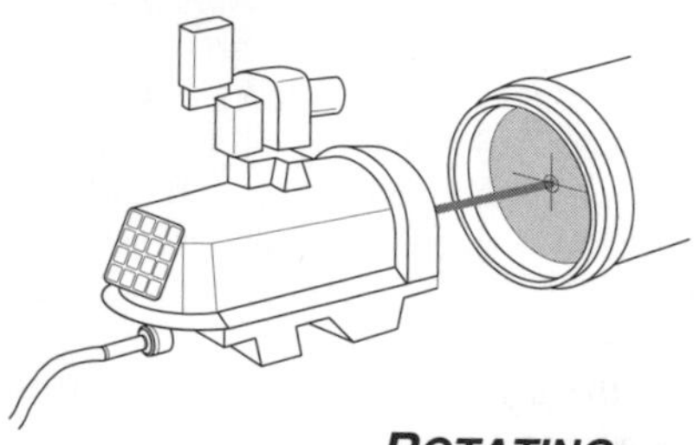

FIXED

The fixed laser projects a single beam of light that can be used as a plumb line, string line, grade line, etc. This laser is used for sewer work, tunnel alignment, plumbing a column, plumbing an elevator shaft, etc. Any layout problem that requires a single line as control should be accomplished with this laser. This is typically considered a visible light laser.

ROTATING

The rotating laser is typically constructed with a rotating prism that projects the laser beam by spinning rapidly. This rapidly spinning beam creates a plane that can be used as a reference. Both visible light lasers and electronic levels can be rotating lasers. Some applications require being able to see the beam while other applications do not. It is up to the contractor to decide which best meets the needs of the work.

The main purpose of the rotating laser is to create a horizontal plane. This emulates the horizontal plane created when an automatic level is set up. Uses of this type of laser are extensive. They include excavating, finishing a subgrade, finishing a floor, installing mechanical systems, installing a drop ceiling, as well as other construction elevation work.

An alternative application of the rotating laser is when it can also be used to project a vertical plane. Some rotating lasers are constructed so they can be turned on their sides to create a vertical plane of reference. It can be used along the side of a building to align forms for the entire length of the building. This type of application is becoming more and more prominent on the jobsite.

UTILITY

The utility laser is versatile. It can project horizontal and vertical lines, horizontal and vertical planes, and slope lines or planes. This type of laser can be used in almost any situation. It is usually much more expensive than the single-function lasers that are available. Utility lasers are typically of the helium-neon class and emit a visible beam that workers can see for line or grade.

USES OF THE LASER BY FIELD ENGINEERS

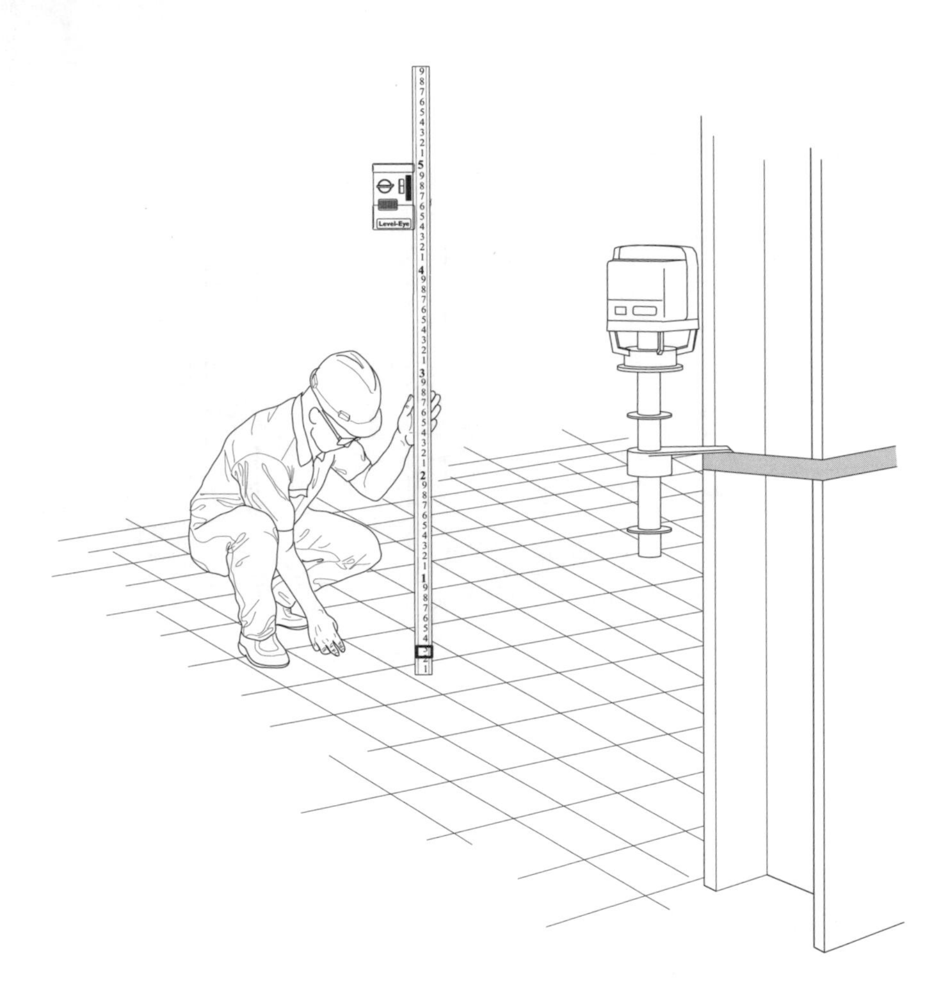

SCOPE

The laser can be one of the field engineer's best friends on the jobsite. It works as long as it has power, and it extends the ability of the field engineer to accomplish more than one task at a time. Anytime a field engineer can do that, success will follow. This section will address some of the uses of the laser and will point out common mistakes that may be encountered.

STANDARD LASER LEVELING PRACTICES

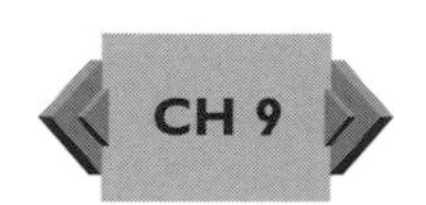

Even though the operation of the laser is simple and elevations are easy to obtain, exact procedures must be used to meet the tolerances required of construction. Basically, standard practices used when performing differential leveling should be followed. These include: balancing sites, plumb rod, solid setups and properly extended rod. Reference *Chapter 9, Leveling.*

APPLICATIONS OF LASERS FOR GRADE

Wherever a constant elevation is needed, a laser is the best tool that can be used. Constant elevation means requiring that the grade is checked at point after point, minute by minute, all day long. Having a person stand behind an instrument constantly reading a rod is not a very productive use of an individual.

Instead, a strategically-placed laser level which is creating a horizontal plane with its beam can be used by one or dozens of workers, needing an elevation, at the same time. Some grade applications include:

Maintaining level on brick or block that is being laid.

Setting cut or fill stakes.

Topographic surveying.

Establishing the design elevation on a foundation.

Leveling concrete forms.

Measuring excavation depths.

Setting grade stakes.

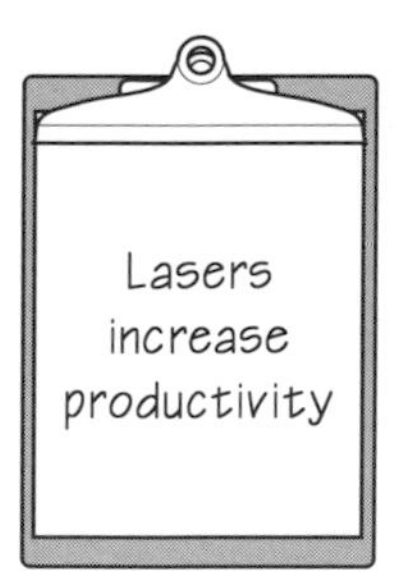

Installing HVAC and electrical duct at the design elevation.

Screeding subgrade.

Setting plumbing drains.

Maintaining finish floor elevation during concrete placement.

Installing sprinkler systems.

Maintaining horizontal drop ceilings.

Landscape terracing.

Installing conveyer systems.

Although this list is long, these are only a few of the many uses for a laser on a construction site. The field engineer must be creative in determining uses of the laser.

USES OF LASERS FOR ALIGNMENT

Utility lasers have the capability to project a horizontal line, vertical line, vertical plane, or slope line. Refer to the owners manual for specific alignment procedures. Uses of these types of lasers for alignment include:

Plumbing a building.

Aligning a wall.

Aligning a form.

Plumbing an elevator shaft.

Aligning a sewer.

Setting centerline of anchor bolts.

Plumbing columns.

Establishing lines on a slab for interior wall construction.

USING A SENSOR

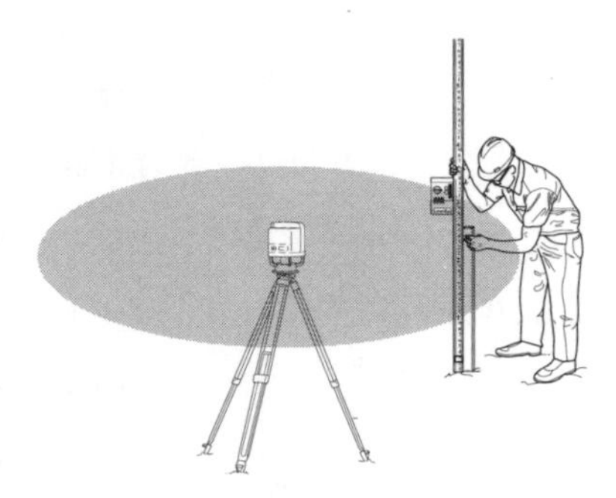

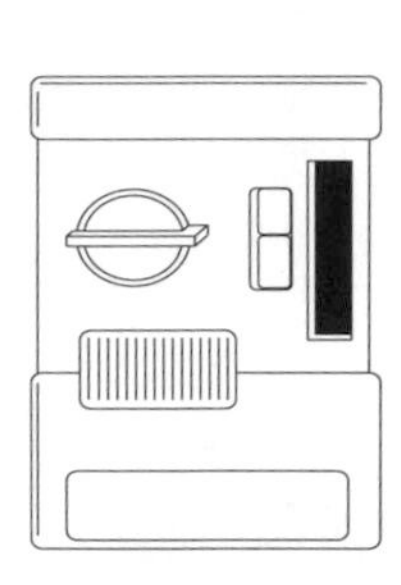

To determine the location of a laser line or plane, an electronic sensor or signal detector is needed with the electronic level, and may be necessary with the visible light laser when it is used in the bright sunlight. These devices are designed to attach to a level rod, a 2"x4", a lath, or just be used alone. They are built to be very rugged and durable for the inevitable abuse that they will encounter by being dropped or getting wet, muddy, etc. They typically have a fine (1/16") and coarse (1/8") mode for detecting the signal. Also, they have an audible signal that can be turned on or off depending on the situation. A visual display indicates to the user whether or not the sensor is above or below the signal line.

STEP-BY-STEP PROCEDURE FOR USING A LENKER ROD

The use of a direct elevation rod (commonly referred to as a Lenker Rod) is ideal for most grade work with a laser. The Lenker rod is designed to directly display the elevation of any point where the rod is held. Below is the step-by-step procedure for using a lenker rod.

STEP 1

Loosen the rod face clamp and set the bottom of the rod on the benchmark

STEP 2

Knowing the elevation of the benchmark, (i.e. 678.35) move the rod face until the instrument line of sight reads elevation 8.35. Remember the numbering is inverted. This will require exact communication between the persons at the instrument and rod.

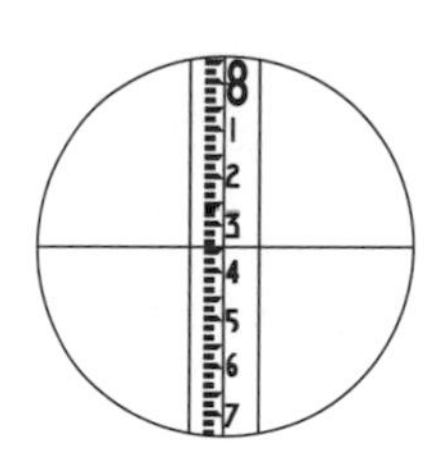

STEP 3

When satisfied that the rod face is set to the elevation of the benchmark, tighten the clamp bracket so that it holds the rod face securely in position.

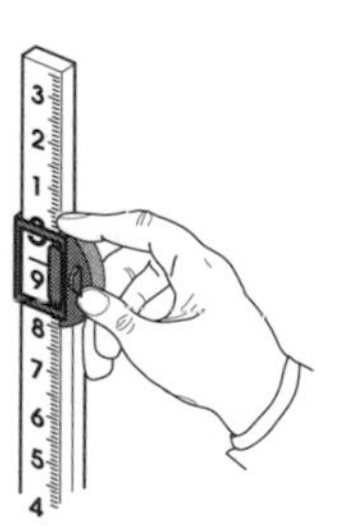

STEP 4

Go to where elevations are needed and hold the rod on a point so the person at the instrument can read it. That reading plus 670.00 is the elevation of the point.

STEP 5

Go to any other point where elevations are needed and repeat step 4 to determine the elevation. Record all elevations in the fieldbook.

STEP 6

If the limits of the lenker rod are reached because of hilly terrain, simply establish a turning point. Hold the lenker rod on the turning point to determine it's elevation, move the instrument, and repeat the above steps to continue the work.

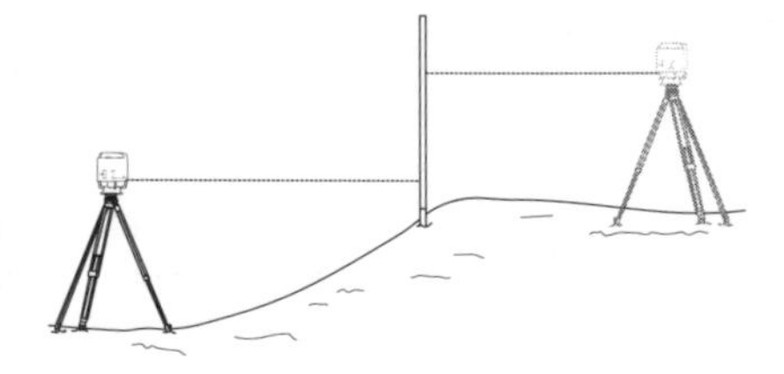

OPERATING CONDITIONS

The laser and sensor are both designed to operate in the environment of a construction site. They will function in the normal operating temperatures expected of a worker. Night work or low light work is possible with a laser when sensors are used.

CALIBRATION OF A LASER

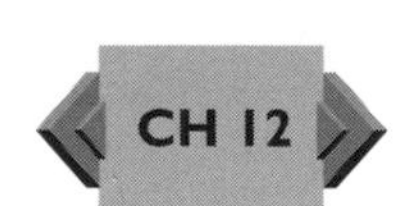

A rotating laser can be checked for calibration in much the same way that a level is checked—by pegging it. Several lasers are made to be easily adjusted in the field if the laser is found not to be projecting a horizontal plane. For more information about instrument calibration, reference *Chapter 12, Equipment Calibration.*

A fixed-line laser used for plumbing is most easily calibrated in the manner that an optical plummet is checked. This is done by rotating it 90º and marking where the beam hits a target. If it hits in the same spot in all four positions, it is projecting a vertical line. If four different spots are marked, the laser is not projecting a vertical line.

COMMON MISTAKES

This list of common mistakes is a short list compared to differential leveling with an automatic level and rod. Because laser leveling is so simple, there just aren't as many opportunities for mistakes. Some that might occur include:

Not checking for calibration.

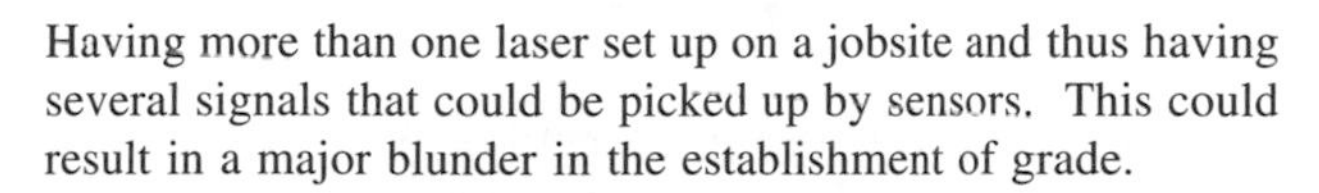

Having more than one laser set up on a jobsite and thus having several signals that could be picked up by sensors. This could result in a major blunder in the establishment of grade.

The out-of-level sensor not working causing the laser to stay on if it is bumped.

Becoming too dependent on the electronic box. Becoming overconfident about the results of the laser, and not following good leveling practices.

Setting the elevation incorrectly with the rod.

SUMMARY OF LASER OPERATION

Field Engineers Operating a Laser Shall	
Read the owner's manual	Read the operations manual thoroughly because it is the understanding and skill of the operator that makes a difference in the ultimate results obtained.
Utilize the proper techniques	Balance sights, if turning points are established during the operation of the laser. Unbalanced BS and FS with the laser, just like normal leveling with a level, can result in incorrect elevations. Locate solid turning points.
	The laser must be set up solidly when on a tripod or attached solidly to a column bracket. Set up tripods taking care to be certain level bubbles are centered.
	Recognize the sensor is attached and the rod must be held plumb to obtain correct results.
	Properly extend the rod.
Calibrate	Check laser calibration at the start of a project and anytime during the project when critical elevations are being established.
Plan	Lay aside the necessary equipment for the day's operations and check to make sure it is in working condition. Check the equipment again at the completion of the day's operations to see that all is complete and accounted for before leaving the jobsite.
Maintain	Utilize protective covers when rainy conditions exist. Dry lasers that get wet lightly with a clean cloth. In cold weather, or when the laser may be moist, leave it out of its case overnight to prevent moisture from penetrating the instrument.
Use good field techniques	Drive grade stakes straight and flush until they are solid. Carry marking pens and spray paint at all times and mark each lath or stake. Clean the bottom of the rod of mud, ice, and snow. Read and reread where the sensor is on the rod to be sure the correct reading is made. Check by tying into another point of known elevation. That is, remember to close the loop. Utilize care when the level is set up in traffic or in busy areas.

EQUIPMENT CALIBRATION

12

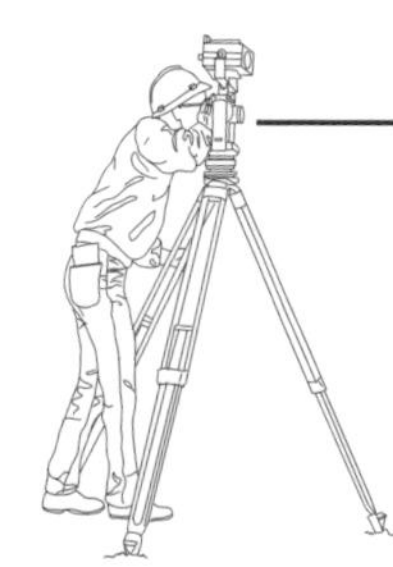

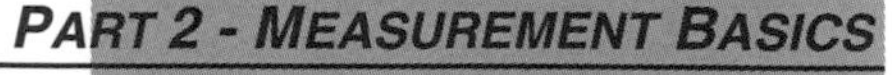

PART 2 - MEASUREMENT BASICS

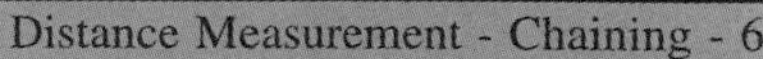

CHAPTER OBJECTIVES:

Describe the general testing requirements that should be followed in preparing to calibrate equipment.
List the equipment and personnel needs to perform chain calibration.
Describe and perform the process of calibrating a chain.
State the objective of chain calibration.
Analyze the results of chain calibration to determine if the chain is usable.
Point out on an angle turning instrument the principle geometric lines.
Illustrate the geometric relationships that should exist in transits and theodolites.
List and describe how to perform the calibration tests for an angle turning instrument.
Describe the principal lines and geometric relationships that are necessary in a typical level.
State the objective of the "Quick Peg" test and describe how it is performed.
Discuss how a single field engineer can perform the quick peg.

INDEX

FIELD CALIBRATION: GENERAL INFORMATION

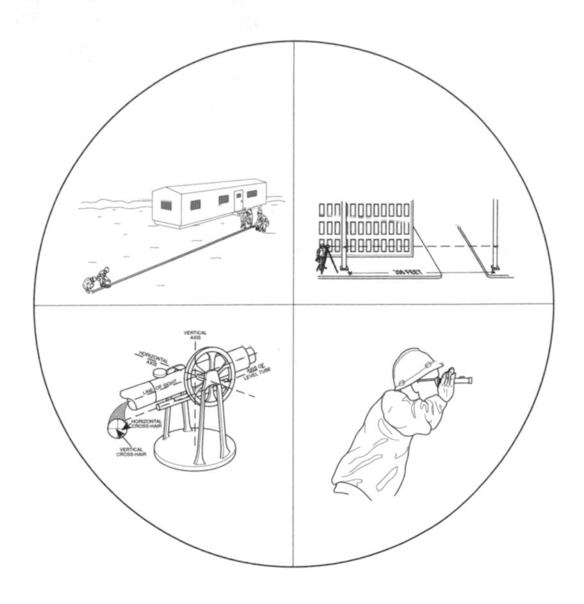

SCOPE

The work of the field engineer depends on having instruments in proper adjustment so results are accurate and precise. When measuring a distance, turning an angle, or setting grade, the field engineer must have no doubt about the reliability of the instruments being used. The field engineer should be able to perform the tests outlined in these sections on calibrating the chain, the transit, the level. The tests are simple; however, they must be performed exactly to ensure good results. If the field engineer is competent at performing these tests, he or she will be more knowledgeable of the limits of the instruments and confident that work will be put in place as specified.

It should be noted that the instrument tests discussed here are to determine if any "adjustments" are necessary. Adjustments refer to the process of bringing the various fixed parts of the instrument into proper relation with one another. Examples include actually moving the adjusting screws on the bubbles, moving the cross-hairs, or moving the horizontal axis, among other things. If, after testing the instrument, it is determined that an adjustment is indeed necessary, the field engineer should have it sent to an adjustment laboratory. At such a laboratory, the instrument would be reliably adjusted while the field engineer would be able to continue with other field engineering activities.

GENERAL

The frequency with which adjustments are required depends on many factors. They include the particular adjustment, the particular instrument, the care of the instrument, and the precision required for the job.

A good rule of thumb to follow is "**Surveying instruments should be tested frequently, but adjusted rarely**." In testing an instrument to determine whether it needs adjustment or not, make certain the instrument is tested at least three consecutive times ("triple tested"). During the triple tests, any error which occurs must be the same error each time. Testing three consecutive times with the same error occurring each time is the only way to know for certain an adjustment is necessary because of the condition of the instrument and not because of faulty testing procedures. Any deviations in the errors indicate faulty testing.

Additionally, in the past, many adjustments were made in the field; and quite often, those adjustments were improper. Today, however, it is encouraged that surveying instruments be sent to an adjustment laboratory rather than taking the chance of making improper adjustments.

INSTRUMENT TEST BASIC PRINCIPLES

Tests can be performed just about anywhere, anytime. However, there are times and places better than others.

AVOID THE HOT SUN

Avoid the hot sun as it will cause bubbles to move erratically as the instrument parts are heated on one side and then the other. Heat waves will make it difficult to sight onto the rod or point. Chains will expand when exposed to the sun. In areas where there is sunshine every day, tests should be performed early in the morning, in the shade, or whenever the sun won't affect the tests.

AVOID INCLEMENT WEATHER

Tests should not be performed in the rain or snow unless absolutely necessary.

SELECT A FLAT AREA

The terrain where the testing is to take place should be relatively flat, and the soil should be firm so solid instrument setups can be made. Avoid wet marshy areas and loose sandy soil. Select a site that has not been disturbed by recent construction.

SELECT AN OPEN AREA

The area should be free of obstacles such as bushes, fences, roads, etc., that would cause plumbing or other additional setups. A distance of approximately 200 feet is needed.

DEVELOPMENT OF A TEST AREA

If the triple testing of an instrument is to be performed often at the same location, it would be wise to take the time to develop a test area that doesn't need to be set up repeatedly. Find a level, open distance of 200 feet and place three concrete monuments in the ground. If time or money do not allow for concrete monuments, just place long wooden hubs firmly into the ground.

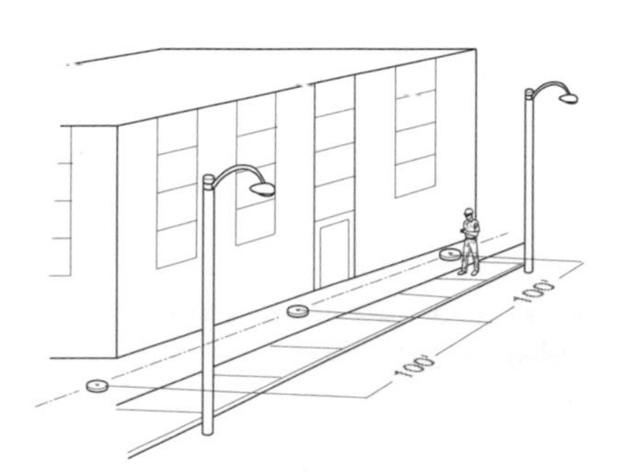

Place a monument at each end of the 200-foot area and one in the middle. A transit/theodolite should be used at one end to position the points exactly "in line" with each other. The monuments should be placed exactly 100 feet apart. Using the finest chaining techniques, a calibrated chain should be used to mark the 100- and 200-foot distances on the monuments.

Once a test area has been developed, it can be used for performing chain calibration, transit/theodolite calibration, and leveling calibration tests.

GENERAL TESTING REQUIREMENTS

To ensure that extraneous factors do not have an adverse effect on testing, the field engineer should check the following items prior to testing. Remember, attempting to perform calibration tests on equipment that has not been maintained and is in poor condition is a waste of time.

CHAINS

Do not use rusty chains. Rust on a chain is a sign of a poorly maintained tool and should not be trusted.

Chains that are kinked cannot give good results.

Chains that are broken should be thrown away and not used unless absolutely necessary.

Leather thongs at the ends of the chain should be strong and long enough to provide a secure hold.

TRIPODS

The head to leg joint should be tight. Often, poor results with an instrument can be attributed to tripods with missing bolts or joints loose because of excessive wear.

The screws which hold adjustable legs in place should prevent any slippage.

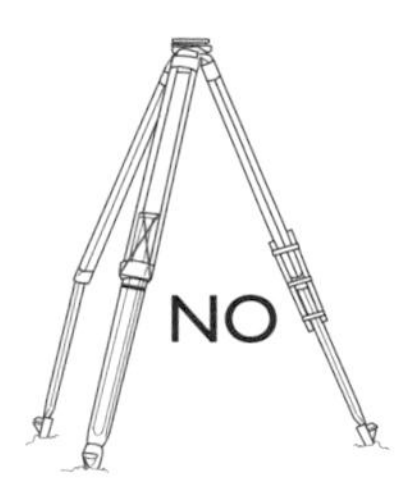

Tripod feet should be securely fastened.

Broken tripod legs should be replaced, not repaired.

Tripod legs should be waxed or oiled to provide smooth operation.

INSTRUMENTS

Parallax should be eliminated at all times.

The objective and the eyepiece lens should be kept clean.

Leveling screws (on four-screw instruments) should always be down on the leveling base.

Clamps should not be over-tightened! Tangent screw use should always end with a clockwise turn to apply pressure to the opposing spring.

Eliminating Parallax		
1	Look through the instrument at a light-colored object: a white house, a piece of paper, someone's shirt, etc. The cross hairs may be fuzzy or almost not visible.	
2	Adjust the eyepiece until the crosshairs are as dark and crisp as they can possibly be. Look from end to end of the crosshair making it the darkest and crispest at the center.	
3	Look at the level rod and move your head slightly, looking at the crosshair. If it stays on the same spot on the rod, parallax has been eliminated for now.	
4	Throughout the day, as your eyes tire, the crosshairs will become fuzzy and parallax will return. Continually repeat the procedure described above to keep parallax in check.	

CALIBRATION TESTING OF A CHAIN

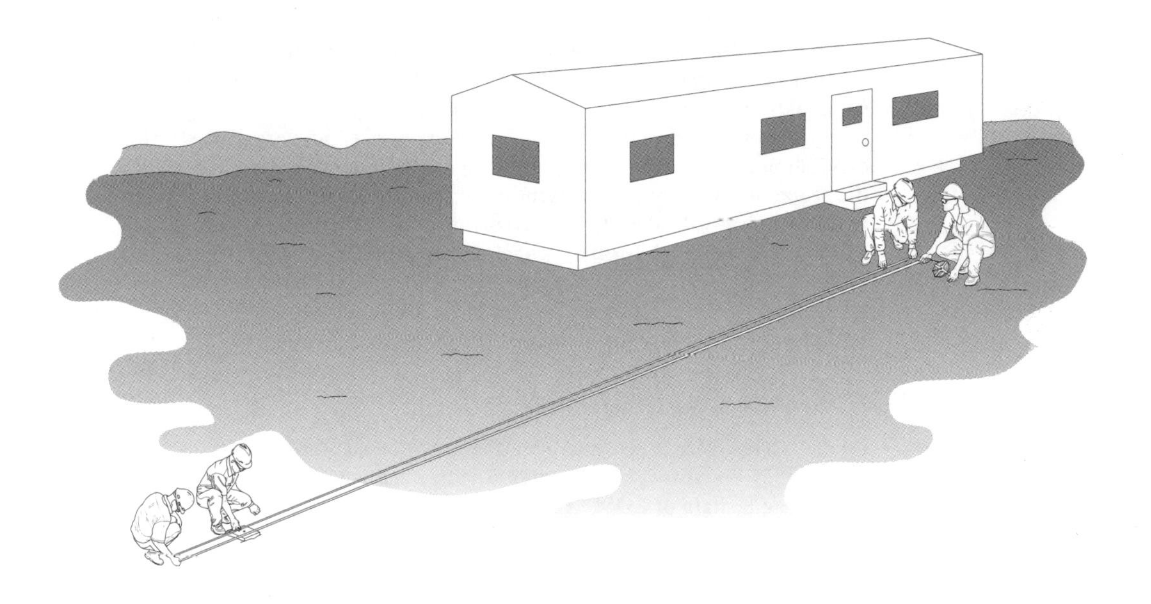

SCOPE

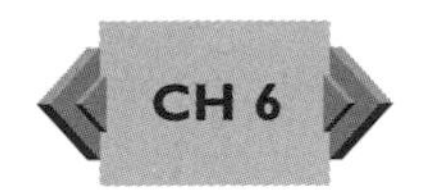

Two factors control the ability of a field engineer to achieve precision and accuracy at the same time. One factor is using refined techniques. This means following very closely the steps involved in measuring a distance with a chain. Refer to *Chapter 6, Distance Measurement - Chaining* for a step-by-step procedure for chaining. The steps should be followed exactly. The second factor is using reliable equipment. Even the best techniques will not produce good results if faulty equipment is being used.

Chain calibration absolutely must be completed before the first jobsite measurement. It should also be accomplished at regular intervals throughout the duration of the project, in order to detect any variation in the chain which would result in errors.

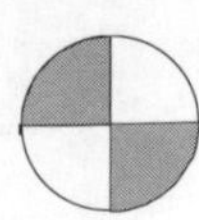

GENERAL INFORMATION

The chain used by the field engineer should be one that has been calibrated to determine the actual distance from the 0 mark to the 100 mark. Field engineers on a recent project started using a brand new chain from a major manufacturer and found it to be 0.6 feet short! They investigated and found the length between 65 and 66 was only four tenths! Needless to say they took it back and received a new one.

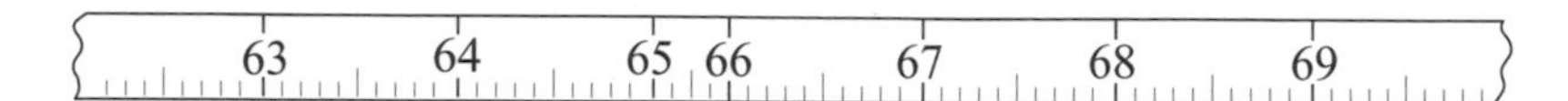

Old chains stretch with use and end up being longer than 100 feet between end marks. Broken chains usually lose length when they are spliced, leaving them too short between end marks. New, old, or spliced chains must be checked so the field engineer knows the actual length of the chain.

Chains can be calibrated by following very rigid, very precise criteria; or, calibration can be achieved by following a more practical approach. It is beyond the scope of this manual to present more than a practical approach. The method presented here will identify discrepancies of less than one hundredth of a foot, which is close enough for most construction projects. If a more exact method is needed, Surveying for Civil Engineers, by Phillip Kissam (McGraw-Hill), should be consulted for recommended precise calibration procedures. A complete reference of surveying texts can be referenced in *Appendix D*.

PLANNING FOR CHAIN CALIBRATION

Look for a standard for comparison. This may be a baseline or an invar chain. In many cities, local surveying professionals establish a baseline for all to use in calibrating a chain or EDM. Call a local surveyor or surveying supply dealer to see if a baseline is available. Invar chains are special chains made of a material that isn't affected very much by the temperature. They have been specially calibrated and are certified to be a certain distance at a specific working temperature. These may be purchased or rented from a surveying supplier. Possibly an invar chain already exists in your company. Contact the support department to see if one is available and arrange to have this calibrated chain sent to your jobsite. Call well in advance to be sure you have it when you need it.

EQUIPMENT NEEDED TO PERFORM CALIBRATION INCLUDES:

Calibrated chain or baseline

Uncalibrated chains

Tension handles

Chain knuckles

Magnifying glasses

Field book

Correction formulas

PERSONNEL NEEDED

One person holding the zero end of the chain

One person holding the 100 end of the chain

One person aligning the zero end of the chain

One person reading the 100 end of the chain

Arrange to have four people available at the time of calibration. This number is necessary to perform the test properly.

CALIBRATION PROCEDURE (COMPARISON TO A CALIBRATED CHAIN)

When a comparison is being made to a calibrated chain, environmental conditions should be controlled as closely as possible. Go indoors into a hallway or large room to avoid sunlight, temperature, wind, etc. If a large room or hallway 100-200 feet long is not available, the comparison can be made outdoors early in the morning.

OBJECTIVE:

To determine the length of the chains being used on the jobsite.

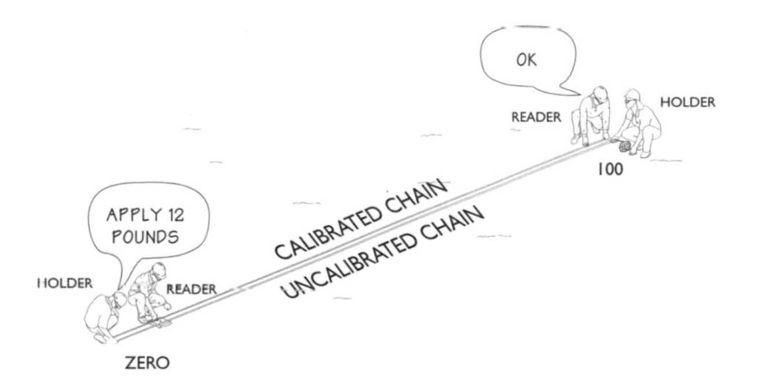

TEST:

STEP 1

Assign specific duties to the individuals assisting in the test: Two persons to hold the ends of the chains and apply the proper tension, one person to align the 0 ends, and one person to make the comparison at the 100- or 200-foot ends. Lay out the chains. Attach the tension handles to the chains. Assuming the chains will be lying on the ground fully supported, a tension of 10-12 pounds will be applied. (Check chain manufacturer for the exact tension required for the particular chain.) The persons doing the comparing should each have a magnifying glass for more precise alignment of the end marks.

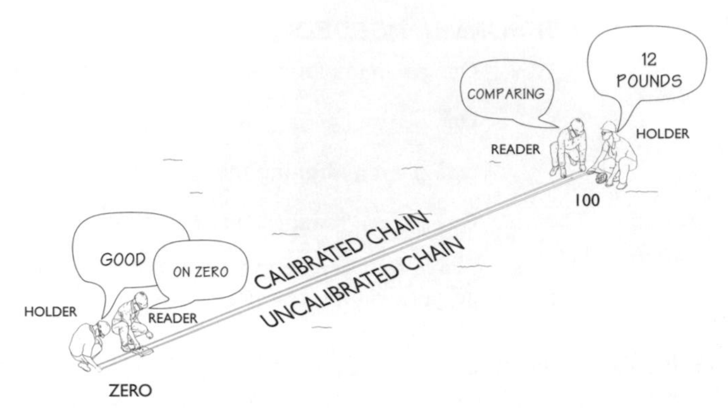

STEP 2

With everyone in position, apply tension to the chains. Have the person at the 0 end communicate to the holders to align the 0's perfectly. When that is ready, communicate to the person measuring the difference between the calibrated chain and the chain being tested. Constant communication between the participants is the key to a successful comparison of the chains.

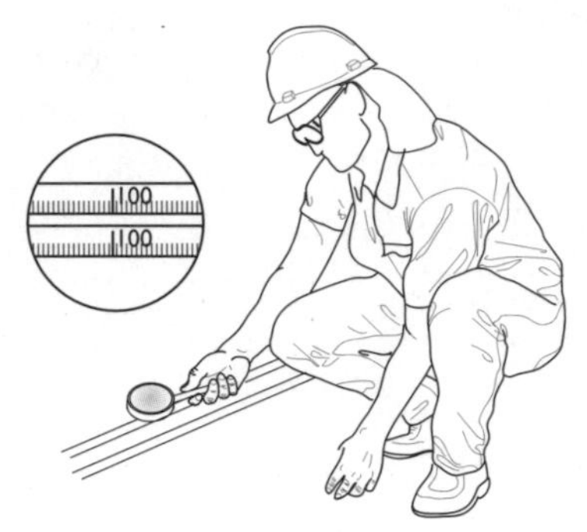

STEP 3

The person doing the comparison should measure and record the difference (estimate to the thousandth of a foot) between the ends of the chain.

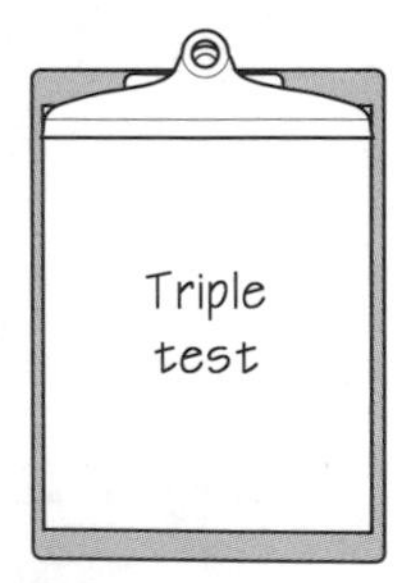

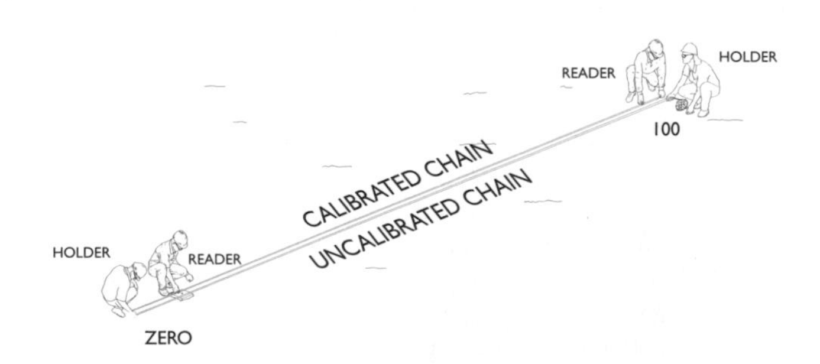

STEP 4

Repeat the test at least three times to get an average of the difference in length. Apply any correction for expansion or contraction of the chain because of temperature. Compare the nominal(what length the chain should be) to the calibrated length.

CALIBRATION PROCEDURE (COMPARISON TO A BASELINE)

The procedure for calibrating a chain when using a baseline is almost identical to comparing the chain with a calibrated chain. Simply make the comparison to the points on the ground and note any difference.

RESULTS

0.00 TO ±0.01 FROM NOMINAL

Chains found to be within 0.01 foot should be marked and used only for critical work.

±0.01 TO ±0.02 FROM NOMINAL

A chain found to be more than 0.01 foot from the actual length should be prominently marked to avoid its use for precise construction layout work.

±0.02 OR MORE FROM NOMINAL

Any chain found to be more than 0.02 foot from the actual length should be discarded.

FIELD TESTING THE TRANSIT AND THEODOLITE

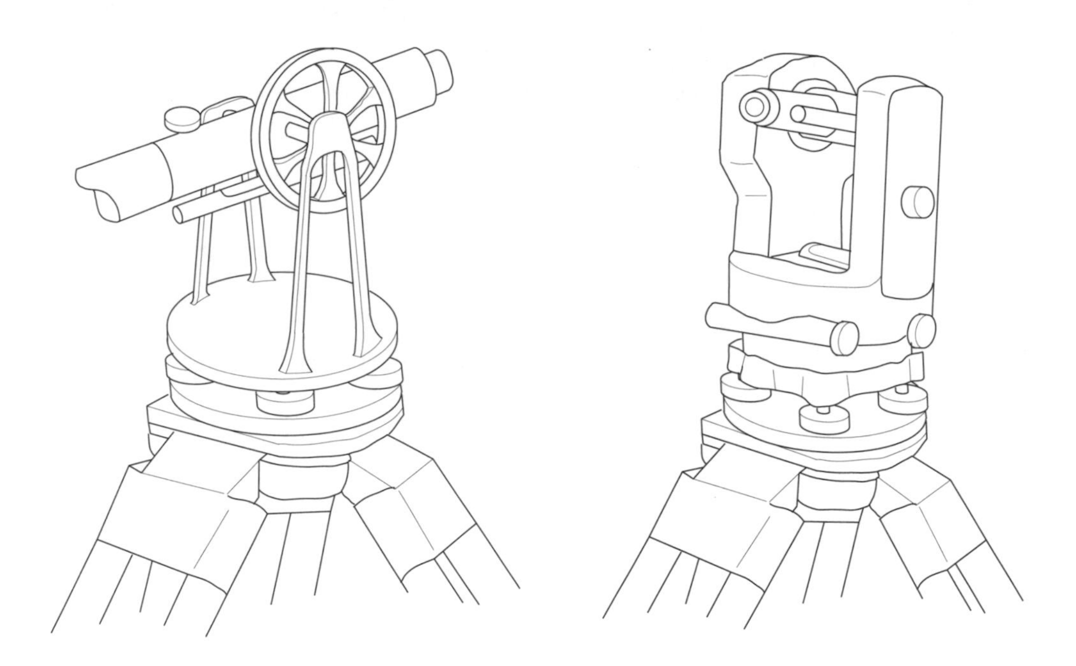

SCOPE

The importance of testing angle measuring instruments in the field cannot be overemphasized. While most errors and misadjustments can be eliminated by proper operation of the instrument (turning direct and then reverse angles for layout and measurement), it is a time-consuming process that is quite often limited to primary control measurements. Therefore, since proper techniques are sometimes neglected, it is very important that the instrument be in proper adjustment for times when only direct angles are turned.

REVIEW OF INSTRUMENT GEOMETRY

THE PRINCIPAL LINES ON A TRANSIT/THEODOLITE ARE SHOWN:

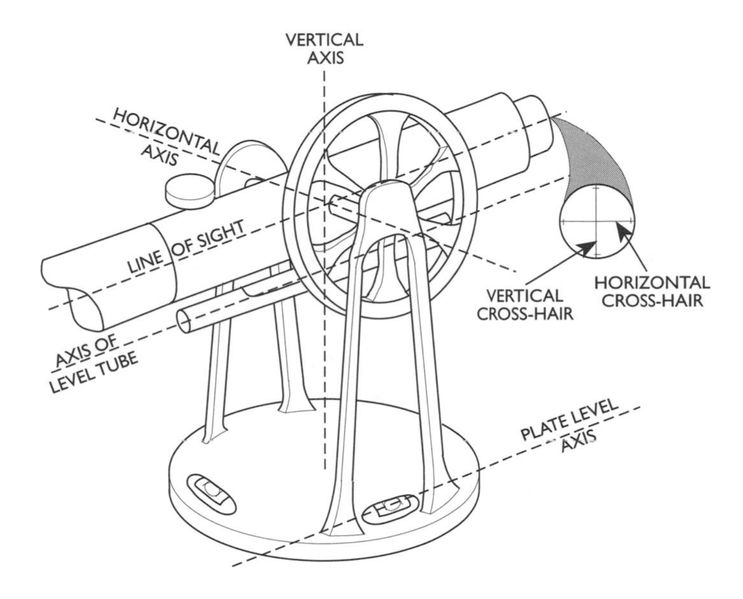

RELATIONSHIPS AND TEST OBJECTIVES

The following are geometric relationships and test objectives for the transit or theodolite.

The vertical axis should be perpendicular to the axis of the plate levels.

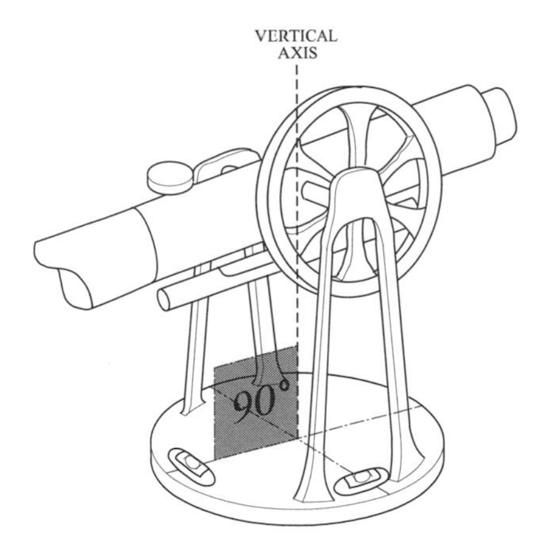

The vertical crosshair should lie in a plane perpendicular to the horizontal axis.

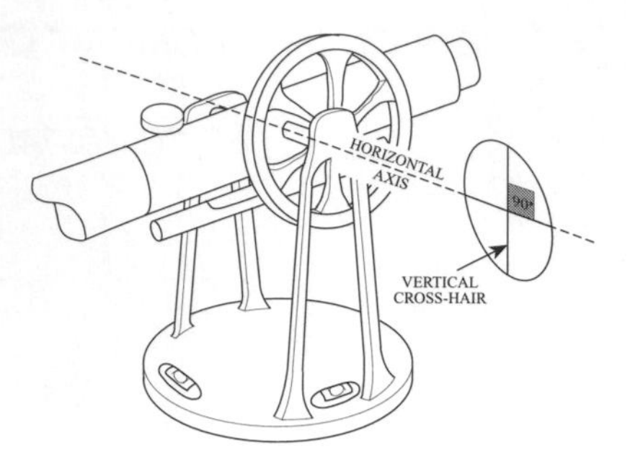

The line of sight should be perpendicular to the horizontal axis.

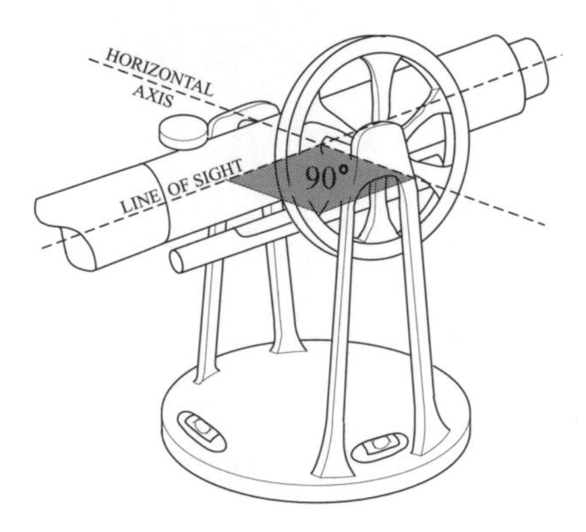

The horizontal axis should be perpendicular to the vertical axis.

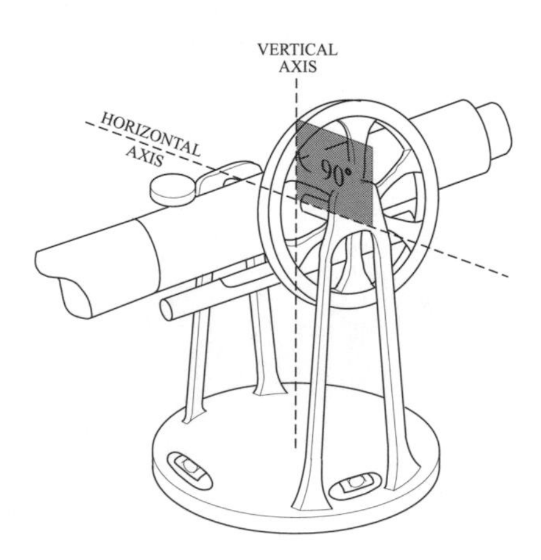

The axis of the telescope level should be parallel to the line of sight.

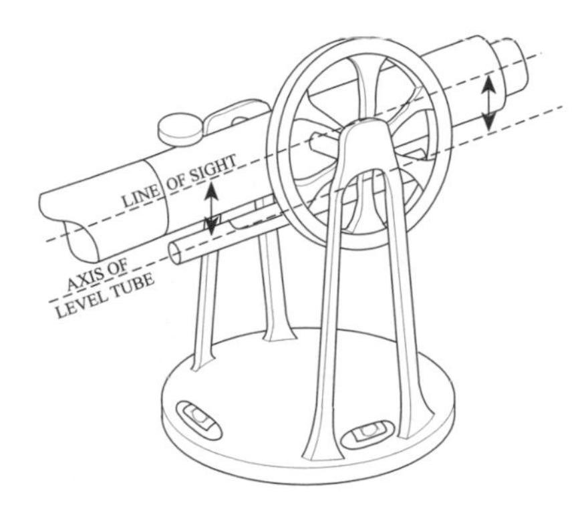

TEST ONE - THE PLATE LEVEL BUBBLE

An off-center bubble indicates the vertical axis is not truly vertical—a fundamental requirement when turning angles.

OBJECTIVE

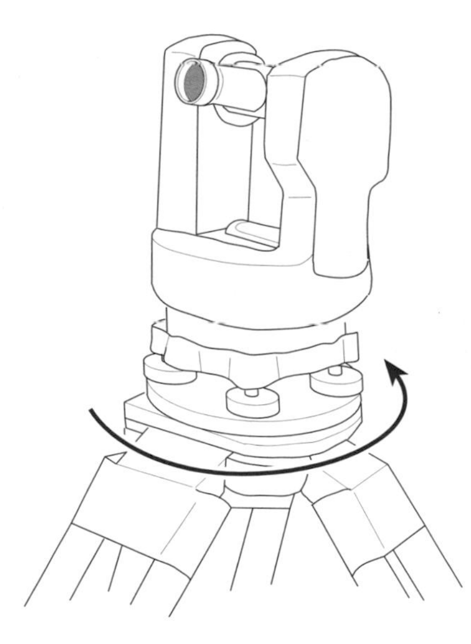

To make the plate bubbles center when the vertical axis is vertical.

TEST

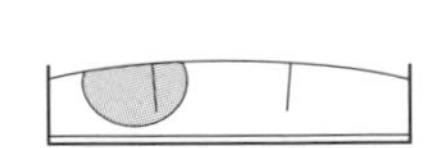

Level the instrument so the bubbles are exactly centered when over two opposite leveling screws.
Rotate the instrument 180 degrees about the vertical axis.
The bubbles should remain exactly centered. If they are, the object has been met.
If the bubble does not remain centered, perform the following adjustment.

ADJUSTMENT

Note the amount the bubble moved from center. This amount is double the total bubble error.
With the proper tool, rotate the capstan screw to move the bubble one half the amount of error towards the center.
Relevel the instrument, rotate 180°, and recheck the bubble.

CHECK ADJUSTMENT

Repeat test and adjustment steps until the bubble remains centered in all locations.

IF NOT ADJUSTING AT THIS TIME

Using the instrument without making the adjustment is possible if the observed total error is balanced. That is, bring the error half way back to center with the leveling screws. It will appear the instrument is out of level, but it is not. Split the difference!

TEST TWO - THE VERTICAL CROSSHAIR

When sighting onto a point, any part of the vertical crosshair should be able to be used. However, if the vertical crosshair isn't properly oriented to the horizontal axis, it will move off of a point as the scope is rotated above the horizontal axis.

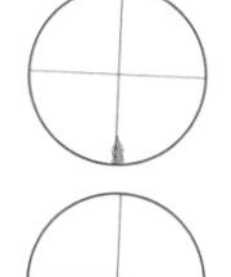

OBJECTIVE

To position the vertical crosshair so it lies in a plane perpendicular to the horizontal axis.

TEST

1. Check the level of the instrument.
2. Sight the telescope on some sharply-defined point approximately 200 feet away.
3. Clamp both horizontal motions when the vertical crosshair is on the point. Use the tangent screw to get it exactly on the point.
4. Using the vertical motion tangent screw, move the line of sight up and down.
5. The vertical crosshair should remain exactly on the point. If so, the objective has been met.
6. If the crosshair appears to move away from the point (see sketch), perform the test two more times obtaining the same results before deciding to send the instrument in for adjustment.

ADJUSTMENT

Loosen two adjacent crosshair capstan screws and rotate the cross-hair.

IF NOT ADJUSTING AT THIS TIME

The instrument can still be used if the actual adjustment is not made. Use only the part of the vertical crosshair which is close to the horizontal crosshair.

TEST THREE - THE LINE OF SIGHT

When turning direct and reverse to set a point on a hub, the line of sight after each turn should hit in the same place on the hub. If instead, two points result from direct and reverse turning, the instrument needs adjusting.

OBJECTIVE

To determine if the line of sight is perpendicular to the horizontal axis.

STEP-BY-STEP TEST

STEP 1

Set up and level the instrument over the center point, "B," of the established test area (see the general calibration information section).

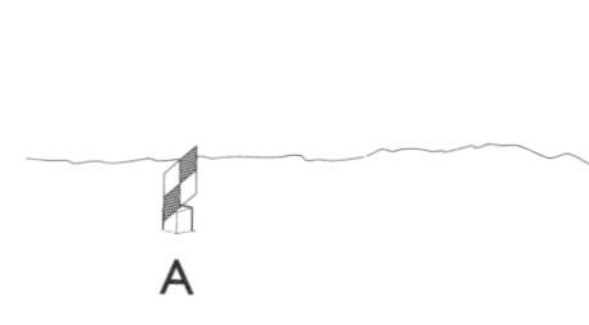

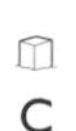

STEP 2

Sight precisely onto point "A" with both horizontal motions clamped and invert (plunge) the telescope towards point "C."

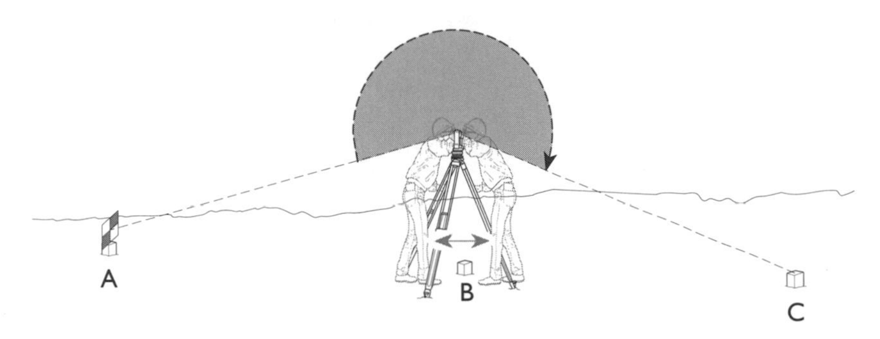

STEP 3

Set a hub at point "C" and very precisely mark the line of sight "#1" on it.

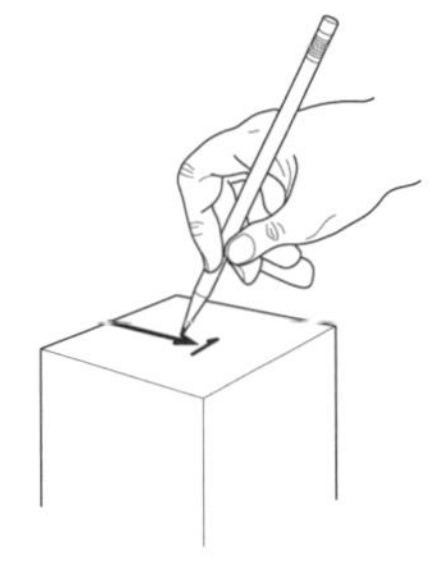

STEP 4

Release the lower motion and turn the instrument 180 ° back to sight onto point "A."

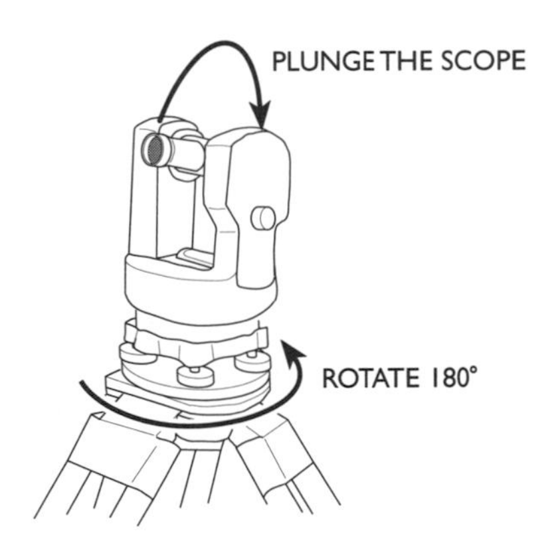

STEP 5

Invert the scope again and sight to the hub "C." If the line of sight "2" hits on the hub, precisely mark it. If the line of sight hits off of the hub, nail a short board to the top of the hub in order to make the mark.

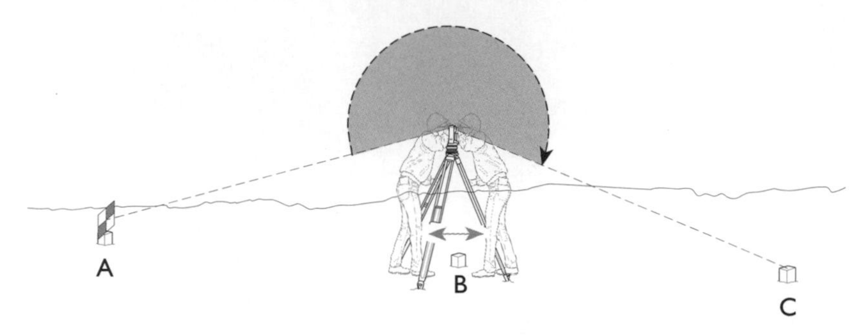

STEP 6

If both line of sight marks, "1"and "2," fall exactly in the same place, the objective has been met.

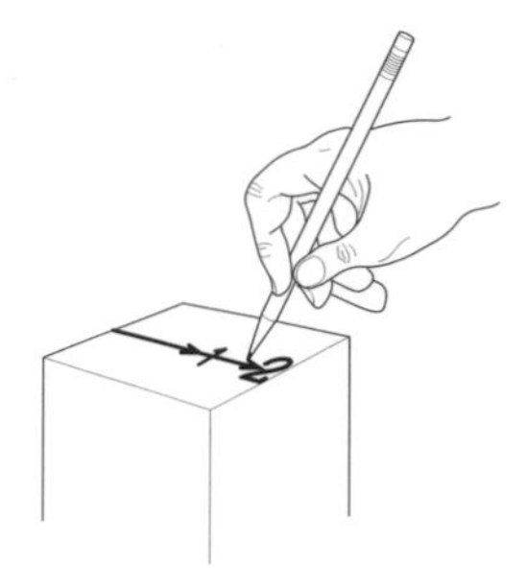

STEP 7

If two separate marks result after performing the test three times, the instrument needs to be sent in for adjustment.

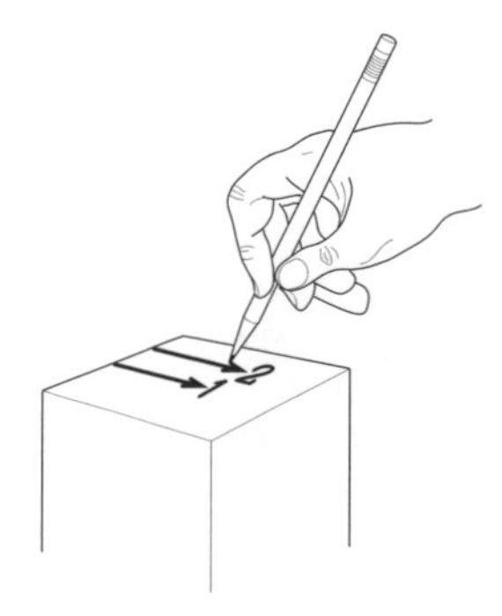

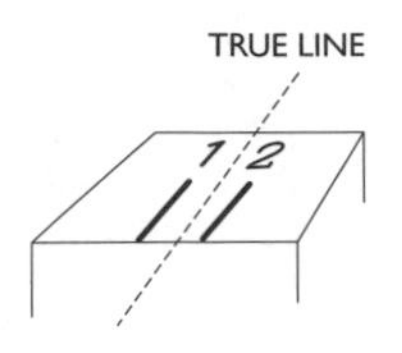

ADJUSTMENT

Alternatively loosen and tighten opposite crosshair capstan screws to move the crosshair in the direction needed. This is best executed by an equipment technician.

IF NOT ADJUSTING AT THIS TIME

The true point is always halfway between "1" and "2." Split the difference to set the actual point.

TEST FOUR - THE HORIZONTAL AXIS

This horizontal axis test is extremely important whenever a transit/theodolite is being used for plumbing high-rise structures. If the horizontal axis isn't truly horizontal, a control line projected upward will not be plumb.

OBJECTIVE

The horizontal axis should be perpendicular to the vertical axis.

TEST

STEP 1

Set up the instrument to within 200 feet of a building with a high, well-defined point, such as a church steeple. Be sure the instrument is level.

STEP 2

Sight onto the high point "A," locking both horizontal motions.

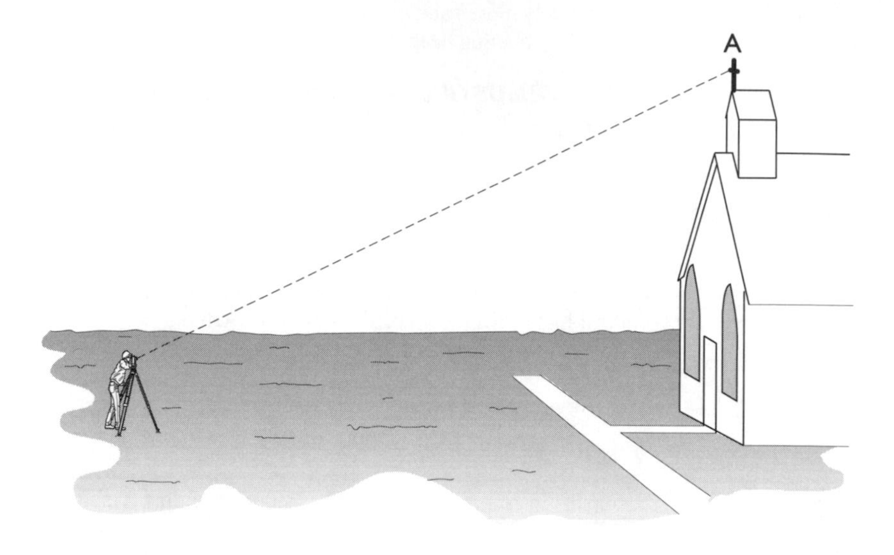

STEP 3

Lower the scope to set a hub at the base of the object sighted on. Place a hub or locate a spot on something solid where a mark can be made. With a sharp pencil, mark the line of sight "B."

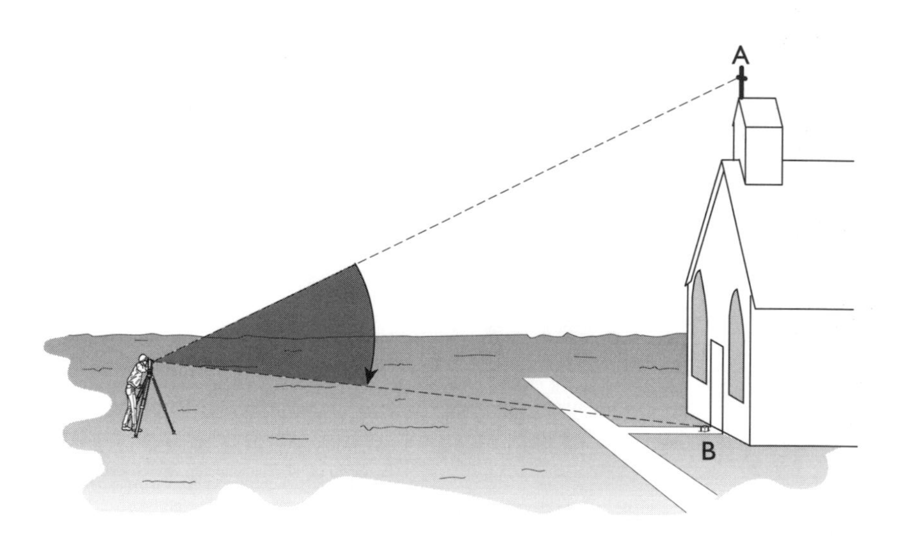

STEP 4

Invert the telescope, release the horizontal motion, rotate the instrument 180º, and sight onto the high point"A" again.

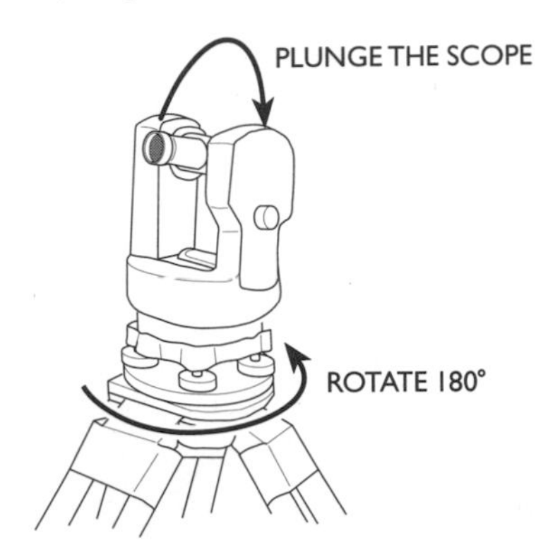

STEP 5

Lower the scope and mark the line of sight "C" on the top of the hub.

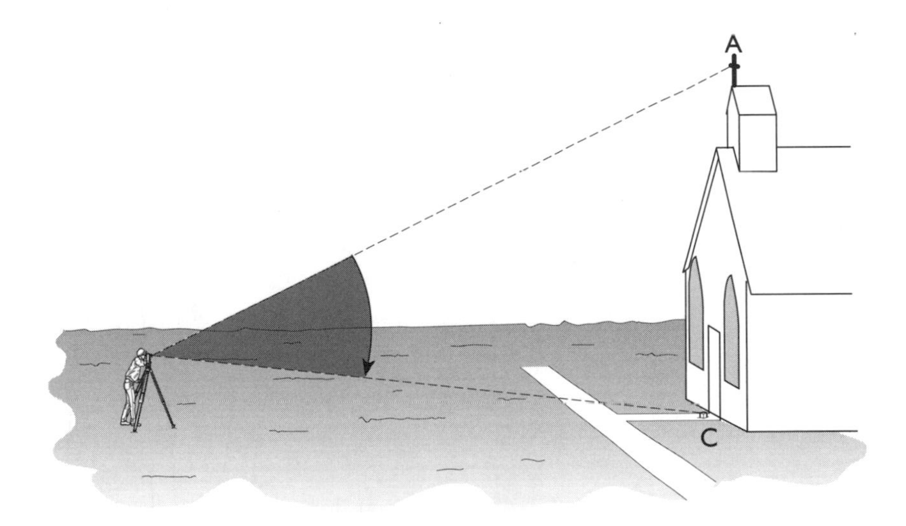

STEP 6

If the two marks "B" and "C" coincide, the objective has been met. If there is a deviation, repeat the test two more times to confirm the results. Then, if two separate marks result, the instrument is in need of adjustment.

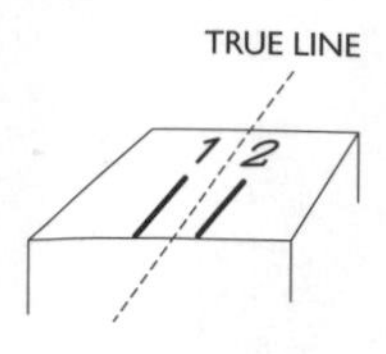

ADJUSTMENT

Locate the capstan screws that raise or lower the horizontal axis and make the necessary adjustment. This is best executed by an equipment technician.

IF NOT ADJUSTING AT THIS TIME

Again, the principle of double centering applies. If plumbing a column, give line direct and then reverse and split the difference to obtain the true line.

TEST FIVE - THE TELESCOPE BUBBLE

Some jobs have only a transit available to turn angles <u>and</u> perform leveling operations. When that is the case, the telescope bubble on the transit should be in good adjustment to give accurate results.

OBJECTIVE

The axis of the telescope level should be parallel to the line of sight.

TEST

This is called the two-peg or quick-peg test. Reference *Field Testing the Level (12-23).*

FIELD TESTING THE LEVEL

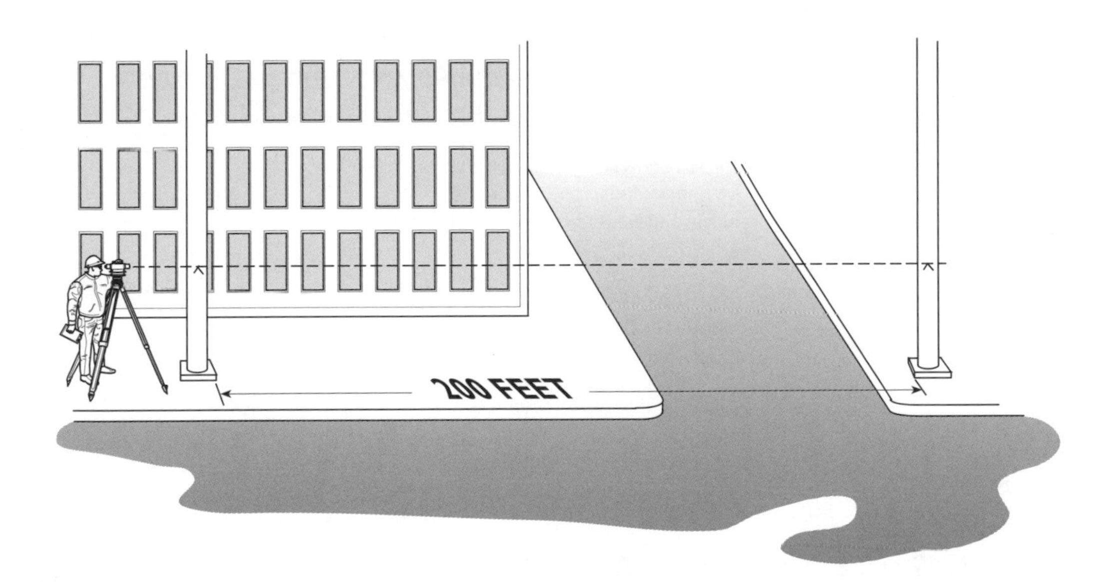

SCOPE

When setting grade, there should be no doubt in the field engineer's mind about the expected results from the level being used. A horizontal line of sight should exist so all points are established from the same reference plane.

Proper procedures must be used to ensure that any error in the instrument is negated. Proper procedures include the field engineer performing the tests outlined in this section whenever the instrument is suspected of being out of adjustment. A thorough understanding of the principle lines and the geometric relations within the level is necessary.

As a reminder, all tests should be performed at least three times (triple test). The results should be the same each time, or the test itself is not being performed properly. The same principles apply to both the Dumpy level and the Automatic level. The sections *Basic Theory of Differential Leveling, and Step-by-Step Procedure for Leveling (Chapter 9)*, should be studied before performing these tests.

REVIEW OF THE GEOMETRY OF A LEVEL

To be able to test the level, a thorough understanding of the principal geometric relationships of a typical level is necessary.

THE PRINCIPAL LINES ON A LEVEL ARE:

Vertical axis

Line of sight

Axis of the level tube (Dumpy level)

Vertical crosshair

Horizontal crosshair

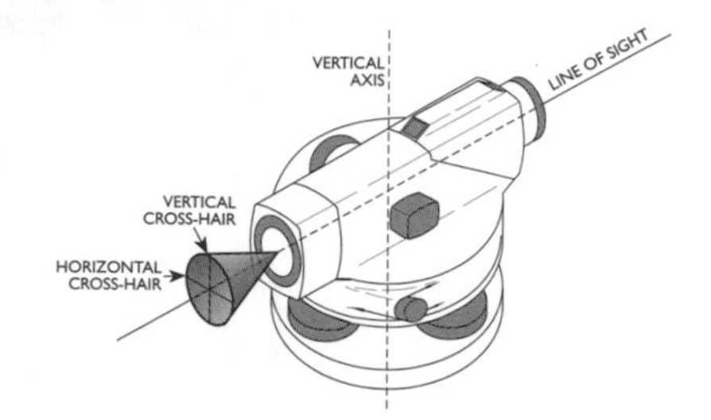

THE GEOMETRIC RELATIONSHIPS AND TEST OBJECTIVES

The vertical axis should be perpendicular to the axis of the bubble (Dumpy level).

The horizontal crosshair should be perpendicular to the vertical axis.

The line of sight should be parallel to the axis of the level tube (Dumpy level).

The vertical axis should be perpendicular to the axis of the compensator (automatic level).

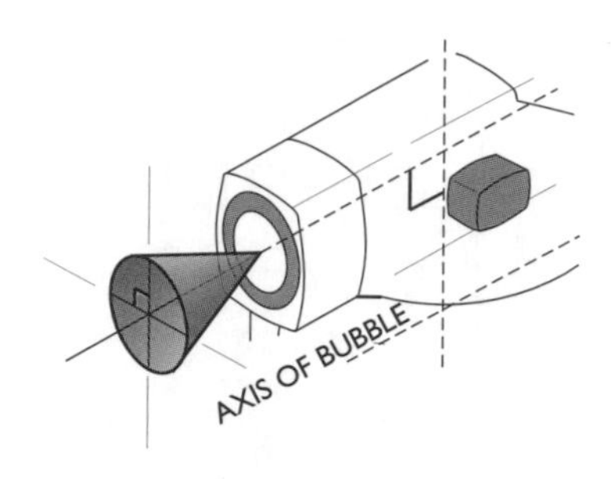

TEST ONE - TELESCOPE BUBBLE (DUMPY)

OBJECTIVE:

Make the telescope bubble center when the vertical axis is vertical.

TEST:

This is identical to test one in the section on Field Testing the Transit or Theodolite. Refer to it for procedure.

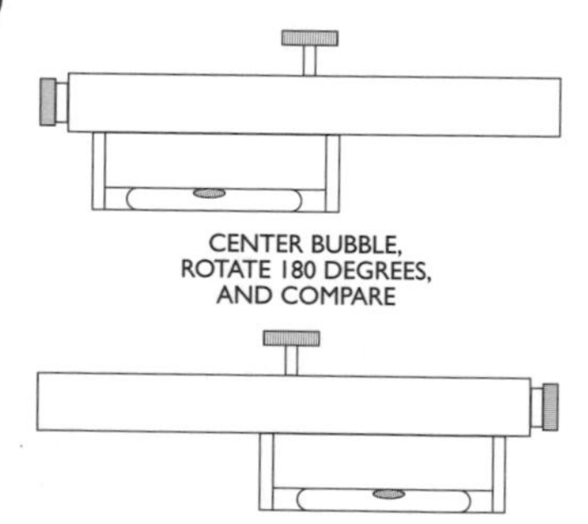

TEST TWO - HORIZONTAL CROSSHAIR

When sighting onto a level rod (story pole), usually the center part of the horizontal crosshair next to the vertical hair is used. However, any part of the horizontal crosshair should be able to be used to read the rod. If the crosshair isn't oriented properly, it may cause error in the rod reading.

OBJECTIVE:

The horizontal crosshair should lie in a plane perpendicular to the vertical axis.

TEST:

Level the instrument and sight the horizontal crosshair towards a well-defined point. A very small pencil mark on a power pole is acceptable.

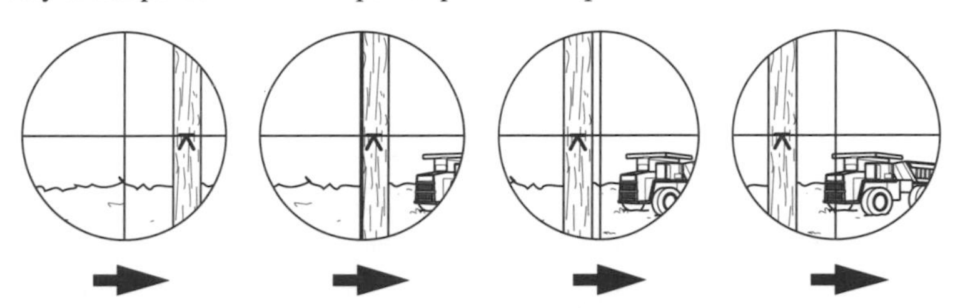

Placing the crosshair on the point, slowly rotate the instrument about its vertical axis, using the tangent screw. The horizontal crosshair should stay on the point as the instrument is rotated.

Repeat the test several times. If the crosshair moves off of the point, the crosshair reticule is in need of adjustment.

ADJUSTMENT

This adjustment cannot be made in the field on many instruments. Take the instruments to a repair facility.

IF NOT ADJUSTING AT THIS TIME

Make it a point to always use the center part of the horizontal crosshair. If this is consistently used, any reading error will be negated.

TEST THREE - LINE OF SIGHT TWO-PEG OR QUICK-PEG TEST

Testing to ensure that the line of sight is horizontal has always been called the two-peg. It requires reading a rod and adding and subtracting the rod readings to determine if the line of sight is horizontal. It works well. However . . .

An alternative method of performing the two-peg is what has appropriately been labeled the Quick-Peg. The same theory applies in setting up in the middle between two points, and then setting up close to one of the points. However, rather than taking readings and having to add and subtract them, NO recorded readings are necessary. Field engineers can even set up this test so it **can be performed by just one person** in less than five minutes!

Even when the instrument bubble is level (Dumpy) or the compensator is working (Automatic) there is the possibility the line of sight is not truly horizontal. The crosshairs may be positioned with the line of sight inclined slightly causing faulty readings.

OBJECTIVE:

The line of sight should be parallel to a horizontal line represented by the axis of the level bubble in a Dumpy Level or the axis through the compensator in an Automatic Level.

QUICK PEG PROCEDURE

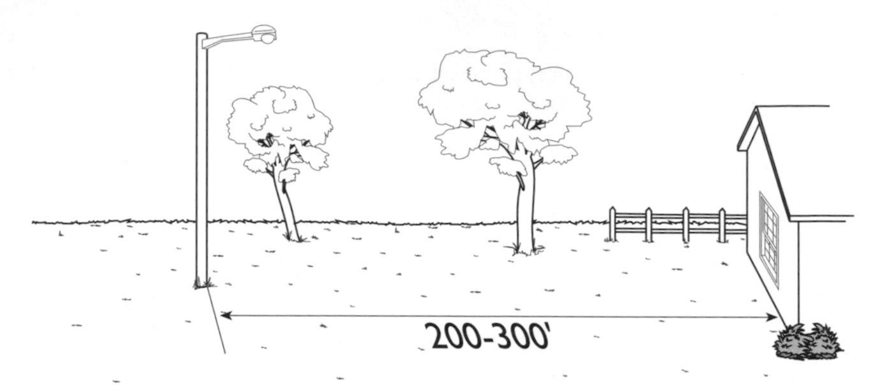

STEP 1

Find two stationary objects (utility pole, fence, building, etc.) that are 200'-300' apart, with a clear line of sight between them, in this case, the light pole and the building.

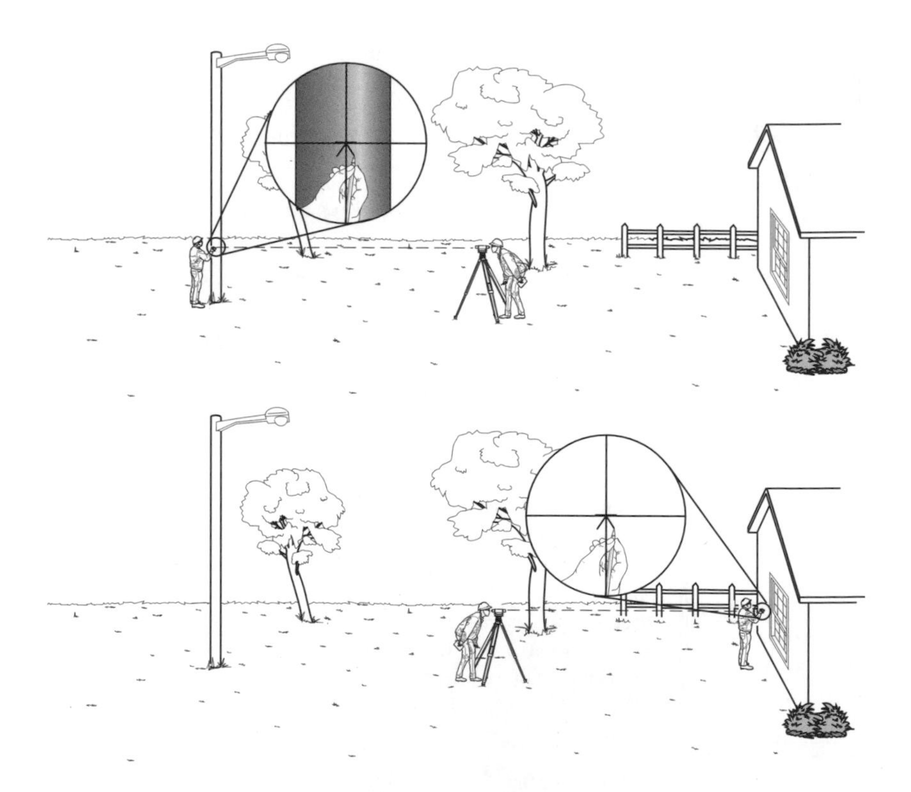

STEP 2

Set up the level at the midpoint between the objects, and mark the line of sight on both objects. The mark should be permanent for later use. Hint; this instrument setup should be a foot or two below your normal setup height.

STEP 3 Move the instrument to the proximity of one of the objects and set up as closely as possible to it. Make sure you can see both points from this setup.

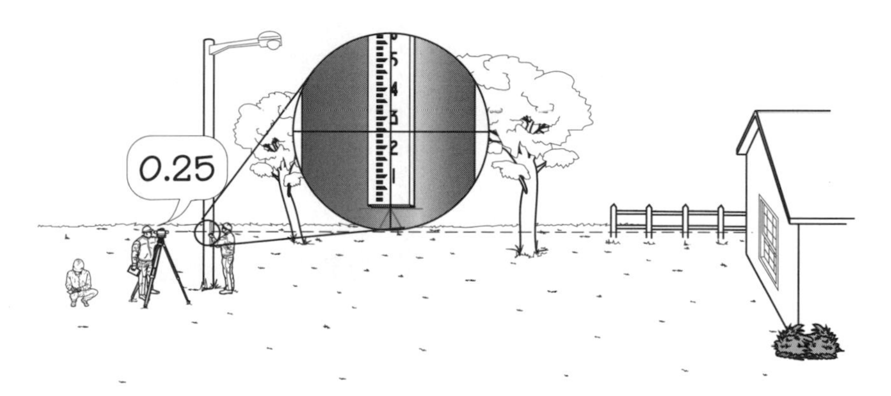

STEP 4 Have a rodperson hold an engineer's rule on the mark on the near object and read the rule.

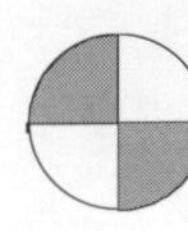

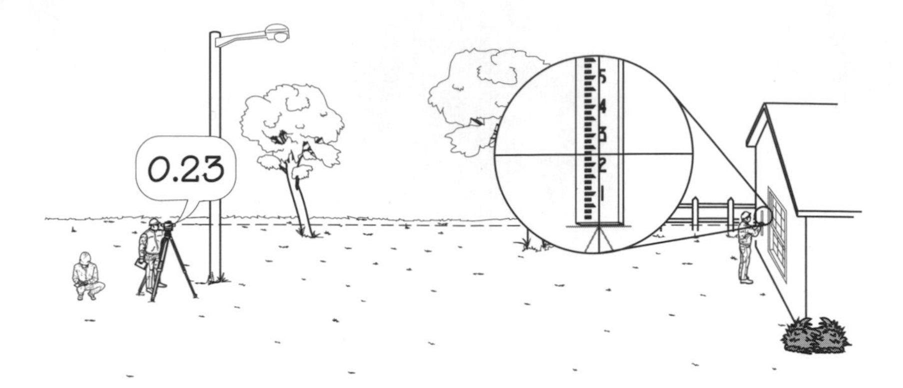

STEP 5

Have the rodperson go to the other object and read the rule there.

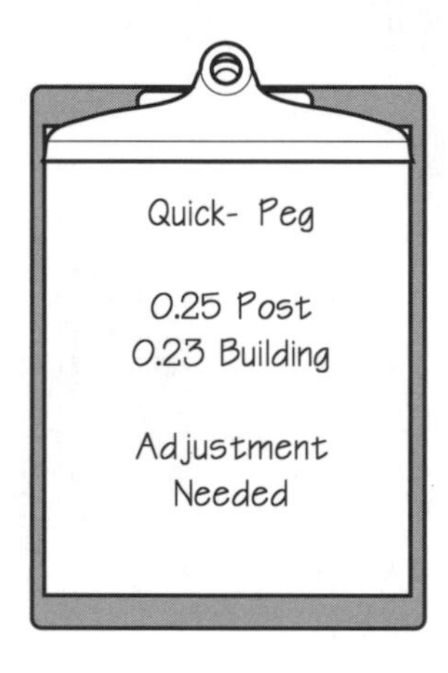

RESULTS

Compare the readings. If the two readings are the same, the instrument is in proper adjustment.

If the readings are different, the line of sight should be adjusted so both readings become the same. The crosshairs are actually adjusted. This is a simple procedure and can be performed by anyone with mechanical skill. Before making any adjustment, be sure to triple test the above procedure to confirm the results.

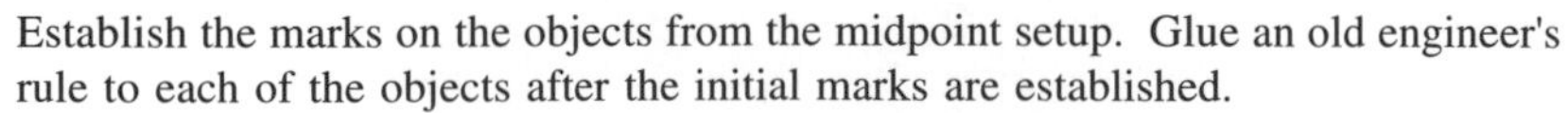

QUICK-PEG BY ONE PERSON:

Establish the marks on the objects from the midpoint setup. Glue an old engineer's rule to each of the objects after the initial marks are established.

From that time on, one person can perform the peg test by making one setup near one of the objects and reading both rules.

That individual can compare the readings and react accordingly.

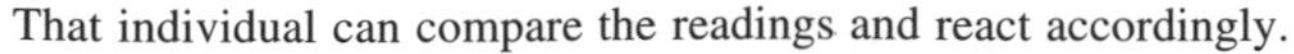

See *Chapter 25, One-Person Surveying Techniques,* for a more detailed description.

CALIBRATION OF THE BUBBLE ON A PRISM POLE

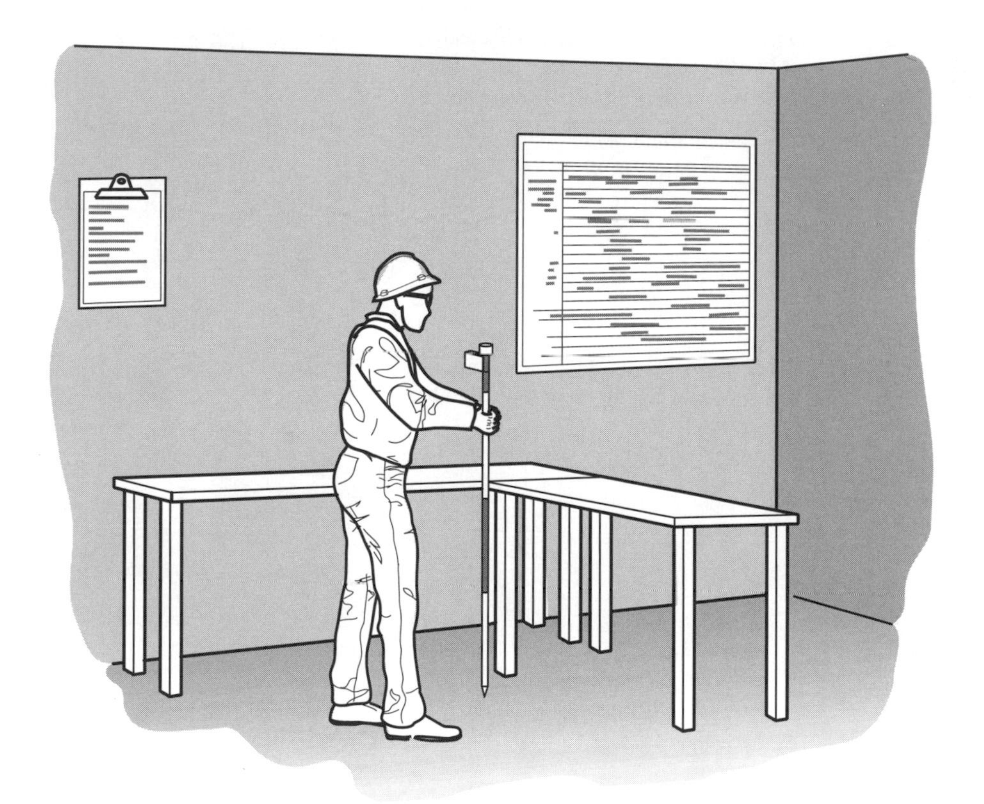

SCOPE

It was stated earlier in Chapter 7 that the prism pole is the greatest source of error in EDM measurement. Two factors cause this. The first factor is the person holding the pole may not be following proper techniques. The second factor is that the bull's-eye bubble on the prism pole is not in proper adjustment. The field engineer should check that the person holding the pole is following proper techniques. In addition, the field engineer should check the bubble frequently to ensure the prism pole is plumb when the bubble is centered. The test to check the bubble is simple and requires no special tools.

GENERAL

The objective is to first determine if the bubble is in adjustment. To determine this, the field engineer simply places the prism pole against a solid object, centers the bubble, and rotates the pole. If the bubble is in proper adjustment, the bubble will stay centered while the pole is rotated. If the bubble is not in proper adjustment, the bubble will drift out of center. Presented here is the procedure for calibrating a bubble that is out of adjustment.

STEP-BY-STEP PROCEDURE: CALIBRATING A BULL'S-EYE BUBBLE

STEP 1

This can be done anywhere there is a solid object - a common place is the intersection of two desks. Preferably the floor will be carpeted or a floormat will be available to prevent the tip of the pole from slipping.

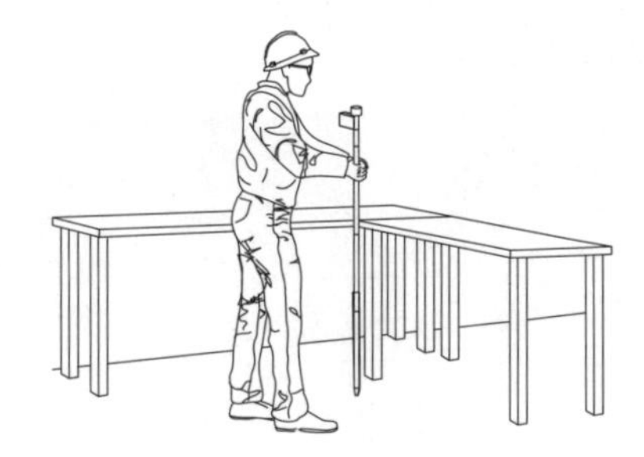

STEP 2

Place the prism pole at the intersection of the desks and position it until the bubble is centered.

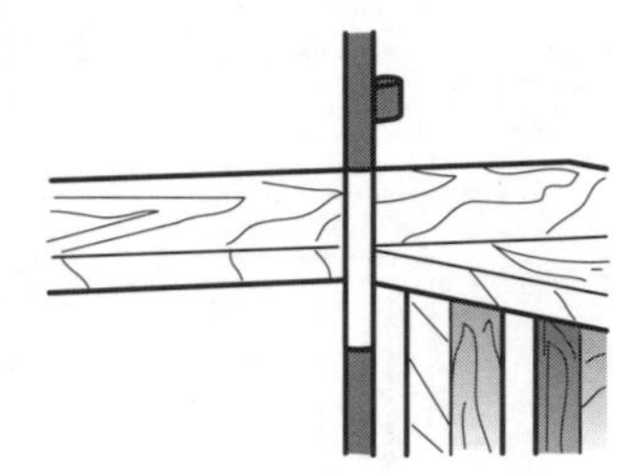

STEP 3

Very slowly rotate the prism pole, watching the position of the bubble.

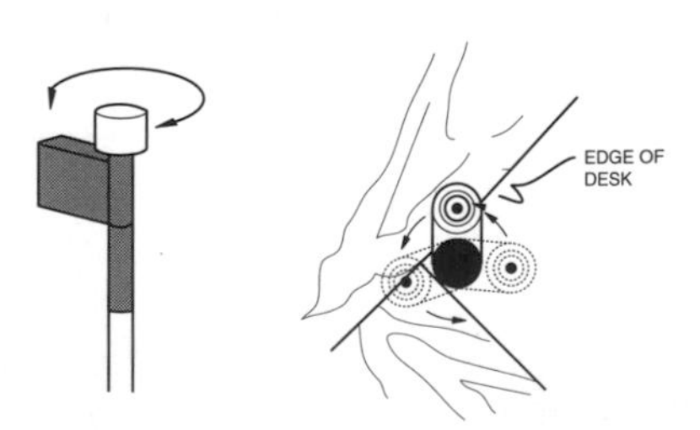

STEP 4

If the bubble is in proper adjustment, it will remain perfectly centered as the prism pole is rotated.

STEP 5

If the bubble is out of adjustment, the bubble will drift slowly off of center as the prism pole is rotated. Note where it is off center the most. That is the location where the bubble must be adjusted from.

STEP 6

Most bubbles are adjusted by using a screwdriver and turning screws on the bottom of the bubble housing. Study yours to determine how to make the adjustment. Use a trial and error approach to get the bubble in proper adjustment.

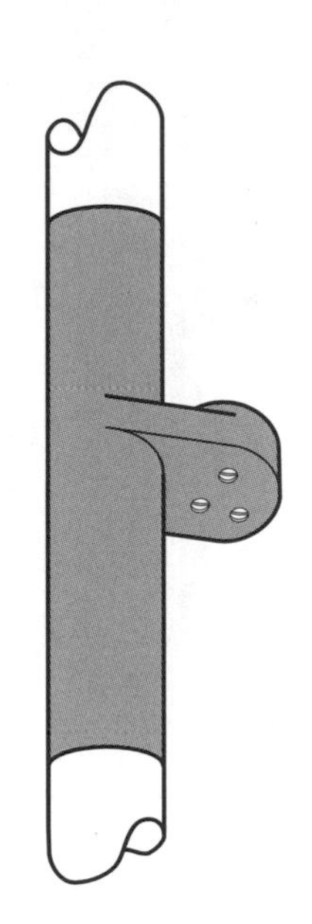

STEP 7

Repeat all steps until the bubble remains centered as the pole is rotated.

CALIBRATION OF A HAND LEVEL

SCOPE

The simple hand level can be one of the most important instruments the field engineer uses. Its uses are practically limitless in establishing rough elevations, determining plumb locations when chaining, setting slope stakes, etc. To perform these many activities within the tolerances required, the hand level must be accurate. The process involved in checking the calibration of a hand level is simple and straightforward.

STEP-BY-STEP PROCEDURE: CALIBRATION OF A HAND LEVEL

STEP 1

Find two stationary objects that are about 20 feet apart. For example a tree and a light pole.

STEP 2

Put a pencil mark on one of the objects at eye level.

STEP 3

Place the hand level exactly on the pencil mark and sight to the other object. Center the bubble in the hand level exactly and communicate where the line of sight hits the object. Mark the line of sight.

STEP 4

Go to the other object and hold the hand level exactly on the pencil mark and sight back toward the original object.

STEP 5

When the bubble is centered exactly, have a helper mark the line of sight. The line of sight should fall exactly on the original sighting mark. If it does, the hand level is in perfect calibration and no further action is needed.

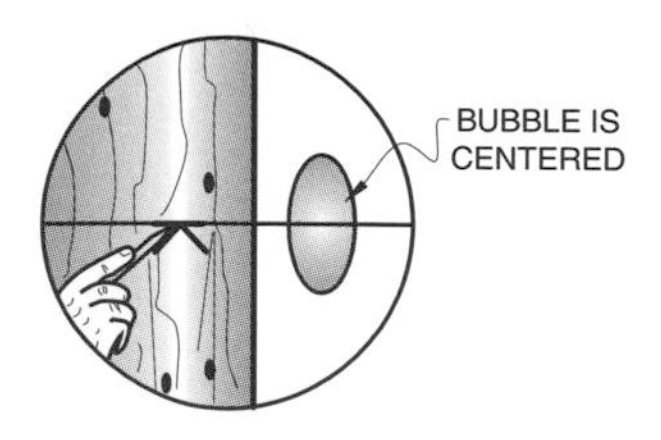

STEP 6

If it doesn't, the hand level is in need of adjustment. Study the prism or mirror mechanism that is used. Look for an adjusting screw. By trial and error make an adjustment that moves the line of sight up or down.

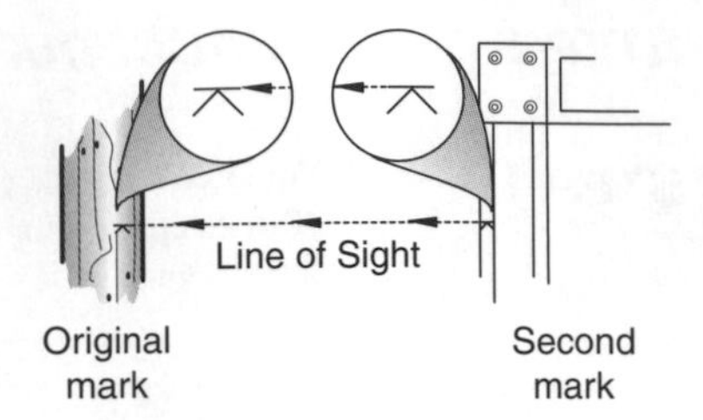

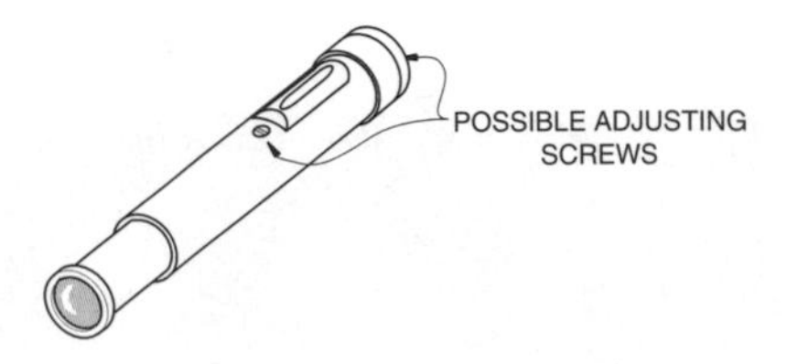

STEP 7

Repeat all steps until the hand level is in adjustment.

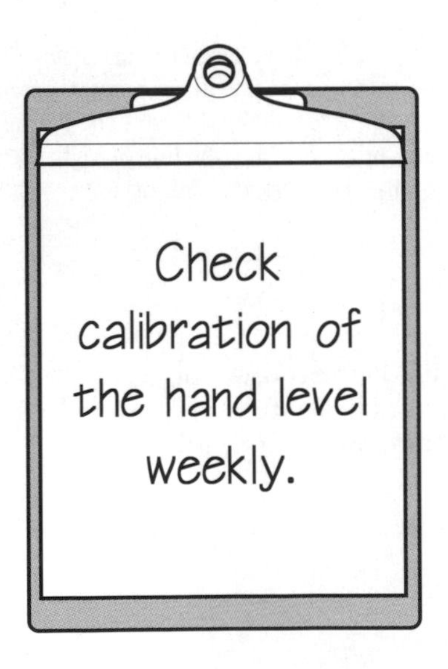

MATH REVIEW AND CONVERSIONS

13

PART 3 - CONSTRUCTION SURVEYING CALCULATIONS

CHAPTER OBJECTIVES:

Define the following geometric terms: point, line, angle, circle, polygon.
Describe the various characteristics of a polygon.
State the principal trigonometric functions based on the right triangle.
Identify the sign (positive or negative) of the sine and cosine functions in the NE, SE, SW, NW quadrants.
State and use the law of sines.
State and use the law of cosines.
Convert feet, inches, and fractions to feet, tenths, and hundredths.
Convert feet, tenths, and hundredths to feet, inches and fractions.
Convert from degrees, minutes, and seconds to degrees and decimal degrees.
Convert from degrees and decimal degrees to degrees, minutes, and seconds.

INDEX

MATH REVIEW

SCOPE

The field engineer should be very aware of the basic languages of mathematics. That is, the field engineer should be fluent in algebra, geometry, and trigonometry. Algebra supplies the language, rules, and methods of calculations for the manipulation of formulas. Geometry develops the methods of reasoning in mathematics and gives the rules and methods for plane and solid measurement. Trigonometry is the detailed study of the properties of the triangle.

Field engineers who fail to take full advantage of the power of mathematics will make more physical work for themselves by measuring when they could be calculating. A competent field engineer will understand mathematical principles and will apply them every day to calculate corrections, traverses, coordinates, curves, and volumes. This section simply lists some common formulas and relationships of geometry and trigonometry the field engineer might need.

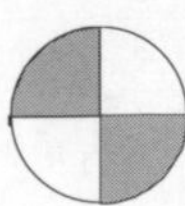

GEOMETRY

POINT

A point is a position. It has no length, width, or height.

LINE

A line connects points. The shortest distance between two points is a straight line. A line of which has no straight portion is considered curved.

ANGLE

An angle is formed by two straight lines that intersect at a point. The unit of measurement for angles in the USA is degrees. A right angle is 90°. A straight angle is 180°. An angle between 0° and 90° is acute. An angle greater than 90° is obtuse. Two angles that add together to total 90° are said to be complementary. Two angles that add together to total 180° are said to be supplementary. If two parallel lines are cut by a transverse line, the alternate interior angles are equal.

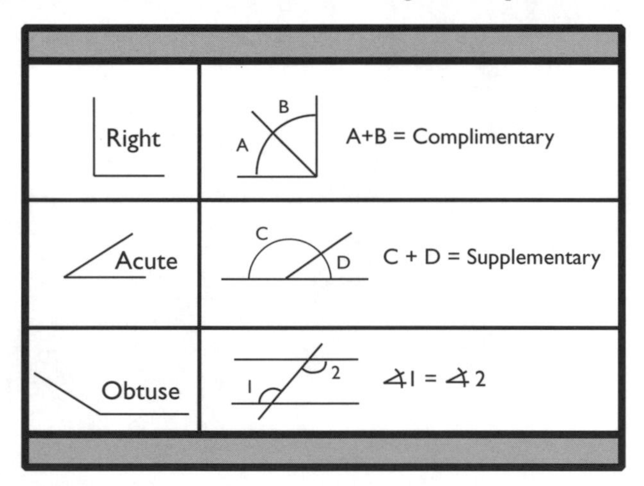

CIRCLE

Circles are formed by a curved line, every point of which is equal distance from the center. The circumference is the length of the curved line forming the circle. The radius is the straight line joining the center and any point on the circle. The diameter is a straight line between two points on the circle passing through the center. A chord is a straight line inside a circle with its ends on the circle. An arc is a portion of the circle. A tangent is a line touching the circle at only one point. A central angle is an angle formed by two radii.

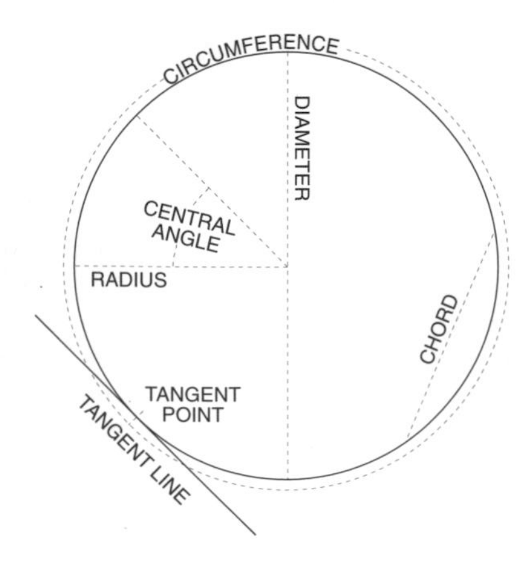

POLYGON

A triangle is a 3-side polygon. A quadrilateral is a 4-sided polygon and can include: parallelograms, rectangles, squares, trapezoids, etc.
The sum of the interior angles of any polygon is (n-2) 180° where n is the number of sides. The sum of the exterior angles of a polygon is 360°.
The square of the hypotenuse of a right triangle equals the sum of the squares of the legs (Pythagorean's Theorem).
The 3/4/5 triangle or any multiple of the sides can be used to establish perpendicular lines.

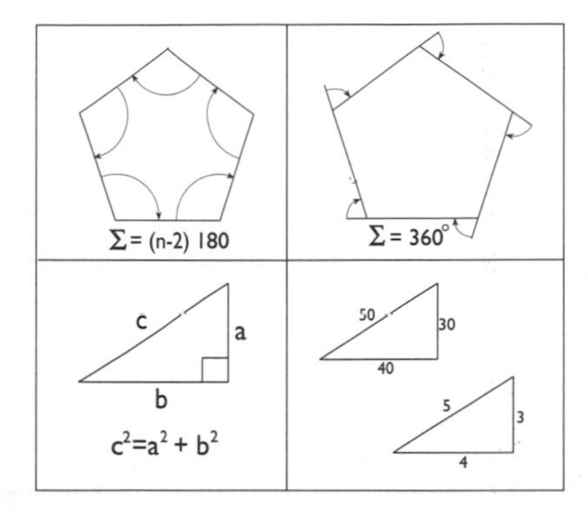

TRIGONOMETRY

RIGHT ANGLE TRIGONOMETRY

Right triangles have one angle that is 90°. The Pythagorean theorem only works if a right angle exists. The longest side of a triangle is called the hypotenuse.

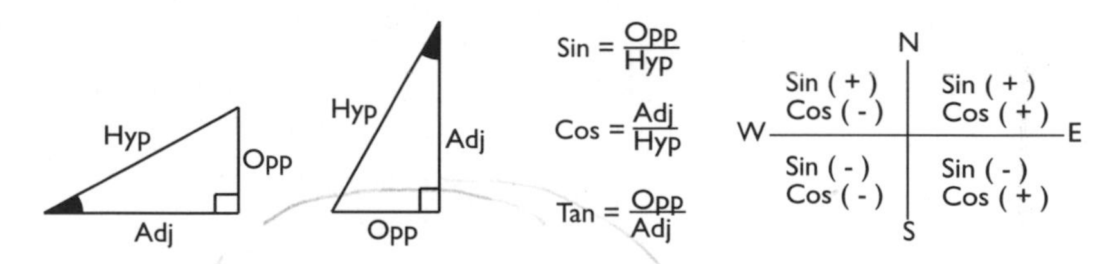

OBLIQUE ANGLE TRIGONOMETRY

No angles in an oblique triangle are equal to 90°. Therefore, none of the formulas or relationships for the right triangle apply.
The oblique triangle can be calculated if at least one distance and any two other parts are known. The other two parts can be either the other two sides, or any two of the angles, or any combination of angles and sides.
The law of sines is used if an angle and its opposite side are known in addition to one other side or one other angle.
The law of cosines is used if two sides and the included angle are known.

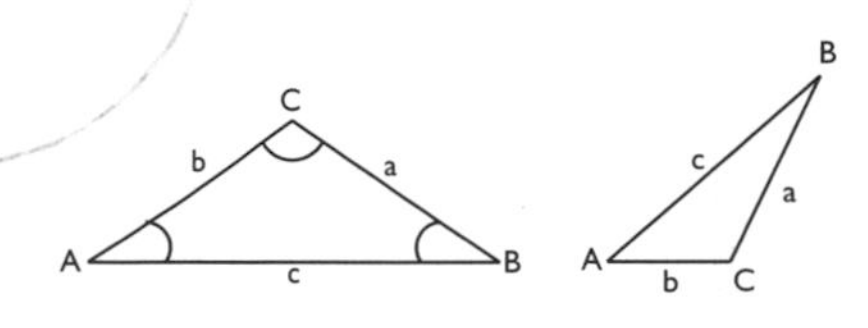

Law of Sines = $\frac{a}{\text{Sin A}} = \frac{b}{\text{Sin B}} = \frac{c}{\text{Sin C}}$

Law of Cosines = $c^2 = a^2 + b^2 - 2ab \text{ Cos}C$

FORMULAS AND RELATIONSHIPS

RIGHT TRIANGLES

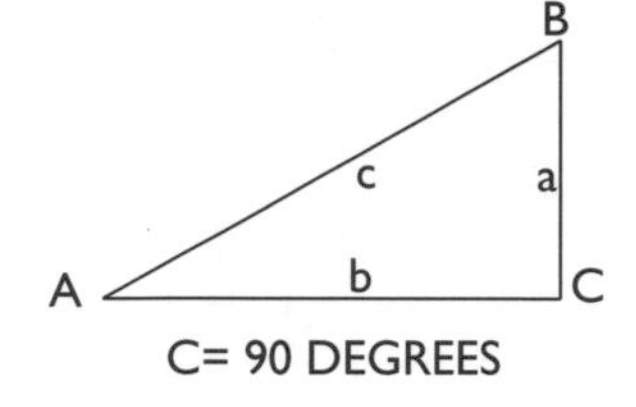

GIVEN	TO FIND	FORMULA
a,c	b	$\sqrt{c^2 - a^2}$
	A	sin A = a/c
	B	cos B = a/c
	Area	$a/2 \sqrt{c^2 - a^2}$
b,c	a	$\sqrt{c^2 - b^2}$
	A	cos A = b/c
	B	sin B = b/c
	Area	$b/2 \sqrt{c^2 - b^2}$
a,b	c	$\sqrt{a^2 + b^2}$
	A	tan A = a/b; cot A = b/a
	B	tan B = b/a; cot B = a/b
	Area	ab/2
A, a	b	a cot A
	c	a/sin A
	B	90 - A
	Area	$\frac{a^2 \cot A}{2}$
A,b	a	b tan A
	c	b/cos A
	B	90 - A
	Area	$\frac{b^2 \tan A}{2}$
A,c	a	c sin A
	b	c cos A
	B	90 - A
	Area	$\frac{c^2 (\sin A)(\cos A)}{2} = \frac{c^2 \sin 2A}{4}$

OBLIQUE TRIANGLES

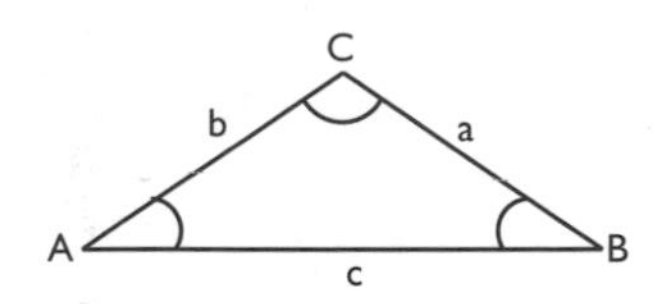

Law of Sines = $\frac{a}{\text{Sin A}} = \frac{b}{\text{Sin B}} = \frac{c}{\text{Sin C}}$

Law of Cosines = $c^2 = a^2 + b^2 - 2ab \text{ Cos} C$

GIVEN	TO FIND	FORMULAS
a,b,c	A,B, & C using "s"	Law of Cosines, $\sin 1/2 A = \sqrt{\frac{(s-b)(s-c)}{bc}}$; $\cos 1/2 A = \sqrt{\frac{s(s-a)}{bc}}$ $\sin A = \frac{2\sqrt{s(s-a)(s-b)(s-c)}}{bc}$ The value "s" = 1/2 (a+b+c) Note: For angles B&C, make appropriate substitutions in these formulas
	Area	$\sqrt{s(s-a)(s-b)(s-c)}$ The value "s" = 1/2 (a+b+c)
a,A,B	b C c	Law of Sines 180 - (A + B) Law of Sines; $\frac{a \sin (A+B)}{\sin A}$
	Area	$\frac{a^2 \sin B \sin (A + B)}{2 \sin A}$
a,b,A	B C c	Law of Sines 180 - (A + B) $\frac{a \sin (A + B)}{\sin A}$
a,b,C	c A B	Law of Cosines $\tan A = \frac{a \sin C}{b - (a \cos C)}$ 180 - (A + C)
	Area	1/2 ab sin C
ABC,a	Area	$\frac{a^2 (\sin B)(\sin C)}{2 \sin A}$

Area Computations

Triangle

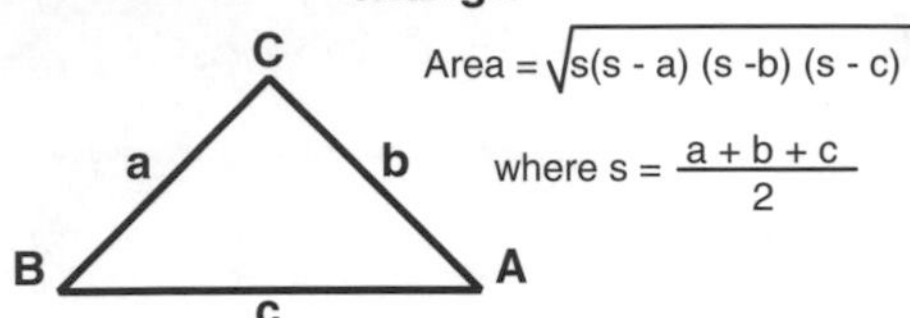

$$\text{Area} = \sqrt{s(s - a)\,(s - b)\,(s - c)}$$

where $s = \frac{a + b + c}{2}$

Circle

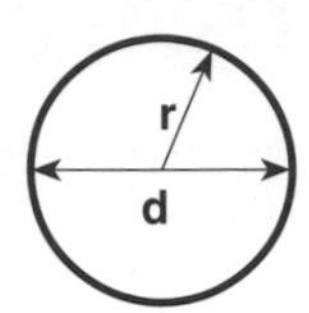

$A = \pi r^2$

$C = \pi d$

Triangle

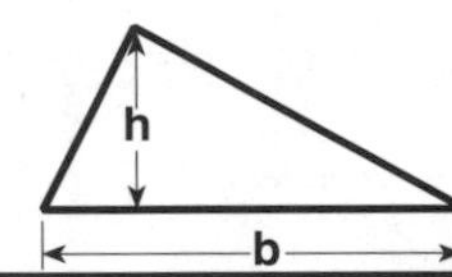

$$A = \frac{bh}{2}$$

where h = the verical height from base to apex

Circular Sector

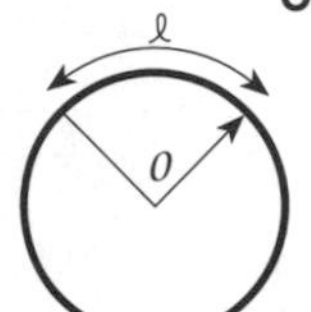

$$A = \frac{r\ell}{2}$$

$$\ell = \frac{\pi r 0}{180}$$

Trapezoid

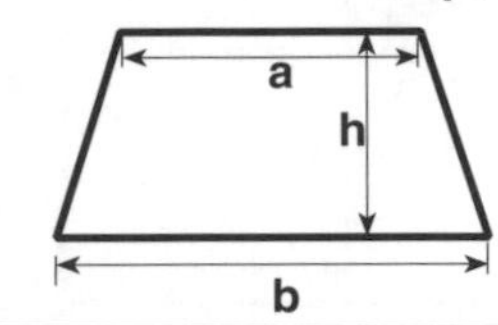

$$A = \frac{h(a + b)}{2}$$

Circular Segment

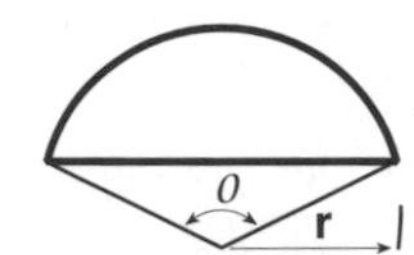

$$A = \pi r^2 \frac{0}{360} - \frac{ab}{2} \sin C$$

Trapezium

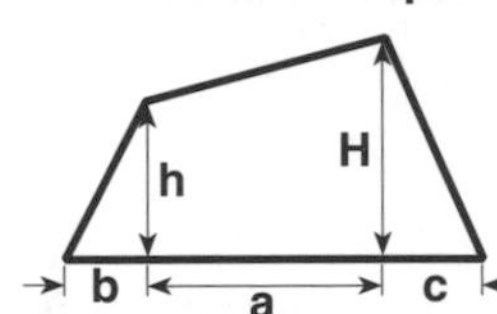

$$A = \frac{(H + h)\,a + bh + cH}{2}$$

Parallelogram

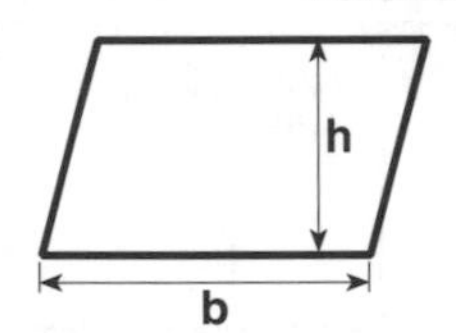

$A = hb$

A = area
π = 3.1414
s = length of side

v = volume
c = circumference

N = no. of sides
r = radius of inscribed circle

CONVERSIONS

SCOPE

Degrees, decimal degrees, feet, tenths, hundredths, inches, fractions, cubic feet, cubic yards, meters, square feet, acres, etc., are all units of measurement or volume the field engineer may deal with on a daily basis. To communicate with all of the individuals on the construction site, the field engineer may find it necessary to convert from one unit to another. Although the information here should be known, it is presented as a reminder of how to convert.

LENGTH

Some craftspersons don't work with and don't understand (or don't want to understand) feet, tenths, and hundredths and will demand that dimensions be given in feet and inches. To be congenial, it is best for the field engineer to go ahead and list dimensions the way the craftspersons wish.

FEET AND INCHES TO FEET, TENTHS, AND HUNDREDTHS

Mathematically, take the feet, inches, and fractions and convert them to decimals. That is, turn the fraction into a decimal, and add it to the inches, divide the inches by 12 to represent the inches in tenths, add that to the feet to obtain feet, tenths and hundredths.
Example: To Convert 25' 8½"

1. ½" is 0.5 inches add that to 8 to obtain 8.5
2. Divide 8.5 by 12 to obtain 0.71'
3. Add .71' to 25' to obtain 25.71'

FEET, TENTHS, AND HUNDREDTHS TO FEET AND INCHES

Mathematically, take the feet and tenths and subtract the feet, multiply the tenths by 12 to obtain the number of inches, subtract the inches and multiply the remainder by 8 to obtain the number of eighths.
Example: To Convert 16.44'

1. Subtracting 16 feet = .44'
2. Multiplying .44 by 12 = 5.28 inches
3. Subtract 5 to obtain 0.28 inches
4. Multiply that by 8 to obtain 2.24 or 2 eighths and ¼ of an inch.
5. Therefore, 16.44' coverts to 16' 5¼"

More practically, take an engineer's rule graduated in feet and tenths on one side and feet and inches on the other and simply observe the conversion by looking at one side and then the other.
Or, the field engineer can take advantage of a chart that lists the common conversions. Such a chart has been provided at the end of this section.

DEGREES

Although angles are listed on plans and are turned on instruments to degrees, minutes, and seconds, the field engineer must convert to degrees and decimal degrees to obtain trigonometric values from calculators and computers. Most calculators have a conversion key for this, but some don't. If needed, here is how to make the conversions.

DEGREES, MINUTES AND SECONDS TO DECIMAL DEGREES

Mathematically, divide the seconds by 3600, divide the minutes by 60, and add both to the degrees.
Example

41° 24' 06" = 41 + 24/60 + 06/3600 = 41.401666667°

DECIMAL DEGREES TO DEGREES, MINUTES, AND SECONDS

Mathematically, subtract the degrees, multiply the remainder by 60, subtract the minutes, multiply the remainder by 60 and subtract the minutes.
Example: To Convert 35.5674932°

1. Subtracting the degrees = .5674932°
2. Multiplying by 60 = 34.049592'
3. Subtracting the minutes = .049592'
4. Multiplying by 60 = 03"
5. Or 35° 34' 03"

VOLUME

Cubic Feet to Cubic Yards
Divide the total number of cubic feet by 27

AREA

The standard surveying unit for area is acres. There are 43,560 square feet in an acre. To convert square feet to acres, divide the total number of square feet by 43,560. To convert acres to square feet, multiply the number of acres by 43,560.

TABLES

Fractions of an Inch	Decimals of a Foot												Fractions of an Inch	Decimals of an Inch
	0"	1"	2"	3"	4"	5"	6"	7"	8"	9"	10"	11"		
0.00	.000	.083	.166	.250	.333	.416	.500	.583	.666	.750	.833	.916	0.00	.000
1/8	.010	.093	.177	.260	.343	.427	.510	.593	.677	.760	.843	.927	1/8	.125
1/4	.020	.104	.187	.270	.354	.437	.520	.604	.687	.770	.854	.937	1/4	.250
3/8	.031	.114	.197	.281	.361	.447	.531	.614	.697	.781	.864	.947	3/8	.375
1/2	.041	.125	.208	.291	.375	.458	.541	.625	.708	.791	.875	.958	1/2	.500
5/8	.052	.135	.218	.302	.385	.468	.552	.635	.718	.802	.885	.968	5/8	.625
3/4	.062	.145	.229	.312	.395	.479	.562	.645	.729	.812	.895	.979	3/4	.750
7/8	.072	.156	.239	.322	.406	.489	.572	.655	.739	.822	.906	.989	7/8	.875
1	.083	.166	.250	.333	.416	.500	.583	.666	.750	.833	.916	1.000	1	1.000

Conversions
Length
1 ft = 0.03048 m exactly (U.S. standard foot)
1 in. = 2.54 cm = 25.4 mm
1 m = 10 decimeters = 100cm = 1000 mm
1 mi = 5280 ft = 1609 m = 1.609 km
1 Km = 1000 m = 0.62137 mi.
Area
1 acre = 43560sq. ft = 4047 sq. m = 10 chains squared
1 hectare = 1000 sq. m = 2.47 acres
Volume
1 cu. m = 35.31 cu. ft
1 cu. yd = 27 cu. ft = 0.7646 cu. m
1 gal (U.S.) = 3.785 litres
1 cu. ft = 7.481 fal. (U.S.) = 28.32 litres
Angles
1 revolution = 360°
1 degree = 60 minutes
1 minute = 60 seconds
1 revolution = 400 grad, also known as grade and as gon
1 right angle = 100.0000 grad
1 revolution = 2 pi radians
1 radian = 57.29578 degrees

DISTANCE CORRECTIONS

14

PART 3 - CONSTRUCTION SURVEYING CALCULATIONS

CHAPTER OBJECTIVES:

Describe why corrections to distances are a necessary activity of the field engineer.
Define the following terms: Chain length, nominal length, too long, too short, recording less, recording more, measuring less, measuring more.
Distinguish between measuring a distance and laying out.
Describe the "per chain length" method of calculating and applying corrections.
Describe how corrections for chain length, temperature, tension, chain not level, alignment, and sag are applied when measuring a distance.
Describe how corrections for chain length, temperature, tension, chain not level, alignment, and sag are applied when laying out a distance.
Calculate and apply corrections for chain length.
Calculate and apply corrections for slope.
Calculate and apply corrections for temperature.
Calculate and combine the corrections for chain length, slope, and temperature and apply them simultaneously to measured distances.
Calculate and combine the corrections for chain length, slope, and temperature and apply them simultaneously to distance to be laid out.

INDEX

OVERVIEW

SCOPE

It was stated earlier that all measurements contain errors and mistakes. Mistakes are eliminated by checking and rechecking the entire measuring process until any mistakes that were made are found. Errors are eliminated by mathematical formula or are reduced in size by refining the measuring techniques. This section deals with the corrections a field engineer should make to critical distances that have been chained.

BACKGROUND

To understand corrections and how to apply them, some basic definitions are necessary. Study them carefully.

CHAIN LENGTH

Chain length is used to express the distance measured in hundreds of feet. For example, when using a 100' chain, 665.32 would be expressed as 6.6532 chain lengths. In addition, if a 50' chain were being used, 665.32 would be expressed as 13.3064 chain lengths. This is used in some of the formulas developed for distance corrections.

NOMINAL LENGTH

Nominal length describes the length the chain should be. That is, a 100-foot chain has a nominal length of 100 even if errors exist and the chain is actually 100.02.

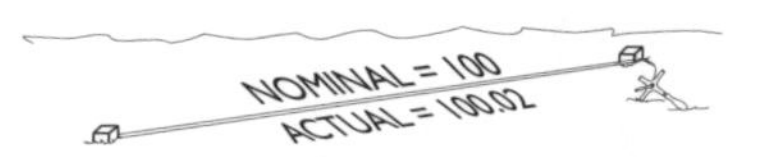

MEASURING CONDITIONS

MEASURING A DISTANCE

When two points are already existing in the ground, such as two control monuments, the field engineer will occasionally check the distance between them. This is referred to as "measuring a distance."

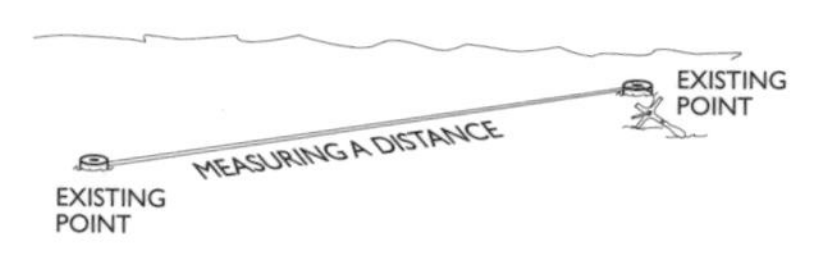

LAYING OUT

When laying out a building, points are going to be set at a prescribed distance taken from the plans or from calculations. This is referred to as "laying out."

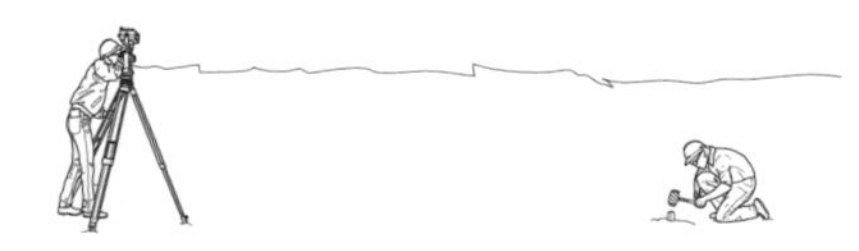

CHAIN CONDITIONS

TOO SHORT

Too short describes the length condition of the chain when a measurement is made. For example, a 100-foot nominal length chain found to be 99.98 feet when calibrated is used to measure a distance. The length of this chain is less than its nominal length. Therefore, the chain is too short.

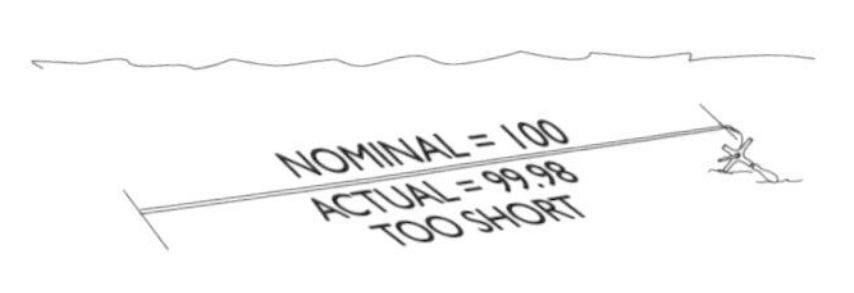

TOO LONG

Too long is just the opposite of too short. It describes the length condition of the chain when a measurement is made. For example, a 100-foot nominal length chain found to be 100.01 feet when calibrated is used to measure a distance. The length of this chain is greater than its nominal length. Therefore, the chain is too long.

CONDITION EFFECTS

RECORDING LESS

When a distance between two points is measured with a chain that is too long, the distance that is recorded will be smaller or less than the actual length of the line.

MEASURING LESS

When a distance is to be laid out with a chain that is too short, the point will be set closer than it should be.

RECORDING MORE

When a distance between two points is measured with a chain that is too short, the distance that is recorded will be larger or more that the actual length of the line.

MEASURING MORE

When a distance is laid out with a chain that is too long, the point will be set farther than it should be.

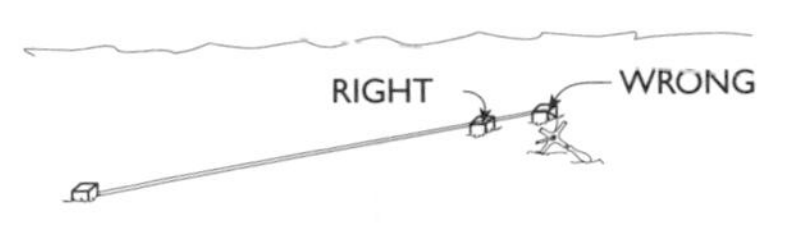

CORRECTION CALCULATION PROCEDURE

There are two distinct methods used to calculate and apply corrections to chained distances. One method applies the correction to the total distance immediately, while the other method applies the correction to chain lengths and then calculates the total correction. The method used depends on the personal preference of the field engineer.

The easiest method to understand is when calculations are corrected in chain lengths. It takes an extra step, but the entire chain correction process is broken down in smaller more understandable parts. This is especially crucial when applying the corrections.

APPLICATION OF CORRECTIONS

Calculation of the amount of correction is easy. Generally, the field engineer uses a formula to obtain the amount of correction that must be applied per chain length. Actually applying the correction is the most difficult part of distance corrections.

> "When the chain is too long, is the correction added or subtracted?"
>
> "When the temperature is cold, is the correction added or subtracted?"
>
> "When measuring on a slope, is the correction added or subtracted?"

These are questions the field engineer must answer in order to be able to apply the corrections effectively. The answer to these questions is a question. Was the field engineer measuring a distance between established points or laying out a prescribed distance from plans?

The following tables are presented to answer this question. Study them carefully because sometimes it seems the correction is applied opposite of what common sense tells you to do.

MEASURING A DISTANCE

Measuring a Distance

Type of Error	Condition	Effect on Distance Measured	Application of Calculated Correction to Recorded Distance
Chain length	Too long	Record less	Add
	Too short	Record more	Subtract
Temperature	Hot	Record less	Add
	Cold	Record more	Subtract
Tension	Less	Record more	Subtract
	More	Record less	Add
Tape not level	Too short	Record more	Subtract
Alignment	Too short	Record more	Subtract
Sag	Too short	Record more	Subtract

LAYING OUT A PRESCRIBED DISTANCE

Laying Out a Prescribed Distance			
Type of Error	**Condition**	**Effect on Laying Out**	**How to Apply Calculated Correction**
Chain length	Too long	Measure more	Subtract
	Too short	Measure less	Add
Temperature	Hot	Measure more	Subtract
	Cold	Measure less	Add
Tension	Less	Measure less	Add
	More	Measure more	Subtract
Tape not level	Too short	Measure less	Add
Alignment	Too short	Measure less	Add
Sag	Too short	Measure less	Add

CORRECTION FOR TENSION AND SAG

The correction for sag and tension can be avoided by simply applying the proper tension to the chain. If field engineers are following the proper procedures for chaining a distance, it is likely these corrections will never need to be made. Therefore, they will not be covered in this manual. Should a field engineer sometime need to perform a sag or tension correction to a chained distance, just about any text written exclusively for surveying will provide the formulas needed. A complete reference of surveying text can be referenced in *Appendix D*.

CORRECTION FOR ALIGNMENT

Alignment corrections are typically applied when chaining long distances through a wooded area when trees and other obstacles have not been removed. The amount the chain was off of line is measured, and a correction is applied to the length recorded, or the Pythagorean Theorem is used for the calculation. Correcting for alignment of the chain is not typically a problem on the construction site because it is not an acceptable procedure. If something is on line, either it is removed or the field engineer moves the entire line one way or the other to avoid the obstacle. The tolerances required for the layout of most points on the construction project are so small it is not acceptable to lay out a point that requires the chain to go around an obstacle.

However, if it would occur that the field engineer needed to set a point around an obstacle, the basic requirement is to measure how much the chain is off of line, read the distance on the chain from the zero end to the object, and measure the distance from the object to the end of the chain. Then, apply the Pythagorean Theorem to the two triangles. By calculating the remaining leg of the triangles and adding them together, the straight through distance can be determined.

CORRECTIONS FOR TEMPERATURE, LENGTH, AND SLOPE

These errors are unavoidable. They cannot be eliminated by any specific procedure, although if all chaining was done at 68° F, then no correction would be needed for the expansion or contraction of the chain. However, this isn't practical, so the field engineer should know how to perform the temperature correction as well as the corrections for length, and slope. These corrections are covered in the following sections.

CHAIN CORRECTIONS FOR LENGTH

Laying Out a Prescribed Distance

Condition of chain	Effect on laying out	How to apply calculated correction to plan distance
Too long	Measure more	Subtract
Too short	Measure less	Add

Measuring a Distance

Condition of chain	Effect on distance measured	Application of calculated correction to recorded distance
Too long	Record less	Add
Too short	Record more	Subtract

SCOPE

Correcting a chain for length is easy to apply by calculation or even while in the field measuring. It is simply the difference between the nominal length and the actual length of the chain. It is that simple--to make an addition or subtraction and have the distance. Or, if the amount of correction is known, the field engineer can apply the correction as the distance is being measured. For example, if after calibrating, a chain is found to actually be 99.99 feet long, the field engineer can simply add 0.01 each time the full 100 feet of the chain is used. If the distance is less than 100, a proportion of the 0.01 can be applied (assuming the error in the chain is uniform throughout its length). Regardless of how it is applied, the field engineer must be familiar with chain length corrections.

REASONS FOR CHAIN LENGTH CORRECTIONS

It would be expected that a chain marked with 0 on one end and 100 on the other end would be 100 feet long exactly. It isn't for a number of reasons:

Errors occur at the time of manufacture which affect the length.

The chain has kinks in it from being bent or stepped on while being used.

The chain has elongated from use and is longer than it was at the time of manufacture.

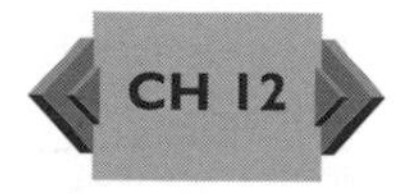

Generally, these errors are very small and may be ignored for most short distance field engineering work. However, if the work is critical, the error should be corrected for distances measured or laid out. The amount the length of a chain varies from its nominal length is determined by calibration testing. See *Chapter 12, Equipment Calibration* for a thorough explanation of this process. When this is done, the amount the chain is long or short becomes known. Sometimes a local university, county surveyor or local surveying association will have a special, calibrated chain that can be used for the comparison. When the error in a chain is known, it can be eliminated from the measurement.

GRAPHICAL SOLUTION FOR A 100' CHAIN

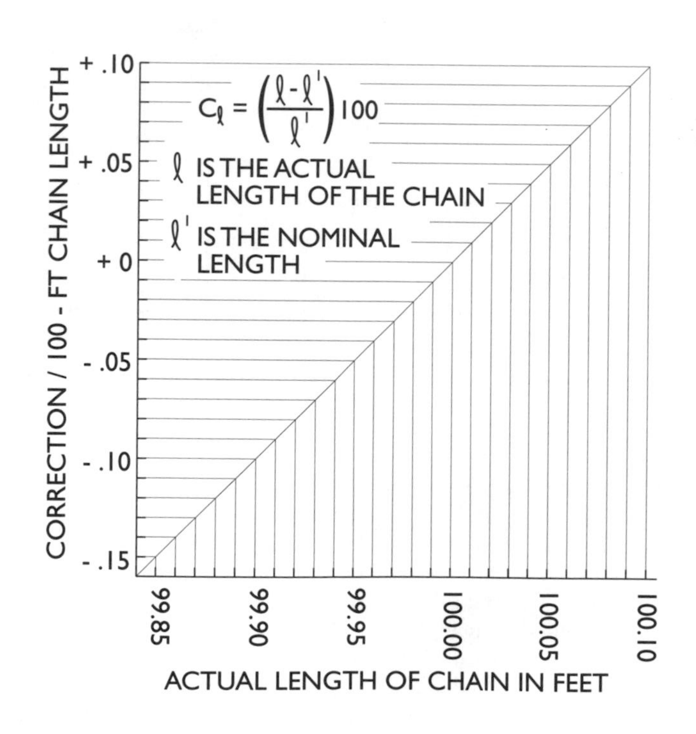

EXAMPLE OF LAYING OUT A POINT

Suppose a chain was used to lay out centerline points on a two-mile length of roadway. After the work was finished, the field engineer decided to check that chain at the local baseline. Upon checking, it was found the chain was actually 100.02 rather than 100.00. What effect will this have on the two miles of centerline stakes? How far off will the stake be at the end of the line (10,560 feet away)?

CORRECTION PER CHAIN LENGTH

The difference between nominal length and actual chain length is 0.02' per chain length. (C.L)

NUMBER OF CHAIN LENGTHS

There are 105.60 chain lengths in 10,560 feet

TOTAL CORRECTION

Multiplying

$$\frac{0.02'}{C.L.} x 105.60 C.L = 2.11'$$

APPLICATION

The stake at the end of the line is 2.11 feet from where it should have been because the chain is 0.02 per chain length longer than nominal. Refer to the *Laying out a Prescribed Distance* table on *14-9.* The end stake will have to be moved 2.11' back towards the beginning of the line. Additionally, every stake along the two miles of line will have to be moved because each of those stakes is also incorrectly placed. It would have been worthwhile to have checked the chain before performing all of that work. The centerline will have to be restaked with a more accurate chain, or each measurement with the existing chain will have to be corrected by the field engineer.

EXAMPLE OF MEASURING ESTABLISHED POINTS

A chain was used to check the distance between two control points on a large building project. The distance was measured and recorded as 2224.40 feet. After the work was finished the field engineer decided to check that chain at the local baseline. Upon checking, it was found that the chain was actually 99.99 feet rather than 100.00. What effect will this have on the recorded distance? What is the actual length of the line?

CORRECTION PER CHAIN LENGTH

The difference between nominal length and actual chain length is 0.01' per chain length.

NUMBER OF CHAIN LENGTHS

There are 22.2440 chain lengths in 2224.40 feet.

TOTAL CORRECTION

Multiplying

$$\frac{0.01'}{C.L.} x22.240 = 0.22'$$

APPLICATION

The recorded distance is off 0.22'. But which way? Is the correction added or subtracted? Refer to *14-9*, the *Measuring A Distance* table. The condition of the chain is too short. Therefore, the effect on the distance measured is to record more. Thus, the 0.22 must be subtracted to correct the recorded distance.

2224.40 - 0.22 = 2224.18 the correct distance between the control points.

CHAIN CORRECTIONS FOR SLOPE

Laying Out a Prescribed Distance on a Slope		
Condition	Effect on distance measured	Application of calculated correction to plan distance
Short	Measure less	Add

Measuring a Distance on a Slope		
Condition	Effect on laying out	How to apply calculated correction to recorded distance
Short	Record more	Subtract

SCOPE

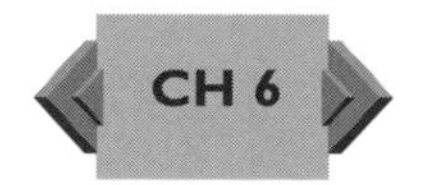

Most construction projects are not level and require the field engineer to occasionally measure distances on slopes. Breaking chain can be performed as discussed in *Chapter 6, Chaining*, or measurements can be made along the slope and then corrected to horizontal. This section deals with the various methods used to correct distances for the effect of slope.

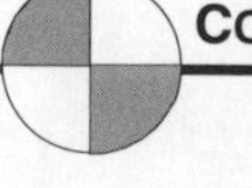

BACKGROUND

MEASURING A DISTANCE

When measurements between two established points are made on a slope, the measurement made is the equivalent of measuring the hypotenuse of a triangle. Remember that the hypotenuce is always longer than the sides, therefore a slope measurement between two existing points will be longer than the corresponding horizontal distance.

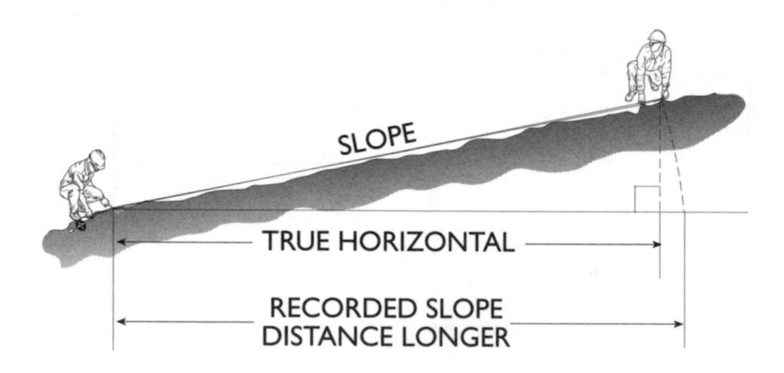

LAYING OUT A POINT

When prescribed distances are laid out on a slope, the effect will be to set points at less than the corresponding distance.

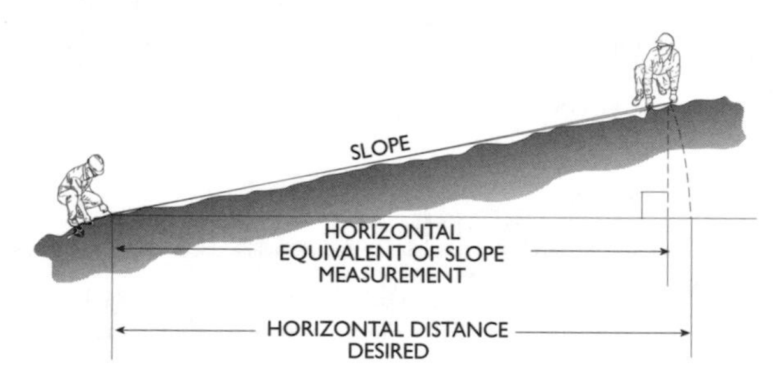

THE CORRECTION FOR SLOPE FORMULAS

There are several methods of correcting distances for the effect of slope. Each yields the same results, but one may be better than another depending on the available data.

PYTHAGOREAN THEOREM

If the slope distance and the change in elevation are known, the recorded or required distance can be determined.

$$Horizontal^2 = Slope^2 - Elevation^2$$

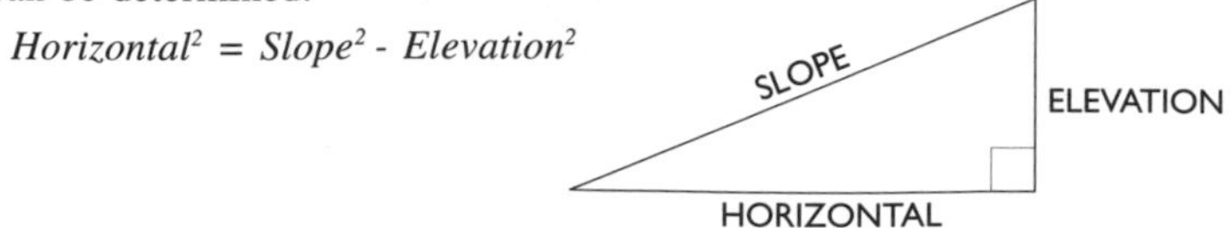

RIGHT ANGLE TRIGONOMETRY

If the slope angle is known, the cosine or sine functions can be used depending on whether the angle is an angle off of the horizon, or is a zenith angle.

For an angle off of the horizon,

Horizontal = Cosine Angle x Slope

SLOPE
COS

For an angle off of the zenith,

Horizontal = Sin Zenith Angle x Slope

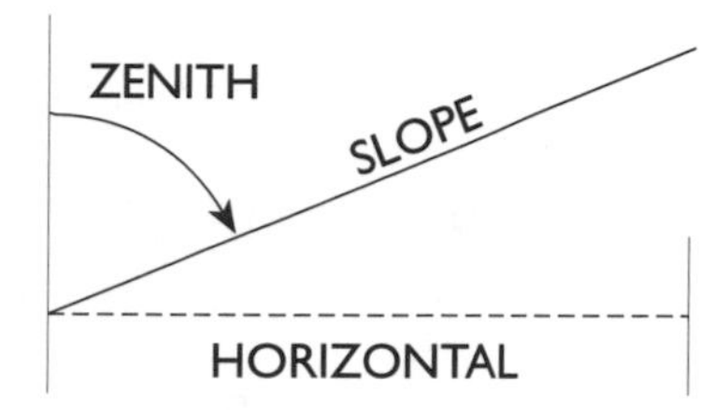

CORRECTION FORMULA

If the change in elevation is known, the correction formula can be used to determine the correction per chain length.

$$Cg = \frac{h^2}{2(L)}$$

GRAPHICAL SOLUTION FOR A 100' CHAIN

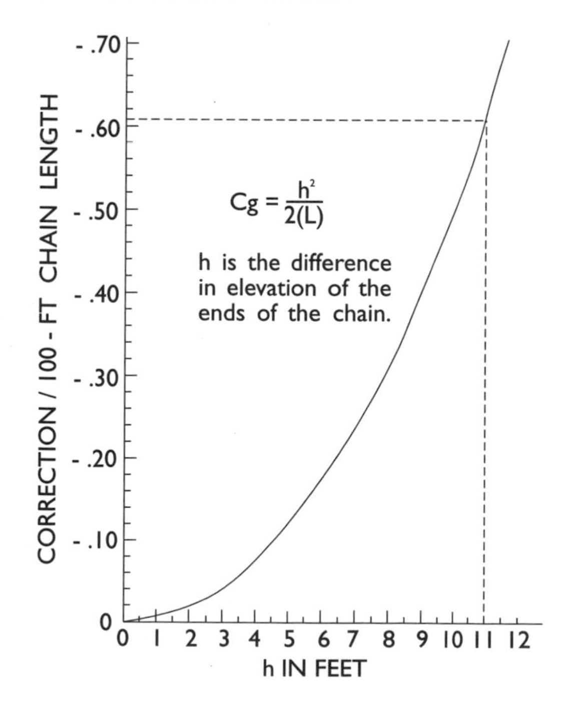

EXAMPLE WHEN LAYING OUT A POINT

A 100-foot chain is used to lay out bridge abutments that are 50 feet apart. The ground is sloping at the rate of 10' per 100 feet. What distance will need to be laid out on the slope to maintain the 50' horizontal spacing?

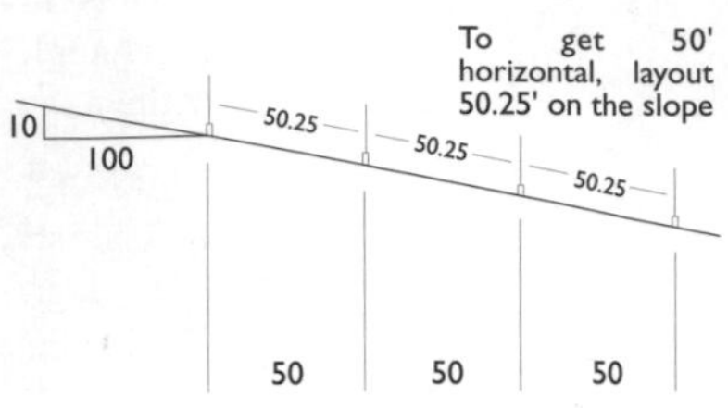

CORRECTION PER CHAIN LENGTH

$C_g = 10^2 / 2\,(100)$

$C_g = 0.5'$ *per chain length*

NUMBER OF CHAIN LENGTHS

There are 0.5 chain lengths in 50 feet.

TOTAL CORRECTION

$0.5'/C.L \times 0.5 = 0.25'$

APPLICATION

Refer to *14-3,* the *Laying Out a Prescribed Distance* table. In this case, the effect will be to measure less than needed to maintain the interval. Therefore, the correction should be added. The field engineer will need to set points at intervals of 50.25 feet on the slope to set the points per the plan distance of 50 feet.
Checking by using the Pythagorean Formula,

$Slope^2 = Horizontal^2 + Elevation^2$

The elevation difference in 50 feet would be 5 feet.

$Slope^2 = 50^2 + 5^2 = 50.25'$

$Slope = 50.25'$

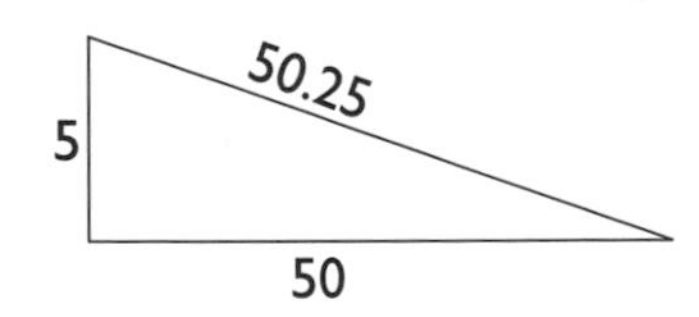

EXAMPLE: MEASURING BETWEEN ESTABLISHED POINTS

A field engineer measured the length between two control monuments with a 200-foot chain. One was at the bottom of a hill and the other was on the side of the hill. A slope measurement of 155.29' was made by holding the zero end of the chain at the center of the instrument and the end of the chain on the other control point. A vertical angle of 15 degrees was measured off of the horizon.

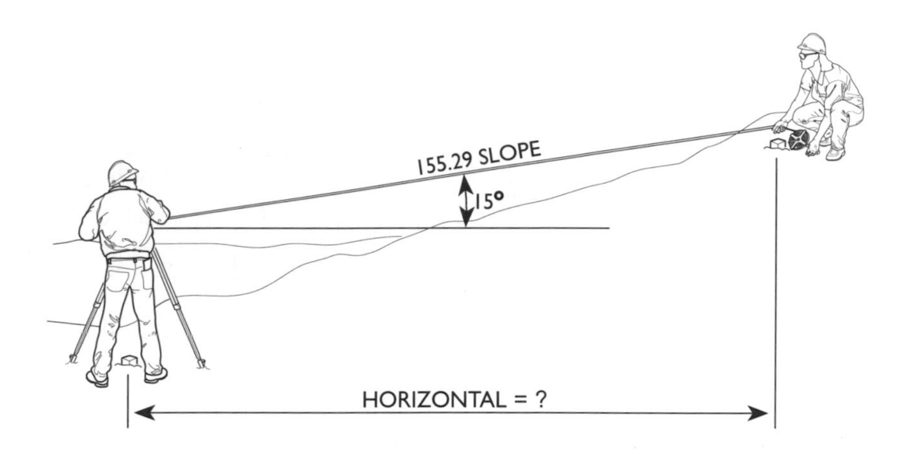

USING TRIGONOMETRY

Horizontal Distance = Cosine Vert. Angle x Slope

= *Cosine 15 x 155.29*

= *150.00'*

APPLICATION

No application is needed in this case as the formula calculates the horizontal distance directly.

CHAIN CORRECTIONS FOR TEMPERATURE

Laying Out a Prescribed Distance		
Condition of chain	Effect on laying out	Application of calculated correction to plan distance
Hot	Measure more	Subtract
Cold	Measure less	Add

Measuring a Distance		
Condition of chain	Effect on distance measured	How to apply calculated correction to recorded distance
Hot	Record less	Add
Cold	Record more	Subtract

SCOPE

Metal expands when it is hot and contracts when it is cold. Most people recognize these facts but never really have to do anything about them. However, any field engineer who will be measuring critical distances with a steel chain will have to do something about it. They will make corrections for the expansion or contraction of the chain due to temperature changes.

BACKGROUND

Standard chains are manufactured to be used at 68° F. Any temperature above 68° will cause the chain to elongate. Any temperature below 68°F will cause the chain to contract. The thermal expansion coefficient for the steel chains are manufactured from is 0.00000645 feet per foot per 1° F .

During measuring, the temperature of the chain itself is required to be known. Not the temperature of the air. Therefore, a thermometer must be attached to and be in direct contact with the steel of the chain.

THE CORRECTION FOR TEMPERATURE FORMULA

Correction $C_t = 0.00000645\ (T_t - 68)\ 100$

Where T_t is the temperature of the steel at the time of measurement.

GRAPHICAL SOLUTION FOR A 100' CHAIN

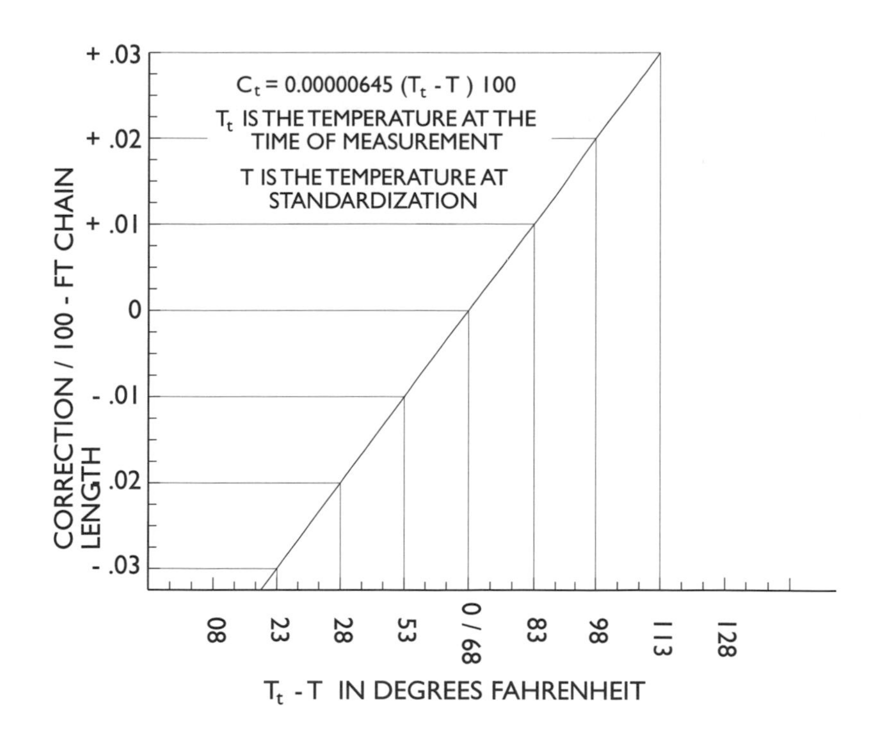

EXAMPLE OF LAYING OUT A POINT

On a cold,blustery day, a 100-foot chain is used to lay out building corners for a 250-foot by 400-foot structure. The temperature of the chain at the time of measurement is 23°. What distance will need to be laid out to set the points at the prescribed distances?

CORRECTION PER CHAIN LENGTH

Correction $C_t = 0.00000645\ (T_t - 68)\ 100$

$= 0.00000645\ (23-68)\ 100$

$= 0.03$

NUMBER OF CHAIN LENGTHS

There are 2.5 chain lengths in 250 feet

There are 4 chain lengths in 400 feet.

400.12

250.075

250 X 400

TOTAL CORRECTION

0.03'/C.L x 2.5 = 0.075'

0.03'/C.L. x 4.0 = 0.120'

APPLICATION

Refer to *14-18,* the *Laying Out a Prescribed Distance* table. In this case, the temperature of the chain is colder than 68° F. Therefore, the effect on the layout will be to measure less than the nominal length. Thus, the calculated correction must be added to the prescribed distances. The building corners must be laid out at distances of 250.075 and 400.120.

EXAMPLE OF MEASURING BETWEEN ESTABLISHED POINTS

On a hot, sunny day, a 100-foot chain is used to measure the distances between two highway centerline monuments. The temperature of the chain at the time of measurement is 108°. The distance was measured and recorded to be 999.74. What is the actual distance?

CORRECTION PER CHAIN LENGTH

Correction $C_t = 0.00000645(108 - 68)\ 100$

$C_t = 0.026$

NUMBER OF CHAIN LENGTHS

There are 9.9974 chain lengths in 999.74 feet

TOTAL CORRECTION

0.026'/C.L x 9.9975 = 0.258' or 0.26

APPLICATION

Refer to *14-18,* the *Measuring a Distance* table. In this case, the temperature of the chain is hotter than 68° F. Therefore, the effect on the layout will be to record less than the nominal length. Thus, the calculated correction must be added to the prescribed distances. The actual distance of the line is 999.74 +.26, or 1000.00 feet.

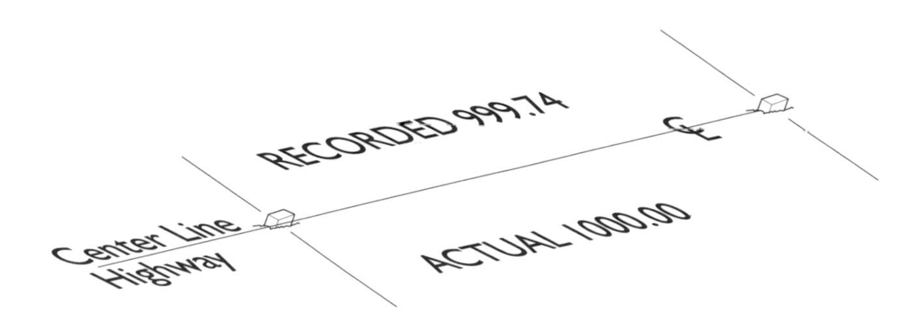

COMBINED CORRECTIONS

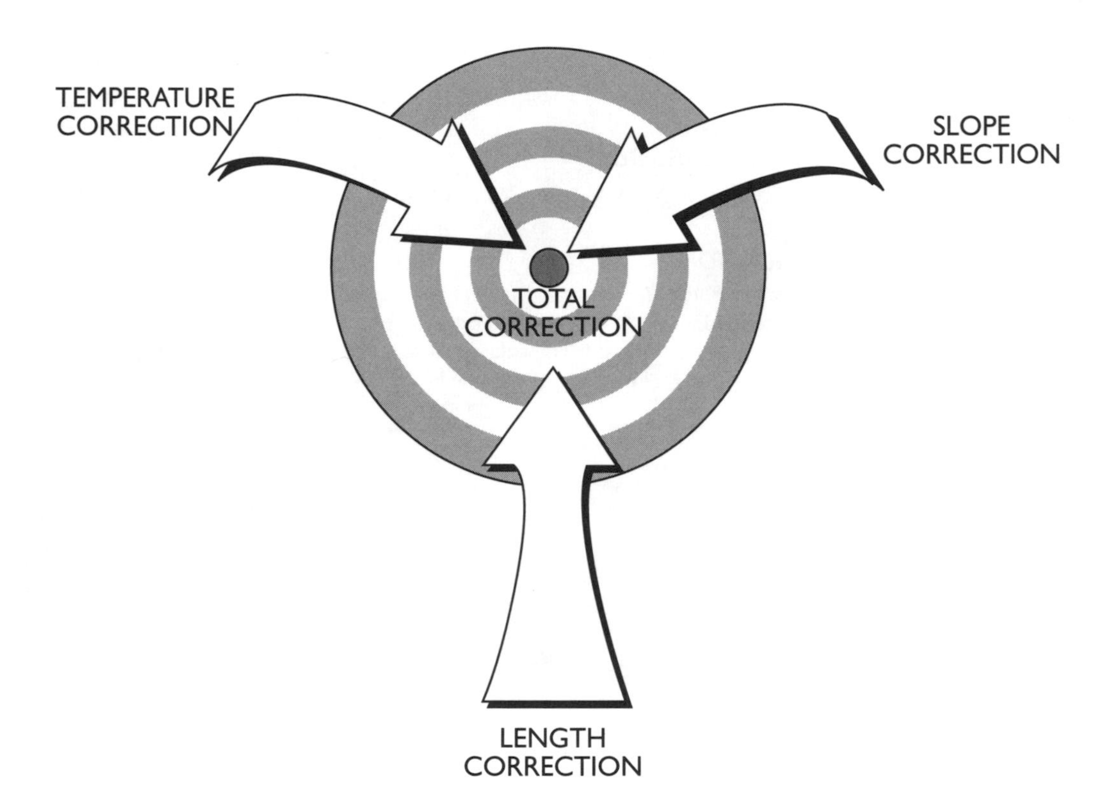

SCOPE

When applying a correction to a distance, it is typical that more than one correction will be necessary. Corrections can be made individually, such as determining the amount of correction necessary, applying it, determining another correction to be made, applying it, and so forth. More efficiently, corrections can be made by determining all of the corrections necessary, totalling them, and applying a single total correction. Applying a total correction is an easy task when all calculations are based upon chain lengths. Discussed here are procedures for combining temperature, length, and slope correction figures, and for applying them as one total correction.

GENERAL

The same conditions and formulas used in the previous sections apply when performing combined corrections. The same thought process of "was it a measurement or a layout, was it too short or too long, did we measure more or less, will we add or subtract," is also used.

PROCEDURE

Combined Corrections Procedure	
1	Calculate each individual correction.
2	Go through the "Add or Subract" thought process to place a sign on the correction.
3	List the corrections and add them using algebra to obtain a combined correction per chain length.
4	Multiply the total correction per chain length by the number of chain lengths to obtain the total correction.
5	Apply the total correction to the distance to obtain the corrected distance.

EXAMPLE OF COMBINING CORRECTIONS WHEN LAYING OUT

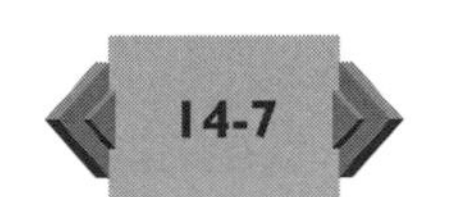

A distance measuring 250 feet to a pier is to be laid out on a bridge project. The terrain from the control point to the pier is sloping at a 7% grade down into the river valley. The temperature of the chain at the time of the measurement is 45° F. The chain used is known to be 100.01 when compared to a standard at 68° F. What is the distance that must be laid out in the field to obtain the desired distance to the pier? Refer to *14-7*, the *Laying out a Prescribed Distance* table, in the *Overview* section of this chapter.

NUMBER OF CHAIN LENGTHS

$$\frac{250}{100} = 2.5 \text{ Chain Lengths}$$

TEMPERATURE CORRECTION PER CHAIN LENGTH

$$\text{Correction}, C_t$$

$$C_t = 0.00000645(T - 68)C.L.$$

$$C_t = 0.00000645(45 - 68)100$$

$$C_t = 0.015 \text{ per Chain Length}$$

In this case, because the temperature is cold, the effect in laying out is to measure less so the calculated temperature correction should be added to obtain the correct distance to be laid out.

$$\text{Temperature Correction, } Ct = +0.015 \text{ per Chain Length}$$

LENGTH CORRECTION PER CHAIN LENGTH

The difference between nominal length and actual chain length is 0.01' per chain length.

APPLICATION

In this case, because the chain length is too long, the effect in laying out is to measure more so the calculated chain length correction should be subtracted to obtain the correct distance to be laid out.

$$\text{Length Correction, } C_l = -0.01 \text{ per Chain Length}$$

SLOPE CORRECTION PER CHAIN LENGTH

A 7% grade means the ground is rising or falling 7 feet in 100 feet. Using the slope correction per chain length formula,

$$\text{Slope Correction, } Cg$$

$$Cg = \frac{h^2}{2(C.L.)}$$

$$Cg = \frac{7^2}{2(100)}$$

$$Cg = 0.24$$

100
7
7%

In this case, because measuring on the slope makes the desired horizontal measurement too short, the effect on laying out is to measure less so the calculated slope correction should be added to obtain the correct distance to be laid out.

$$\text{Slope Correction, } Cg = +0.24 \text{ per Chain Length}$$

COMBINING THE CORRECTIONS

Combined Correction = $C_l + C_g + C_t$
Length Correction, C_l = *-0.01 per chain length*
Slope correction, Cg, = *0.24 per chain length*
Temperature Correction C_t = *+0.015 per Chain Length*
Combined Correction = *(-0.01) + (+0.24) + (+0.015)*
Combined Correction = *+0.245 per chain length*

TOTAL CORRECTION

Multiply the number of chain lengths by the combined corrections.

Total Correction = (+0.245)/C.L. X 2.5 C.L. = +0.613'

APPLICATION

By combining the corrections as shown and watching the signs, the method of applying the correction to the desired distance is automatic. In this case, the correction of +0.613 is added algebraically to the plan distance of 250 feet to obtain the distance to be laid out in the field.

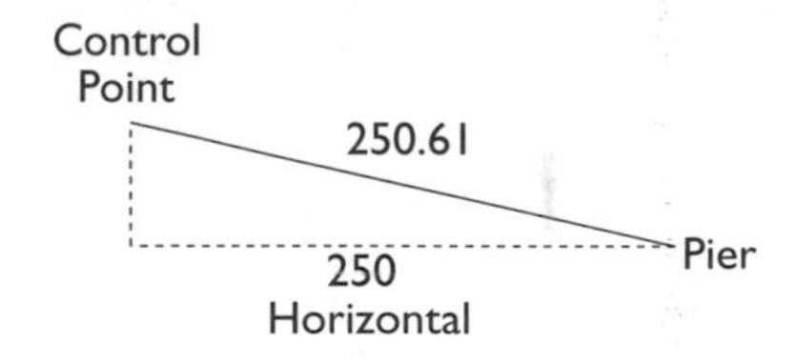

Distance to be laid out = 250.00 + 0.613 = 250.613 or a practical measurement of 250.61.

EXAMPLE OF MEASURING BETWEEN TWO POINTS

In this example of a start-up of a project, two field engineers are measuring distances between control points (CP) that had been set by a professional surveyor. The distance they measured and recorded, between CP 1004 and CP 1005, was found to be 500.17. Upon checking the "report of survey" from the professional surveyor, the distance between CP 1004 and CP 1005 is supposed to be 500.00'. They aren't concerned though because they haven't made the necessary corrections. At the time of the measurement, the temperature of the chain was a scorching 103° F. The difference in elevation between the two control points is 20 feet. The calibrated length of chain is 100.02'. What is the actual distance between the control points that must be laid out in the field to obtain the true distance between the control points? Reference *14-6,* the *Measuring a Distance* table, in the *Overview* section of this chapter.

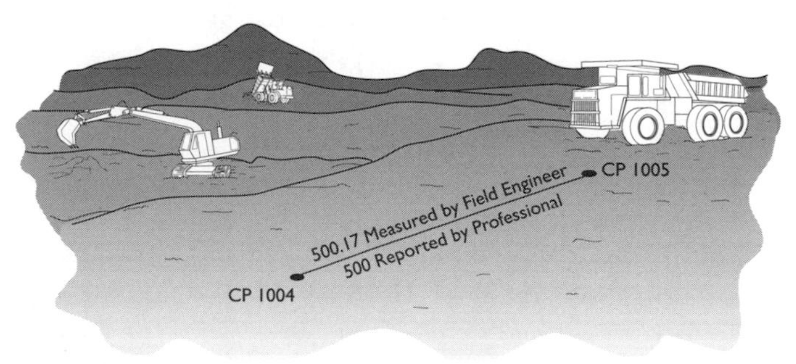

NUMBER OF CHAIN LENGTHS

$$\frac{500.17}{100} = 5.0017 \textit{ Chain Lengths}$$

TEMPERATURE CORRECTION PER CHAIN LENGTH

$$\begin{aligned} &\textit{Temperature Correction, Ct} \\ Ct &= 0.00000645(T-68)100 \\ Ct &= 0.00000645(103-68)100 \\ Ct &= 0.02 \textit{ per Chain Length} \end{aligned}$$

In this case, because the temperature is hot, the effect in measuring is to record less, so the calculated temperature correction should be added to obtain the correct distance between the control points.

$$\textit{Temperature Correction, } Ct = +\ 0.02 \textit{ per Chain Length}$$

LENGTH CORRECTION PER CHAIN LENGTH

The difference between nominal length and actual chain length is 0.02' per chain length.

APPLICATION

In this case, because the chain length is too long, the effect in measuring is to record less, so the calculated chain length correction should be added to obtain the correct distance between the control points.

$$\textit{Length Correction, } C_l = +\ 0.02 \textit{ per Chain Length}$$

SLOPE CORRECTION PER CHAIN LENGTH

To use the correction-per-chain-length method when a 20-foot difference in elevation between control points approximately 500 feet apart is known, the difference in elevation per 100 feet must be determined. A simple proportion can be used to calculate the amount.

$$\frac{20}{500} = \frac{x}{100}$$

solving for x = 4 feet elevation difference per 100 feet. Therefore,

$$\begin{aligned} &\textit{Slope Correction, Cg} \\ Cg &= \frac{h^2}{2(C.L.)} \\ Cg &= \frac{4^2}{2(100)} \\ Cg &= 0.08 \end{aligned}$$

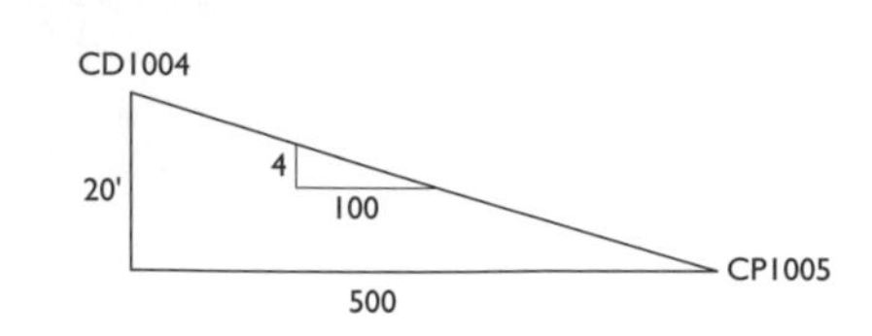

In this case, measuring on the slope has the effect of making the distance too short, the effect is to record more, so the calculated slope correction should be subtracted to obtain the correct distance between the control points.

$$Slope\ Correction,\ Cg = -\ 0.08\ per\ Chain\ Length$$

COMBINING THE CORRECTIONS

Combined Correction = $C_l + C_g + C_t$
Length Correction, C_l = *+0.02 per chain length*
Slope correction, Cg, = *-0.08 per chain length*
Temperature Correction C_t = *+0.02 per Chain Length*
Combined Correction = *(+0.02) + (-0.08) + (+0.02)*
Combined Correction = *- 0.04 per chain length*

TOTAL CORRECTION

Multiply the number of chain lengths by the combined corrections.

(-.04)/C.L. X 5.0017 C.L. = -0.2' total correction

APPLICATION

By combining the corrections as shown and watching the signs, the method of applying the correction to the desired distance is automatic. In this case, the correction of -0.20 must be subtracted from the recorded distance of 500.17 feet to obtain the distance.
Actual distance = 500.17 + (-0.20)
Actual distance = 499.97
Uh-oh, the field engineers measured 499.97 and the professional surveyor shows the distance to be 500.00. Who is right? The field engineers went to the field and remeasured the distance. They still obtained the same results. The next step is to call the professional surveyor to confirm his results.

SUMMARY

It has been presented in this section that corrections can easily be calculated, combined and applied at one time to obtain a distance to be laid out or a distance that has been measured. Again, there are other methods of calculating corrections. This author has shown one simple method. Each individual determines which method to use. Refer to another surveying text for additional examples of chain corrections. For a listing of surveying texts, reference *Appendix D*.

TRAVERSE COMPUTATIONS

15

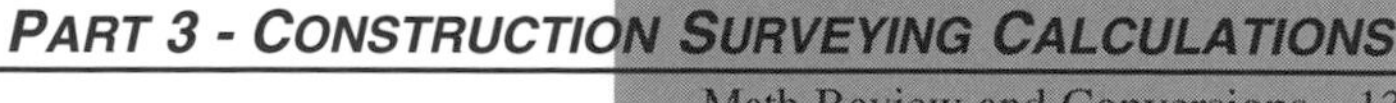

PART 3 - CONSTRUCTION SURVEYING CALCULATIONS

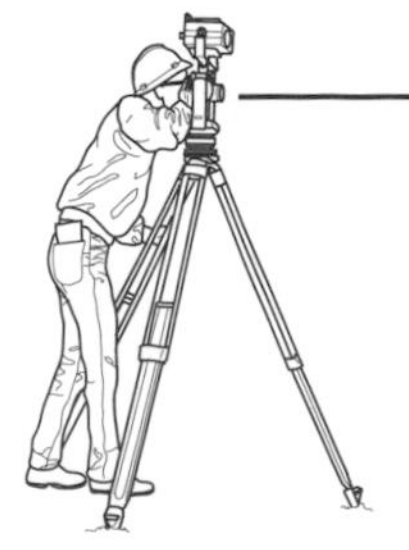

CHAPTER OBJECTIVES:

Define traverse.
State the ultimate objective of traverse computations.
Outline the step-by-step procedure for traverse computations.
Describe what type of traverse should be used to confirm the viability of the work when establishing control on a jobsite.
Describe when angles should be adjusted as part of traverse computations and list the basic rule of geometry that must be met.
Compare azimuth to bearings and use adjusted angles to calculate the directions of lines.
Calculate latitudes and departures for a line.
Define error of closure and precision and calculate them for a traverse.
State why latitudes and departures are adjusted, calculate the adjustment, and apply the adjustment for a traverse.
Describe and calculate coordinates for a traverse using adjusted latitudes and departures.
Describe and perform the process of calculating area using coordinates.

INDEX

OVERVIEW

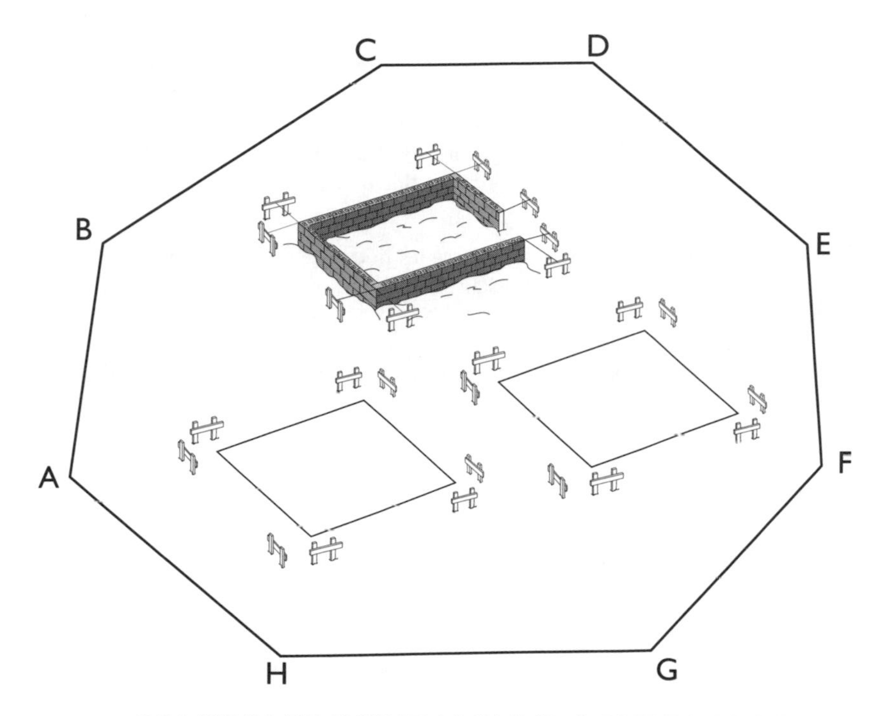

CONTROL TRAVERSE AROUND A BUILDING SITE

SCOPE

A traverse is a series of points connected by lines in which the distance and direction is measured by the field engineer. By knowing the distance and direction of each line for the traverse, the relative position of the points to each other can be determined. This is the objective of traverse computations.

The traverse is the most important geometric tool the field engineer has available. It is a means by which the quality of the field measurements can be checked and compared to a standard. Without the ability to perform traverse computation, the field engineer would never know if the precision of the control was good enough to be used for the layout.

STEP-BY STEP PROCEDURE FOR TRAVERSE COMPUTATIONS

Traverse computations are very systematic and very exact. You can't go to step two until you have performed step one and so on. You can't make a mistake at any stage or the end result will be wrong. Field engineers performing traverse computations must know the procedure by heart and must be able to perform the calculations with 100% accuracy.

The systematic approach to traverse computations is simplified by using a standard form to record the results of the calculations. Traverse computations use basic formulas from geometry and trigonometry. The computation of the formulas can be performed easily with a hand-held scientific calculator, by using a standard computer spreadsheet, or by writing computer programs.

STEP 1

Measure the distances and angles in the field. Use good measurement techniques and record angles and distances clearly in the field book.

STEP 2

Adjust the angles. After angles have been measured, check them to see if they represent a closed geometric figure. If they do, no adjustment is needed. If not, they can be adjusted randomly, systematically or from notes in the field book

STEP 3

Determine starting direction for one of the lines. A starting direction can be assumed, determined from magnetic North, taken from plans, etc. If the traverse isn't part of a large network, any direction can be used.

STEP 4

Calculate the direction for all the lines. Direction of lines can be represented in either azimuths or bearings. Calculation in azimuths is the preferred method as it is done by following very simple rules. Bearings are the preferred method of representing directions on drawings. All direction calculations must be checked and confirmed to be completely correct or all of the following calculations will be incorrect.

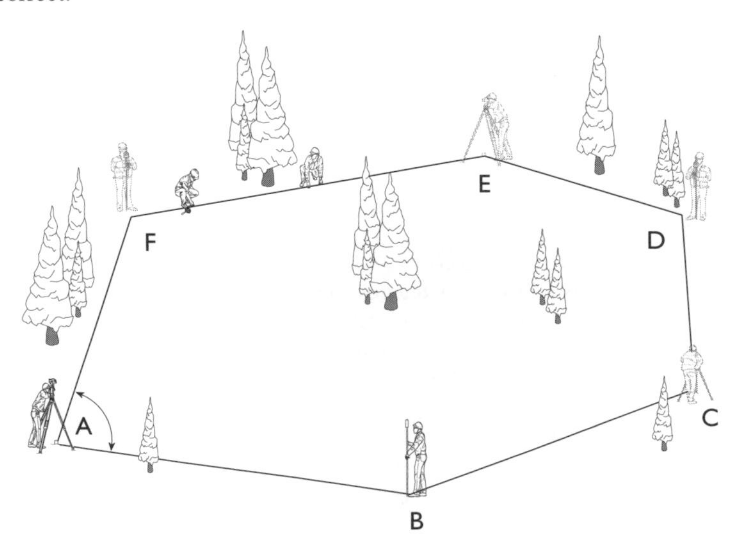

STEP 5

Determine trig functions of directions. The sine and cosine of the directions are determined by inputting the angle into a scientific calculator or a computer and requesting the trigonometric value.

STEP 6

Calculate latitudes and departures. Having the distance and direction of each line, the rectangular components of the line can be calculated. These are the latitude and departure. The formulas for this are:

Latitude = distance X cosine of direction

Departure = distance X sine of direction

Based on the direction of the line, North and South latitudes and East and West departures are calculated for each line.

STEP 7

Calculate linear error of closure. By summing the North Latitudes, South Latitudes, East Departures, and West Departures and determining the difference in latitudes and difference in departures and applying these differences to the Pythagorean theorem, the linear error of closure (LEOC) for the traverse can be determined. When calculated, the LEOC represents the amount the distance measurement failed to close for the traverse.

STEP 8

Calculate precision. Determining the precision of a traverse gives the field engineer a ratio to compare to specified standards. If the precision of the field measurements meets the standards, the field work is complete. If it doesn't, the field engineer will need to return to the field to remeasure angles and/or distances to determine the erroneous measurement.

STEP 9

Adjust the latitudes and departures. If the precision meets standards, it is standard practice to perform an adjustment to the latitudes and departures to eliminate any small error that exists in the measurements. Several methods have been developed to calculate this adjustment.

STEP 10

Determine starting coordinate. Starting coordinates can be assumed, determined from the work of a registered surveyor, taken from plans, etc. If the traverse isn't part of a large network, any starting coordinates can be used. Typically, the starting coordinates are chosen large enough so there can never be a negative coordinate.

STEP 11

Calculate coordinates. Using the adjusted latitudes and departures and the starting coordinates, the coordinates of each point are calculated. After calculating around the traverse, the field engineer is careful to return to the starting coordinate as a check.

STEP 12

Apply coordinates to construction. With today's technology (electronic theodolites, EDMs, calculators and computers) the most powerful and versatile method of surveying layout is using rectangular coordinates to determine field layout data. If coordinates of structures and control points are known, the layout data of the field engineer can be easily calculated using what is commonly called coordinate geometry. Coordinates represent power!

ANGLE ADJUSTMENT

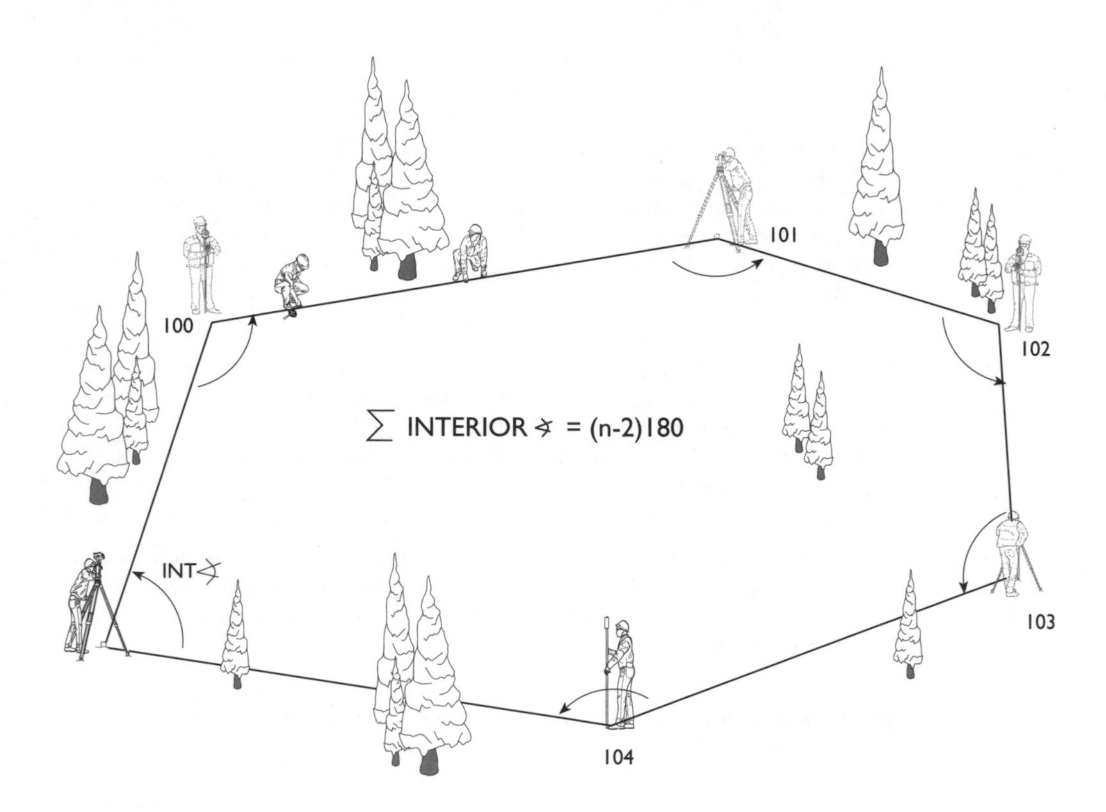

SCOPE

When angles are turned on a traverse, it is expected there will be small errors in each of the angles that are measured and recorded. Sometimes these errors cancel themselves out, and other times they still exist even though exact procedures were used by the field engineer. In traverse computations, it is common practice to adjust those errors out of the angles so the angles will represent a closed geometric figure.

GENERAL

WHEN TO ADJUST

Adjustment of angles should only be performed if the angles turned do not meet the expected results attainable from the instrument. See the instrument specifications for this value. In other words, do not adjust out mistakes. Turn the angles again if mistakes exist. However, if mistakes have been eliminated and the amount of error is small, it is common practice to apply an adjustment to the angles. this adjustment will force a geometric closure of the angles of the figure which was measured.

A RULE FROM GEOMETRY

After the interior angles of traverse have been measured, they should be added together. The sum of the interior angles should be compared to the basic rule from geometry which states :

The sum of the interior angles of a closed figure will equal (n-2)180

"n" represents the number of sides in the figure.

For a three-sided figure, (3-2)180 = 180 degrees

For a four-sided figure, (4-2)180 = 360 degrees

For a six-sided figure, (6-2)180 = 720 degrees

and so on...

ADJUSTMENT OPTIONS

RANDOMLY ELIMINATE THE ERROR

If the error is small and it would not make a great deal of difference exactly how the error was eliminated, do whatever will make the numbers work out evenly. For example, if the error was 1' 30"apply 45" to two angles, or 30" to three angles, or any combination to make numbers work out easily.

FOLLOW THE NOTES IN THE FIELD BOOK

If the field notes say "the setup at point 4 was on soft ground and the instrument wouldn't stay level," it might be a good idea to apply all of the error to the angle measured at point 4. Review the field notes before making any adjustment.

EQUALLY ADJUST ALL ANGLES

If there is no information which indicates the adjustment should be applied to one or more angles, the error can be distributed equally.

STEP-BY-STEP PROCEDURE FOR ADJUSTING ANGLES

STEP 1 **Sum the interior angles of the traverse.**

STEP 2 **Compare the sum of the angles**, that were measured and recorded to the (n-2)180 rule. Example: For a five-sided figure, the sum of the angles should be 540 degrees. From the field book, the angles added to 540° 02' 30", the two are not equal.

STEP 3 **Determine the difference between the two**. In the above case, 2' 30"

STEP 4 **Perform the adjustment**. Adjust randomly, equally, or follow field notes. In this case adjust randomly to eliminate seconds from the angles.

STEP 5 **Sum the angles.** As a final check of the adjustment made, add the angles again. If everything was executed properly, the interior angle total should equal the total obtained from the angle summation rule from geometry.

EXAMPLE TRAVERSE COMPUTATION

SKETCH OF FIELD ANGLES

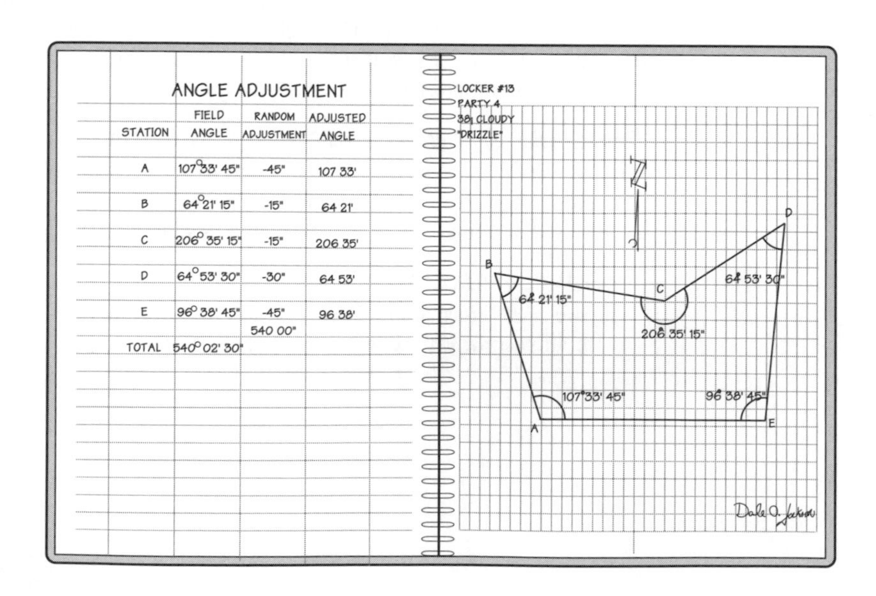

ANGLE ADJUSTMENT

STATION	FIELD ANGLE	RANDOM ADJUSTMENT	ADJUSTED ANGLE
A	107° 33' 45"	-45"	107 33'
B	64° 21' 15"	-15"	64 21'
C	206° 35' 15"	-15"	206 35'
D	64° 53' 30"	-30"	64 53'
E	96° 38' 45"	-45"	96 38'
		540 00"	
TOTAL	540° 02' 30"		

Angle Adjustment			
Station	***Field Angle***	***Random Adjustment***	***Adjusted Angle***
A	107° 33' 45"	-45"	**107° 33'**
B	64° 21' 15"	-15"	**64° 21'**
C	206° 35' 15"	-15"	**206° 35'**
D	64° 53' 30"	-30"	**64° 53'**
E	96° 38' 45"	-45"	**96° 38'**
Total	540° 02' 30"		**540° 00'**

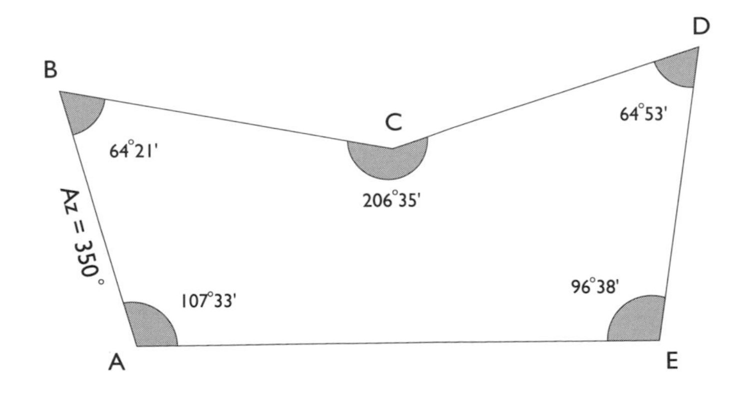

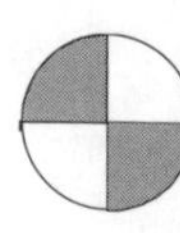

DIRECTION CALCULATONS

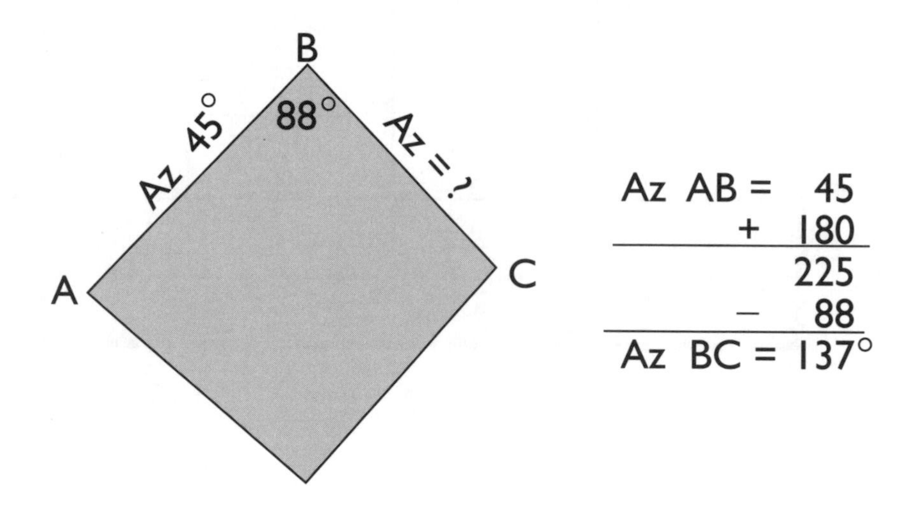

SCOPE

After traverse angles have been adjusted to represent a closed geometric figure and a starting direction has been determined or assumed, the direction of each line can be calculated. This is accomplished using either azimuths or bearings. Some field engineers prefer to calculate azimuths and others prefer to calculate bearings. It is purely a matter of personal preference because they both represent the same thing—a direction. However, azimuths are much easier to calculate. Since they are easier fewer mistakes in calculations generally occur. Therefore, although the properties of both bearings and azimuths will be discussed, only the calculation of azimuths will be presented in this section. It will also be shown how bearings can easily be obtained by converting from azimuths.

GENERAL INFORMATION

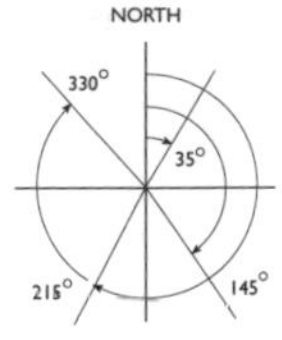

NORTH
N 30° W
N 60° E
WEST
EAST
S 45° E
S 30° W
SOUTH

COMPARISON BETWEEN AZIMUTHS AND BEARINGS

Azimuths are measured clockwise from North and vary from 0 to 360°. They require only a numerical value (35°, 145°, 215°, etc.). Azimuths may be referenced to true North, to magnetic North, to an azimuth taken from plans, or from an assumed direction. In addition, azimuths may be forward or backward.

Bearings are measured either clockwise or counterclockwise from North and South. They vary from 0 to 90° and require two letters and a numerical value (N 60° E, S 30° W, etc.). Bearings may be referenced to true North, magnetic North, to an azimuth taken from plans, or from an assumed direction. Bearings may also be forward or backward.

DETERMINING THE QUADRANT

By observation of the value, the field engineer should be able to instantly visualize where the bearing or azimuth is located. Whether it is in the Northeast quadrant, Southeast, Southwest, or Northwest.

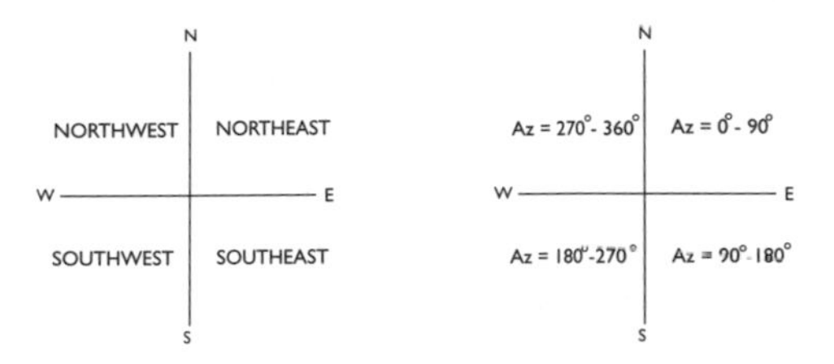

Converting from Azimuths to Bearings		
Azimuth	**To Bearing**	N 360° - Az / NONE W / E Az - 180° / 180° - Az S
0° to 90°	none	
90° to 180°	180° - Azimuth	
180° to 270°	Azimuth - 180°	
270° to 360°	360° - Azimuth	
	Example	
Azimuth	**Calculation**	**Bearing**
54°	no action needed	N 54° E
112°	180° - 112° = 68°	S 68° E
231°	231° - 180° = 51°	S 51° W
345°	360° - 345° = 15°	N 15° W

Converting from Bearings to Azimuths		
Bearing	**Azimuth**	
N 0° E to N 90° E	none	
S 0° E to S 90° E	180° - bearing	
S 0° W to S 90° W	180° + bearing	
N 0° W to N 90° W	360° - bearing	
Example		
Bearing	**Calculation**	**Azimuth**
N 54° E	no action needed	54°
S 68° E	180° - 68° = 112°	112°
S 51° W	180° + 51° = 231°	231°
N 15° W	360° - 15° = 345°	345°

N
360° - Bearing
NONE
W
E
180°+ Bearing
180° - Bearing
S

OBTAINING A BACK AZIMUTH

To calculate azimuths about a traverse, it is necessary to obtain the back azimuth of a line. To calculate a back azimuth, simply add 180° to the azimuth of the line. In formula form:

Back Azimuth = Forward Azimuth + 180°

For example, if a line has an azimuth of 75° it's back azimuth would be 255°.

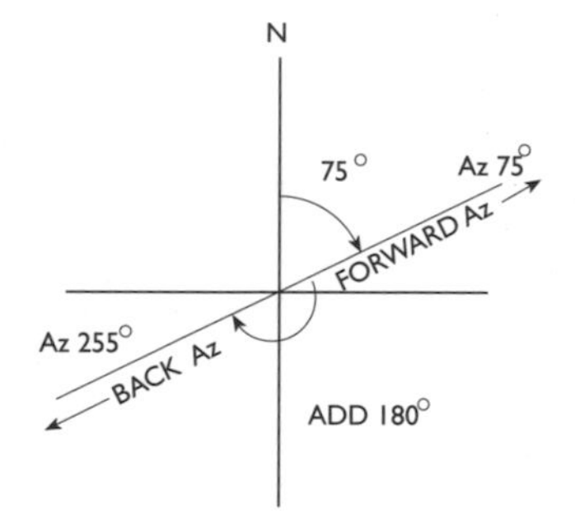

OBTAINING A BACK BEARING

If back bearings are ever needed, all the field engineer needs to do to change the direction is reverse the letters. That is, N 45° E has a back bearing of S 45° W and S 82° 20' E has a back bearing of N82° 20' W.

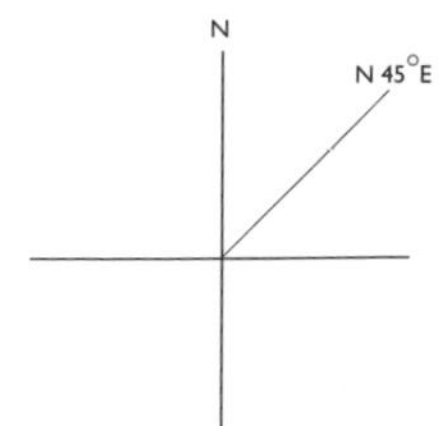

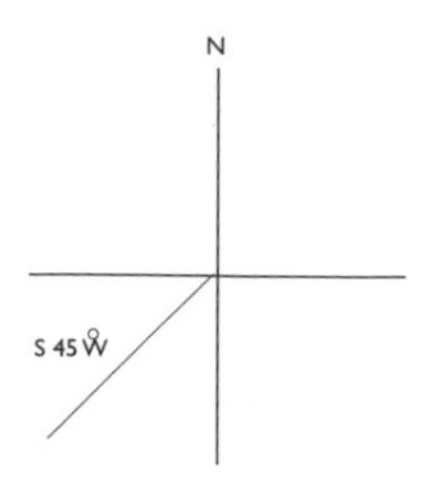

CALCULATE DIRECTIONS CLOCKWISE OR COUNTER-CLOCKWISE

Direction calculations can be performed in either a clockwise or counter-clockwise manner about a traverse. There are specific rules for calculating in either direction.

To calculate azimuths clockwise around a traverse: Subtract the adjusted interior angle from the back azimuth of the preceding line.

To calculate azimuths counter-clockwise around a traverse: Add the adjusted interior angle to the back azimuth of the preceding line.

STEP-BY-STEP PROCEDURE FOR CALCULATING DIRECTIONS

The following is the step-by-step procedure for calculating azimuths clockwise. See the table on the following page.

STEP 1

Plan and prepare. Determine a known azimuth, andthe direction that the calculation will proceed (clockwise or counterclockwise). List the adjusted interior angles of the traverse and draw a sketch of the entire traverse. Be sure it is reasonably accurate (that is, angles and distances should be close to scale). Label the points, the starting direction, and the interior angles. Orient the drawing properly to North.

STEP 2

Perform the calculation. Start by writing down the starting azimuth. Add 180° to obtain the back azimuth. Subtract the adjusted interior angle to obtain the azimuth of the next line. If the result is greater than 360, subtract 360. Write down the azimuth on the sketch.

STEP 3

Repeat the calculation. For each line of the traverse the same calculation should be repeated. That is, add 180° and subtract the interior angle.

STEP 4

Check the calculations by using the last interior angle to recalculate the starting azimuth.

EXAMPLE

Calculate the direction of each line and provide a check. (clockwise)

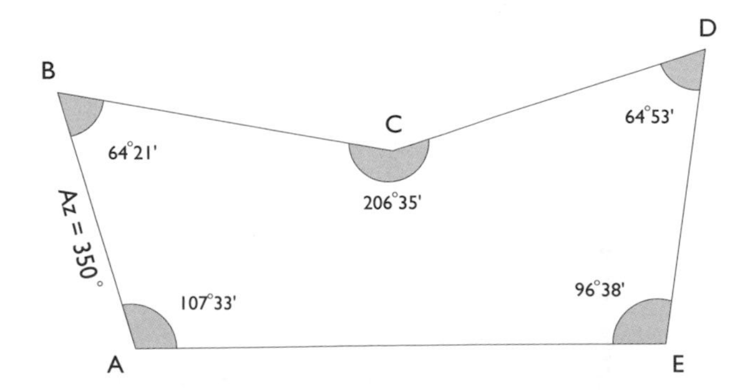

EXAMPLE SOLUTION

Clockwise Azimuth Calculation	
Proceeding Clockwise about the traverse.	
Azimuth AB = 350°	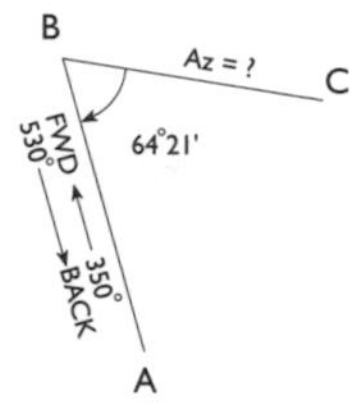
Add 180°	
Back Azimuth = 530°	
Subtract interior Angle at B - 64° 21'	
Azimuth BC = 105° 39'	
Add 180°	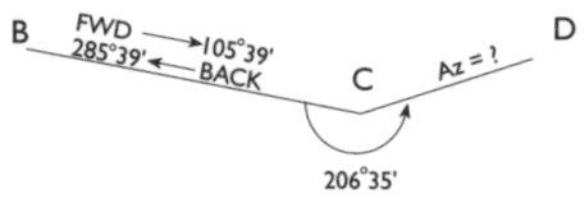
Back azimuth = 285° 39'	
Subtract interior angle at C - 206° 35'	
Azimuth CD = 79° 04'	
Add 180°	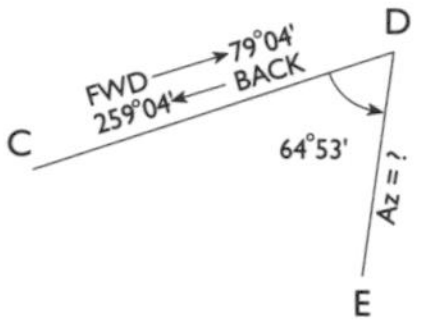
Back azimuth = 259° 04'	
Subtract interior angle at D - 64° 53'	
Azimuth DE = 194° 11'	
Add 180°	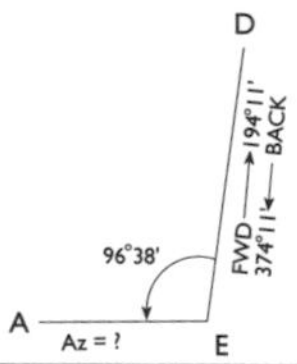
Back azimuth = 374° 11'	
Subtract interior angle at E - 96° 38'	
Azimuth EA = 277° 33'	
Add 180°	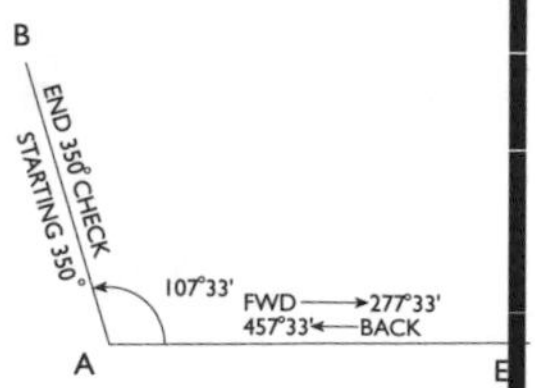
Back azimuth = 457° 33'	
Subtract interior angle at A - 107° 33'	
Azimuth AB = 350° 00'	
This checks with the starting azimuth.	

SUMMARY OF RESULTS

Directions for Example Traverse			
Station Point	**Adjusted Angle**	**Line**	**Azimuth**
A	107° 33'		
		AB	350°
B	64° 21'		
		BC	105° 39'
C	206° 35'		
		CD	79° 04'
D	64° 53'		
		DE	194° 11'
E	96° 38'		
		EA	277° 33'
Total	540° 00"		

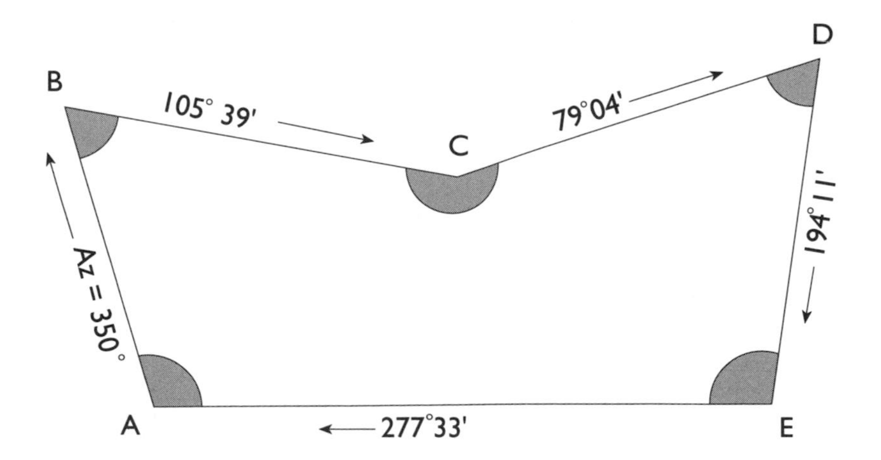

LATITUDES, DEPARTURES, PRECISION, AND ADJUSTMENTS

SCOPE

One of the immediate goals of traverse computations is to determine the precision of fieldwork measurements. This is accomplished by breaking each line into its component parts so a mathematical closure of the fieldwork can be determined. With a mathematical closure the field engineer can calculate a precision ratio to compare to established standard and make a decision about the acceptability of the work.

LATITUDES AND DEPARTURES

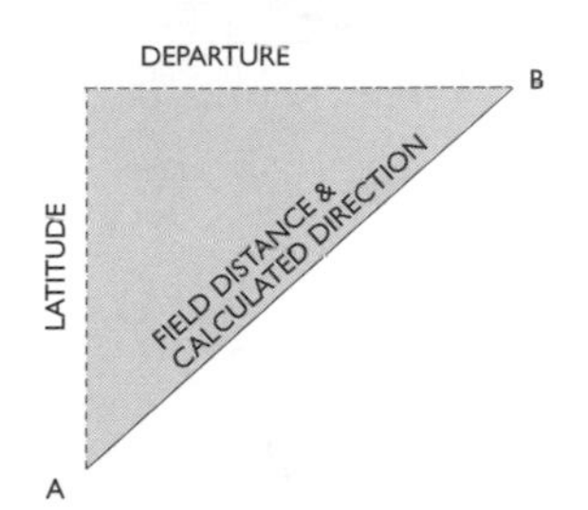

Using simple formulas which are based on right angle trigonometry and utilizing the field distance and calculated direction from point to point, the field engineer can calculate the North and East components of any line. These components are also referred to in surveying terms as the latitude and departure of the line. The North component of a line is the latitude, and the East component of a line is the departure. Simple sine and cosine values of the direction of the line are multiplied with the distance to determine these values.

CALCULATING LATITUDES AND DEPARTURES

LATITUDE

To calculate the North/South component (Latitude) of a line, determine the trigonometric value of the cosine of the direction and multiply it by the distance.

Latitude = Distance x Cosine Direction

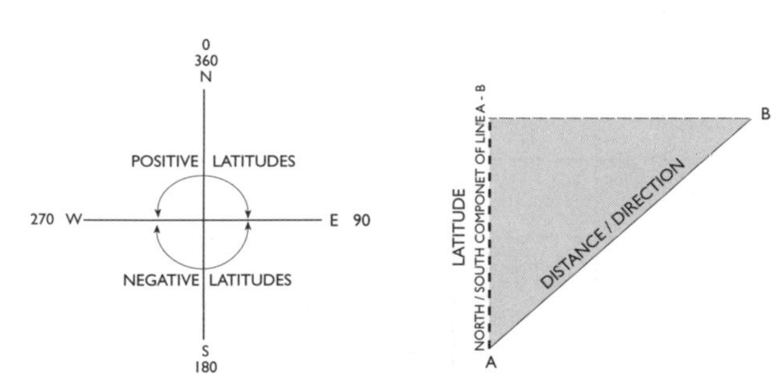

The direction of the line will determine the sign of the latitude. Positive indicates a North latitude and negative indicates a South latitude. If using azimuths, an azimuth between 0 and 90 or between 270 and 360 will yield a positive latitude and an azimuth between 90 and 270 will yield a negative latitude. If using bearings, the sign is determined from the stated direction, North or South.

DEPARTURE

To calculate the East/West component (Departure) of a line, determine the trigonometric value of the sine of the direction and multiply it by the distance.

Departure = Distance x Sine (Direction)

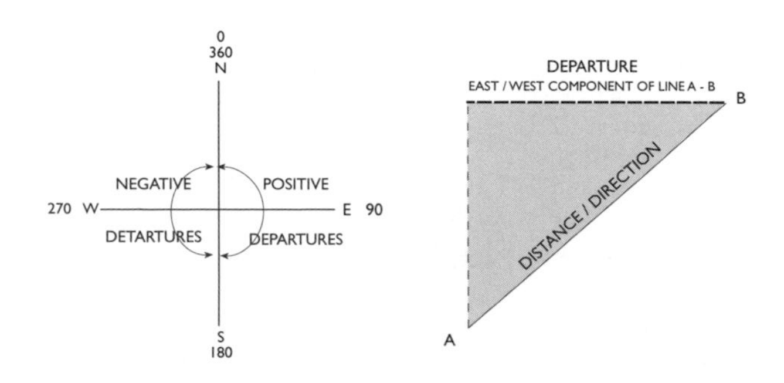

The direction of the line will determine the sign of the departure. Positive indicates an East departure, and negative indicates a West departure. If using azimuths, an azimuth between 0 and 180 will yield a positive departure, and an azimuth between 180 and 360 will yield a negative departure. If using bearings, the sign is determined from the stated direction, East or West.

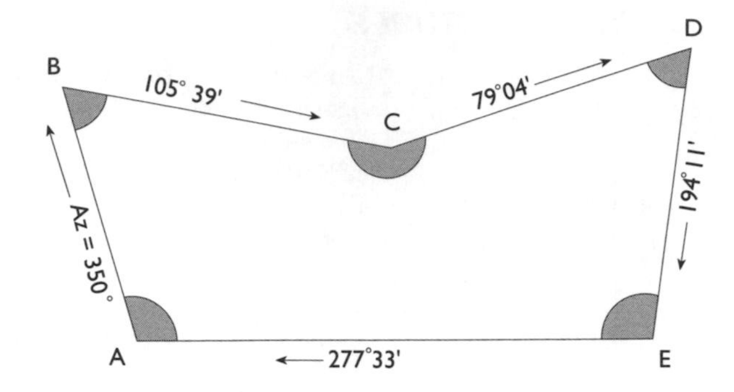

EXAMPLE FOR LINE AB

$Latitude_{AB}$ = Cos 350° x 677.97' = 667.6701

$Departure_{AB}$ = Sin 350° x 677.97' = -117.7283

Follow the same procedure for each line. See the table below for a summary of the latitude and departure computations.

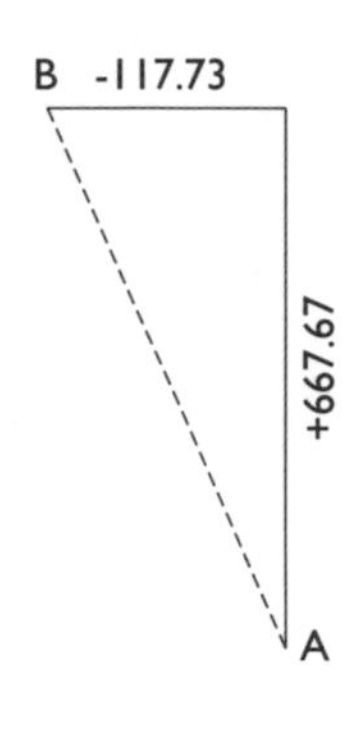

Line	Azimuth	Distance	Cosine	Lat	Sine	Dep
AB	350°	677.97	0.9848	**667.6701**	-0.1736	**-117.7283**
BC	105°39'	616.05	-0.2698	**-166.1858**	0.9629	**593.2115**
CD	79°04'	690.88	0.1897	**131.0369**	0.9818	**678.3395**
DE	194°11'	783.32	-0.9695	**-759.4418**	-0.245	**-191.9333**
EA	277°33'	970.26	0.1314	**127.4838**	-0.9913	**-961.8484**

LINEAR ERROR OF CLOSURE AND PRECISION

Finally, after measuring distances and angles in the field, adjusting the angles, calculating directions of lines, and calculating latitudes and departures, the field engineer is at the important step of determining the linear closure and precision of the fieldwork. These calculations will tell the field engineer if the fieldwork meets acceptable standards. If it does not, the field engineer will return to the field to remeasure angles and/or distances. If it does meet acceptable standards, calculations can proceed to the next step - the calculation of adjustment of the latitudes and departures.

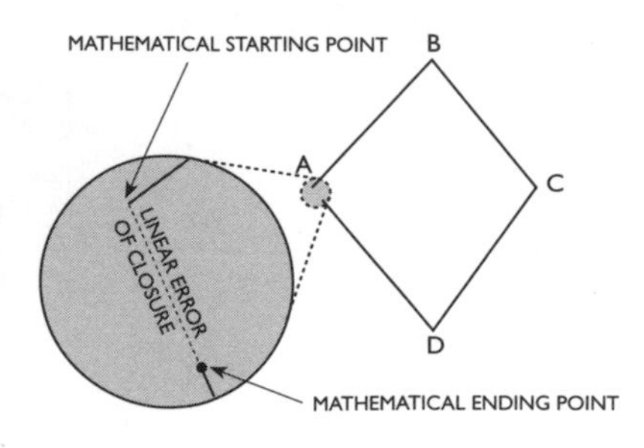

FORMULAS

The formula used in the calculation of the linear error of closure is simply an application for the Pythagorean theorem. The process of determining the closure in the North/South Latitudes and the closure in the East/West departures, yields values which can be substituted as follows: Reference *Chapter 13, Math Review* for more information about formulas.

$$c^2 = a^2 + b^2$$

$$LEOC^2 = (Closure\ in\ Latitudes)^2 + (Closure\ in\ Departure)^2$$

To determine the precision, a ratio of the LEOC to the total distance measured (perimeter) is calculated. The formula is:

$$\frac{LEOC}{Perimeter} = \frac{1}{X}$$

STEP-BY-STEP PROCEDURE

The following is the step-by-step procedure for calculating LEOC and precision. It allows the field engineer to methodically use the calculated latitudes and departures to determine their mathematical closure that is referred to as the Linear Error of Closure (LEOC). Using the LEOC and the distance measured, a ratio called the precision is calculated and used as a comparison to published standards.

STEP 1 — **Sum the distances** to obtain the total distance around the perimeter of the traverse. Sum the North Latitude column. Sum the South Latitude column. Sum the East Departure column. Sum the West Departure column.

STEP 2 — **Determine the difference** between the summations of the North and South Latitudes (Closure in Latitude). Determine the difference between the summations of the East and West Departures (Closure in Departure).

STEP 3 — **Insert the differences** into the Pythagorean Theorem to calculate the Linear Error of Closure (LEOC) for the traverse.

STEP 4 — **Divide** the LEOC by the perimeter to calculate the precision. Represent the precision as a ratio. For example 1:6622

CONTINUING EXAMPLE

		Latitudes		Departures	
Line	**Distance**	**N. Lat.**	**S. Lat.**	**E. Dep.**	**W. Dep.**
AB	677.97	667.6701			117.7283
BC	616.05		166.1858	593.2115	
CD	690.88	131.0369		678.3395	
DE	783.32		759.4418		191.9333
EA	970.26	127.4838			961.8484
Totals	**3739.48**	**926.1908**	**925.6276**	**1271.551**	**1271.51**
		926.1908 - 925.6276 Closure in Lat.=0.5632		1271.5510 - 1271.51 Closure in Dep.=0.0410	

LINEAR ERROR OF CLOSURE

$LEOC^2 = (Closure\ in\ Latitude)^2 + (Closure\ in\ Departure)^2$

$\text{LEOC}^2 = 0.5632^2 + 0.0410^2$

LEOC = 0.5647

PRECISION

$LEOC\ /\ Perimeter = 1\ /\ X$

0.5647 / 3739.48 = 1 / X

Precision = 1 / 6622

After measuring 3739 feet, the field engineers had a LEOC of 0.5647 feet. Expressed as a ratio, 1 foot in 6622 feet meaning that if the field engineers had measured 6622 feet, using the same techniques and precision, they would have been off 1 foot. For ordinary chaining this is an acceptable precision. When using an EDM for control work, a precision of 1/20,000 or more would be common.

ADJUSTMENT OF LATITUDES AND DEPARTURES

The computation of the LEOC indicates there are small errors that exist in the linear measurements of the traverse. It is common practice to perform an adjustment to the latitudes and departures to force the traverse into a closed geometric figure. In other words, the sum of the latitudes and the sum of the departures should be zero.

GENERAL

Although there are several different methods of adjustment available to the field engineer, the most common method used is the compass rule. Application of the compass rule assumes the angles and distances measured on the traverse were measured with equal precision. Thus, this rule is appropriate when similar precision equipment is used together. If a transit and chain are used, the compass rule applies. If a theodolite and EDM are used together, the compass rule is a still good choice. If a theodolite and chain are used together, perhaps a different method of adjustment should be considered because of the inequity in the precision of the two measuring devices.

The very best method of adjustment is by "least squares." This is a statistical adjustment that allows for all the variables that might affect the measuring process to be written into the equations. It is beyond the scope of this manual to discuss the development of the least squares condition equations. If a field engineer ever has a need for this type of adjustment, entire books are available that have been written about the process. See *Appendix D* for a listing of several textbooks that are specifically written for surveying computations.

COMPASS RULE

The compass rule calculates an adjustment that is applied to each latitude and each departure individually. A proportion is established that uses the length of the line, the perimeter, and the closure. In generic form the formula is:

$$\frac{Correction}{Closure} = \frac{Distance}{Perimeter}$$

THE FORMULA TO CORRECT FOR LATITUDES

$$\frac{Latitude\ Correction}{Closure\ in\ Latitudes} = \frac{Distance}{Perimeter}$$

$$Latitude\ Correction = \frac{(Closure\ in\ Latitude)(Dist.)}{Perimeter}$$

THE FORMULA TO CORRECT FOR DEPARTURES

$$\frac{Departure\ Correction}{Closure\ in\ Departures} = \frac{Distance}{Perimeter}$$

$$DepartureCorrection = \frac{(Closure\ in\ Departure)(Dist.)}{Perimeter}$$

CALCULATE THE CORRECTIONS

STEP 1

Calculate the latitude correction for each line. Use the distance for the line, the perimeter, and the closure in latitude, and substitute into the latitude correction formula. Write the value of the correction in the column between the latitudes.

STEP 2

Calculate the departure correction for each line. Use the distance for the line, the perimeter, and the closure in departure, to substitute into the departure correction formula. Write the value of the correction in the column between the departures.

SAMPLE CALCULATIONS

$$Correction\ Lat_{AB} = \frac{(0.5632)(677.97)}{3739.48} = 0.1021$$

$$CorrectionDep_{AB} = \frac{(0.0410)(677.97)}{3739.48} = 0.0074$$

Follow the same procedure for each line. See the table below for a summary of the calculated correction for each line.

Calculated Corrections for Latitude							
Line	**Distance**	**N. Lat**	**Correction**	**S. Lat**	**E. Dep**	**Correction**	**W. Dep**
AB	677.97	667.6701	-0.1021			+0.0074	117.7283
BC	616.05		+0.0928	166.1858	593.2115	-0.0068	
CD	690.88	131.0369	-0.1041		678.3395	-0.0076	
DE	783.32		+0.1180	759.4418		+0.0086	191.9333
EA	970.26	127.4838	-0.1461			+0.0106	961.8484
Totals	3739.48	926.1908	<Compare>	925.6276	1271.551	<Compare>	1271.51
		Larger: subtract correction		Smaller: add correction	Larger: subtract correction		Smaller: add correction

APPLY THE CORRECTIONS

The objective is to balance the latitudes and departures so the closure in both is zero. **Basic Rule: Add to the smaller and subtract from the larger.**

LATITUDE CORRECTIONS

Review the summations of the North latitudes and the South latitudes. To balance these numbers, one must be increased and the other decreased so they are the same. Therefore, the corrections must be added to the smallest latitude and subtracted from the largest latitude.

In the example, the total of the North latitudes = 926.1908 and the total of the South latitudes = 925.6276. Therefore, the North latitudes need to be decreased (subtract from the larger) and the South latitudes need to be increased (add to the smaller).

DEPARTURE CORRECTIONS

Review the summations of the East departures and the West departures. To balance these numbers, one must be increased and the other decreased so they are the same. Therefore, the corrections must be added to the smallest departure and subtracted from the largest departure.

In the example, the total of the East departures = 1271.5510 and the total of the West departures = 1271.5100. Therefore, the East departures need to be decreased (subtract from the larger) and the West departures need to be increased (add to the smaller).

SUMMARY OF ADJUSTED LATITUDE AND DEPARTURES

Adjusted Latitudes and Departures				
Line	*N. Lat.*	*S. Lat.*	*E. Dep.*	*W. Dep.*
AB	667.568			117.7357
BC		166.2786	593.2047	
CD	130.9328		678.3319	
DE		759.5598		191.9419
EA	127.3377			961.859
Total	**925.8385**	**925.8384**	**1271.5366**	**1271.5366**
	Closure in Lat. = 0.00		**Closure in Dep. = 0.00**	

FINAL CHECK

It can be seen by observation that the summation of the North and South latitudes is equal (925.8385 & 925.8384 The 0.0001 represents a rounding error) and the summation of the East and West departure is equal (1271.5366). Therefore the adjustment was carried out successfully.

CALCULATING RECTANGULAR COORDINATES

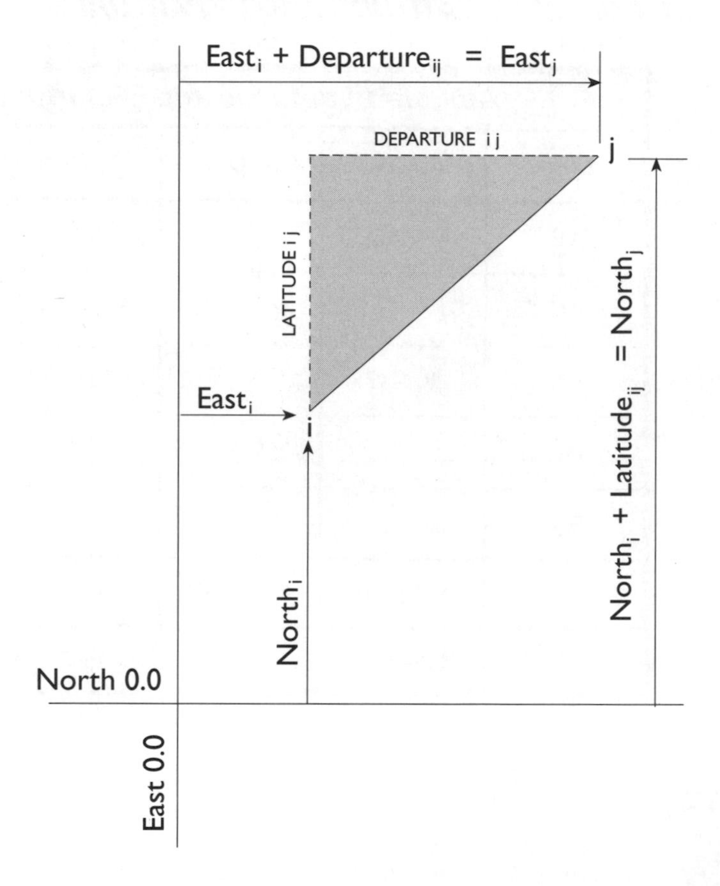

SCOPE

Coordinates are numbers. They represent a method of defining the location of a point by referencing to North and East axes of a rectangular coordinate system. The North and East coordinates for a point represent the distance to the point from the origin of each axis. Every point on a construction site, the control, the structure, utilities, etc., will have a unique North and East coordinate that can be used by the field engineer for layout purposes.

GENERAL

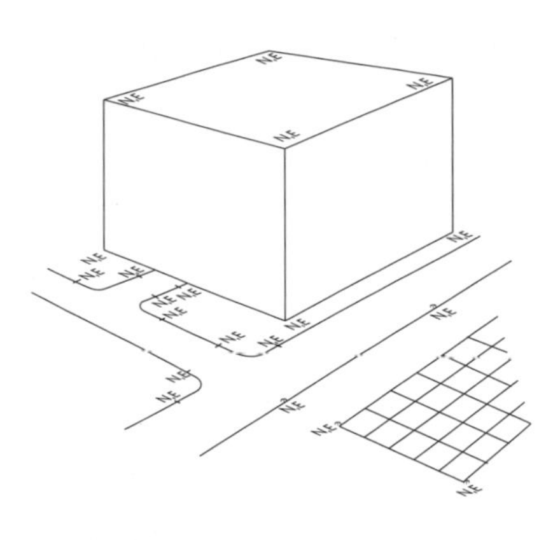

Coordinate systems on most building construction sites are typically only used on that particular site. That is, rather than having one coordinate system for all construction, each building project will have its own coordinate system. This hasn't been a problem because over the years there wasn't a real need to reference a project on one side of a town to a project on the other side of the town. Additionally, there hasn't been a cost effective method of keeping each project on the same system or making the measurements required to do it.

However, this is changing with the availability of GPS equipment. Soon, this equipment will easily allow projects to be designed using a state or even a national coordinate system. See *Appendix C* for more information on GPS and construction. The use of coordinates by the field engineer will be explained more thoroughly in the next chapter, *Coordinates in Construction*.

CALCULATING COORDINATES ON A TRAVERSE

Calculating coordinates for a traverse is very systematic. A starting coordinate for a point must be known, and the adjusted latitudes and departures for each line must be known.

STARTING COORDINATE

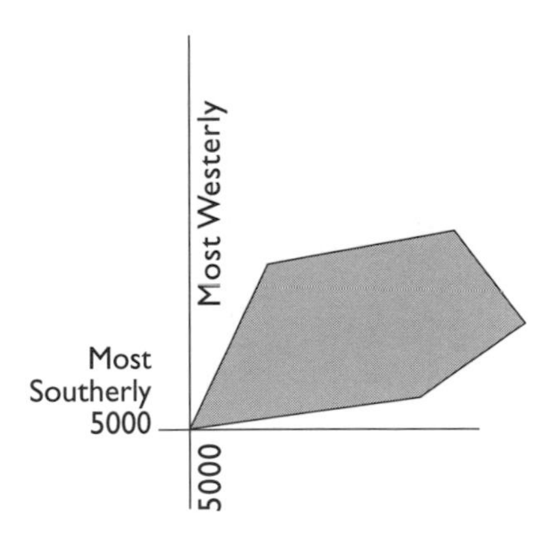

Use the most Westerly and most Southerly Point.

The field engineer needs to have starting coordinates for one of the points of the traverse. This starting coordinate may come from the plans, from the report of survey by the professional surveyor, or by just assuming some numbers for one of the points. Often times the starting coordinates are assumed. If they are, it is a good idea to start the coordinates with the most southwesterly point of the traverse and to start with fairly large initial coordinates so there will be no possibility of negative coordinates on the traverse. Having negative coordinates just introduces an additional source of error in the calculations. See *Chapter 16, Coordinates in Construction (16-5)* for a more complete discussion of this topic.

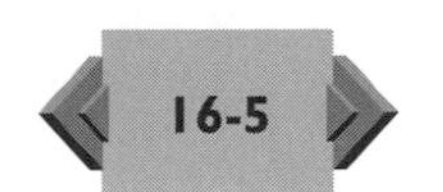

ADJUSTED LATITUDES AND DEPARTURES

At this point in the discussion of traverse computations, adjusted latitudes and departures are available for the continuing example problem.

THE FORMULAS

The calculation of coordinates at this stage only requires addition and subtraction. In generic form, the formulas to calculate the coordinates for line (i-j) are:

$$North_{(i)} + Latitude_{(ij)} = North_{(j)}$$
$$East_{(i)} + Departure_{(ij)} = East_{(j)}$$

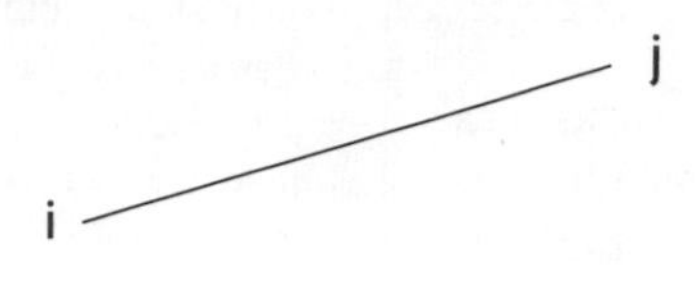

More specifically for a traverse ABCDEA the North formulas would be:

$$North_{(A)} + Latitude_{(AB)} = North_{(B)}$$
$$North_{(B)} + Latitude_{(BC)} = North_{(C)}$$
$$North_{(C)} + Latitude_{(CD)} = North_{(D)}$$
$$North_{(D)} + Latitude_{(DE)} = North_{(E)}$$
$$North_{(E)} + Latitude_{(EA)} = North_{(A)}$$

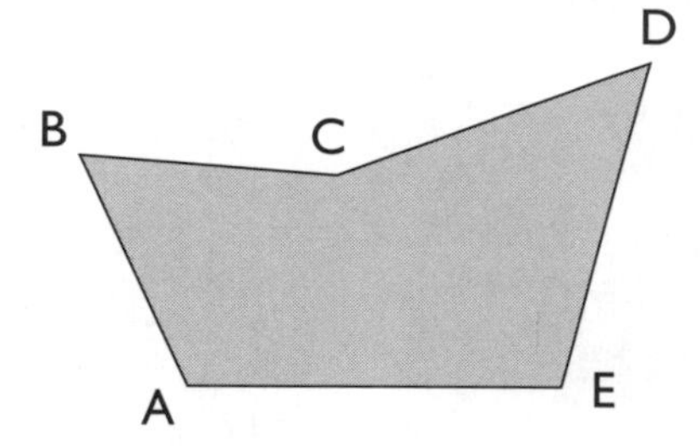

and the East formulas would be:

$$East_{(A)} + Departure_{(AB)} = East_{(B)}$$
$$East_{(B)} + Departure_{(BC)} = East_{(C)}$$
$$East_{(C)} + Departure_{(CD)} = East_{(D)}$$
$$East_{(D)} + Departure_{(DE)} = East_{(E)}$$
$$East_{(E)} + Departure_{(EA)} = East_{(A)}$$

It should be evident that the formulas for calculating coordinates for a traverse involve simply moving sequentially from one point to the next to the next and so forth until returning to the original starting coordinate.

WATCH THE SIGNS

The sign of the latitude and departure must be strictly followed to get the correct coordinates. Recall that North latitudes are postive and Southlatitudes are negative and East departures are postive and West departures are negative.

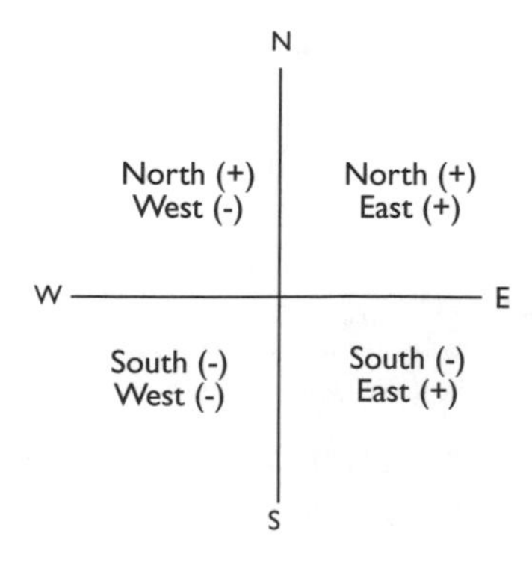

CONTINUING EXAMPLE

Given: Traverse ABCDEA with starting coordinates of N1000 and E1000 for point A and the adjusted latitudes and departures.

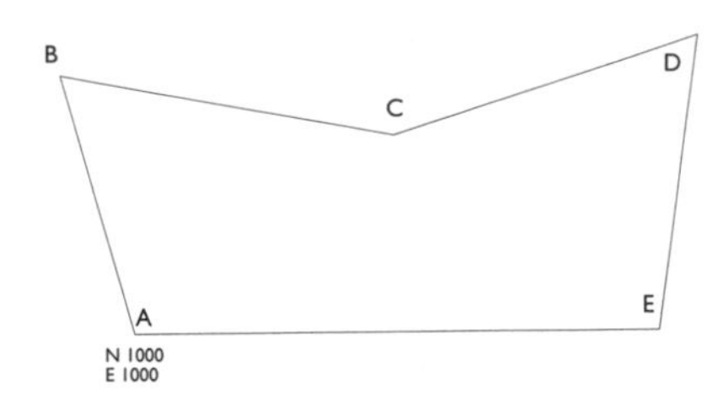

For point B

$North_{(A)} + Latitude_{(AB)} = North_{(B)}$
$1000 + 667.5680 = 1667.5680$

$East_{(A)} + Departure_{(AB)} = East_{(B)}$
$1000 + (-117.7357) = 882.2643$

For the remaining points, see the summary in the table below.

Calculation of Coordinates

Line	Latitudes	Departure	Point	North	East
			A	1000	1000
AB	N667.5680	W117.7357			
			B	1667.568	882.2643
BC	S166.2786	E593.2047			
			C	1501.2894	1475.469
CD	N130.9328	E678.3319			
			D	1632.2222	2153.8009
DE	S759.5598	W191.9419			
			E	872.6624	1961.859
EA	N127.3377	W961.8590			
			A	1000.0001	1000.0000

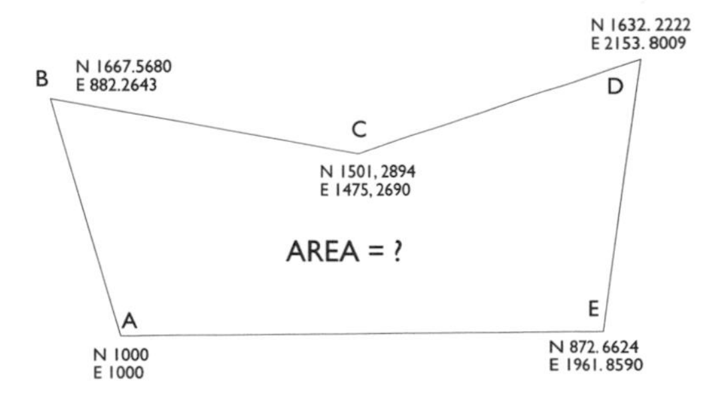

The 1000.0001 North Coordinate represents a rounding error when the latitudes were adjusted. It is not measurable, therefore it can be ignored.

AREAS BY COORDINATES

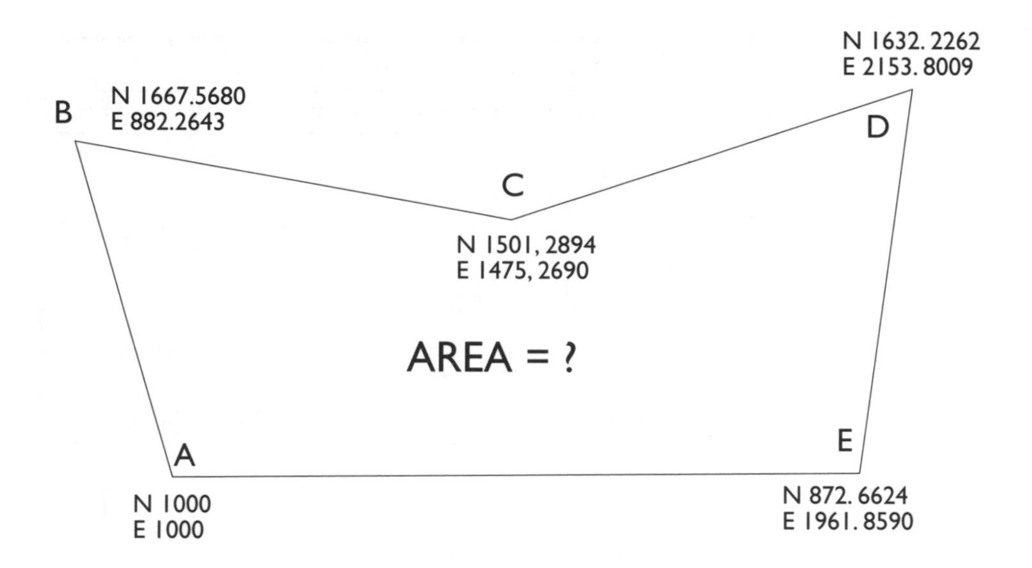

SCOPE

Now that coordinates have been calculated, the field engineer can begin to use their power. One of these uses is the determination of area. The area of traverses, cross sections, or any other irregular shape can easily be calculated if coordinates around the figure are known. This process involves using a very specific method of listing the coordinates and a systematic method of computations. For additional uses of Coordinates see *Chapter 16, Coordinates in Construction.*

GENERAL

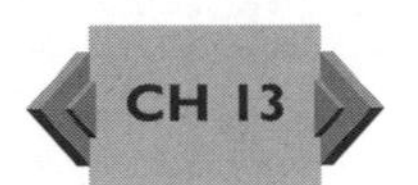

Calculating area is fairly easy if the shape of the figure is one in which geometric formulas are known. See *Chapter 13, Math Review and Conversions* for a summary of formulas for standard shapes. However, if the shape is irregular it becomes much more difficult to obtain an area. Sometimes an irregular shape can be broken down into smaller figures that conform to known shapes and area formulas. But, that is a very time consuming and laborious task. There are several methods that have been developed for calculating area by coordinates. All are based on the mathematics of breaking the irregular shape into a series of trapezoids and systematically determining the area. Some of the methods involve a seemingly complex listing of the points and keeping track of numerous multiplications and subtractions. A simple method of determining areas with coordinates is presented here.

PROCEDURE

This method is best explained through the use of an example problem. The figure at right shows a simple shape that the area is easily determined by the common geometric formulas. Area one is a triangle and area two is a square. The total area of this figure is determined by calculating each area and then adding the areas together.

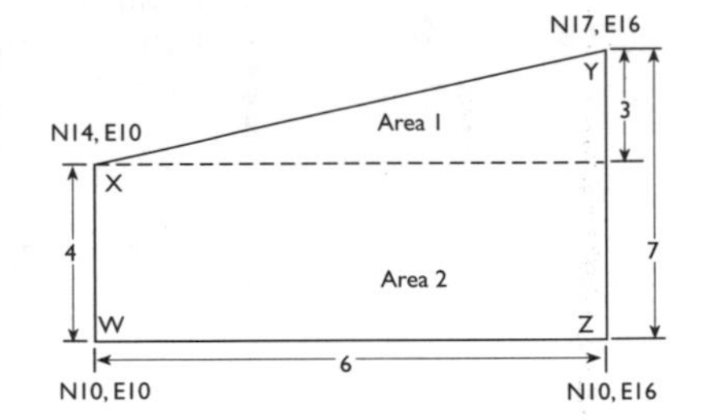

$$Area1 = \frac{bh}{2} = \frac{6x3}{2} = 9$$

$$Area2 = bh = 6x4 = 24$$

$$Total\ Area = 33$$

This certainly isn't a difficult problem to solve geometrically, but it can easily be shown how to determine the same area using coordinates.

Always repeat the first coordinate

Area Calculation by Coordinates

List the coordinates in consecutive order around the figure making sure to repeat the first coordinate.

Pt	N	E
W	10	10
X	14	10
Y	17	16
Z	10	16
W	10	10

Cross multiply all of the corrdinates from left down to the right and sum these values.

Pt	N	E	
W	10	10	
X	14	10	= *10x10 = 100*
Y	17	16	= *14x16 = 224*
Z	10	16	= *17x16 = 272*
W	10	10	=*10x10 = 100*
			Total = 696

Cross multiply all of the coordinates from right down to the left and sum these values.

	Pt	N	E
	W	10	10
10x14 = 140	X	14	10
10x17 = 170	Y	17	16
16x10 = 160	Z	10	16
16x10 = 160	W	10	10
Total = 630			

Determine the algebraic difference between these sums.

$$696 - 630 = 66$$

Divide this difference by 2 to get the area.

$$Area = \frac{66}{2} = 33$$

which is the same area that was obtained by geometric formula

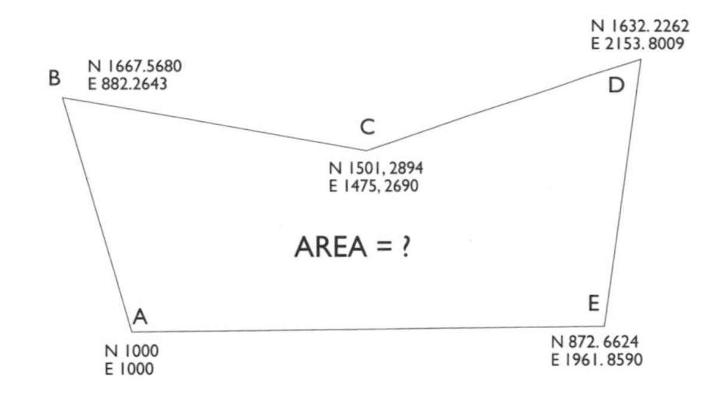

Point	North	East
A	1000	1000
B	1667.568	882.2643
C	1501.2894	1475.2690
D	1632.2262	2153.8009
E	872.6624	1961.859
A	1000.000	1000.0000

CONTINUING EXAMPLE

By following the area by coordinates procedure, to calculate the area of this figure:

Cross multiply left down to right and summed = 10,650,714.20

Cross multiply right down to the left and summed = 9,241,474.82

Difference between the summations = 1,409,239.38

Divide by 2 to get area = 704,619.69 Sq. Ft.

Divide by 43560 to get acreage = 16.1758 acres

SUMMARY

This method of calculating area by coordinates is very quick and easy. It can be calculated using a hand held calculator or can be programed easily into an electronic spreadsheet. The field engineer should always perform this type of computation twice to ascertain that the correct result has been obtained.

TRAVERSE COMPUTATION SHEET

The following computation sheet can be used effectively by the field engineer to keep all of the calculations together and organized.

TRAVERSE COMPUTATIONS

Field Notes

Book No._____ Page _____

					Latitudes		Departures		Adjusted		Point	Coordinates	
Line	Direction	Distance	Cosine	Sine	N (+)	S (-)	E (+)	W (-)	Lat	Dep		North	East
	Σ			Σ									

Computation, Date __________

Computed by __________

Checked by __________

Closure = __________

Precision = 1/__________

Area = ______ Sq. Ft. _____

Area = ______ Acres _____

COORDINATES IN CONSTRUCTION

16

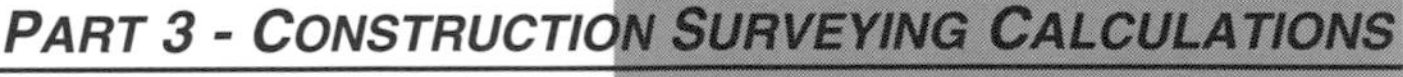

PART 3 - CONSTRUCTION SURVEYING CALCULATIONS

CHAPTER OBJECTIVES:

Describe how coordinates represent power to the field engineer.
Compare the rectangular coordinate system terms for mathematics and for surveying.
State the advantages and disadvantages of coordinates in construction.
List and state the uses of coordinates on construction sites.
Describe the process of inversing coordinates for distance and calculate an inverse distance.
Describe the process of inversing coordinates for direction and calculate an inverse direction.
Distinguish between distance/distance, distance/direction, and direction/direction intersection situations and be able to calculate for the given intersection situation.

INDEX

OVERVIEW

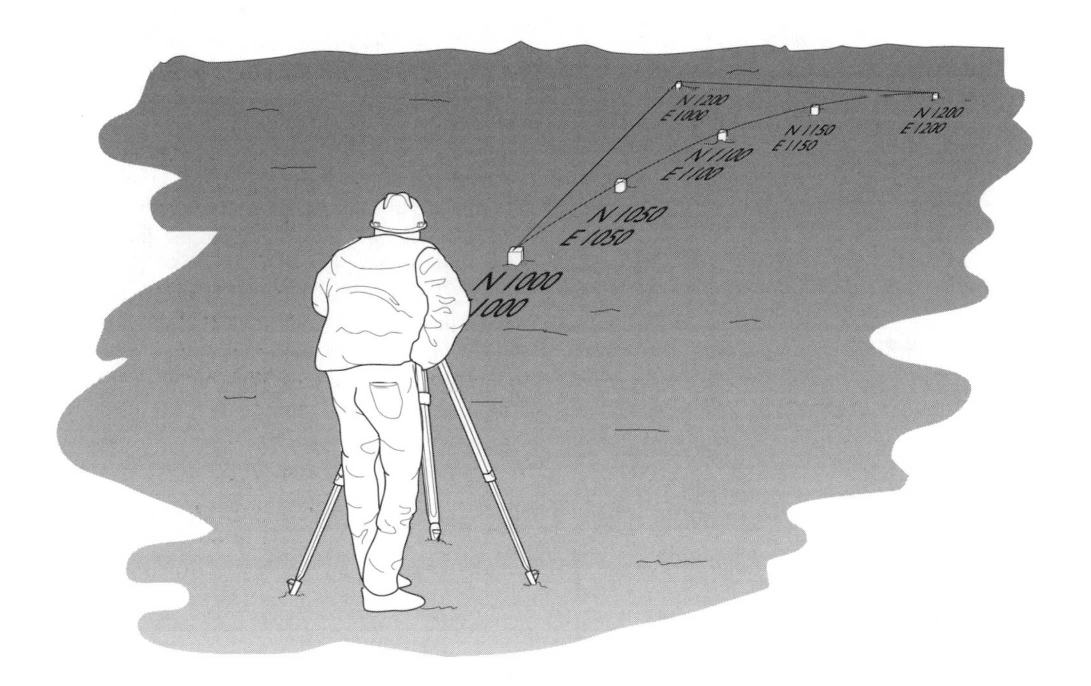

SCOPE

To have power is to control. It is the responsibility of the field engineer to exercise the power of surveying measurement to control the layout of structures on the jobsite. The most powerful and versatile method of surveying layout is utilizing rectangular coordinates.

Coordinates can be readily applied to construction. Many of today's projects are designed on CADD systems which utilize coordinates as their bases of operation. If coordinates of the structures and control points are known, the layout data of the field engineer can be easily calculated. Having coordinate values for all points on a project allows the field engineer the convenience of using any two points on a construction site to lay out any other point. That is power. That is control.

Using coordinates makes the job of the field engineer both easier and harder. Easier because of the power of coordinates; harder because of the chance for calculation and layout errors. Any field engineer using coordinates must thoroughly understand the coordinate system, coordinate geometry, radial layout techniques, and redundant checking of calculations and field layout. If proper procedures are used, near perfection in the layout of simple to complex projects can be achieved.

BACKGROUND (PAST, PRESENT, AND FUTURE)

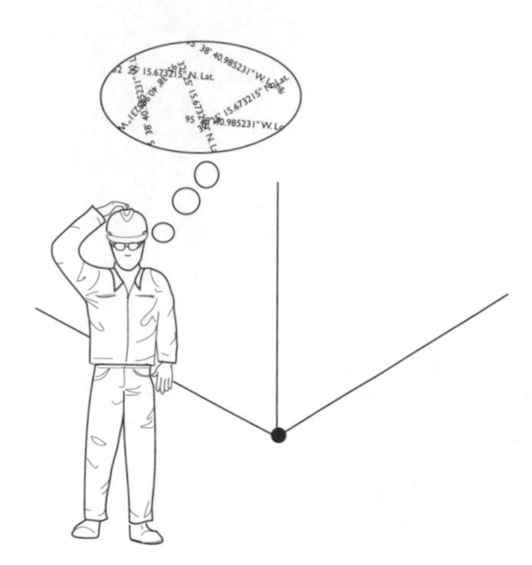

The use of coordinates in construction is not new. Coordinates have been used since man first determined a need to describe the location of a point. In ancient Egypt, along the Nile river, a crude system of coordinates was used to describe the location of property corners so they could easily be located after the yearly floods.

Coordinates were used by explorers to determine their relative location on the surface of the earth. Columbus and others used coordinates to determine the location of land they discovered. The system used then and now is a geodetic coordinate system called latitude and longitude. The equator represents 0° latitude and Greenwich, England, represents 0° longitude. Every point on earth has a unique latitude and longitude. That system was also used by Lewis and Clark in their exploration of the West. Their daily astronomical observations gave them data which allowed them to develop maps of the Missouri and Columbia Rivers.

We could use geodetic coordinates to describe points on our jobsites, but we haven't because it is difficult to determine latitude and longitude, and because it is based on a sphere. To obtain the detail needed, we would have to have something like 32°25'15.673215" N Lat., 95°38'40.985231" W Long. to describe the location of a column. Which is not quite practical for the construction site. The Global Positioning Systems (GPS) that are at the forefront of today's technology do give readouts in Latitude and Longitude. GPS is changing the use of geodetic coordinates in the everyday world, it still is not practical for typical construction although great advances are occurring everyday. To learn more about this subject, reference *Appendix C*.

RECTANGULAR COORDINATE SYSTEM REVIEW

Because most construction sites are limited to a relatively small area, the coordinate system that we utilize is based on a plane. That is, we assume the area of the earth where the project is located is theoretically flat. Even though we all know the shape of the earth is spherical, for all practical purposes, we assume we are working on a plane. We work from a rectangular coordinate system rather than the geodetic one that explorers and travelers on GPS uses.

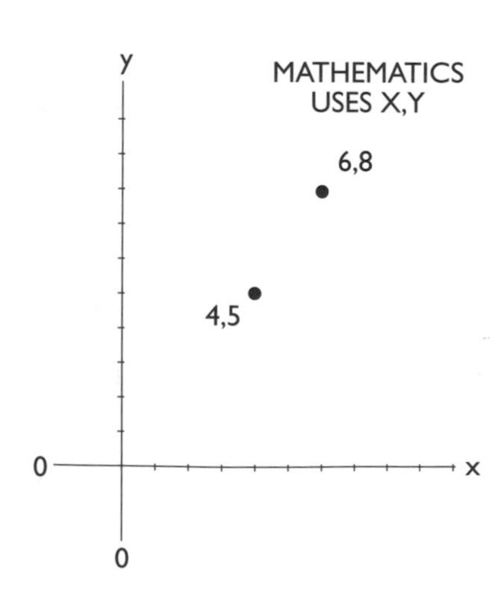

The rectangular coordinate system as we know it was first developed by a French mathematician in the 17th century. He devised a system which consisted of two geometric axes perpendicular to each other on a plane to locate a point. In mathematical terms, a horizontal axis and a vertical axis were used to describe the system. Again mathematically, coordinates of points are represented by listing first the horizontal component and then the vertical component. Coordinate pairs of 4,5 indicate a point that is 4 units to the right of the origin and 5 units above the origin. If another coordinate pair such as 6,8 is known, equations of geometry and trigonometry can then be applied to determine distance and direction between the points.

In surveying, one axis is oriented to North and the other to East. The origin is where the two axes intersect. The origin is represented by North 0.00 and East 0.00, respectively. Typically the N0.00, E0.00 origin is only a theoretical location and is never physically occupied or used in the field. If coordinates are to be assigned to a project, large numbers such as N5000, E5000 or larger, are used as starting coordinates to ensure that negative coordinates will never exist.

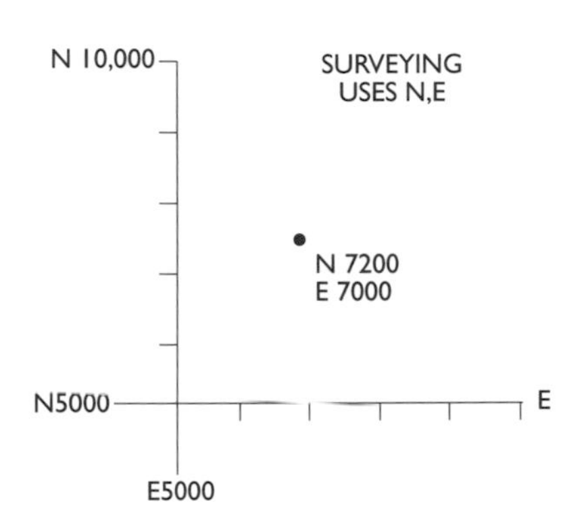

The main reason negative coordinates are avoided is to eliminate another source of potential errors. If negative coordinates do exist when using coordinates and performing coordinate geometry calculations, there must be strict adherence to the rules of algebra regarding negative signs. Although this isn't difficult, it is just one more step in the calculation process that could result in adding instead of subtracting or vice versa. It is good practice to avoid sources of mistakes in the calculation of fieldwork whenever possible.

ADVANTAGES

The advantages of rectangular coordinates center around their ease of use. They only require simple, straightforward math. They can be used and easily adapted to any project. In addition, horizontal coordinates transfer vertically. That is, the coordinate values of a point are the same at the bottom of a column as they are at the top of the column.

DISADVANTAGES

A disadvantage of rectangular coordinates is that they are not entirely understood by all jobsite personnel. It can also be difficult to detect mistakes if specific checks are not used.

WHERE TO USE COORDINATES

Anywhere. They can be used on any jobsite, building, roadway, bridge, etc. Coordinates can be used on large or small projects. However, they should only be used when they are the best method for the job. Large complex jobs spread over a large area lend themselves very well to the use of coordinate control and layout to tie all of the projects together. Small bridges may also use coordinates very well. Small buildings such as garages, small houses, etc. may not need coordinates. It is the responsibility of the field engineer to select the best method of layout for the site conditions.

CONSIDERATIONS IN COORDINATE USE

The size of the project

The complexity of the project

The project schedule

The types of surveying instruments available

The types of computational equipment available

SETUP OF A COORDINATE SYSTEM

DESIGNED INTO PLANS

The designer will provide coordinates of main building points. The field engineer will determine coordinates for other points as needed. See *Calculating Coordinates* in *Chapter 15, Traverse Computations*. The illustration below shows coordinates used for locating corners and columns of a building.

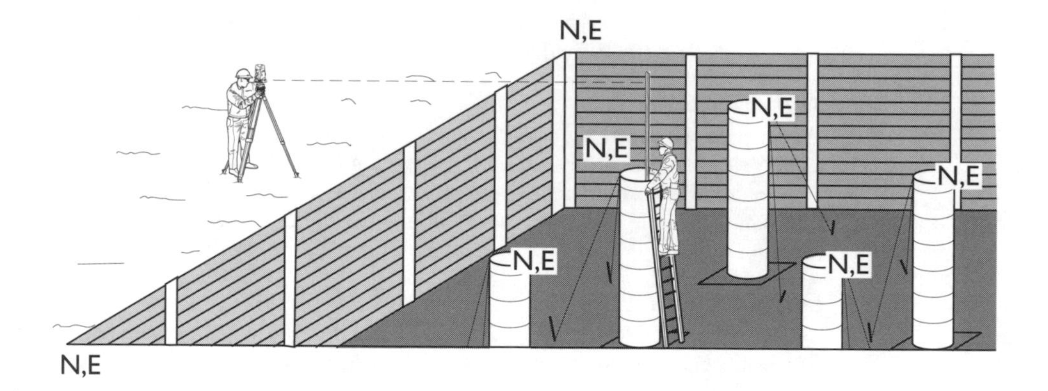

NOT ON PLANS

If coordinates are not used on the plans and the field engineer plans to use them for layout purposes, it will be necessary to design a coordinate system to fit the project. A primary control traverse must be ran and an arbitrary coordinate system must be chosen. The field engineer will have to calculate the traverse and coordinates of all control points as well as coordinates of all points on the structure. This can be a monumental job if the project is large. A computer with coordinate geometry software is very helpful.

USES OF COORDINATES

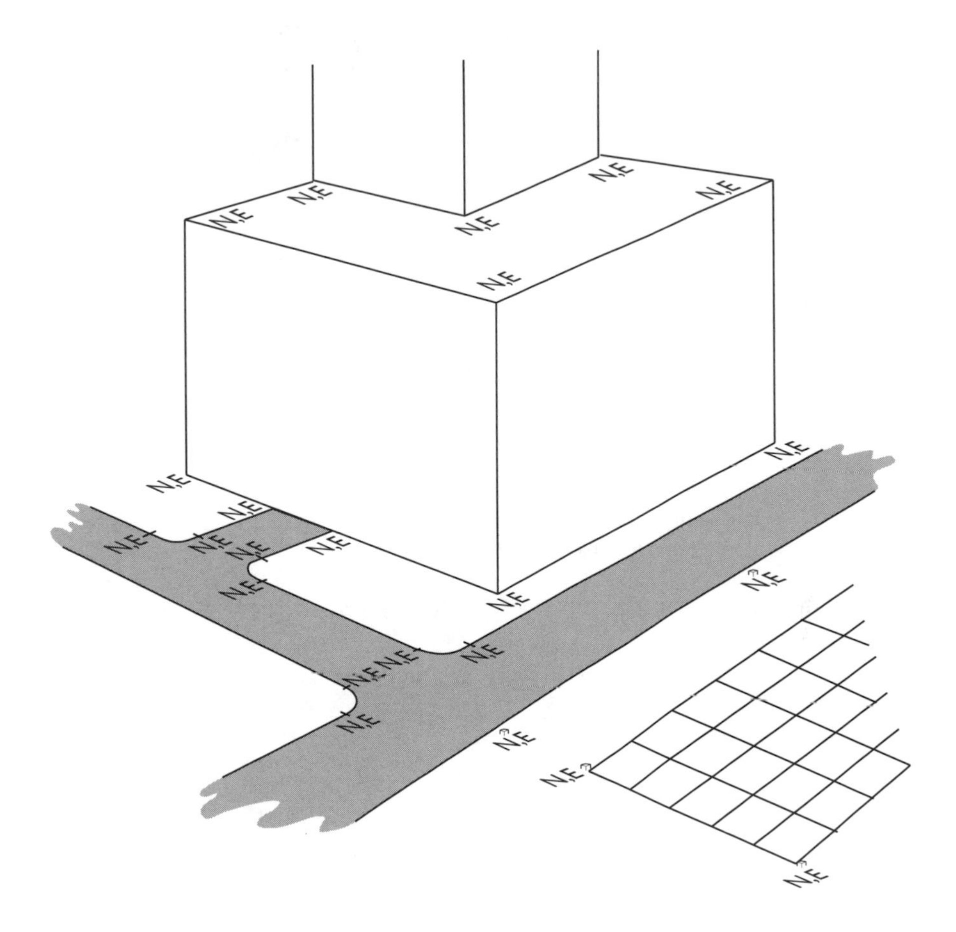

SCOPE

The uses of coordinates in construction are too numerous to attempt to list. Just about any layout activity can be performed by using coordinates. It is up to the field engineer to determine whether coordinates are the best method to use or not. Some examples are shown in this section.

CONSTRUCTION

BUILDINGS

Rectangular buildings, irregularly shaped buildings, buildings with curved surfaces, etc., are all adaptable to the use of coordinates.

BRIDGES

Bridges come in all sizes and shapes. Some are practically on the surface of the water and some are tall enough for large ships to go under. Some span small creeks or ravines and are less than 50 feet long, while other bridges go across large bodies of water and are miles long. All bridges are unique, but at the same time, they all are easily adaptable to being controlled by a coordinate system.

ROUTES

Highway projects sometimes extend for miles and contain curves and bridges. Coordinates can be used to define every point on the route project.

UTILITY

Utilities are part of building construction, route construction, sitework, etc. Coordinate systems developed for the overall project can be used to locate utilities from any control point on the project.

SITEWORK

Sidewalks, curbs, gutters, catch basins, etc., are all part of the overall construction project. Again, they can be easily described by coordinates and located by coordinate layout data by the field engineer.

INVERSE

$$\text{Distance}_{12} = \sqrt{(N_2 - N_1)^2 + (E_2 - E_1)^2}$$

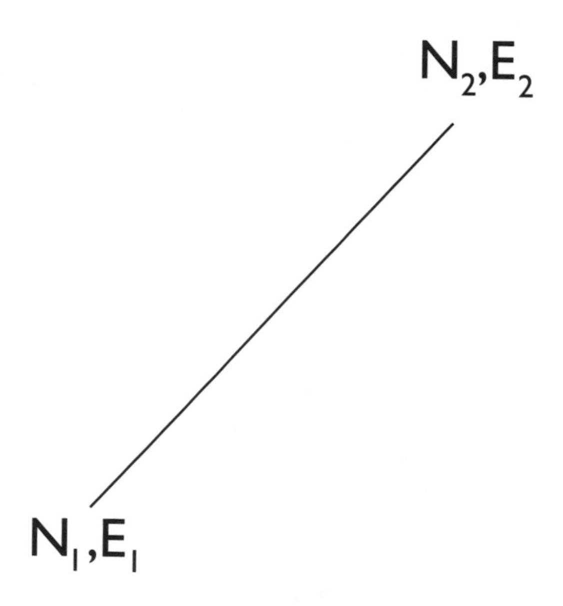

$$\text{Direction} = \text{Arctan}\,\frac{E_2 - E_1}{N_2 - N_1}$$

SCOPE

Very likely the most calculated aspect of coordinates will be determining the distance and direction between two pairs of existing coordinates. This process is called inversing. Field engineers use this constantly to obtain data for radial layout.

INVERSE DIRECTION

Part of the power of coordinates is being able to determine the direction between one point with coordinates and any other point with coordinates. Differences in coordinate pairs can be applied to right angle trigonometry to determine the direction simply and quickly. Field engineers performing radial layout of points will calculate directions from coordinates daily.

$$Direction = \arctan\frac{E_2 - E_1}{N_2 - N_1}$$

$$Direction = \arctan\frac{Departure}{Latitude}$$

FROM MATHEMATICS

In mathematics, the primary axis from which calculations are based is the x-axis. The coordinates are used to determine rectangular differences between the respective coordinates. These rectangular differences in the x and y directions correspond to the opposite and adjacent sides of a right triangle. This calculation is based on the right angle trigonometry formula where the opposite and adjacent sides of the triangle are used to determine the value for the tangent function of an angle. When that is known, the angle off of the x axis in degrees, minutes, and seconds can be determined. This is illustrated below.

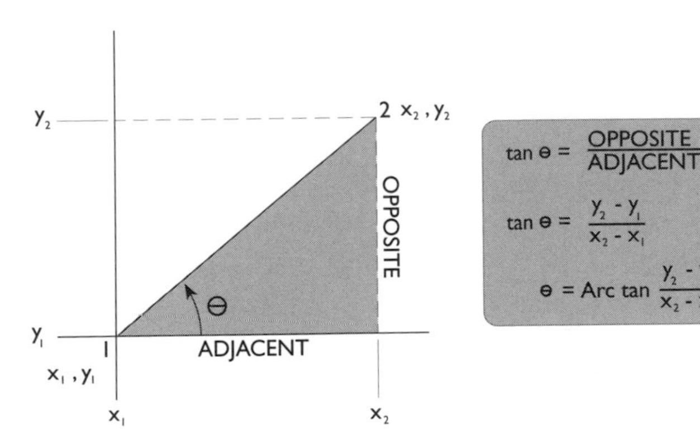

EXAMPLE

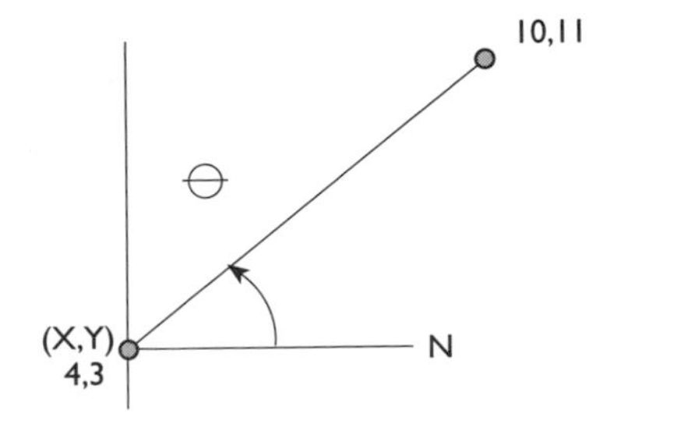

$$\Theta = \arctan\frac{11-3}{10-4}$$

$$= 53^\circ 07' 48''$$

IN SURVEYING

The same formula applies in surveying calculations, but the axis that is used is different and the names of the components are different. The "y" or North axis is the reference direction in surveying so the calculations are based off of it. Again, the coordinates are used to determine rectangular differences (latitudes and departures) between the respective starting and ending points of the line. These rectangular differences in the North and East directions correspond to the opposite and adjacent sides of a right triangle. The same right angle trigonometry formula is used where the opposite and adjacent sides of the triangle are used to determine the value for the tangent function of an angle. When that is known, the direction in degrees, minutes, and seconds can be determined. This is illustrated below.

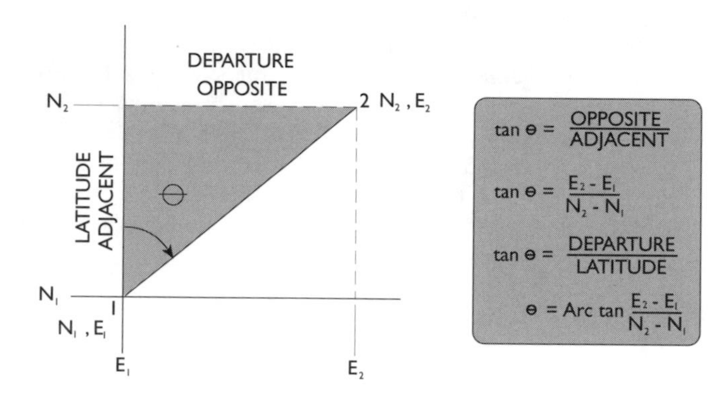

EXAMPLE

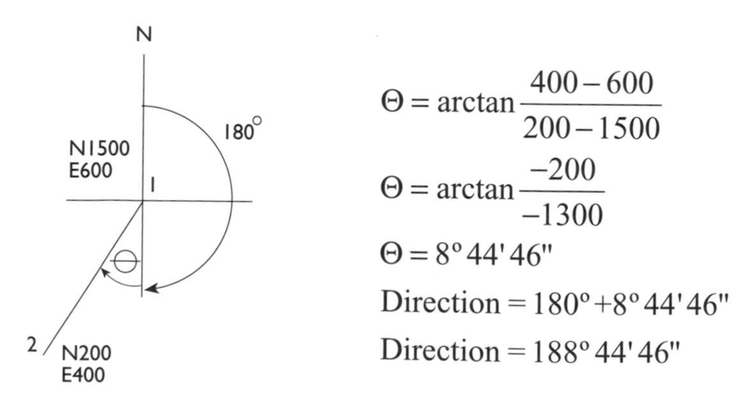

$$\Theta = \arctan \frac{400 - 600}{200 - 1500}$$

$$\Theta = \arctan \frac{-200}{-1300}$$

$$\Theta = 8^\circ 44' 46''$$

$$\text{Direction} = 180^\circ + 8^\circ 44' 46''$$

$$\text{Direction} = 188^\circ 44' 46''$$

DETERMINING THE DIRECTION

Rule of Thumb: To be able to determine the correct quadrant the direction is in, always subtract the starting coordinate from the ending coordinate.

$$E_2 - E_1$$
$$N_2 - N_1$$

The field engineer must be very careful to observe the quadrant in which they have calculated direction. The quadrant can be determined by observing the signs of the results of E_2-E_1 and N_2-N_1. For example, if E_2-E_1 is positive, and N_2-N_1 is negitive, the direction is SE or an azimuth between 90° and 180°. Many calculators automatically display the correct directions for the field engineer. The illustration and table below will be helpful in determining the correct result.

N

$E_2 - E_1 = (-)$ $N_2 - N_1 = (+)$	$E_2 - E_1 = (+)$ $N_2 - N_1 = (+)$
W	**E**
$E_2 - E_1 = (-)$ $N_2 - N_1 = (-)$	$E_2 - E_1 = (+)$ $N_2 - N_1 = (-)$

S

Direction When Inversing			
Bearing	Azimuth	$E_2 - E_1$	$N_2 - N_1$
NE	0°-90°	+	+
SE	90°-180°	+	-
SW	180°-270°	-	-
NW	270°-360°	-	+

INVERSE DISTANCE

Part of the power of coordinates is being able to determine distance between one point with coordinates and any other point with coordinates. Differences in coordinate pairs can be applied to the basic principle of the Pythagorean Theorem to determine distances quickly. Field engineers performing radial layout of points will calculate distances from coordinates constantly.

$$\text{Distance} = \sqrt{(N_2 - N_1)^2 + (E_2 - E_1)^2}$$

$$\text{Distance} = \sqrt{(Latitude)^2 + (Departure)^2}$$

FROM MATHEMATICS

This calculation is based on the distance formula from mathematics. The distance formula is used with a rectangular coordinate system to determine rectangular differences between the respective coordinates. Since, by definition, the axes of rectangular coordinate systems come together at a right angle, the Pythagorean Theorem is used to determine the distance between points. This is illustrated below.

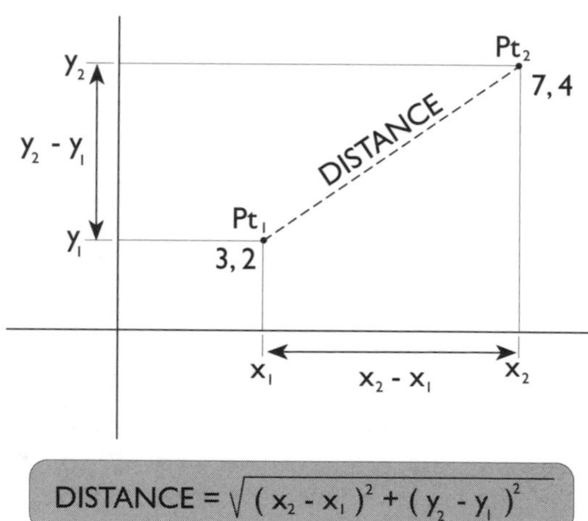

$$\text{DISTANCE} = \sqrt{(x_2 - x_1)^2 + (y_2 - y_1)^2}$$

$$= \sqrt{(7 - 3)^2 + (4 - 2)^2}$$

$$= \sqrt{16 + 4}$$

$$= 4.47$$

Remember, the coordinate at the beginning of the line is always subtracted from the coordinate at the end of the line.

IN SURVEYING

The same formula applies in surveying calculations, but the axis that is used is different and the names of the components are different. The "y" or North axis is the reference direction in surveying so the calculations are based on the axis. The coordinates are used to determine rectangular differences (latitudes and departures) between the respective starting and ending points of the line. These rectangular differences in the North and East directions correspond to the opposite and adjacent sides of the triangle. Thc same Pythagorean Theorem is applied where the opposite and adjacent sides of the triangle are used to determine the distance between the end points of the line. This is illustrated below.

EXAMPLE

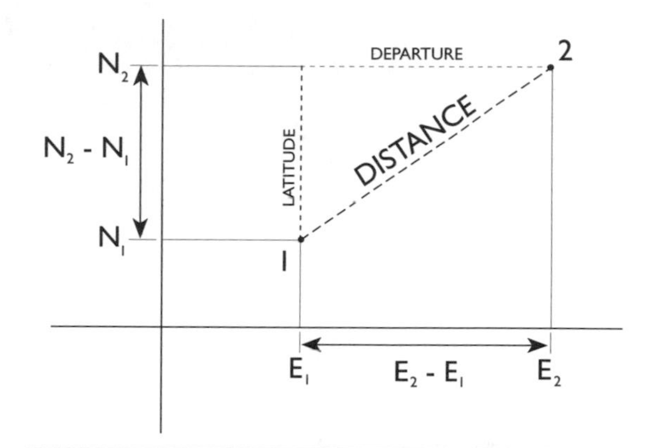

$$\text{INVERSE DISTANCE} = \sqrt{(N_2 - N_1)^2 + (E_2 - E_1)^2}$$

$$\text{INVERSE DISTANCE} = \sqrt{(\text{Latitude})^2 + (\text{Departure})^2}$$

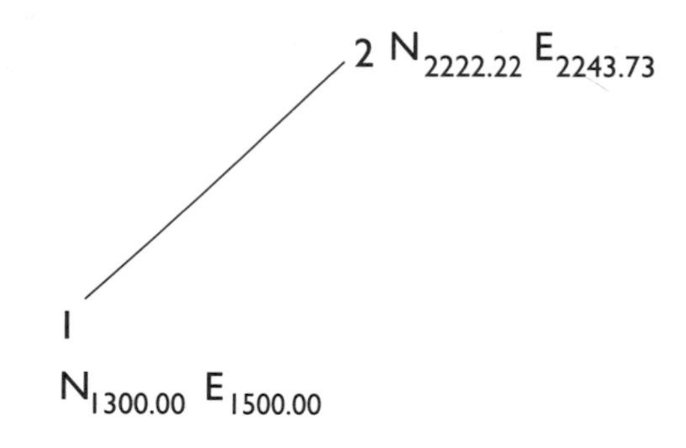

$$\text{DISTANCE}_{2-1} = \sqrt{(2222.22 - 1300)^2 + (2243.73 - 1500)^2}$$

$$= 1184.75'$$

INTERSECTION CALCULATIONS

SCOPE

While calculating layout data, the field engineer will occasionally encounter a situation where needed distances or directions to a point are not known. Often times these situations can be solved by the use of intersections. Three types of intersection problems are possible: distance/distance, distance/direction, and direction/direction. Using the available information, the Law of Sines or the Law of Cosines, and traverse computation techniques, the unknown distances and/or direction and the coordinates can be determined.

DISTANCE/DISTANCE INTERSECTION

This situation may be encountered if the field engineer has two known points on the plans and distances to a third point but no directions in which to calculate the coordinates. See the illustration below for the known and unknown information. Because there are no directions, two possible solutions are possible. Be careful when performing these calculations to obtain the desired results.

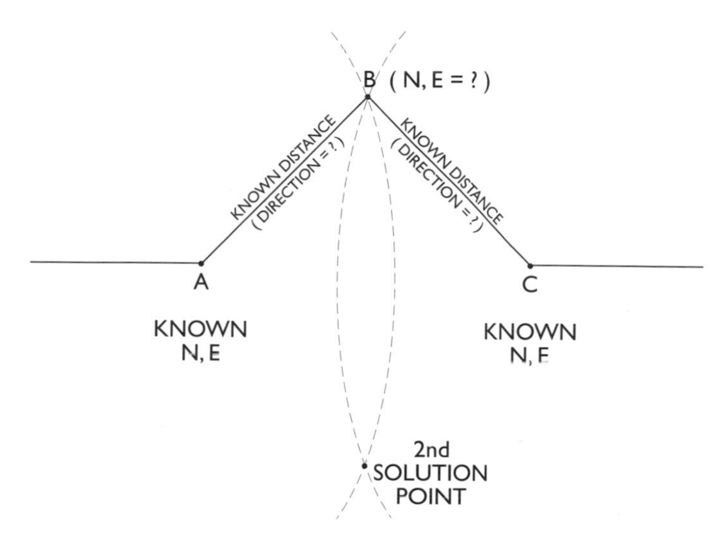

STEP-BY-STEP DISTANCE/DISTANCE EXAMPLE

Calculate the coordinates of point B for the given situation.

STEP 1

Sketch the problem and list what is known and unknown. Determine which intersection solving method should be used and proceed with the solution.

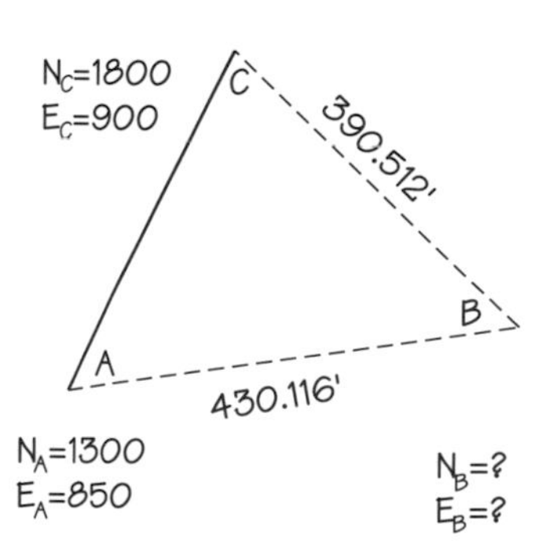

STEP 2

Inverse between the known coordinates for A and C.

$$\overline{AC} = \sqrt{(1800-1300)^2 + (900-850)^2}$$

$$\text{Distance}_{AC} = 502.494$$

$$Azimuth_{AC} = \arctan\frac{(900-850)}{(1800-1300)}$$

$$Azimuth_{AC} = 5^{\circ}\,42'38''$$

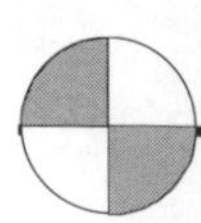

STEP 3

Use the Law of Cosine to solve for Angle A

$$Cos\frac{A}{2} = \sqrt{\frac{s(s-a)}{bc}}$$

$$S = \frac{a+b+c}{2}$$

$$Angle\ A = 48^{o}45'06"$$

STEP 4

Use the direction of the line AC and the calculated angle at A to determine the direction from A to B.

For this example, adding the direction and angle together will obtain the direction from A to B. That isn't always the case. Refer to *Chapter 15, Traverse Computations* for the process to use.

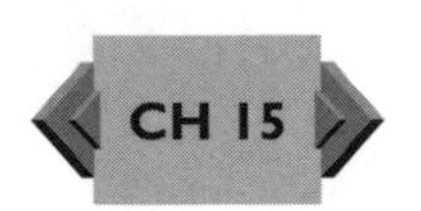

$$5^{o}42'38" \ + \ 48^{o}45'06"$$

$$\text{Azimuth}_{AB} = 54^{o}27'44"$$

STEP 5

Using the known distance and direction of line AB, calculate the latitudes and departures for the line.

$$Latitude_{AB} = 430.116Cos54^{o}27'44"$$

$$Latitude_{AB} = 250.00$$

$$Departure_{AB} = 430.116Sin54^{o}27'44"$$

$$Departure_{AB} = 350.00$$

STEP 6

Using the coordinates of point A and the latitude and departure for line AB, calculate the coordinates of Point B.

$$North_{A} \ + \ Latitude_{AB} \ = \ North_{B}$$

$$North_{B} \ = \ 1550.00$$

$$East_{A} \ + \ Departure_{AB} \ = \ East_{B}$$

$$East_{B} \ = \ 1200.00$$

STEP 7

Check the coordinates of Point B by performing the same calculations from point C. Identical results should be obtained

North B = 1550.00
East B = 1200.00

DISTANCE/DIRECTION INTERSECTION

This situation may be encountered if the field engineer has two known points on the plans and a distance from one point and a direction from another point to a third point but lacks enough information to calculate the coordinates. See the illustration below for the known and unknown information. Be careful when performing these calculations to obtain the desired results.

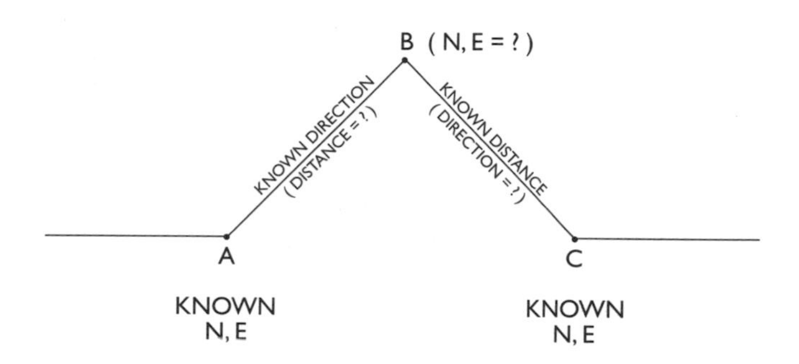

STEP-BY-STEP EXAMPLE

Calculate the coordinates of point B for the given situation.

STEP 1

Sketch the problem and list what is known and unknown. Determine which intersection solving method should be used and proceed with the solution.

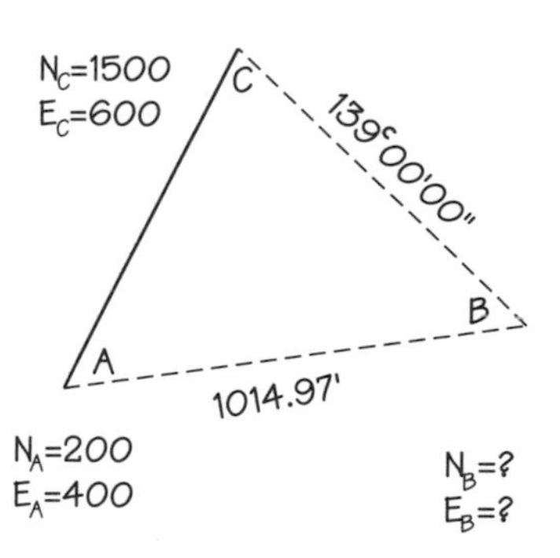

STEP 2

Inverse between the known coordinates for A and C.

Distance$_{AC}$ = 1315.29 Direction AC = 8°44'46"

STEP 3

CH 15

Use the back azimuth of line AC and the direction of line CB to calculate the interior angle at C. For this example, the direction of CB is subtracted from the direction of AC. Again, that isn't always the case. Refer to *Chapter 15, Traverse Computations*, for the process to use.

188°44'46" - 139°00'00" = Angle C = 49°44'46"

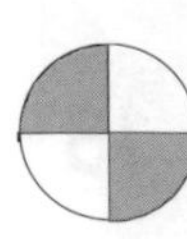

STEP 4 With the angle as C, the distance AB and AC, use the Law of Sines to solve for the angle at B. *Angle B = 81°29'51"*

STEP 5 Add the angle at C and the angle at B and subtract from 180 to calculate the remaining angle for the triangle. *Angle A = 48°45'23"*

STEP 6 Use the remaining angle and the Law of Sines to determine the distance CB. $Distance_{CB} = 1000.00$

STEP 7 Knowing the distance and direction of line CB, calculate the latitudes and departures for the line. $Latitude_{CB} = -754.71$ $Departure_{CB} = 656.06$

STEP 8 Using the coordinates of point C and the latitude and departure for line CB, calculate the coordinates of Point B. $North_B = 745.29$ $East_B = 1256.06$

STEP 9 Check the coordinates of Point B by performing the same calculations from point A. *Check calculations*

DIRECTION/DIRECTION

This situation might occur when the field engineer is in the field and wants to determine the coordinates of a point that cannot be occupied. Using a theodolite the field engineer turns angles to the point from two known points. See the illustration below for the known and unknown information. Be careful when performing these calculations to obtain the desired results.

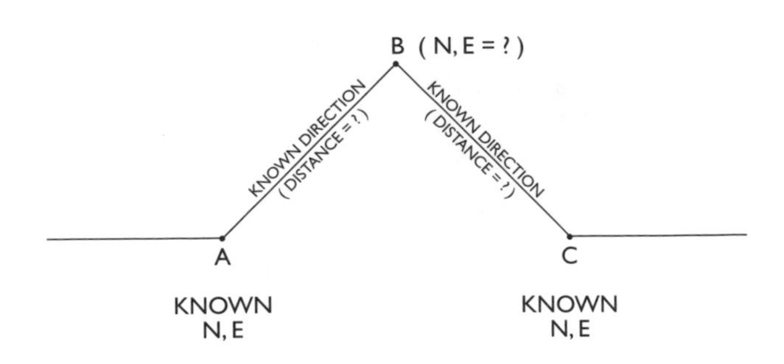

STEP-BY-STEP DIRECTION/DIRECTION EXAMPLE

Calculate the coordinates of point B for the given situation.

STEP 1

Sketch the problem and list what is known and unknown. Determine which intersection solving method should be used and proceed with the solution.

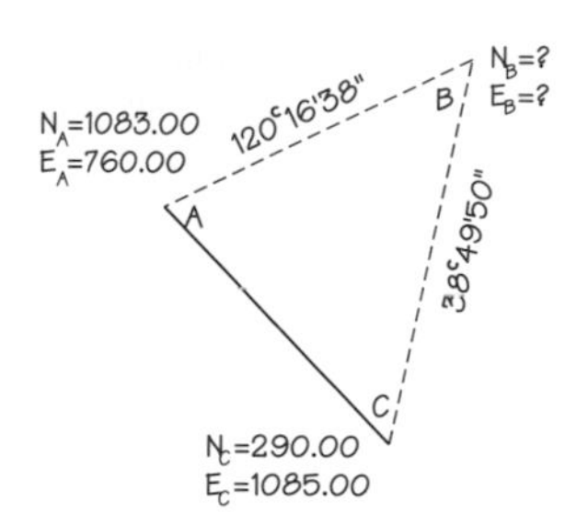

STEP 2

Inverse between the known coordinates for A and C.

Distance $_{AC}$ = *857.01*
Azimuth $_{AC}$ = *157°42'52"*

STEP 3

Use the calculated direction of AC and the known directions to determine the interior angles of the triangle.

Angle A = *37°26'14"*
Angle B = *81°26'48"*
Angle C = *61°06'58"*

STEP 4

With all three angles known and one distance, use the Law of Sines to solve for distances AB and CB.

$$\frac{857.01}{Sin81°26'48"} = \frac{AB}{Sin61°06'58"}$$

Distance AB = *758.84*
Similarly,
Distance CB = *526.83*

STEP 5

Using the coordinates of point A and the latitude and departure for line AB.

$Latitude_{AB}$ = *-382.595*
$Departure_{AB}$ = *655.33*

STEP 6

Calculate the coordinates of Point B.

$North_B$ = *700.40*
$East_B$ = *1415.33*

STEP 7

Check the coordinates of Point B by performing the same calculations from point C. If the calculations are performed correctly, the same values should be obtained.

$North_B$ = *700.40*
$East_B$ = *1415.33*

SUMMARY

It has been shown that seemingly unattainable information can be determined if a step-by-step problem solving approach using coordinates is used. The ability to calculate intersections gives the field engineer additional options in selecting the best layout method for the work.

HORIZONTAL CURVES

17

PART 3 - CONSTRUCTION SURVEYING CALCULATIONS

CHAPTER OBJECTIVES:

State the purpose of horizontal curves for modern transportation systems.
Describe the theory of the horizontal curve.
Describe, define and illustrate the parts of a simple horizontal curve.
Describe the basic geometric relationships that must exist in a horizontal curve.
Define degree of curve and its relationship to the radius of the curve.
Develop the formulas for the parts of a curve and use the formulas to calculate the parts of a curve.
Describe how deflections are used to layout a horizontal curve.
Explain the need for short and long chords in the layout of a curve.
List the step-by-step layout process of laying out a horizontal curve.

INDEX

THEORY

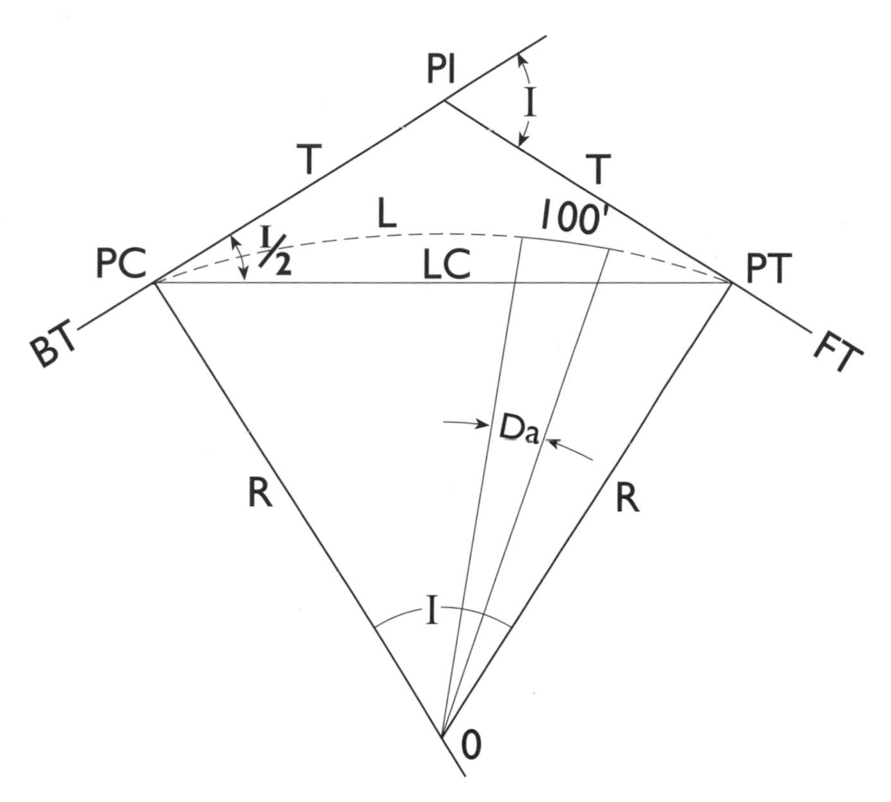

SCOPE

Have you ever traveled a country road in hill country? If you have, you know the country road followed the contour of the hills and valleys. This caused your travel to be very slow, and you sometimes seemed to meet yourself coming around turns. These country roads were usually built along ancient animal trails that followed the easiest and most comfortable route for the slow travel of wagons and carts. Early automobile travel was also very slow so road builders built the country road along the same route. Modern automobile travel is now high speed, and our patience is tested by the lack of designed curves on country roads.

The horizontal curve is used to provide a smooth transition between straight stretches and curves. Modern highways have curves designed to accommodate high speed travel in all types of terrain. The curves are designed to provide the motorist with the safest and most comfortable travel possible.

Curves do not occur just on highway construction projects. Most buildings will have roads on the site that will contain curves. More and more buildings themselves have curved sides for aesthetic purposes. No matter where the field engineer works, it is likely that horizontal curves will be encountered.

HORIZONALCURVE BASIS

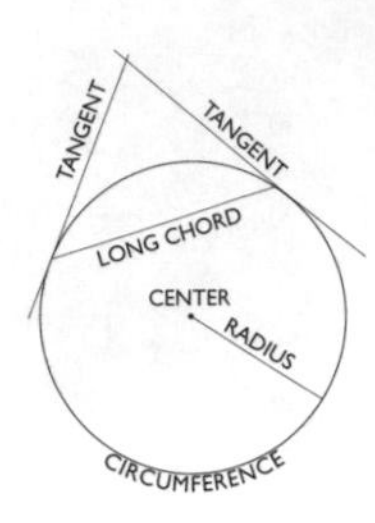

The simplest circle is what is used as the basis of the horizontal curves in highway design. In highway design, a small part of the circle is inserted where the two tangent lines intersect. When the circle is placed so it touches the tangent, it is done so that a line drawn from that point to the center of the circle is perpendicular to, or at right angles to, the tangent line. This line also represents the radius of the circle. The illustration shows the basic theory used to define the relationship of the simple circle to the two tangents. All curves will be perpendicular to the tangents. Now that the basic relationship can be seen, some additional parts of the curve can be defined and abbreviated.

PARTS OF A HORIZONTAL CURVE

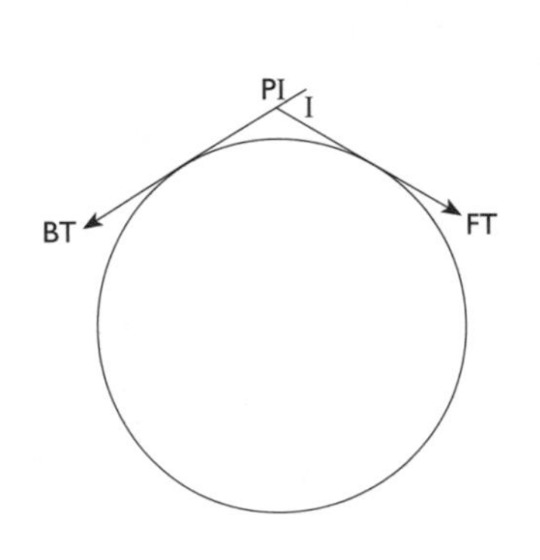

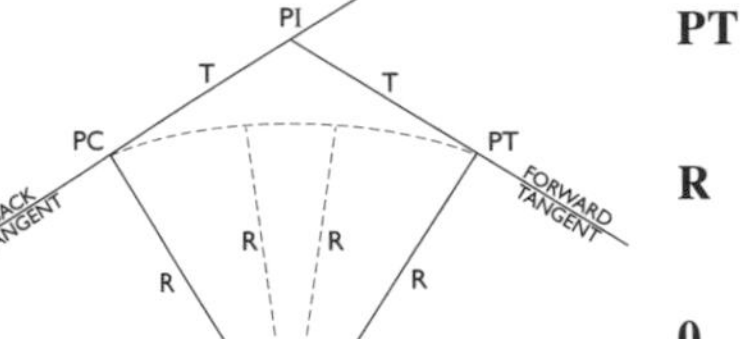

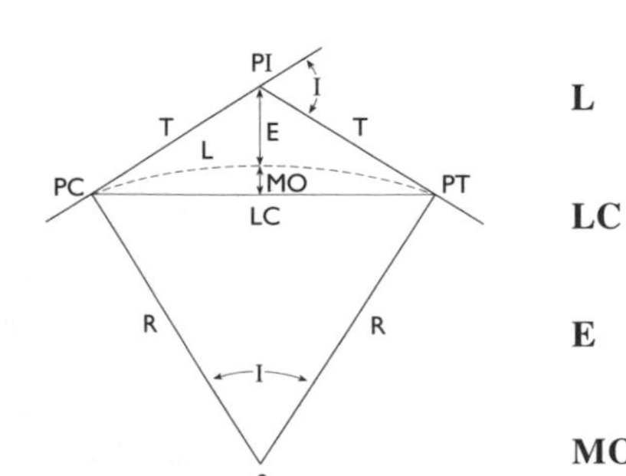

PI — **Point of Intersection** - That point where the two tangents intersect.

I — **Angle at the PI** - This is the deflection angle measured off of a prolongation of the back tangent to the forward tangent. This is also called "Delta."

BT — **Back Tangent** - The straight tangent extending towards the beginning of the project.

FT — **Forward Tangent** - The straight tangent extending towards the end of the project.

PC — **Point of Curvature** - That point where the tangent ends and the curve begins.

PT — **Point of Tangency** - That point where the curve ends and the tangent begins.

R — **Radius** - The distance from any point on the curve to the center of the circle.

0 — **Center of the Circle** - That point which is equal distance from all parts of the curve.

T — **Tangent Distance** - The distance along the tangent from the PC to the PI, or from the PT to the PI.

L — **Length along the Curve** - The distance along the curve between the PC and the PT.

LC — **Long Chord** - The straight line distance between the PC and the PT.

E — **External** - The distance from the PI to the center point on the curve.

MO — **Middle Ordinate** - The distance from the center point on the curve to the center point on the long chord.

GEOMETRIC RELATIONSHIPS

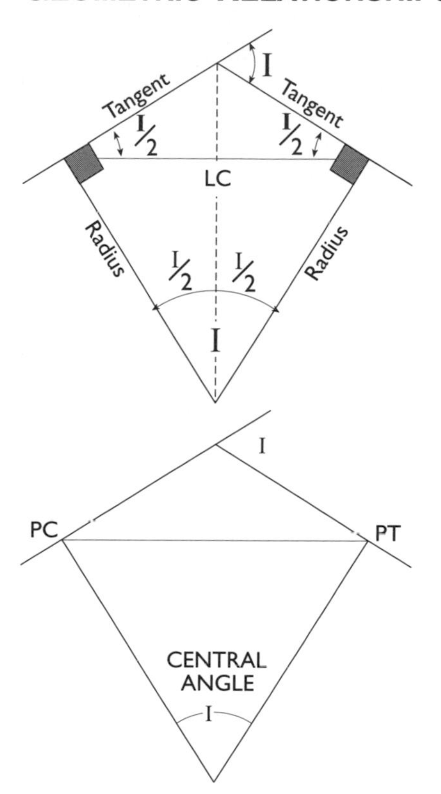

At the PC and PT, the radius and tangent are perpendicular.

The angle between the tangent and the long chord originating at the PC or PT is equal to one half of "I".

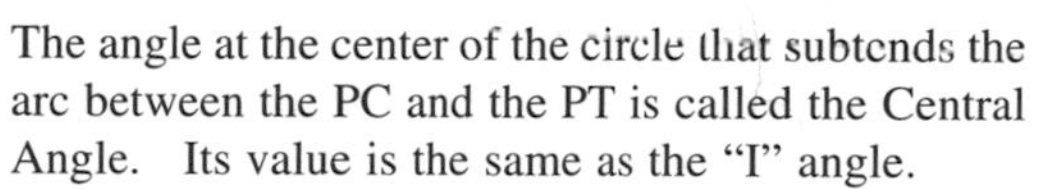

The angle at the center of the circle that subtends the arc between the PC and the PT is called the Central Angle. Its value is the same as the "I" angle.

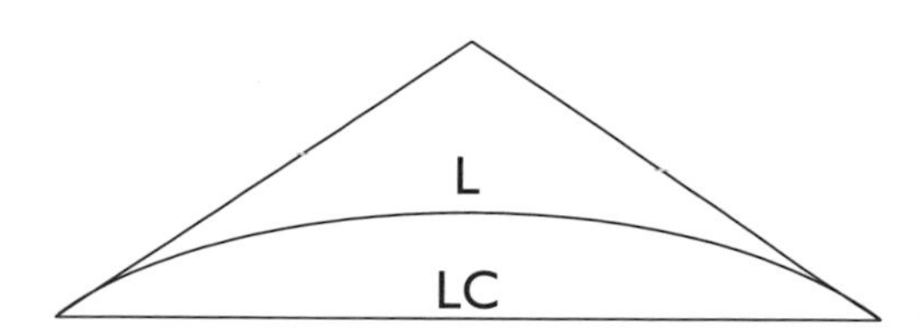

L is always greater than LC

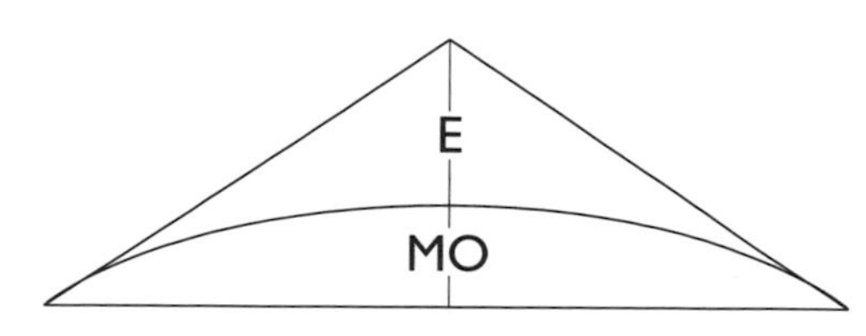

E is always greater than MO

DEGREE OF CURVE D_A

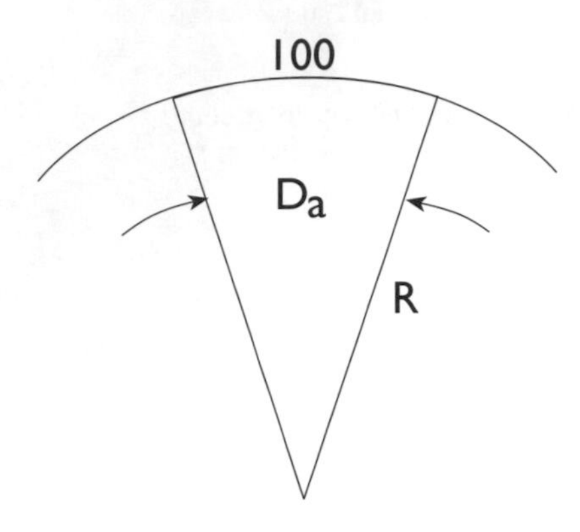

D_a Degree of Curve (Highway) is defineable as the angle at the center of the circle subtended by an arc of 100 feet on the curve.
Curves are obviously different as you travel on them. Some you can travel very fast with ease while others require that you slow down considerably in order to stay on the pavement. The degree of curve **D_a** and the radius **(R)** are used interchangeably to describe an important property of the curve, the sharpness.

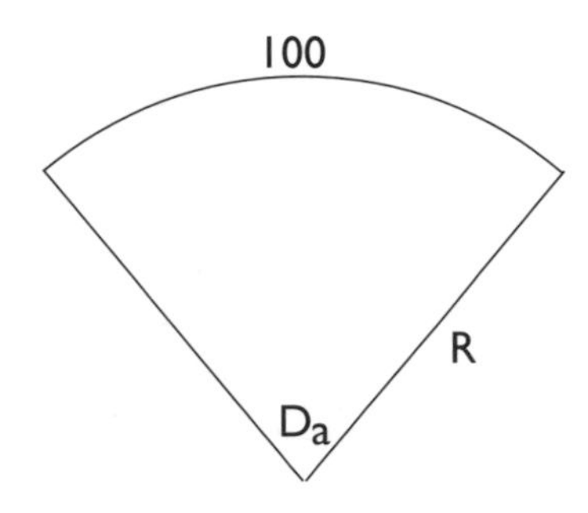

The **sharpness of a curve** can be described by the Degree of Curve (D_a) or the radius (R). That is, the sharper the curve, the larger the D_a and the smaller the radius.

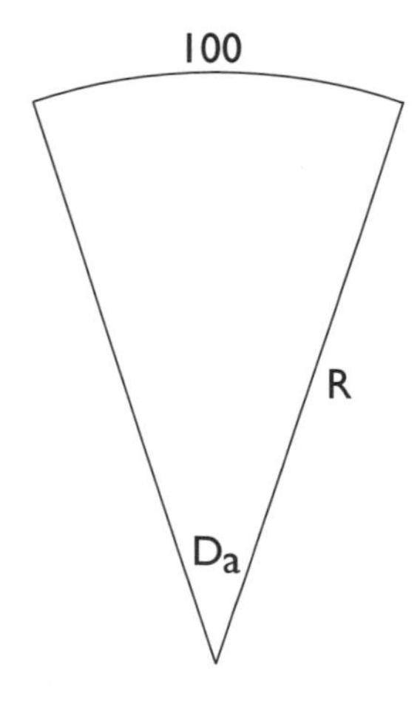

The flatter the curve, the smaller the D_a and the larger the radius.

FORMULA DEVELOPMENT

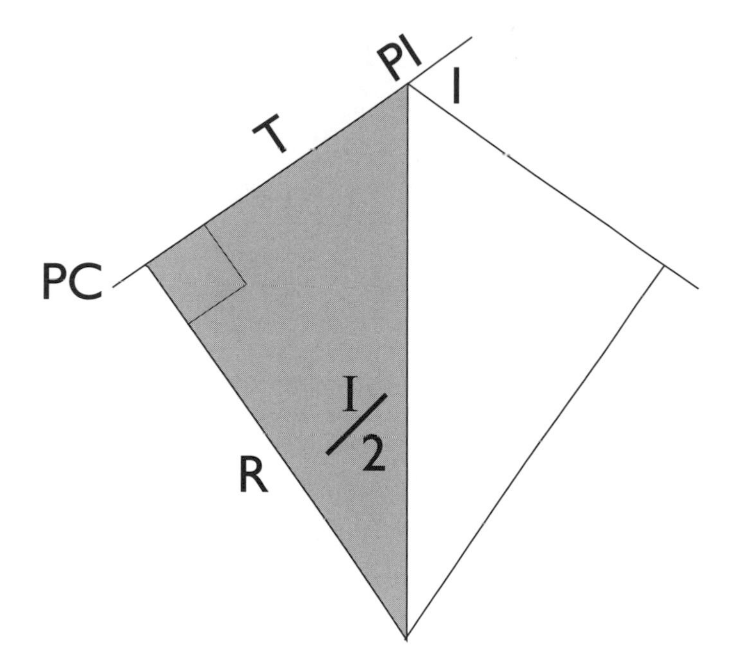

SCOPE

The formulas of the horizontal curve are based on the simple circle and the right triangle. By observing the geometric relationships and using standard right angle trigonometry, the formulas used to calculate the parts of the horizontal curve can be easily developed by the field engineer.

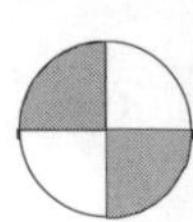

TANGENT

On the illustration, it is shown that the radius is perpendicular to the tangent, and that a line can be drawn from the center of the circle to the PI. This creates an angle that is one-half the **I** angle. Refer to the shaded triangle.

Using basic right angle trigonometry, the triangle we have created has the following known and unknown elements.

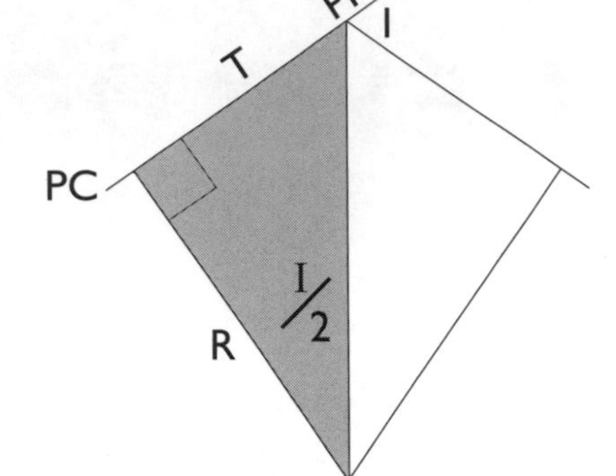

KNOWN

1. A right triangle.
2. One-half of the central angle, I/2
3. The radius, R, which is the adjacent side for the triangle.

UNKNOWN

"T" which is the opposite side.

DEVELOPING THE FORMULA FOR T

$$\tan\Theta = \frac{opp}{adj}$$

Substituting known terms

$$\tan\frac{\mathrm{I}}{2} = \frac{T}{R}$$

Solving for T

$$T = R\tan\frac{I}{2}$$

LONG CHORD

On the illustration, the long chord is bisected and a right triangle is shown. Solving for the parts of the triangle results in halving of the bisected long chord. Simply multiply by 2 to get the long chord length.

KNOWN

1. A right triangle
2. One-half the central angle, I/2
3. The radius, R, which is the hypotenuse of the triangle.

UNKNOWN

LC/2 which is the opposite side of the triangle.

DEVELOPING THE FORMULA FOR LC

$$Sin\Theta = \frac{opp}{hyp}$$

Substituting known terms

$$\text{Sin}\,\frac{\text{I}}{2} = \frac{LC/2}{R}$$

Solving for LC

$$LC = 2R\sin\frac{I}{2}$$

LENGTH OF CURVE, L

The length of curve is determined by a simple proportion.

$$\frac{Da}{100} = \frac{I}{L}$$

Solving for L

$$L = 100\frac{I}{Da}$$

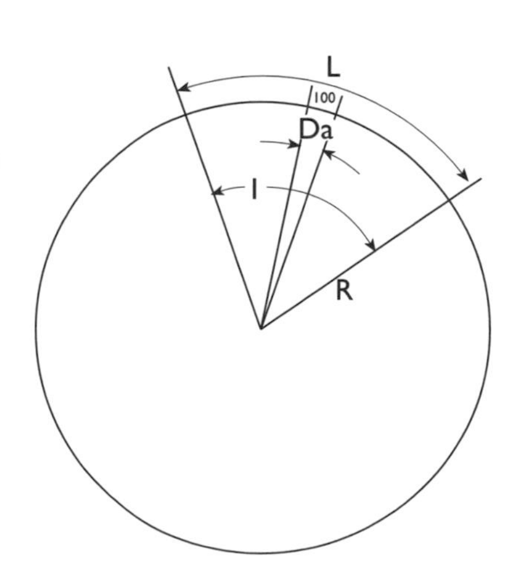

EXTERNAL, E

On the illustration on the right, it is shown that a right triangle exists from the PI to the center of the circle. Using the known parts, the hypotenuse can be calculated. Subtracting the radius yields the external.

KNOWN

1. Right triangle
2. One-half the central angle
3. Radius

UNKNOWN

Hypotenuse

DEVELOPING THE FORMULA FOR E

$$Cos = \frac{adj}{hyp}$$

Substituting

$$Cos\frac{I}{2} = \frac{R}{R+E}$$

Solving for E

$$E = \frac{R}{\cos I/2} - R$$

MIDDLE ORDINATE, M.O.

On the illustration at the right it is shown that a right triangle exist to the long chord. Using the know parts the long-leg of the triangle can be calculated. Subtracting the leg of the triangle from the radius yields the middle ordinate.

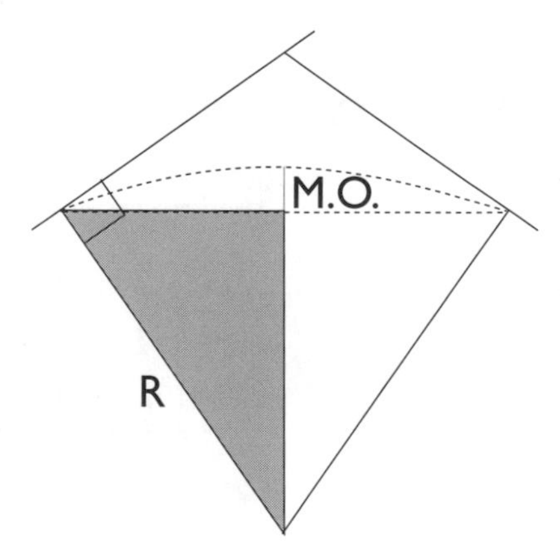

KNOWN

1. Right triangle
2. One-half the central angle
3. Radius

UNKNOWN

The leg of the triangle

DEVELOPING THE FORMULA FOR MO

$$Cos\frac{I}{2}=\frac{R-MO}{R}$$

Solving for MO

$$MO=R-R\cos\frac{I}{2}$$

DEGREE OF CURVE

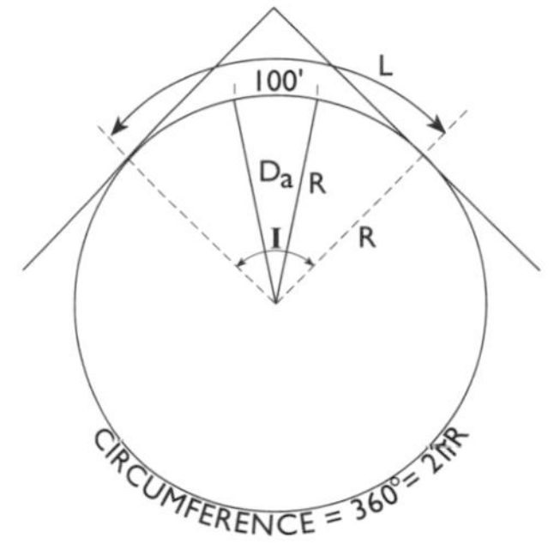

The degree of curve is determined by setting up a simple proportional relationship. By definition, the degree of curve can be represented as Da/100. The total arc of a circle is represented by $2\pi r$ with a total angle of 360°. Similarly this can be represented as 360°/$2\pi r$.

DEVELOPING THE FORMULA FOR DA & R

$$\frac{Da}{100}=\frac{360}{2\pi r}$$

Solving for Da

$$Da=\frac{5729.58}{R}$$

Solving for R

$$R=\frac{5729.58}{Da}$$

STATIONING

Now that formulas are available to calculate the parts of the curve, the stationing for the PC and PT can be obtained. When the PI was originally established, a station was calculated for it from the measured distances. This station can now be used to calculate the stations of the PC and PT. The method used is to work backwards from the PI to the PC and forwards from the PC along the curve to the PT. This sounds simple and it is!

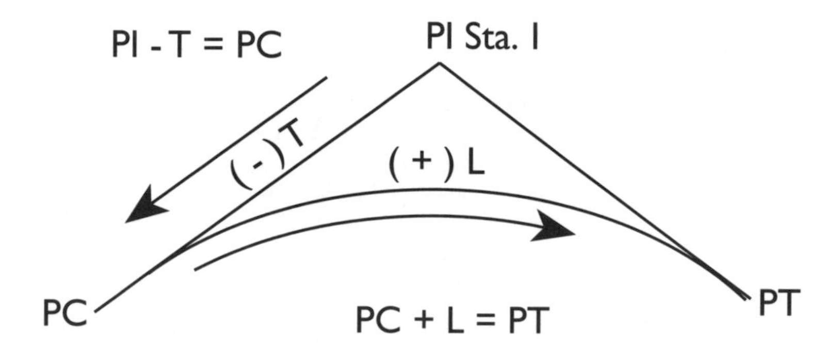

$$PI - T = PC$$
$$PC + L = PT$$

CALCULATION OF CURVE PARTS

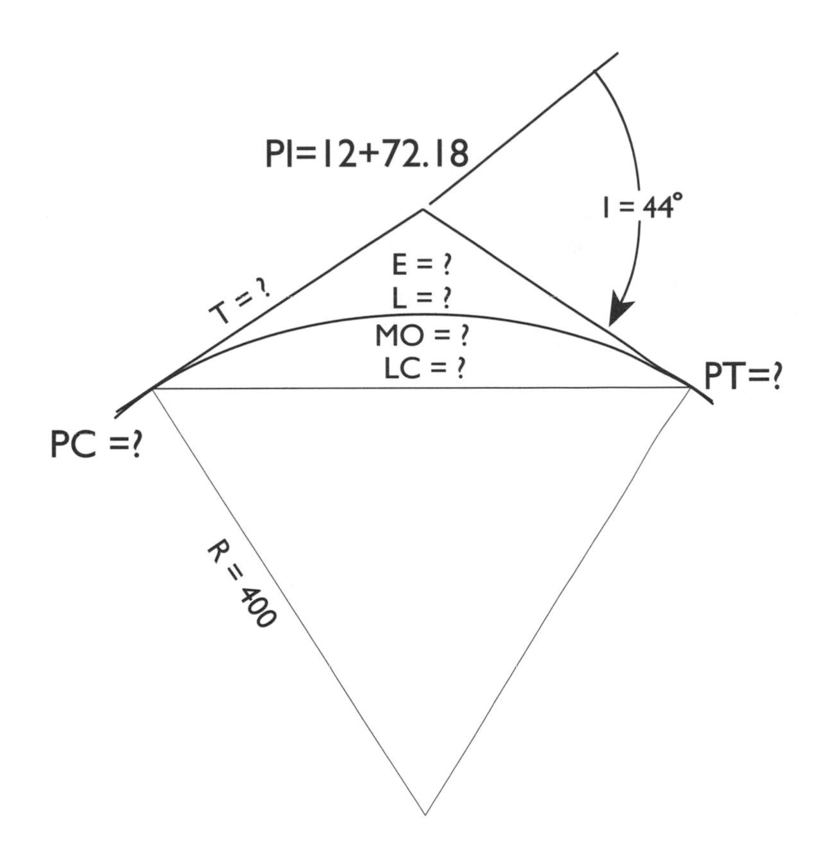

SCOPE

Given any two parts of a horizontal curve and the formulas for the curve, all remaining parts of the curve can be determined. That seems like a pretty strong statement. It is, but it is true. It may take some manipulation of formulas and graphical solutions to get answers in some situations, but it can be done. Generally, though, from the project plans, the field engineer is given the **I** angle and the **R**adius, he/she needs to calculate the **T**angent, **L**ength, **L**ong **C**hord, **E**xternal, **M**iddle **O**rdinate, and the **D**egree of Curve.

EXAMPLE

Calculation of Horizontal Curve Parts		
Given I = 44 R = 400 PI Sta. = 12 + 72.18		
Step 1	**Compute the Tangent**	
	T = R tan 1/2 I = 400 tan 44°/2 = 161.61'	
Step 2	**Compute the Da**	
	Da = 5729.58/R Da = 5729.58/400 = 14.324°	
Step 3	**Compute the Length**	
	L = 100 I/Da = 100(44° /14.324°) = 307.18'	
Step 4	**Compute the Long Chord**	
	LC = 2R Sin I/2 = 2(400) Sin 44°/2 = 299.69'	
Step 5	**Compute the External**	
	E = (R/Cos I/2) - R = (400/Cos 44°/2) - 400 = 31.41'	
Step 6	**Compute the Middle Ordinate**	
	M.O. = R - (RCos I/2) = 400 - (400Cos 44°/2) = 29.12'	
Step 7	**Compute the Stations of the PC and PT**	
	PI Station - T = PC Station	PC Station + L = PT Station
	1272.18 - 161.61 = 11 + 10.57	1110.57 + 307.18' = 14 + 17.75

SUMMARY

At this point, the parts of the curve have been calculated. The next section will discuss how to calculate information to lay out a curve in the field.

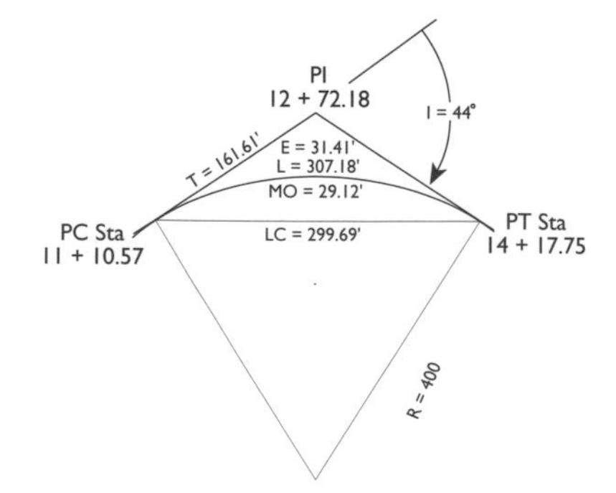

DEFLECTIONS AND SHORT AND LONG CHORDS

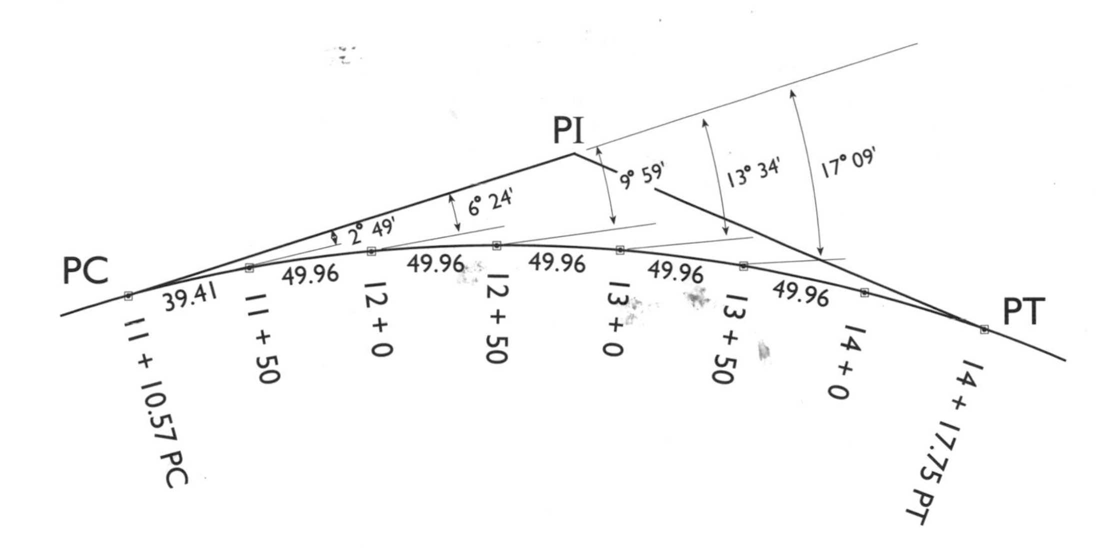

SCOPE

Field engineers are often required to stake out horizontal curves on building or highway projects. Although there are several methods of laying out a curve, the most common method is to use deflection angles and short or long chords. The actual layout procedure will be explained in the next section. This section will deal with the calculation of the deflection angles and the short and long chords.

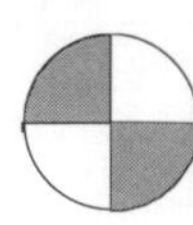

GENERAL

Calculation of deflections and chords for the layout of a curve uses very simple formulas and a straightforward procedure. Some basic terms and an understanding of where numbers are coming from is needed before describing the actual calculation process.

STATION INTERVAL

The first step in the process is for the field engineer to decide by consulting with the superintendent the interval that the curve needs to be staked. That is, 25', 50' or 100'. Once the interval is determined, the stations where stakes will be placed can be listed on the calculation sheet.

ARC LENGTH

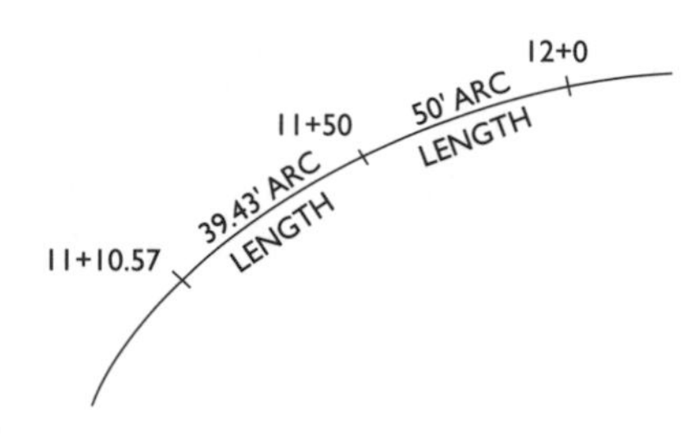

Calculations of the deflections are based on a formula that requires inputting the arc length to each station. Arc length is the distance along the curve. For example, from station 11+10.57 to station 11+50 the arc length is 39.43 feet and from station 11+50 to 12+00 is an arc length of 50 feet. Obtaining arc length is determined by simple subtraction of stations. Arc length values between the individual stations on the curve will be listed on the calculation sheet as well as the total arc length from the PC of the curve.

DEFLECTION PER FOOT OF ARC

Preliminary calculation of the deflection for one foot of arc creates a constant that the field engineer can use to easily and quickly calculate the deflection for any arc length. In the deflection angle formula that will be presented later, simply insert one foot in for arc length to determine this value.

CHORD CALCULATIONS

Calculation of the short or long chord is based on the same formula that was used to calculate the long chord for the entire curve. The angles used in the equation are the deflection angles that are calculated for each station. The equation is first used to determine the short chords for the curve. . The equation is then used to determine the long chords from the PC to each station on the curve. Long chords will especially be needed if an EDM is being used to layout the curve.

DEFLECTION INCREMENT

Deflection increment is a term that is used in the calculation process to describe the amount of deflection between each station. There will be an increment for the arc length from the PC to the first station on the curve, for the standard station interval on the curve (50 Feet), and for the closing distance into the PT at the end of the curve.

When calculating a curve, if the deflection increment for 50 feet is known, it is simply added to each successive station to get the total deflection of the next station. For example if the deflection for station 12+00 is 6°24'17", adding the deflection increment for 50 feet of 3° 34' 52" will obtain the deflection for station 12+50 of 9°59'09".

CALCULATE AND RECORD KNOWN DATA

Set up a field book or notepad for listing the calculated curve data. See the headings on the example below. Defl. = Deflections.

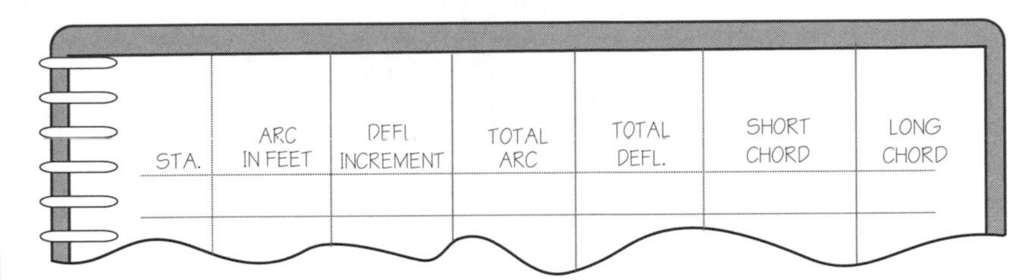

Start with the PC station at the bottom of the page and list the stations up the page from it. Listing route surveying information this way is a common field practice. The field engineer using the notes can then stand at the PC and look ahead at increasing stations in the fieldbook as well as looking ahead at increasing stations on the ground. This is done for practicality as well as convenience. List on the calculation page the parts of the curve if they have been previously calculated.

THE DEFLECTION ANGLE EQUATIONS

There are two equations which can be used to calculate the deflection. Use depends on the information that is available. Both equations yield the deflection angle in minutes of arc which must be converted to degrees, minutes, and seconds for turning with a theodolite or transit.

$$d_{(MINUTES)} = (1718.873 / R)(Arc\ Length)$$

$$d_{(MINUTES)} = (.3)(Arc\ Length)(Degree\ of\ Curve)$$

THE SHORT AND LONG CHORD EQUATIONS

Since the length of each chord is slightly less than the length of the arc it subtends, the chord lengths must be computed.

$$Short\ Chord = 2R\sin(\text{Incremental deflection})$$

$$Long\ Chord = 2R\sin(\text{Total deflection at a curve point})$$

STEP-BY-STEP PROCEDURE

Use the following procedure to compute the deflections and chord lengths for a curve.

Step	Calculate the:	Process
1	Individual Arc Lengths	Subtract each successive station on the curve to determine the arc lengths.
2	Total Arc Lengths	Calculate the total arc length from the PC to each station on the curve
3	Deflection per foot	Substitute one foot into the deflection angle equation to obtain this constant.
4	Deflection increment for the first station beyond the PC	Insert the arc length from the PC to the first station on the curve into the deflection angle equation.
5	Standard deflection increment per station interval	If the interval between stations is 50 feet, input an arc length of 50 into the deflection angle equation.
6	Deflection closing into the PT.	Insert the arc length from the last station to the PT into the deflection angle equation
7	Total deflection for each station from the PC	Individually, insert the arc length from the PC to each station on the curve into the deflection angle equation.
8	Short Chords	Use the short chord equation and the deflection increment to the first station, the standard interval, and the last station to determine the short chords.
9	Long Chords	Use the long chord equation and the total deflection from the PC to each station to calculate each long chord.

CHECKS

The calculation for the last deflection from the PC to the PT will use the length of curve and if no mistakes have been made, will equal one half the central angle. The calculation of the last long chord to be used for the layout will equal the long chord from the PC to the PT.

CONTINUING EXAMPLE

List the Known Information
Given I = 44 R = 400 PI Sta. = 12 + 72.18
T = 161.61'
Da = 14.324°
L = 307.18'
LC = 299.69'
E = 31.41'
M.O. = 29.12
PC Station = 11 + 10.57
PT Station = 14 + 17.75

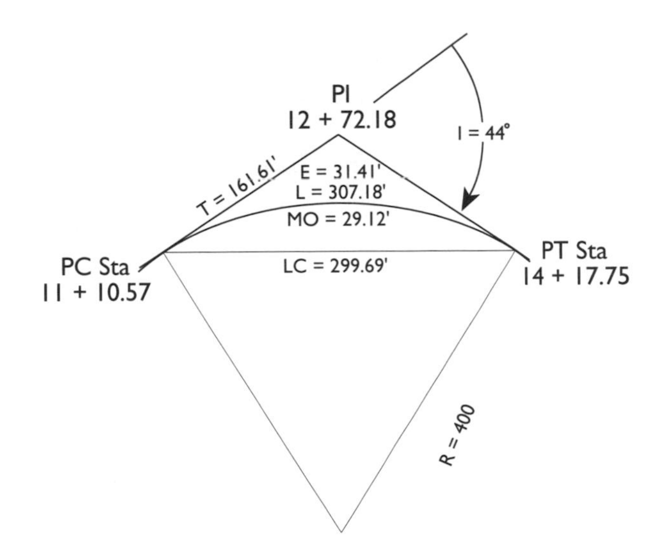

CALCULATIONS:

The deflection per foot, deflection increments, the total deflections, the short chord, and long chord for even 50 foot stations on the curve.

DEFLECTION PER FOOT

$d_{(minutes)} = (1718.873/R)(\text{Arc Length})$

$d = (1718.873/400)\ (1)$

$d = 4.297183$ minutes per foot of arc

DEFLECTION INCREMENTS

To first station from PC - Sta 11 + 50

$d_{(minutes)} = (4.297183)(39.43) = 169.437906 \text{ minutes} = 2° 49' 26''$

At 50 foot intervals

$d_{(minutes)} = (4.297183)(50) = 214.859125 \text{ minutes} = 3° 34' 52''$

For the closing distance into the PT

$d_{(minutes)} = (4.297183)(17.75) = 76.274989 \text{ minutes} = 1° 16' 16''$

TOTAL DEFLECTIONS

@ 11 + 50 $d_{(minutes)} = (4.297183)(39.43) = 2° 49' 26''$

@ 12 + 0 $d_{(minutes)} = (4.297183)(89.43) = 6° 24' 17''$

@ 12 + 50 $d_{(minutes)} = (4.297183)(139.43) = 9° 59' 09''$

etc. to the end of the curve.

THE SHORT CHORD LENGTHS.

To first station from PC, 11+50

Short Chord = 2R sin (2° 49' 26") = 39.41"

At 50 foot intervals

Short Chord = 2R sin (3° 34' 52") = 49.96"

For the closing distance into the PT

Short Chord = 2R sin (1° 16' 16") = 17.74"

THE LONG CHORDS

@ 12 + 0 Long Chord = 2R sin (6° 24' 17") = 89.24'

@ 12 + 50 Long Chord = 2R sin (9° 59' 09") =138.72'

etc. to the end of the curve.

TABLE OF RESULTS

Station	Arc in Feet	Deflection Increment	Total Arc	Total Deflection	Short Chord	Long Chord
14 + 17.75 PT	17.75	1° 16' 16"	307.18	22° 00' 00"	17.74'	299.68'
14 + 00	50	3° 34' 52"	289.43	20° 43' 44"	49.96'	283.16'
13 + 50	50	3° 34' 52"	239.43	17° 08' 52"	49.96'	235.87'
13 + 00	50	3° 34' 52"	189.43	13° 34' 01"	49.96'	187.66'
12 + 50	50	3° 34' 52"	139.43	9° 59' 09"	49.96'	138.72'
12 + 00	50	3° 34' 52"	89.43	6° 24' 17"	49.96'	89.24'
11 + 50	39.43	2° 49' 26"	39.43	2° 49' 26"	39.41'	39.41'
11 + 10.57 PC	0	0	0	0	0	0

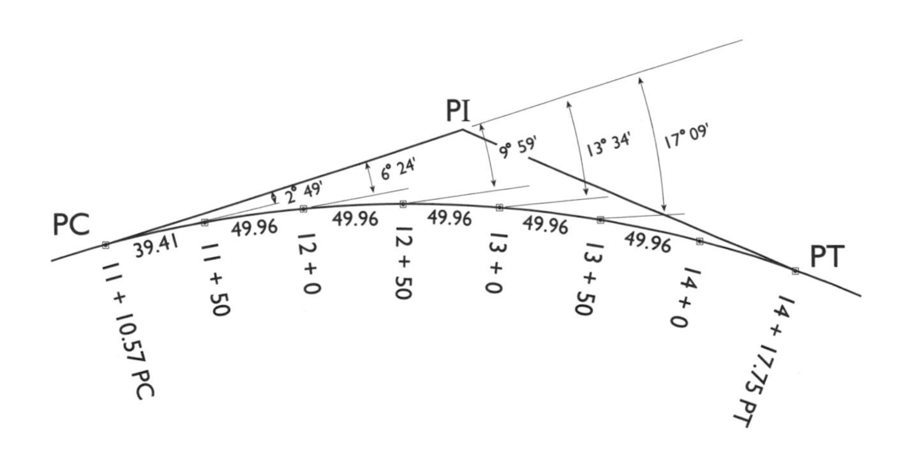

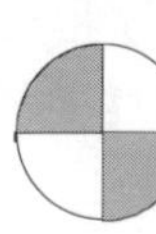

LAYOUT

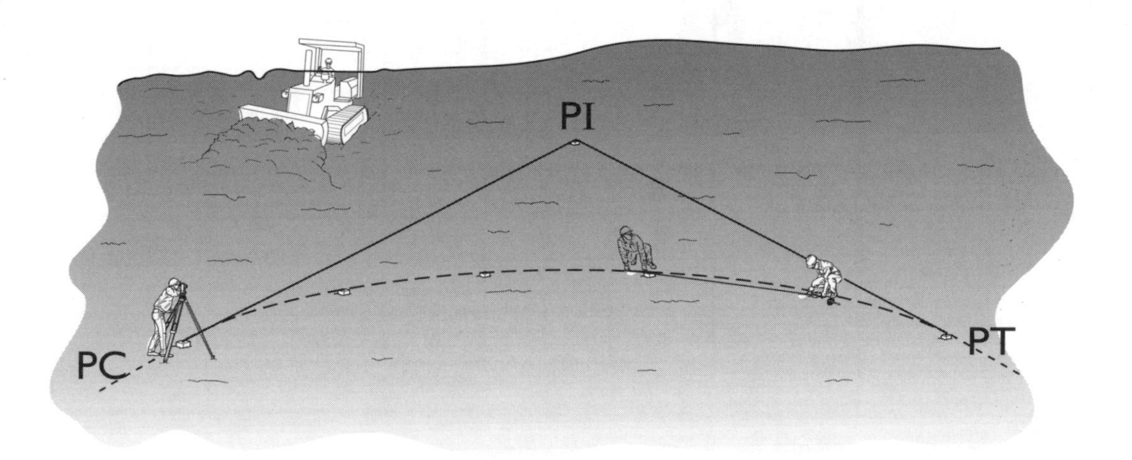

SCOPE

One of the most satisfying aspects of construction route surveying is the development and layout of a circular curve. This section will introduce the field engineer to the actual field process necessary to stake a horizontal curve. All standard layout practices should be followed to reduce the possibility of errors in the measurements.

EQUIPMENT

Horizontal Curve Layout Equipment	
Per crew	**Per person**
Transit or theodolite	Field book
Range pole	Plumb bub
Tripod	Straight edge
	4H pencil

STEP-BY-STEP CURVE LAYOUT PROCEDURE (SHORT CHORD)

The following is the step-by-step procedure for Deflection Angle Layout (Short Chord Method). Refer to the table on 17-17 for the curve data used in this procedure.

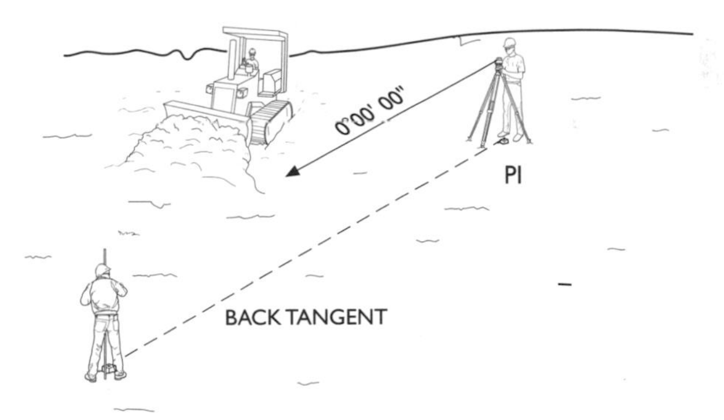

STEP 1

Set up the transit over the point of intersection (P.I.). Sight at a point on the back tangent with "0" on the instrument.

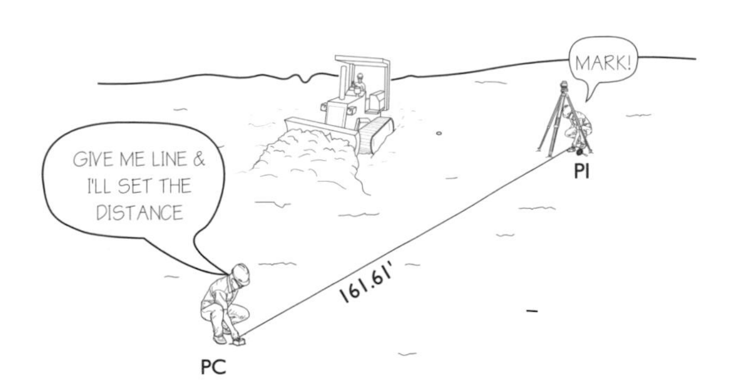

STEP 2

Measure the tangent distance along the back tangent to set a hub and tack at the point of curvature (P.C.).

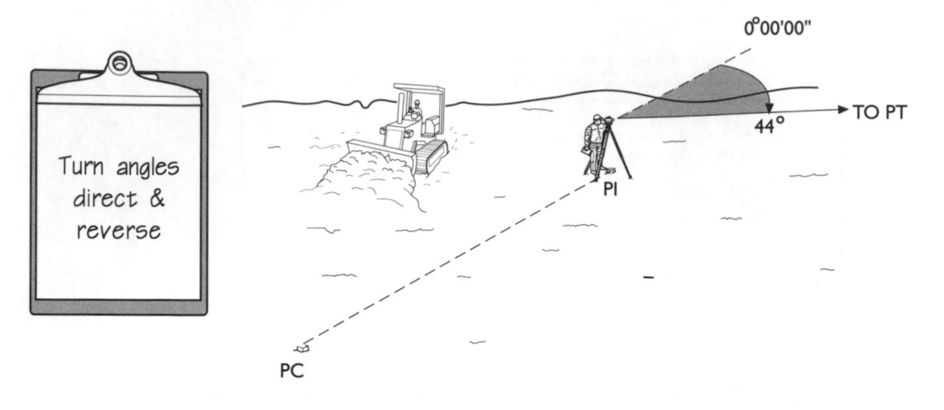

STEP 3

Turn the "I" Angle. Sighting on the point of curvature, invert the scope, and turn the required "I" angle toward the P.T. (Direct & Reverse).

STEP 4

Set a hub at the P.T. Along the line of sight, again measure the tangent distance to set a hub at the point of tangency (P.T.).

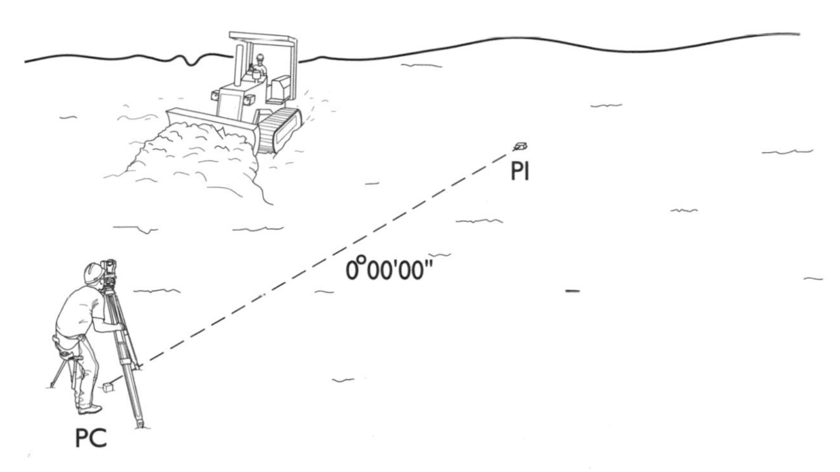

STEP 5

Set up the transit over the P.C. and sight on the P.I. with 0° set on the vernier.

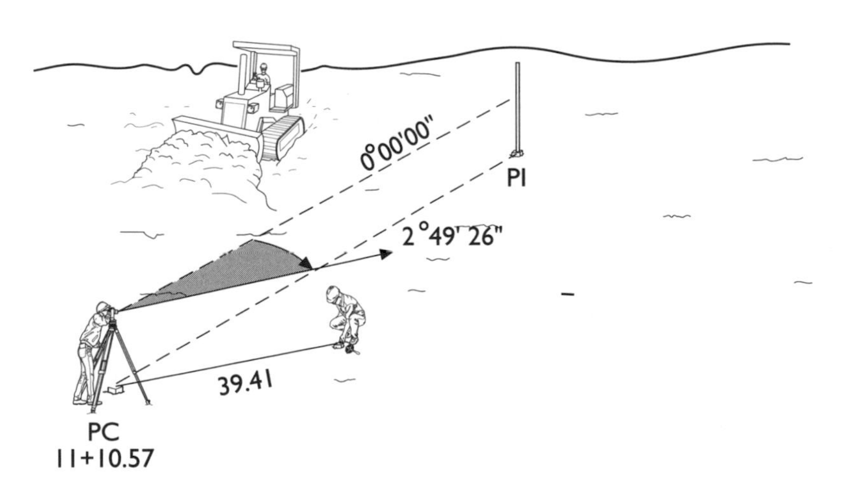

STEP 6

Measure the deflection short chord distance. Turn the calculated deflection angle and measure the short chord distance from the P.C. along the line of sight to the first point on the curve.

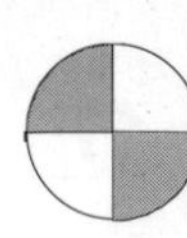

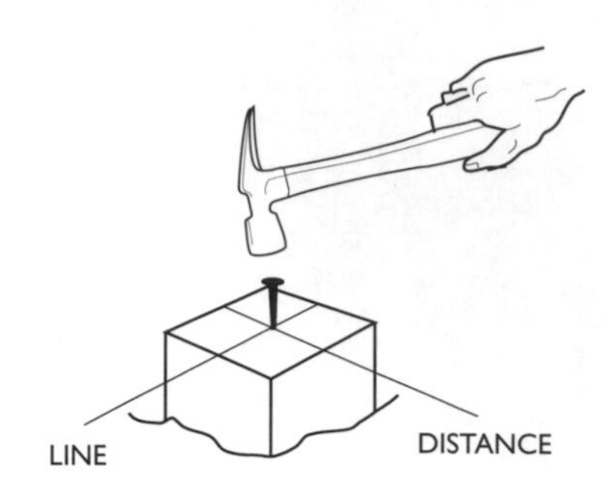

STEP 7

Set the first point. Place a hub at the curve point. Place line on the hub and then measure the distance. Place a tack at the point.

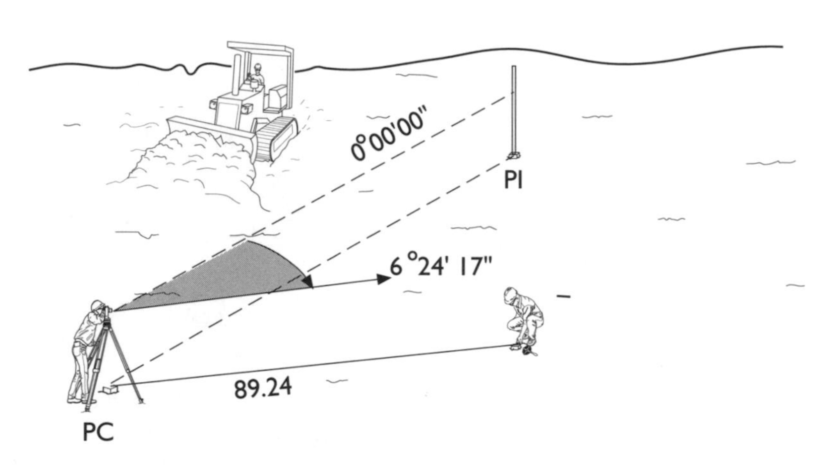

STEP 8

Set additional points. Refer to the data sheet and turn the next deflection angle. Measure the short chord distance and set a hub and tack as before.

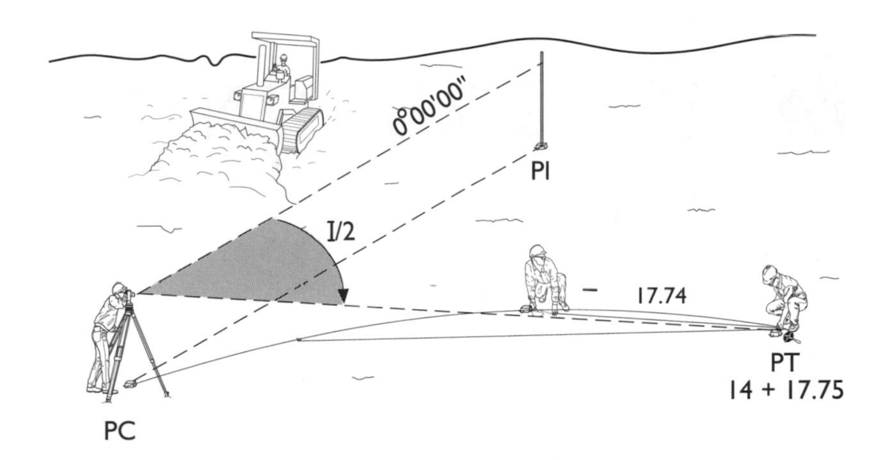

STEP 9

Repeat to the P.T. Continue this process until you reach the PT. Your last deflection angle and measured distance should tie into the P.T. within acceptable tolerances.

MOVING UP ON A CURVE

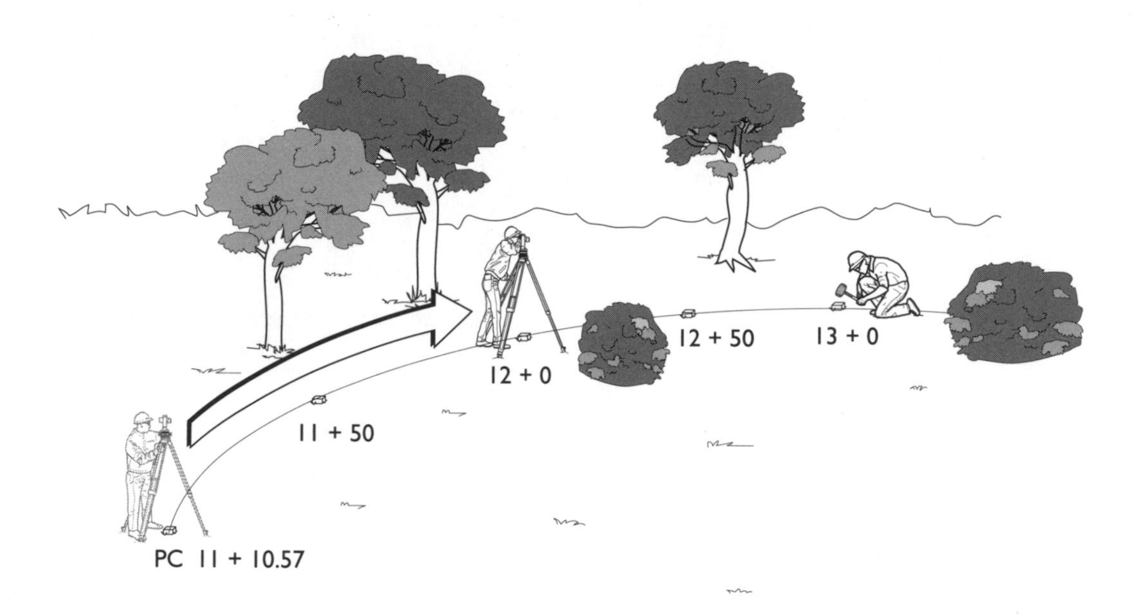

SCOPE

In the process of staking out a curve, it is common for the field engineer to be unable to layout the entire curve from one instrument setup at the PC. This may occur for any number of reasons including the length of curve being very long or obstacles that prevent seeing from the PC to the PT. The method used to move up on a curve is a simple one that allows the field engineer to continue to use the original layout notes. That is, no additional calculations are needed to move up on a simple curve.

BASICS

The following is a description of how to move up on curve and continue the original notes.

SETUP

Set up on any point on the curve. (Usually the last point established.)

BACKSIGHT

Backsight onto any other point on the curve with the deflection angle of the point that is sighted on set on the horizontal circle. (Follow the rule of thumb: long backsight and short foresight to reduce the effect of sighting errors, and sight on the point farthest away on the curve.)

PLUNGE

Plunge the telescope to face in the direction the layout of the curve is proceeding.

TURN THE ANGLE TO THE NEXT POINT ON THE CURVE

Refer to the original curve layout notes and turn the instrument to read the angle for the next point to be established on the curve.

CONTINUE THE LAYOUT

Continue the layout until the end of the curve is reached or another obstacle is encountered.

REPEAT IF NECESSARY

If another move-up is required, repeat the process.

NOTE:

The original notes can be continued "as is" if short chords are being used to layout the distances. If long chords are being used, additional calculations of new long chords from the new set-up point will be required.

STEP-BY-STEP EXAMPLE

Refer to the notes for the curve calculated in the previous section.

Station	Arc in Feet	Deflection Increment	Total Arc	Total Deflectio	Short Chord	Long Chord
14 + 17.75 PT	17.75	1° 16' 16"	307.18	22° 00' 00"	17.74"	299.68
14 + 00	50	3° 34' 52"	289.43	20° 43' 44"	49.96"	283.16
13 + 50	50	3° 34' 52"	239.43	17° 08' 52"	49.96"	235.87
13 + 00	50	3° 34' 52"	189.43	13° 34' 01"	49.96"	187.66
12 + 50	50	3° 34' 52"	139.43	9° 59' 09"	49.96"	138.72'
12 + 00	50	3° 34' 52"	89.43	6° 24' 17"	49.96"	89.24'
11 + 50	39.43	2° 49' 26"	39.43	2° 49' 26"	39.41'	39.41'
11 + 10.57 PC	0	0	0	0	0	0

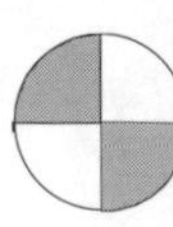

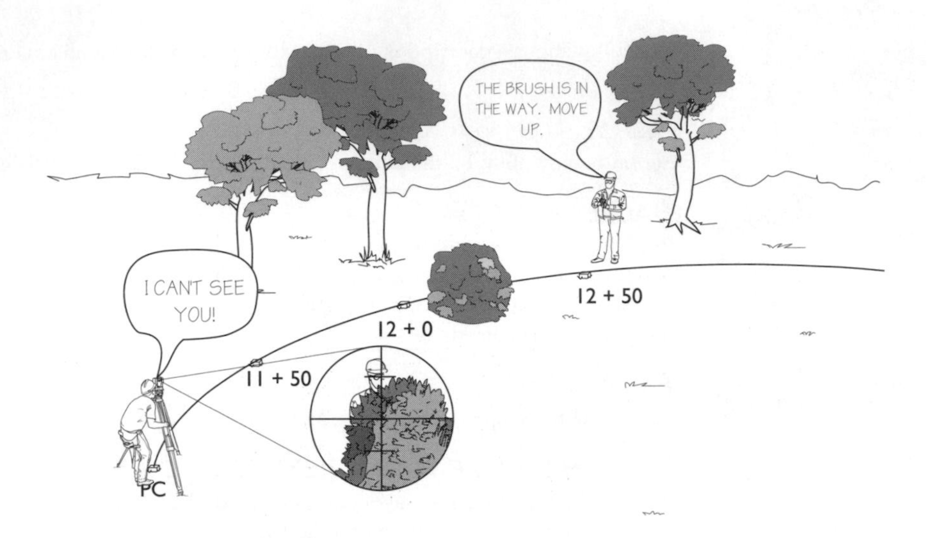

STEP 1 After laying out stations 11+50 and 12+00 from the PC, a large bush prevents the field engineer from establishing station 12+50 and the rest of the curve.

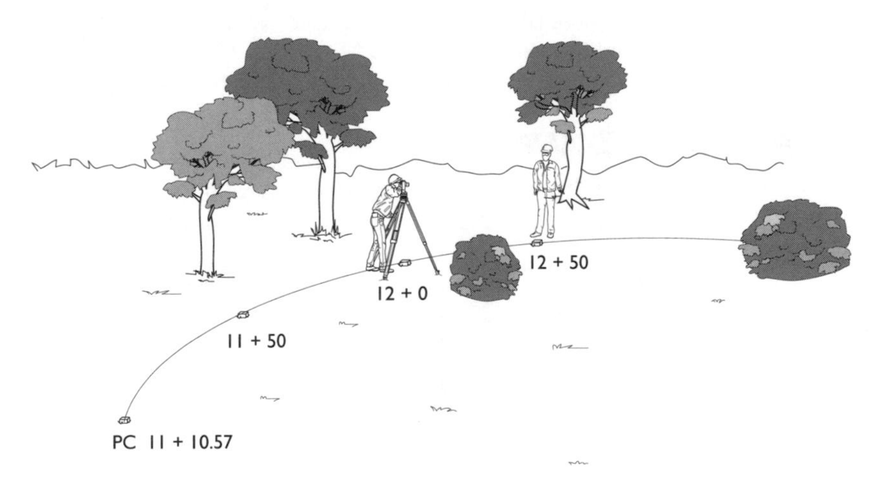

STEP 2 The field engineer moves the instrument to 12+00 and sets up over the point.

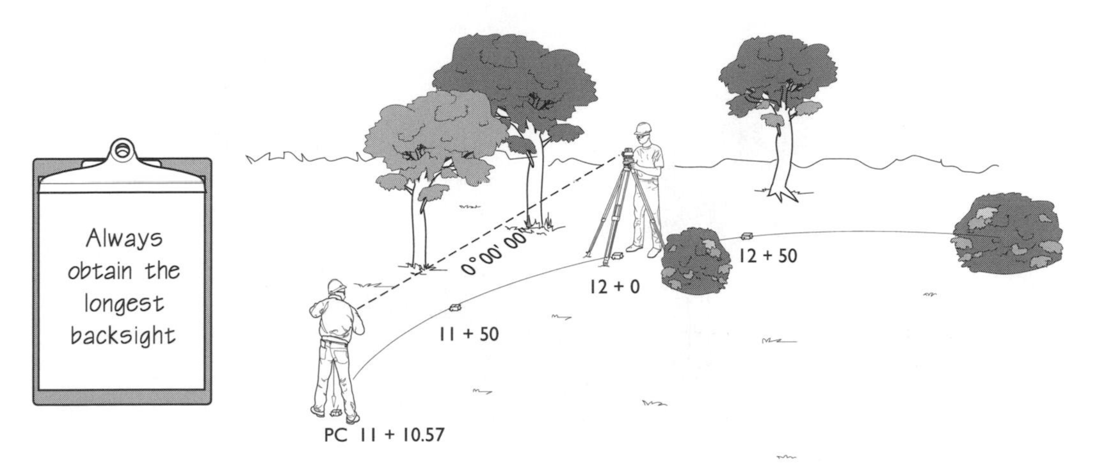

STEP 3

The field engineer decides to backsight on the PC (Longest Backsight) and sets 0° 00'00" on the instrument.

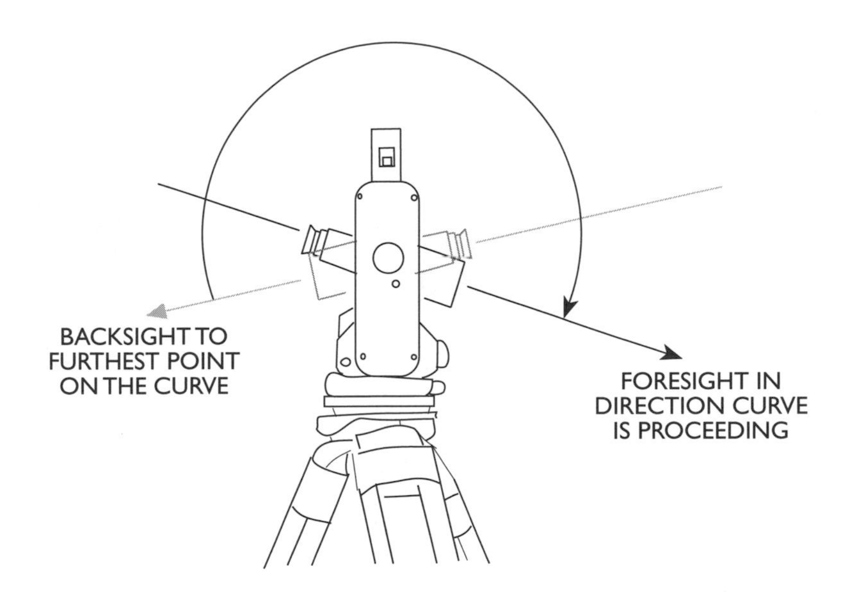

STEP 4

After backsighting, the field engineer plunges the telescope to sight in the direction the curve is proceeding.

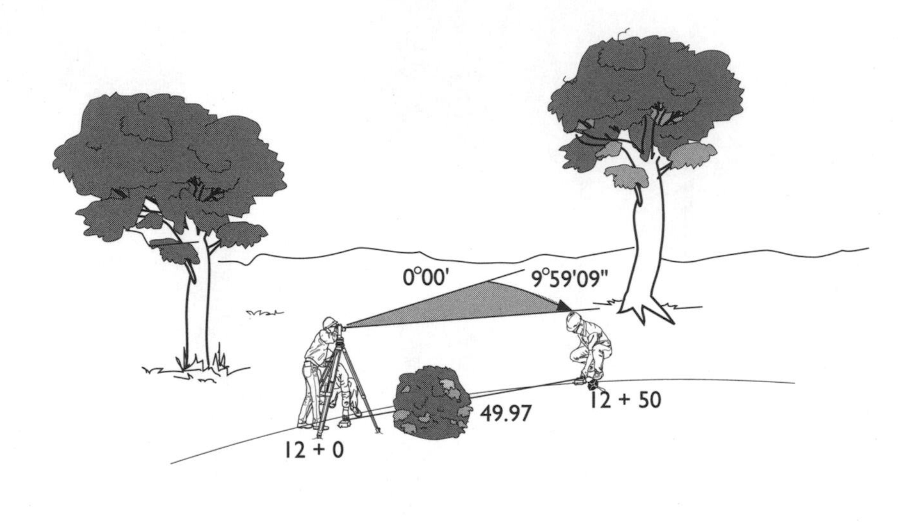

STEP 5

Referring to the original notes, the field engineer turns the instrument to 9° 59' 09" and directs the head chain to measure the short chord of 49.97 to establish point 12+50.

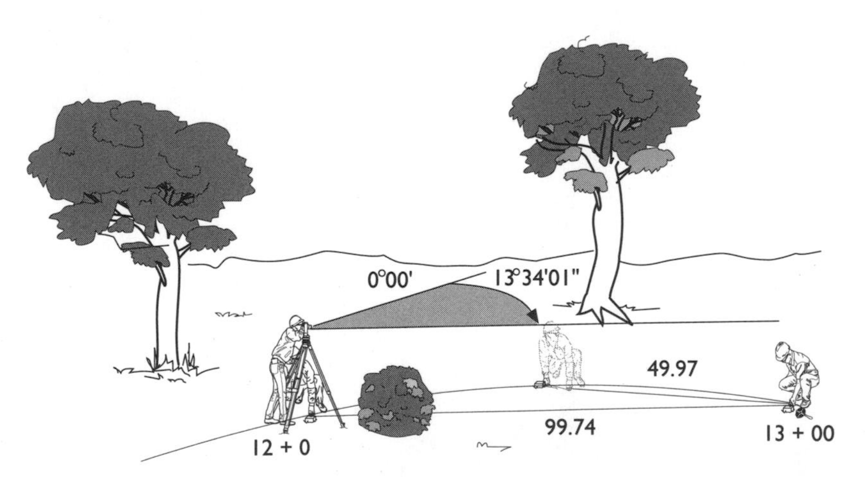

STEP 6

Referring to the original notes, the field engineer turns the instrument to 13° 34' 01" and directs the head chain to measure the short chord of 49.97 to establish point 13+00. If long chords are being used, the field engineer uses the chord formula and calculates a long chord of 99.74 to measure from the instrument setup.

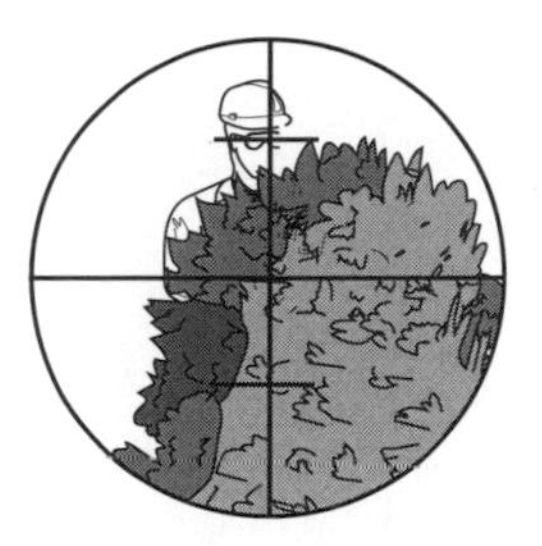

STEP 7 After setting point 13+50, another bush is in the line of sight and prevents laying out additional points from this setup. The field engineer moves the instrument up to 13+50 and sets up over the point.

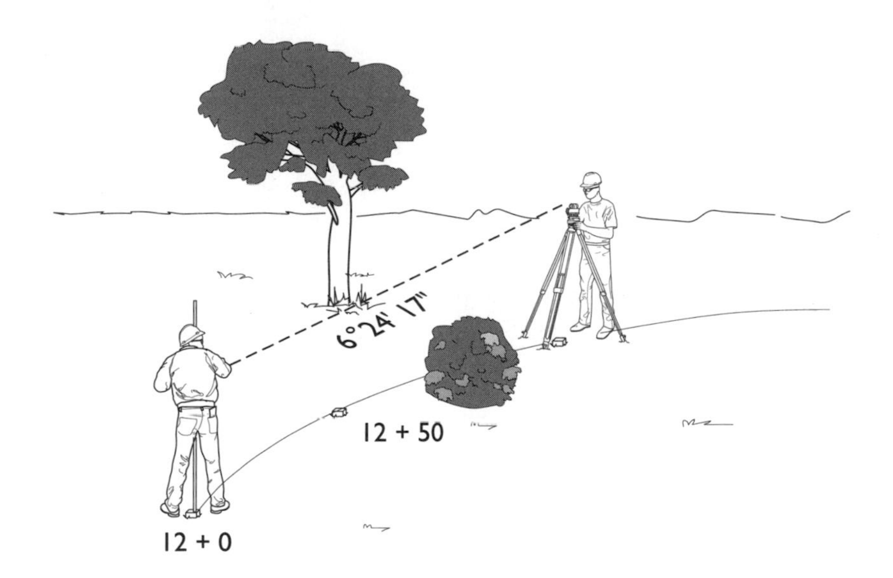

STEP 8 The field engineer backsights on the previous set-up point at station 12+00 (Longest Backsight) with the deflection of 6° 24' 17" on the instrument (The deflection of the point sighted on!).

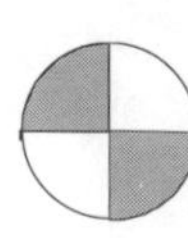

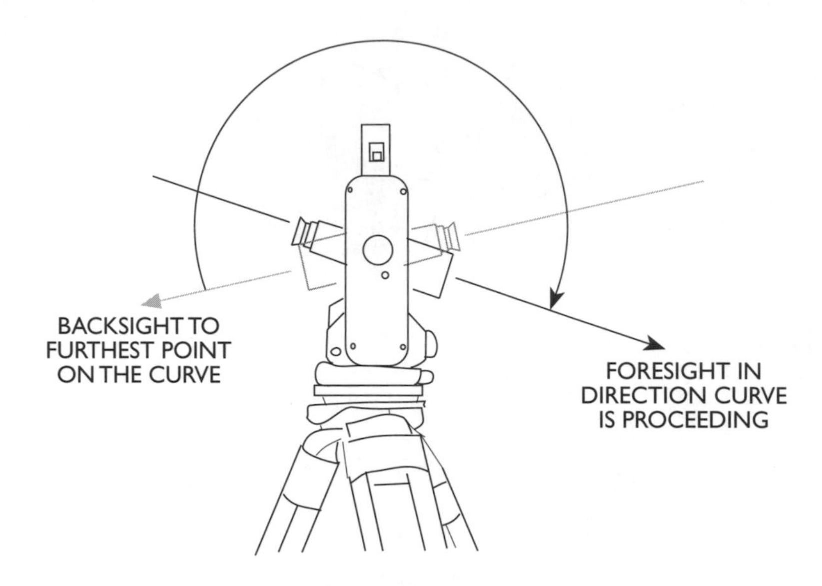

STEP 9

After backsighting, the field engineer plunges the telescope to sight in the direction the curve is proceeding.

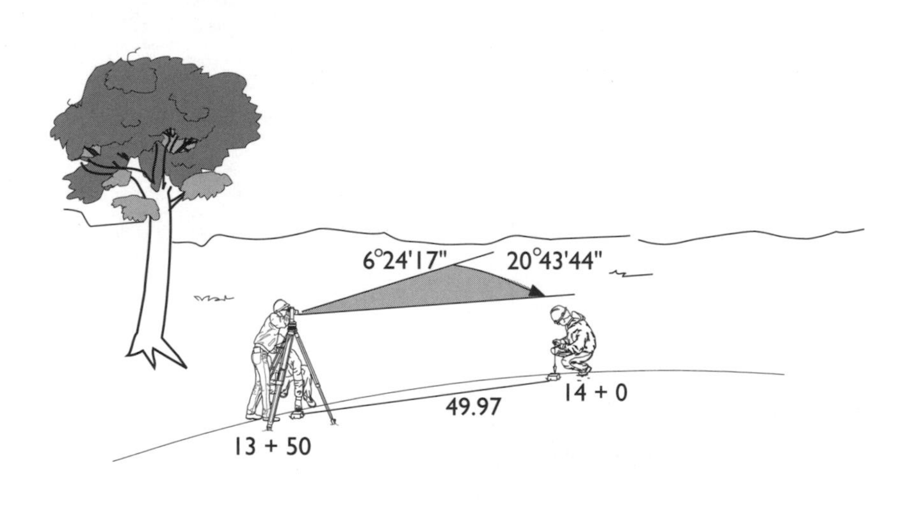

STEP 10

Referring to the original notes, the field engineer turns the instrument to 20° 43' 44" and directs the head chain to measure the short chord of 49.97 to establish point 14+00.

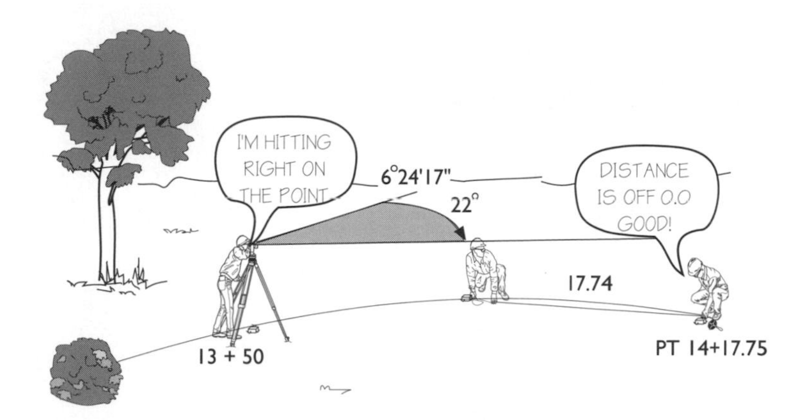

STEP 11

Referring to the original notes, the field engineer turns the instrument to 22° 00' 00" and directs the head chain to measure the short chord of 17.74' to establish the PT at station 14+17.74 or check into the PT if it had been established previously. The curve layout is complete.

SUMMARY

It has been shown and described that moving up on a curve is very methodical. Any field engineer should be able to perform this activity if needed on the jobsite.

OFFSET CURVE CALCULATION AND LAYOUT

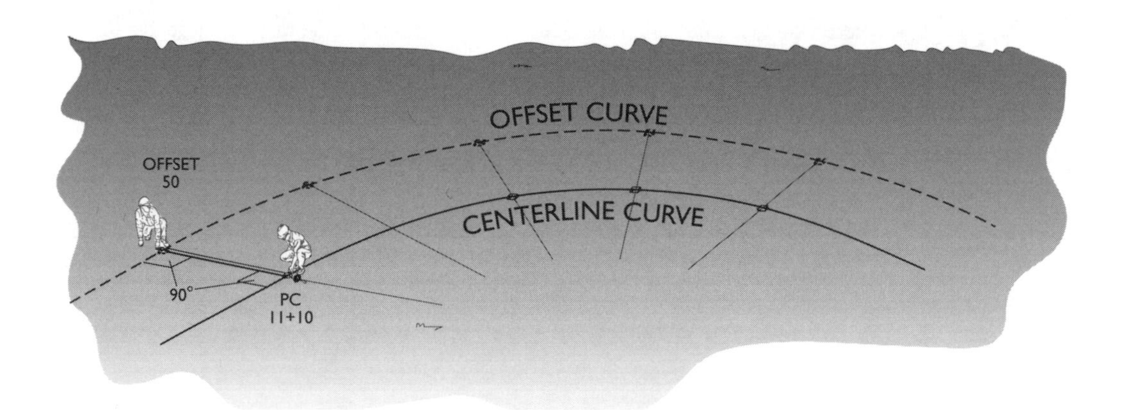

SCOPE

When a field engineer is laying out a route it is good practice to locate the control points outside the construction area to avoid the points becoming obliterated as the clearing, cut and fill, rough grade, and final grade work is performed. Offsetting right or left of the limits of construction can provide a safe location for this control. Along the straight areas of the route, offsetting is a simple process of running a line parallel to the centerline. When a curve is encountered, it is again a simple process of running an offset curve line that is parallel to the centerline. With the exception of calculating and measuring different chord lengths, the same procedure described for laying out a curve along the centerline is followed.

TERMINOLOGY

To describe the direction the curve is being offset, several terms can be used. As the route increases in stations, ***offset right*** is used to describe establishing control to the right of the centerline.

As the route increases in stations, **offset left** is used to describe establishing control to the left of the centerline. Other terms used to describe the direction of the offset are "in" and "out."

"Offset in" occurs when the curve is being offset toward the radius point of the curve and the length of the radius is shortened. As a result, the smaller radius used in the chord formula has the effect of decreasing the chord lengths used in the layout.

"Offset out" occurs when the curve is being offset away from the radius point of the curve and the length of the radius is lengthened. As a result, the larger radius used in the chord formula has the effect of increasing the chord lengths used in the layout.

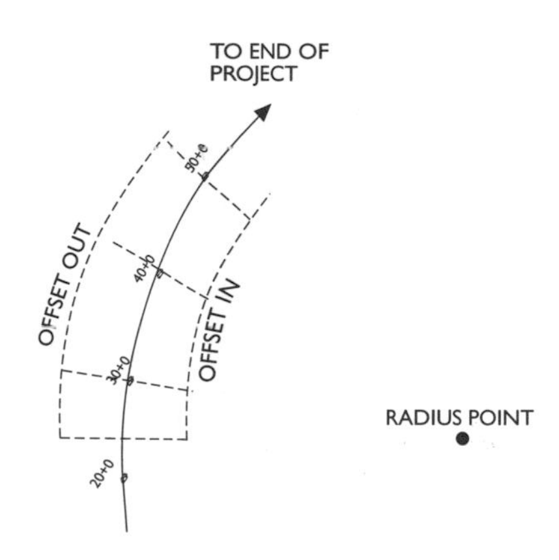

THEORY AND OBSERVATIONS

THE FUNDAMENTAL CHANGE IS THE RADIUS

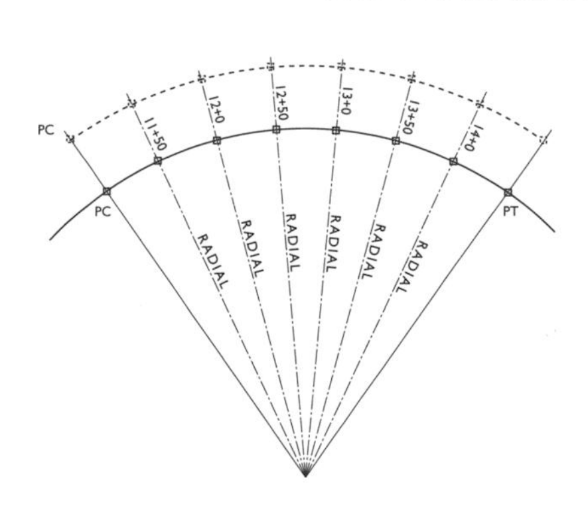

The only real change from the standard horizontal curve is that the radius and the chord lengths change for the offset curve. The deflections stay the same, and the stationing stays the same. Calculations are simple and only require the calculation of new short and long chord lengths.

THE OFFSET CURVE IS RADIAL

In route layout, stations on the offset line are used just as they were on the centerline. That is, when a curve is offset, the original stations are used even though the new arc lengths on the offset line are larger or smaller than they were originally. This occurs because the curve points are being offset radially from the radius point of the curve. Refer to the illustration of these relationships.

THE STATIONING STAYS THE SAME

Because the offset curve is considered to be radial from the centerline, the field engineer can use the same stationing that is used on the centerline when performing the layout of the offset curve.

THE DEFLECTION ANGLES STAY THE SAME

Recall that the formula used to calculate curve layout deflection angles uses the radius and the arc length. When a curve is offset, the radius and the arc length will change proportionately. The deflection angles used to layout an offset curve remain the same as they were for the centerline layout.

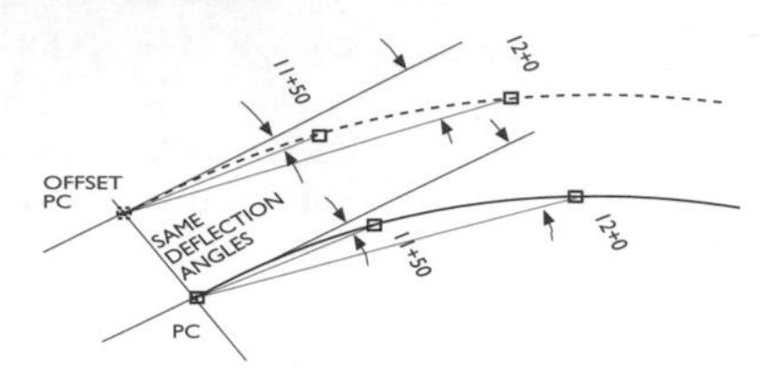

CONTINUE THE ORIGINAL NOTE FORM

Because the stationing and the deflection angles stay the same for the offset curve, the original note form can be used by simply adding columns that list the new short and long chord lengths to be measured.

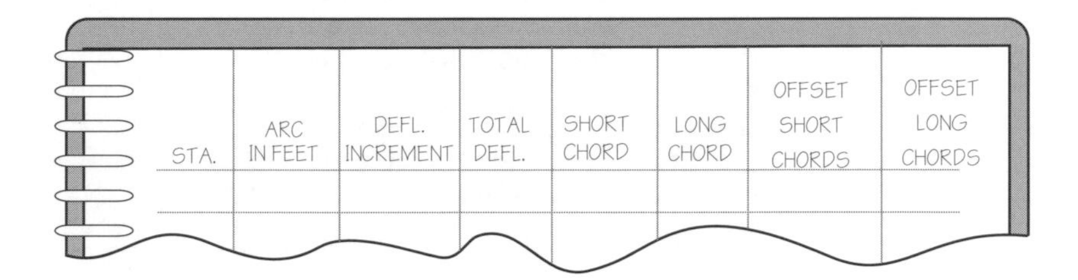

THE SHORT AND LONG CHORD EQUATIONS

The same formula that was used to calculate the short and long chords is used to determine the offset chord lengths. The only variable that changes is the radius.

Short Chord = 2R Sin (Incremental Deflection)

Long Chord = 2R Sin (Total Deflection at a Curve Point)

PROCEDURE

PREPARE

Determine the direction that the curve must be offset as "in" or "out."

Determine the amount of offsetting that is needed to locate the route outside of the construction work area.

Gather a calculator, notepad, and fieldbook.

Set up a fieldbook page or notepad page for listing the calculated data. See the headings on the example on the previous page.

CALCULATE THE NEW RADIUS

Add or subtract the amount of offset depending on whether or not the curve is being offset "in" or "out".

CALCULATE SHORT AND LONG CHORDS

Use the equation for the chord length and the deflection angles for the first station, the standard interval, and the last station to calculate the short chords.

Use the equation for the chord length and the cumulative deflection angles to calculate the long chords.

EXAMPLE SOLUTION:

CONTINUING OUR SAMPLE PROBLEM

Given: I = 44° R = 400 PI Sta. = 12 + 72.18

Calculate a 50' out offset curve (R = 450')

COMPUTE THE SHORT CHORD LENGTHS

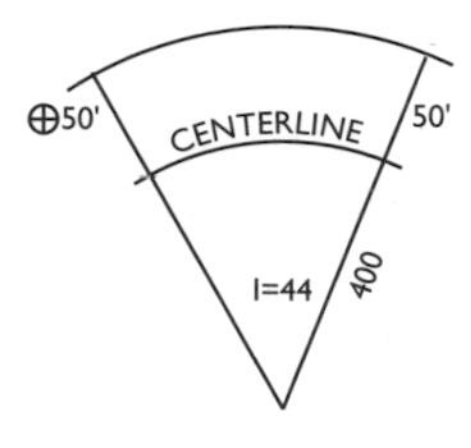

Recall, Short chord = 2R sin (deflection)

To first station from PC

Short chord = 2(450) sin (2° 49' 26") = 44.34

At 50 foot intervals

Short chord = 2(450) sin (3° 34' 52") = 56.22

For the closing distance into the PT

Short chord = 2(450) sin (1° 16' 16") = 19.96

LONG CHORDS

Long chord to station 12+00 = 2(450) sin (6° 24' 17") =100.40'

Long chord to station 12+50 = 2(450) sin (9° 59' 09") = 156.06

Long chord to station 13+00 = 2(450) sin (13° 34' 01") = 211.12

etc.

TABLE OF RESULTS

Station	Deflection Increment	Total Deflection	Short Chord	Long Chord	50' Offset Short Chord	50' Offset Long Chord
14 + 17.75 PT	1° 16' 16"	22° 00' 00"	17.74"	299.68	19.96	337.14
14 + 00	3° 34' 52"	20° 43' 44"	49.96"	283.16	56.22	318.55
13 + 50	3° 34' 52"	17° 08' 52"	49.96"	235.87	56.22	265.35
13 + 00	3° 34' 52"	13° 34' 01"	49.96"	187.66	56.22	211.12
12 + 50	3° 34' 52"	9° 59' 09"	49.96"	138.72'	56.22	156.06
12 + 00	3° 34' 52"	6° 24' 17"	49.96"	89.24'	56.22	100.4
11 + 50	2° 49' 26"	2° 49' 26"	39.41'	39.41'	44.34	44.34
11 + 10.57 PC	0	0	0	0	0	0

STEP-BY-STEP LAYOUT PROCEDURE

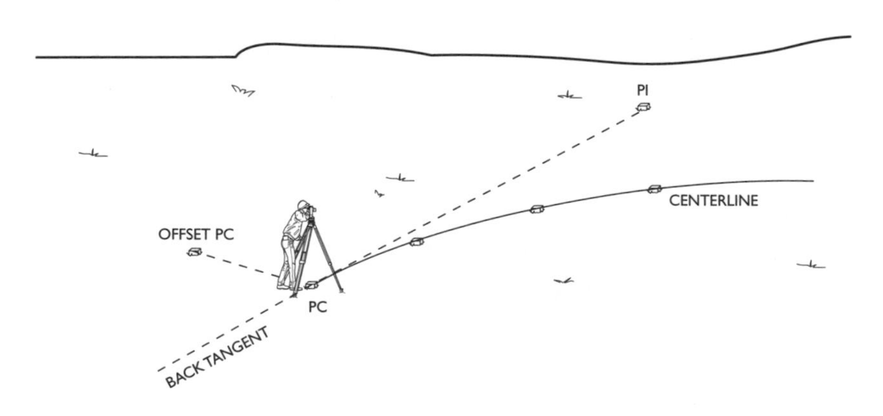

STEP 1

Set on the PC of the Centerline curve and sight on the PI or back tangent, and turn a 90 degree angle in the direction of the proposed offset curve.

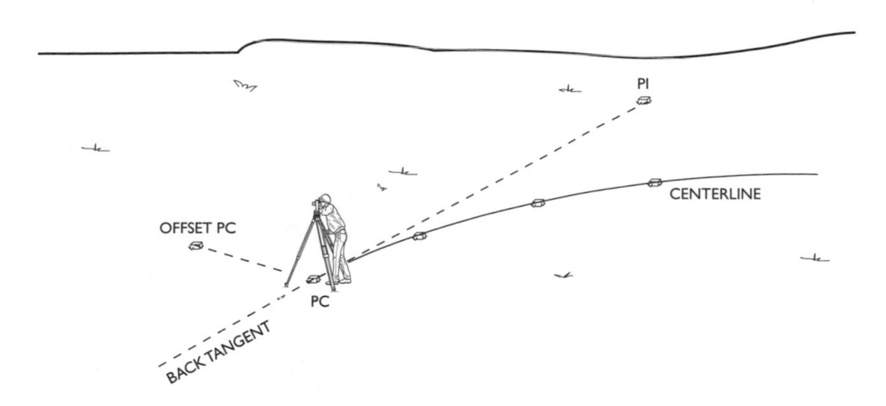

STEP 2

Along the line of sight, measure by chaining or EDM the distance to the offset PC. Set a hub and tack at that point.

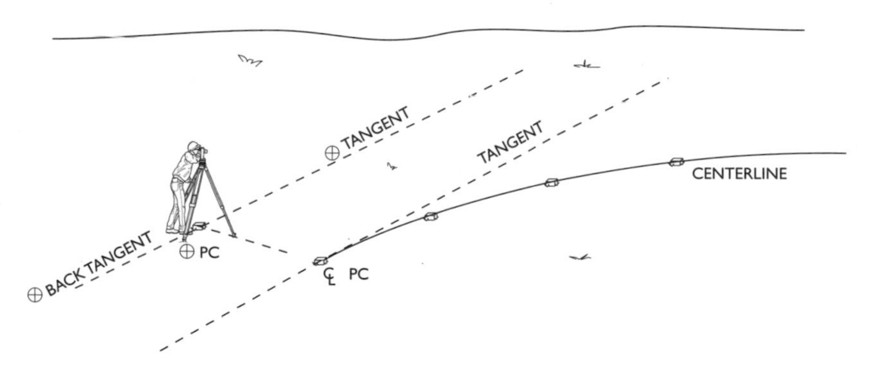

STEP 3

Move the instrument to the offset PC, set up, and level over the point. Sight onto the centerline PC with 90 degrees on the instrument, and turn to 0 degrees to create an offset line parallel to the back tangent of the centerline. Or, preferably, go as far as visible on the back tangent and set an offset point. Sight onto that point from the offset PC. This follows the principle of "long backsight and short foresight".

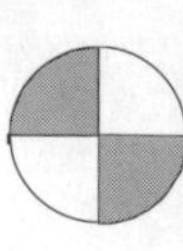

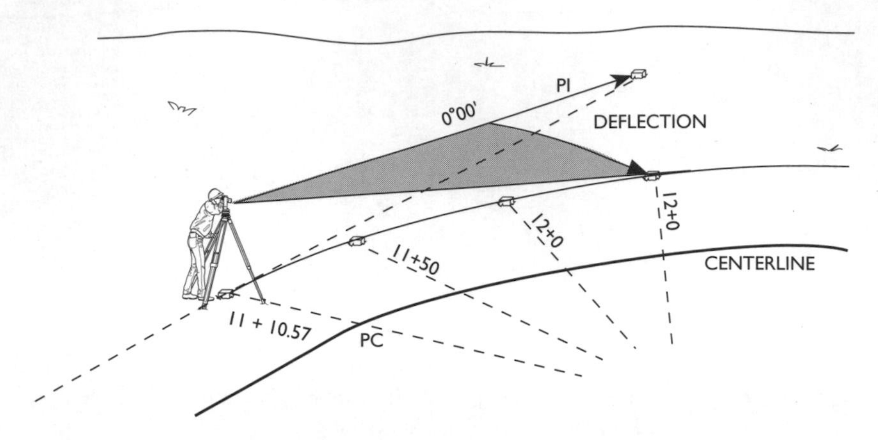

STEP 4

Refer to the offset curve calculation table and follow the procedure for laying out a horizontal curve. Turn the same deflection angles and measure the new chord distances.

VERTICAL CURVES

18

PART 3 - CONSTRUCTION SURVEYING CALCULATIONS

CHAPTER OBJECTIVE:

Describe why vertical curves are necessary for modern transportation.
Describe the following terms related to vertical curve computations: slope, gradient, tangent.
Calculate the elevations of stations on a gradient.
Describe, define and illustrate the parts and elements of a vertical curve.
Describe the properties of the vertical curve.
State the equations of the vertical curve and define the terms.
Calculate tangent offsets and apply them to tangent elevations to determine vertical curve elevations.
Describe and calculate the low or high point of a vertical curve.

INDEX

OVERVIEW

SCOPE

Vertical curves are used in design to provide a smooth transition between two slopes. They are used extensively in highway design to provide a comfortable and roadway for today's high-speed travel. Vertical curves are also used on site work for access roads, as well as for aesthetic purposes when designing an elaborate landscaped site entrance. The field engineer should be familiar with vertical curves and how they are calculated as well as staked in the field.

GENERAL

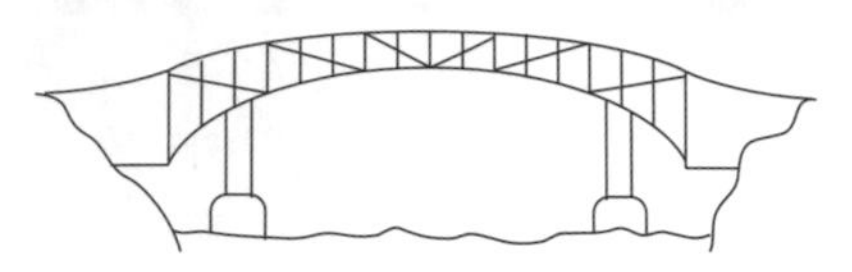

Vertical curves are entirely different from horizontal curves. Vertical curves lie in a vertical plane whereas horizontal curves are in the horizontal plane. Vertical curves usually exist in the same area of a highway as a horizontal curve. When trying to visualize a vertical curve, think about looking at a bridge that is constructed high over a river to allow tall sail boats to pass under it. The bridge starts out at ground level and curves upward towards the top. Approaching the top, it begins to curve downward. As the bottom is reached on the other side of the river, it gently transitions into the roadway . In this scenario, three vertical curves were used. One at the beginning, one at the top, and one at the end.

VERTICAL CURVES ARE ELEVATIONS

A vertical curve is represented by elevations. These elevations are determined by calculations that are based on gradients (slope) on both sides of a vertical intersection, the elevation at the intersection, and the designed length of the vertical curve. Unlike horizontal curves, there are no deflections or chord distances to measure to locate a vertical curve. In fact, at the time the vertical curve is established in the field, there should be no measurements required. Simply subtract the profile ground elevation from the calculated vertical curve elevation to get a cut or fill to write on the stake.

Summary of Calculating a Vertical Curve	
1.	Obtain vertical curve design data from the plans.
2.	Calculate the elevations of the design gradient
3.	Calculate the rate of change "r" of the vertical curve in percent per station
4.	Calculate the tangent offset - "y"
5.	Calculate the Curve Elevation.

This chapter will introduce one of several different methods that are used to calculate a vertical curve. If more information is necessary, see *Appendix D* for a list of route surveying textbooks.

GRADIENT AND ELEVATIONS ALONG THE TANGENT

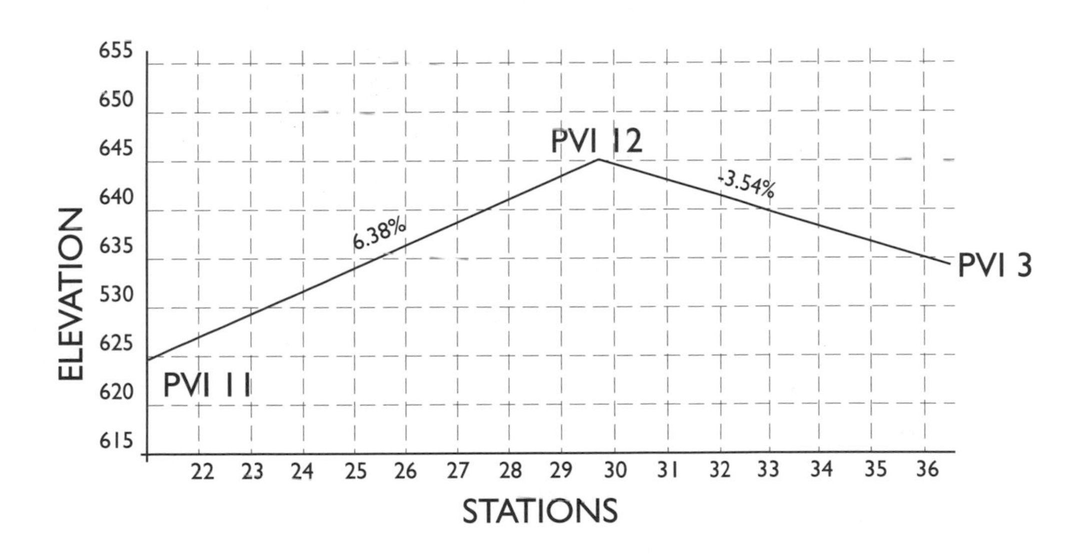

SCOPE

Before actually calculating the vertical curve elevation, the field engineer will first need to calculate elevations along the tangent. Even before that the field engineer will have to understand and be able to calculate gradient. This section offers an overview of gradient calculations and provides an example of the calculations needed to determine elevations along the tangent.

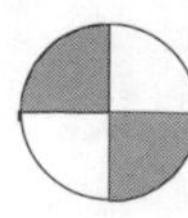

DEFINITIONS

SLOPE

The term **slope** is used to describe the steepness of a line. Mathematically, it is the ratio of the rise in elevation to the run in horizontal distance. This can be expressed, and shown by:

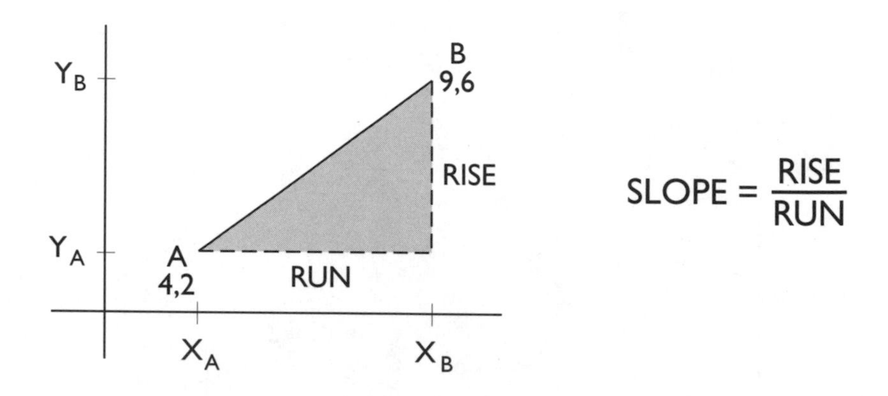

Using coordinates, the slope of the line is defined by:

$$\text{SLOPE OF AB} = \frac{\text{RISE}}{\text{RUN}} = \frac{Y_B - Y_A}{X_B - X_A} = \frac{6-2}{9-4} = \frac{4}{5}$$

GRADIENT OR GRADE

In road design, the term **gradient** is used instead of slope to describe the profile of the centerline. In practice, gradient is expressed two ways - either as a ratio of feet per foot, or as a percent.

For example, a gradient that has been calculated as + 0.0453 ft/ft can be converted to a percent by multiplying by 100.

+ 0.0453 ft/ft x 100 = 4.53%

or 4.53 feet per 100 feet or per station, and converting back:

4.53/100 = 0.0453

Most highway plans will express the gradient as a percentage. There are several additional properties associated with gradient that must be understood.

POSITIVE GRADIENT

A positive gradient is a line that rises as stations increase.

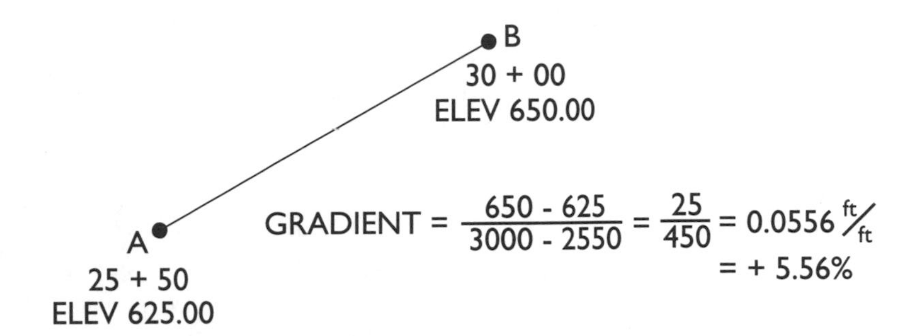

NEGATIVE GRADIENT

A negative gradient is a line that falls as stations increase.

B
30 + 00
ELEV 650.00

C
34 + 25
ELEV 642.50

$$\text{GRADIENT} = \frac{642.5 - 650}{3425 - 3000} = \frac{-7.5}{425} = -0.0176\ \text{ft}/\text{ft}$$

$$= -1.76\%$$

ZERO GRADIENT

A line that neither rises nor falls as stations increase.

C
34 + 25
ELEV 642.5

D
38 + 75
ELEV 642.5

$$\text{GRADIENT} = \frac{642.5 - 642.5}{3875 - 3425} = \frac{0}{450} = 0\ \text{ft}/\text{ft}$$

POINT OF VERTICAL INTERSECTION

A point of vertical intersection (PVI) occurs where grade lines intersect. From one PVI to another PVI the gradient lines are straight, and the grade change is constant. When design conditions are known, a vertical curve is inserted at each PVI on the centerline to provide a smooth transition between tangent lines.

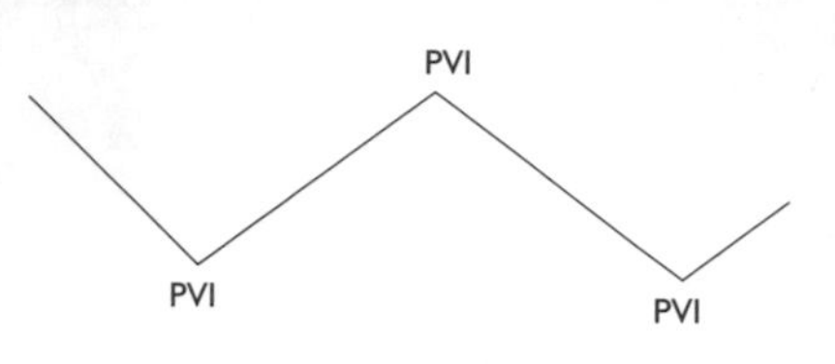

CALCULATION OF THE GRADIENT FROM PVI TO PVI

The following is a road centerline from PVI 1 through PVI 2 to PVI 3. The first step in vertical curve calculations is to determine the gradient between the PVI's. The sample calculations show the method to determine the gradient from PVI 1, station 21+50, elevation 626.34, to PVI 2, station 26+00, elevation 655.06; and from PVI 2 to PVI 3 station 29+25, elevation 643.54.

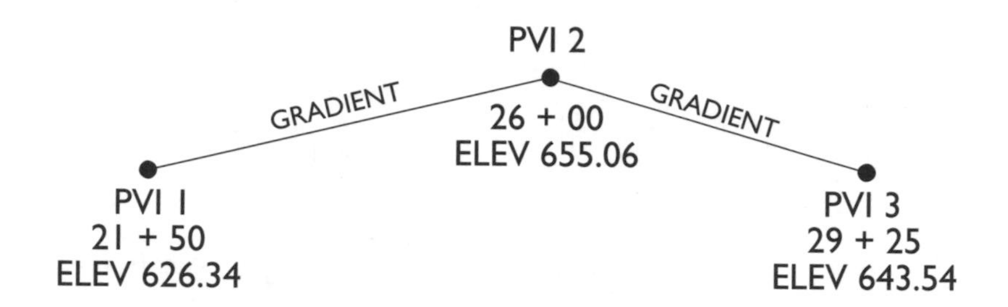

SOLUTION PVI 1 TO PVI 2

$$\text{GRADIENT} = \frac{655.06 - 626.34}{2600.00 - 2150.00} = \frac{28.72}{450}$$

$= 0.0638\ ^{ft}/_{ft}$ of constant change from PVI 1 to PVI 2

$= 6.38\%$

SOLUTION PVI 2 TO PVI 3

$$\text{GRADIENT} = \frac{643.54 - 655.06}{2925.00 - 2600.00} = \frac{-11.52}{325}$$

$= -0.0354\ ^{ft}/_{ft}$ of constant change from PVI 2 to PVI 3

$= -3.54\%$

TANGENT ELEVATIONS

When the gradient is known, or has been calculated, elevations along the tangent from one PVI to another PVI need to be determined before elevations on the vertical curve can be obtained. Because the gradient is constant, the elevation for any station between PVIs can readily be calculated as follows:

Calculation of Tangent Elevations	
Step 1	Determine the distance from the PVI to the first full station.
Step 2	Multiply the distance in stations times the gradient to obtain the change in elevation.
Step 3	Apply the change in elevation to the known elevation at the PVI to obtain the elevation for the first full station. Add for positive gradient; subtract for negative gradient.
Step 4	Repeat for each full, half, or quarter station, or any other point until the next PVI is reached.

EXAMPLE

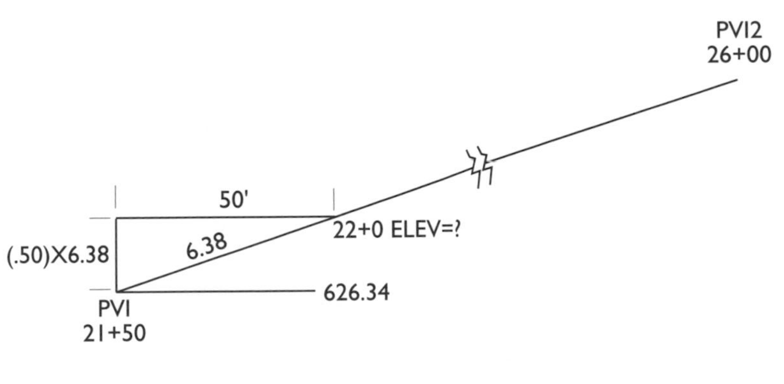

626.34 + (0.50) X(6.38) = ELEV 22+0

626.34 + 3.19 = ELEV 22+0

629.53 = ELEV 22+0

SUMMARY TABLE

The following table summarizes the calculations required to determine the elevations of the half stations between the PVIs.

Point	Station	Computation Elev. + Distance Sta. X Gradient	Tangent Elevation
PVI 1	21+50		626.34
	22+00	626.34+(0.50)(6.38)=	629.53
	22+50	626.34+(1.00)(6.38)=	632.72
	2300	626.34+(1.50)(6.38)=	635.91
	23+50	626.34+(2.00)(6.38)=	639.10
	24+00	626.34+(2.50)(6.38)=	642.29
	24+50	626.34+(3.00)(6.38)=	645.49
	25+00	626.34+(3.50)(6.38)=	648.68
	25+50	626.34+(4.00)(6.38)=	651.87
PVI 2	26+00	626.34+(4.50)(6.38)=	655.06
	26+50	655.06+(0.50)(-3.54)=	653.28
	27+00	655.06+(1.00)(-3.54)=	651.52
	27+50	655.06+(1.50)(-3.54)=	649.74
	28+00	655.06+(2.00)(-3.54)=	647.97
	28+50	655.06+(2.50)(-3.54)=	646.20
	29+00	655.06+(3.00)(-3.54)=	644.42
PVI 3	29+25	655.06+(3.25)(-3.54)=	643.54

Although the elevation of the PVIs was already known, it is good practice to recalculate it as shown to check for possible errors in gradient calculations.

ELEMENTS AND PROPERTIES

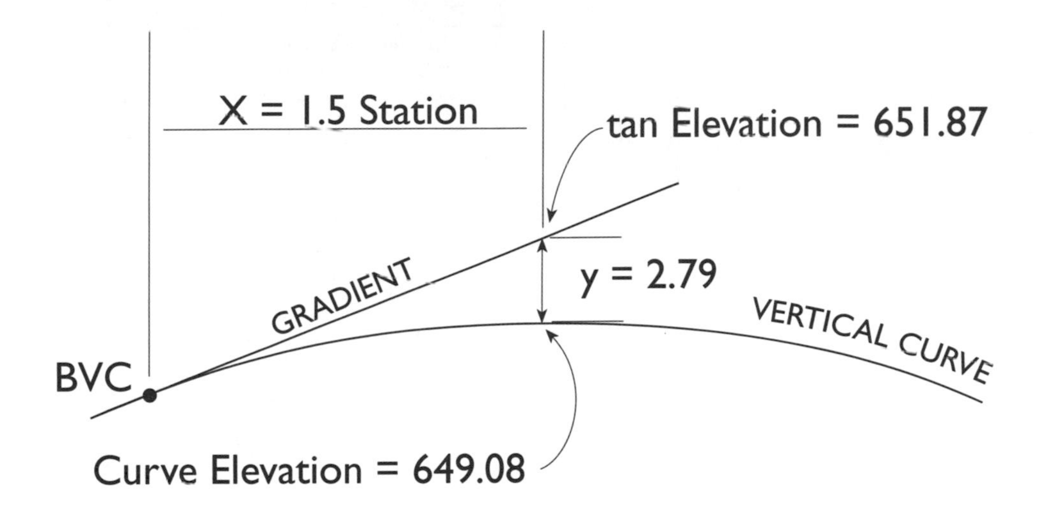

SCOPE

The vertical curve is completely different from the horizontal curve. Recall the horizontal curve is based on the simple circle, and there are numerous formulas and calculations required to calculate the parts and layout information for it. The vertical curve is based on the simple parabola which has one very basic formula. It must be visualized that vertical curves lie in the vertical plane. Surveying in the vertical plane relies on elevations; therefore, vertical curves are based on elevation calculations.

PARTS OF THE VERTICAL CURVE

Specific terminology is used to describe the parts of the vertical curve. Abbreviations have also been developed for each term. Care must be taken to distinguish between the vertical curve terminology and the horizontal curve terminology (some of the terms are similar).

Elements of the Vertical Curve	
PVI	Point of vertical intersection. The point where the two tangents meet.
g1	Back gradient. This represents the slope of the connecting tangent from the PVI back towards the beginning of the project (ft/ft or %).
g2	Forward gradient. This represents slope of the connecting tangent from the PVI towards the end of the project (ft/ft or %).
BVC	Beginning of vertical curve. That point where the tangent ends and the vertical curve begins.
EVC	End of vertical curve. That point where the vertical curve ends and the forward tangent begins.
L	Length of vertical curve (in stations; in all formulas!). This is a horizontal distance.
L/2	One half the length or distance from the BVC to the PVI, or EVC to the PVI
e	Ordinate from the P.V.I. to the vertical curve (feet).
x1	Horizontal distance from the BVC (stations).
y1	Ordinate distance at any station between the BVC and the PVI (feet) (between the gradient and curve.)
x2	Horizontal distance from the EVC (stations).
y2	Ordinate distance at any station between the EVC and the PVI (feet).

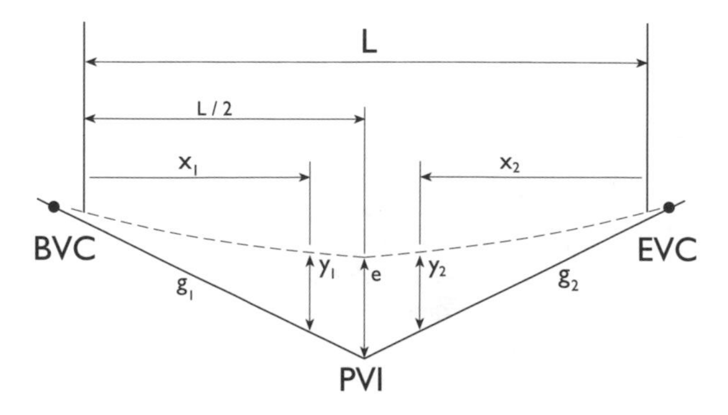

PROPERTIES OF THE VERTICAL CURVE

As stated previously, the vertical curve is based on the parabola. The curve is not based on a whole parabola but rather a small part of it - just as we only use a small part of the simple circle for horizontal curves. See the illustration below:

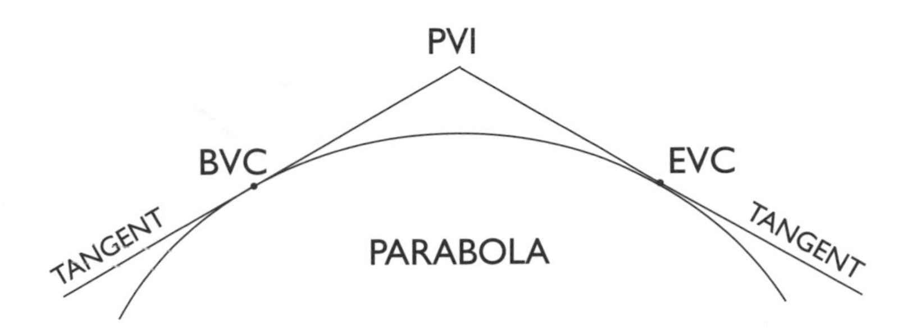

SYMMETRY

The use of the parabola for our calculations could become rather difficult if the theory was not set up so that the curve is symmetric. That is the vertical curve has equal tangents, or, half of the vertical curve falls before the PVI, and half of it falls after the P.V.I. This property makes vertical curve calculations simple. NOTE: Sometimes unequal tangent or non-symmetric vertical curves are encountered. They are simply calculated by creating two equal tangent curves. See Appendix D, for a route surveying textbook for this procedure.

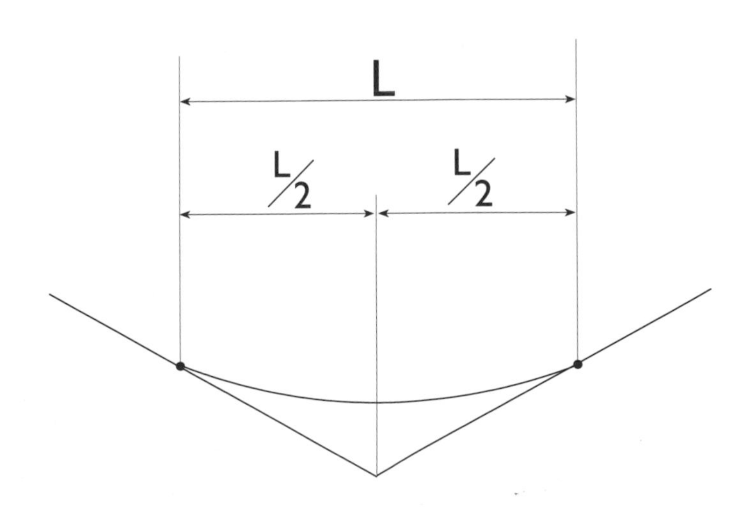

SAG CURVE

A sag curve is a vertical curve that occurs when a negative gradient is followed by a positive gradient.

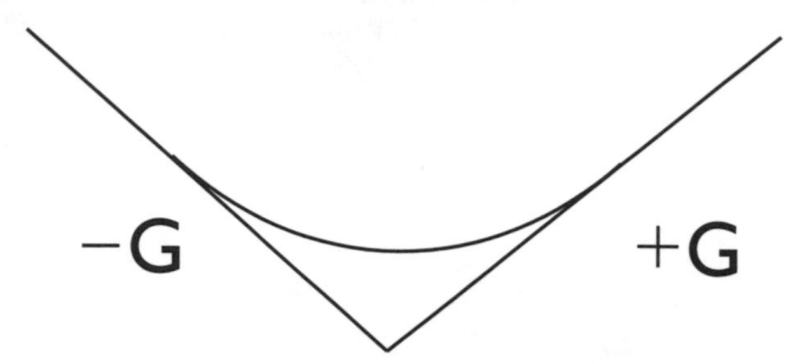

CREST CURVE

A crest curve is a vertical curve that occurs when a positive gradient is followed by a negative gradient.

CALCULATION OF CURVE ELEVATIONS

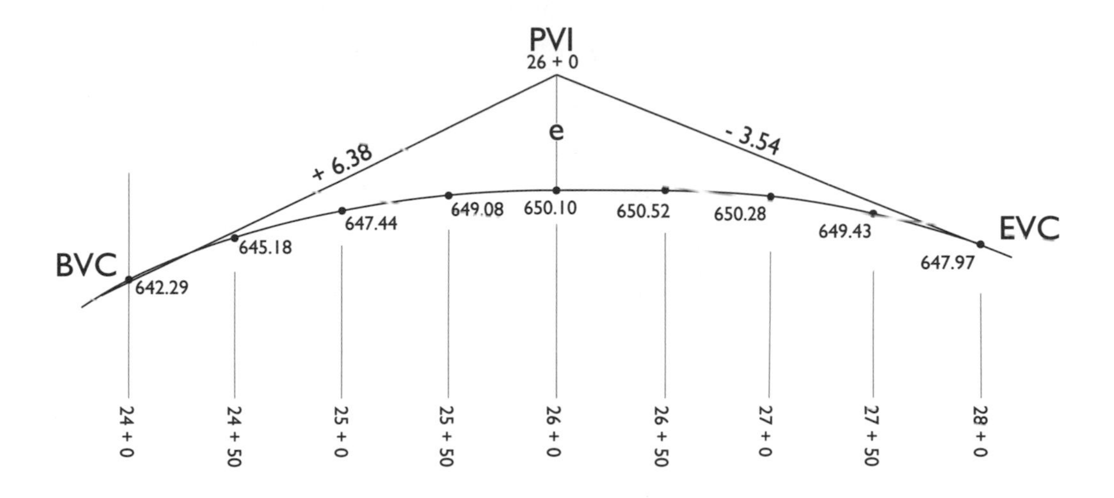

SCOPE

Just like horizontal curves, vertical curves are designed into the roads we travel to provide us with comfort and safety. Without vertical curves, there would be abrupt up-and-down changes as the highway followed the terrain. This would not be practical for our high speed travel.

Many factors must be considered when vertical curves are designed. These include: vehicle speed, height of vehicles, passing-sight distance, stopping distance, rider comfort, drainage, and appearance. These factors as well as the difference in gradient of the intersecting tangents control the calculated values of each vertical curve.

We are not concerned with these design factors. An expert in road design will provide the vertical curve information on the plans. We will take the plans and calculate the data needed to stake the vertical curve in the field. In the field a vertical curve is defined as a series of elevations along a center line.

As stated earlier, the simple parabola is the geometric figure used to describe vertical curves used in road design. The parabola is used because it provides a gradual change in direction, is convenient to use, and is simple to calculate.

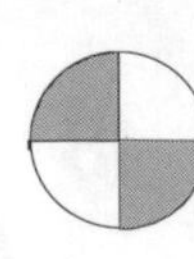

EQUATIONS OF THE VERTICAL CURVE

The parabola defines the vertical curve because it provides gradual changes in direction. The equation for the parabola that will be used has been derived from its simplest form: It is represented by:

$$y = rx^2$$

where

y = the vertical distance (feet)

x = the horizontal distance (stations)

r = rate of change of the gradient in percent per station

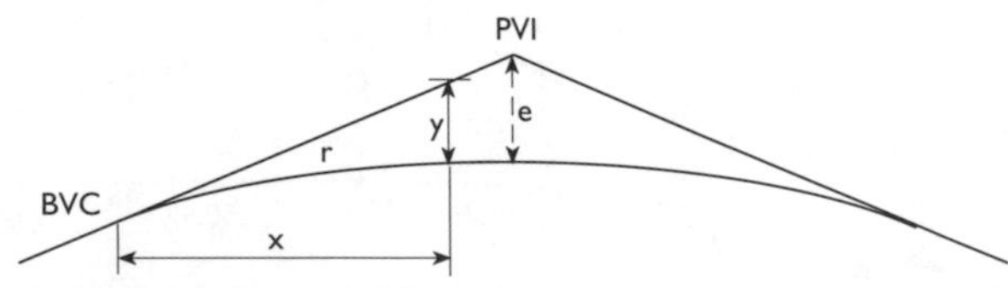

From this simple formula, useful formulas for use in vertical curve calculations have been derived by substituting elements of the vertical curve. These equations are:

$$r = \frac{(g_1 - g_2)}{2L}$$

$$e = \frac{(g_1 - g_2)L}{8}$$

$e =$ vertical distance from PVI to curve

The calculation of "r" will give results that can be used to calculate elevations which define the vertical curve. The calculation of "e" can be used as a check in the vertical curve calculation process.

EXAMPLE

PVI=26+0
EL 655.06

g_1 =6.38 g_2 =-3.54

L=4

$$e = \frac{(g_1 - g_2)L}{8} = \frac{(6.38 - (-3.54))4}{8} = \frac{(9.92)4}{8} = 4.96$$

$$r = \frac{(g_1 - g_2)}{2L} = \frac{(6.38 - (-3.54))}{2(4)} = 1.24$$

CALCULATION PROCESS

Calculating a Vertical Curve	
1	Setup a table to record thc calculations.
2	Input the known information: Stations, PVI, BVC, EVC, and the previously calculated tangent elevations.
3	Calculate "r": the rate of change of the slope in percent per station.
4	Compute "y" using the formula: $y = rx^2$ where y is the vertical distance between the tangent and curve and x is the horizontal distance from the BVC or the EVC to the stations on the curve. This method calculates from the BVC to the PVI and from the EVC to the PVI.
5	Apply the vertical distance "y" to the tangent elevation to obtain the curve elevation. If the curve is a crest curve, subtract "y" from the tangent. If the curve is a sag curve, add "y" to the tangent elevation.

SAMPLE CALCULATION

$y = rx^2$

at 24+50

$y=1.24(.5^2)$
$y=0.31$

EXAMPLE VERTICAL CURVE ELEVATION CALCULATIONS

With the information covered, it is a simple matter of applying what is known to obtain the elevations which define the vertical curve. The final computation is to obtain the ordinate value, y, for each of the stations along the curve. This calculation can be accomplished by using the fundamental formula for the parabola: $y = rx^2$.

This information is taken from the plans:

PVI = 26+00 Elevation = 655.06
g_1 = +6.38% g_2 = - 3.54%
L = 4 stations

SUMMARY TABLE

Vertical Curve Elevations					
Point	**Station**	**Gradient Elevation**	**Computation ($y = r(x^2)$)**	**y**	**Curve Elevation (elev - y)**
PVI 1	21 + 50	626.00			
	22 + 00	629.53			
	22 + 50	632.72			
	23 + 00	635.91			
	23 + 50	639.1			
BVC	24 + 00	642.29	1.24 (0^2)	0	642.29
	24 + 50	645.49	1.24 (0.50^2)	0.31	645.18
	25 + 00	648.68	1.24 (1.0^2)	1.24	647.44
	25 + 50	651.87	1.24 (1.50^2)	2.79	649.08
PVI 2	26 + 00	655.06	e	4.96	650.1
	26 + 50	653.28	1.24 (1.50^2)	2.79	650.52
	27 + 00	651.52	1.24 (1.0^2)	1.24	650.28
	27 + 50	649.74	1.24 (0.50^2)	0.31	649.43
EVC	28 + 00	647.97	1.24 (0^2)	0	647.97
	28 + 50	646.2			
	29 + 00	644.42			
	29 + 50	643.42			

SUMMARY

In this case, the vertical curve was a crest curve. Therefore, the "y" was subtracted from the gradient elevation.

ELEVATION OF THE HIGH OR LOW POINT

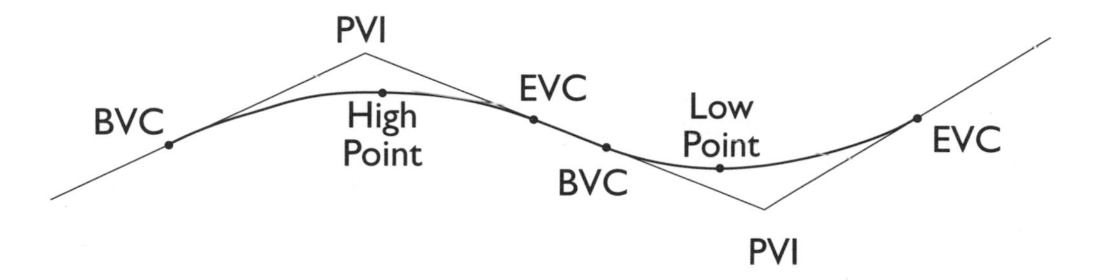

SCOPE

When traveling down a highway after a rainstorm, it is easy to determine where catch basins were not located properly. Water is puddled up in the low spot of the road. The catch basin isn't draining water away because it is located higher than the low spot. The field engineer who is working for a paving contractor, on highway construction projects or even on building sites may be responsible for calculating where the low point is to locate drainage structures. This is a very simple calculation and is based, again, on the formula for the parabola.

GENERAL

The determination of the location of a high or low point is generally performed at the design stage of a project by the design engineer. Calculations are performed so that the location of drainage structures can be determined and located on the plans. However, in some instances, the location of structures has been left to a note on the plans that says "exact location to be determined by field calculation and measurement."

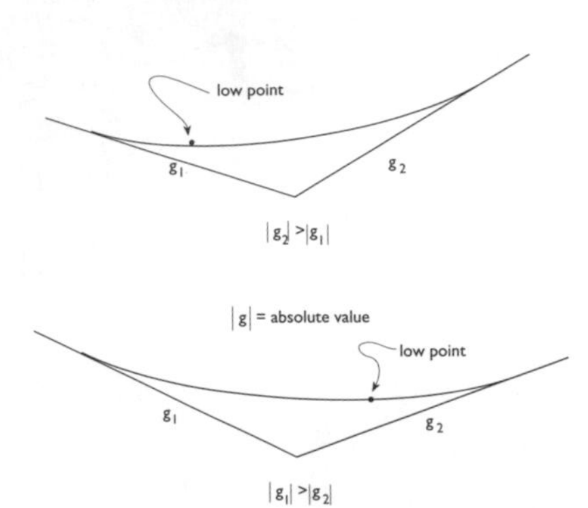

If this note is encountered when locating the low point on a curve, the field engineer should be prepared to perform the necessary calculations.

By observation of the gradients of the vertical curve, the general location of the low point being before or after the PVI can be estimated. Using the absolute values of the gradients, it should be noted that the high or low point of a vertical curve occurs on the side of the smallest gradient. If g_1 is larger than g_2, the low point occurs after the PVI. If they are the same, the low point occurs at the PVI.

EQUATION FOR THE HIGH OR LOW POINT

The formula given below was derived from the basic equation for the parabola. The way it is set up, the calculation of the distance to the high or low point is determined from the BVC of the vertical curve. Therefore, "x" can fall before the PVI or after it.

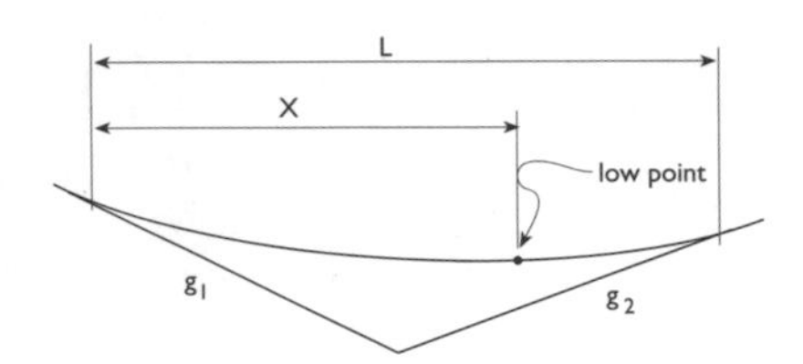

$$x = \frac{g_1 L}{g_1 - g_2}$$

x = Distance from BVC to the high or low point
L = Length of Vertical Curve in Stations
g_1 = back gradient
g_2 = forward gradient

When performing the calculations necessary, be sure to watch algebraic signs to obtain the correct results.

CALCULATING A LOW POINT LOCATION AND ELEVATION

STEP 1

Determine what is given and draw a sketch labeling the known values.
g_1 = -2%,
g_2 = +1.6%,
L = 8 Stations
PVI @ 87+00 PVI Elev. = 743.00

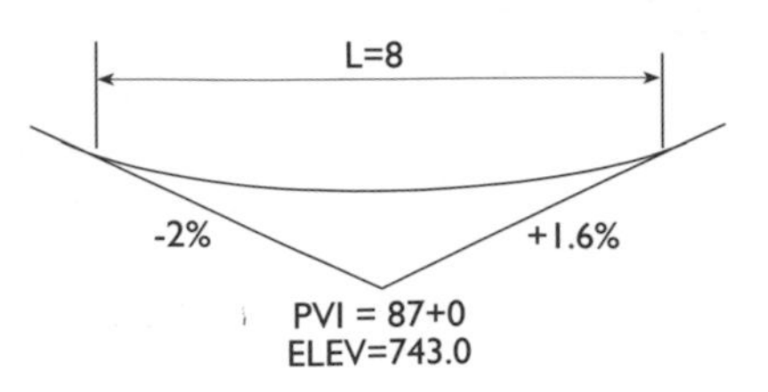

STEP 2

Observe and mentally note if the low point is before or after the PVI.

STEP 3

Using the low point formula, substitute known values and solve for the distance from the BVC to the low point "x"

$$x = \frac{-2x8}{-2-(+1.6)}$$

$$x = 4.444 stations$$

STEP 4

Determine the station of the low point.

BVC Station - 83+00
83+00 + 4.4444 = Station of Low Point
Station of low point = 87+44.44

STEP 5

Determine the gradient elevation at the low point.

In this case, work from the PVI at 87+00 for 44.44 feet to the low point location.
743.00 + 1.6(.4444 stations)
Elevation of gradient at low point = 743.71

STEP 6

Determine the "y" tangent offset at the low point.

Calculate needed information to solve for y. Distance from the EVC to the low point is 3.56 stations.
Solving for "r" = -0.225

$$y = rx^2$$

$$y = -0.225(3.56)^2$$

$$y = -2.85$$

STEP 7

Calculate the elevation of the vertical curve at the low point.

Low Point Elevation = Gradient Elevation -y
Low Point El. = 743.71 - (-2.85)
Low Point El. = 746.55

SUMMARY

The low point for this vertical curve is at station 87+44.44 and at elevation 746. 55. The field engineer should locate any drainage structure at that station and elevation.

LAYOUT IN THE FIELD

SCOPE

The calculated curve elevation represents the elevation of the finished grade. The field engineer will have run profile levels and the existing ground elevation along the centerline. Stakes will be set at the centerline with the difference (cut or fill) written on them. This provides the contractor with the information necessary to perform the work on the project.

GENERAL

To "layout" a vertical curve in the field the field engineer needs curve elevations and ground elevations. The curve elevation are calculated using the procedures outlined in this chapter. The ground elevations are obtained by profile leveling techniques. By determining the difference in elevation between "ground" and "Curve" elevations, the field engineer can provide a cut or fill amount on each stake or centerline. This grade information may also be used for setting slope stakes if the typical section is level.

STEP BY STEP PROCEDURE

The following is the procedure to be followed for Layout and Staking a Vertical Curve.

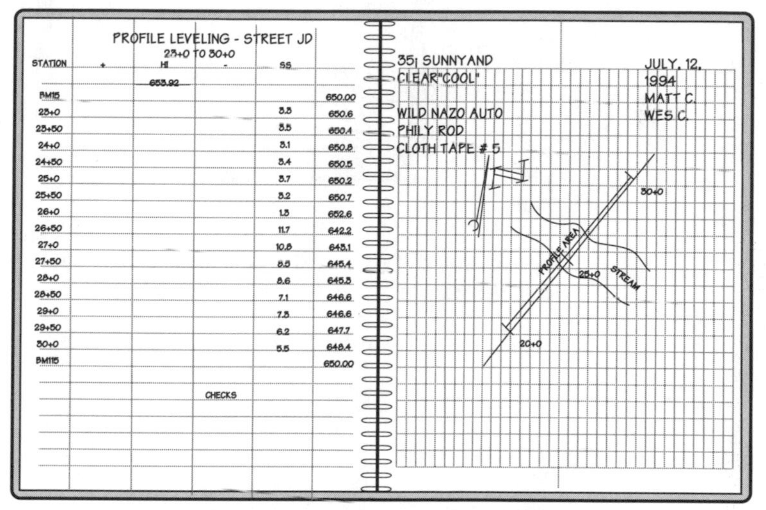

PROFILE LEVELING - STREET JD
23+0 TO 30+0

STATION	+	HI	-	SS	
		653.92			
BM15					650.00
23+0				3.3	650.6
23+50				3.5	650.4
24+0				3.1	650.8
24+50				3.4	650.5
25+0				3.7	650.2
25+50				3.2	650.7
26+0				1.3	652.6
26+50				11.7	642.2
27+0				10.8	643.1
27+50				8.5	645.4
28+0				8.6	645.3
28+50				7.1	646.6
29+0				7.3	646.6
29+50				6.2	647.7
30+0				5.5	648.4
BM115					650.00

CHECKS

35| SUNNYAND
CLEAR"COOL"
WILD NAZO AUTO
PHILY ROD
CLOTH TAPE # 5

JULY. 12.
1994
MATT C.
WES C.

STEP 1

Obtain the fieldnotes for the centerline profile ground elevations.

Vertical Curve Elevations		
Point	**Station**	**Curve Elevation**
BVC	24 + 00	642.29
	24 + 50	645.18
	25 + 00	647.44
	25 + 50	649.08
PVI 2	26 + 00	650.1
	26 + 50	650.52
	27 + 00	650.28
	27 + 50	649.43
EVC	28 + 00	647.97

STEP 2

Obtain the calculated curve elevations for the vertical curve.

Station	Gradient Elevation	Cut / Fill
24 + 00	642.29	C 8.5
24 + 50	645.49	C 5.3
25 + 00	648.68	C 2.8
25 + 50	651.87	C 1.6
26 + 00	655.06	C 2.5
26 + 50	653.28	F 8.3
27 + 00	651.52	F 7.2
27 + 50	649.74	F 4.0
28 + 00	647.97	F 2.7

STEP 3 Develop a cut/fill sheet and determine the difference between the profile ground elevations and the designed curve elevations

STEP 4 Go to the field and mark the cut or fill on the stakes.

EXCAVATIONS

19

PART 3 - CONSTRUCTION SURVEYING CALCULATIONS

CHAPTER OBJECTIVES:

Describe why a field engineer must understand volume computations.
List and explain the various methods of collecting data for volume computations.
Identify the key to volume computations.
List and describe the various method used to compute area.
Discuss why accurate volume computations are important to the owner and to the contractor.
State and use the formula for computing volumes by average end areas.
State and use the formula for computing volumes when borrow pit leveling data is available.

INDEX

VOLUME

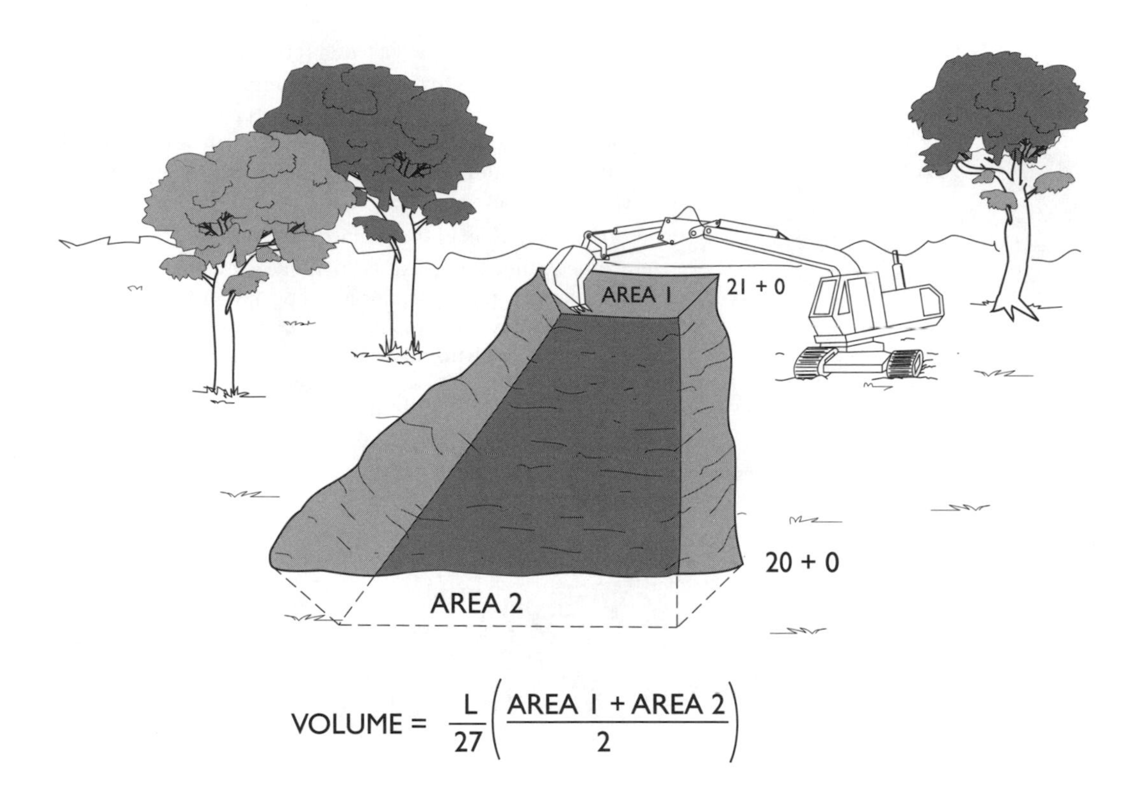

SCOPE

The determination of volume is necessary before a project begins, throughout the project, and at the end of the project. In the planning stages, volumes are used to estimate project costs. After the project is started, volumes are determined so the contractor can receive partial payment for work completed. At the end, volumes are calculated to determine final quantities that have been removed or put in place to make final payment. The field engineer is often the person who performs the field measurements and calculations to determine these volumes. Discussed here are the fundamental methods used by field engineers.

GENERAL

To compute volumes, field measurements must be made. This typically involves determining the elevations of points in the field by using a systematic approach to collect the needed data. If the project is a roadway, **cross sectioning** is used to collect the data that is needed to calculate volume.

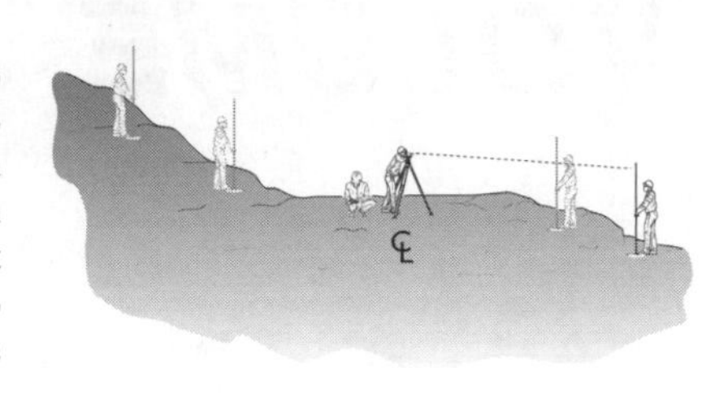

If the project is an excavation for a building, **borrow pit** leveling will be used to determine elevations of grid points to calculate the volume. Whatever the type of project, the elevation, and the location of points will need to be determined. It is the responsibility of the field engineer to determine the most efficient method of field measurement to collect the data.

However, it should be mentioned that sometimes volumes can be determined by using no field measurements at all. In some situations, the contractor may be paid for the number of truckloads removed. Keeping track of the number of trucks leaving the site is all that may be necessary. However, this isn't a particularly accurate method since the soil that is removed expands or swells and takes up a larger amount of space than the undisturbed soil. Depending on how the project is bid, it is sometimes accurate enough. For more information on cross sectioning and borrow pit leveling reference *Chapter 22, Leveling.*

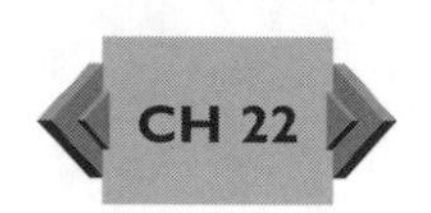
CH 22

FIELD MEASUREMENTS FOR VOLUME COMPUTATIONS

Measurements for volume are nothing more than applying basic distance and elevation measurements to determine the locations and elevations of points where the volume is to be determined. It usually is not practical to take the time to collect data everywhere there is a slight change in elevation. Therefore, it must be understood that volume calculations do not give exact answers. Typically, approximations must be made and averages determined. The field engineer will analyze the data and make decisions that result in the best estimate of the volume.

THE KEY TO VOLUME IS AREA

The key to volume calculation is the determination of area. Most volume calculation formulas contain within them the formula for an area, which is simply multiplied by the height to determine the volume. For instance, the area of a circle is pi times the radius squared. The volume of a cylinder is the area of the circle times the height of the cylinder. If an area can be determined, it is generally easy to determine the volume.

METHODS OF DETERMINING AREA

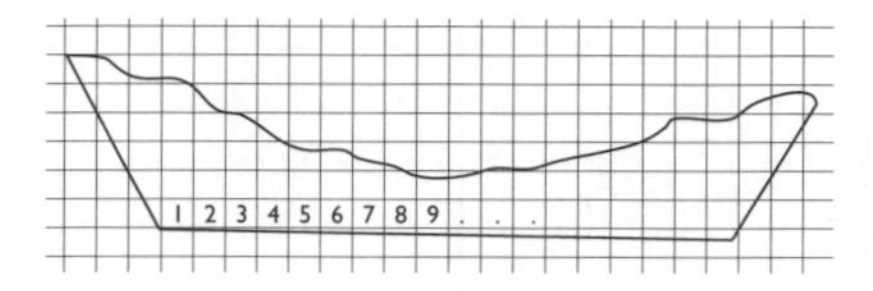

COUNTING SQUARES

Approximation of area can be determined by plotting the figure to scale on cross-sectional paper and counting the squares. Each square represents "x" number of square feet. Incomplete squares along the edges of the cross-section are visually combined and averaged.

PLANIMETER

The electromechanical digital planimeter is a quick method of determining the area of irregular shaped figures. The irregular shape is drawn to scale and the planimeter is used to trace the outline of the shape. Inputting a scale factor into the planimeter results in a digital readout of the area.

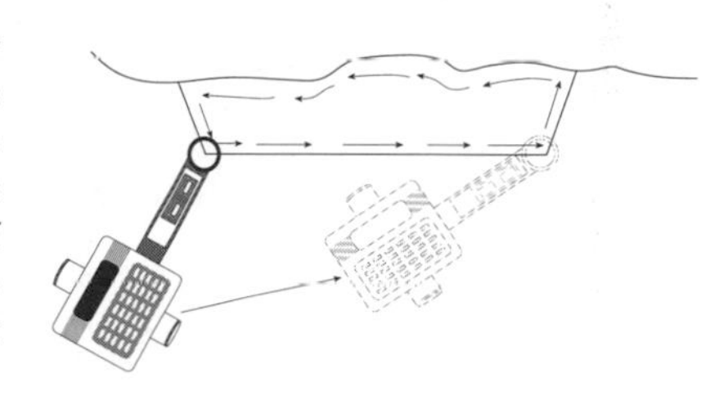

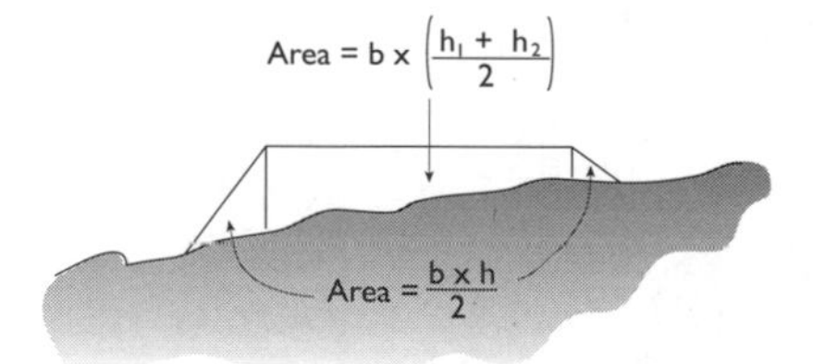

GEOMETRIC FORMULA

Although a shape at first may seem irregular, it is often possible to break it into smaller regular shapes such as squares, rectangles, triangle, trapezoids, etc. that will allow the use of standard geometric formulas to determine the area. This method may be cumbersome because of all the shapes that may need to be calculated.

CROSS-SECTION COORDINATES

If cross-sectional field data is available, this is the recommended method of calculating volume. Once understood, it is fast and the most accurate way of determining area. Cross-section data collected on a project represents elevation and location information for points on the ground. These points can be used as coordinates to determine area. To determine the area follow the procedure described in *Chapter 15, Traverse Computations*.

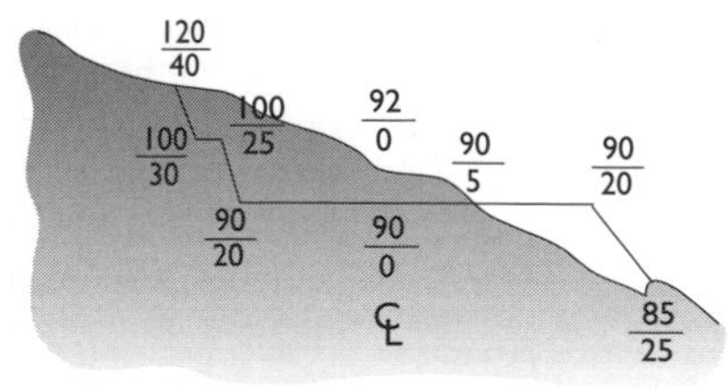

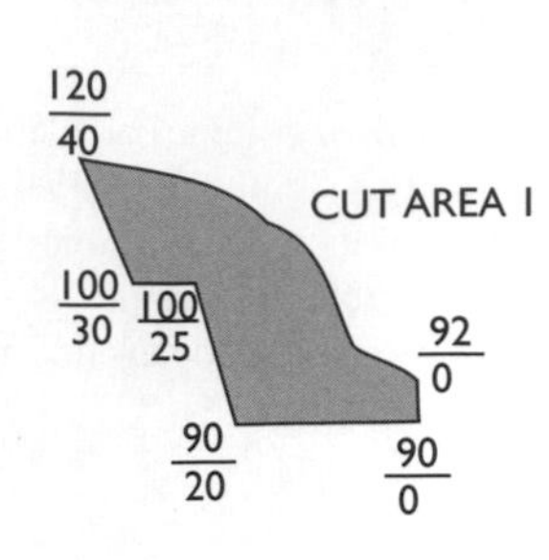

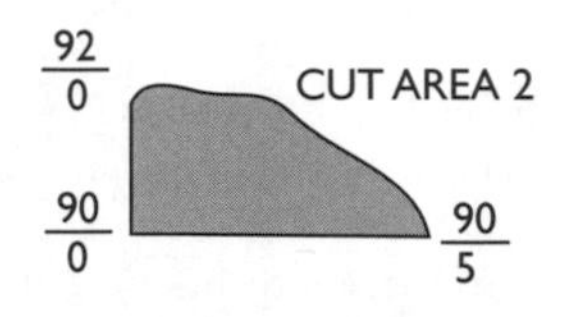

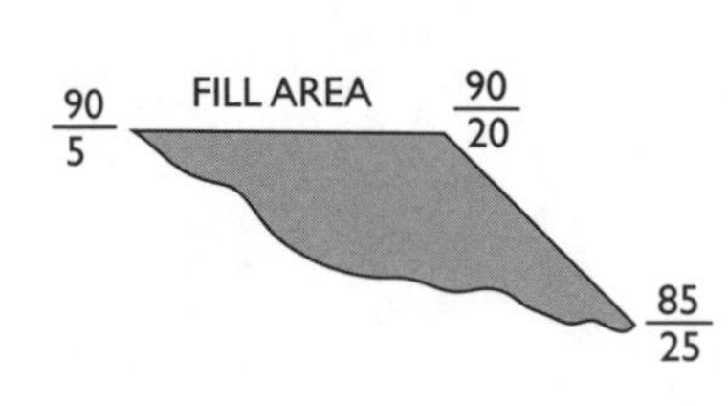

Area by Coordinates

Left of Centerline Cut Area 1	Right of Centerline Cut Area 2	Right of Centerline Fill Area
92 0	92 0	90 5
90 0	90 5	90 20
90 20	90 0	85 25
100 25	92 0	90 5
100 30		
120 40		
92 0		
Σ ↙ 11780	Σ ↙ 450	Σ ↙ 4400
Σ ↘ 11050	Σ ↘ 460	Σ ↘ 4475
Difference = 730	Difference = 10	Difference = 75
Divide by 2	Divide by 2	Divide by 2
Area = 365 Sq. Ft.	Area = 5 Sq. Ft.	Area = 37.5 Sq. Ft. of Fill
Total Cut Area = 370 Sq. Ft.		

VOLUME COMPUTATIONS - ROAD CONSTRUCTION

In road construction, the shape of the ground must be changed to remove the ups and downs of the hills and valleys for the planned roadway. Often mountains of dirt are moved to create a gentle grade for the roadway. Payment for the removal and placement of dirt is typically on a unit cost basis. That is, the contractor will be paid per cubic yard of soil moved. There will be a separate price per cubic yard of rock. It can be seen that accurate determination of the volume moved is critical to the owner and contractor. Each wants an accurate volume so payment for the work is correct.

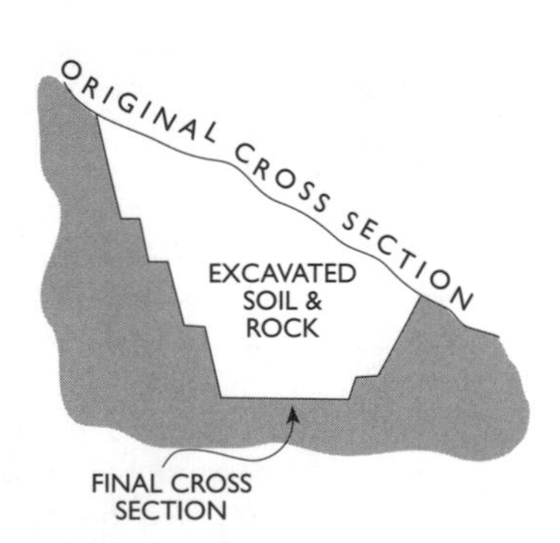

For road projects, cross-sections of the ground elevations are measured at the beginning of the project, during the project, and at the end of the project. Comparisons of the final cross-sections to the original cross-sections are used to determine the volume moved. Areas of the cross-sections are most easily determined by using the elevations of the points and their locations from the centerline (coordinates).

CALCULATION OF CROSS SECTION VOLUMES

AVERAGE END AREA METHOD

The average end area method uses the end areas of adjacent stations along a route and averages them. This average is then multiplied by the distance between the two end areas to obtain the volume between them. In formula form:

$$Volume = \frac{L}{27}\left(\frac{Area1 + Area2}{2}\right)$$

Where "L" represents the distance between the cross sectional end areas being used in the formula. "27" represents the number of cubic feet in one cubic yard. Dividing cubic feet by 27 converts to cubic yards.

EXAMPLE:

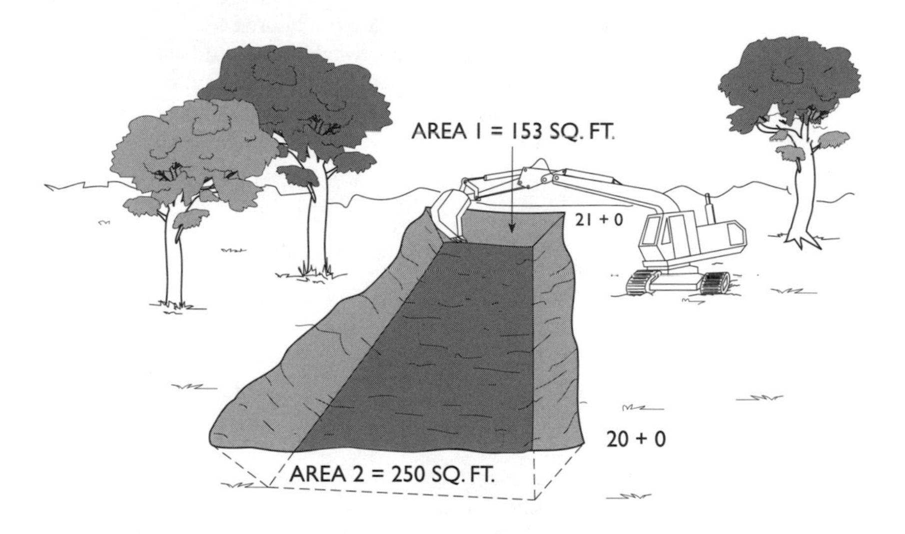

$$Volume = \frac{100}{27}\left(\frac{153 + 250}{2}\right)$$

$$Volume = \frac{100}{27}(201.50)$$

$$Volume = 746.3 cu. yd.$$

The average end area can be used in any situation where areas can be determined.

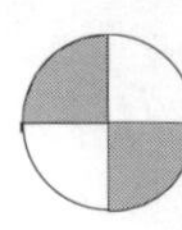

VOLUME COMPUTATIONS - BUILDING EXCAVATION

A method known as borrow-pit leveling can be used effectively to determine volume on building projects. A grid is established by the field engineer and elevations on the grid points are determined before the excavation begins and when the work is complete.

If the fill or excavation shape conforms to a common solid shape such as a cube or a prism, common geometric formulas can be used. The best method to use depends on the specific project conditions where the volume is to be determined, and on the measurement information available. The engineer must analyze the situation and select the method that will yield the most accurate results in the shortest time.

BORROW-PIT

The borrow pit method uses a grid and the average depth of the excavation to determine the volume. Before the excavating begins, the field engineer creates a grid over the entire area where the excavation is to occur. Elevation data is collected at each of the grid points and recorded for future reference. At any time during the excavating, the field engineer can reestablish the grid and determine new elevations for each of the field points. Using the average height formula shown below, the volume of soil removed from each grid area can be determined. The smaller the grid interval, the more accurate the volume. In formula form:

$$Volume = \frac{Area}{27}\left(\frac{h_1 + h_2 + h_3 + h_4}{4}\right)$$

An example is demonstated on the following page.

SUMMARY

Only two general methods of calculating volumes have been presented here. There are many others that are very specific for the particular situation. For example, when determining volumes along a roadway, there is a constant transition from cut to fill to cut to fill, etc. To more accurately compute the volume, a prismoidal formula is used. Like this situation, there are many others in which formulas are used. The field engineer should check with textbooks that discuss in detail route surveying and earthwork for additional information. See *Appendix D* for a listing of several textbooks that are specifically written for route surveying.

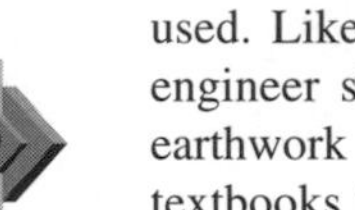

EXAMPLE

$$Area = 50x\left(\frac{30+40}{2}\right) = 1750$$

$$Volume = \frac{1750}{27}x\left(\frac{0+0+(93.2-83.1)+(96.7-87.2)}{4}\right)$$

$$Volume = \frac{1750}{27}x\left(\frac{0+0+10.1+9.5}{4}\right)$$

$$Volume = \frac{1750}{27}x(4.9)$$

$$Volume = 317.59cu.yds.$$

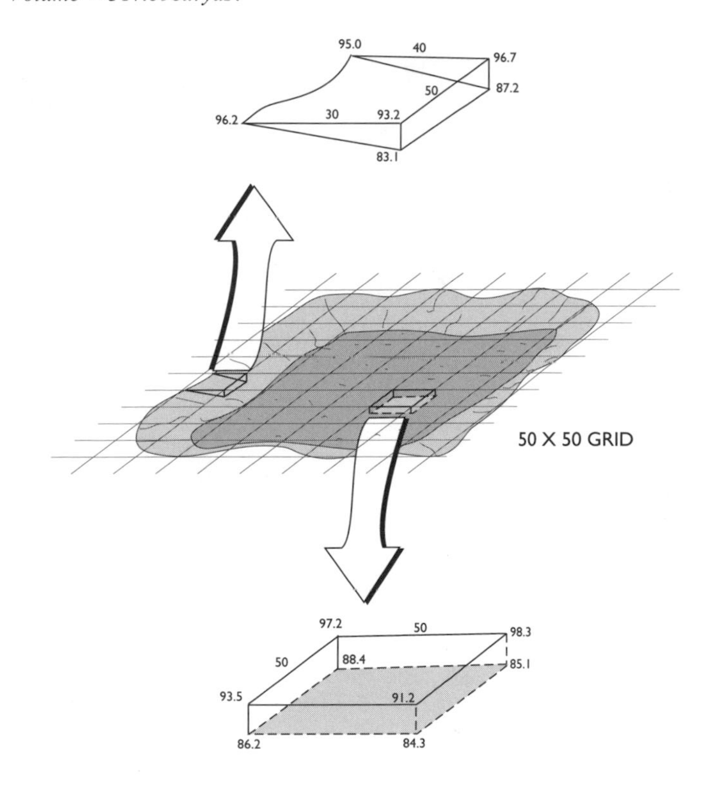

$$Area = 50x50 = 2500$$

$$Volume = \frac{2500}{27}\left(\frac{(93.5-86.2)+(97.2-88.4)+(98.3-85.1)+(91.2-84.3)}{4}\right)$$

$$Volume = \frac{2500}{27}\left(\frac{7.3+8.8+13.2+6.9}{4}\right)$$

$$Volume = \frac{2500}{27}(9.05)$$

$$Volume = 838cu.yds.$$

SLOPE STAKING

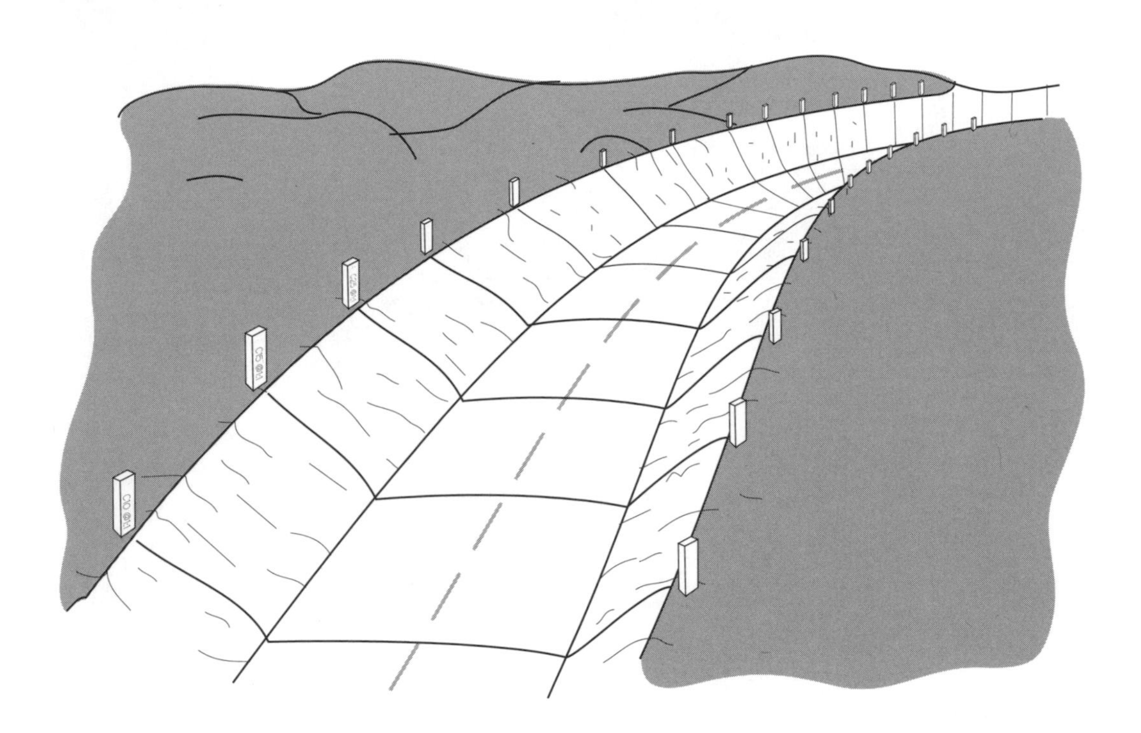

SCOPE

When excavating for a building foundation or a route of transportation, the excavation contractor's goal is to remove or place only the amount of soil or rock required in the contract. If more rock is removed than designed, the contractor will not be paid for the excess removed. If less is removed, the owner will require the contractor to return to the site to remove what was required. Either of these situations will cost the contractor time and money. The field engineer for an excavation contractor has the responsibility of placing stakes that locate the limits of the excavation so that only the volume of earthwork required is moved. Slope stakes are the communication tools used by the field engineer to inform the equipment operators where to work, how much to cut or fill, and what slope angle must be maintained by the operator.

OBJECTIVE

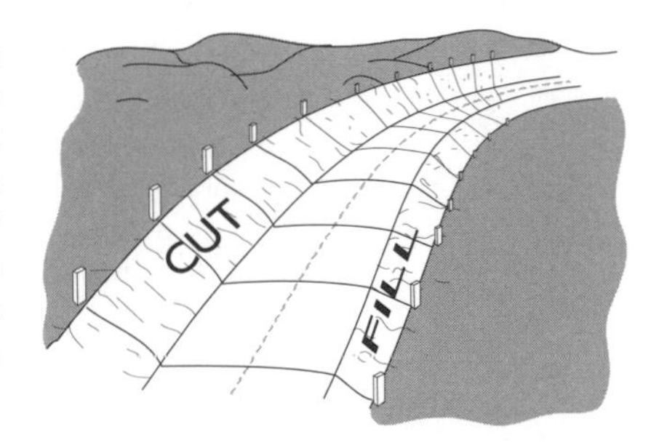

The objective of slope staking is to locate where the design slope intersects the original ground. There is only one spot where this can occur. To locate that spot will require that the field engineer combine field measurements of distances and elevations with the typical section design information that is obtained from the plans and specifications.

HIGHWAY PROJECT

For a typical highway project, there will be a spot to the right and to the left of centerline to mark where the slopes intersect the ground. In technical terms, the objective is:

Distance measured must equal Distance calculated, or,
Distance Measured, $d_m = d_c$ *Distance Calculated*

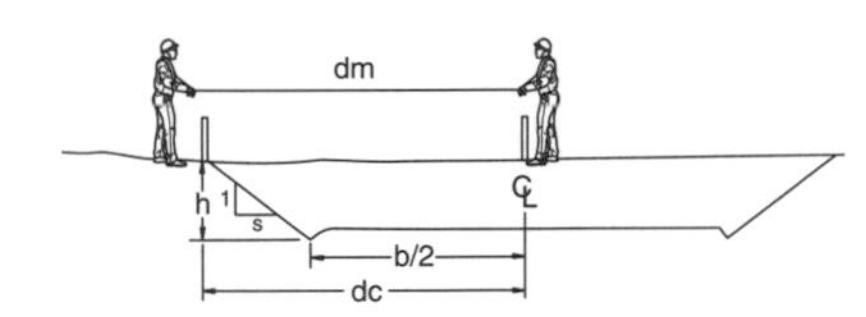

Where d_m is the field measurement from centerline to the location of the slope stake, and d_c is the calculated distance from the centerline. d_c is determined by a basic formula that is based on design dimensions and slope, and the field measurement of the height of cut or fill. In an example for a highway, the formula is:

$$d_c = \frac{b}{2} + sh$$

"b" is the base width of the road, "s" is the slope, and "h" is the height of the cut or fill. These terms will be defined in detail later.

BUILDING SITE

For a building, the same approach is taken, but the formula must be adapted for the given situation. It may be that the reference line for setting the slope stakes will be the edge of the building and that the slope will be cut from the bottom of the foundation. Assuming a working space between the wall and the slope, the formula might be:

$$d_c = Work\ space + sh$$

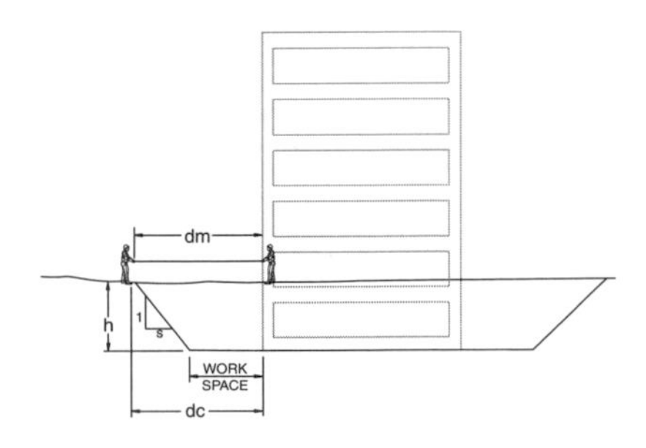

In this case, the field engineer would measure d_m and compare it to d_c. Are they equal? If they are, set the stake. If they aren't, determine a new "h" and measure a new d_m and perform the computation and comparison until they are equal. Once the field engineer understands the basic concept of setting a slope stake, writing a formula to determine d_c for any situation will be easy.

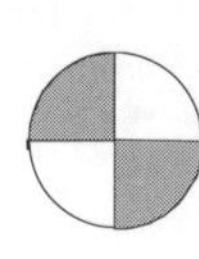

TERMS AND DEFINITIONS

In the plans of a highway or a building project are elevation drawings that illustrate to the builder what the project will look like from a side or end view, In route construction, these drawings are called the sections. The highway designer uses standard specifications that are based on the general purpose of the highway to design the typical sections that are to be followed during the construction. Terms used in the section include:

BASE, "B"

"B" is the base which is the width of the bottom of a cut section or the top of the fill section.

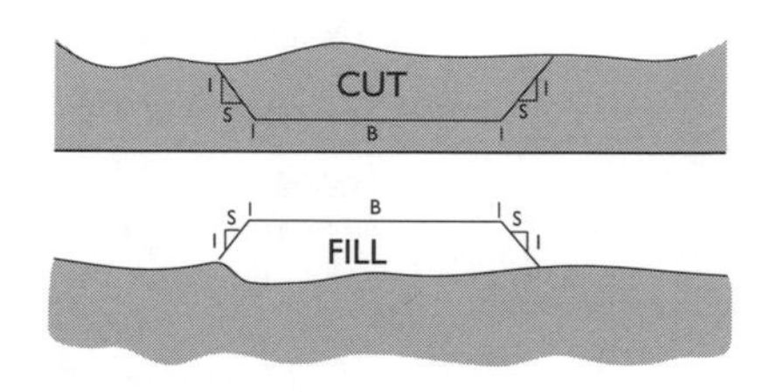

SLOPE, "S"

"S" is the horizontal component of the slope. When representing slope, there are two components to a sloped line, the vertical component and the horizontal component. Standard notation in route surveying is for the vertical component to be listed as "one." When communicating slope, the horizontal component is typically listed first and the vertical component is listed last. Some examples follow:

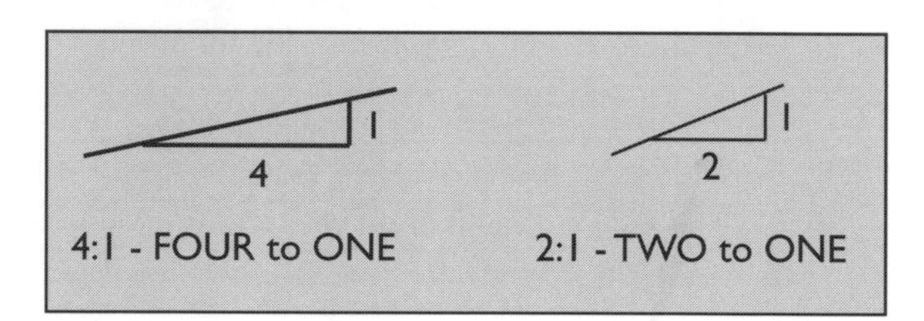

GRADE ELEVATION, "GRADE"

"Grade" is the elevation that comes from the design elevation.

GROUND ELEVATION, "GROUND"

"Ground" is the actual elevation of the ground obtained by leveling techniques in the general area where the slope stake will be located. Several ground elevations may be determined in the trial and error process of setting a slope stake.

HEIGHT, "h"

The height is the difference between the ground elevation and the grade elevation. In equation form it would be represented as:

$$h = \text{ground elevation} - \text{grade elevation}$$

BENCH

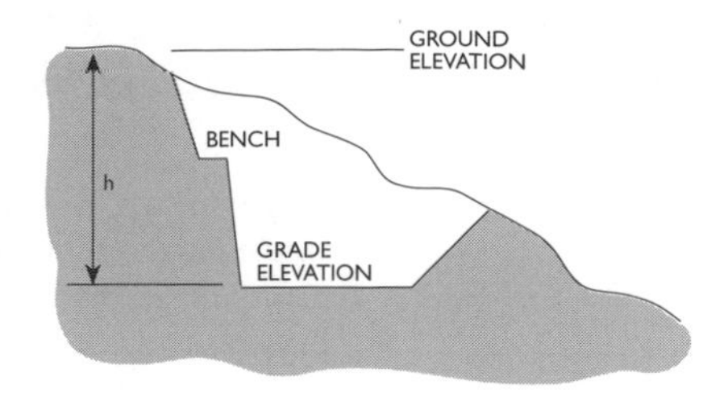

A bench is a level area on the slope that is used to stop falling rocks and soil. It also is used to divert water from flowing down the slope. Several benches may be used on a very deep cut. See the illustration for a complex cut section that follows.

EXAMPLES OF HIGHWAY SECTIONS

Typical sections can be simple or can become complicated, depending on the type of terrain where the project is being built.

SIMPLE CUT AND FILL SECTIONS

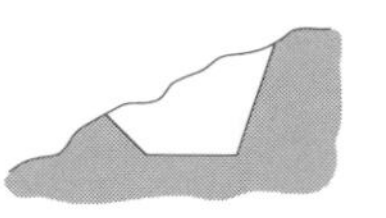

If there isn't much cut, the typical roadway section in a cut area may appear similar to this.

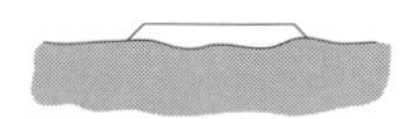

If the roadway is being built on a relatively flat area of ground in a fill area the section appear similar to this.

COMPLEX CUT SECTION

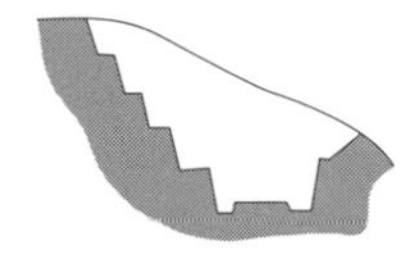

When a roadway is being built through a mountainous area, cuts on the side of the mountain or even through a mountain are not uncommon. Several slope angles and benches may be used depending on the conditions of the soil or the rock that is encountered.

PREPARING TO SET HIGHWAY SLOPE STAKES

Typically when staking a highway, the field engineer will initially set stakes at the centerline of the road to determine its location. The centerline will then be profile leveled so that an initial ground elevation will be known. To set highway slope stakes, the field engineer must have the following information.

NECESSARY FIELD INFORMATION

A centerline staked in the field is needed along with elevations of the ground at the centerline.

INFORMATION FROM THE PLANS AND SPECIFICATIONS

The finish grade elevation at centerline, and the typical section that is being used must be taken from the plans and specifications. In addition, the base width of the highway, the ditch width information, side slope information, and bench width is also needed.

FIELD MEASUREMENTS

Distance measurement from the reference line to where it is estimated the design slope intersects the ground.

Leveling measurements to determine the elevation of the ground where it is estimated the design slope intersects the ground.

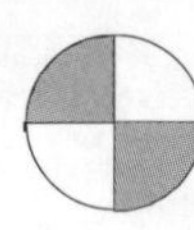

PROCEDURE FOR SETTING A HIGHWAY SLOPE STAKE

Slope staking is best understood when actual numbers are used. The following data will be used in the step-by-step illustrated description.

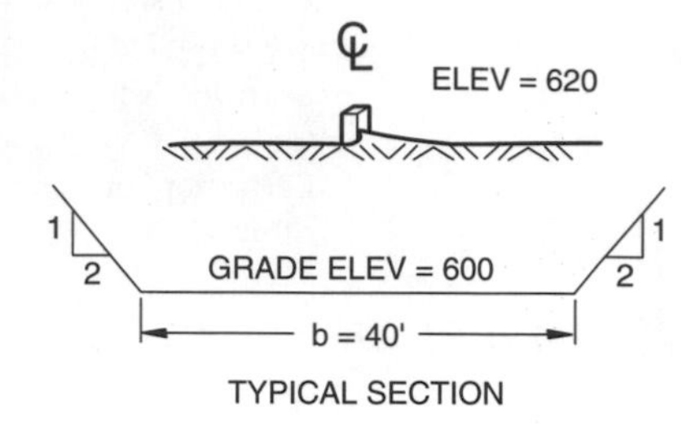

STEP 1

Gather all necessary equipment and tools, design information, and the fieldbook with the centerline profile elevations. Stand at the centerline and review the typical section details. Attempt to visualize the section and the lay of the ground and formulate a guess for the distance from centerline that the slope stake will be placed.

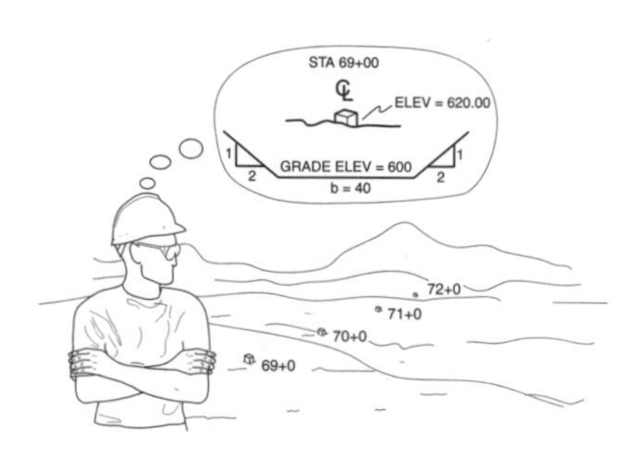

STEP 2

Standing at centerline, face in the direction to set the slope stake. Hold the zero end of the cloth tape and send the person with the rod out, perpendicular to the centerline, a measured distance equal to your estimate. This distance will be d_m.

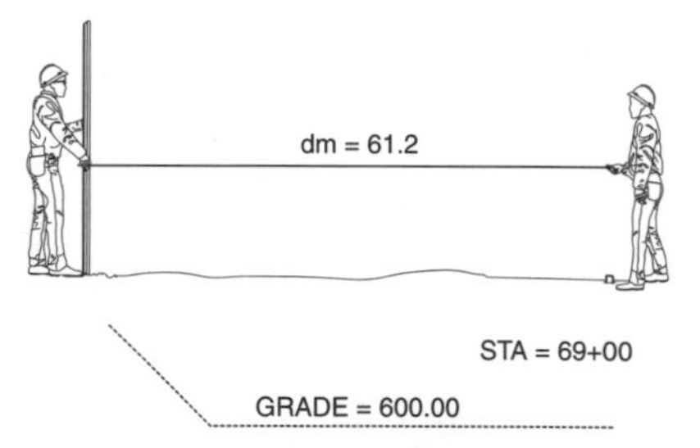

STEP 3

Using a hand level and knowing the distance from your eye to the ground, take a reading on the rod at the estimated point. Add the profile elevation and your eye height to obtain the h.i. Subtract the rod reading to obtain the elevation of the ground at the point.

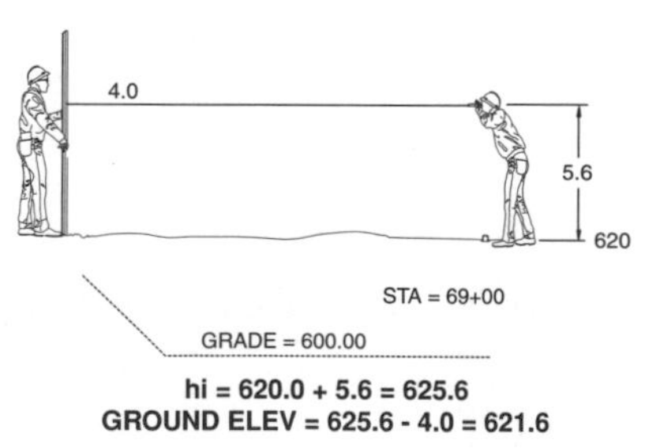

hi = 620.0 + 5.6 = 625.6
GROUND ELEV = 625.6 - 4.0 = 621.6

STEP 4

Obtain the grade elevation from the design information. Subtract ground elevation and grade elevation to obtain "h". Using the design information, determine "b" and "s" and insert them into the formula to determine d_c.

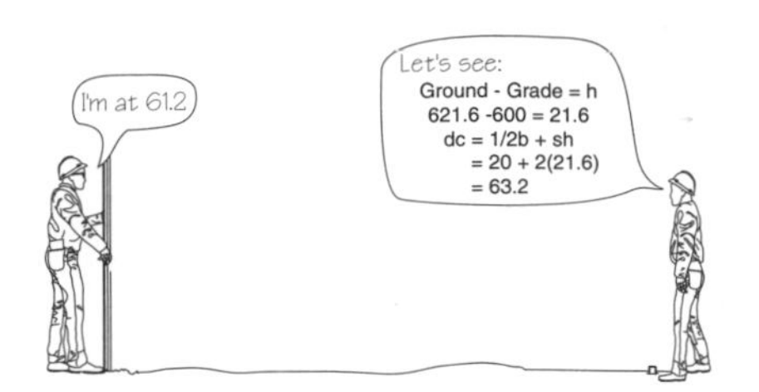

STEP 5

Compare d_m and d_c. If they are within a tenth, set the slope stake and the cut and slope information for on the stake. If d_m and d_c are greater than a tenth, determine if another elevation reading is needed.

STEP 6

Measure a new d_m, take a new rod reading, calculate a new ground elevation, calculate a new h, solve for d_c.

STEP 7

Compare d_c to d_m. If they are within a tenth, set the slope stake. If not, repeat the process again.

SETTING ANY TYPE OF SLOPE STAKE

The process illustrated and described above can be used in any slope staking situation: be it slope stakes in fill areas, complex cut sections, building excavations, etc. The process of measuring a distance and comparing it to a calculated distance is the gist of slope staking. In all slope staking situations, the key is being able to determine "h." Once "h" is known, d_c can be calculated and compared to the easily measured d_m.

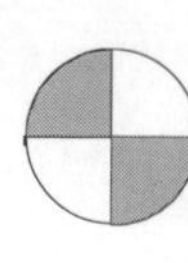

FIELDNOTES

All observations are to be recorded in the field notebook at the time they are observed. Refer to the following illustration for an example slope stake noteform.

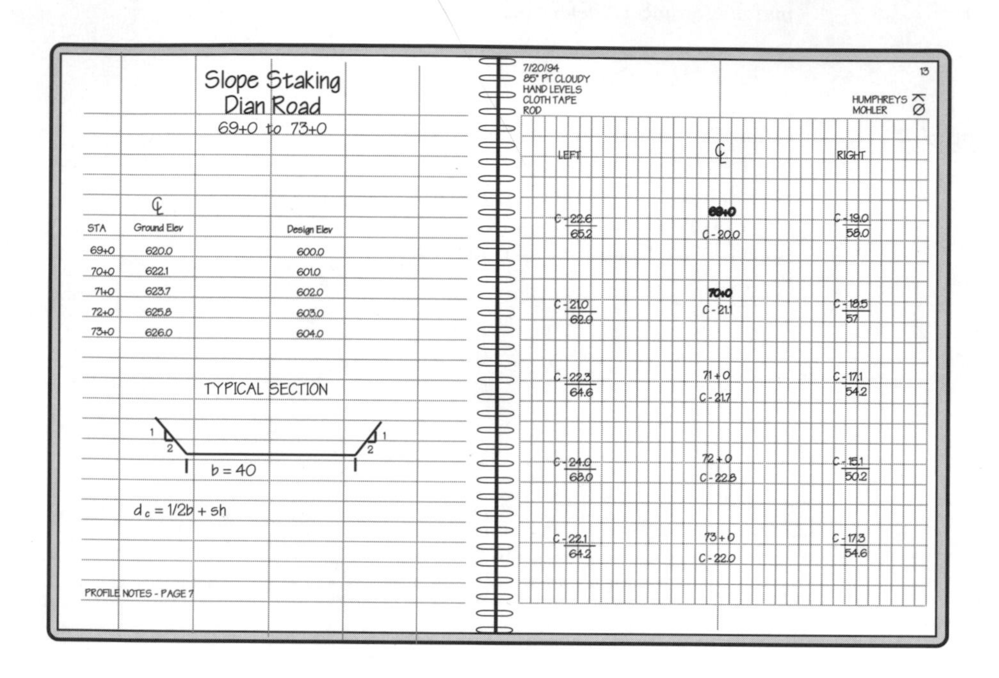

DISTANCE APPLICATIONS

20

PART 4 - CONSTRUCTION LAYOUT APPLICATIONS

CHAPTER OBJECTIVES:

Describe how pacing is a useful tool of the field engineer.
Determine your pace length and use it to measure distances.
Describe the process used for chaining the distances on a traverse.
List the equipment used when chaining a traverse.
State the rule of thumb for construction precision in chaining.
Describe the purpose of referencing points.
List the equipment, tools, and supplies that are required when referencing points.
Explain the procedure used for referencing.
List and describe common points used when referencing.
State the rule of thumb for precision when chaining.

INDEX

CHECKING DISTANCES BY PACING

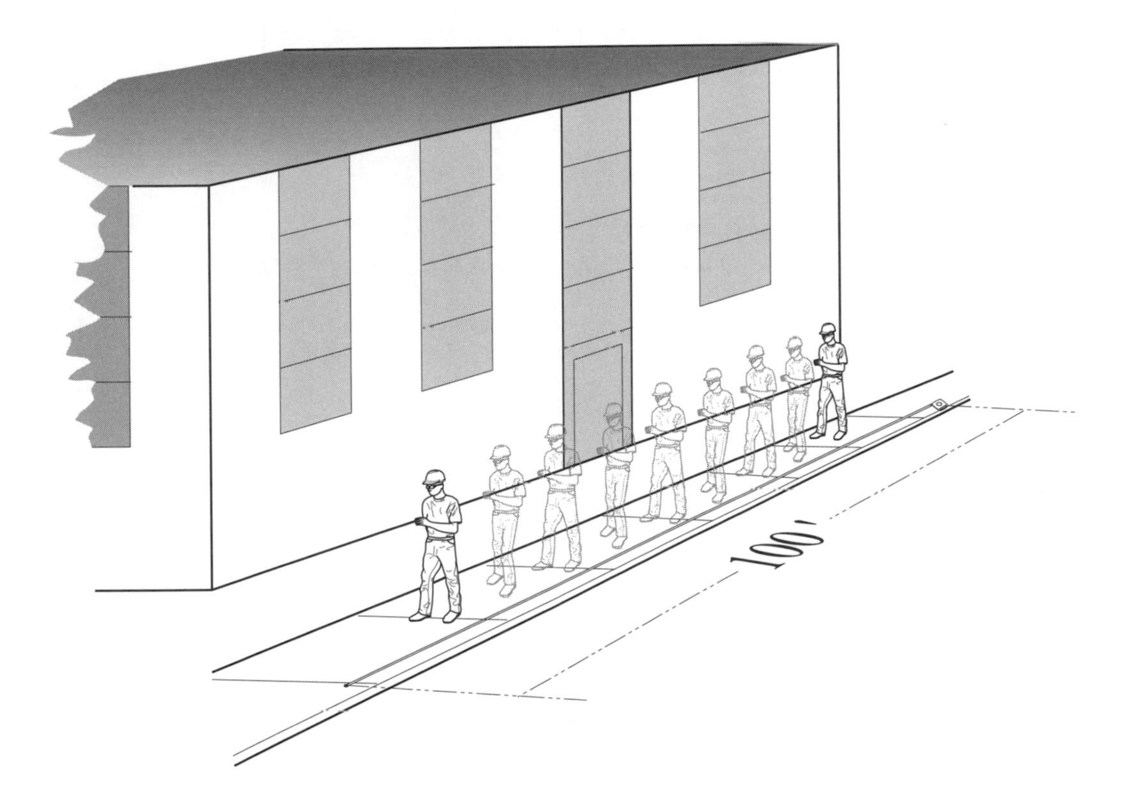

SCOPE

The ability to pace distances with reasonable precision is very useful to almost anyone. The field engineer, in particular, can use pacing to quickly check measurements made by more precise means. By so doing, he/she will often be able to detect large mistakes.

A person's average pace length can be determined by counting the paces necessary to walk a distance which has previously been measured more precisely (i.e., with a steel tape). For most people, pacing is accomplished most satisfactorily when taking natural steps. Some others like to try to take paces of certain lengths (i.e., a length of three feet), but this method is tiring for long distances and usually gives results of lower precision.

As horizontal distances are necessary in construcion, some adjustments should be made when pacing on a sloping ground. With a little practice, a person can pace distances with a precision of roughly 1:50 to 1:200 depending upon the ground conditions.

EQUIPMENT

Distance Checking Equipment	
Per crew	***Per person***
100-foot cloth tape	Field book 4H pencil Straight edge

PROCEDURE

PLAN AND PREPARE

Select a location that is level and has a clear distance of 100 feet. Lay out a 100-foot cloth tape on the ground.

Some individuals count each step as a pace and others only count when their right or left foot steps. It really doesn't matter which method is used. Just be consistent.

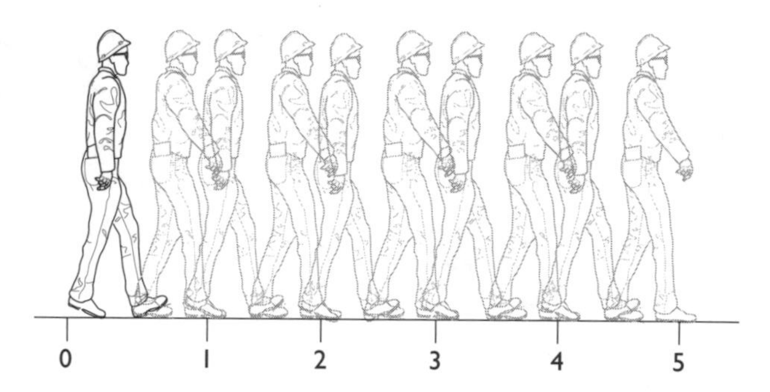

One pace counted every time the right foot steps down.

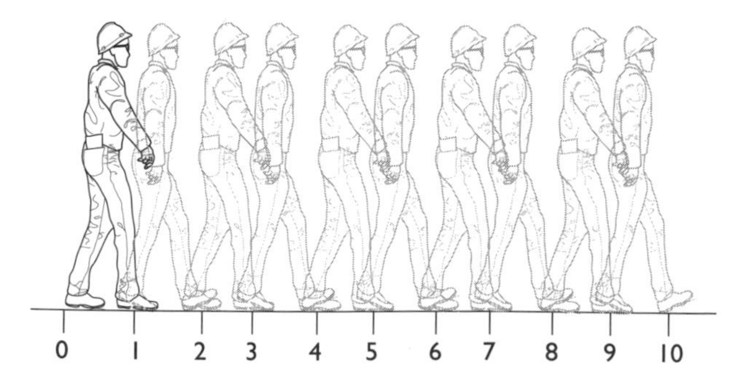

One pace counted for every step taken.

DETERMINE PACE LENGTH

Walking naturally, pace between the end points of the chain at least four times. Record 1/2 or even 1/4 paces if it is possible to determine your pace that closely. Record the number of paces in your field book.

Determine the average number of paces and divide it into the 100 foot distance to determine a pace length. Record this distance in the field book.

CALCULATIONS

Distance divided by # of Paces = Length of Pace

Length of Pace x Average Pace = Distance

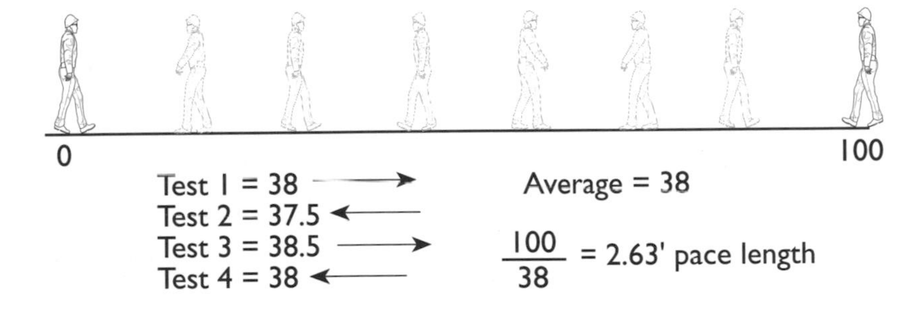

DEVELOP CONFIDENCE IN PACING

To gain confidence in your pacing ability, pace along the sides of a building and calculate the building dimensions by multiplying your pace length by the number of paces you recorded. Use a chain to measure the dimensions of the building to check your paced distance. Repeat by pacing lengths you can later measure. Continue this practice until you can pace distances to a precision of one foot in 50 feet. With some practice, you'll be able to pace quite accurately.

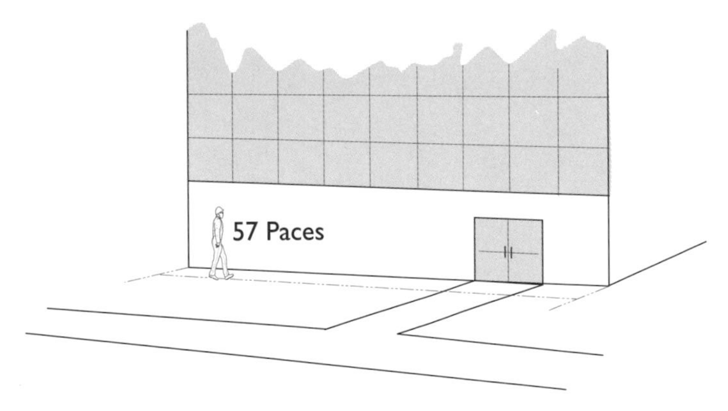

57 x 2.63 = 150' Building

REMEMBER YOUR PACE LENGTH

Commit your pace length, when walking naturally, to memory, so you can use it anytime you need a quick check of a distance.

FIELD NOTES

Note keeping for determining one's pace should utilize a combination of tabulation, sketches, and descriptions.

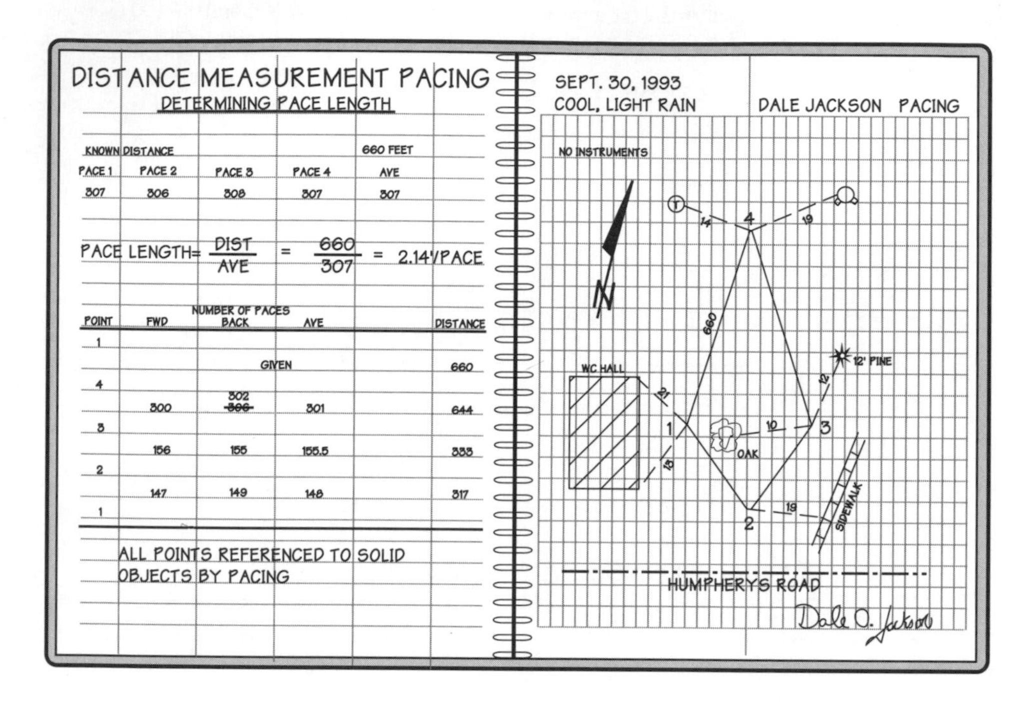

DISTANCE MEASUREMENT PACING

DETERMINING PACE LENGTH

KNOWN DISTANCE				660 FEET
PACE 1	PACE 2	PACE 3	PACE 4	AVE
307	306	308	307	307

$$\text{PACE LENGTH} = \frac{\text{DIST}}{\text{AVE}} = \frac{660}{307} = 2.14'/\text{PACE}$$

POINT	NUMBER OF PACES FWD	BACK	AVE	DISTANCE
1				
		GIVEN		660
4				
	300	302 ~~306~~	301	644
3				
	156	155	155.5	333
2				
	147	149	148	317
1				

ALL POINTS REFERENCED TO SOLID OBJECTS BY PACING

SEPT. 30, 1993
COOL, LIGHT RAIN

DALE JACKSON PACING

NO INSTRUMENTS

CHAINING A TRAVERSE

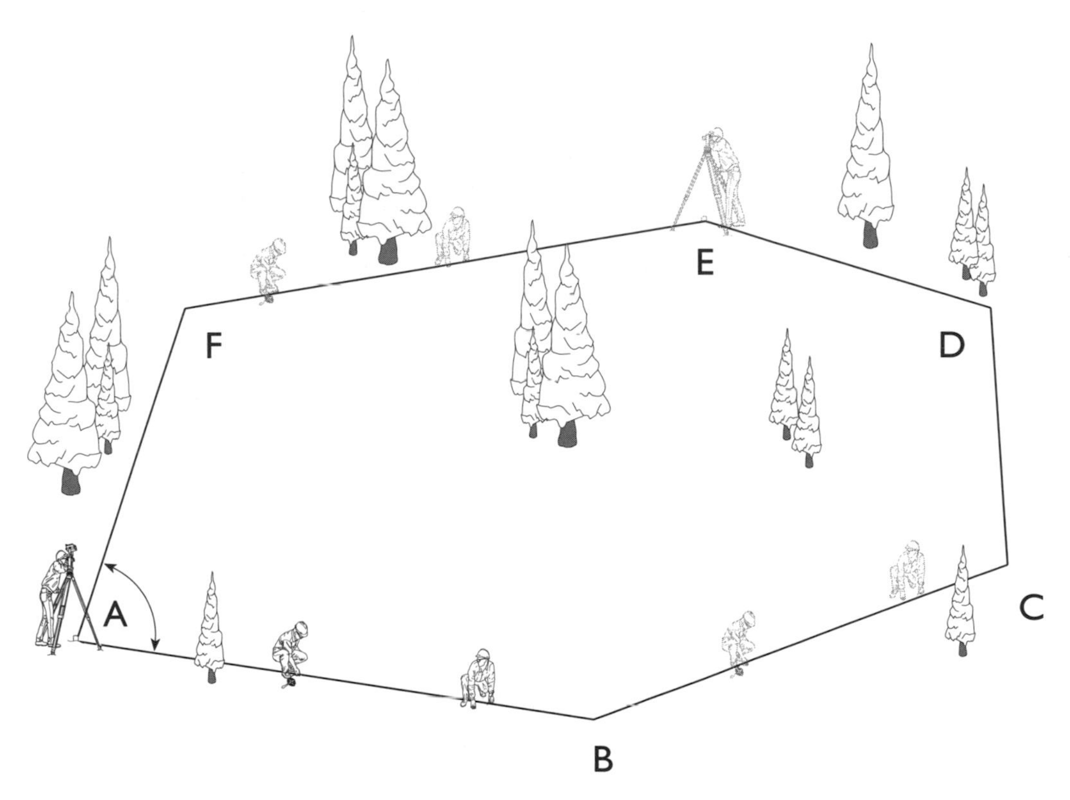

SCOPE

As linear measurement is the basis of all field engineering operations, traverse chaining is the way the field engineer begins to determine the relative location of points. By also measuring interior angles at the points, the field engineer can determine the precision of the work which is a necessity when establishing control networks on construction projects.

The actual process of chaining a traverse differs from measuring a single line only in the number of lines to be measured. The same procedure of locating or establishing the points, setting range poles, aligning, applying tension, plumbing, marking, reading, and recording the forward and back distance is simply repeated for every line to be measured. This process is used time and again on all types of field engineering applications of horizontal distance measurement. Therefore, the field engineer should learn inside and out the steps and techniques of chaining.

BACKGROUND AND REFERENCES

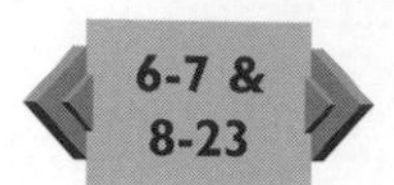

Chapter 6, Distance Measurement-Chaining, Step-by-Step Procedure (6-7)
Chapter 8, Angle Measurement, Measuring Traverse Angles (8-23)

EQUIPMENT

Traverse Chaining Equipment	
Per crew	**Per person**
Chain	4H pencil
Chaining pins or nails	Plumb bob
Two range poles	Field book
	Straight edge

PROCEDURE

FOLLOW CHAINING FUNDAMENTALS

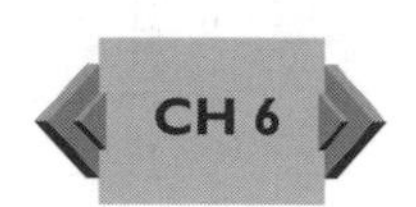

You will be measuring from traverse point to traverse point. Identify the points that define the traverse. Determine who will be head chain and who will be rear chain. Prepare a sketch of the traverse in the field book. Set range poles and begin. Chain from point to point utilizing the chaining procedure specified in *Chapter 6, Distance Measurement - Chaining*. Repeat the process for every line of the traverse.

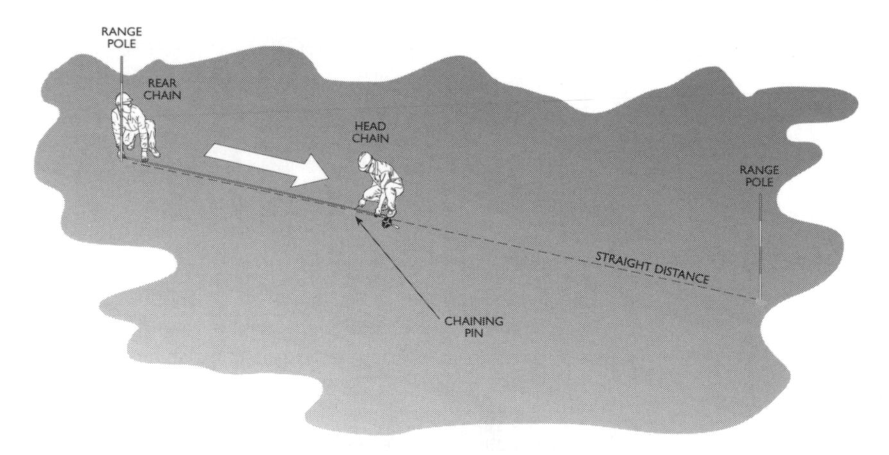

Review of Chaining Fundamentals	
Measure both ways	All distances must be measured forward and back as a check.
Chain horizontally	Plumb to ensure all measurements are horizontal distances. If the terrain is sloping at more than 5 feet of elevation per 100-foot chain length, it will be necessary to utilize the procedures indicated for breaking chain.
Know your equipment	A variety of types of graduated chains may be encountered. Study your chain to be sure you understand its graduations and use (feet, tenths, and hundredths, or feet and inches).
Pull hard	Improper tension is the greatest factor affecting accuracy in chaining. It is better to place too much tension on the chain rather than too little. So, PULL HARD!
Maintain good alignment	Make sure good alignment is maintained while chaining distances greater than one chain length for straight line measurement.
Set solid points	Drive stakes until they are solid and flush. Tacks will be placed exactly on line and exactly at the measured distance.
Protect points	Place guard lath around stakes to warn others of the importance of the point. Lath will be marked clearly with information that describes the use of the point.
Maintain equipment	Clean, dry, and oil chains regularly. Watch out for loops.

MAINTAIN PRECISION

The rule of thumb for construction is that forward and back measurements should agree within 0.01 per 100 feet measured. Thus, for a distance of 200 feet, the difference between the recorded forward and back measurement should be 0.02 or less. For a distance of 500 feet, the difference should be 0.05 or less. Actually, any distances over 200 feet should be measured with an EDM.

CALCULATIONS

The forward and back distances are averaged by adding them together and dividing by 2. Some persons chaining also calculate a discrepancy ratio for each line measured. The calculated ratio is compared to an established standard to determine if the distance is acceptable or needs re-measured.

Forward - Back = Discrepancy

$$\frac{Discrepancy}{Mean} = \frac{1}{x}$$

For example, in the notes below for line 19-13.

$$\frac{168.29 - 168.25}{168.27} = \frac{1}{4200}$$

FIELD NOTES

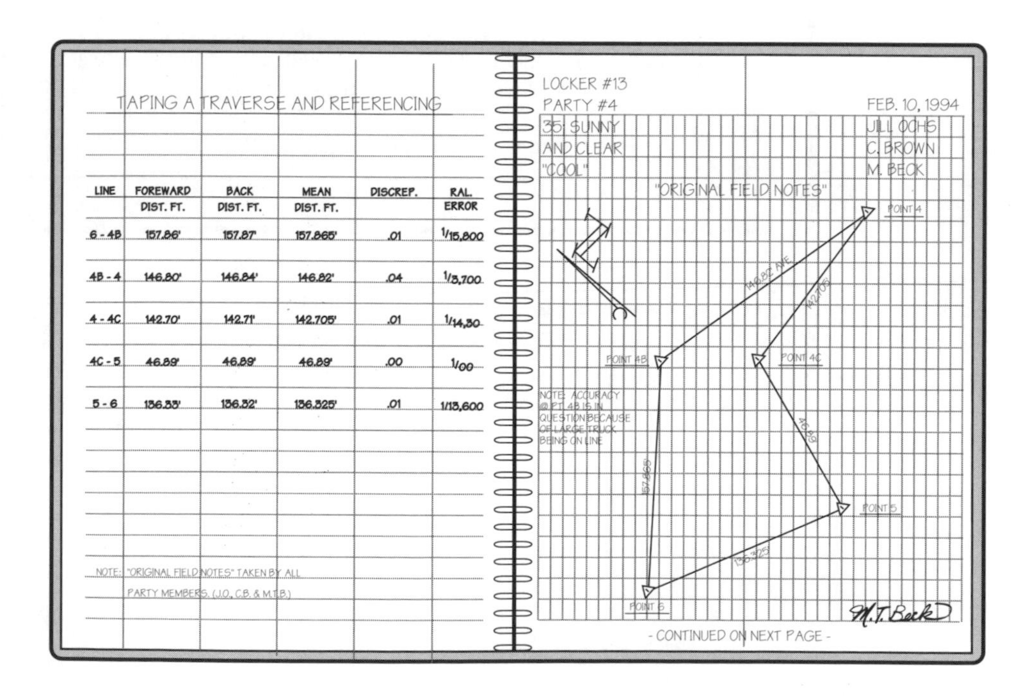

TAPING A TRAVERSE AND REFERENCING

LINE	FOREWARD DIST. FT.	BACK DIST. FT.	MEAN DIST. FT.	DISCREP.	RAL. ERROR
6 - 4B	157.86'	157.87'	157.865'	.01	1/15,800
4B - 4	146.80'	146.84'	146.82'	.04	1/3,700
4 - 4C	142.70'	142.71'	142.705'	.01	1/14,30
4C - 5	46.89'	46.89'	46.89'	.00	1/00
5 - 6	136.33'	136.32'	136.325'	.01	1/13,600

NOTE: "ORIGINAL FIELD NOTES" TAKEN BY ALL PARTY MEMBERS. (J.O., C.B. & M.T.B.)

REFERENCING A POINT

SCOPE

In field engineering, points such as benchmarks, control points, and other types of points will need to be used at a later date. To ensure they can be found by anyone who might want to use them, it is necessary to reference points to surrounding permanent objects. Doing this precisely will allow obliterated points to be relocated. This is one of the small tasks a person can do to become an effective field engineer.

Referencing is an easily understood activity. Simply stand at the point to be referenced, look around for objects that are permanent, and measure distances from the point to the objects. Prepare a clear sketch of the location and distances to the referenced objects, and store the reference field book in a safe place. If you have referenced properly, anyone should be able to use your field notes to find or establish a point. If they cannot, it is your fault. Reference carefully!

EQUIPMENT

Much of the equipment listed below is optional. Its use depends on the conditions that allow the field engineer to use it.

Optional Equipment	
1	Standard chaining equipment
2	Shiny metal washers
3	Marking paint
4	PK nails
5	20 penny nails
6	Railroad spikes
7	Surveying ribbon
8	Stakes
9	Hubs
10	Plastic targets

PROCEDURE

PLAN AND PREPARE

Look around. What object is available for referencing to? Is there a fire hydrant, power pole, manhole cover, sidewalk, curb, tree, street sign, etc.? Any number of objects that are solid and permanent can be used as reference points. Make sure you have the equipment required for referencing.

ESTABLISH THREE REFERENCES

Each point should be referenced to at least three but preferably four permanent or semi-permanent objects such as trees, large rocks, fence posts, fire plugs, etc.

REFERENCE IN ALL DIRECTIONS FROM THE POINT

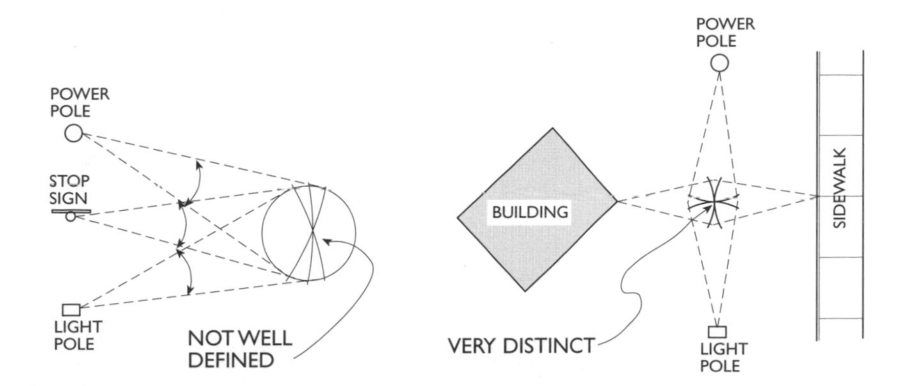

Same direction When arcs are swung from references in the same direction, there is a large area where the actual point can be. It would be difficult to relocate a point with any certainty if it was referenced like this.

All directions When arcs are swung from references in all directions, there is a very small, distinct area where the point could be. An obliterated point could be reestablished with a high degree of confidence in this situation.

REFERENCE TO DEFINABLE POINTS

There are many common references that are used by field engineers to reference points.

Fire hydrant

If referencing to a fire hydrant, select the bonnet bolt that is nearest the point and mark it. Field engineers generally spray paint the particular bolt on the hydrant they are using. (Don't violate local ordinances by defacing property.) In your notes say you have measured to the center of the bonnet bolt painted red.

Wooden pole

If measuring to a wooden power pole, drive a nail in it. Some field engineers will put a washer or nut or something on the nail so it can be distinguished from any other nails in the pole.

Sidewalk

If measuring to a sidewalk, consider driving a PK nail in a crack and using it as the reference point.

Tree

Measuring to a tree might seem to be a good reference. But was the measurement to the center or the face of a tree? Be sure to be very specific in your field notes. Trees are not good references because there is not a definable point on the tree. There could be if you drive a nail in them, but, many people frown on that. Thus trees are generally a second choice as far as using them as a reference object.

Fence
Fences are good references, but care must be taken to properly describe the particular post being used. Identify it with markings and describe it in your notes by saying, "the 5th post from the north end."

Building
Corners of buildings are excellent references. Be sure to describe which corner of the building is being used. Ask permission to make an unobtrusive identifiable mark if needed. Don't spray paint a big "X" on a building.

Sign post
There are many sign posts along highways. Be sure to clearly identify the one that is being used.

STAY WITHIN ONE CHAIN LENGTH

If possible, stay within one chain length of the point when referencing. More than one tape length introduces the possibility of additional errors in the measurement.

MEASURE TO THE HUNDREDTH

Measure and record the distance to reference points to the nearest 0.01 ft. so the point can be relocated accurately if necessary.

DRAW A CLEAR, COMPLETE SKETCH

The best-referenced point in the world will not be a good one if the sketch of the referencing is poor. See the example in the field notes below.

FIELD NOTES

The field notes should show a detailed sketch of each point indicating the type of point, and the distance and direction to the objects being referenced. Describe all pertinent information which would allow another field engineer to use your reference notes to locate the point.

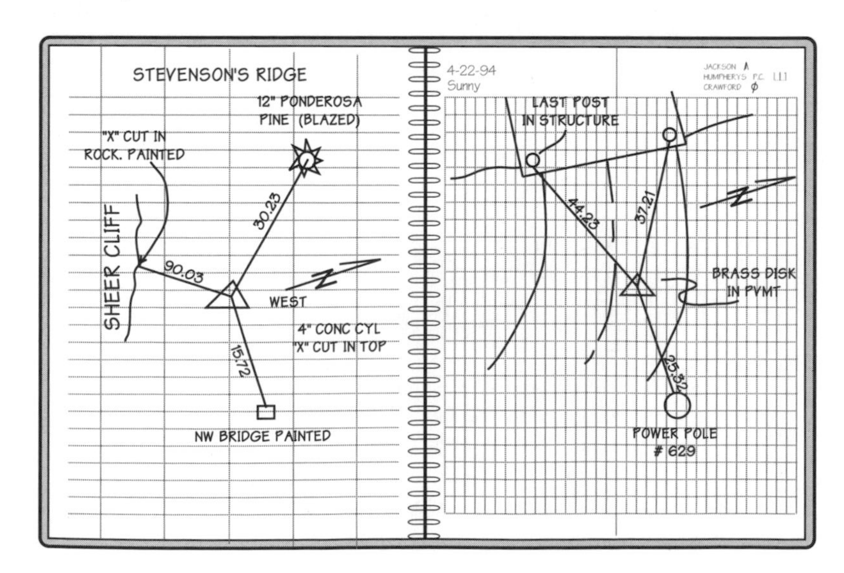

MEASURING AND LAYING OFF ANGLES WITH A CHAIN

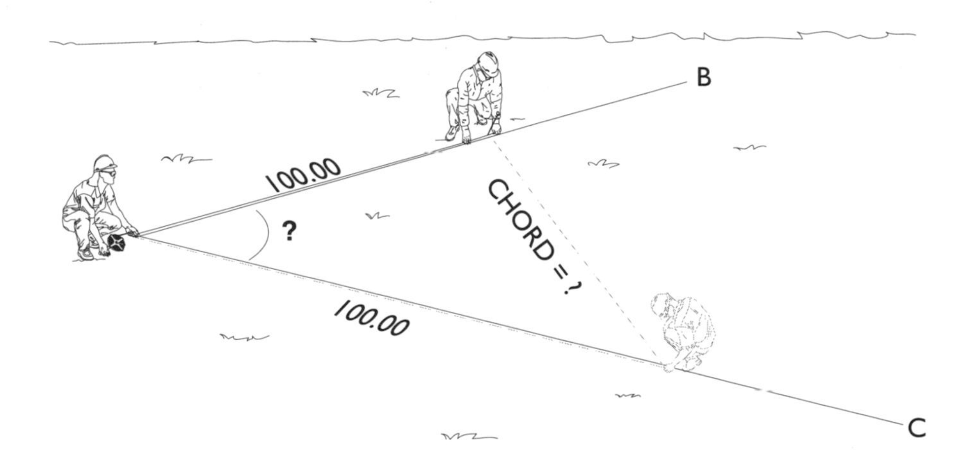

SCOPE

When we talk of measuring an angle, the first thing that comes to mind is a protractor, or in surveying, a transit or theodolite. It is possible to determine the value of an angle or to lay out a specific angle by means of linear measurement. Relatively simple mathematics is all that is required. Field engineers should be aware of this method of angle measurement. In the unlikely event that they do not have access to a theodolite, this method can be used. This method could also be used if an angle must be measured to the nearest second and the theodolite will only measure to the nearest 20".

MEASURING AN ANGLE

Measuring angles with a chain can best be accomplished by using oblique angle trigonometry. This versatile method used is straightforward and simple to calculate. Exact measurement of the sides of the triangle must be made to obtain an accurate value for the calculated angle. Being off even a hundredth will affect the angle considerably.

USING OBLIQUE ANGLE TRIGONOMETRY

This method involves measuring defined distances along the legs of a triangle and then measuring the closing length. This is shown in the illustration. The formula used in this case is based on a trigonometric identity. The formula used is:

$$Sin\frac{A}{2}=\frac{Chord\ Length}{2L}$$

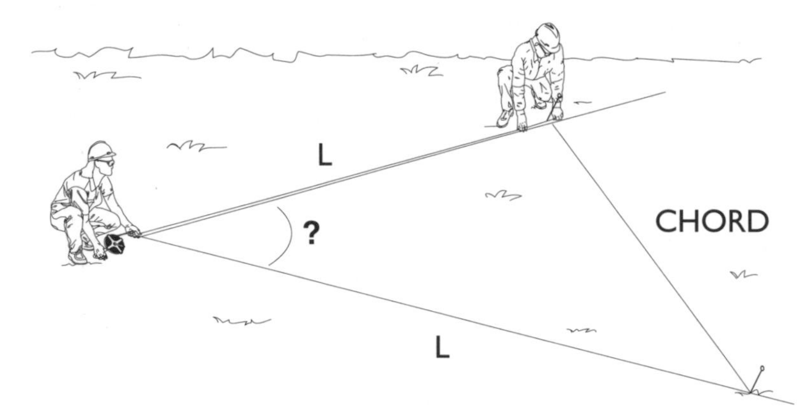

"L" is the length of the two sides of the triangle used. L can be any length that is practical for the situation. Fifty or one hundred are commonly used. If only a 100-foot chain is available, using side lengths of 50 or less is a good idea so the chord length is less than 100.

The field engineer will have to accurately define the lines where the angle is to be measured. This can be accomplished by using string lines or by careful alignment by sighting across a plumb bob string. Whatever method is used, the points marking the legs of the triangle must be located accurately.

AN EXAMPLE

$$Sin\frac{1}{2}A=\frac{55.4}{200}$$

$$Sin\frac{1}{2}A=0.277000$$

$$Arc\ \sin=32^{\circ}09'45''$$

By using this method, any angle between intersecting lines can be determined. For convenience, 100-foot horizontal distances were used. However, any length could have been used.

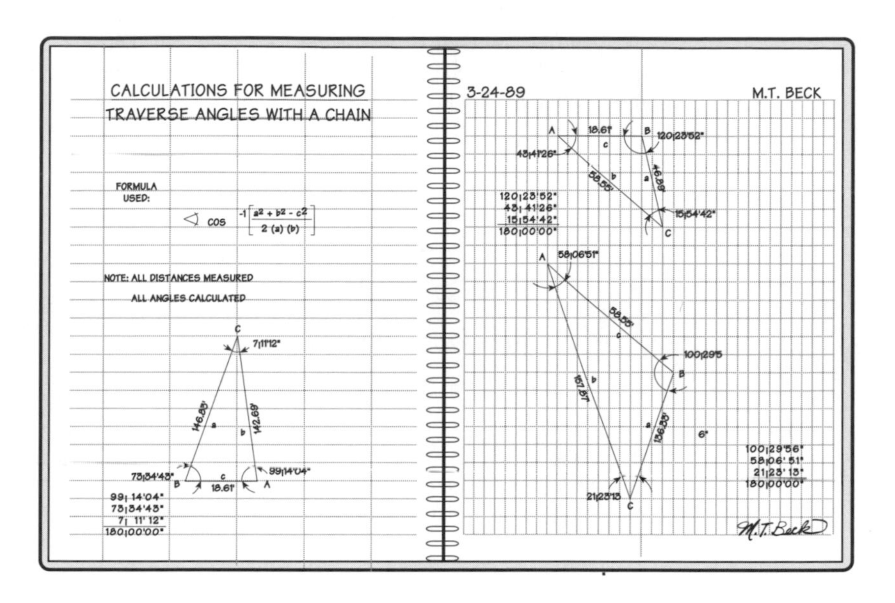

IMPROVING ANGLE PRECISION

Suppose that all a field engineer has available at a remote construction site is a transit that measures to the nearest one minute of arc. If the construction plans call for the critical layout to an equipment pad 1,000 feet away at an angle of 15° 27'40" what can be done to measure to the 40"?

It could be estimated on the vernier, but that doesn't provide much comfort to the field engineer who knows the layout must be exact. The field engineer could try to get an instrument shipped in and use it for the one measurement, but that may not be practical. Or, the field engineer could use trigonometry for a quick practical solution.

This involves exactly measuring the angle of 15° 27' and establishing it onto a temporary point perpendicular to the point that must be established. Using right angle trigonometry a perpendicular offset distance representing the 40" can be calculated and measured to establish the exact 15° 27' 40" angle.

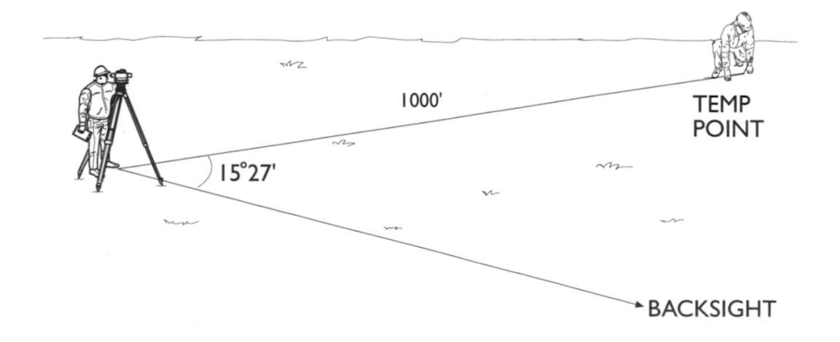

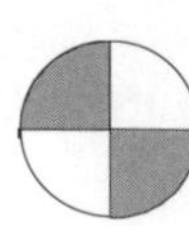

EXAMPLE PROCEDURE

Set up the 1-minute transit over the point from where the angle is to be measured. Backsight and turn the angle to the nearest minute to set a temporary point perpendicular to where the point is to be established. Turn the angle Direct and Reverse several times to get as close to the true angle as possible. Using the tangent function, calculate the perpendicular offset (opposite side of the triangle) using the 40" and the distance of 1000 feet.

$$Tan\,A = \frac{opposite}{adjacent} \quad or,$$

$$opposite = (adjacent)\tan A$$

$$therefore,\ opposite = 1000\tan 40''$$

$$offset = 0.19\,feet$$

From the temporary point, measure 0.19 feet perpendicular to the line back to the transit and set the permanent layout point. It should be noted that the perpendicular offset direction at the temporary point is generally estimated.

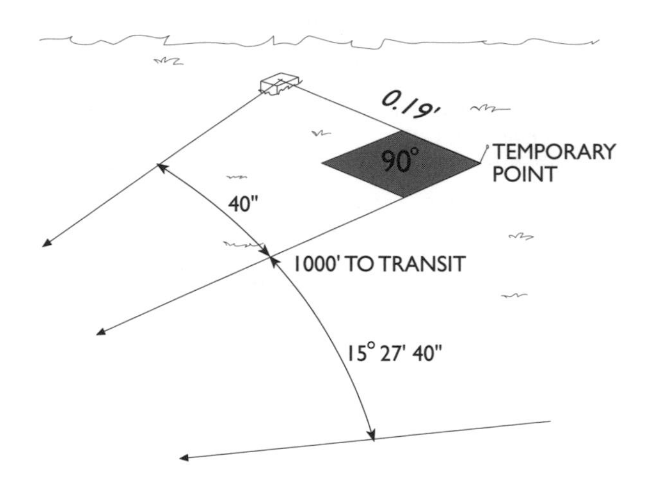

SUMMARY

The methods for measuring an angle or laying off an angle with a chain may not be used often, but understanding the process is a valuable knowledge for the field engineer.

ANGLES AND DIRECTION

21

PART 4 - CONSTRUCTION LAYOUT APPLICATIONS

CHAPTER OBJECTIVES:

State the purpose of bucking in on line on the construction site.
Describe why the process of bucking in on line is basically a trial and error process.
List the equipment used when bucking in on line.
Describe the procedure to be followed when bucking in on line.
State the objective of bucking in on line.
Describe the purpose of prolonging a line by double centering.
Describe the procedure to be followed when prolonging a line.
Explain where the actual line is when two distinct lines are marked in the double centering process.

INDEX

BUCKING IN ON LINE

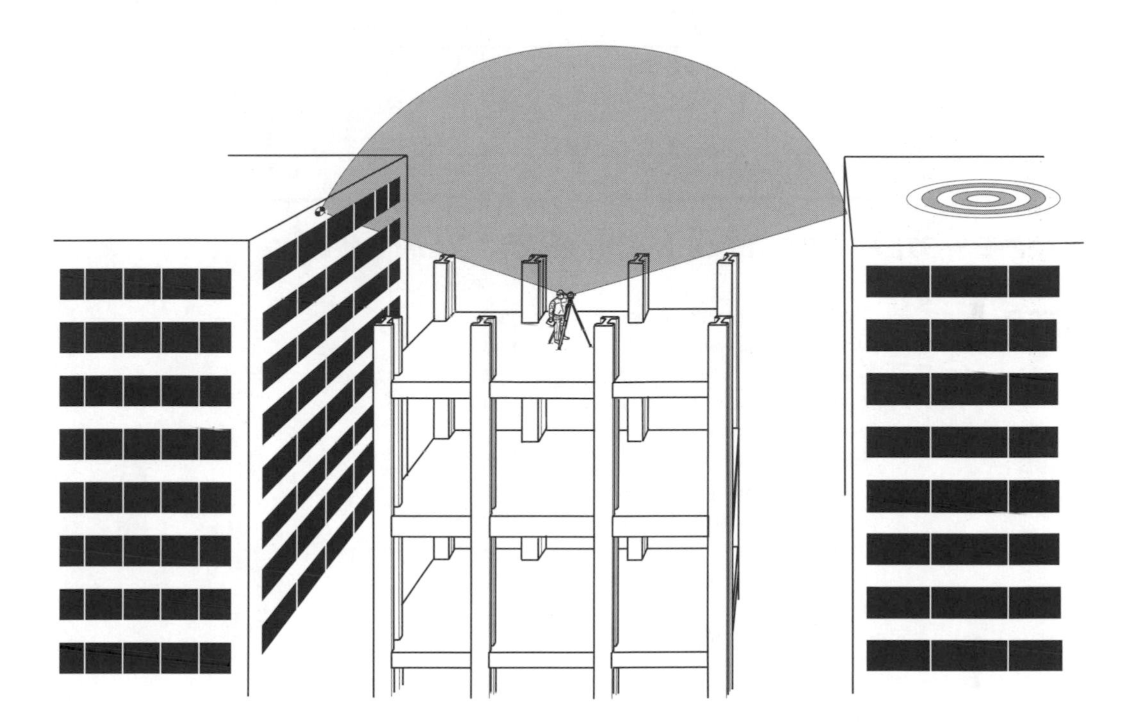

SCOPE

In field engineering, there are times when it is necessary to establish a line between two points that are not intervisible, yet both can be seen from a point between the two points. For example, points on both sides of a hill are not visible to each other, but each would be visible from a single point on top of the hill. This can occur often in high-rise construction when line is needed on a floor and the control is on targets that have been established on adjacent buildings. The field engineer will not be able to set on one of the control points to establish line so bucking in (also called wiggling in) on line is necessary to establish a line on the floor of the building between the control points.

Bucking in is basically a trial and error process. The field engineer guesses where the line between control points is and sets up the instrument and goes through the process to see how close the guess was. Seeing the results of the first try, the field engineer again makes an educated guess and repeats the process. This is repeated until the instrument is on line between the control points.

BACKGROUND REFERENCE

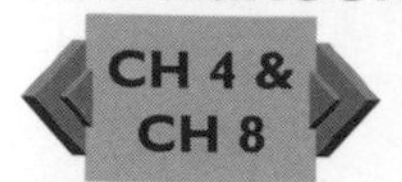

Chapter 4 Fieldwork Basics, Setting Up, and Leveling an Instrument
Chapter 8 Angle Measurement

EQUIPMENT

Bucking in Equipment	
Per crew	**Per person**
Transit or theodolite	Field book
6-foot rule	Straight edge
Hub & hammer	4H pencil

STEP-BY-STEP PROCEDURE FOR BUCKING IN ON LINE

STEP 1

Review the points between where you want to establish a line. Think about where to set up the instrument for the bucking in process. To reduce instrument error, it is best to set up the instrument halfway between the points, if possible. Stand at the selected instrument location. Be sure you can see clearly both of the points you are establishing a point between. If the points are hubs, consider placing a target on them. If possible, make them visible so the bucking in process can be accomplished by one individual at the instrument, rather than three people where one is at the instrument and one is at each of the two control points.

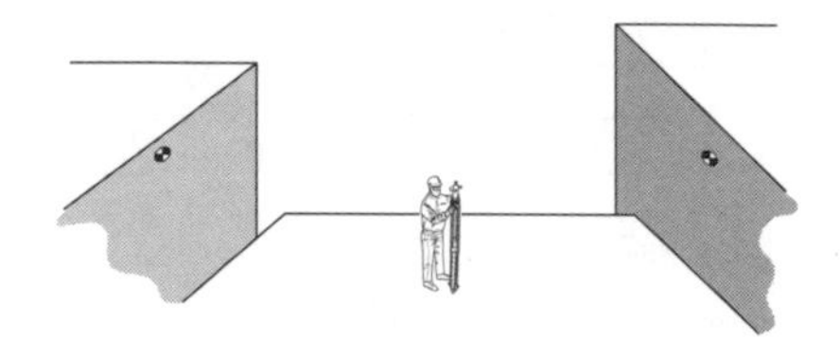

STEP 2

Set up the instrument at a location which you estimate is half way between and on line with the two points. Level it precisely.

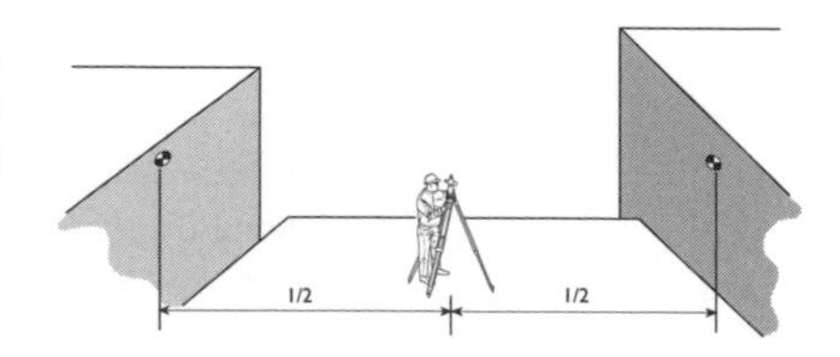

STEP 3

Sight exactly onto the first point, plunge the scope, and see how closely you hit the second point. More than likely, you won't hit the other point. However, if you are good at guessing and you hit exactly on the point, you are on line between the points and can establish the required line. If you did not hit the point, continue through this procedure.

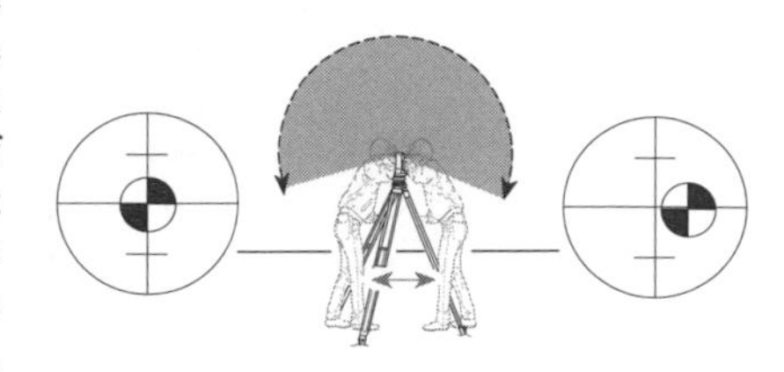

STEP 4

Estimate how much the instrument is off of line. If you are set up halfway between the two points, you can closely guess how much the transit needs to be moved by estimating or measuring how far the line of sight missed the point. If the line of sight was about 1 foot from the target, move the instrument one half foot. (If you aren't set up halfway, a proportional relationship can be set up. If you don't know where you are set up, move the instrument wherever you think.)

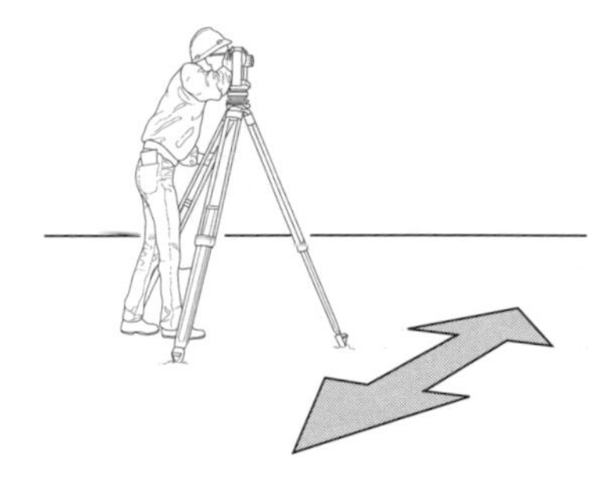

STEP 5

Move the instrument, set up, sight, plunge, and observe. Repeat the above steps over and over, getting closer and closer to being on line between the points. Each movement of the instrument should be smaller and smaller as you get closer to being on line. When you plunge and the line of sight is within a few hundredths of the second point, stop this process and go on to the next step.

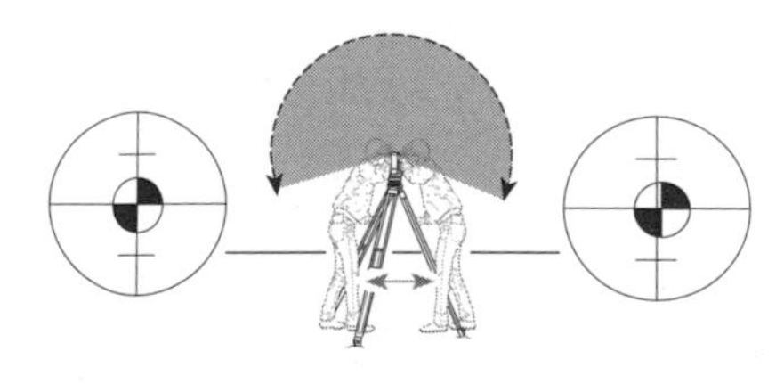

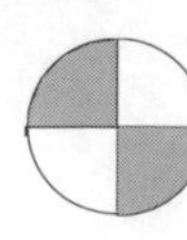

STEP 6

Shift the head of the instrument on the tripod.When you get to within a few hundredths after plunging, you should be able to shift the transit on the leveling head enough to get exactly on line. Try this then sight, plunge, and observe where the line of sight hits the second point.

STEP 7

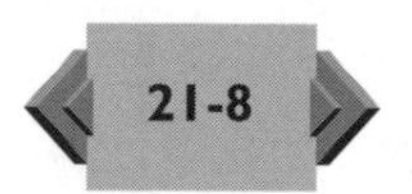

Double-center as a check. When you hit the point exactly after plunging, use the procedure of double centering to be sure of your work. Reference the next section, *Prolonging a Line by Double Centering,* for a complete description.

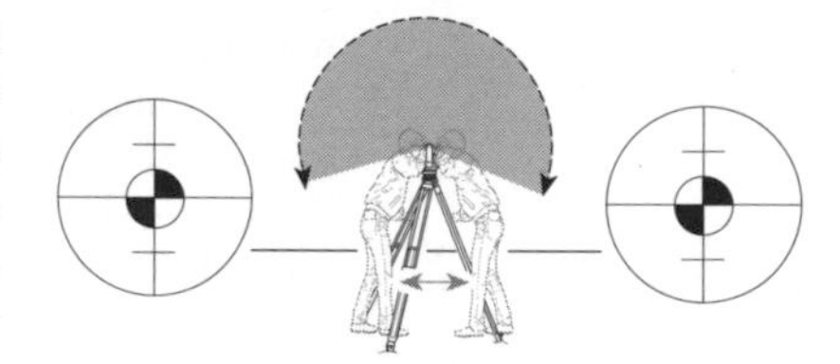

STEP 8

Mark the final instrument location.Do whatever is necessary to mark the point directly under the instrument by sighting through the optical plummet. Drive a hub if the setup is on the ground; etch a point if the setup is on concrete, etc.

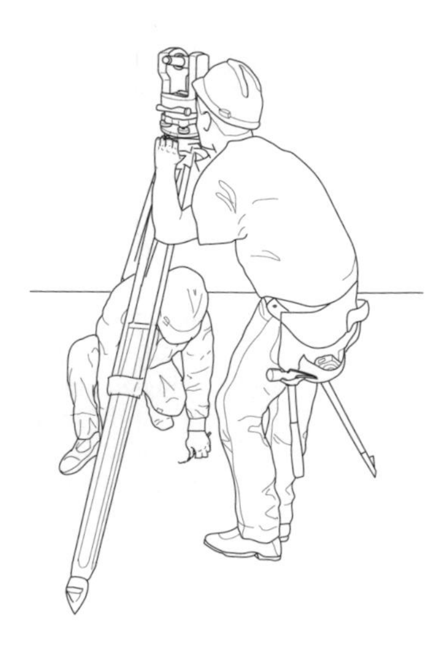

STEP 9

Establish the required line.Using the final instrument location and line of sight, mark other points on the line for future use as a control line.

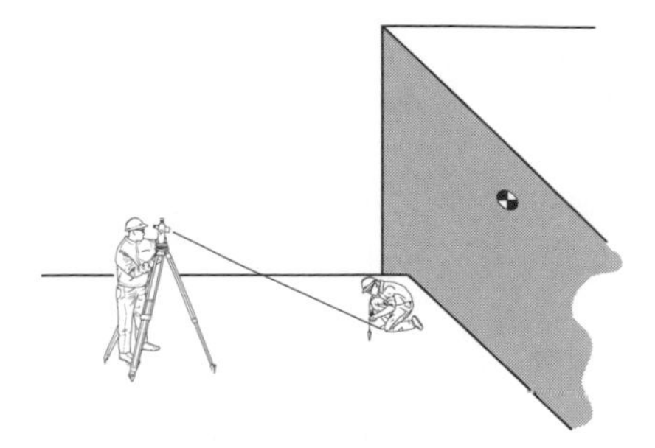

STEP 10

Record your work in the field book.For future reference and as a record of your work, prepare a sketch of the site, the points and the lines you established.

FIELD NOTES

A sketch of the area, the points used, the points established, and the line established are needed for this activity. Include all information provided, and anything else you feel is necessary.

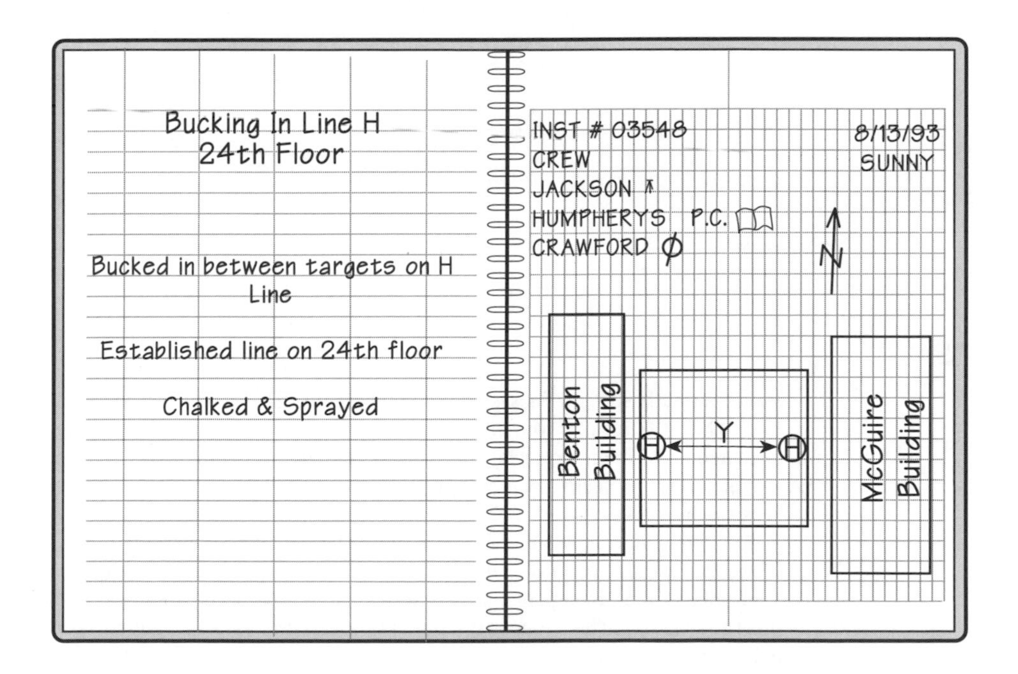

PROLONGING A LINE BY DOUBLE CENTERING

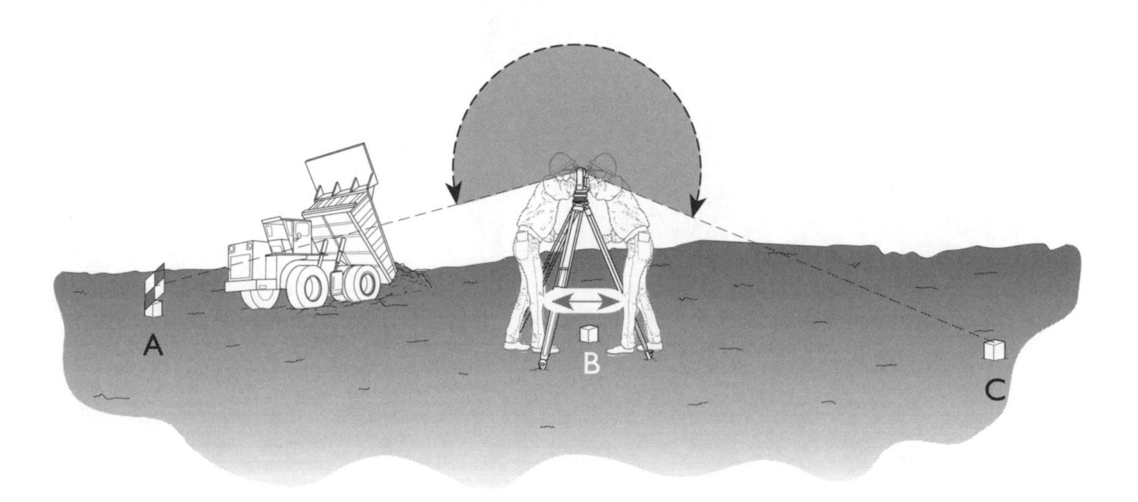

SCOPE

Projecting perfectly straight lines is often a daily activity of the field engineer on some projects. This is especially true in the layout of route projects such as highways or pipelines and on building projects where lines are being projected into the building. The best method of projecting or prolonging lines is one which eliminates instrumental errors in the process.

Double-centering should be used in all instances of prolonging a line by the field engineer. However when using double-centering, the field engineer must remember one of the fundamental principles of layout: always have long backsights and short foresights. This reduces the effect that sighting errors might have on the line being prolonged.

EQUIPMENT

Double Centering Equipment	
Per crew	**Per person**
Transit or theodolite	Plumb bob
2 range poles	Field book
Targets	Straight edge
	4H pencil

STEP-BY-STEP PROCEDURE FOR DOUBLE CENTERING

Using two points extend a line beyond them by the following method:

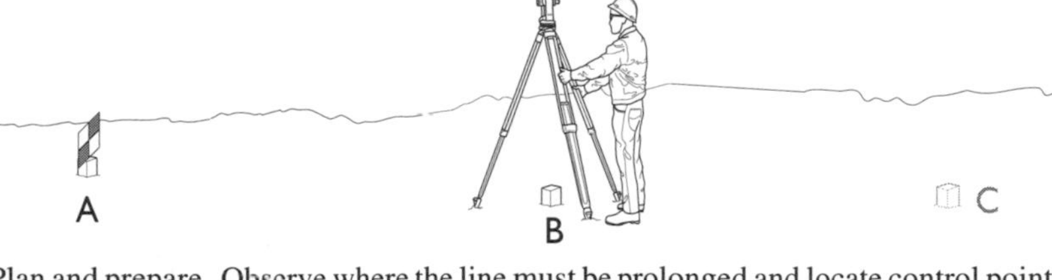

Plan and prepare. Observe where the line must be prolonged and locate control points that can be used. Typically, a point that is closest to the work will be used for the instrument to be set up on, and a point that is farthest from the work will be used as a backsight. Determine where to set a point that will define the prolonged line and use the following procedure.

STEP 1

Set up on point B. Set up and level the instrument exactly over the point.

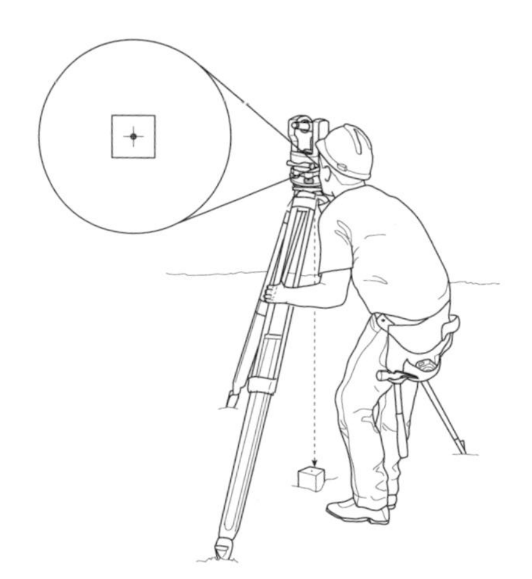

STEP 2

Backsight on the rear station A with the telescope in the direct position.

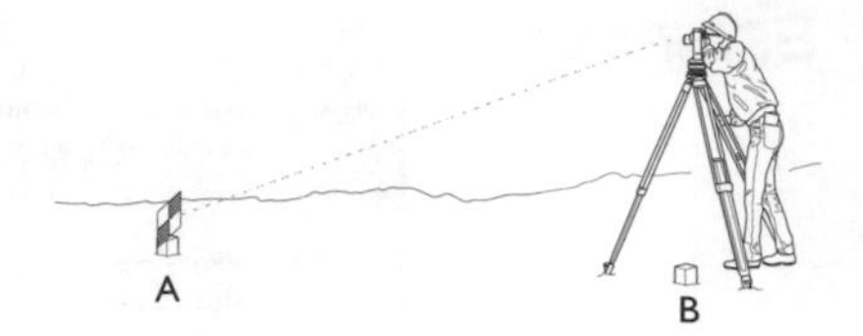

STEP 3

Plunge the telescope to the reverse position, and direct the field engineer to set Point C.

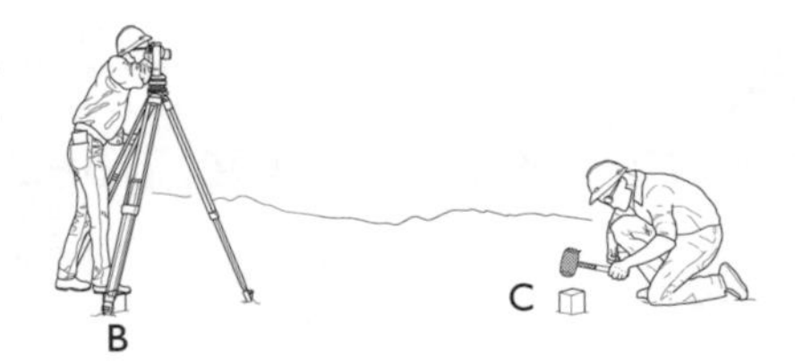

STEP 4

Put line exactly on point C.

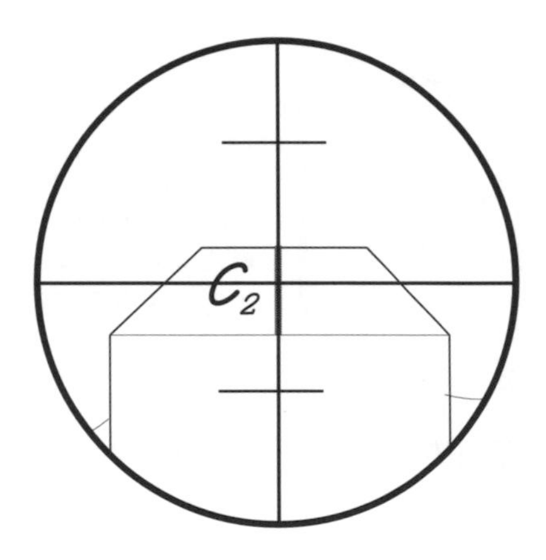

STEP 5

Rotate the instrument 180º.

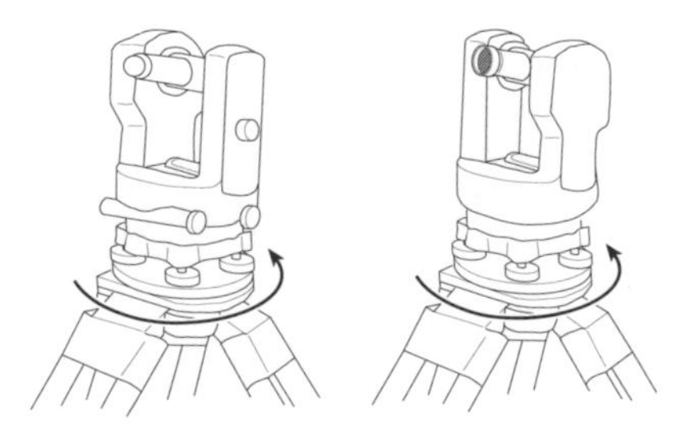

STEP 6

Backsight on point A with the telescope in the reversed position.

STEP 7

Plunge the telescope to the direct position, and sight onto point C.

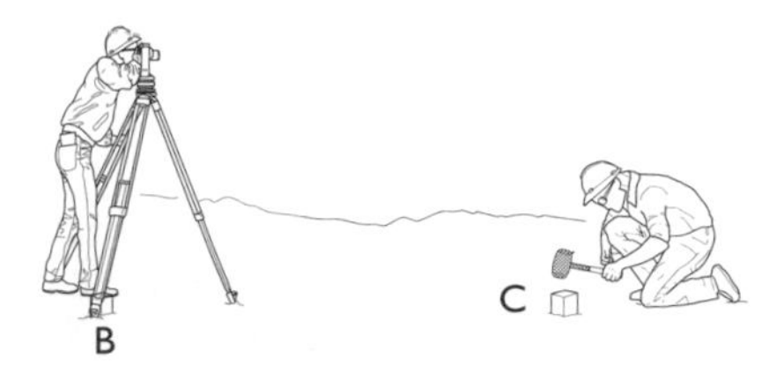

STEP 8

Analyze the results. If both points C_1 and C_2 hit exactly in the same line, they represent a true prolongation of the line ABC, and you have accomplished the task of prolonging the line. If they don't fall on the same line, and actually represent two distinct lines, repeat the above process to confirm your work.

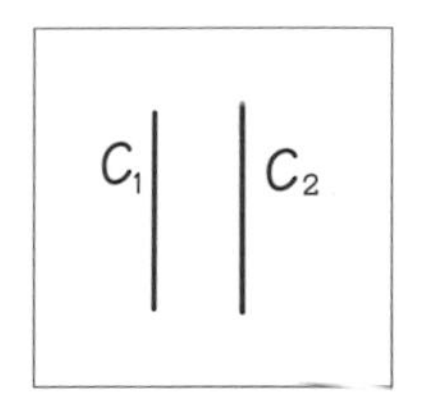

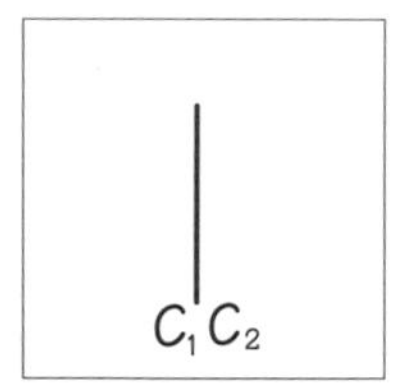

STEP 9

Split the difference and set point C. When two marks result from the above process, it is standard practice to split the distance between point C_1 and point C_2 and set point C which represents the straight line through ABC. When two distinct points result every time, it indicates that the instrument is out of adjustment and it should be sent in for calibration.

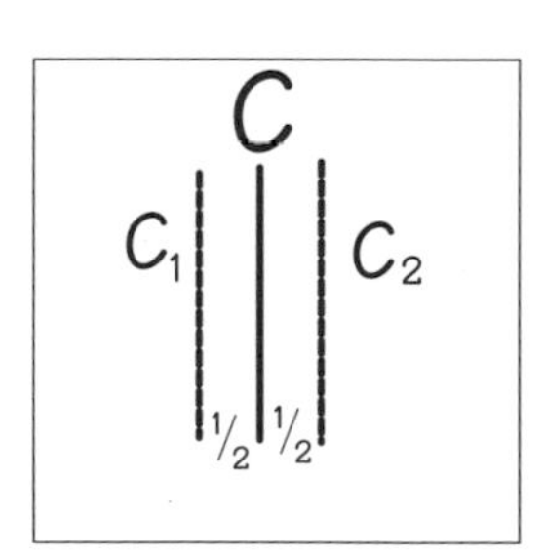

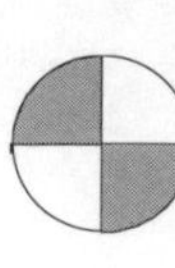

STEP 10

If possible, set on C, sight on B and check any intermediate points back towards A. Set the instrument on Point C and check your work by sighting on point B and check any intermediate line points that might be visible towards point A. They should be on line.

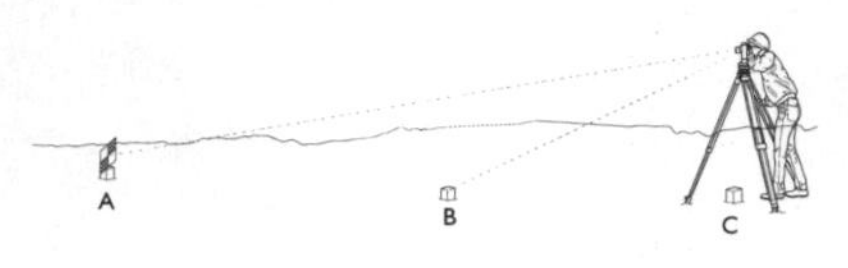

SUMMARY

Double centering is one of the most fundamental of all operations with a theodolite. It should be performed every time a point is set regardless of the type of instrument used, from a transit to a total stations. (Note: some newer total stations have dual-axis compensators which theoretically eliminate intrumental errors that are eliminated by double centering. However, they do not compensate for sighting errors which should be averaged by this method.)

Leveling Applications

22

Part 4 - Construction Layout Applications

CHAPTER OBJECTIVES:

Describe why it is important to plan the location of benchmarks on the construction site.
List the supplies and tools needed to establish benchmarks.
Describe the procedure used to set benchmarks.
Describe what should be considered when selecting a benchmark from existing objects.
Explain the rationale for setting benchmarks within one instrument set up of needed use.
Describe the process of profile leveling and identify its use on the construction site.
Explain the criteria used for deciding on the proper location to set the instrument when profile leveling.
Identify the precision required for ground shots when profile leveling.
Describe the notekeeping process used when profile leveling.
Explain how the profile is plotted on paper.
Describe the purpose of borrow pit leveling.
Describe the process used to perform borrow pit leveling on the jobsite.
Explain the need for transferring elevations up a structure.
List the equipment used to transfer elevations up a structure.
Describe the step-by-step procedure used for transferring elevations up a structure.

INDEX

SETTING BENCHMARKS

SCOPE

Setting benchmarks on the jobsite can be a very rewarding task for the field engineer if it is done correctly. Having benchmarks readily available anywhere on the jobsite throughout the project's duration will save the field engineer time in setting grade for the crafts. Consequently, it is important that benchmarks be set in a permanent manner so they are not easily disturbed by activities on the construction site or by weather conditions. It is important to locate benchmarks outside of the construction area so they will not need to be relocated during construction.

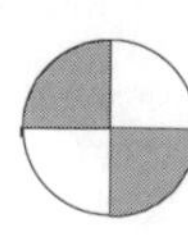

EQUIPMENT AND SUPPLIES

The supplies and tools listed may or may not be used depending on the types of benchmarks being established. Carry the appropriate tools and supplies in the field engineering vehicle at all times so they are available for use.

Benchmark Equipment	
Supplies	***Tools***
Paint	Hammer
Railroad or boat spikes	Sledge hammer
Small nails	Brush-hook or axe
Wood lathes	Cold chisel
Blue keel	Scribe
Survey ribbon	

BASIC PRINCIPLES

PLAN AND PREPARE

The field engineer should establish benchmarks at convenient intervals. For the construction of buildings, the rule of thumb is at least two benchmarks should be visible from an instrument setup anywhere on the jobsite. For route construction, benchmarks should be established at a maximum of 1000' to 1200' intervals. For bridge construction, the rule of thumb for building construction applies. There should be numerous benchmarks available around a bridge project.

Locate benchmarks so they can be seen from the project site. That is, avoid setting benchmarks on the backsides of trees and poles where it will be difficult to use them.

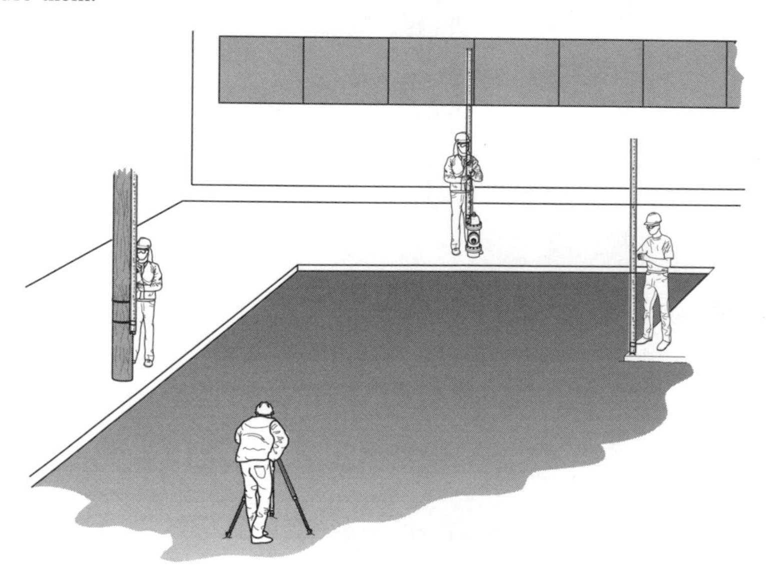

SELECTING A BENCHMARK FROM EXISTING OBJECTS

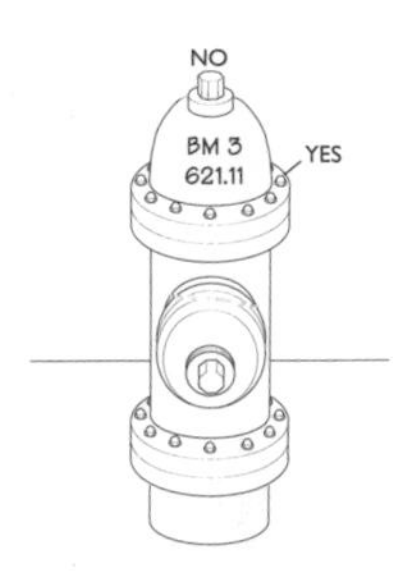

Select permanent and solid objects, such a large trees, culvert headwalls, and bridge abutments.

When using trees, select a location for the railroad or boat spike on a side where the level rod can be fully plumbed without obstruction. Drive the spike at a slightly downward angle leaving it protruding 1-1/2" to 2".

When using culvert headwalls of bridge abutments, select a suitable corner, and chisel an "**X**" on the spot you are using as a benchmark. Don't be overly destructive; just mark it so it can be found if the weather washes paint or other marks away.

When using fire-hydrants, do not use the top nut as it is a valve which moves every time the hydrant is used. Use a bonnet bolt instead. Mark the bonnet bolt used.

Do not use manhole covers, small rocks, wobbly posts or any other objects that can be moved or shifted in position.

BUILDING BENCHMARKS

When good existing objects are not available on which to establish benchmarks, it may be necessary to make your own. Follow the same criteria as for existing objects: make it permanent and solid. Generally, this means digging a hole and pouring it full of concrete to ensure that it is secure and will last. If you do dig a hole and pour concrete, be sure to determine the depth of the frost line for the area. Make the hole deeper than the frost line by at least a foot; otherwise, the winter freezing and thawing cycle will move the benchmark.

Be sure to put a distinct high point on the benchmark. Rounded brass caps can be embedded into the concrete or a piece of rebar can extend out of the concrete to serve as the high point.

PLACING BENCHMARKS

Select objects that are near the elevation where they will be needed. It doesn't make sense to have benchmarks that are more than one or two setups away.

MARKING BENCHMARKS

The field engineer should mark the benchmarks so everyone on the jobsite is aware of their importance. A specific color of surveying ribbon should be designated for benchmarks. To mark a benchmark on a pole or tree, tie two bands of survey ribbon around the tree or pole high enough so it is easily visible. Nail a lath vertically and write the number and elevation of the benchmark on it with a permanent marker. Paint it with fluorescent paint. Everyone on the site should know it is a benchmark.

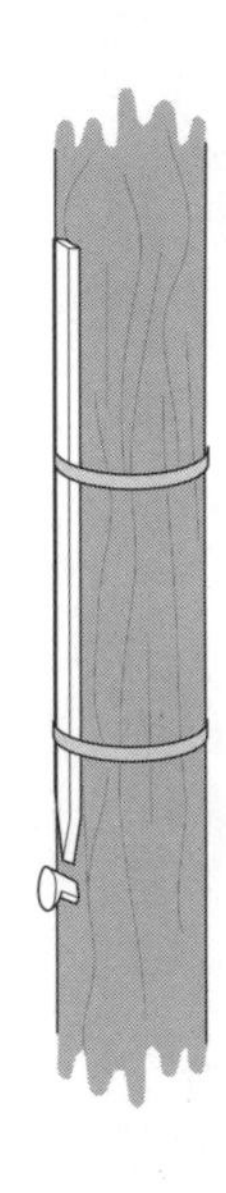

FIELD NOTES

In the field book, completely describe all benchmarks set. If a telephone pole or power pole is used, record the pole numbers. Describe the type of material, size, and type of tree, etc. Show on a sketch the location of the benchmark and surrounding features such as buildings, roads, etc. Benchmarks should be referenced so they can be located. See the section on referencing in this chapter.

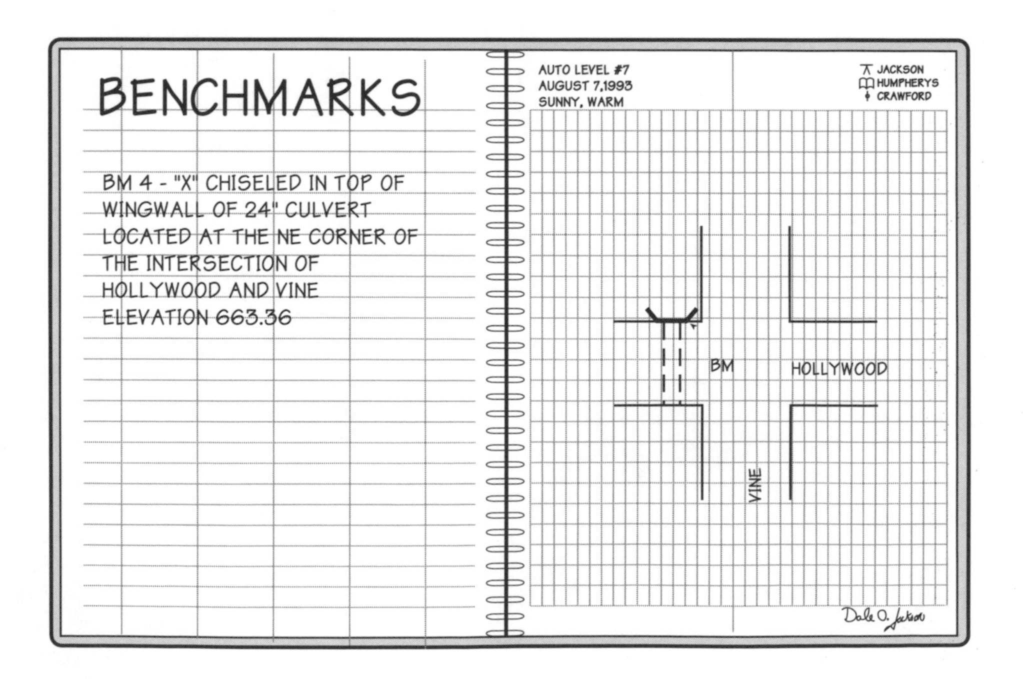

PROFILE LEVELING

SCOPE

Determining the elevation of the ground along a defined line is a constant activity of some field engineers. Field engineers who work on route projects such as highways will constantly be determining the profile of the ground along the centerline of the route. Profile leveling is an extension of the basic differential leveling process. The main difference is that numerous foresights (sideshots) are taken along a defined line. Otherwise, the process and calculations are the same. The notekeeping format for profile leveling is very similar to differential leveling with the addition of subtracting the sideshots from the H.I. to obtain the desired ground elevations. When leveling along a defined line, elevations must be taken so the exact profile is determined. This requires the careful attention of the field engineer to take elevations at all places along the line where the slope changes.

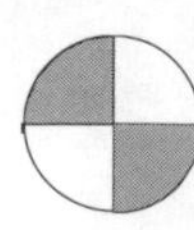

REFERENCES

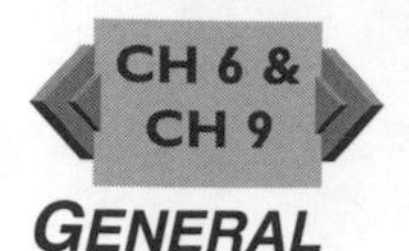

Chapter 6, Distance Measurement - Chaining, Proper Use of a Chain
Chapter 9, Leveling, Differential Leveling Procedure

GENERAL

The saying, "A picture is worth a thousand words." applies to profiles. Civil engineers, architects, landscapers, contractors and, of course, field engineers, use profiles to visually examine what the ground looks like on paper. A plotted profile allows the user to observe the general slope of the ground, determine where drainage structures should be located, and calculate amounts of cut or fill from the existing profile to a proposed profile, etc. Profiles are extremely important in all aspects of construction.

As stated earlier, profile leveling is very similar to differential leveling except that sideshots onto the profile are now part of the process. In fact, differential leveling is performed as part of the profile leveling process. Benchmarks, backsights, foresights, height of instrument, turning points, and balanced readings are all part of profile leveling. The difference is that after an H.I is determined, readings are taken onto the profile points and recorded systematically in the field book as sideshots.

The process involves using a theodolite and chain to define the line to be profiled, locating a benchmark, performing differential leveling to proceed from the benchmarks to the profile line, taking readings on the profile, using the differential leveling process to move the instrument along the profile as required by the terrain or length of the profile, and using differential leveling to return to a benchmark to close the loop. So, if differential leveling is understood, learning profile leveling is an easy transition.

STEP-BY-STEP PROCEDURE FOR PROFILE LEVELING

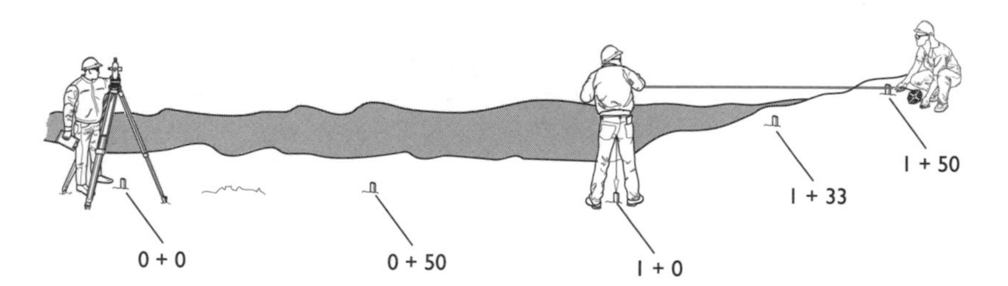

STEP 1

Define the line to be profiled by determining the starting and ending points of the line and setting a transit or theodolite on one end of the line. Sight the other end of the line. Using a chain or EDM to measure distances and the theodolite to stay on line between the end points, establish profile points on the line. The profile points should be taken at regular intervals such as every 25, 50, or 100 feet, and at any other locations on the line where there is an observable change in the slope of the ground. Mark each profile point with a stake, or lath, or chaining pin with paper, and write on it information to define its position on the profile. The method of stationing (0+00, 1+00, 11+73, 12+50, etc.) is typically used on profiles to mark the positions of points.

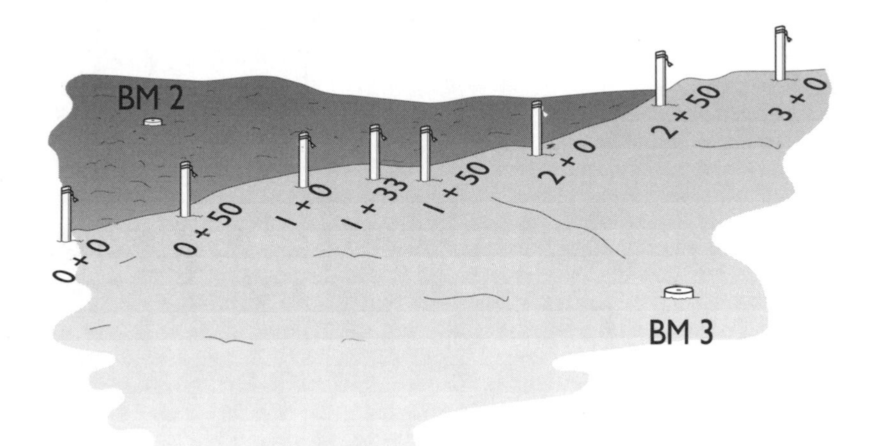

STEP 2

Locate benchmarks that can be used to obtain a starting elevation and to tie into at the end of the loop. The elevation of each profile station and the necessary intermediate turning points will be determined. Positions of known elevations are to be determined. They should be approximately equal distance from instrument to foresight station and instrument to backsight station. Assume a maximum sight distance (determined by pacing) of 150 feet in order to minimize errors in reading the level rod. Examine the slope of the terrain and ensure that the instrument person can observe the backsight, foresight, and profile stations with the instrument leveled.

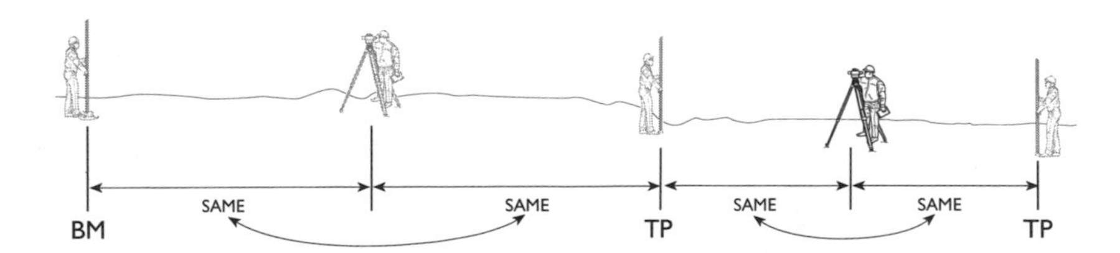

STEP 3

Read and record to the nearest 0.01 ft. (hundredth) the backsight reading with the rodperson holding the level rod plumb over the backsight station.

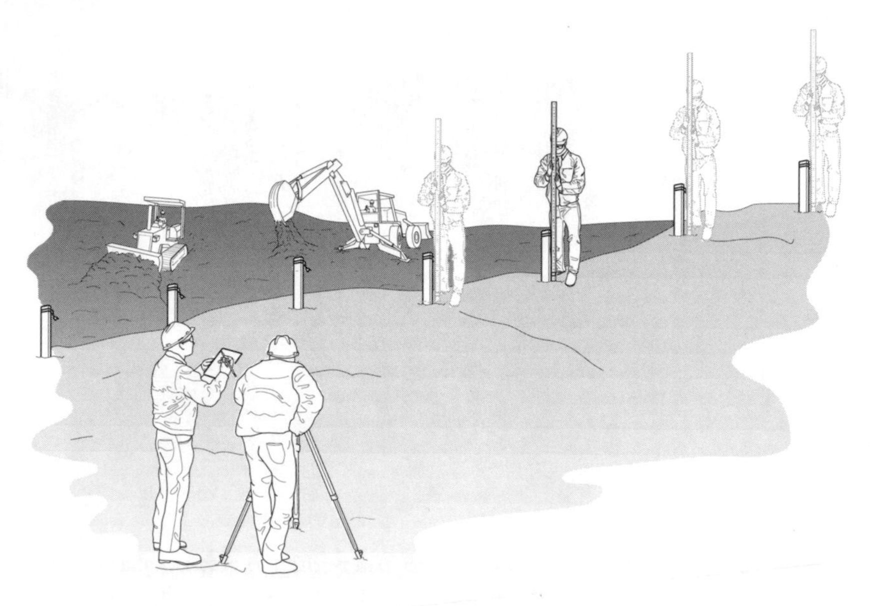

STEP 4

Advance the rodperson to the previously determined profile points; and read and record the sideshot reading on the points to the nearest 0.1 ft. (tenth). Repeat this process for as many stations as possible until a turning point is needed.

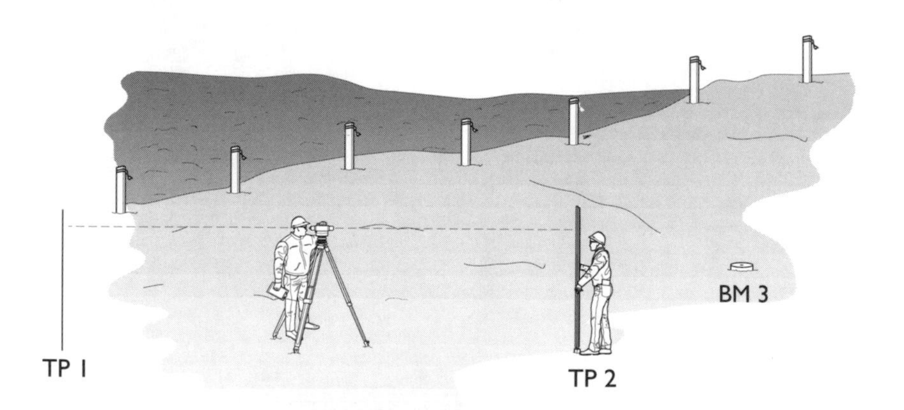

STEP 5

The person holding the rod should select a turning point that is an equal distance from the instrument as was the backsight reading. Read and record the foresight reading.

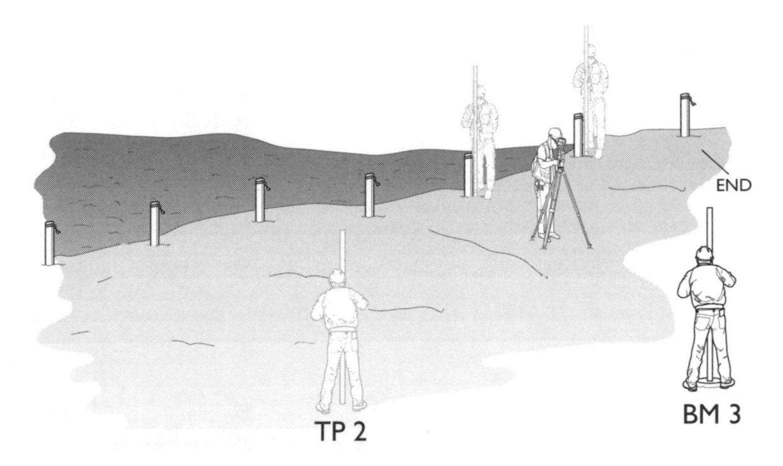

STEP 6

Repeat the steps until the elevation of each designated profile point has been determined and the level circuit closes back on a known benchmark.

CALCULATIONS

The level circuit data should be checked for numerical mistakes of addition and subtraction in the field book by performing an arithmetic check on each page of notes. Initial Elevation plus Sum of B. S. minus Sum of F. S. = Final Computed Elevation. Reminder: the sideshots are not used in the arithmetic check process.

FIELD NOTES

See *Fieldnotes for Leveling, Chapter 9*, for an explanation of notekeeping, observe that each sideshot is subtracted from the preceeding HI.

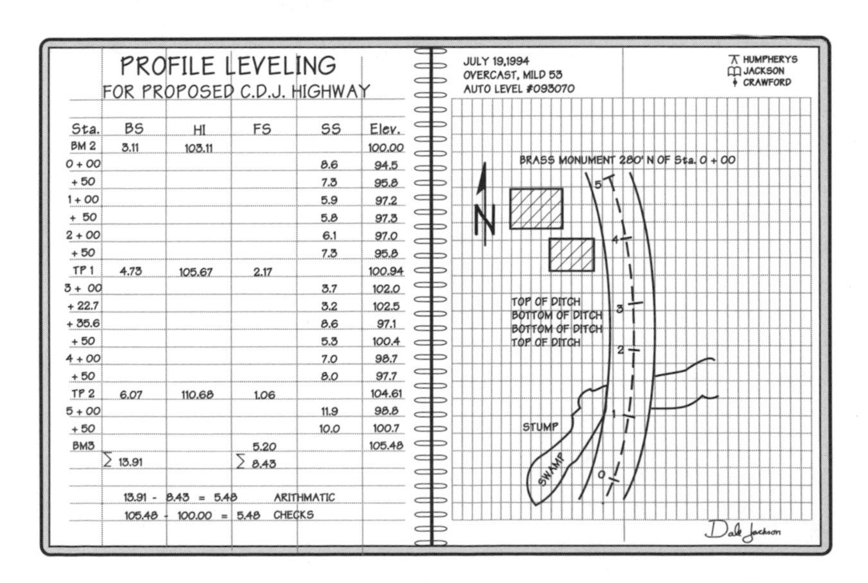

PROFILE LEVELING
FOR PROPOSED C.D.J. HIGHWAY

Sta.	BS	HI	FS	SS	Elev.
BM 2	3.11	103.11			100.00
0 + 00				8.6	94.5
+ 50				7.3	95.8
1 + 00				5.9	97.2
+ 50				5.8	97.3
2 + 00				6.1	97.0
+ 50				7.3	95.8
TP 1	4.73	105.67	2.17		100.94
3 + 00				3.7	102.0
+ 22.7				3.2	102.5
+ 35.6				8.6	97.1
+ 50				5.3	100.4
4 + 00				7.0	98.7
+ 50				8.0	97.7
TP 2	6.07	110.68	1.06		104.61
5 + 00				11.9	98.8
+ 50				10.0	100.7
BM3			5.20		105.48
	Σ 13.91		Σ 8.43		

13.91 - 8.43 = 5.48 ARITHMATIC
105.48 - 100.00 = 5.48 CHECKS

PLOTTING

After the fieldwork has been completed and checked, the fieldnotes can be used for plotting the profile. Determine the use of the profile so that a proper horizontal and vertical scale can be selected. For example, in an area where there is not much change in elevation, the vertical scale may need to be exaggerated by as much as ten or more times the horizontal scale in order to show the ground changes. See the illustration below. If the profile is through a hilly area, no difference between the horizontal and vertical scale may be necessary to show ground changes.

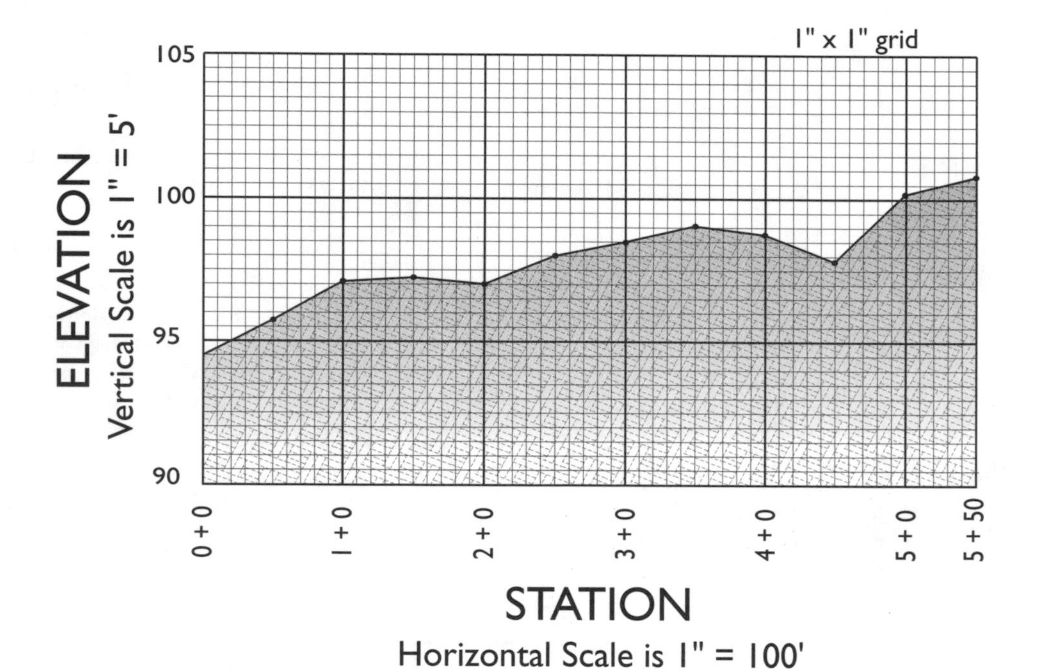

GRID (OR BORROW-PIT) LEVELING

SCOPE

Determining the elevation of the ground in an excavation or on a mound is a common activity of some field engineers. Field engineers who work on building or route projects may be asked to calculate the volume of material that has been excavated or placed. Grid leveling is an excellent method used to determine volumes of irregularly shaped excavations. Grid leveling is another extension of the basic differential leveling process. Grid leveling is different from basic differential leveling because numerous foresights (sideshots) are taken on the grid. Otherwise the process and calculations are the same. The notekeeping format for grid leveling is very similar to profile leveling except it is also necessary to subtract the sideshots from the H.I. to obtain the desired ground elevations. When leveling a grid, elevations at each grid point must be taken so the exact contour of the ground is determined. If the grid interval is too large and high and low points are being missed, the field engineer should consider decreasing the interval and setting more grid points.

GENERAL

Grid leveling is typically a redundant activity. That is, it is performed over and over on the same points as volume calculations are needed. For an excavation, a grid is established in the area where the volume will be needed. The leveling process is performed and elevations of the grid points are recorded. After excavating, the elevations of the grid points are reestablished and leveling is performed again. The difference in the original elevation and the final elevation is used in a volume computation formula to determine the amount of material excavated. Volumes are extremely important to the contractor since payment for work performed is often based on the volume of material moved. See *Chapter 19, Excavation,* for a discussion of volume calculations.

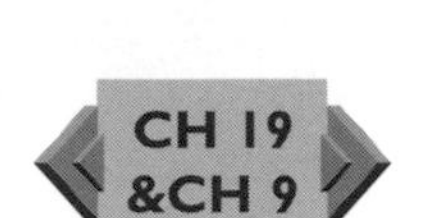

As stated earlier, grid leveling is very similar to differential leveling except that sideshots onto the grid are now a part of the process. In fact, differential leveling is performed as part of the grid leveling process. Refer to *Chapter 9, Leveling* for the step-by-step differential leveling procedure. Benchmarks, backsights, foresights, height of instrument, turning points, and balanced readings are all part of grid leveling. The difference is that after an H.I is determined readings are taken onto the grid points and recorded systematically in the field book as sideshots, just like profile leveling.

The process involves measuring and defining the grid using a chain, locating a benchmark, performing differential leveling to proceed from the benchmark to the grid, taking readings on the grid, using the differential leveling process to move the instrument around the grid as required by the terrain or size of the grid, and using differential leveling to return to a benchmark to close the loop. If differential leveling is understood, learning grid leveling is an easy transition.

STEP-BY-STEP PROCEDURE FOR GRID LEVELING

Define the grid to be leveled by determining the limits of the excavation or pit. Randomly establish the size of the grid so it completely surrounds the site. Check with the superintendent and the specifications to determine the grid interval that will best determine the volume to the accuracy required. Typically, regular intervals such as every 25, 50, or 100 feet are used.

STEP 1

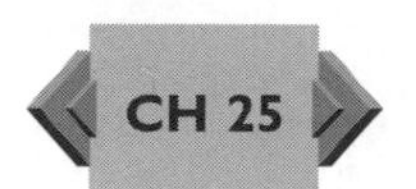

Use one side of the grid as a baseline and begin measuring and marking the intervals. After that side is marked, use a right angle prism or use the 3/4/5 triangle method of establishing a right angle to determine the direction of one of the other sides of the grid. See *Chapter 25, One Person Surveying Techniques*. Begin measuring and marking the intervals on that side. Repeat the process for the remaining sides of the grid. Use the method of stationing on one side of the grid to label the points and use letters to mark the points in the other direction, or some other notation which avoids confusion. With the intervals established on the outsides of the grid, the actual grid points can be located fairly accurately by having two individuals sighting across the grid from one side to the other while a third person sets the interior points of the grid.

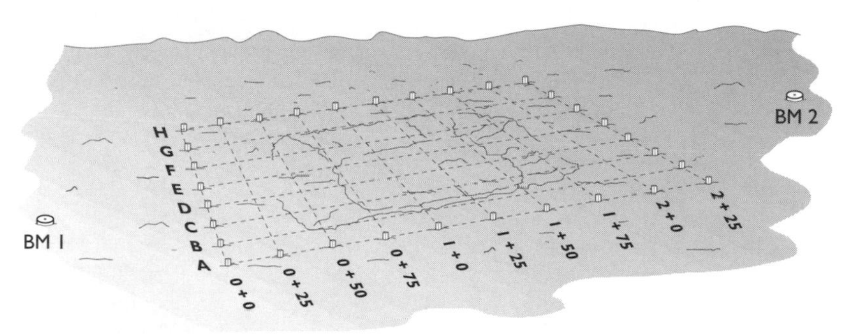

STEP 2

Locate benchmarks that can be used to obtain a starting elevation and to tie into at the end of the loop. The elevation of each grid point and the necessary intermediate turning points will be determined. Assume a maximum sight distance (determined by pacing) of 150 feet in order to minimize errors in reading the level rod. Examine the slope of terrain to ensure that the instrument person can observe the backsight, foresight, and grid points where the instrument will be set up.

Turning Point

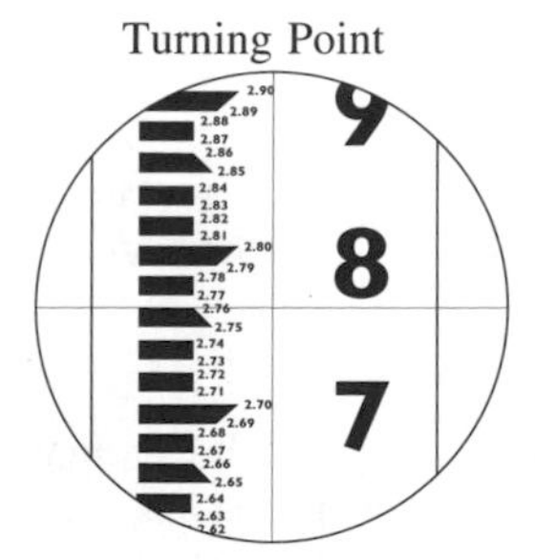

Side shot

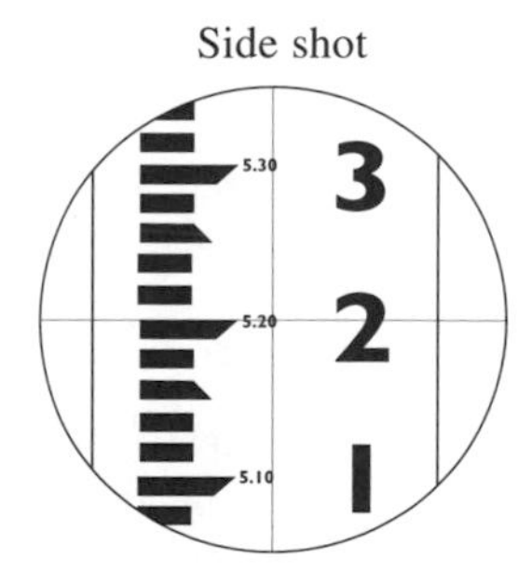

STEP 3

Read and record to the nearest 0.01 (hundredth) foot the turning point reading with the rod person holding the level rod plumb over the turning point station. Advance the rod person to the previously determined grid points and read and record the sideshot reading on the points to the nearest 0.1 (tenth) foot. Repeat for as many points as possible until a new turning point is needed.

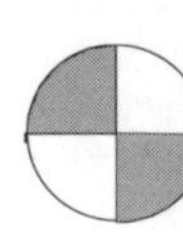

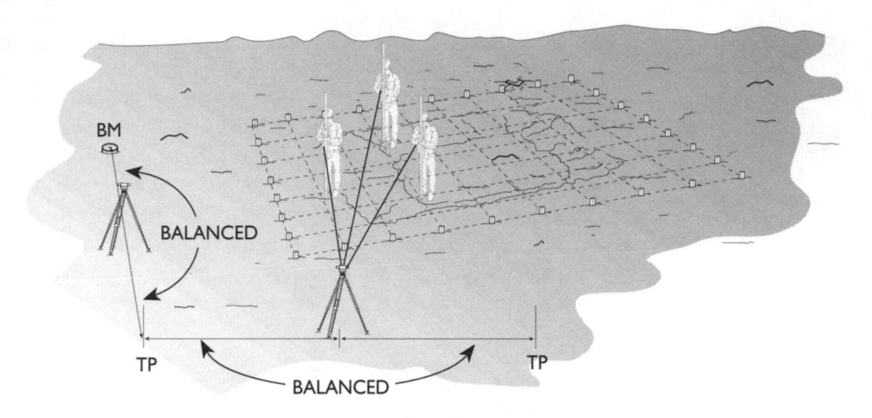

STEP 4

Balance BS and FS between turning points. The person holding the rod should select a turning point that is an equal distance from the instrument as was the backsight reading. Read and record the foresight reading.

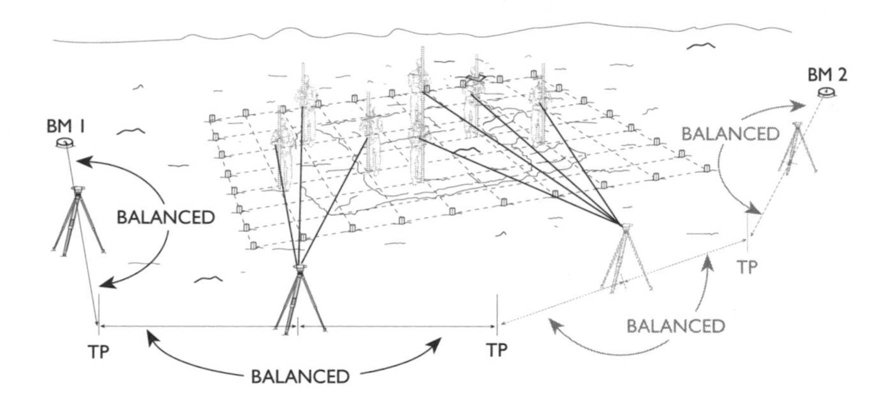

STEP 5

Repeat the steps until the elevation of each grid point has been determined and the level circuit closes back on the known benchmark.

CALCULATIONS

If the grid leveling process required several turning points, the differential level circuit data should be checked for numerical mistakes of addition and subtraction in the field book by performing an arithmetic check on each page of notes.

Reminder: the sideshots are not used in the arithmetic check process.

FIELDNOTES

The following is a representative example of fieldnotes which could be used for grid leveling

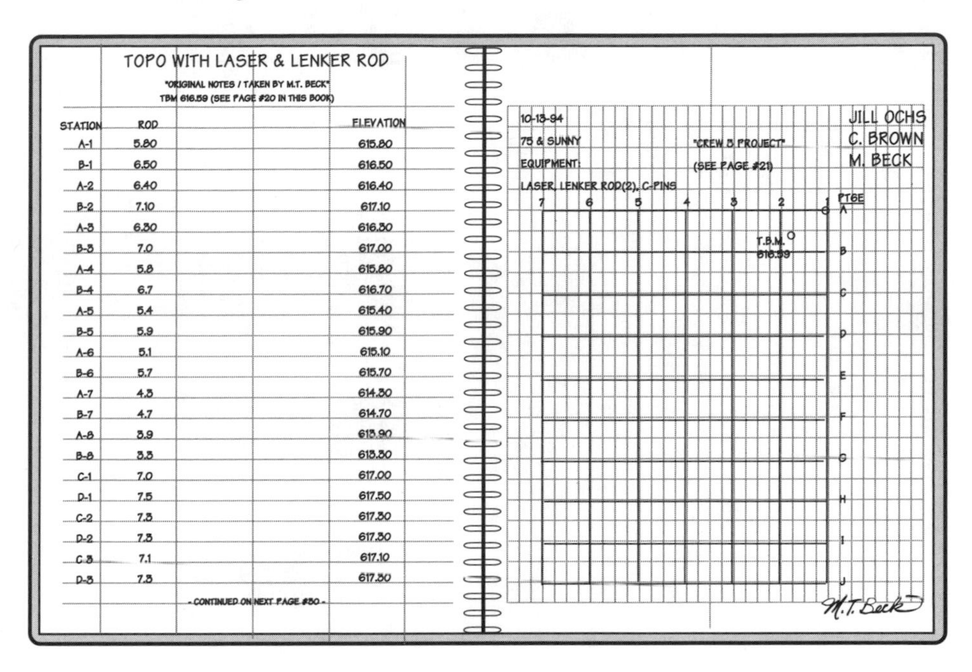

TOPO WITH LASER & LENKER ROD

"ORIGINAL NOTES / TAKEN BY M.T. BECK"
TBM 616.59 (SEE PAGE #20 IN THIS BOOK)

STATION	ROD			ELEVATION
A-1	5.80			615.80
B-1	6.50			616.50
A-2	6.40			616.40
B-2	7.10			617.10
A-3	6.30			616.30
B-3	7.0			617.00
A-4	5.8			615.80
B-4	6.7			616.70
A-5	5.4			615.40
B-5	5.9			615.90
A-6	5.1			615.10
B-6	5.7			615.70
A-7	4.3			614.30
B-7	4.7			614.70
A-8	3.9			613.90
B-8	3.3			613.30
C-1	7.0			617.00
D-1	7.5			617.50
C-2	7.3			617.30
D-2	7.3			617.30
C-3	7.1			617.10
D-3	7.3			617.30

- CONTINUED ON NEXT PAGE #30 -

TRANSFERRING ELEVATIONS UP A STRUCTURE

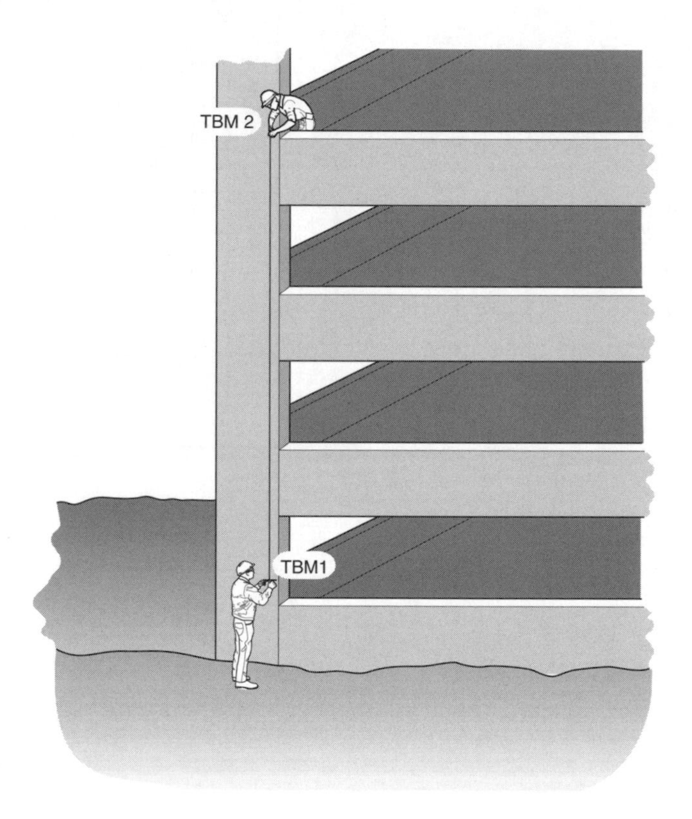

SCOPE

Multi-story building construction and tall column construction on a bridge project require that ground elevations be transferred vertically as the structure progresses upward to maintain the design grades. Occasionally, this can be accomplished by normal differential leveling procedures of backsights and foresights when there is an adjacent structure or high ground. However, the most common method of transferring elevation up into a structure is to use a good calibrated chain and measure vertically along a column, the elevator/stair core, or any other part of the structure which is solid and progressing vertically. It is a simple process, but one that requires following exact procedures to obtain accurate results.

REFERENCES

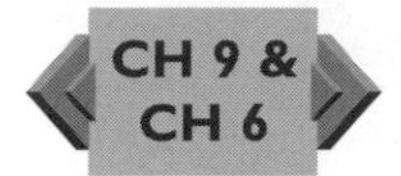

Differential Leveling Procedure, Chapter 9, Leveling
Proper Use of a Chain, Chapter 6, Distance Measurement with a Chain

EQUIPMENT

Transfering Elevation Equipment	
Per crew	**Per person**
Leveling instrument	Plumb bob
Level rod	Field book
Turning pin	Hand level
Chain	Straight edge
Hammer	4H pencil

STEP-BY-STEP PROCEDURE

STEP 1

Plan and prepare. Discuss a plan of attack for the efficient and accurate completion of the task. Obtain the necessary equipment. Locate a nearby starting benchmark and determine its elevation.

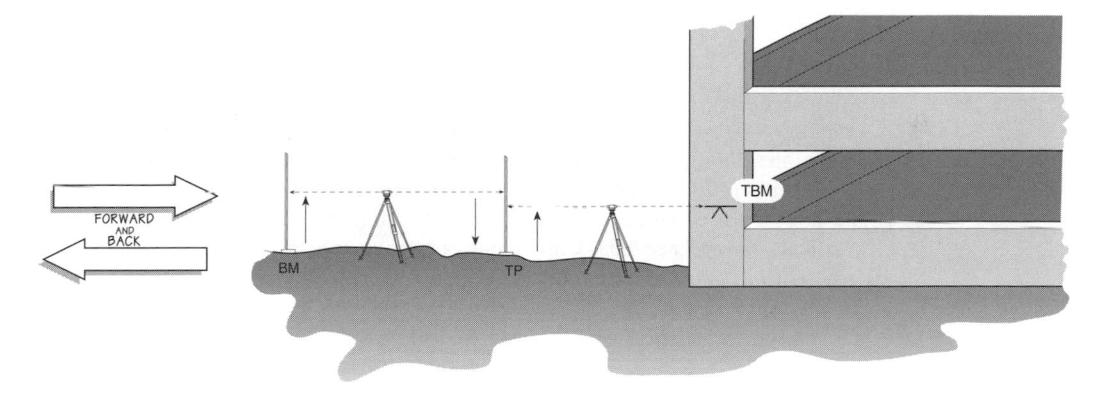

STEP 2

Using the techniques of differential leveling, run a level loop to the structure and establish a temporary benchmark (TBM) at the bottom of the structure. Typically, TBMs are simply a mark on the wall with the elevation written there. Remember to describe and identify the point chosen in your field notes.

STEP 3

Complete the level loop by returning to the starting benchmark. Check your results. If your work is acceptable (i.e., it meets accuracy standards for the project), proceed with the transfer. If your work is unacceptable, re-run your level loop.

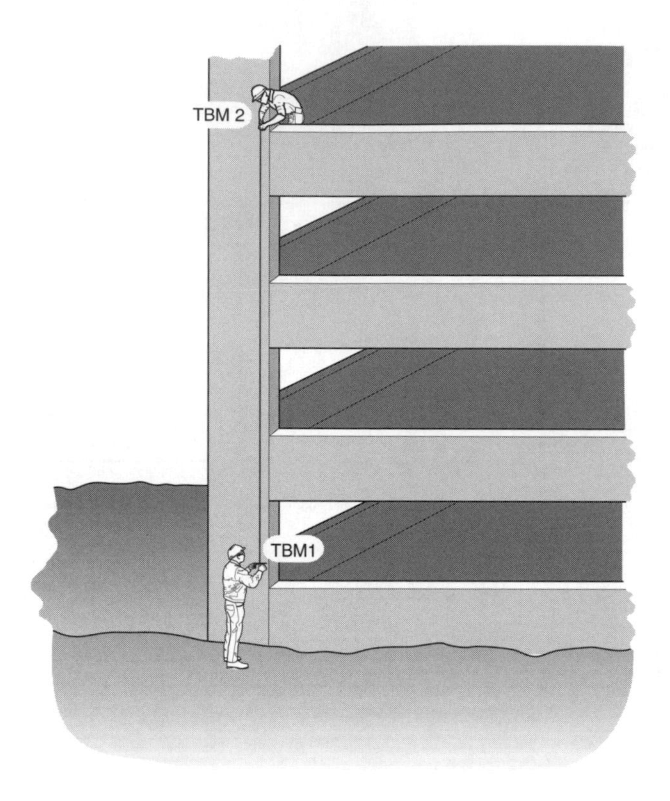

STEP 4

Using the chain, transfer the elevation up the side of the structure to the floor on which the elevation is needed. Have one field engineer accurately hold the zero end of the chain on the TBM just established at the bottom and have another field engineer carefully establish a new TBM up on the floor of the structure. (For buildings, this is typically four feet above finish floor.) Measure the appropriate vertical distance needed. Care should be taken to ensure the TBM is located where it will be visible from the instrument. No extra tension needs to be applied to the chain in this instance. It's own weight should keep it straight and taut.

STEP 5

To use the TBM up on the structure, set up a leveling instrument and establish your HI by backsighting a rod held on the TBM. Perform leveling tasks as if you were on the ground.

Note: Often, points that need elevations up on structures may be above the line of sight of the instrument. If this is the case, those which are on the ceiling or on the bottom of a beam will require the level rod to be used upside down. This will result in a positive foresight that must be added to the HI rather than subtracted! Be careful in your notekeeping.

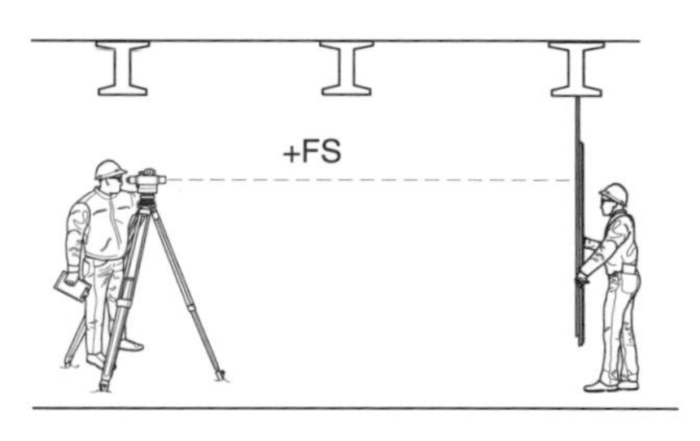

CALCULATIONS

Differential leveling calculations:
HI = BS + Elevation
Elevation = HI - FS (except with positive foresight, then: *Elevation = HI + FS*)

SAMPLE FIELD NOTES

The following is a representative sample of fieldnotes which could be used for transferring an elevation up a structure

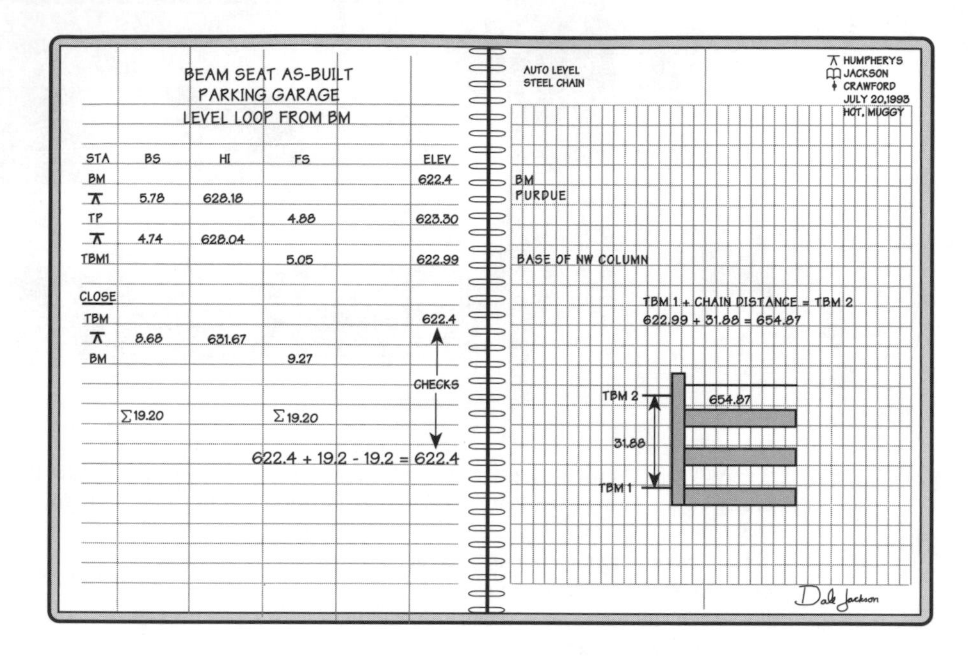

CROSS-SECTION LEVELING

SCOPE

Cross-section leveling is important to the contractor and to the owner because the data is used in the computation of volumes of earth that is cut or filled. Cross sectioning itself is a very simple task. All it requires is that the elevation and location of ground points be determined and recorded along a line that is perpendicular to the centerline. The type of instrument used to perform cross sectioning varies from a hand level to a total station. The field engineer should choose an instrument that can be used efficiently to collect the data that meets the accuracy requirements of cross sectioning.

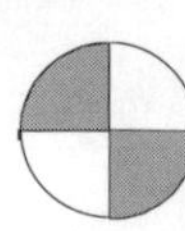

GENERAL

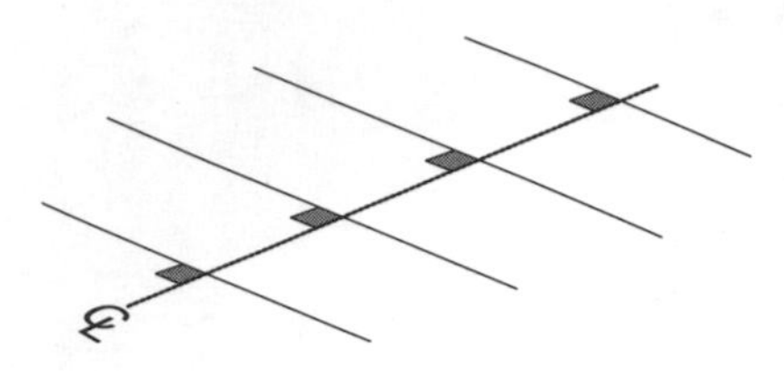

Basically, the cross-sectioning process is the same as profile leveling. Rather than determining elevations along the centerline, cross-sectioning determines elevations for lines that are perpendicular to the centerline. There is only one centerline profile for a project but there may be hundreds or thousands of cross-sections. For the entire length of a route project, cross-sections are typically required at every full station and at every half station.

Cross-sectioning is a constant activity of the field engineer on a highway project. After the centerline for a highway project has been established, the first activity of the field engineer will be cross sectioning, and after the roadway has been constructed the last activity of the field engineer will be cross sectioning. Additionally, cross-sections are often required during the project for partial payment of work performed.

PLANNING

The field engineer preparing to cross-section needs to review the scope of the work to evaluate the length of the project, the required interval between cross-sections, the elevation differences that will be encountered, the accuracy requirements, and the equipment available.

CROSS-SECTION INTERVAL

Most often the specifications from the owner (State Department of Transportation) will specifically indicate the required interval for cross-sectioning. Usually this distance is 50 feet but may be 100 feet if the terrain is not changing much. The closer the interval, the more accurate the volumes that will be calculated.

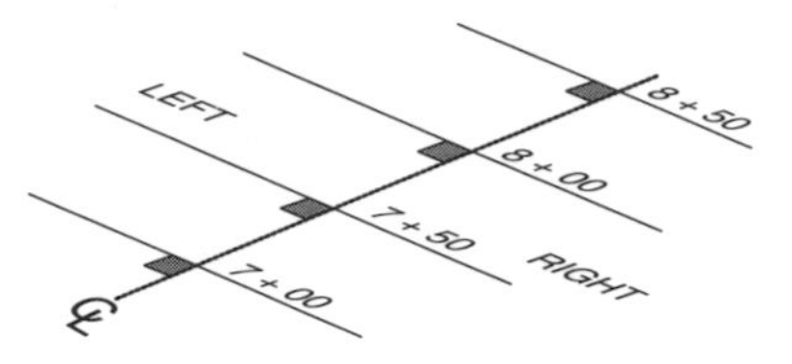

ELEVATION SPOTS

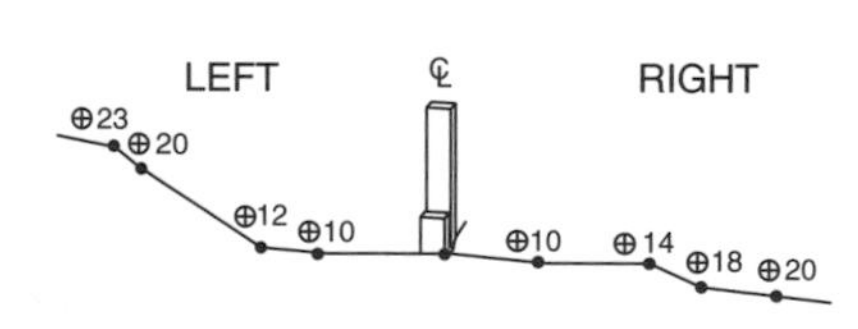

Elevations on the cross-section are typically determined wherever there is a change in the slope of the ground. That is, if the ground is flat and all of a sudden starts sloping downhill, the spot where that change took place is where an elevation needs to be determined. Additionally, it may be required that elevations be taken at specified offsets from centerline such as 10, 20, 30, 40, etc. feet from centerline.

ACCURACY REQUIREMENT

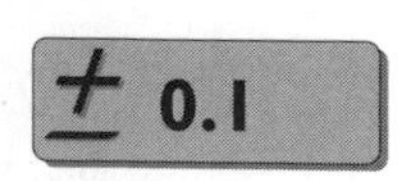

Typically, cross-section data for readings taken onto the ground are read to the nearest tenth of a foot. Distances from centerline are also measured to the nearest tenth. Confirm in the specifications or with the owner the required accuracy for the work.

MAINTAIN PERPENDICULARITY

When cross-sectioning it is essential that subsequent cross-sections be performed where the original cross-section was measured. In other words, all of the cross-sections taken at a particular station along the centerline must be along the same perpendicular line that the original cross-section followed. If much deviation from this line occurs, the data collected will yield an incorrect volume. Most often when cross-sectioning, no instrument is used to determine the perpendicular line that is required. A double-right angle prism or a "Right Angle Wingding" is most often used by the field engineer. Refer to *Chapter 25, One-person Surveying Techniques* for an explanation of how to obtain a right angle with these methods.

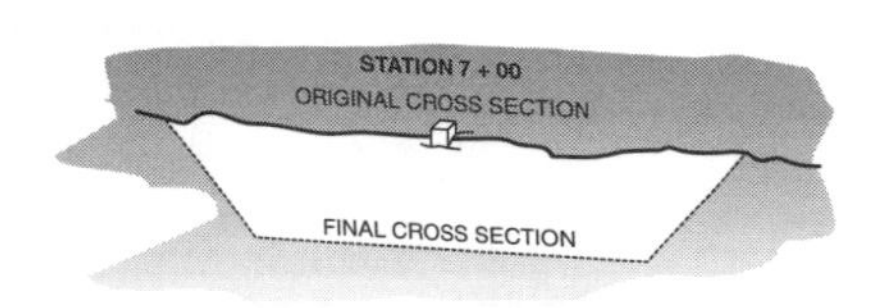

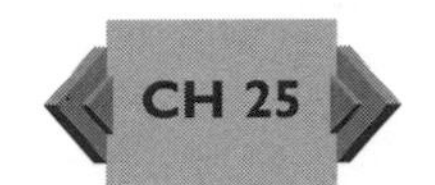

EQUIPMENT

To cross-section a field engineer can use a hand level, a Rhodes Reducing Arc, a dumpy or automatic level, a reducing tachometer, or a total station. The choice of instrument depends on what is best suited for the terrain that will be encountered, and what type of equipment is available within the company, or what the budget is for purchasing equipment. Everyone would like a total station with a data collector for cross-sectioning, but that isn't always the best or the most practical. Once the scope of the work is understood, the field engineer may want to check with an equipment supplier for a recommendation.

STEP-BY-STEP PROCEDURE FOR CROSS-SECTIONING

As indicated earlier, cross-sectioning can be performed with a wide variety of equipment. To understand the basic concept, cross-sectioning with a hand level, level rod and cloth tape will be described and illustrated in detail. Later, a total station will be shown being used for measuring cross-sections.

STEP 1

Gather the equipment and proceed to the area that needs cross-sectioning. Locate the previously set centerline profile stakes. Assign the tasks of holding the rod along the section, of being the person at centerline using the hand level, and notekeeping.

STEP 2

At the starting station, determine a cross-section line that is a perpendicular line to centerline by the use of a double right angle prism or a right angle wingding.

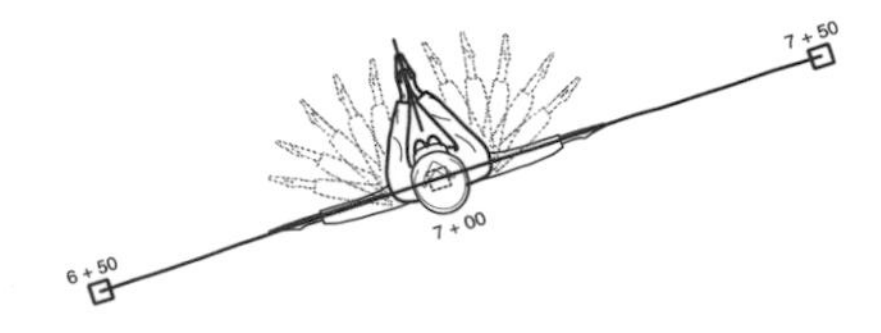

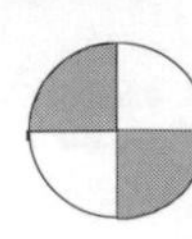

STEP 3

The person at centerline directs the person holding the rod to proceed along the cross-section line with the cloth tape. It is the responsibility of the person holding the rod to select the spots where the slope of the ground changes and an elevation is needed. At the first spot where a break in grade occurs, stops to take a rod reading. Reads the distance on the cloth tape and yells it back to the notekeeper.

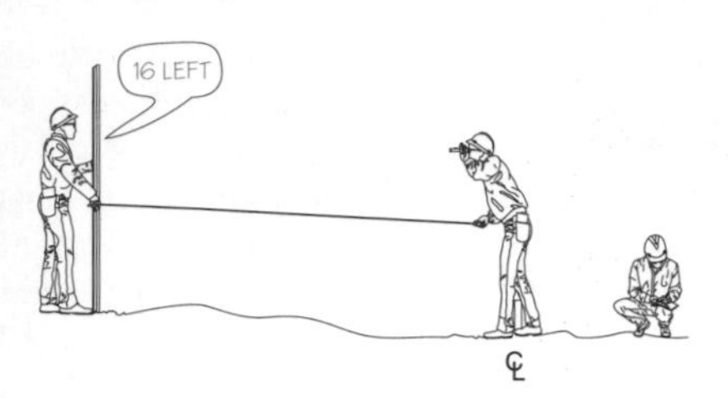

STEP 4

The person at centerline, using a hand level, centers the bubble while looking through the instrument and reads the level rod to the nearest tenth. If a notekeeper is available, states the reading loudly and the notekeeper repeats the reading for confirmation and then records it in the field book.

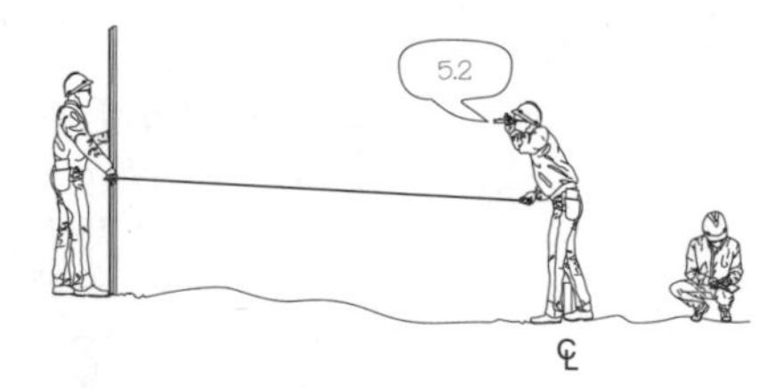

STEP 5

The person holding the rod turns and begins walking on the cross-section line looking for the next change in grade. When it is located, stops, reads the distance out loud to the notekeeper and hold the rod for the reading. The person with the hand level observes the level rod and communicates the rod reading to the notekeeper.

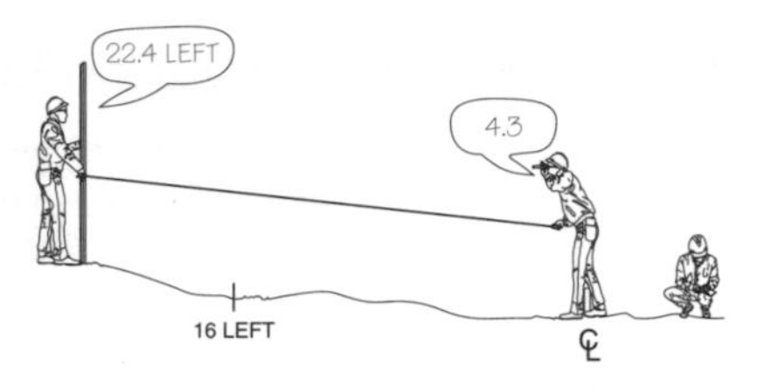

STEP 6

This process is repeated until the required distance out from centerline is met. The person holding the rod then walks back to centerline and proceeds to the other side of centerline to cross-section it. At each spot where the ground changes, a distance is measured, a rod reading is taken, and the data is recorded.

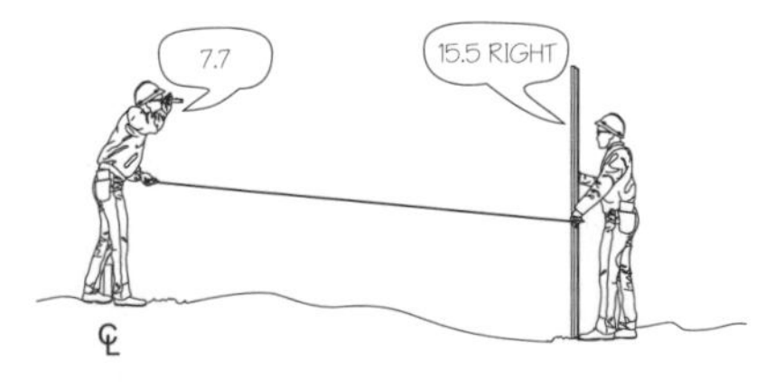

STEP 7

This process is repeated over and over until all centerline stations have been cross-sectioned.

STEP 8

The cross-sections are then taken to the office for plotting and used in the determination of volumes.

PROCEDURE WITH A TOTAL STATION

Using a total station and a data collector for cross-sectioning provides the ultimate in speed of measurement and the ability to almost instantly plot the cross-section as well as determine volumes.. The following illustration shows how the power of the total station can be used to cross-section several stations on the centerline from one setup of the total station. Read the owner's manual for the total station to determine how it can most efficiently be used for collecting cross-section data.

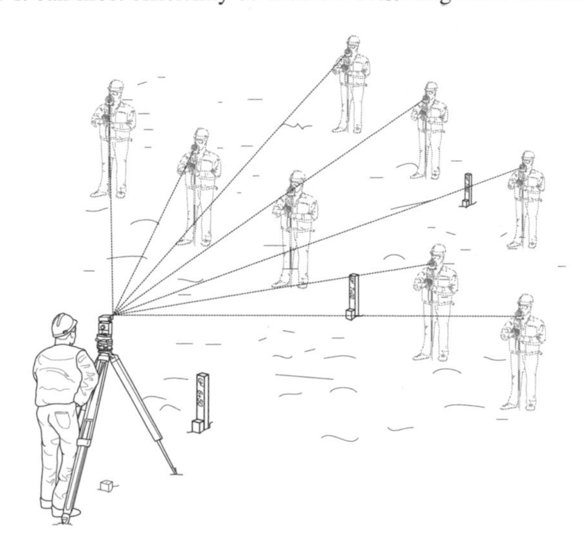

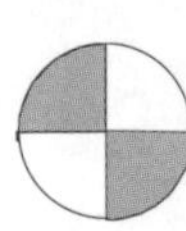

FIELDNOTES

The field notes listed below represent two of many methods of notekeeping that is used for cross-sectioning. See one of the other surveying textbooks listed in *Appendix D* for additional samples.

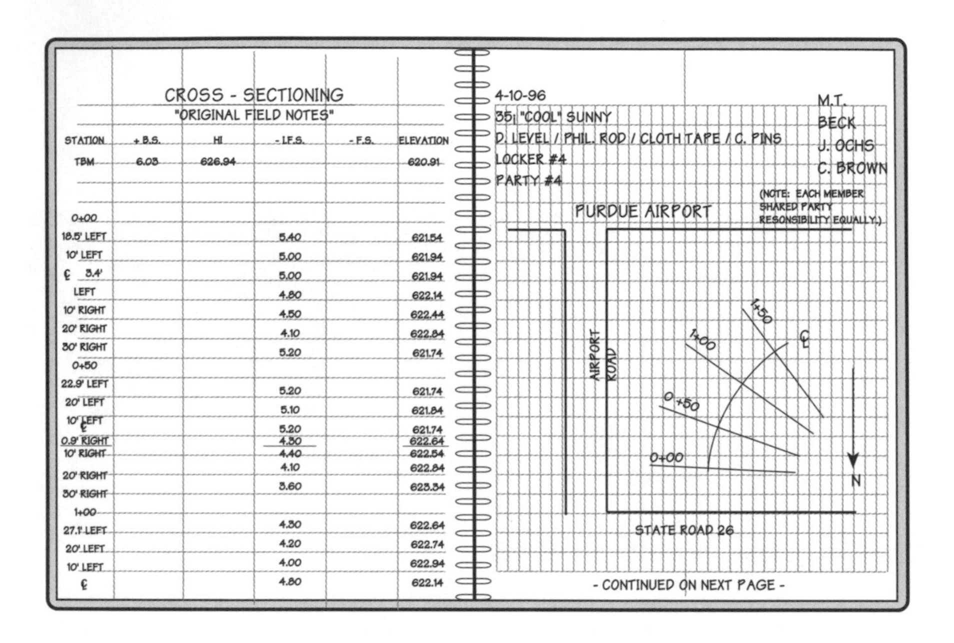

CROSS - SECTIONING
"ORIGINAL FIELD NOTES"

STATION	+ B.S.	HI	- I.F.S.	- F.S.	ELEVATION
TBM	6.03	626.94			620.91
0+00					
18.5' LEFT			5.40		621.54
10' LEFT			5.00		621.94
℄ 3.4'			5.00		621.94
LEFT			4.80		622.14
10' RIGHT			4.50		622.44
20' RIGHT			4.10		622.84
30' RIGHT			5.20		621.74
0+50					
22.9' LEFT			5.20		621.74
20' LEFT			5.10		621.84
10' LEFT ℄			5.20		621.74
0.9' RIGHT			4.30		622.64
10' RIGHT			4.40		622.54
20' RIGHT			4.10		622.84
30' RIGHT			3.60		623.34
1+00					
27.1' LEFT			4.30		622.64
20' LEFT			4.20		622.74
10' LEFT			4.00		622.94
℄			4.80		622.14

- CONTINUED ON NEXT PAGE -

CROSS - SECTIONING
OF RAVINE

STATION	B.S.	HI	F.S.	S.S.	ELEVATION
TBM A	5.6	105.6			100.00

5/15/96
COOL 56°
HAND LEVEL, CLOTH TAPE

W.G CRAWFORD
MATT C.

Station	LEFT			℄	RIGHT		
0+0	101.3 / 4.3 / 25	99.8 / 5.8 / 19	100.4 / 5.2 / 10	100.6 / 5.0 / 0	100.5 / 5.1 / 10	100.9 / 4.7 / 18	101.5 / 4.1 / 30
0+50	101.3 / 4.3 / 25	97.9 / 7.7 / 22	97.0 / 8.6 / 15	97.1 / 8.5 / 0	97.9 / 7.7 / 10	99.3 / 6.3 / 17	101.2 / 4.4 / 29
1+0	100.3 / 5.3 / 28	94.1 / 11.5 / 21	93.0 / 12.6 / 15	92.9 / 12.7 / 0	94.3 / 11.3 / 15	96.5 / 9.1 / 19	100.8 / 4.8 / 28
1+50	99.1 / 6.5 / 25	91.3 / 14.3 / 15	90.1 / 15.5 / 10	89.9 / 15.7 / 0	92.9 / 12.7 / 10	95.3 / 10.3 / 15	99.5 / 6.1 / 25

- CONTINUED ON NEXT PAGE -

CONSTRUCTION LAYOUT TECHNIQUES

23

PART 4 - CONSTRUCTION LAYOUT APPLICATIONS

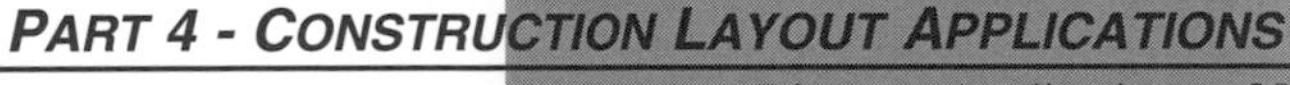

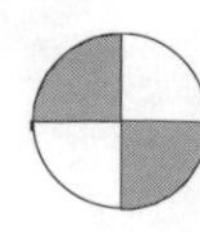

CHAPTER OBJECTIVES:

Outline the activites that must be considered when planning to perform construction layout.
Identify and discuss the various methods of laying out that can be used by the field engineer.
Explain how the field engineer can establish points that are permanent.
Define radial layout.
Describe the step-by-step process of laying out points with a top-mount EDM using radial layout techniques.
Describe the most effective method of checking radial layout.

INDEX

OVERVIEW

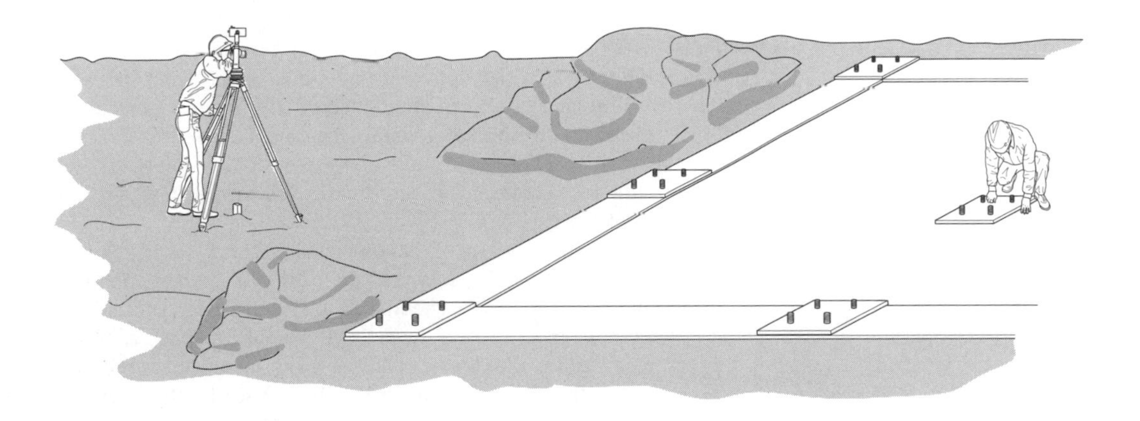

SCOPE

Generally speaking, there are no procedure patterns for how layout is to be accomplished. For each project, you will need to plan how the job is to be carried out. Layout of a residential structure is different yet similar to the layout of a commercial structure. Layout of a waste water treatment plant is different yet similar to the layout of a bridge. Field engineers should be familiar with the various methods of layout and be able to choose the best method for the job. This chapter will introduce fundamental principles of building layout that can be applied to all types of projects.

APPX. D

The material in the overview has been written in reference to Measuring Practice on the Building Site. See the bibliography in *Appendix D* for more information.

BEFORE THE LAYOUT

EXAMINE DRAWINGS

When preparing for laying out a new job it is necessary to thoroughly study all drawings. While doing this, you can form an idea of the kind of project which you will be laying out. Considerations include type of project, size of project, number of buildings, method of construction, etc.

Check the dimensions stated on the drawings and points for which there are coordinates, e.g., the lengths of the buildings between the corners and the relative positions of the buildings.

CHECK RESOURCES

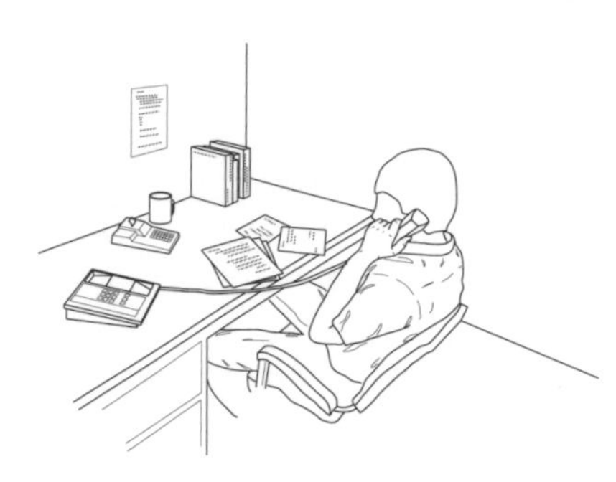

Find out what subcontractors and others will lay out and what will be laid out by field engineers. If you are to locate the position of the building, obtain the data necessary such as the coordinates for the corners and the reference points. Check with the architect for coordinate information that may be available on CAD.

Contact the surveyor who performed the property survey of the site to obtain a report of the survey, if available. Perhaps the surveyor already established a coordinate system that you can use for your work. Investigate whether there are any local restrictions about establishing monuments on streets or sidewalks. Check standards which specify the accuracy requirements for the layout.

EXAMINE THE SITE CONDITIONS

By reviewing the drawings and checking with resources, you will now have learned so much about the project you can now go and look at the site. You can form an idea of the natural obstacles on the site and the kinds of monuments and targets that can be used, etc. Make sketches and notes of all that you see. If the scope of the project calls for the layout of buildings, storage areas, transport roads, etc., you should think ahead about how you would like to place your control so you can prevent unnecessary obstacles to layout.

After you have collected sufficient information concerning the project and the required field engineering work, you can assess the project's scope. Decide how the layout is to be performed, the instruments required for the work, and the number of people required for the field crews, etc.

PLAN THE CONTROL

Grids, baselines, or control networks can be established as the control system for the project. The nature and scope of the project are main considerations on how to proceed. If only a single building or bridge has to be laid out, it is possible that only a baseline will be needed. If the project consists of multiple buildings or multiple bridges, a primary control network interconnecting the control points will be necessary. The primary control traverse can then be measured and computations performed to determine the quality of the fieldwork.

By deciding at an early stage in the project how you will lay out the structural grid, base lines, or primary control, you will be prepared for any eventual field engineering concerns during the construction of the project.

ACCURACY CONSIDERATIONS

Before any lay out is undertaken, instruments to be used must be checked for proper calibration. Follow the step-by-step procedures outlined in *Chapter 12, Equipment Calibration*, to check your equipment. The field engineer should always use correct instrument techniques which minimize the effects of instrument errors. Do not forget to read the operating manual for the instrument which is being used. Layout should always be arranged to be self-checking as much as possible. For example, leveling should always be checked into a reliable benchmark and layouts should be checked against other control points on the site. In addition, it is always advisable to give the layout a visual check to be sure it "looks right."

SCHEDULE THE LAYOUT WORK

Develop a layout schedule and determine the construction method which will be used during the various stages of the layout. It is vital that points should be ready when needed. The control points-building corners, etc.-must be set before actual construction begins. Once the excavators are on the site and begin work, it may be difficult to perform the layout. When you have drawn up the layout schedule, you must make a detailed plan of each principal stage in this schedule. As a field engineer, you must be flexible and be prepared to alter these plans as construction proceeds and when you are in a better position to decide what kind of layout will be most suitable.

LAY OUT METHODS

Layout work usually will vary from project to project. The order in which layout work is described here is only an example. If, for instance, the buildings are laid out from primary control, then it may be best to determine the corner points by radial layout, intersection, or baseline offset.

RADIAL

7-16

Radial layout is based on a known point and a known direction. With the theodolite over the known point, the calculated direction and distance are measured and the point is laid out and staked. For more information about Radial Layout reference *Chapter 7, Distance Measurement - EDM.*

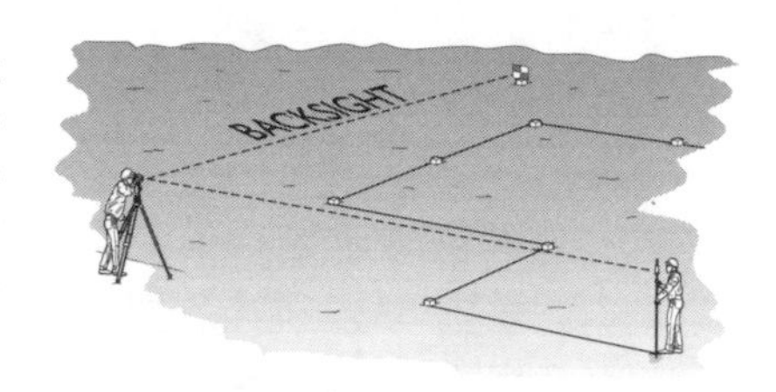

INTERSECTION

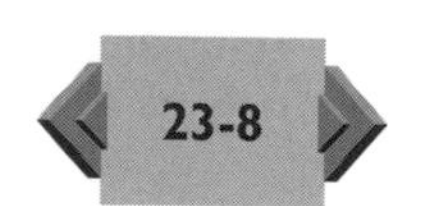

When the method of intersection is used, it is best to use two theodolites, each set up over a suitable known point. The directions towards the structure point are laid out from these points, and the position of the point is determined by the intersection of the two lines of sight. Reference the next section, *Setting Points by Intersecting Lines (23-8).*

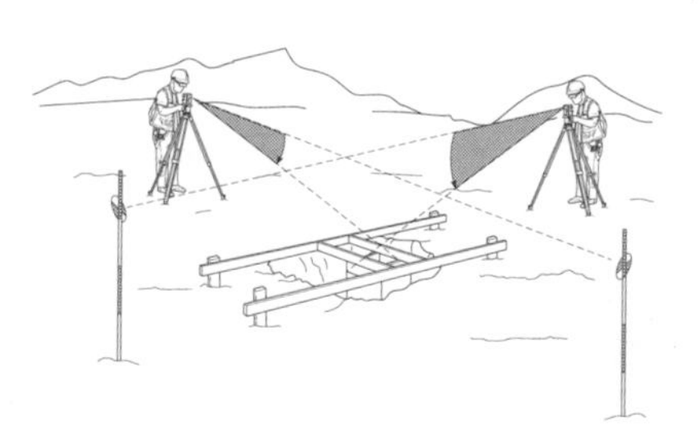

BASELINE / OFFSET

When the baseline offset method is used, work is based on a straight line between two known points, i.e., two points on the baseline or two boundary posts, where the line of sight is unobstructed. The distance along the baseline or boundary line from one of the points to a point at a right angle to the required corner point is determined, and this distance is laid off.

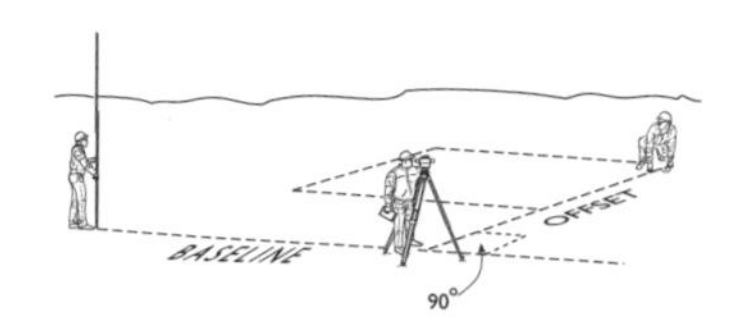

A theodolite should be used to measure the perpendicular angle and the straight line to the required point. Reference *Chapter 23, Baseline Offset* (23-12) for more information.

When the corner points have been laid out, check the work by measuring the lengths of the sides, the diagonal between the corner points, and the relation to other important reference points! Never forget the necessary chain corrections! The initial layout for site clearance and excavation can now be done. Base this work on the corner points which have already been laid out.

ESTABLISHING PERMANENT CONTROL POINTS

Since the corner points will disappear in the course of excavation make sure their positions are tied in so they can be located again. Consider using batterboards to reference the lines and to work from. Position permanent control points in such a way that they always can be checked and used throughout the whole building process. Always provide a redundant number of permanent points. The points must be of a permanent nature. Where rock is not available, the marks must be constructed in concrete at a frost-free depth.

ESTABLISHING BENCHMARKS

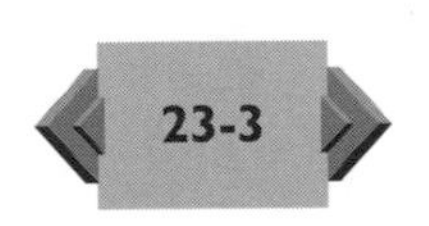

Establish some benchmarks for use in leveling later on. Keep the layout plan in mind when selecting the positions of these marks and remember that the distance between the level and the rod should not exceed 150 feet. Reference *Chapter 22, Leveling Applications, Setting Benchmarks (23-3).*

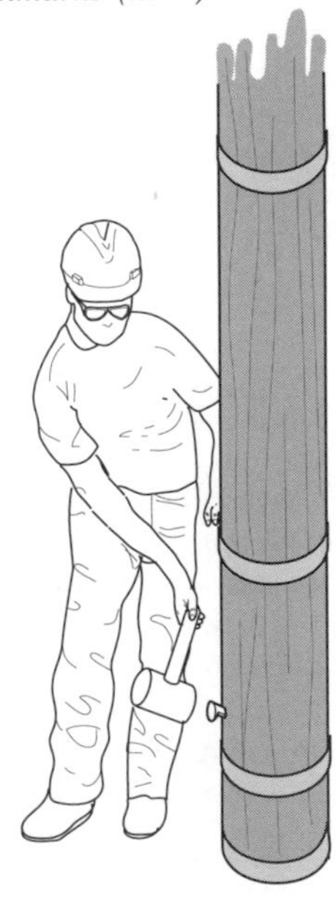

SETTING POINTS BY INTERSECTING LINES

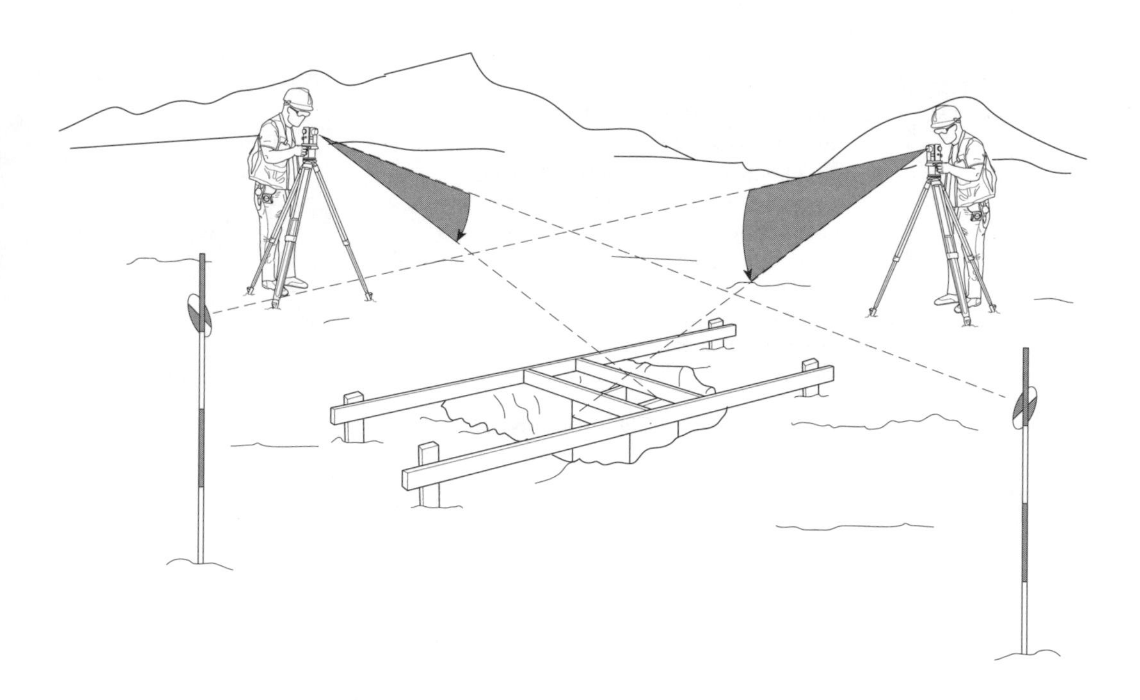

SCOPE

In building construction, setting points can be accomplished by measuring distances with a chain, measuring distances with an EDM, or by intersecting the line of sight of theodolites. Many field engineers favor the last method because it is quick and accurate and is preferable over stretching a chain which may require temperature, length, or slope corrections. It is even preferred over the EDM because no measurements are taken that need to be checked. If the field engineer carefully designs control points and targets so they surround the project site, intersecting lines with a theodolite works very well in controlling the layout of the structure.

PLAN THE CONTROL

Planning is the key to being able to locate points by intersecting lines. A building site must have control completely around it and targets must be placed to create an efficient method of creating the line of sight. It takes time in the initial layout of the project to accomplish this, but the efficiency of layout will be worth the effort.

REVIEW THE PLANS

Determine first if intersecting lines will work for the project. Projects most suitable for this method are those that are rectangular or square and have columns that basically line up. Projects that have numerous walls and unusual dimensions to columns, or have curved lines, etc., are adaptable to this method of layout, but will require more initial layout work.

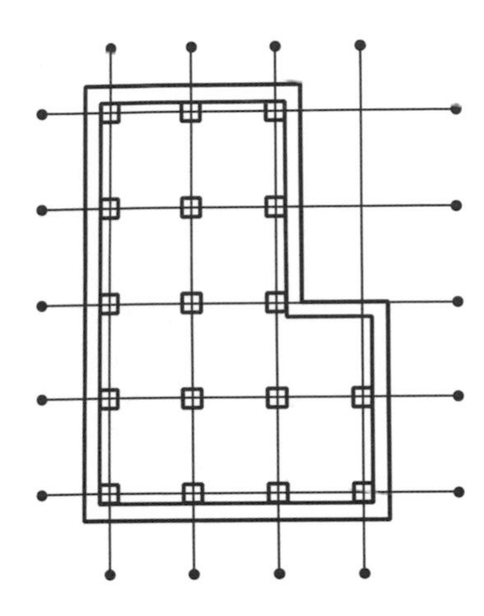

LOCATE CONTROL POINTS

Surrounding the site and locating control points on every column line will require that more points be placed and monumented. The field engineer will have to communicate closely with the superintendent and the crafts to ensure that points are located outside the limits of construction and material storage areas.

ESTABLISH TARGETS

Having a backsight available on each line is the key to efficient use of intersection lines layout. Place targets on hubs, adjacent buildings, sidewalks, etc. Paint or use sticker targets. Improvise targets by placing a nail in the top of a hub and placing surveying ribbon behind it.

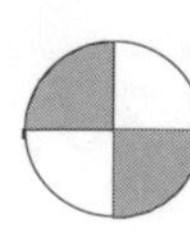

PROCEDURE

Assuming that control has been established around the site and that targets are available, use the following method to locate a needed point.

EXAMPLE

The illustration at the right is for a simple building. Control points and target locations have been established and are indicated. Exact location of an anchor bolt template at B2 has been requested by the superintendent. Two instruments and two field engineers are available. A carpenter has located the template approximately but needs the exact location from the engineers.

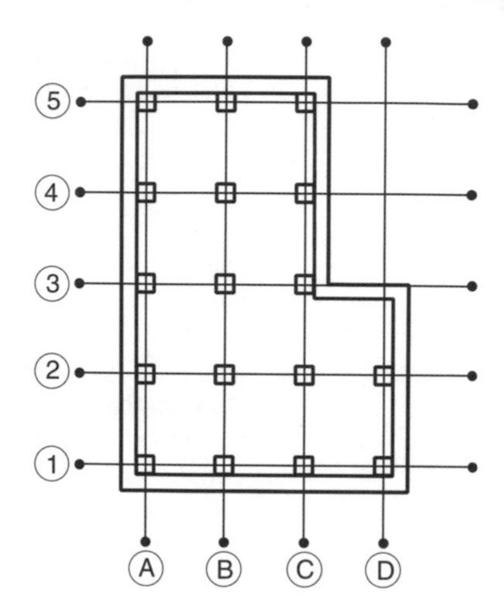

LAYOUT BY INTERSECTING LINES

STEP 1

One field engineer sets up an instrument on "B" line. Backsights onto a "B" line target across the building site. Signals to the carpenter at the anchor bolt location to mark on the template the "b" line. A point is marked on one side and then the other, and a straight edge is used to connect the points with a line.

STEP 2

The other field engineer sets up an instrument on "2" line. Backsights onto a "2" line target across the building site. Signals to the carpenter at the anchor bolt location to mark on the template the "2" line. A point is marked on one side and then the other and a straight edge is used to connect the points with a line

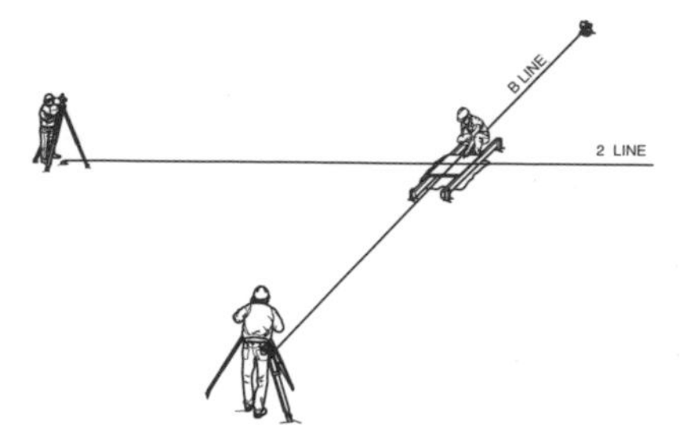

STEP 3

The intersection of "B" line and "2" line marks the center of the template and is indicated on the template. These lines can now be used by the carpenter from which to measure and locate the position of the anchor bolts.

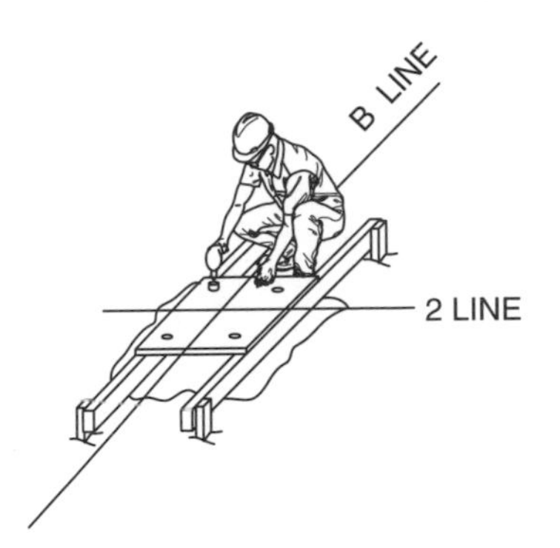

STEP 4

Repeat this procedure for any point that needs to be located within the structure.

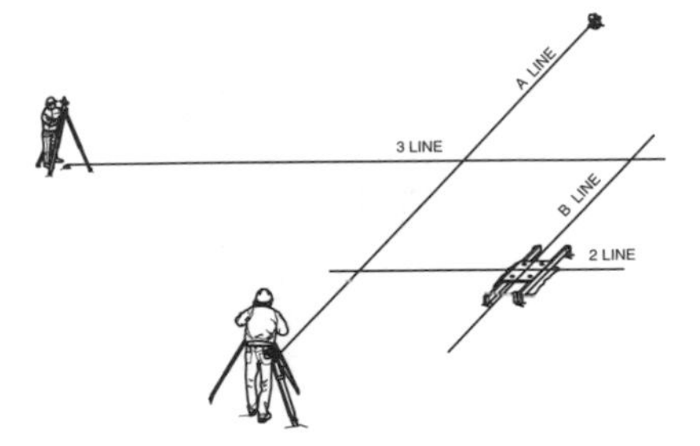

STEP 5

Check the layout of the anchor bolts before the concrete is placed.

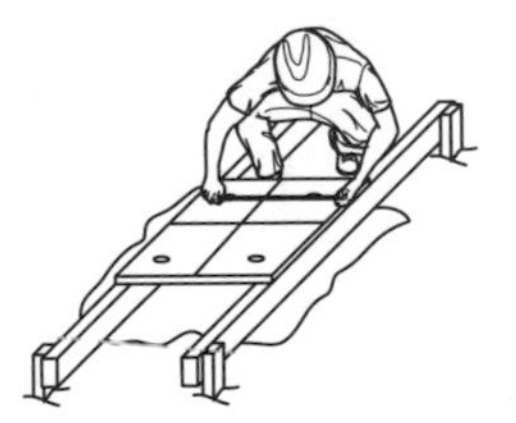

SUMMARY

It can be seen that intersecting lines is a simple process. If a field engineer takes the time to establish control points and targets that surround the site, it can be the best method of locating points.

BASELINE OFFSET

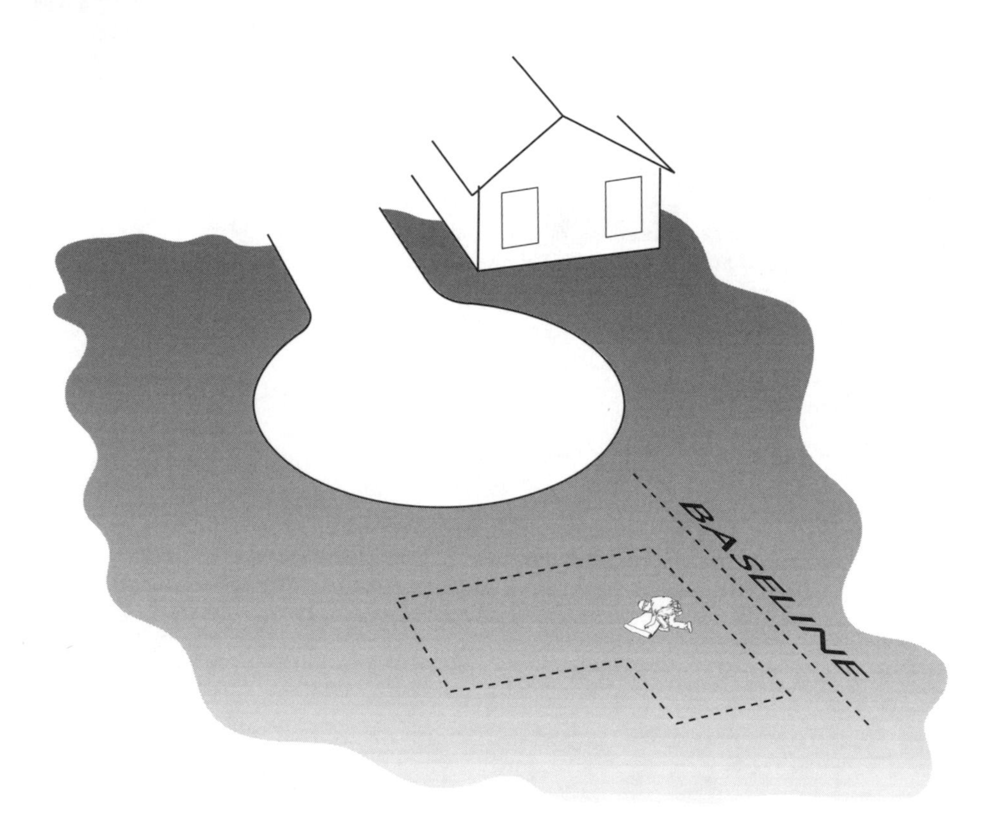

SCOPE

This method of layout is used for residences and commercial and industrial structures. It doesn't matter what the size or shape of the structure, the method of establishing a baseline and turning a 90° angle and measuring distances to locate points is exactly the same. More points have to be located with the larger and more complex structures. Usually several instrument setups will be required to complete the layout and to check the points. Care must be taken in turning the angles and measuring the distances so the building will be square and the proper dimensions. The instrument used may be a builder's transit/level, a transit, an optical theodolite, a digital theodolite, or a total station. If an EDM is used, simply measure the distances with it, rather than using a chain as described on the following pages. However, if an EDM is used, it expected that radial layout would be the method of choice.

BASELINE OFFSET WITH AN INSTRUMENT AND A CHAIN

STEP 1

Place a hub and tack at one of the front corners of the building to use as a starting point of the layout baseline. Set up an instrument over the point and sight in the direction of the other front corner of the building.

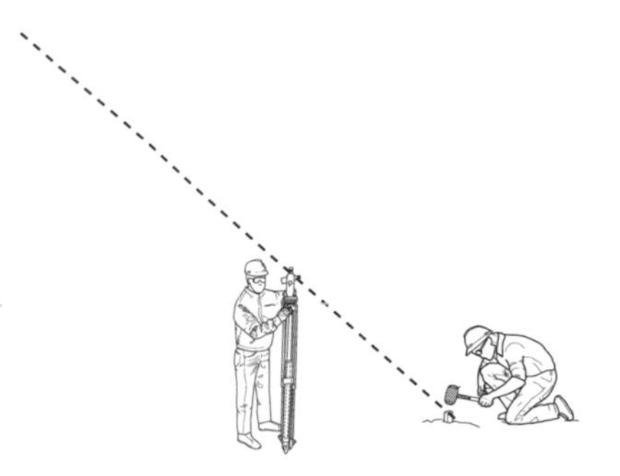

STEP 2

Chain the prescribed distance to the other front corner and set a hub and tack, This represents the other end of the baseline.

STEP 3

Set zero degrees on the instrument and turn 90° in the direction of the adjacent corner of the building. Measure the prescribed distance from the plans and set a hub and tack exactly on line and distance.

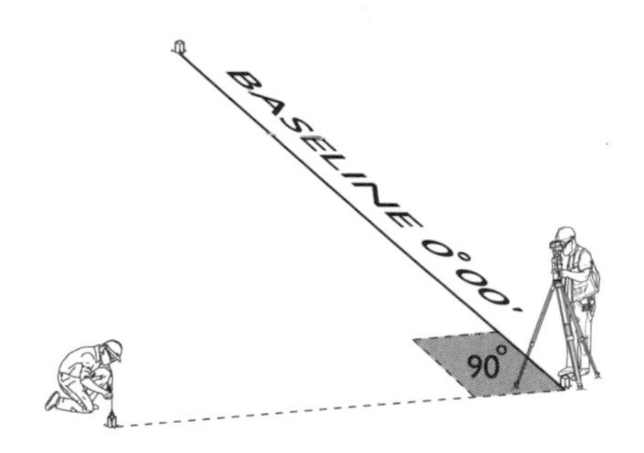

STEP 4

Move the instrument to the other end of the baseline and set it up over the point. Sight the other end of the baseline with zero on the instrument and turn 90° toward the adjacent corner of the building. Measure the prescribed plan distance and set a hub and tack.

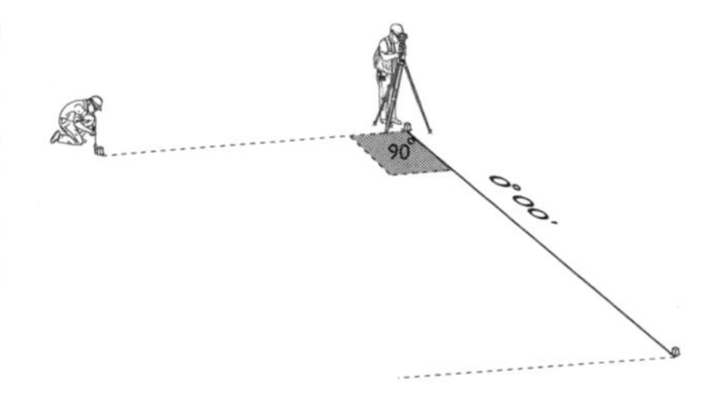

STEP 5

Measure along the baseline whatever distance is needed for setting the instrument up to turn an angle and set any interior points or corners on the structure.

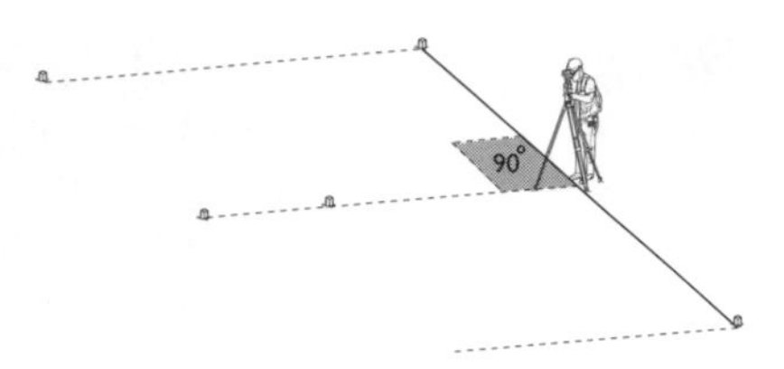

STEP 6

Measure the diagonals to determine if the building is square. If the diagonals are equal and the planned dimensions have been established, the building is square. If the diagonals are not equal, the angles can be turned again or the 3/4/5 can be used at each of the corners to determine which is not square.

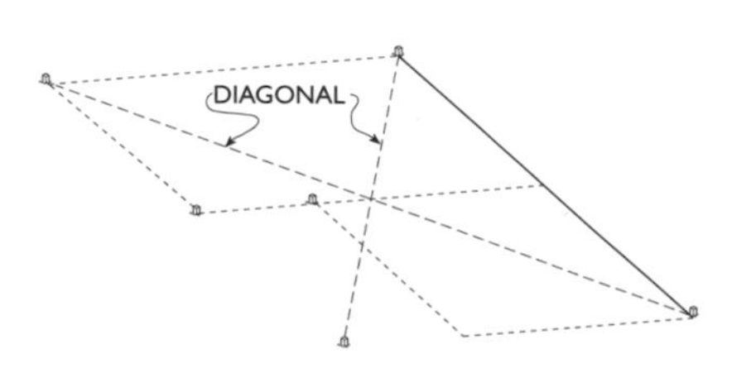

STEP 7

Set batterboards at each corner and stretch strings or wire. Measure the distances and diagonals between the string intersections as a final confirmation that the building is the proper size and is square.

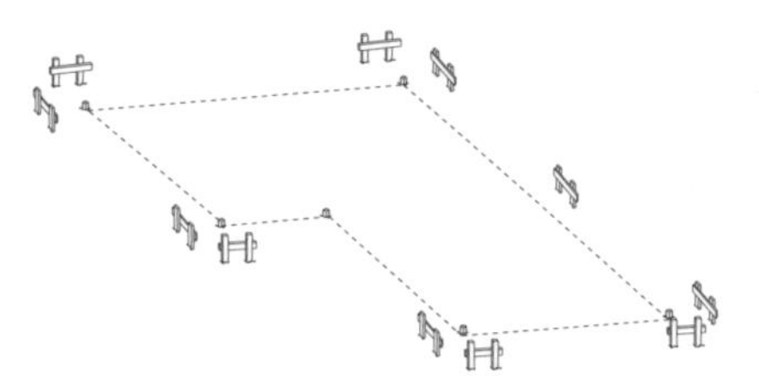

STEP 8

Inform the construction crews the building is ready for excavation.

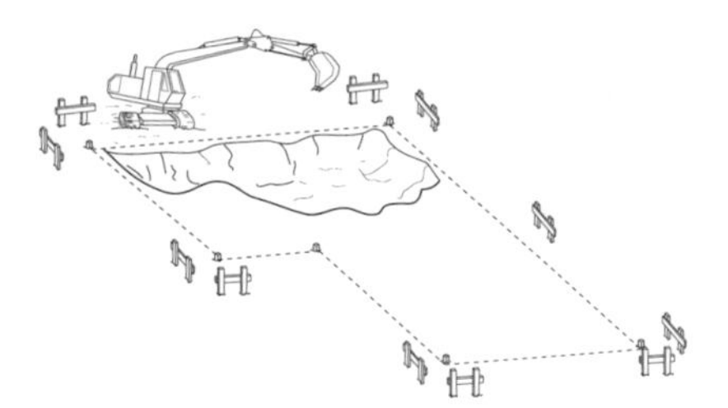

SUMMARY

It has been shown that the baseline offset method of layout is simple and straightforward. If the field engineer can establish the baseline in a location that is safe from construction activity, it will be able to be used throughout the project. For long-term use, baselines should be well monumented and defined by at least four points on the baseline. The field engineer should reference all baseline points so they can be re-established if necessary.

BATTERBOARDS

SCOPE

Anyone who has seen a construction site has undoubtedly seen batterboards. They are often the first sign that construction is going to take place; they also indicate exactly where the project is to be built. Batterboards are used by a variety of crafts to control the excavation of the foundation, locate the footings, align the masonry block, or build the walls. Although location or alignment is the primary function of batterboards, a reference elevation to the top of a footing or the finish floor of the building is also often placed on the batterboard. Batterboards may be set by field engineers, professional surveyors, carpenters, masonry crews, excavators, etc. Regardless of who sets the batterboards, certain principles should be followed.

GENERAL PURPOSE

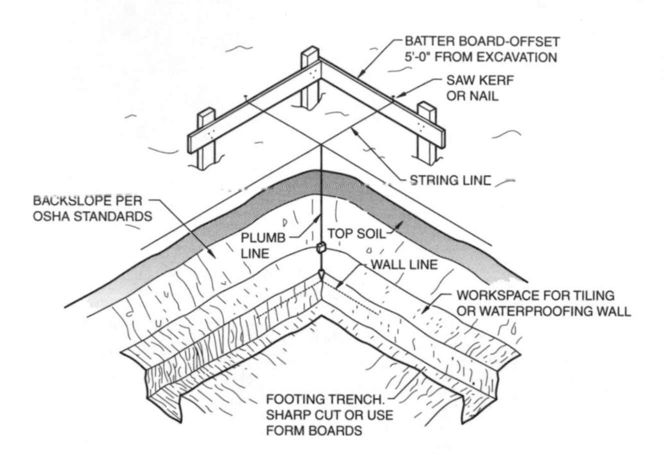

The primary function of batterboards is to establish a reference line. For a residential structure the reference line is typically the outside face of the foundation. For pipelines, the center of the pipe is located. For column footings, the center of the footing is generally located. No matter what the batterboard is referenced to, it is serving the same purpose in all situations. The batterboard is built so a semi-permanent control line will be available as needed during construction.

Batterboards must come in pairs. That is, where there is one there must be another. Two batterboards are required to stretch a string or wire between or to sight between and create the reference line.

TYPES OF BATTERBOARDS

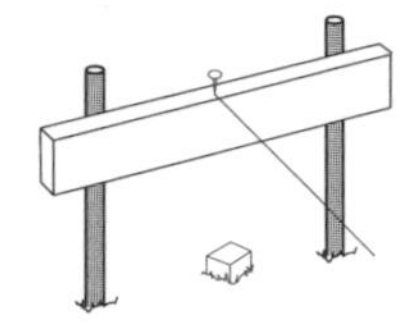

Batterboards are built in many ways. Most often, the construction of the batterboard is an afterthought and the nearest thing available is used. Batterboards can be built from new or scrap lumber, rebar, wooden posts; just about anything will work. The following illustrations show a few examples of the construction of batterboards.

LUMBER

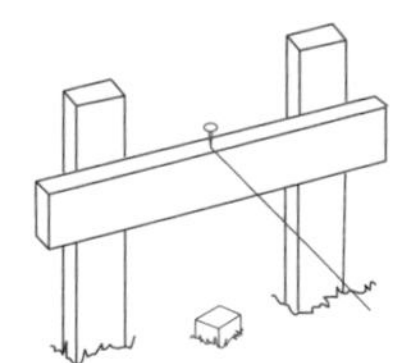

When using lumber, 2 x 4s are usually used as the stakes that are driven into the ground and a 2x4 or 1x4 is used as the batterboard. If the soil is soft and the stakes are not firm in the ground, it is a good practice to drive additional 2x4s to support the initial batterboard. It is not uncommon for the stakes to pull out of the ground when wire is being used and it is pulled tight.

REBAR AND LUMBER

Sometimes rebar is used as the stakes to support the batterboard. The batterboard is attached to the rebar by wire or by drilling a hole in the rebar and using nails or screws. When the ground is very hard, rebar is much easier to drive into the ground than a 2x4.

PORTABLE AND REUSABLE

In an effort to cut wood costs and to be able to recycle the wood, a field engineer might choose to build batterboards as shown in the illustration. To use these batterboards, simply use form stakes or rebar and solidly attach them to the ground. Drive a nail on line and attach a string or wire and they are ready for use. Since moving and storing them is a consideration, they might be the best option when a series of houses is being built in a subdivision.

USES OF BATTERBOARDS

BUILDINGS

The predominant use of batterboards is in building construction. They are used when building garages, houses, gas stations, restaurants, schools, factories, and many other types of structures. Any time constant alignment is needed, using batterboards is the method of choice by most builders. For a simple building, batterboards might be installed as shown in the illustration.

HIGHWAY

Batterboards are of limited use on highway projects. The alignment of the roadway is typically maintained from stakes in the ground. However, anytime a culvert, abutment, pier, etc. is constructed, batterboards may be used constantly by the crafts.

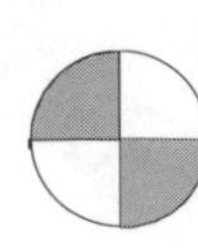

PIPELINE

Although pipeline lasers are the primary method of controlling line and grade of pipelines, batterboards are still used in some instances. The illustration shows a typical use of batterboards for pipeline construction.

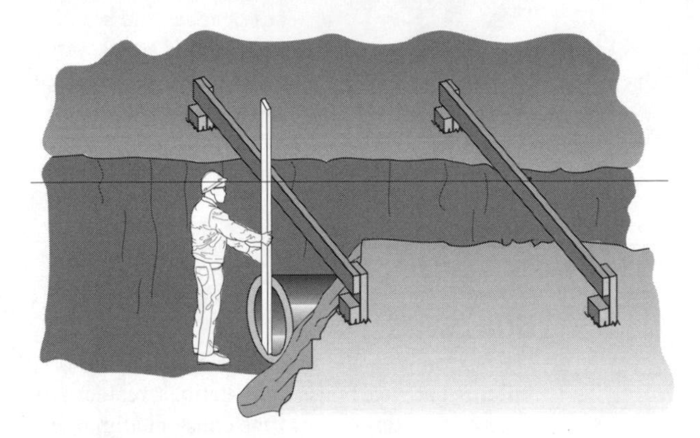

ESTABLISHING LINE ON BATTERBOARDS

Line on batterboards is generally established after the building line has been determined. For instance, if the corners of a building have been set, an instrument is set at one of the corners and a sight is taken on the other corner. The telescope is sighted onto one batterboard and a nail is driven on line or a saw kerf is cut. The telescope is then plunged and a nail or kerf is placed on the other batterboard. The instrument is then moved from the corner point and a string is stretched between the batterboards.

SETTING AN ELEVATION ON BATTERBOARDS

In building construction it is often desirable to reference the height of the batterboard to a reference elevation on the structure, usually the elevation of finished floor. The batterboard is then set at that elevation.

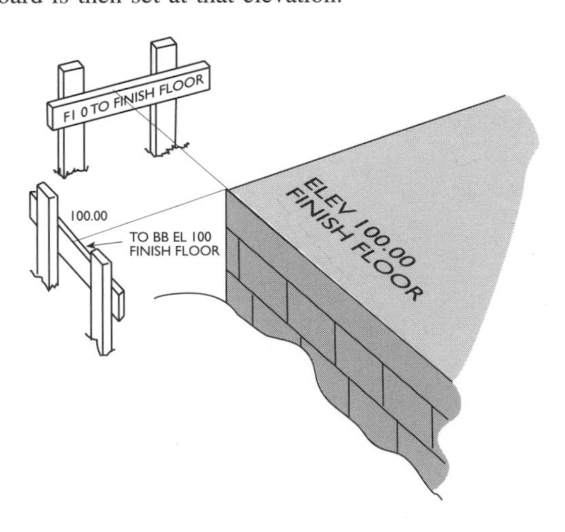

USING THE 3/4/5

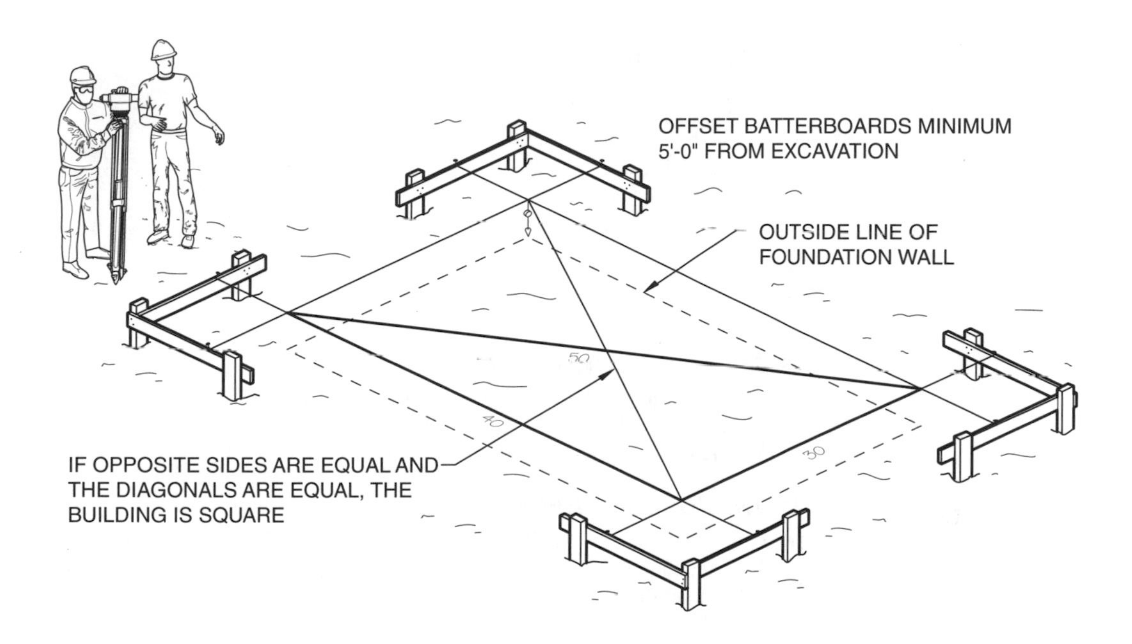

SCOPE

Since the days of the ancient builders of Greece and Egypt, the 3/4/5 relationship has been used on construction sites to establish or to check perpendicular lines. Today's field engineer follows in the footsteps of the ancient builders whenever this relationship is used. Field engineers in all types of construction-residential, commercial, industrial, etc.-find the 3/4/5 useful for quick checks to confirm that a true 90° angle has been established even when using the most precise surveying equipment. The carpenters on projects use this relationship constantly to square walls and even to plumb walls. In construction, the 3/4/5 is probably the most used of all geometry principles.

GENERAL

By using common geometric and trigonometric formulas, the fact that the 3/4/5 is a right angle can be proven. A mathematics text should be referenced for more information.

MULTIPLES OF THE 3/4/5

To use other triangles similar to the 3/4/5, simply multiply each side by the same constant. Some examples are shown

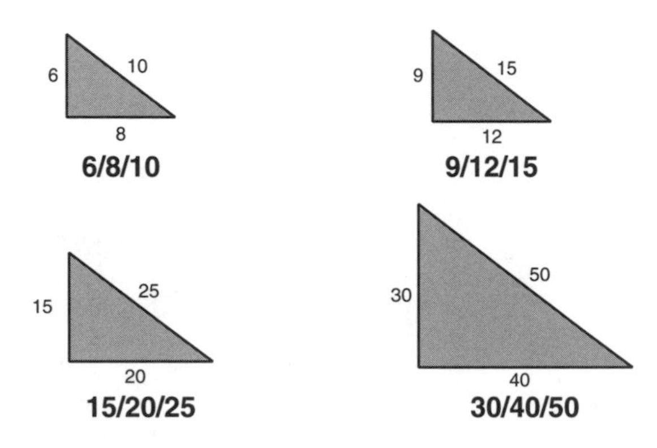

OTHER TRIANGLE COMBINATIONS

If circumstances warrant it, there are other combinations of integers that can be used to obtain a right angle triangle. They include: 5/12/13, 8/15/17, 10/24/26, etc. Undoubtedly, there are many more.

THE PYTHAGOREAN THEOREM

If the field engineer wants to be able to use any distances to layout or check a right angle, the Pythagorean theorem is the tool to use. If the sides of a building are 26.5 and 32.5, inputting these numbers into the Pythagorean formula will yield an hypotenuse of 41.93. The field engineer can use this distance to be sure a right angle exists. Recall:

$$c^2 = a^2 + b^2$$

c b a

3/4/5 MEASUREMENT REQUIREMENT

Using the 3/4/5 is a very simple process. However, to achieve the expected results the measurements must be exact. That is, follow proper chaining techniques carefully. Any error in the measurement of the distances will result in not establishing a right angle as desired. Or if an established 90° angle is being checked, inaccurate measurement may cause the field engineer to make an adjustment that is unnecessary.

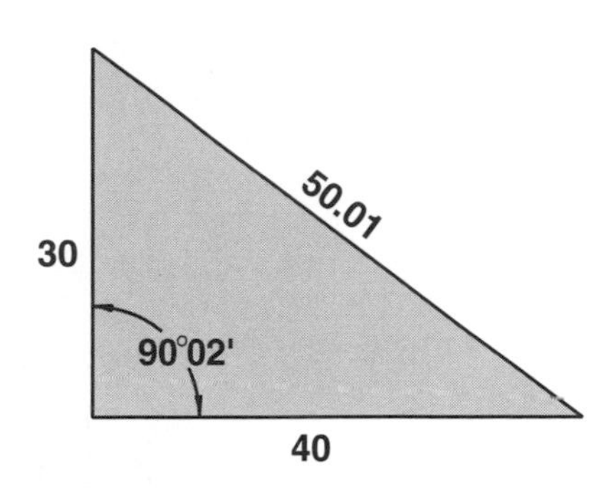

If using the 30/40/50 feet triangle, calculations show that being off just 0.01 in the measurement of the 50' hypotenuse results in an angular error of over two minutes of arc. In other words, the planned right angle will be 89°58' or 90°02' depending on whether the distance is 49.99 or 50.01. If shorter distances such as 3, 4, and 5 feet are used, an error of 0.01 when measuring the 5 foot hypotenuse results in an angular error of almost 15 minutes! It cannot be stressed enough that exact measurements are necessary to get desired results when using the 3/4/5 method of laying out or checking a 90° angle.

COMMON USES OF THE 3/4/5 IN CONSTRUCTION

The field engineer can use the 3/4/5 anytime a 90° angle is to be laid out or a 90° angle is being checked.

LAYOUT

The 3/4/5 can be used when laying out the corners of a building, offsetting a point at 90° to a highway centerline, and squaring up footing formwork.

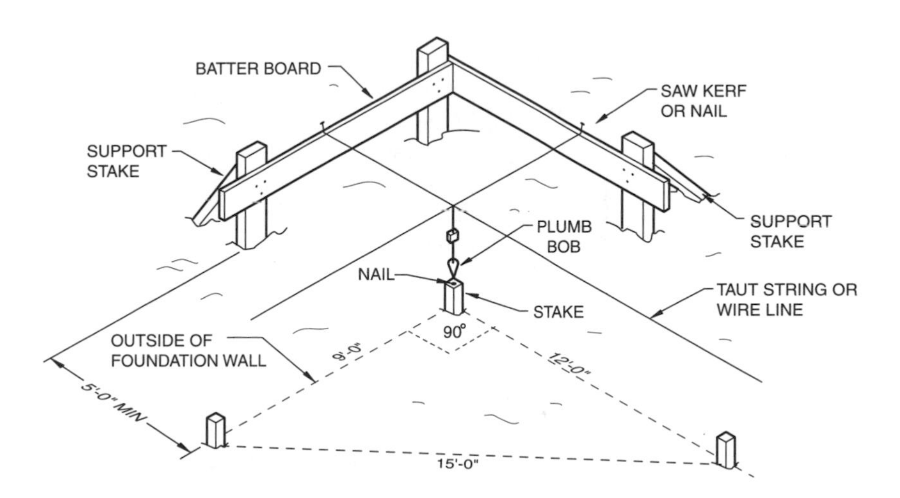

Laying out the corners of a 30' x 40' building can be done using 3/4/5. The method shown requires 3 persons to hold a single 100' chain at the points.

Establishing a 90 degree using the 3/4/5		
1	Determine the location of the point and the baseline off of which a 90° line is to be established.	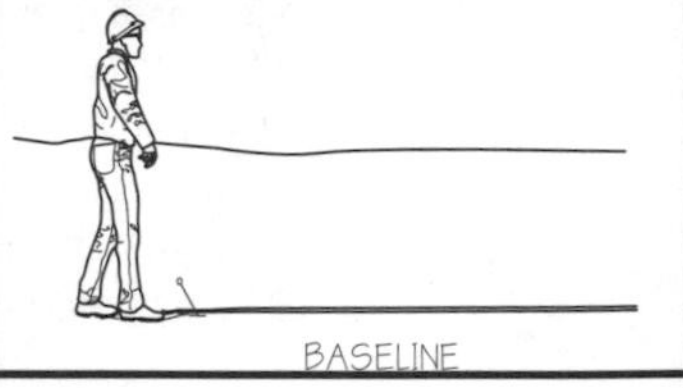 BASELINE
2	Measure 30 feet along this line, putting a hub and tack exactly on the line and distance.	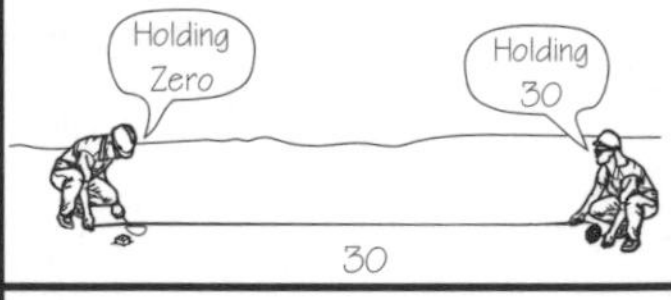 Holding Zero Holding 30 30
3	One person holds zero on the original point while another holds 100 on the point 30 feet away. The remaining field engineer will take the chain and will loop and hold 40' and 50' together. The chain must be pulled tight in two directions at the same time. The exact intersection of the 40' and 50' is 90° to the baseline.	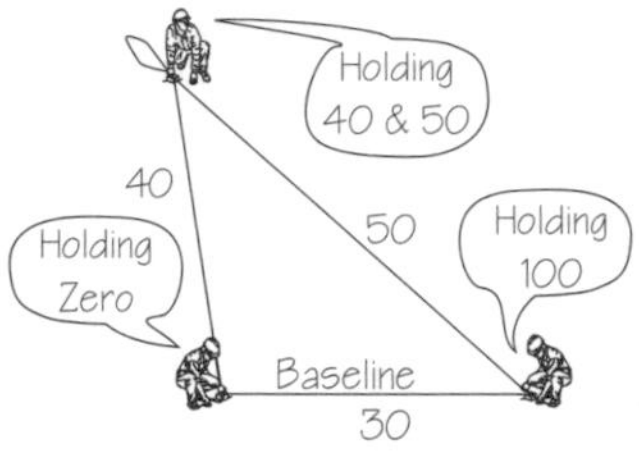 Holding 40 & 50 40 50 Holding Zero Holding 100 Baseline 30
4	A stake and tack are set and the measurements are checked to confirm that a right angle was established.	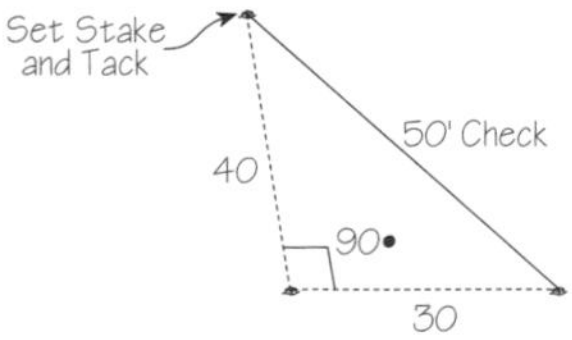 Set Stake and Tack 50' Check 40 90• 30
5	Repeat this process for the other corners of the building.	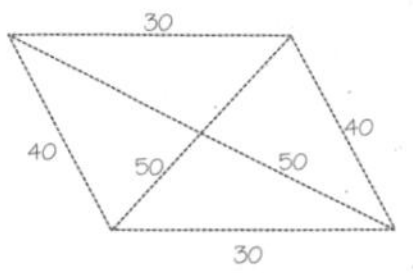 30 40 50 50 40 30

CHECKING

Checking to see if the corners of a building are square can be done by following this method. It requires 2 people to perform.

Checking a 90 degree using 3/4/5		
1	Determine the location of the point where the 90° is to be checked	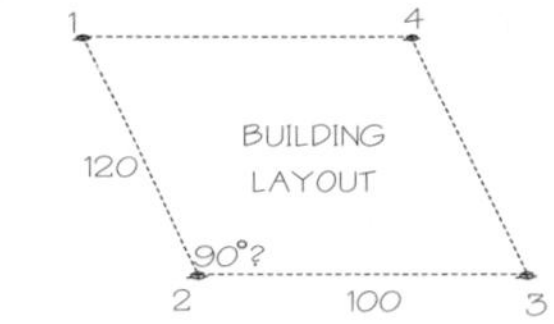 1 4 BUILDING LAYOUT 120 90°? 2 100 3
2	Measure 30 feet along one of the lines, putting a hub and tack on the line at exactly 30 feet.	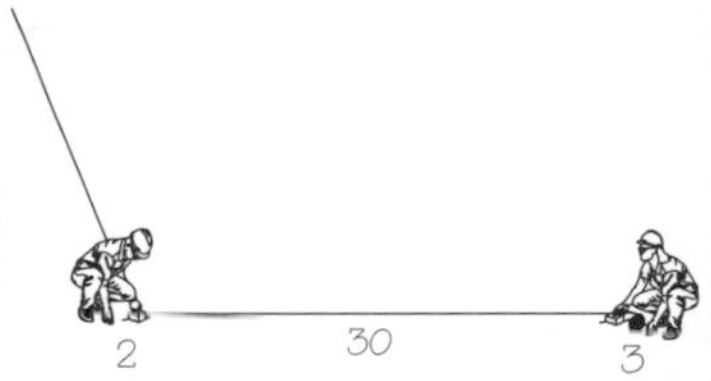 2 30 3
3	Measure 40 feet along the other line and put a hub and tack exactly on the line at 40 feet.	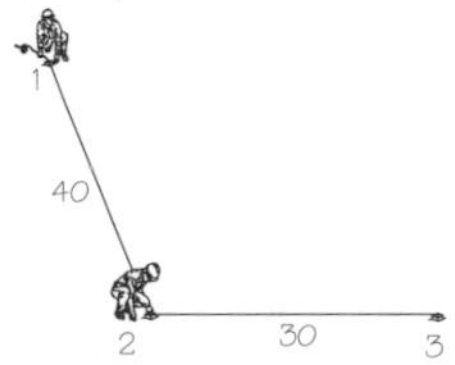 1 40 2 30 3
4	Measure the distance between the two hubs. If a true 90° angle exists, the measurement will be exactly 50 feet.	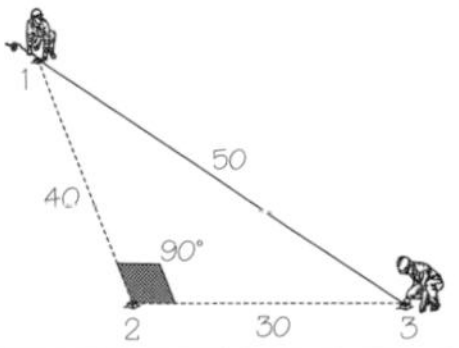 1 50 40 90° 2 30 3
5	If the measurement is not exactly 50 feet, the angle is not 90°. The direction of one of the lines or the corner point must be adjusted until a right angle exists.	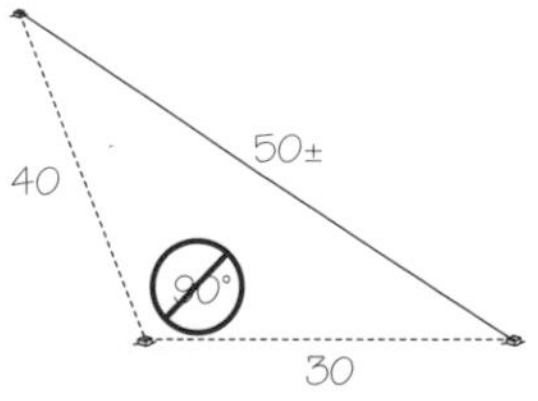 40 50± 90° 30

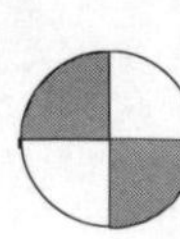

Construction Control and Layout

24

Part 4 - Construction Layout Applications

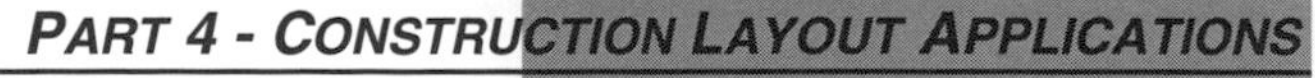

CHAPTER OBJECTIVES:

Describe the role of the field engineer in establishing primary and secondary control.
Identify the steps involved in planning project control.
Distinguish between primary, secondary, and working control.
Discuss how points are set and kept on the construction site.
Discuss the establishment of baselines for primary horizontal control.
Discuss the establishment of vertical control on the construction site.
Describe the methods that are used to layout and control sitework, retaining systems, caissons, footings and foundations, anchor bolts, structural steel, structural concrete, embedded items, concrete columns, elevator and stair cores, and sidewalks, curbs, and gutters.

INDEX

OVERVIEW

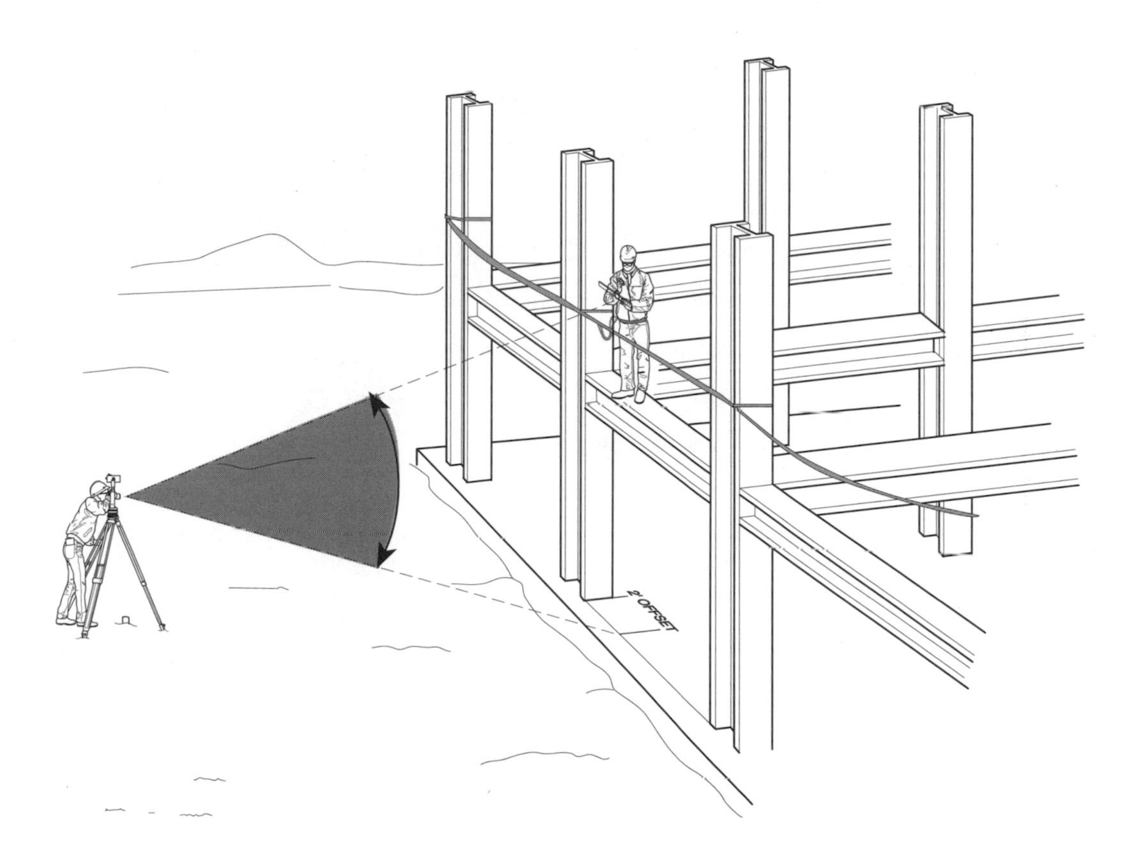

SCOPE

All chapters in this book have been written to prepare the field engineer to perform surveying and layout tasks on the jobsite. This chapter is different in that it assumes the field engineer has mastered those skills and is ready to apply them to construction control and layout. Although the principles and practices presented here are directed toward building construction, they can be applied to any type of project; bridge construction, route surveying, water treatment plants, etc.

CHAPTER ORGANIZATION

This chapter is presented in the order that a project might be surveyed and laid out by the field engineer. Section topics include:

Establishing Primary and Secondary Control

Sitework

Retaining Systems

Caissons

Footings and Foundations

Anchor Bolts

Structural Steel

Structural Concrete and Metal Decks

Embedded Items, Sleeves, and Blockouts

Concrete Columns

Elevator and Stair Cores

Sidewalks, Curbs and Gutters

Each section starts with planning as this is the key to the success of a field engineer. Planning control, planning the equipment, or planning the layout. The field engineer must realize that planning is the first step in any activity. The layout considerations of the particular topic are then discussed and illustrated. It is intended that this material will assist the field engineer in visualizing the task and as a result being better prepared to perform in the field. The control of the elevations on a project is a major part of this chapter. Every task that requires horizontal layout generally requires establishing an elevation. What elevations are needed and how to maintain them are discussed. In some of the topics there are discussions about the actual construction methods. Things that the field engineer must consider when developing the control for the project as well as controlling the specific construction activity, such as checking anchor bolts during concrete placement.

REFERENCE

Some of the basic measurement material presented has been discussed in earlier chapters. It is repeated here because of its importance to the broad scope of construction control and layout. Anytime this occurs, a reference icon is used to direct the reader to where additional information is available.

ESTABLISHING PRIMARY AND SECONDARY CONTROL

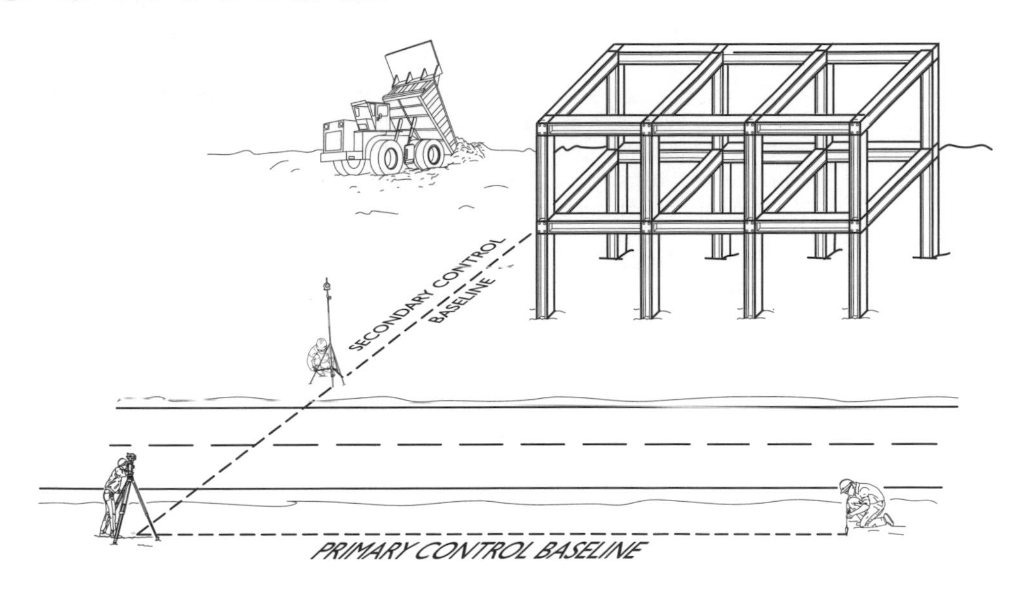

SCOPE

One of the most important responsibilities of the field engineer is establishing project controls. How well the initial control work is planned and laid out will be a major factor in the progress of the project. The field engineer will use the primary control to give line and grade information to the crafts on a moment's notice. How well this is done will be a direct reflection on the perceived competence of the field engineer.

Project controls are the horizontal and vertical planes which form a three-dimensional grid or network around the project. These grid lines are usually the dimensioning points or column lines described in the construction documents by the architect or engineer. Before this network may be utilized, a thorough understanding of its origin and of the relationship to the property and the project must be developed.

GENERAL PLANNING

Planning is the key to success when establishing primary and secondary control. Plan, plan, plan; follow your plan; update your plan. Planning helps to ensure the success of your layout, your projects, and your company. The ability to plan well separates the average field engineer from the excellent one.

PLANNING PROJECT CONTROL

When planning project control, the field engineer must consider the type of project that is being built. Is it a multi-story building? Is it a one-story building? Are there multiple buildings? Is it a bridge or a wastewater treatment plant? This list could go on forever. Identify the type of project and determine, by reviewing the plans, the best method for establishing and then controlling the structure.

SURVEY THE SITE

Is it urban or rural? If it is an urban location with limited and congested access, it might be difficult to find a spot in the vicinity of the site that will be undisturbed during the construction. If it is rural, it is likely there will be unlimited space to place control away from the construction site so it is protected.

The terrain must be considered when locating control. If the project is a bridge, consideration must be made about locating control close to the work, but avoiding the flood plain if the bridge is over a river. Other considerations include soil conditions, utility locations, site access, frost line depth, intervisibility between points, excavation limits, etc.

SCHEDULE YOUR WORK

If the project includes many repetitions of the same duties, such as setting control on every floor of a 40-story high-rise, then a daily or weekly schedule of the field engineer's work should be prepared and given out to all foremen. That way, everyone will have a clear idea of when they may expect layout of control points that they need.

CONTACT THE PROFESSIONAL SURVEYOR

Contact the licensed surveyor who performed the property survey of the construction site. Request:

Documentation of the property monuments placed at each corner of the site. Monuments should be permanent, solid, easily identified, and well marked. Each monument should be referenced to three other permanent objects, with distances and descriptions of the objects shown on a reference report.

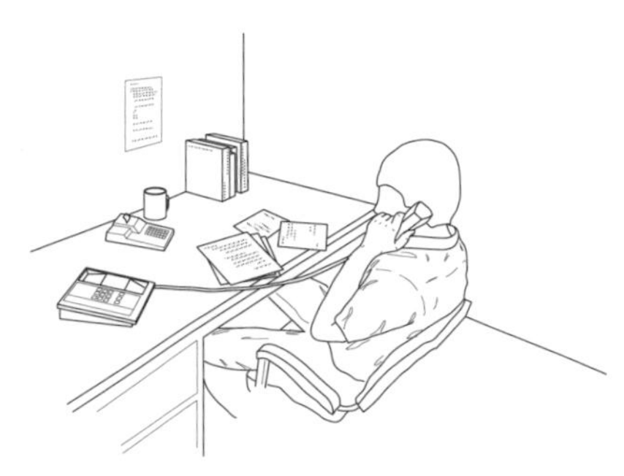

Copies of the legal Survey that was prepared.

A "Report of Survey" that describes in detail how the survey was performed, types of equipment used, problems encountered, etc.

PLAN FOR CONTROL MONUMENTS

When it is time to establish a monument control, have necessary materials and equipment available. Including:

Materials for monuments. These can be lengths of rebar, PK nails, or brass caps. If setting a monument in soil, dig a hole below local frost line and fill with concrete. Place a rebar or brass cap to mark the point.

Equipment for placing monuments. This could include a post hole digger, shovel, backhoe, concrete drill, chisel, sledge, etc. Have this equipment available for when you need it.

DEVELOP YOUR SURVEYING EQUIPMENT MAINTENANCE PLAN

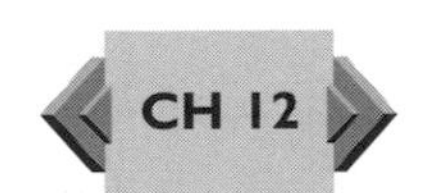

At this stage in the project, the equipment that is available may be brand new, or it may be from other projects. Regardless of its origin, before any work is conducted, a thorough check of all the desired geometric relationships of the transit or theodolite should be performed. Chains (even new ones) should be taken to a local calibration baseline and tested. Levels should be pegged. Level rods should be checked. Perform all necessary tests before the primary control fieldwork begins. Reference *Chapter 12, Equipment Calibration.*

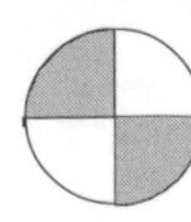

VERIFYING THE SITE

Before the project begins, a licensed survey team will have established the boundaries and geometry of the project site. Do not proceed without site information from the licensed surveyor. Be sure the surveyors have provided enough information for you to physically recreate the site survey. Read the report of the survey to determine what was accomplished on the site by the licensed surveyor and crew. Review the legal survey that was provided for additional notes about the site.

The control monuments should be well marked and easily identifiable. If there is a question about a monument, have the licensed surveyor or a representative come to the site to explain. Your traverse closure in the field must match the site dimensioning and description found on the site survey. Use your best standard surveying techniques to measure all distances and angles. Most survey firms will use an EDM for long distances. If possible, you should also. If chaining, correct for temperature variations. Be sure to double all angles. It may be necessary to contact the licensed surveyor to compare results due to the mathematical closure adjustments typically performed.

The checking of the licensed surveyor's work is especially critical when the project's building line is on or is close to the property line. Note: Even if a member of your construction firm is a licensed surveyor, that person should not perform the site survey. A lawsuit could seriously damage the integrity of the company, its ability to operate, and it's financial status. Always hire someone else to assume liability for the site survey!

DESIGNING THE CONTROL

Control around and within the site must be planned to provide the field engineer with readily available working lines. Too many engineers have simply plotted out the column lines shown on the architectural drawings to be base control, without considering how these lines will be used later. Think ahead to actual use during the construction phase. Ask questions like: What control lines are needed for the type of formwork that will be used? Where will haul roads be located? Where are job trailers going to be? What control will be needed after the frame is erected? etc.

Once the project controls have been closed, it is time to plan ahead and design the control which is going to be utilized throughout the project. It is more important to have a few strong, usable, and constantly checkable lines throughout the project, than it is to try to provide every column line.

You must also consider the accessibility of this information once the job is in progress. Every project will provide different obstacles and challenges which need to be addressed when designing control. However, the following is a partial list of things to consider when planning your project control.

Traffic obstacles such as the location of trucks, equipment, roads, etc.

Stockpiling of materials on location or in the area of control points.

Job progress changes in the structure throughout the project.

Weather related snow piles, ice, flood-plain.

Target or batterboard stability, removal, tampering or shifting of targets, such as when highway crews pave over points or demolition crews destroy targets on adjacent properties.

Construction interference such as formwork, scaffolding, tool boxes, trailers, temporary structures.

Safety on walkways, tie-cables, and barricades.

Effective prejob planning will make an incredible difference in the accuracy and timeliness of your work in the field.

KINDS OF HORIZONTAL CONTROL AND TOLERANCES

± 0.00

PRIMARY CONTROL

Primary means permanent. It is at the limit of the site or even off of the site. It should still exist at the end of the job. It is something you can always depend on.
No Tolerance. Measure until you can't get it any closer.

SECONDARY CONTROL

Secondary means semipermanent. It is established within a tape length of the work. It is well planned. An attempt is made to make it last for weeks or months or even the entire job if lucky. Measure to the 100th.

WORKING CONTROL

Working means very temporary! It may be removed the instant you leave it. It is often covered up by the work as it is used. Close the control by traverse and mathematical closure to confirm that the overall network meets specification.
Rule of Thumb: 1/2 of what is coming behind you.
Example. Wall Forms. Need to be to the nearest 1/4? Then, set working control to the nearest 1/8th.

SETTING PERMANENT CONTROL POINTS

A good field engineer will very carefully establish control points so they will last as long as possible. Hopefully, until the end of the project. To ensure that points are permanent requires using methods that are permanent.

ULTIMATE PRIMARY CONTROL

Set tripod legs in a concrete pillar.

PRIMARY CONTROL

Use concrete monuments that extend below the frost line.

Set-up concrete barriers around the points for ultimate protection.

SECONDARY CONTROL

Scratch and paint points on sidewalks, completed structures, etc.

Set semipermanent batterboards or 4"x4"'s.

WORKING CONTROL

Chalk and spray with varnish.

Set hubs as close as necessary to the work.

Use string lines between hubs.

Set batterboards.

PRESERVING POINTS

To be able to use control throughout the duration of the project, a field engineer should follow some basic principles and standard practices.

PROTECTION

Protect the points by locating them away from the construction crew.

Set rebar in the ground next to the points.

Flag the rebar for high visibility.

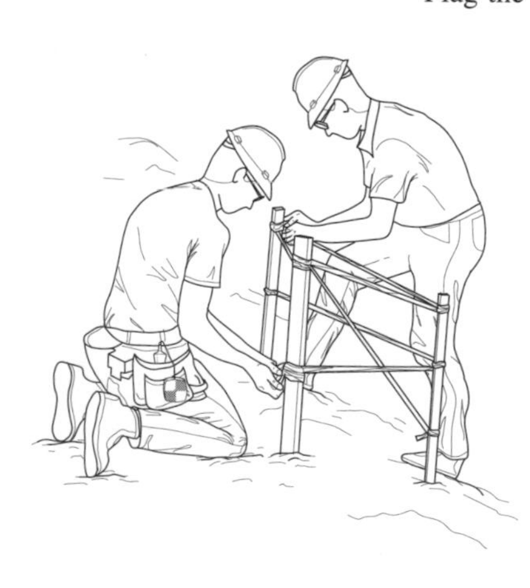

REFERENCE THE CONTROL POINTS

All primary and some secondary points should be referenced.

Place at least three (preferably four) reference ties per point.

Record references in a field book.

KEEP RECORDS OF THE CONTROL POINTS

Set aside a field book for primary control and for referencing.

Copy the control field book and store the copies at the main office.

EVALUATE

Control will be used at different times during the construction process. Evaluate when control points need to go into the building. It may be necessary to establish temporary control points knowing that they will be destroyed during certain phases of the construction work. Move control into the building as soon as possible.

LINE OF SIGHT CONSIDERATIONS

Always use long backsights and short foresights.

Surround the site with control so you are always sighting into the site. This should ensure that BS will be longer than FS.

Always establish four points on a line. If one is lost, there will still be three to use in assuring the line is good.

SHORT FORESIGHT

LONG BACKSIGHT

ESTABLISHING BASELINES FOR PRIMARY HORIZONTAL CONTROL

Baselines are often the foundation on which layout work will be based. Therefore, the utmost in precision must be used to obtain the best accuracy. The best—that summarizes everything about establishing baselines—the best in planning, the best equipment, the best methods of measuring distances, angles, and elevations, the best in checks, and so on.

PLAN THE BASELINE

There should always be one or two lines to refer back to in case of a discrepancy anywhere on the project. Baselines will be the foundation of the primary and secondary control network.

Baselines should be located in areas which allow the greatest amount of undisturbed access and lines of sight. Often, the very best place for baselines is off-site. Contact the owner of any off-site property where you are planning the baseline. Obtain permission!

The plans should be reviewed and the location of the structure on the site determined. The baselines should be placed so they are perpendicular and parallel to the major axis of the structure.

Every project will provide different obstacles and challenges which need to be addressed when planning controls. Look ahead!

STATIONING

Stationing along baselines represents the intersection of perpendicular control lines. Stationing may begin at the point of intersection of property lines or at other primary control lines which will serve as the starting point for stationing. Whether the stationing goes in two directions from the point of intersection or only continues in one direction, always close the dimensioning into at least one other point.

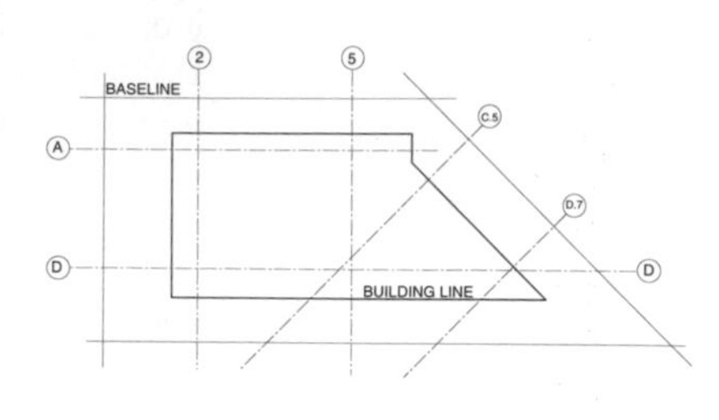

PERPENDICULAR BASELINES

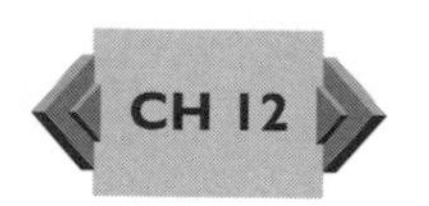

If setting two baselines perpendicular to each other, measure the angle at least six times. Average the angles to obtain a better determination of the true value. Again, use a different instrument, if available, to check the angle turned. If any discrepancies exist, refer to *Chapter 12, Equipment Calibration.*

USING THE BASELINE

Again, don't forget to use long backsights and short foresights. Regularly check the instrument setup for level and ensure it has remained over the point. Secondary and working controls are the offsets or non-column lines which are established and utilized during construction. These lines should be created and used with the same attention to accuracy as with any control line.

When developing working controls, be sure to consider how, where, and when they are to be utilized. Avoid conflicts with sight obstructions and work activities which may interfere with the line of sight on layout work. Be sure working lines are clearly labeled and explained to all crafts so there is no confusion with adjacent column lines.

ESTABLISHING VERTICAL CONTROL (BENCHMARKS)

A plan should be followed in establishing a system of benchmarks on site. Often, the elevation on the plans is some distance from the site. Or, there are two elevations listed. Follow the best procedures for differential leveling while setting the benchmarks.

POSITION BENCHMARKS CAREFULLY

Plan to locate benchmarks no more than 200 feet apart in order to have two benchmarks visible from any instrument setup.

BENCHMARKS SHOULD BE PERMANENT, SOLID OBJECTS

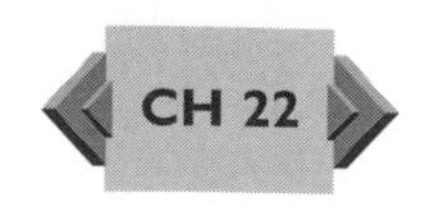

Benchmarks may be the tops of stakes, fire hydrant bolts, rebar, anchor bolts, high points on sidewalks, etc. Review the site plans and walk the site to locate potential benchmarks. See *Chapter 22, Leveling.*

DOCUMENT ALL BENCHMARKS

Plan a notebook that is just to be used for describing benchmarks.

SET AND MARK THE BENCHMARKS

Paint the objects that were located during the planning phase. Be careful to avoid wild, careless markings. Respect the property of others. Dig holes and pour concrete around rebar if solid, permanent objects are not available in the immediate area. Place a guard stake to warn others of their existence.

Always, always, always close a level line onto a known benchmark elevation. There is no excuse for leaving a loop open-ended. There are too many possibilities for mistakes and blunders that could remain unnoticed if the loop is left open.

Use a standard note form that is easily understood. Perform all available checks to ensure accuracy of notes.

CHECK, CHECK, AND RECHECK

Primary control must have no mistakes. While establishing primary control, repeat every operation at least three times. Check differently each time. The subconscious mind will tend to make the same mistake if similar circumstances are encountered. Use different routes. Use different chains. Turn angles to the left instead of to the right. Do whatever is necessary to eliminate all errors!

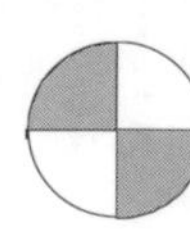

SITEWORK

SCOPE

Sitework is the first and last operation on a construction project. Often the site must be prepared for the work by clearing and demolition. A thorough check for utilities must be made or temporary rerouting may be necessary. Earthwork, mostly excavation, will prepare the site to the construction sub-grades specified on the plans. Throughout the project, elevations of surrounding buildings must be checked for possible settlement. At the end of the project, stakes will be placed to control the final grade. All stakes set for sitework should be referenced back to the primary horizontal and vertical control or laid out to follow the finish contour lines.

CLEARING AND DEMOLITION

After the site has been verified, the limits of usable space must be marked. Fences should be located to enclose all work and storage areas accessible to the project. Fences should be set just inside the limits so they don't encroach onto other property.

Careful study of the site plan will reveal what is to be removed. Often, attempts are made to save mature trees. Any natural or man-made object that is to be saved should be well marked with a surrounding fence of 4' painted lath and colorful surveying tape. Operators should be personally contacted and taken over the site pointing out objects to be preserved.

Demolition is dangerous! Falling debris, unknown utilities, poor walking conditions, etc. comprise just a few of the dangers. Keep bystanders off of the site at all times. A company has an obligation to protect employees and the public during all phases of the project. Safety tool box talks should address this issue.

UTILITIES

Check plans for location of utilities. Contact utility companies to confirm locations and to search for unspecified lines. Utility services may need to be rerouted during the course of the project on a temporary or permanent basis. Follow through and expect the unexpected!

EARTHWORK

SHALLOW EXCAVATION

Shallow excavations include roads, parking lots, shallow buildings, etc.

Determine limits of excavation. Excavate a larger area than needed in order to accommodate formwork. Set stakes at a specified offset from the edge of the pavement, center of footing, or wherever the carpenter foreman desires (5.0' offset is typical).

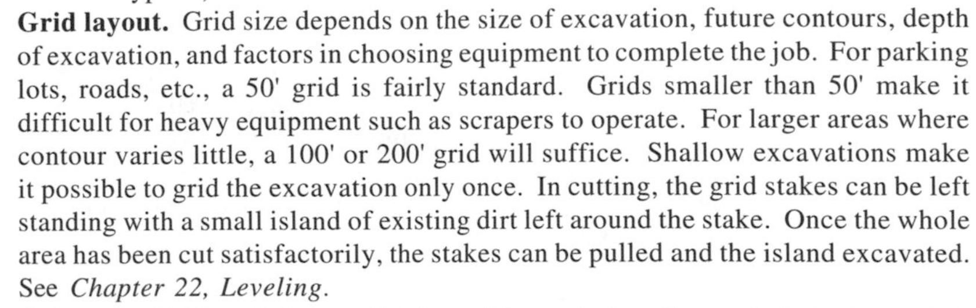

Grid layout. Grid size depends on the size of excavation, future contours, depth of excavation, and factors in choosing equipment to complete the job. For parking lots, roads, etc., a 50' grid is fairly standard. Grids smaller than 50' make it difficult for heavy equipment such as scrapers to operate. For larger areas where contour varies little, a 100' or 200' grid will suffice. Shallow excavations make it possible to grid the excavation only once. In cutting, the grid stakes can be left standing with a small island of existing dirt left around the stake. Once the whole area has been cut satisfactorily, the stakes can be pulled and the island excavated. See *Chapter 22, Leveling*.

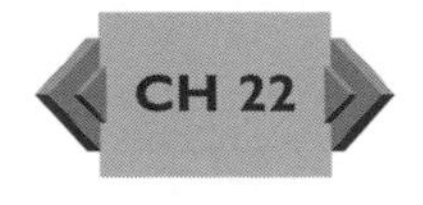

Reference Grid to Primary Control. The main baseline and stationing points should be referenced. Measure the angles and distances. This information is a necessary safeguard in case the grid is inadvertently lost during the construction operation.

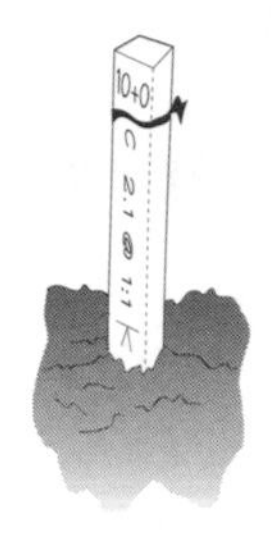

Staking. Use 48" lath, typically, for cut/fill stakes. For clarity, always face the stakes in one direction. Mark both sides with a cut or fill (ex. C2.21) and flag the tops (typically red for cuts and blue for fills). Also mark the bottom of the stake with a line indicating existing grade. See the illustration.

DEEP EXCAVATIONS

Determine the limits of the excavation. Have the soils engineer determine the angle of repose for the soil to slope from the toe of excavation up to the existing ground elevation.

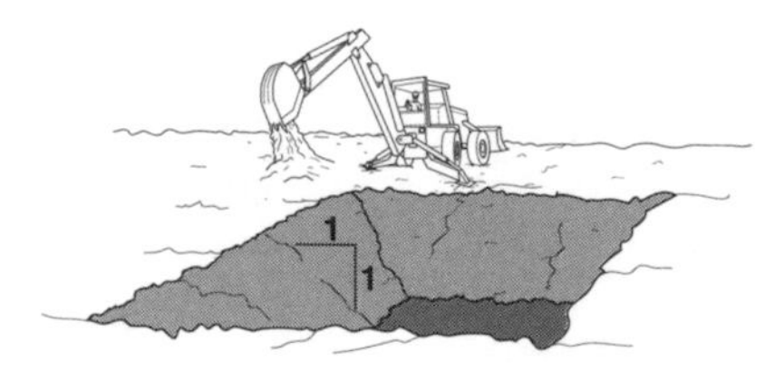

If the excavation will have sheet piling or lagging, determine the limits of excavation necessary for the system used. Maintain an up-to-date schedule of the operations which will be needed to keep the excavator and the pile placers working together placing shoring where needed.

The room which is necessary behind foundation walls (5' typical) depends on the formwork and lagging systems used. Refer to formwork drawings and shoring details and talk to the crafts.

When determining rough-grade elevation, allow additional depth for waste material obtained from caisson placement, subexcavations, etc. If the excavation is to be sloped, one must locate the limits of excavation by accounting for several variables:

the location of the outermost perimeter walls or footings.

the rough grade elevations desired in all portions of the excavation. Generally, the elevation will be taken down to the level of the bottom of the slab on grade.

the amount of working room needed by the crafts.

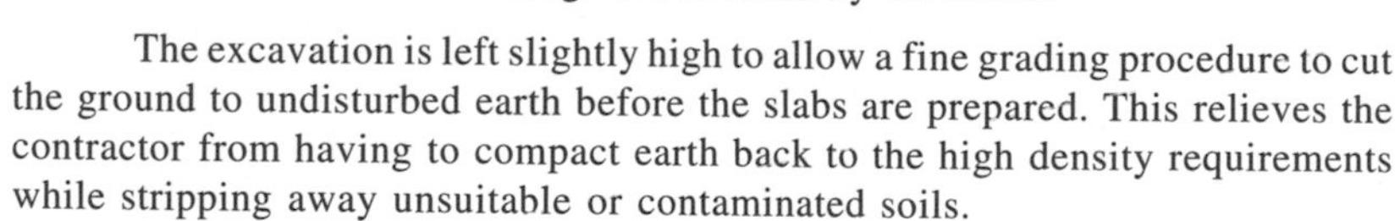

The excavation is left slightly high to allow a fine grading procedure to cut the ground to undisturbed earth before the slabs are prepared. This relieves the contractor from having to compact earth back to the high density requirements while stripping away unsuitable or contaminated soils.

Sometimes the area is over-excavated to allow additional room for inherent spillage from caissons or excavating footings. This may be done when the sub-grade material is not structural in nature.

Determine the horizontal distance which will be necessary to accommodate the slope. Follow the requirements of OSHA. An angle of repose for the type of soil encountered is used to determine how the slope must be cut. Have a registered engineer or surveyor check the slope angle, since liability over a possible mishap could become an issue.

Don't forget to leave room for the formwork behind the perimeter walls or footing when determining the limits of excavation. Nothing slows a formwork crew more than having to work in unnecessarily crowded conditions. Additionally, leaving too narrow of a space behind the wall can create a dangerous trench-like situation. Extra excavation is inexpensive when compared to the labor cost associated with poor productivity or costs related to injury.

STAKING

Excavating or filling at a predetermined slope will make it necessary to locate grade stakes away from the top of the slope. This is properly done by utilizing the method of setting slope stakes found in most surveying texts. Basically, the slope is multiplied by the amount of cut or fill to determine the distance from the structure to where the top or toe of slope should occur. Offset slope stakes 5 feet back from this catch point. See *Chapter 19, Excavations,* for a discussion of slope staking

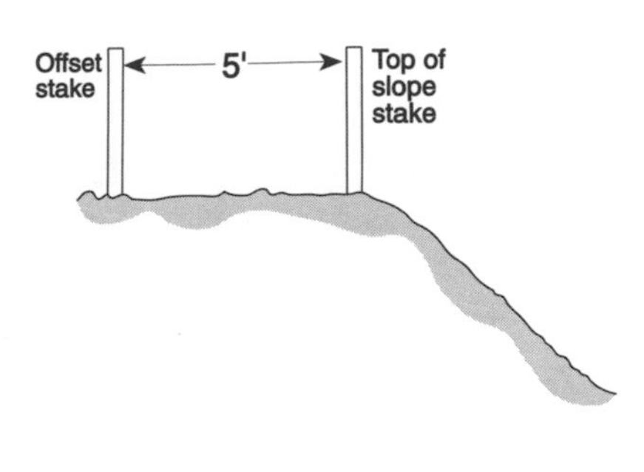

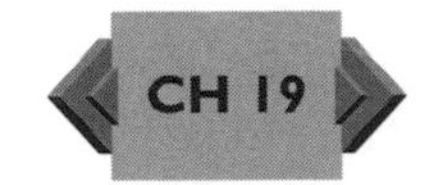

VERTICAL CONTROL

Establish temporary benchmarks (TBM) around the jobsite, both at original ground elevation and in the excavation.

TBM's should be placed in out-of-the-way locations so they are not disturbed during normal construction operations.

Locate a sturdy point such as an edge of footing, top of anchor bolt, or small angle iron welded to H-pile, scribe mark on the wall, etc.

TBM's should be located evenly around the site and at elevation intervals so an HI of the leveling instrument can be determined from just one setup.

For every instrument setup, check your HI by backsighting onto two TBM's, if possible. Then if one is disturbed, it will be readily apparent.

Plan ahead for all vertical control operations.

RETAINING SYSTEM (PILES & LAGGING)

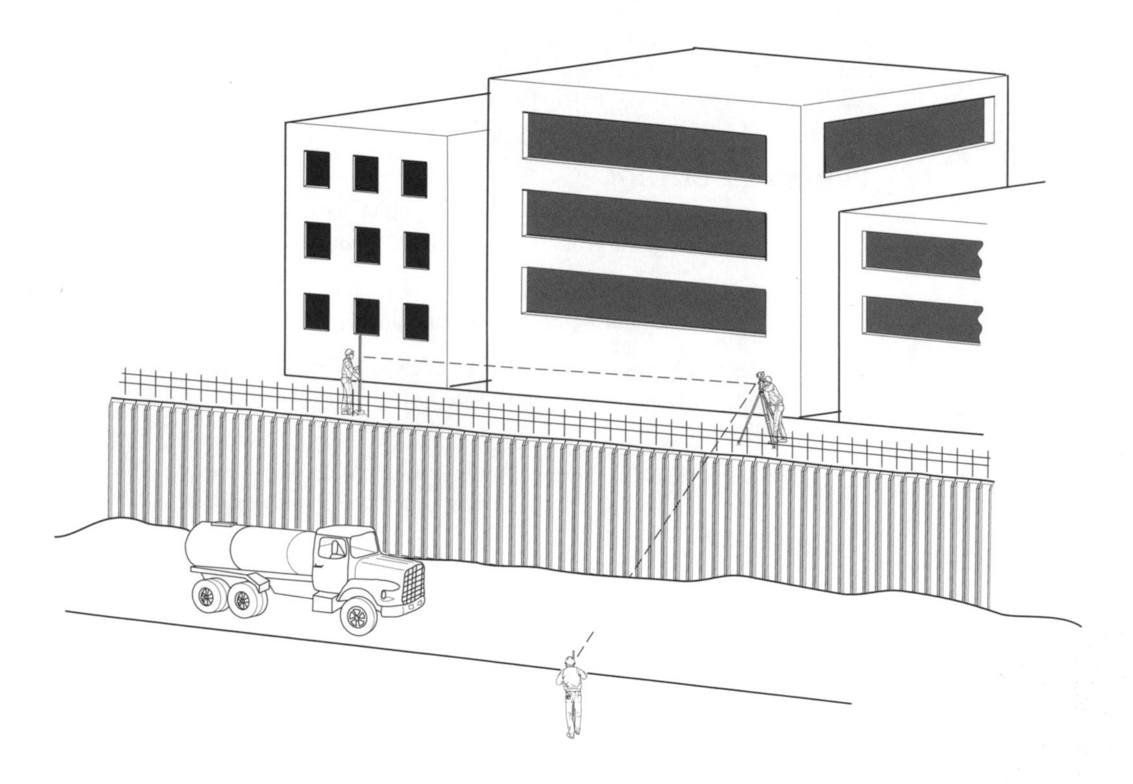

SCOPE

When excavating for a project, several conditions may exist that require holding the soil from tumbling into the excavation site. The most common is a deep excavation through loose soil. Various forms of piling may be driven into the soil to hold it back while excavation and foundation construction is taking place. Another is excavation on a limited-size site. Although the excavation may not be deep and the soil may be able to hold a slope, there might not be room for the required slope. Retaining systems allow work to progress when unstable soil conditions exist, when building on the property line, or when other site limitations exist.

PLANNING

If pile or lagging is required, check drawings provided by the designer. Refer to the drawing for layout and tolerance information.

If it has not been determined whether to use piles or lagging, carefully study the site plan, foundation elevations, and the soil bore sheets. These will reveal the soil conditions, the excavation depth, and the limits of the site. With this information, a registered engineer can determine if sloping of the soil can be used or if a retaining system must be used. **Contact a registered engineer or shoring contractor to design the system.**

RETAINING SYSTEM LAYOUT

REFERENCE TO PRIMARY CONTROL

All measurements for layout of the retaining system should be referenced to primary horizontal and vertical control points.

LOCATE CENTERLINE

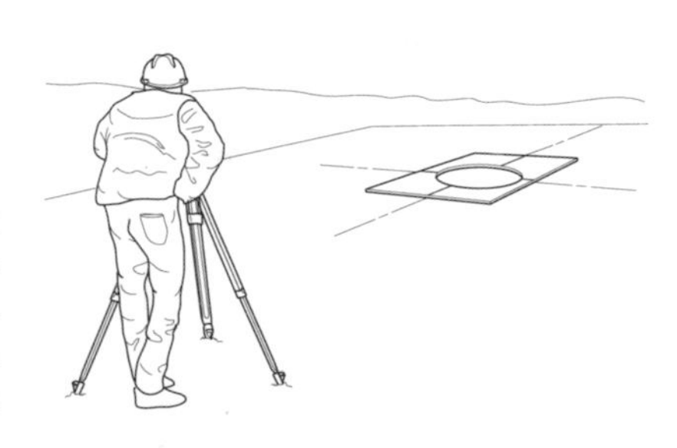

Refer to plans and locate the centerline of the retaining system as noted. Usually this is parallel to the foundation wall.

To locate the centerline of piling in both directions, the lagging and pile thickness plus formwork information must be evaluated. Establish layout from structural drawings for the wall location and for pile centerlines.

OFFSET

Stakes are offset from the actual location to allow for equipment movement. Check with the operator to locate offset stakes where they are not likely to be disturbed. Offset your point in two directions and have a benchmark available for grade.

LOCATION

Plumb and secure pile. See plumbing information in caisson section.

Check location from offset points. Locate within tolerance per specifications.

USING FORM ATTACHMENT BRACKETS

Form attachment brackets should be used when shoring is used as a form face.

Lay out vertical centerline at top of pile. Typically this is the centerline of the form bracket.

Establish horizontal centerline of form bracket. This is an elevation derived from the form drawings.

Double check from primary control and offset stakes.

LAGGING

Offset the wall line. A reminder: lagging may not protrude into the finish wall.

Weld a bracket to the face of the pile and sight the outside face of the wall with a transit to check it for plumbness. This can be plumb bobbed down and used to measure lagging for plumbness from the base of the system.

Double check the wall line for plumb.

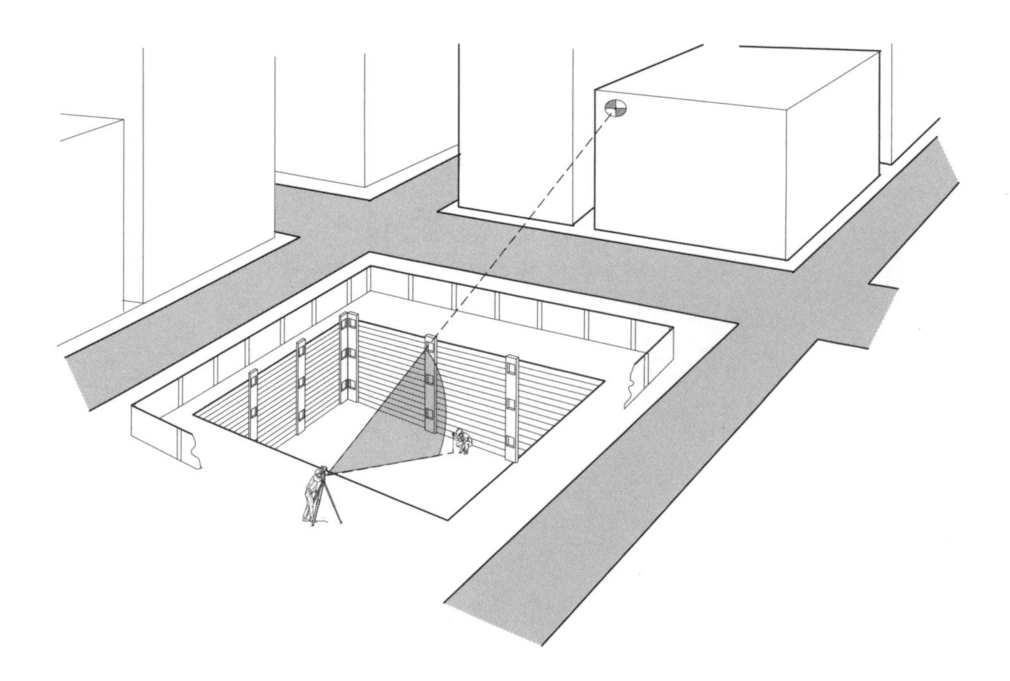

CAISSONS

SCOPE

"BUILDING SINKING, CAISSON WORK TO BLAME?" was the title of an article published in a leading construction magazine. There were several problems with the caisson work which caused the building to settle more than allowed. The subcontractor was trying to increase profits, so he cut the cost of his work. He didn't place the caissons in the exact locations specified; he didn't drill deep enough to obtain the necessary soil strength; he didn't drill the holes plumb, etc. He just did shoddy work. He is now out of business.

Sometimes it might appear that once the foundation floor of any structure is in, what can't be seen will soon be forgotten. That obviously wasn't true in the summary of the above article. Caissons are very important as they are the foundation of the job. Specifications must be followed carefully.

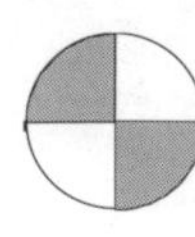

PLANNING

Develop field drawings showing the location of caissons and all pertinent information. Have the project engineer check the drawings for omissions or corrections. Use these field drawings as your guide during layout.

Read, study and know the caisson drawings!

LAYOUT

Use primary control to locate caissons

Primary control should be used as the basis for all caisson locations. If secondary control must be used, it should be checked and double checked during the caisson work.

Typically, column center is equal to the caisson center. However, there are occasions when this is not true. Any unusual caisson locations should be red lined on the plans.

From caisson shop drawings, determine at least two approaches to locate a caisson from primary or secondary control. Both should be used!

Conventionally located caissons are laid out by intersecting the necessary working lines which represent the caisson center. Care should be taken when chaining along working lines for caisson stationing. Be aware that caisson centerlines may not be on gridlines. Always check layout from two different means as a double check.

Once the double check closes, offset caisson in two directions at least 90° apart. Offset stakes or monuments should be located where they will be out of the way of equipment travel and caisson spoils. Five to ten feet is a typical offset built with hubs or sturdy monuments.

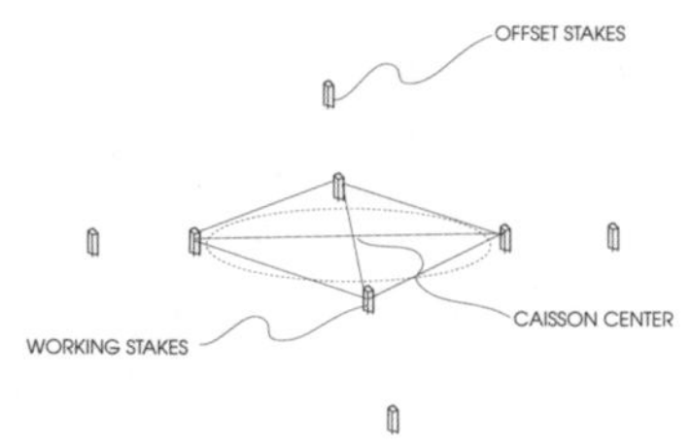

DRILLING AND HOLE PREPARATION

During rig setup, talk to the operators about the importance of the offset stakes. Make sure they understand the manner in which the stakes will be established and how they will be used.

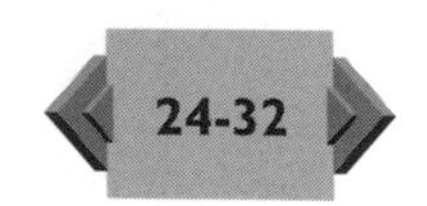

If anchor bolts are to be placed in the top of the caisson, templates should be used for accurate placement. Refer to the *Anchor Bolt section in this Chapter* for placement and alignment procedure.

EXAMPLE COMPUTATION OF AMOUNT CAISSON IS OUT OF PLUMB

Given:	**5'-0" Offset From Edge** **16 oz. Plumb Bob**
Find:	Plumbness in 20' Location Top of Case Location Top of Caisson Total Caisson Out of Plumb

Solution:

Plumbness = 1.5" - .75" = North +.75"

Location Top of Case = +2" North (5'-2" - 5'0" = 2")

Location Top of Caisson = Top Case - Top Caisson (208' - 199' = 9')

Out of Plumb Per Foot = .75"/20' = .0375" per ft.

Out of Plumb to Top of Caisson = 9' x .0375" = .3375" North 2" (north) + .3375" = 2.3375"* (of case Plumb Off at Top of Caisson)

Total Out of Plumb .0375"/Foot x (199-139) = 2.25"

*Watch your signs (±) (North/South)

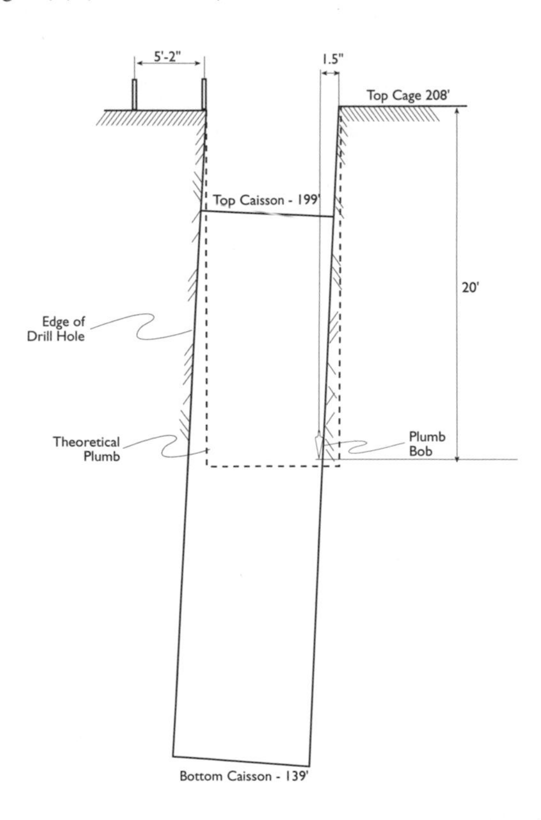

FOOTING AND FOUNDATIONS

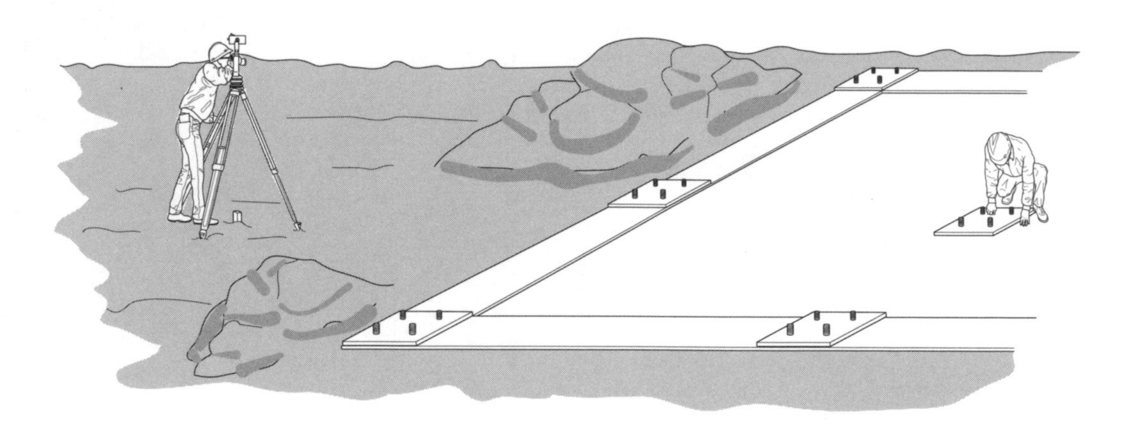

SCOPE

During and after excavation, it will be necessary to give line and grade to the carpenter foreman so crews can begin placing formwork for the concrete work that will support the columns and walls of a structure. This will include spread footings, grade beams, pier caps, slabs, or caisson tops. Although good layout and alignment techniques must be used for initial formwork line and grade, for excavation, and for rough locating purposes, tolerances for this type of work are not as precise as for primary or secondary control work. Normal referencing, double checking, closing level circuits, and information gathering steps should continue to be followed.

PLANNING

Primary control should be used as the starting point for all footing and foundation work. If the work area cannot be seen from primary control locations, care should be taken to establish secondary control locations that can be directly used while line and grade are being placed on formwork. Be sure to check secondary control back into primary control on a regular basis.

Be sure to describe line and grade information to the excavator foreman and operators.

Contact the carpentry crew foreman well in advance of the forming and placing operations to assure that a mutual understanding exists about how forms are being set, how line and grade are to be established, and what means of corrective alignment may be used in the event of a problem. Think ahead!

Line and grade should be easily identified by all concerned. Forms will be oily and layout information will be hard to see, so mark all points well. Use sharp pencils to establish well-defined lines; use permanent markers to label needed line or grade information; circle information with paint.

HORIZONTAL LAYOUT

Layout for excavation may be accomplished by defining the area of cut with hubs, lath, lime, etc. Be sure to consider the means of forming before digging. If the earth is too soft to act as a form face and is to be "high walled," be sure to communicate that to the excavator. If reusable formwork is used, be sure to provide ample work space for the crafts.

Wooden bearing pads may be set at subgrade elevation to facilitate firm bearing for formwork. Form line can then be snapped directly onto wooden pads. Batterboards may be used around the perimeter of subgrade foundations to reestablish working lines. This can be useful on opposite sides of grade beam trenches and pier cap excavations to provide reusable points throughout the forming operation. String lines may be used from point to point to recreate lines of intersection and alignment.

Monitor location during concrete placement

Always monitor formwork during placing operations. Forms may shift or deflect if not properly secured. Formwork which is kicked and stacked with turnbuckles will provide good alignment flexibility. Forms kicked to the trench banks provide a less secure bearing and little means of adjustment.

When using instruments or previously set batterboards and string to monitor footer alignment and location during concrete placing, be sure they are set where the placing operation will not disturb them or the line of sight.

Foundation walls may bear on grade beams, pile caps, slabs, or caisson tops. Be sure the surface on which you will provide layout information is firm and free of obstructions before you begin. Before providing layout information, check with the crew foreman to determine what information is needed. The foreman may want actual concrete line or may ask for a continuous line which represents back of form, stiff back, or whale line.

When possible, set layout line directly onto form foundation tops. Avoid working from an offset line. This is simply one more chance to make a mistake. However, if you are working from an offset and reading directly to a rod or a stiff rule, always be sure to have the person holding the rod swing an arch with the rule, then "read the lowest value." This same principle applies when reading a target or rod during alignment operations while occupying an offset.

Once the first form side is in place and secure, you will need to supply any control layouts and information so blockouts, sleeves, points of change, etc. can be established.

Once the second form side is "buttoned and secure," check form alignment by setting the instrument on the working offset and setting the bottom, middle, and top of the formwork throughout the length of the pour. Work closely with carpenters during alignment operation.

ELEVATIONS

Temporary benchmarks should be established in the area of the formwork as soon as possible. The tops of caissons, protected stakes, etc., may be used. Set several in the area to be checked into frequently.

Once formwork and bulkheads have been set, grade information needs to be provided. Grade may be marked as top of form or a grade line with nails may need to be snapped on the form face.

Place a target on the level rod at the correct reading off of the formwork. This will help to eliminate the human error when reading a rod. Be sure to consider form thickness when placing a target on the level rod.

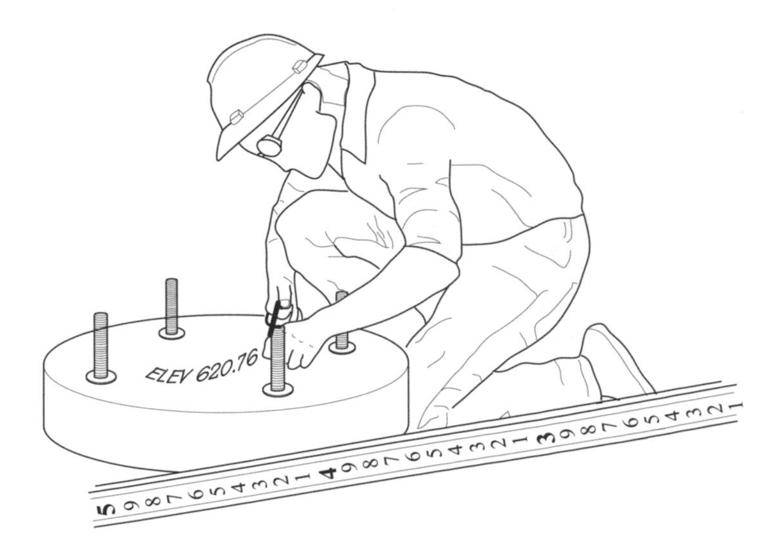

BATTERBOARDS AND OFFSETS

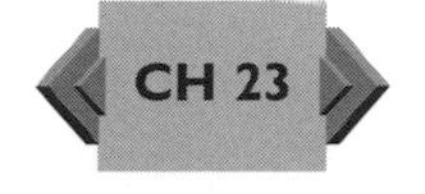

Be sure batterboards are firm, well protected, and clearly labeled. Tie wire may be used to add diagonal support to batterboards. Avoid placing working controls directly on bulkheads and formwork until they have been set, kicked, plumbed, and angled.

When designing an offset distance, be sure to consider formwork, safety equipment, and the means of concrete placement to avoid conflict and to ensure that control information will be available throughout the operation. See *Chapter 23, Construction Layout Techniques,* on setting benchmarks.

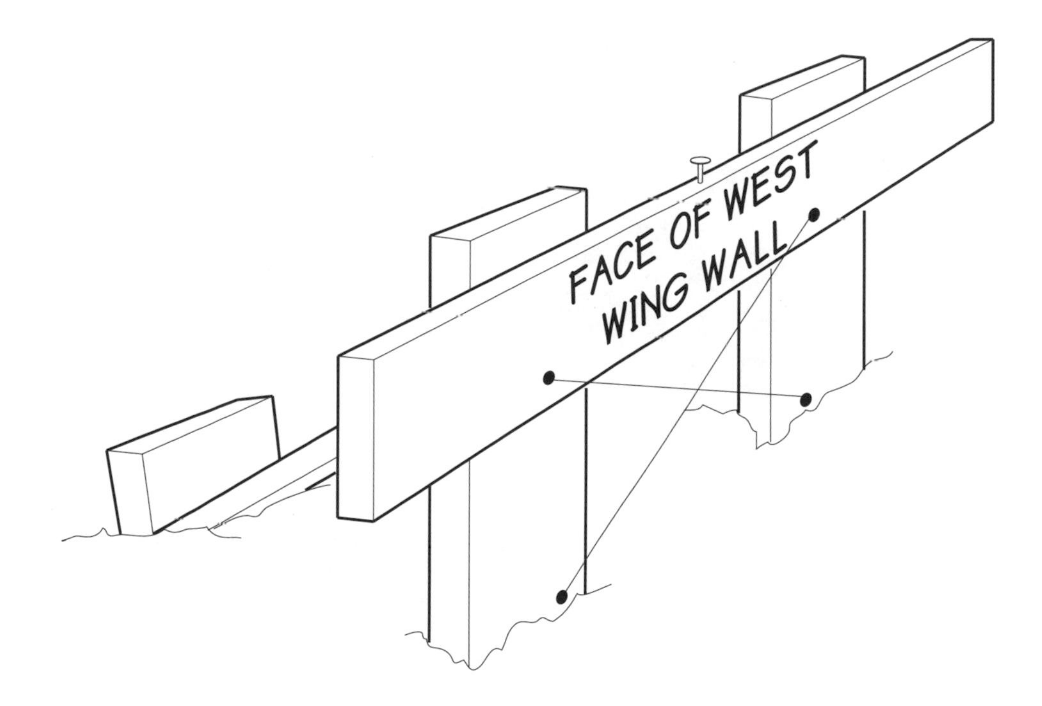

ANCHOR BOLTS

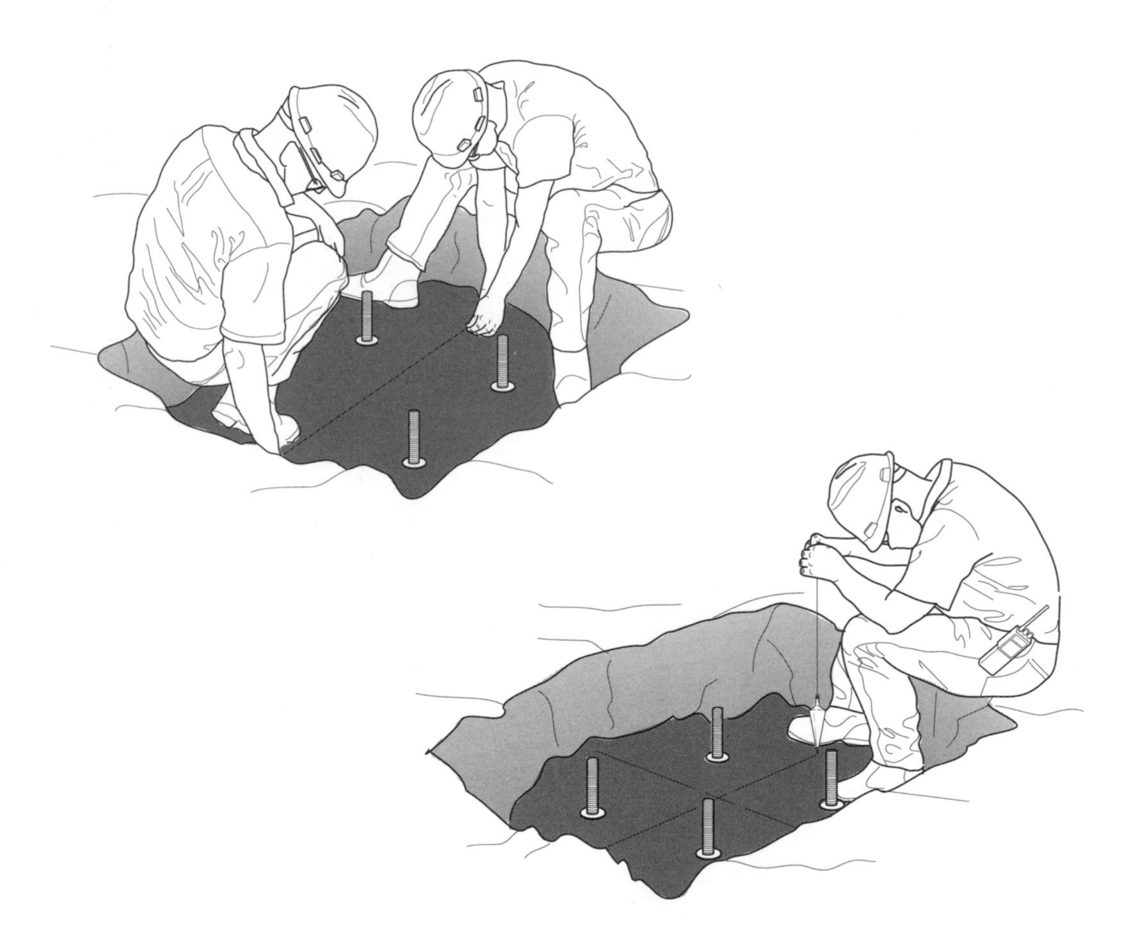

SCOPE

Anchor bolts are used to hold structural members to the foundation. There is little room for error in location or spacing. Steel or precast is being fabricated and brought onto the site with the expectation of being easily slipped into place. The field engineer must use surveying knowledge and skills to ensure that the bolts are located according to plan.

PLANNING

Dimensions between anchor bolt locations, the size of anchor bolts, and the spacing of anchor bolts will vary widely throughout the project. Many will be the same. However, it is the ones that are different that are usually improperly placed. Red-line on the plans unusual anchor bolt situations so they will not be overlooked during layout.

When planning a control network for placing anchor bolts, try to establish points for every grid line on which anchor bolts lie. These points should be part of the primary control network.

HORIZONTAL LAYOUT

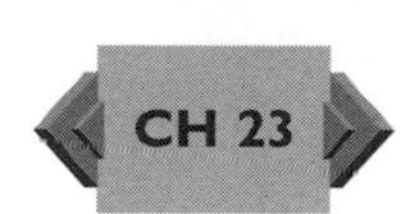

Try to locate control points so anchor bolt templates can be sighted directly with the transit/theodolite. If possible, two instruments should be used simultaneously to set the template by intersecting the two lines of sight. This will increase the accuracy of the location of the anchor bolts. In situations which do not allow for direct sighting from primary control points, it is best to establish secondary control points at a location that will permit direct sighting. Reference *Chapter 23, Construction Layout Techniques,* for information on intersecting lines.

Such practices as plumbing with a plumb bob or hand level or chaining from other points to locate bolts are common practices but should be avoided because of inherent errors in the procedure.

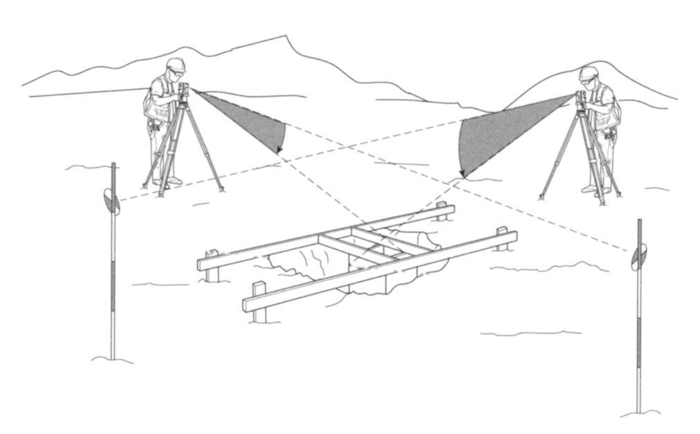

Caution must be taken when deciding to use secondary control points. Secondary points must be located so they will not be disturbed by drilling, concrete placement, etc. and will still allow for checks into primary control.

When adequate protection cannot be provided or good checks into primary control cannot be made from secondary points, it is best to go ahead and work from secondary control with extreme care.

ANCHOR BOLT TEMPLATE PLACEMENT

Before placing, always check the anchor bolt pattern, which is shown on the anchor bolt drawing against the base or structural steel piece drawing to ensure that the two agree.

Whenever you have more than one anchor bolt pattern, always prepare a drawing that clearly identifies the different patterns. Color code the various patterns on the drawing. Then paint the corresponding templates the same color.

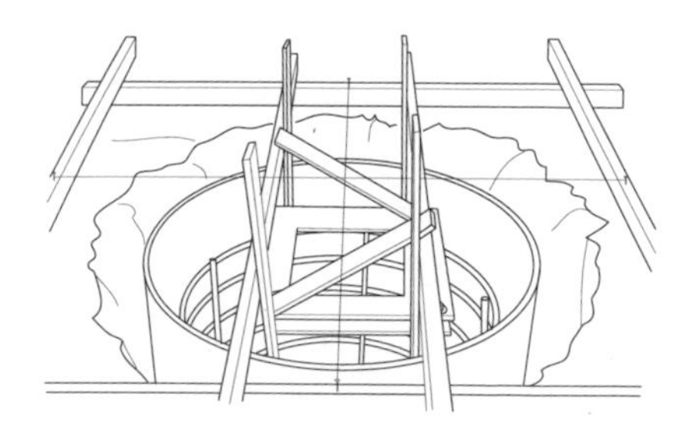

TOLERANCES

Unless otherwise specified in contract documents, anchor bolts should be set within tolerances outlined by the Commentary on the Code of Standard Practice, Section 7, A.I.S.C. Steel Construction Manual.

CONCRETE PLACEMENT

Anchor bolts must be held securely in place while concrete is being placed. Generally, this is accomplished by attaching the template to 2 x 4's and nailing them securely to the formwork.

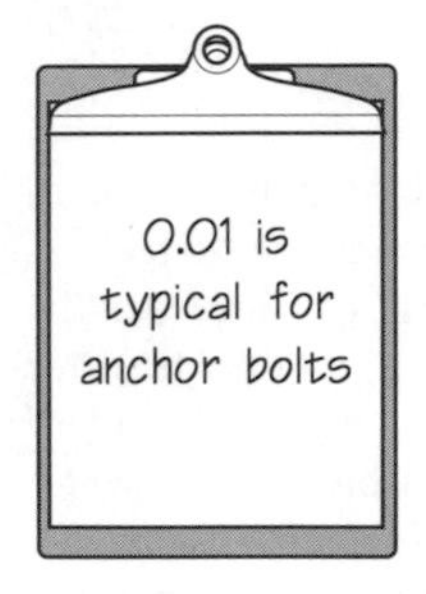

Care must be taken during the template placement to properly orient the template. Rotating the template 90° or 180° from the correct position could have serious results.

Check and double check that instrument sights before, during, and after line are given to the crew member aligning and securing the template.

If possible, set templates as close to pour time as your schedule allows to ensure that the template or formwork doesn't move prior to the pour. Recheck the position of the template during concrete placement.

It is good practice to make sure the labor crew puts a coating of grease or duct tape on the anchor bolt threads prior to concrete placement. This will prevent the threads from filling with concrete. This eliminates requiring a laborer to spend considerable time cleaning the threads.

CHECKING ANCHOR BOLTS AFTER CONCRETE PLACEMENT

After anchor bolts have been set and concrete is placed, grid lines will need to be established on the concrete surface for checking and plumbing the bolts. Concrete nails can be placed in the top of the caisson while concrete is still "green." The heads of the concrete nails will provide durable points that can survive sand blasting during caisson cleaning. The nail heads also provide good secondary control points.

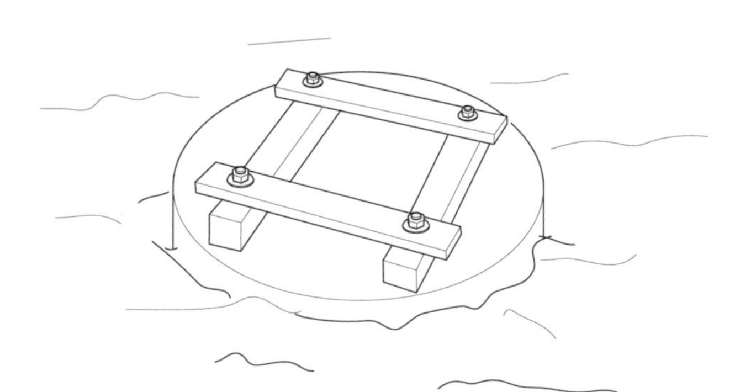

When checking to see if anchor bolts are within tolerance, be sure to check not only the location of the bolts but also the plumbness. Bolts can be within tolerance at the top of the caisson, and out of tolerance several inches up from the caisson top where the base will sit.

To protect bolts after the template is removed, sleeves made of 2 x 4's can be cut and placed on the bolts. This will prevent them from becoming damaged from materials being laid on them or vehicles running over them.

ELEVATIONS

Always check top of anchor bolt elevation against top of base elevation. Make sure anchor bolt elevation is high enough to allow for all required washers and a fully threaded nut.

When setting anchor bolts to elevation, shoot the top of the bolt for elevation, then check the top of concrete elevation, then check the bolt projection.

It is always best to set anchor bolts just a little high rather than to try to raise them as the concrete is setting.

FINAL CHECK

Always check anchor bolt location and elevation when the concrete is "wet". After it drys it will be too late to make any adjustments.

STRUCTURAL STEEL

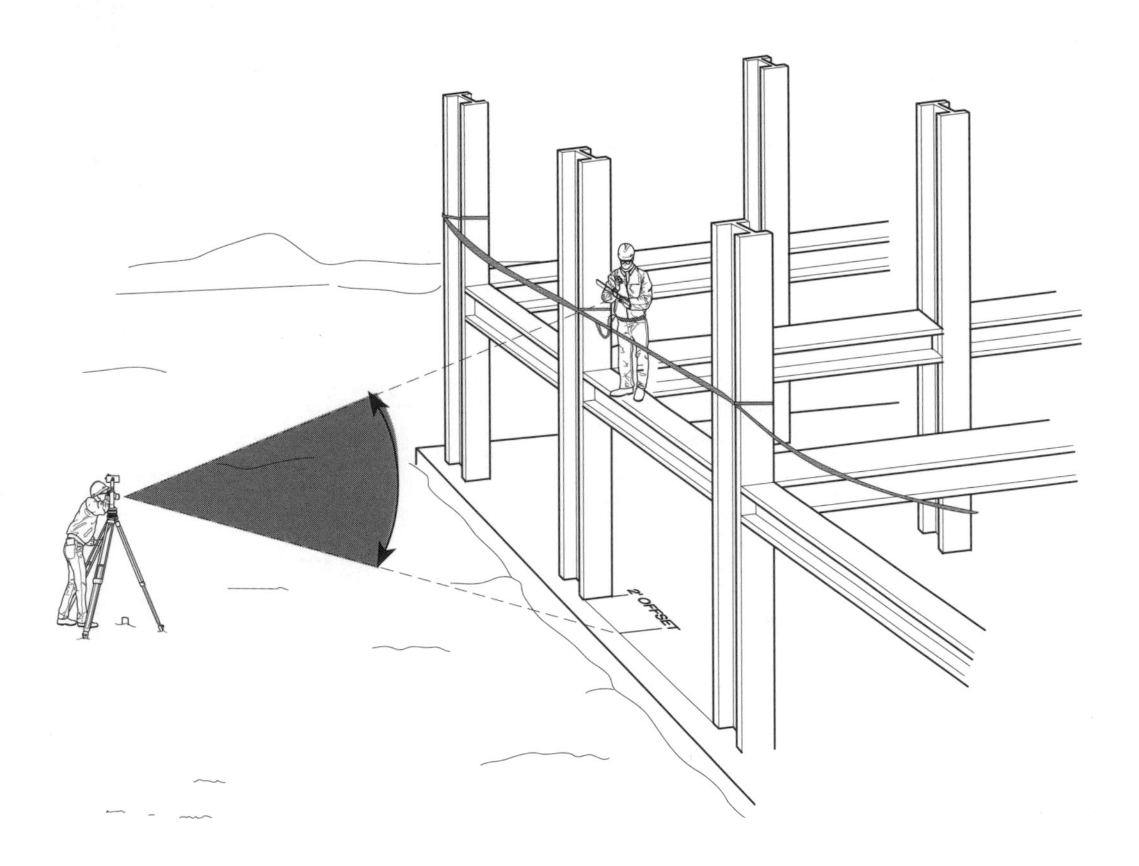

SCOPE

The arrival of the structural steel is a major milestone during the construction of a project. The overall shape of the structure will finally begin to occur. That is, if the steel has been fabricated correctly, if layout and the plumbness of the structure are maintained correctly during erection, and if everyone has been following the plans, the pieces should fit together quickly and easily. However, if the plans are not read or layout and plumbness are not correct, problem after problem will occur creating unnecessary rework and lots of headaches.

PLANNING

USE PRIMARY CONTROL

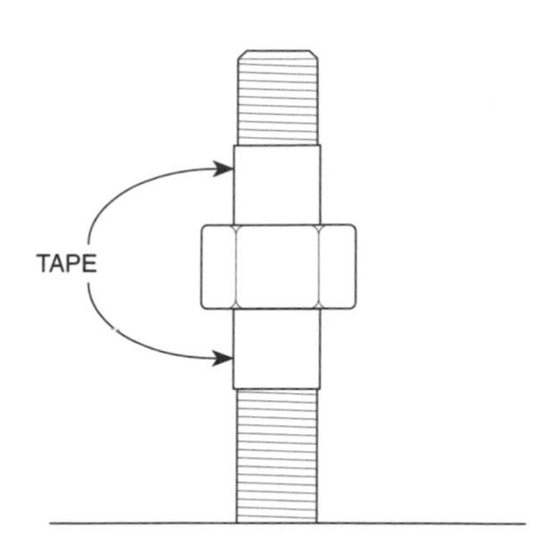

Primary control points used during the location of the anchor bolts should be rechecked for accuracy prior to use during the erection process. If primary control points are established on column lines, it may be necessary to establish secondary control points which are offset a few feet from the column line to be used during the plumbing process.

BASE PLATE ELEVATION

Erection cannot begin until the base plates or columns with base plates are set at the proper elevation. This is done with shims or with leveling nuts threaded onto the anchor bolts.

When shims are used, shim packs can be made in advance. Shim packs are made by stacking shims on top of the bearing surface to get the necessary thickness to establish the elevation of the bottom of the base plate. The shim pack is then taped together and labeled for use when the erection begins.

If leveling nuts are used, they are threaded onto the anchor bolts to the correct elevation. Tape is then placed just above and just below the nut to keep it stationary.

CHECK EQUIPMENT FOR PROPER ADJUSTMENT

TRANSIT AND THEODOLITE

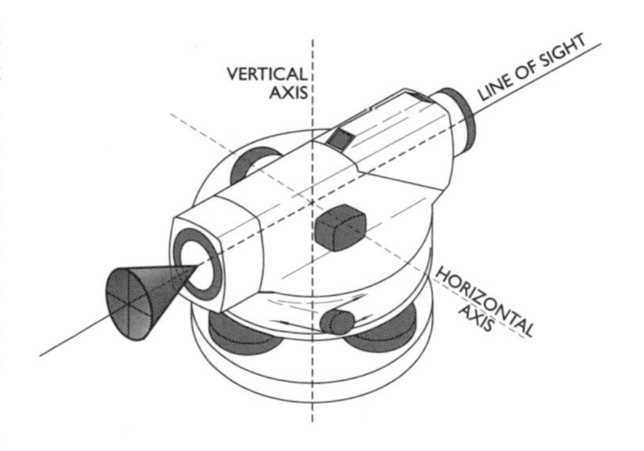

If a transit or theodolite is to be used during the plumbing process, care must be taken to ensure proper adjustment. Establishing plumb at high elevations opens the possibility of errors not normally encountered. Transits and theodolites must be checked to be sure the correct geometric relationship exists. The horizontal axis is perpendicular to the vertical axis. See *Chapter 12, Equipment Calibration,* for an explanation of how to perform the tests.

LASER

If a laser is used, tests should also be conducted to see that the laser line of sight is being projected as vertical. Rotate the laser about its vertical axis and check to be sure the line of sight stays in the same position. If it doesn't, a repair service should be contacted and the laser should be adjusted by qualified persons. Refer to *Chapter 11, Construction Lasers for Line and Grade.*

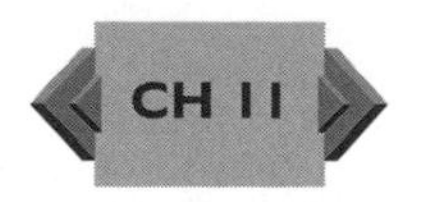

OPTICAL PLUMMET

If an optional vertical alignment instrument is used, it too must undergo tests to ensure it is in proper adjustment.

PLUMBING STRUCTURAL STEEL

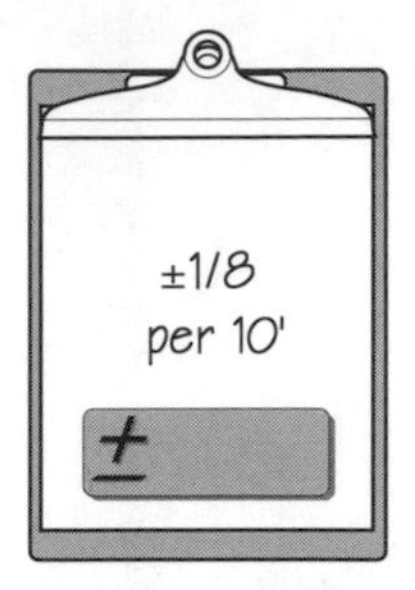

TOLERANCE

Unless otherwise specified by contract documents, structural steel should be plumbed within tolerances outlined by the *Commentary on the Code of Standard Practice, Section 7, A.I.S.C. Steel Construction Manual.*

SET UP OVER CONTROL POINT

The instrument is set up over a control point and a backsight is taken.

READ THE ROD

A rod or story pole is read from the offset line of the column.

EXTEND THE LINE OF SIGHT

The line of sight is extended to the area of work and the craftspersons are given instructions orally or with hand signals about the location of the column.

ATTACH TURNBUCKLES

Turnbuckles and cables will be used to move the steel in the direction needed.

TIGHTEN THE BOLTS AND CHECK

After aligning and tightening the bolts, a recheck of the backsight and foresight should be taken by inverting the scope about its axis and repeating the process. As is all operations with an instrument, everything must be rechecked twice.

OUT OF TOLERANCE TROUBLESHOOTING

If one of the perimeter or interior columns is out of tolerance, check the center to center spacing of the bolt holes from the beam framing into the column. Do not simply check the beam length. Column elevation should be checked to monitor any shortening or fabrication errors. Often, structural columns will be engineered to compensate for the compression experienced after loading. Be sure to consider this factor when determining column elevations.

STRUCTURAL CONCRETE AND METAL DECKS

SCOPE

Structural concrete and metal deck work consist of two major categories: Establishing line and grade and supervising the work of the crafts by checking their interpretation of the plans and specs. Line and grade must be completed quickly on each floor. As work progresses from floor to floor, the crafts will expect line and grade to be available as they need it. Contact should be made with each of the foremen to develop a schedule for establishing line and grade. All line and grade should be established off of primary or secondary control.

After the concrete is poured is not the time to closely read the plans. Many embeds, dowels, conduits, etc. will have to be placed in the concrete. Lift and blockout drawings should be developed for each floor and used to check off items to be installed or openings to be formed. A double check of the lift drawings should be completed just prior to the pour to ensure items have not been forgotten.

PLANNING

All concrete and deck control should be directly established from the primary control system. If existing primary control is not available to set the points, care should be taken to set secondary control points that can be referred to during the deck work.

Coordinate with the craft foreman or the general foreman to determine the control lines they want to use. Communicate to them the method of marking the line and grade information so there is no confusion.

Usually, giving more line and grade information than is necessary is good practice in case it is destroyed. However, like the deck work, many crafts will be using your control. Therefore, to avoid the confusion of having control marked everywhere, only mark control information that is to be specifically used by the crafts. Keep deck controls well labeled and consistent from floor to floor if possible.

As always, be sure that the concrete mix design is appropriate for the application. Check the specifications and confirm that the concrete delivered is correct.

Determine if floors are to be poured level or with a slope to compensate for column compression or differential settlement.

HORIZONTAL CONTROL (LINE)

Be sure to place control on decks as soon as possible so layout of edge forms, beams, blockouts, embeds, etc. can proceed quickly. Check plans for all slab penetrations. Confirm location from control lines and make sure blockouts are of proper size. Finally, check and recheck control during the slab/deck construction as movements may occur within the structure.

ELEVATIONS (GRADE)

Set up a level or laser in an area that will not interfere with the work of the crafts. Preplan this location. Setting up next to a column is best because the floor is more stable. Setting up on a metal deck is very unstable.

When marking slab elevations on adjoining walls and columns, mark grade 3-1/2" above concrete elevations. This is the common height of a finisher's screed. In this manner, the finisher can place concrete to grade without obscuring the grade line.

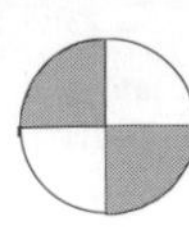

Be sure to check all pre-established concrete grades which need to be matched. Previously poured slab sections, door and elevator sills, etc. may need to be matched in elevation. If discrepancies arise, be sure to consider future impact of your corrective decision. It may be necessary to remove the already placed, but incorrect, material. Read the plans!

The embed plates at the edge of slab or deck that need specific coverage will dictate grade. Care must be taken to ensure that these plates have been located properly.

Door jamb locations should carry the same elevation to ensure consistent door head to ceiling height.

Carefully monitor deck deflection.

Excessive quantities of concrete being used to maintain a level surface may cause an overload condition that will impair structural integrity.

Most slabs are to be poured level. However, in some high-rise work, slabs may need to be placed at a slope to compensate for differential compression at the perimeter or core.

GRADING WITH A LASER

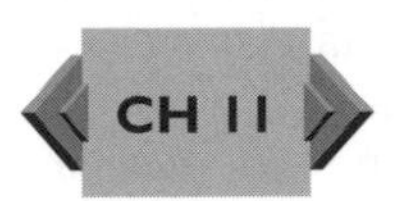

This instrument, although common, is still looked upon with awe, and as infallible. It is simply a level. See *Chapter 11, Lasers in Construction.* The same care is given to it that is used with normal leveling instruments. Always check your work with a calibrated instrument.

Set up out of the way of the work. Plan the setup to be able to maximize the readings without setting up again.

It is sometimes very convenient and acceptable to attach the laser to a column or wall to get it out of the way of work. Special mounts can be purchased (or made) to accomplish this.

As soon as the laser is set up and operating, a mark indicating the line of sight should be made on a nearby column. Throughout the day, the line of sight should be checked on this mark to determine if any instrument movement has occurred.

Use laser sensors to detect the beam if necessary. The sensors can be mounted to a level rod or to a 2"x 4". Care should be taken in initially setting the sensor and also in periodically checking its location on the rod.

CONCRETE PLACEMENT

Care should be exercised when planning the concrete placement sequence on decks to eliminate overloading. Shoring below should be constantly checked and monitored before and during concrete placement. Decks on structural steel or precast beams should be loaded center span first.

Field Engineer's Concrete Placement Checklist	
Develop Checklist	To be sure all materials are in place before pouring.
Check off the work of others	For location and grade of subcontractors.
Check	All reinforcing is in and properly placed: clearance, additional threaded dowels, etc.
Check	Electrical conduit of proper size and outlets at proper locations.
Check	Unit and duct hangers for HVAC.
Check	Embeds for ceiling grid hangers.
Check	Curb steel.
Check	If steel studs are used, make sure they have been inspected and meet specifications.
Check	To see that all slab penetrations are blocked out and checked for location and grade.
Check	Embedded plates for proper location and for correct embed designation and count.
Check	To ensure that bulkhead in pour are properly placed according to structural criteria, and are doweled and keyed. Also check that chamfer strips are in place.
Final Check	Make sure all mesh, conduit and other embedded items are below top of concrete with proper coverage.

On form decks which are carefully leveled to grade from below, finishers may "Pogo" off of the form surface — building a small metal stick which indicates the desired deck thickness and probing it through the concrete to the surface below.

After the concrete is placed and cured, it is too late to read the plans and find out that something is missing. So, check, check, check, and re-check your work!!

LAYOUT OF EMBEDDED ITEMS, SLEEVES, AND BLOCKOUTS

SCOPE

To construct a concrete structure with all of the embeds, sleeves, and block-outs in proper location requires a very methodical, systematic approach. There are countless details that must not be overlooked. Concrete embedded items include, but are not limited to, anchor bolts and base plates, rebar dowels, blockouts, beam pockets, rustication strips, sleeves (for mechanical, electrical, fire protection, etc.), miscellaneous metals, and hollow metal frames. Forgetting to locate just one of these will cost the company time, money, and headaches.

PLANNING

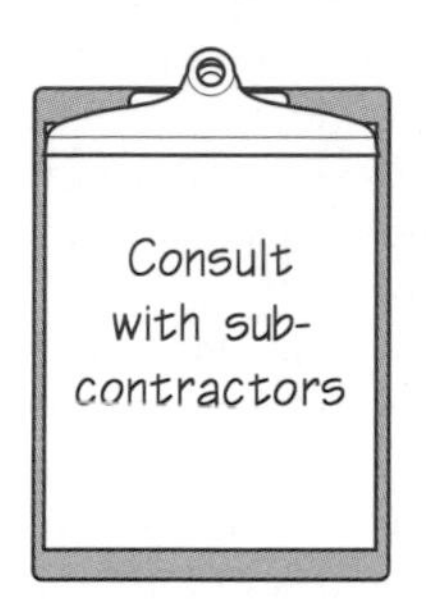

"As always, good planning is a key to success." A familiar statement and never so true as here. It is too late to easily place an embedded item after the concrete has been poured. The work may have to be redone. To avoid encountering that something that has been left out, plan well!

Begin your planning by thoroughly researching the contract documents to determine the embedded items that will be included in the concrete pour on which you are working.

Consult with the key subcontractor and carpenter foreman for their input.

After you have collected data on all embedded items, organize this information on a concrete lift drawing.

LIFT DRAWING - PREPARATION AND USE

Establish an elevation line on the lift drawing that will be the same elevation you will eventually establish on the formwork. Dimension all embedded items off of this elevation line.

Build a system to double check the elevations of all embeds either off of a separate elevation line or in relation to other embeds. Show these relationships on the lift drawing!

Establish one or more working control lines and/or offsets. Once again, the lines established on your lift drawing should be the same lines you intend to establish on the formwork.

Dimension all embeds from a minimum of 2 working control lines to allow for easy layout and checking. Label all embedded items as to their type, size, and use, for easy reference in the field. Watch for these problem areas when preparing the lift drawing:

Between embedded items--Are there overlapping conflicts?

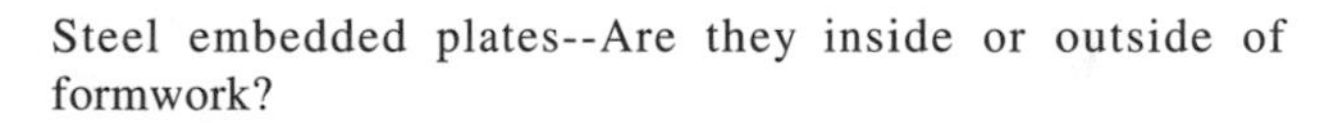

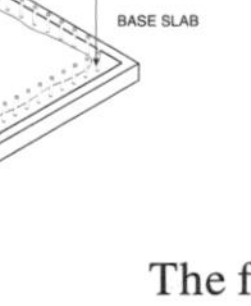

Steel embedded plates--Are they inside or outside of formwork?

Reinforcing steel--Are there heavily reinforced areas and conflicts with embed or blockouts? (Additional trim bars may be required around blockouts.)

Architectural concrete walls--Are proper blockouts attached? (Normal methods of nailing or bolting a block-out to the wall form or using conventional rebar tie wire usually result in rust stains on the architectural concrete face. A better method is using coated tie wire.)

Blockouts and Pipe sleeves -- Centerline of edge or flowline?

The final step in the preparation of your lift drawing is to double check your work for accuracy and completeness. It is always helpful to have a separate party double check your work to avoid overlooking mistakes.

LAYOUT

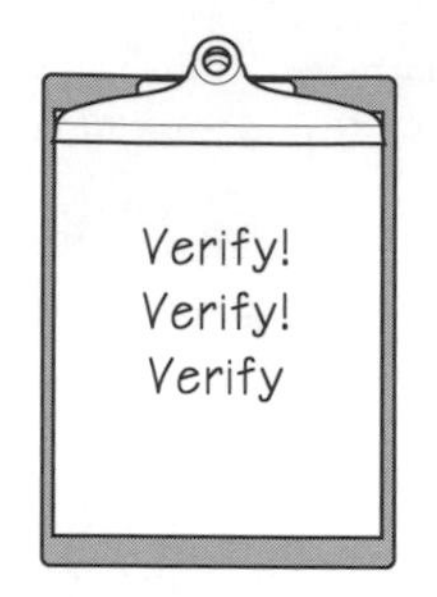

As soon as one face of the wall form is in place and securely fastened, establish your predetermined elevation lines and grid lines. If this information is established prior to any trades working on the wall form, it will allow all parties to work off of the same lines. It is much quicker and easier to perform layout prior to installation of block-outs and reinforcing in a wall pour.

Once the required lines have been established, lay out all embedded items from the lift drawing. All embeds should be clearly labeled as to their type and size from the information shown on the lift drawing.

Contact your field foreman and/or superintendent to discuss methods of marking on the formwork. Several possible materials include chalk (a different color than is being used by the carpenters), magic marker, paint, nails, etc. Extra precautions must be taken on architectural concrete walls to insure that the layout information does not appear on the finished product.

Double check all layout for accuracy and request that the carpenter foreman does the same.

Prior to setting the second form face, verify that all embedded items have been installed per your layout and lift drawing. If you do not verify, any errors will result in delays and additional cost.

CHECK, CHECK, AND RECHECK

While checking the final installation of the embedded items, the following things should be investigated:

Blockouts are securely fastened to the form surface.

Rectangle blockouts have square corners, and have sufficient kickers so they will not change shape under the force of fluid concrete.

Embedded sleeves, electrical boxes, etc. should have tape over all openings to prevent concrete from flowing in.

Verify that blockouts have chamfered corners if required by the contract documents.

Verify that reinforcing has the specified clearance for embedded items and form surfaces.

Failure to check and recheck may result in a jackhammer correcting your mistake.

CONCRETE COLUMN LAYOUT

SCOPE

As an integral part of a concrete structure, concrete columns serve as the load supporting members of beams and decks. The columns may be located in the base of the structure, never to be seen again, or they may be part of the architectural front, seen by thousands each day. Regardless of the location within the structure and the differences in specifications, the same care must be taken in establishing the locations of all columns.

A variety of forming systems may be used to construct concrete columns. The field engineer should become familiar with the type to be used and plan the layout accordingly. Close communication with the carpenter foreman is necessary to agree on the most suitable method of layout and marking the location.

PLANNING

Work on concrete columns should be based on primary control. During the planning of points on primary control, remember that column control points established on grid lines become useless as the columns typically sit on the grid line. More useful are offset lines that can be used for all columns on the grid line.

Before construction begins, the carpentry foreman should be contacted for input into the layout line locations. Are 2 foot offsets from centerline required? Does this forming system require the face of the concrete to be marked? Will templates be constructed to locate forming lines and dowel locations? Communication is the key!

Columns vary in location, size, elevation, mix design, rebar placement, etc. Read the plans to become familiar with what is being constructed. Highlight unusual columns on your plans so you do not overlook them during construction. A good method of rapidly determining everything needed concerning the columns is to prepare a column lift drawing. Reference *Chapter 5, Officework Basics, (5-11)* for a discussion on lift drawings. This drawings should show:

Column layouts that are repetitive from floor to floor, and the dimensioning of the columns from the grid or offset lines. This will provide a quick reference for layout.

Top of column elevation for each column.

Column size.

Concrete mix design for each column.

HORIZONTAL CONTROL AND LAYOUT

Concrete columns may sit on the top of a pier, caisson, slab, or deck; and they may be rectangular, circular, or irregular in shape. Regardless of where they are encountered or the design shape the fundamental principles of control and layout are the same.

COLUMN DOWEL LAYOUT

The first step to ensure accurate location of a concrete column is the layout of the rebar dowels which will penetrate the slab on which the column sits.

It is imperative that the dowels be embedded in a location where the rebar cage for the column will fit over the dowels and allow for proper location of the column form.

For dowels protruding from a concrete slab or pier cap, begin layout by establishing control adjacent to the column (from offset control lines).

Locate center of column in both directions. This can be done with string lines attached to batterboards.

Have the carpenter foreman build a rebar template for the maximum size the rebar dowel cage can encompass. Mark the template with column grid lines to allow the crafts easy identification.

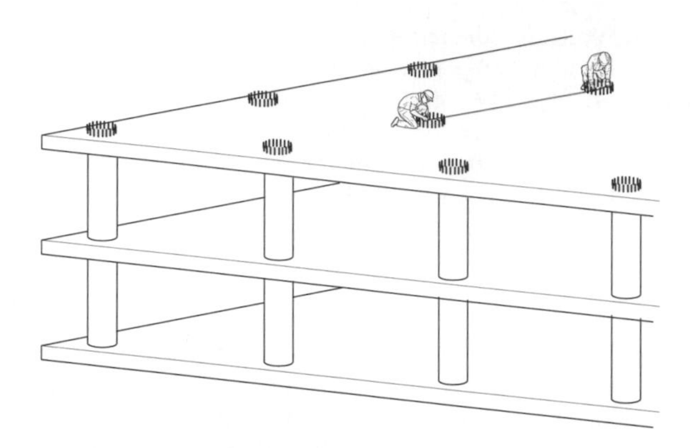

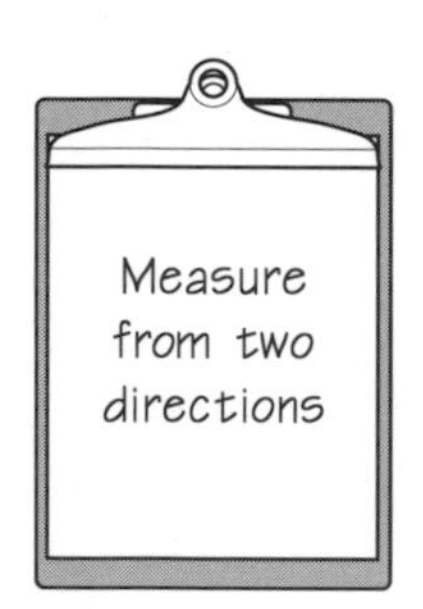

COLUMN LAYOUT ON CONCRETE SLAB

Begin column layout by establishing building control on the concrete slab. A minimum of 2 grid lines in each direction is recommended to allow for minimum chaining distances and double checking.

A good method for establishing grid lines on a concrete slab is a chalk line covered with a clear lacquer paint. This will protect the line for future use in layout of exterior skin, interior partitions, etc.

COLUMN LAYOUT ON RECTANGULAR COLUMNS

Establish the outside perimeter of the column on the concrete slab. This will be measured directly from the offset lines. This will allow the carpenters to set their column base plates directly from the column layout.

A wood template is often useful for ensuring that the top and bottom of the column remains square. Often, two faces of a column could be perfectly plumb while the column could be out of square resulting in an unacceptable product.

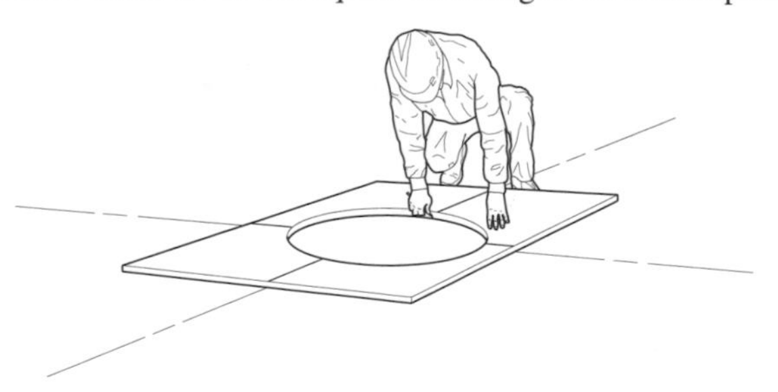

COLUMN LAYOUT ON CIRCULAR COLUMNS

Establish the grid line center of the column and use a circular template to lay out the base of the column on the concrete slab. This layout will allow the carpenter to attach the column base plate directly.

Another method for laying out the column bases is to establish the outside of the column base plate. This can be done for both rectangular and circular columns. To aviod confusion, it is recommended that this be discussed in advance with your carpenter foreman to reach an agreement on the most suitable method.

PLUMBING THE COLUMN

Once the carpenter crew has erected a column form, the form must be plumbed. Use a heavy plumb bob or shoot it with an instrument. Both rectangular and circular columns must be plumbed in two directions.

PLUMBING WITH A PLUMB BOB

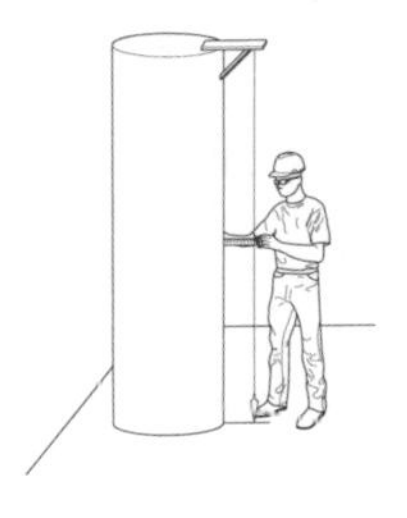

Close attention must be paid to several aspects of the plumbing.

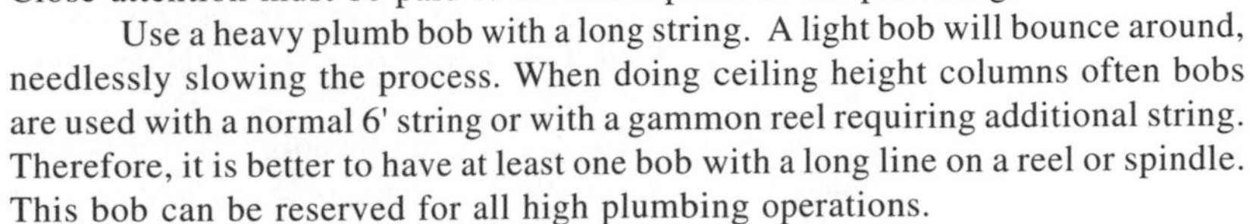

Use a heavy plumb bob with a long string. A light bob will bounce around, needlessly slowing the process. When doing ceiling height columns often bobs are used with a normal 6' string or with a gammon reel requiring additional string. Therefore, it is better to have at least one bob with a long line on a reel or spindle. This bob can be reserved for all high plumbing operations.

Watch for irregular offsets in the forms themselves. The forms may be constructed with 2" x 6"'s at the bottom and with 2" x 4"'s at the top. Vertically aligning the edges of those members would cause the inside of the form to be out of plumb.

PLUMBING WITH AN INSTRUMENT

There are also some aspects of plumbing with an instrument that require careful attention.

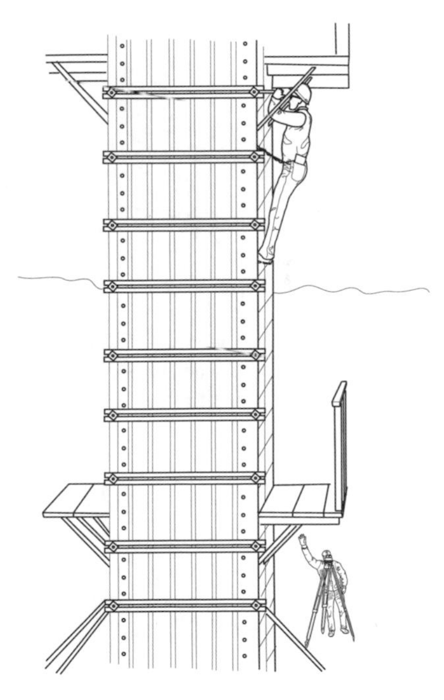

Make sure the instrument is level. Improper leveling of the level plate will result in the vertical crosshair not being truly vertical possibly resulting in a column which could be out of plumb.

Have the person holding the story pole or engineer's rule swing an arc to obtain a reading when the rod is exactly perpendicular to the line of sight.

Again, beware of irregular offsets in the forms. Watch the rodperson carefully to be sure offsets are all being measured from the same plane.

ELEVATIONS

Prior to placement of the concrete, top of column elevation must be established.

On plywood forms, top of concrete is normally established by nails driven to grade. This can also be made with a chalk line, but this line is difficult to find during concrete placement.

Another method would be to identify the distance from the top of form to top of column so the pouring crew can simply measure down. The distance would be clearly marked on the outside face of the form so errors do not result. This method allows for more human error than other methods and, therefore, is not strongly recommended.

The top location of concrete columns is often critical. Investigate the contract drawings for the connection of columns to the structure above to determine the allowable tolerance. Also, double and triple check before and after placement of concrete to ensure the proper elevation is achieved.

When placing columns to be used with cast-in-place deck systems, it is often advisable to place the columns 1/2" above the bottom of deck elevation. This makes the deck forming easier and eliminates concrete leakage during deck placement.

CHECK, CHECK, AND RE-CHECK

One useful double check when several columns are formed in a straight line is to string line or eyeball down the row of columns to ensure a true line of columns. However, be careful eyeballing round columns because their alignment can be deceiving. Check all columns during pouring to observe any movement and make adjustments accordingly.

CONTROL OF ELEVATOR AND STAIR CORES

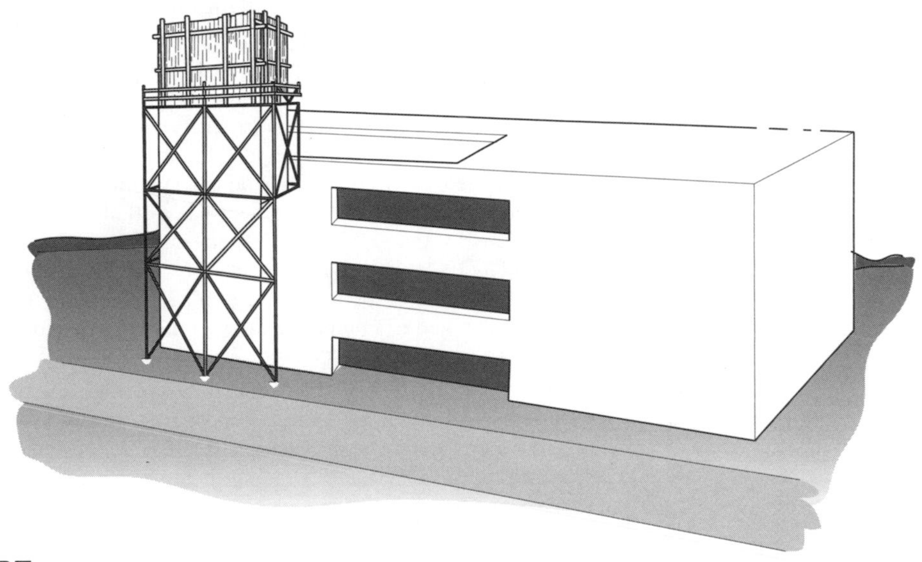

SCOPE

Controlling elevator and stair cores follows very closely with the field engineering work associated with structural concrete and structural steel. The core is an integral part of the structure and must be given more attention than other phases of the construction project. Core location is particularly critical in high-rise work. Structural steel often has to tie into a concrete core, fitting prefabricated steel members (beams, columns, trusses) into the concrete core structure. Any mislocation of the core can easily make the members too long or too short to fit and out of plumb. (Refer to AISC Manual)

This section lists several ways to control the core location as the construction occurs. As with any type of construction, the foundation is very critical; extreme precision is absolutely necessary. The entire core work depends on the accuracy of the base of the core.

Throughout the height of the core work, vertical alignment and horizontal control of twist must meet exacting tolerances. The elevator subcontractor will base work on how well the core is controlled. With elevators and stairs being part of the critical path of most schedules, the field engineer must work well so others can work well.

PLANNING

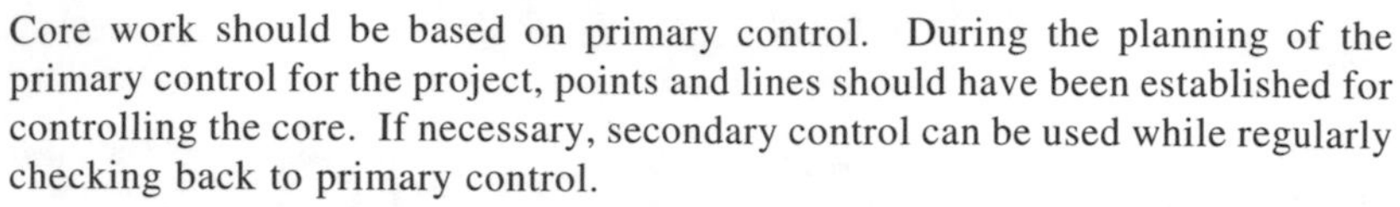

Core work should be based on primary control. During the planning of the primary control for the project, points and lines should have been established for controlling the core. If necessary, secondary control can be used while regularly checking back to primary control.

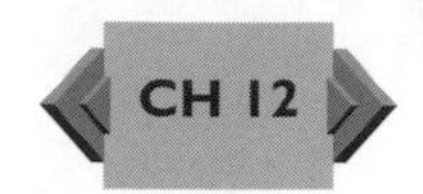

Check equipment for proper adjustment. Refer to *Chapter 12, Equipment Calibration.*

Before construction occurs, one should establish the procedure to be used for all core work. Look at all aspects of the job to gain insight into what access will be available to perform alignment work. Will the core be slipformed the entire height before construction starts on the main structure? Will the core precede the decks? Will there exist any shafts or mechanical cases that can remain empty during the entire core construction time to allow for visual access of the core's plumbness? Can the core be visually checked from the ground without obstructions? How can checks be performed to be absolutely sure your location system is correct? These procedure questions and others must be answered before decisions on a proper system to locate the core can be made.

The shape of each core is usually unique. Some are rectangular, some trapezoidal, some circular, some irregular, and some unsymmetrical. Because of this uniqueness, no one method of control can be used in all cases. Varying the offsets, using coordinate systems, establishing several planes to check the concrete face, plumbing by lasers, etc., may be necessary. Careful thought should go into the method of control prior to the beginning of construction. The following are suggestions when encountering the various types of cores:

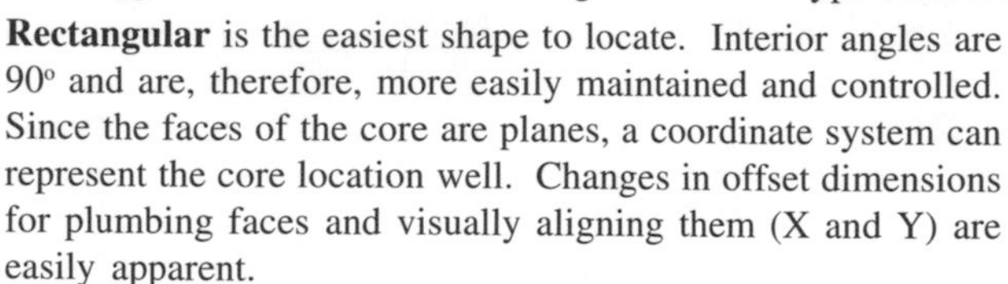

Rectangular is the easiest shape to locate. Interior angles are 90° and are, therefore, more easily maintained and controlled. Since the faces of the core are planes, a coordinate system can represent the core location well. Changes in offset dimensions for plumbing faces and visually aligning them (X and Y) are easily apparent.

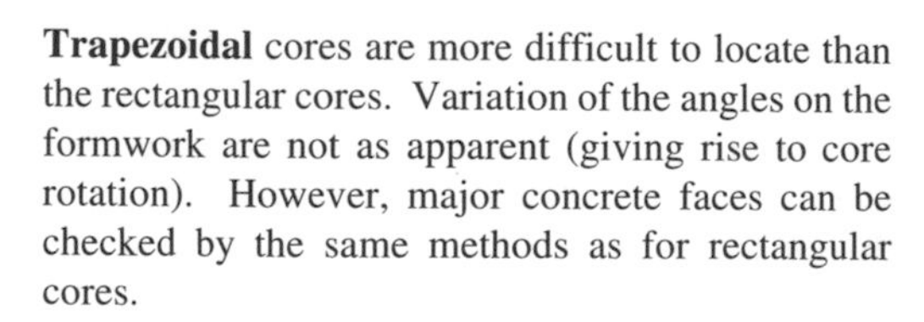

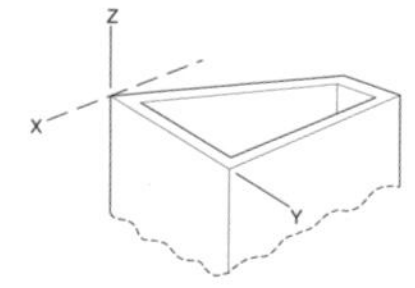

Trapezoidal cores are more difficult to locate than the rectangular cores. Variation of the angles on the formwork are not as apparent (giving rise to core rotation). However, major concrete faces can be checked by the same methods as for rectangular cores.

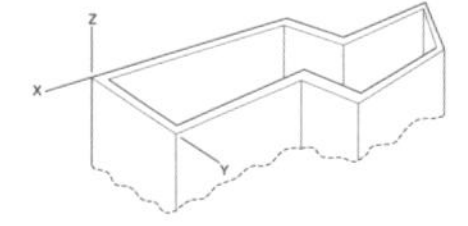

Irregular, unsymmetrical cores are the most difficult to locate. Use of a Cartesian coordinate system is difficult—although it is the most effective method available. Planes can be established from which to check the concrete faces and for rotation of the core.

If deck work is close behind the core, the deck controls can be used as the principal control lines for the core. (Don't forget to check into original primary control.) The close proximity of the deck controls will provide generally easy access to the core.

VERTICAL ALIGNMENT

There are a variety of methods of plumbing the core, including using the plumb bob, a theodolite, or with the laser. The method will depend on the equipment available and the access provided to the core.

USE OF THE PLUMB BOB

A plumb line off of formwork works very well and maintains core location. While ensuring that formwork dimensions and angles remain constant, drop a plumb line from the top of the core, perpendicular to the X and Y faces. The plumb bob works well when used properly. Some points to remember are:

The wind affects the stability of the line. It is best to use piano wire to reduce wind resistance.

A heavy plumb bob, 2-3 pounds, should be used. To reduce movement, the bob should be immersed into a barrel of oil.

The line creates physical limitations to the work. Because of its existence, other work must avoid disturbing the line.

Magnetic forces created by electric motors can measurably affect the line. Precise plumbing should take place when nearby motors are not operating.

USE OF THE LASER

The laser works extremely well as a vertical aligning tool. A major advantage is there is not a physical line to interfere with other work. An initial disadvantage is cost. However, this can be quickly recovered in increased productivity.

In core work, there are situations that adversely affect the laser's precision. BE CAREFUL:

When the laser beam passes through a cold shaft into a heated room, it will have a tendency to be displaced by refraction through these temperature layers. This temperature shear will be apparent as the laser beam appears to bounce around. It is a problem in very large cores or when the laser is used in pipelines. This can be avoided by using the laser at night or in early morning hours when these temperature conditions are not present.

Pressure differentials also exist throughout the core and can cause problems similar to the temperature shear. Pressure differentials are caused by many things including drafting through the core. Nighttime or early morning work will minimize the pressure differentials.

Accuracy of lasers is generally taken for granted. However, lasers are tools—they are not calibrated instruments. This becomes very apparent when used over long distances. The automatic electronic plumbing device on the laser is never absolutely correct and will give slightly incorrect results over long distances. This problem is easily detected and can be avoided by rotating the laser about its base. The average of the points marked during the rotation is the actual position of the point. Reference *Chapter 11, Construction Lasers for Line and Grade.*

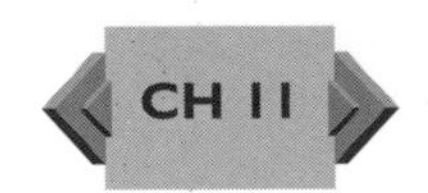

USE OF THE TRANSIT OR THEODOLITE

Use of the theodolite or transit is another acceptable procedure for aligning a core. As stated before, care must be taken when using any instrument for extreme vertical operations. There are a few additional considerations when using transits or theodolites to visually align the core.

One can establish an offset off of the concrete faces and shoot the formwork. However, access for a clear sighting can be difficult. When shooting up to the core, quite often it is difficult to see around the formwork and supports. Also, a Philadelphia rod may be difficult to carry or to hold horizontally over the side for a shot. A smaller, stiff engineer's rule is usable and is readable up to about 80'. When using a rod or rule to shoot the core offset, always swing an arc with the rule to be sure the reading is perpendicular to the line of sight. As with normal leveling, always read the smallest reading throughout the arc.

A better line of sight is available when checking concrete rather than formwork. You are too late to affect the location of that part of the core. However, future lifts can be corrected. By checking both ends of the wall, core rotation (twist, skew) will be apparent.

ROTATION

As with structural steel, if a plumb line or laser line of sight is being used to plumb the structure or core, at least two lines must be used. If not, the structure will be able to rotate or twist without detection. Always use at least two plumb lines or laser lines of sight when vertically aligning a structure!

EXTERNAL FACTORS WHICH AFFECT PLUMBING

Vertical alignment work should take place early in the day before the following factors can have an adverse effect on the plumbing:

Temperature: The sun causes the structure to expand as the temperature rises. Precise plumbing in necessary early in the day to avoid this effect.

Tower Crane: If the tower crane is attached to the structure, the tall structure will move in the direction of the load on the crane. This makes plumbing difficult. Precise plumbing is necessary early in the day or when the crane is not busy to avoid this effect.

Wind: When it is extremely windy, work elsewhere. Even a moderate wind will cause a tall structure to sway. Precise plumbing is should be done when the air is still.

ELEVATIONS

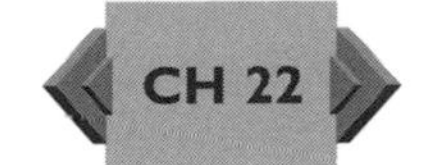

Elevations necessary for core construction are first established by a temporary benchmark near the foundation of the core during the footing and foundation work. The elevations are then transferred to several positions along the side of the core by normal leveling techniques. Next, they are transferred up the side of the core by using an engineer's chain at several locations around the core. See *Chapter 22, Leveling,* for a discussion of transferring an elevation. Then they are transferred around the formwork using an engineer's level, water level, or rotating laser.

Most normal chaining procedures can be used. However, there are a few variations. One variation is that the chain must be held vertically. A plumb line should be used to insure that the chain is not out of plumb. A second variation is that the weight of the chain should supply all of the tension necessary. Do not pull hard on the tape in this position as it will cause stretching and inaccurate readings.

SETTLEMENT

Don't forget to constantly check, check, and re-check the TBM at the foundation level to detect any settlement that might occur during the construction process. If settlement does occur, it may be necessary to rechain up the core to compensate for this shrinkage and displacement. This is most important when the building is of a composite nature with the core being of a different material than the superstructure. Ignoring this possibility could cause the floors to be sloped unacceptably.

Total settlement can be estimated using factors such as soil conditions, foundation design, and total loads. Compressive strain can be estimated by analyzing the combination of all live loads, dead loads, and external loads (structural steel connections). One can then build the core higher to account for the total displacement. Contact the structural engineer for confirmation of any adjustments that are found to be necessary.

LOCATION OF EMBEDS, BLOCKOUTS, ETC.

Care should be exercised in placing embeds and blockouts off of primary control. Checks of the specifications and shop drawings must be made of the tolerances for these items. Metal frames and closures for doors should be compared to blockout sizes. Core location will directly affect the embed location and variances. Typically, the embeds are bolted to the formwork in the same position following the core exactly. However, do not assume this to be the case. Always check to verify there are not subtle changes from lift to lift. Typically, in complex high-rise structures these embeds must fall within ½ inch or less of the correct location. This illustrates the importance of the core location.

SIDEWALKS, CURBS, AND GUTTERS

SCOPE

Construction of the sidewalks, curbs, and gutters generally indicates that the job is winding up and the layout work of the field engineer is almost complete. After having been involved with the critical layout of items that directly affect the structure, sidewalk work may seem insignificant. That isn't the case at all!

Sidewalk, curb, and gutter work represent some of the most visible indications of the quality of the field engineers work. No one likes to open a car door and step into a puddle that has not drained away or trip on a crack of a poorly laid sidewalk. The same quality of work must be maintained for these activities as was maintained for the rest of the project. Just because the end is in sight is no reason to sacrifice quality.

PLANNING

Depending on the project, sidewalk, curb, and gutter work may take no more than a day or, if it is extensive, it could last months. Regardless of the time involved, good planning is a must to finish the job on time and within cost. Not only must the equipment, material, and labor be well scheduled, but there are various inspections that must be arranged. There must be close coordination with the affected traffic management departments.

Sidewalk, curb, and gutter work cannot start until the site is cleaned and the major landscaping is complete. Cleanup should be a topic of concern throughout the project. However, towards the end of the project, it becomes a major activity. At weekly meetings, the subcontractors and foremen should be advised of the upcoming sidewalk, curb, and gutter work, and the need for cleanup in their areas so the work will progress efficiently.

The location of sidewalks, curbs, and gutters along the property lines is very critical. The backs of curbs on both sides of a street are often used by licensed surveyors to locate the center of the street, and ultimately to assist in the location of the property lines. Therefore, the location of the sidewalks, curbs, and gutters must be accurate. Property corners used in the design and layout of the primary control system, should again be located and used to perform the layout of sidewalk, curb, and gutter work. If in doubt about the location of a property corner, call the licensed surveyor who established the original civil drawings used at the beginning of the project.

CURB AND GUTTER CONSTRUCTION

If the exact location isn't given on the architectural or civil drawings, have the licensed surveyor who prepared the property survey indicate where the back of curb should be located. (The City Engineer's office may also be able to provide this location.)

LAYOUT

Establish working control lines, typically 1'0", off back of curb towards gutter, allowing easy control transferring after curb and gutter are placed. Cross reference architectural and civil drawings for flow line location, then adjust for curb configuration with 1'0" off back of curb.

Lay out section of curb and gutter and stake, typically 2'0" off back of curb.

If 2'0" offset is used, place a nail in the hub exactly 2'0" off back of curb and shoot grade. Write the + elevation difference from top of the nail to the top of the curb on the hub. See civil drawings and check architectural drawings for elevations.

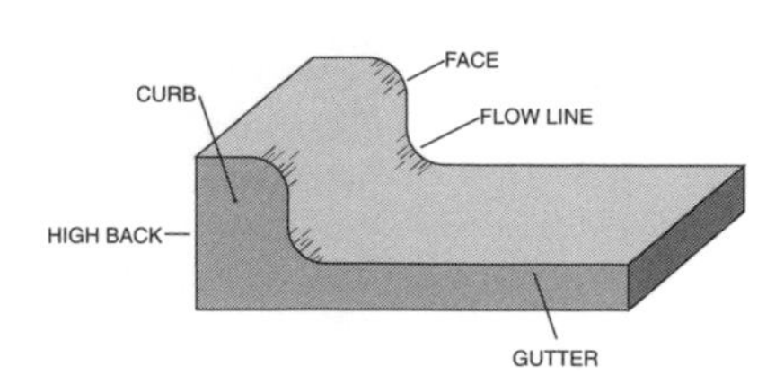

SIDEWALK CONSTRUCTION

Sidewalks may be tight against the back of curb or may be located in the open with nothing around them. They are not as critical for property location as curbs. However, surveyors do use them if the curb and gutter have been destroyed. Ask the licensed surveyor about their location.

LAYOUT

Check civil and architectural drawings for construction joint layout. These construction 0' expansion joints are critical if stone work or pavers are placed on top of the sidewalk.

Lay out construction joints and check subgrade for correct elevations. Check contract documents for minimum and maximum slopes. Double check local ordinances.

Lay out light boxes and street lights Schedule electric utility or government agency for setting any precast bases.

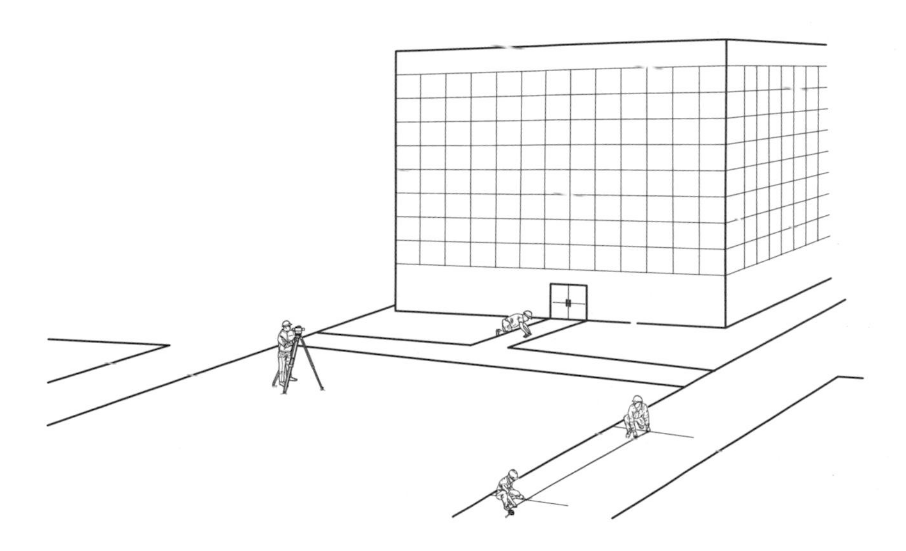

ONE PERSON SURVEYING TECHNIQUES

25

PART 4 - CONSTRUCTION LAYOUT APPLICATIONS

CHAPTER OBJECTIVES:

Explain why one person surveying techniques are important to the field engineer.
Describe the various methods of establishing a backsight.
Describe how the level rod can be used as a tool for establishing line.
Explain how the heighth of instrument can be determined when working alone.
Describe how an old tripod can be used as a tool for checking grade.
Explain how a field engineer can chain a distance alone.
Define wingding and explain its use on the construction site.
Describe the use of the double right-angle prism as a layout tool.
Identify a process that can be used to establish a permanent instrument set up on a construction site.

INDEX

ONE-PERSON SURVEYING TECHNIQUES

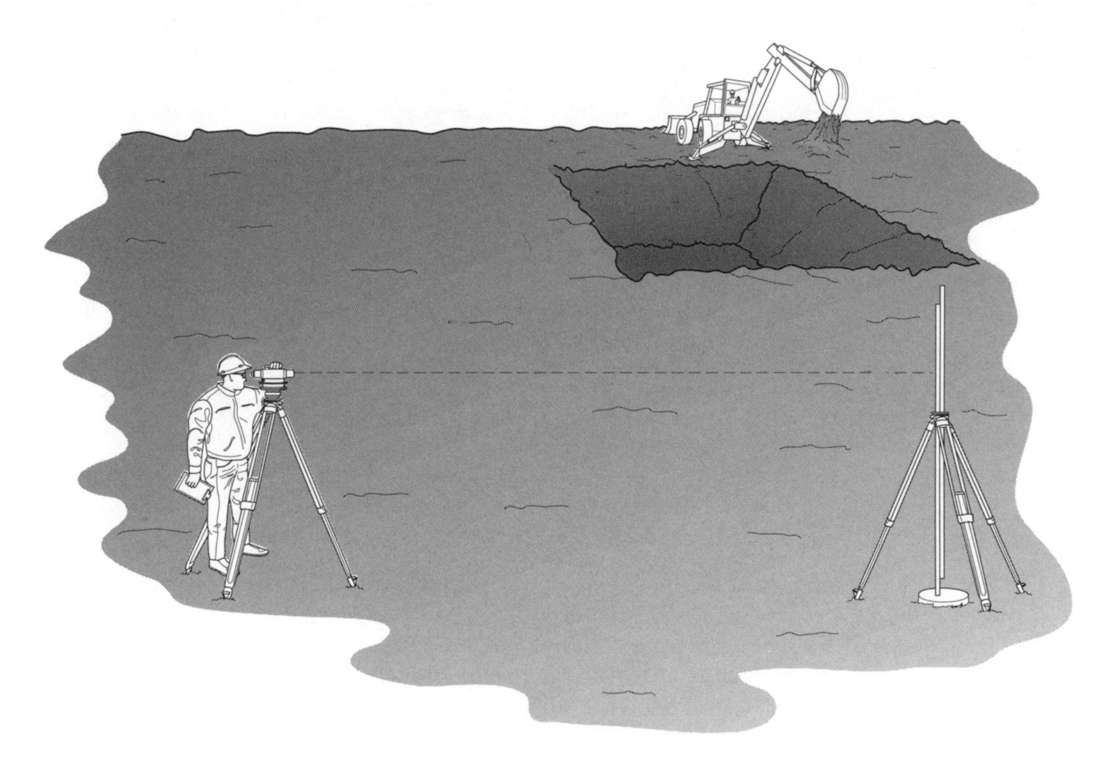

SCOPE

Many times, only one field engineer is assigned to a small job. How is it possible for only one person to perform construction layout? How can lines be measured? How can elevations be determined? How can alignment be given? It isn't as easy as when two people are available, but all of these activities, and many more, are completed every day by field engineers working alone on a project. A lone surveyor is like a woodworker. An efficient woodworker figures out what is necessary to build a jig to do a repetitive task. The jig then acts as a second person in the activity. A field engineer has to figure out how to create "jigs" in the field to act as a second person. It may be frustrating to constantly create unique methods to survey alone; however, the accomplishments can be very satisfying.

GENERAL

A single field engineer on a project can either perform the measurement and layout tasks alone, or ask someone else on the site to help. Probably the field engineer will do both. This section will cover one-person techniques in layout which are used when working alone, or when working with a laborer or carpenter. The field engineer will have to teach the individual helper how to hold the chain or rod, etc. Remember, sometimes the needed measurement will require a lot of planning and creativity, but where there's a will, there's a way.

ESTABLISHING A BACKSIGHT

Obtaining a backsight by yourself requires some planning and a little extra effort when setting a point, but it is a simple task that should be used all of the time.

ON A HUB

When setting a point, take the time to drive a nail directly on line behind the cupped tack that marks the point. This nail can then be used to sight on (if there aren't any obstacles in the way) from an instrument set up on the same line. To make the nail more visible, put a piece of surveying ribbon on it. If the hub is behind grass or dirt, consider hanging a plumb bob from a stake or stick.

ON A WALL

Often, when setting a point on a line, there are buildings directly behind the point where a small target could be placed. Of course, you should ask for permission to do so. Explain that it is temporary and that you will remove it after the project is complete. Generally, a small plastic target can be glued to the wall and will not leave any noticeable marks when removed.

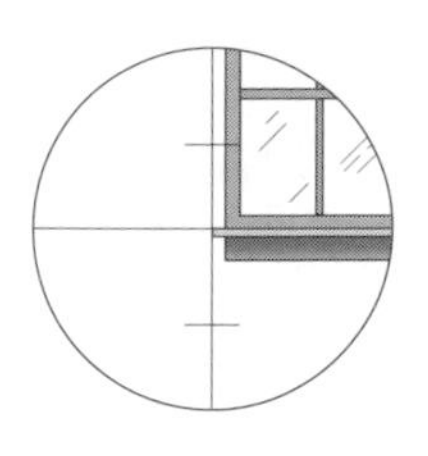

The owner of a building may refuse to give you permission to place a target. While your instrument is sighting on the point, move your line of sight to the building. Look for any distinguishable mark on the building you can come back to anytime you need a backsight on the line you are setting. Possibly the line of sight will hit a crack in the wall just above a window or will hit the edge of a joint, or something similar that can be described in your field book and be used by anyone using the line.

By having a nail in a hub, a target on a wall, or some other type of backsight available at all times, one person can perform many of the layout activities on the jobsite.

ESTABLISHING A LINE

Marking a line can easily be accomplished alone. For example, here is a method that can be used to mark the center line for a column form on a concrete slab.

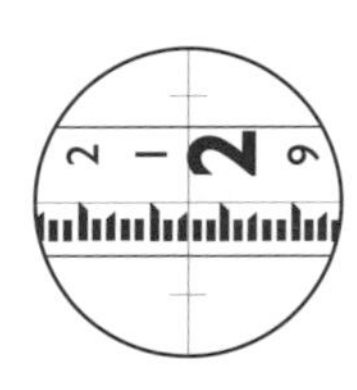

After an instrument has been set up and a backsight has been taken on the nail or target described above, turn the angle required for the layout and sight to where the line is required. Since you are operating alone, no one is there to mark the point. You will have to walk there yourself and mark it. Take a level rod or something else you may have that has graduations on it. Knowing about where the line of sight is, lay the rod down facing the instrument. Go back to the instrument and read the rod with the vertical crosshair carefully. Return to the rod and mark the reading on the surface. That mark represents the point you wanted to establish.

This procedure can be expanded to set points on both sides of a column so a straight edge can be used to establish the centerline of the column. Furthermore, the instrument can be moved to a control line 90 degrees away to establish the centerline of the column in the other direction using the same procedure.

This method of using a rod or something similar that is graduated can be used time and again to establish line when working alone.

SETTING TEMPLATES

A similar procedure to the one described before can be used to set templates in place. Typically, templates are used to hold anchor bolts in place during concrete placement.

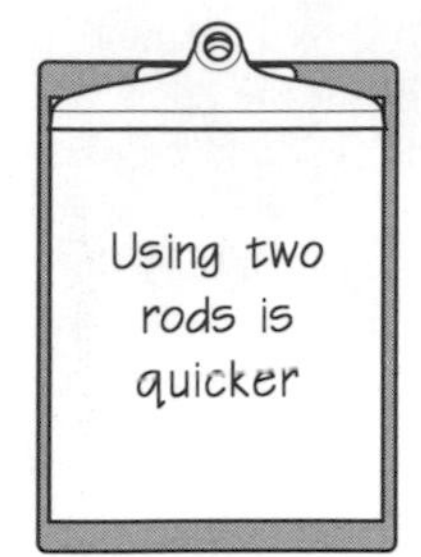

After the carpenters have drilled the holes for the anchor bolts and have established the centerline of the template in both directions, hammer a nail into the template at the end of each centerline. Place the template in the approximate location and put your graduated rod on one side of it. Go back to the instrument and carefully read the rod—go to the template and move it until the nail is at the rod reading. Repeat this process for both sides of the template in both directions until the template is correctly positioned. Secure the template to the form.

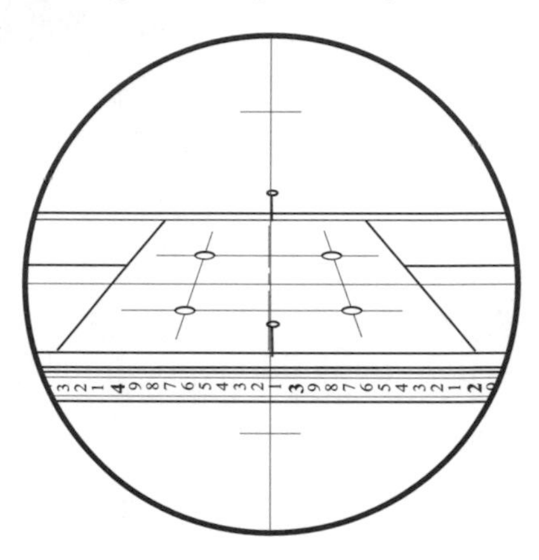

OBTAINING AN H.I.

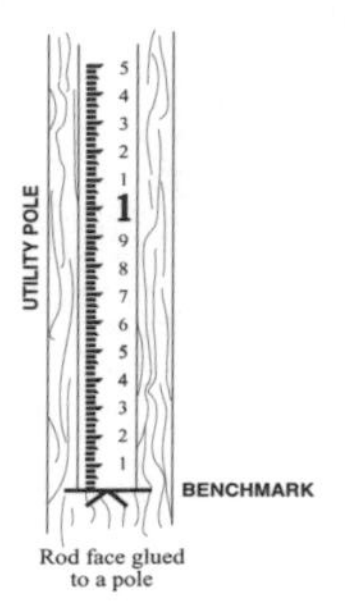

Rod face glued to a pole

Obtaining an H.I. to begin leveling requires some preliminary work. When a benchmark is established, it must be located near something that is vertical such as a power pole, a building, etc. Locate a place to attach the face of a rod on the object. Put it on so the bottom is at an even foot and attach it firmly. Now, whenever you need an H.I., set up the instrument so the face of the rod can be seen, read the rod, and add the reading to the even foot elevation. Now, you have an H.I. for the work you are ready to do.

CHECKING GRADE

How can you stand at the instrument and hold the rod at the same time to do one-person grade checking? It isn't always easy, but it can be done.

An old spare tripod might be used as a support for the level rod. Modify it so that the rod can be held in place while you walk back to the instrument to take a reading. Large rubber bands can be stretched to hold the rod firmly. Or, a C-clamp can be used to hold the rod to the tripod. Equipment suppliers sell a device for just this activity. If you have an old level rod that is still usable, consider attaching 1"x2"'s to it with screws to serve as tripod legs.

MEASURING A DISTANCE

Measuring a long line can be rather difficult alone, but pulling a single chain length can easily be accomplished.

When measuring from a point, see if a piece of rebar can be driven into the ground behind the point. If this is possible, it can be used to attach the chain. The rebar will replace the rear chain person in the leveling process. Use tie wire to attach it, trying to line up zero over the point. If it becomes time consuming to line up zero, just determine what the reading is and add or subtract it from whatever is read or laid out.

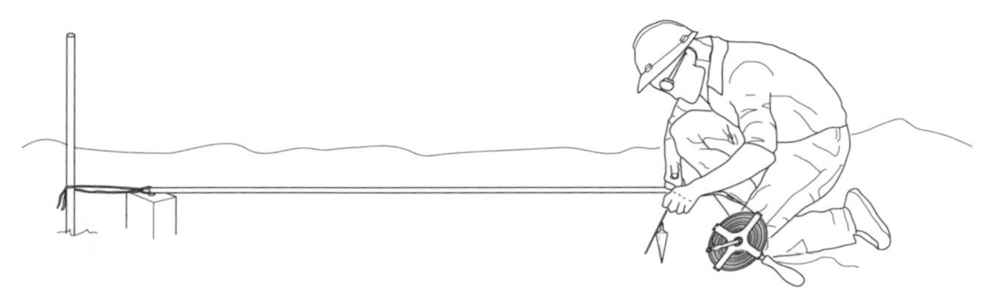

CHECKING PLUMB

Checking a column for plumb is as easy as setting up an instrument. Simply set up the instrument approximately parallel to the side you want to plumb. Level the instrument exactly and sight at the edge of the column. Move the vertical crosshair up and down on the edge of the column to determine if the edge of the column is parallel to the vertical crosshair. If it is, the column is plumb in that direction. You will have to move the instrument to be able to check the column in the other direction. If the column isn't plumb, contact the work foreman for action.

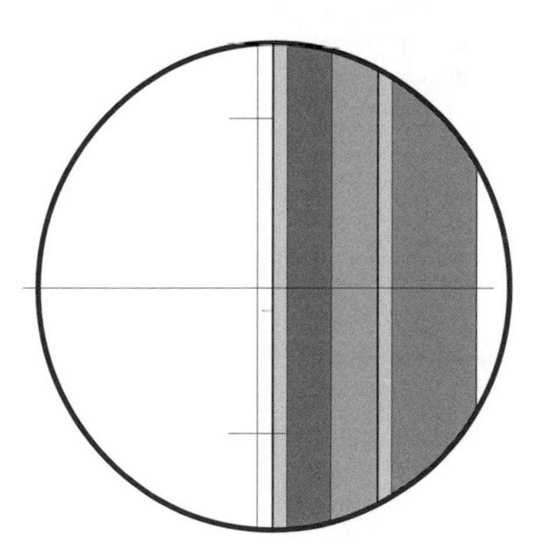

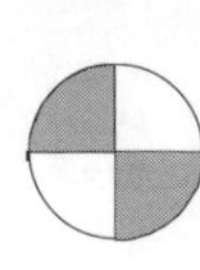

PERMANENT SETUP

Sometimes a job can be laid out from one instrument setup. If the job is going to last awhile, it may be to the advantage of the field engineer to consider creating a permanent setup point.

To establish a permanent setup, the field engineer must locate a point which will be used throughout the project. This point will probably be part of the control network. Have a hole drilled (10 to 30 feet deep) at the point, place a 24" dia. tube form in the hole and fill it with concrete. Place an old set of tripod legs in the concrete at the exact location of the point. After the concrete hardens, the tripod is ready at all times for a setup to lay out a structure. In an instant, the field engineer can be set up and ready to turn angles and give line, or set up the level and be ready to shoot grade.

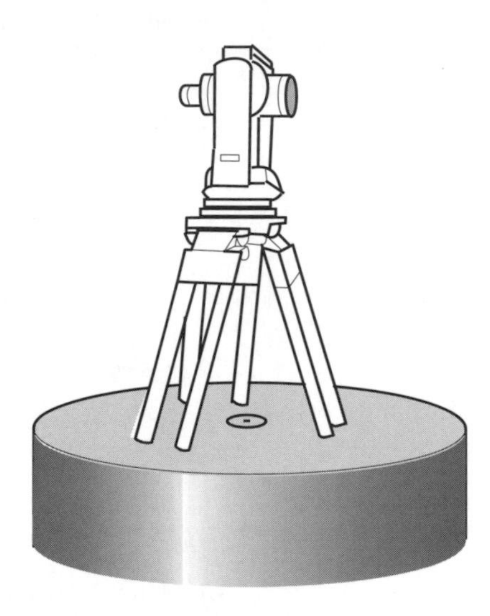

This method of establishing a permanent setup was especially helpful on a US Naval Station project in Texas. Such a point was established in the shifting sand on the centerline of a 1500 foot pier. It was a very effective method of controlling the project.

OBTAINING AN APPROXIMATE RIGHT ANGLE

Often when checking work or setting rough clearing stakes, it is useful to obtain an approximate right angle. Obtaining an approximate right angle can be used when an exact right angle with an instrument isn't necessary, or a right angle prism isn't available. The method that has been used since the beginning of time is commonly called the "wingding." It consists of simply using your arms to measure the 90 degrees. This process is illustrated and explained in the table on the following page.

The Wingding Right Angle		
1	At the point on a line where a ninety degree angle is needed, carefully position yourself so that you are standing exactly on the line.	
2	With your feet aligned 90 degrees to the line, hold one arm out and sight down the line, moving your arm until it is point exactly on the line. Hold it steady on the line.	
3	Rotate your head to look in the opposite direction and raise your other arm and sight on the line, moving your arm until it is pointing exactly on the line. Be careful not to move your other arm.	
4	Rotate your head right and left confirming that both arms are sighted on the line.	
5	Close your eyes and quickly bring your arms together in front of you. Match up your fingertips and your hands should be pointing in a direction that is approximatly 90 degree to the line.	90°

ESTABLISHING A PERPENDICULAR LINE WITH A PRISM

If an angle turning instrument isn't available and a perpendicular line is needed, the double right-angle prism can be effectively used in many instances. The double-right angle prism is often the tool of choice when measuring cross-sections or setting slope stakes. This tool consists of two pentagonal prisms for line of sight to the right and left. Using a double right-angle prism isn't difficult. The following illustrates how to use the double right-angle prism for establishing lines perpendicular to a centerline.

STEP 1

Attach a plumb bob and string to the bottom of the double right-angle prism and hold the prism horizontal and exactly over the point on the centerline where a perpendicular is needed.

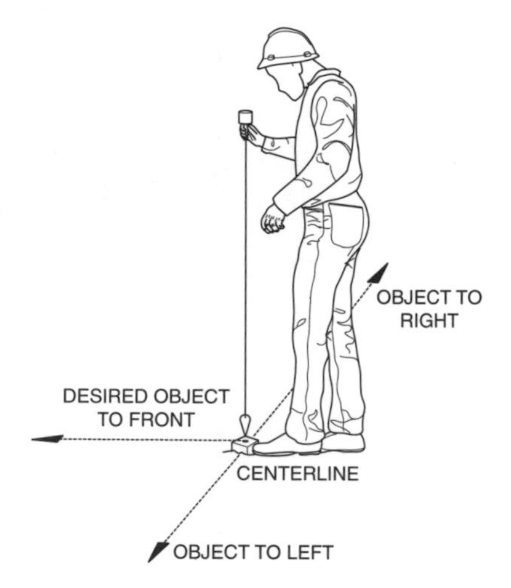

STEP 2

Open up the protective lens cover on the prism. Looking directly at the exposed glass, peer through the middle glass at an object directly ahead.

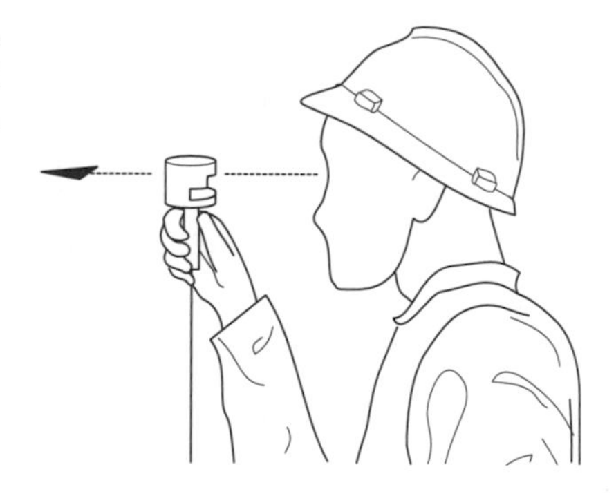

STEP 3

Look at the top prism and observe features that are at a 90 angle to the right and to the left of the prism. Look at the bottom prism and observe features that are at a 90 angle to the left of the prism.

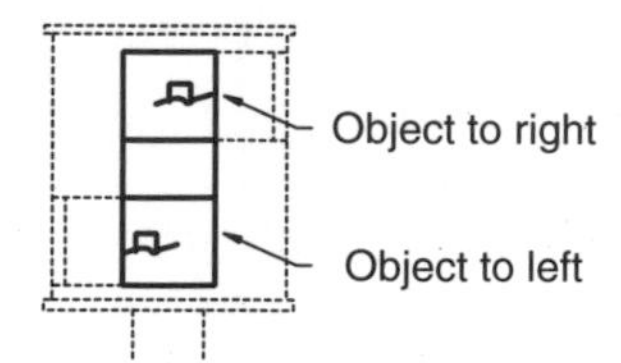

STEP 4

Align the images in the top prism, and the bottom prism with the view straight ahead.

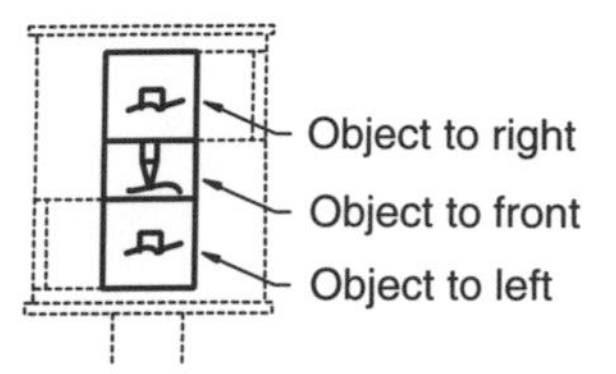

STEP 5

When all three images are in a vertical line, the point straight ahead is perpendicular to the line.

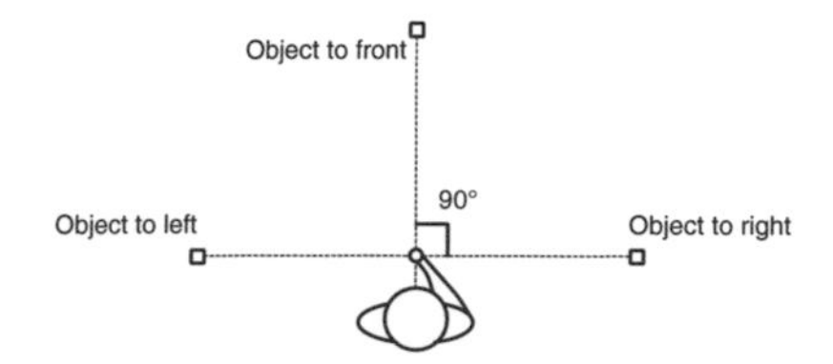

SUMMARY

This chapter lists only a few of the activities that have been accomplished by one-person working alone. Undoubtedly there are many more.

THE PUNCH LIST

26

CHAPTER OBJECTIVES

Describe the common sense approach to checking construction layout work.
State the rules of thumb for eliminating blunders in construction measurements.
List the common mistakes that should be avoided in chaining a distance.
List the common mistakes that should be avoided in leveling.
List the common mistakes that should be avoided in angle measurement.
List the common mistakes that should be avoided in layout.
List the common mistakes that should be avoided in layout calculations.
List the common mistakes that should be avoided in notekeeping.
List the common mistakes that should be avoided in plan reading.
State the ten commandments for field engineers.

INDEX

COMMON-SENSE CHECKS OF CONSTRUCTION LAYOUT

SCOPE

Is it laid out correctly? Does it look right? These are fundamental questions that all field engineers should ask themselves after establishing any points in the field. This section provides some helpful hints that have been suggested by experienced superintendents on what they looked for when they were field engineers and what they continue to look for as they walk the jobsite now. As can be expected, most of their suggestions are common-sense approaches to reviewing the work.

STEP BACK, LOOK AT IT, AND QUESTION IT!

After laying something out, walk 50 or more feet away from it and look it over. Eyeball things for proper alignment, grade, etc. Does it look like you visualized it? Do the distances look right? Do the columns line up? Are they supposed to?

Has a common-sense approach to layout been used? Could a different approach be taken? Were there violations of standard practices?

Watch for too much control in a small area. Several lines or stakes within a few feet of each other is confusing. Look for improper markings on stakes. In haste, someone may have marked a stake as being offset 5 feet when it is actually 10 feet.

REMEASURE IF NECESSARY

Come in from a different direction to check critical points. This is a method that must be used frequently, especially when coordinates are being used as the basis of the layout system. The only way to discover some mistakes is to do it differently.

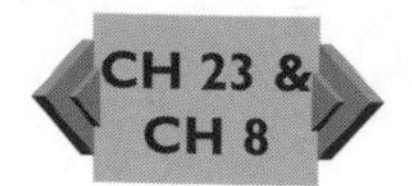

Cross-check measured distances. Measure diagonals if possible. Use the 3/4/5 method to check 90° angles *(Chapter 23, Construction Layout Techniques)*. When measuring angles, use the Close-the-Horizon method of summing the angles to check their total *(Chapter 8, Angle Measurment)*.

Check dimensions of work already in place. Check old work to see if they meet the tolerances established. Check new work to see if you are obtaining the same results. Any deviation indicates a problem in procedures or instrumentation.

CALIBRATE IT FIRST

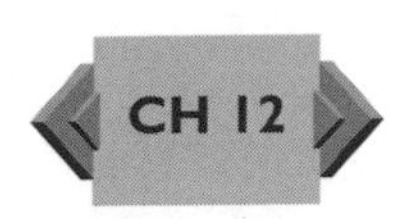

New equipment should always go through a calibration check. Calibration should be preformed for transits and theodolites, levels, site levels, rods, and especially chains. Theodolites should be checked for proper internal geometric relationships. Levels should be pegged. Reference *Chapter 12, Equipment Calibration.*

Only after each new piece of equipment has been personally checked, should it be used for layout work.

GET ANOTHER OPINION

If the points are very critical and the outcome of the project depends on the points being absolutely correct, it may be necessary to ask someone else to perform the check another crew or even an outside consultant. If others do come in, they should check using independent methods and their own equipment.

If you don't have the luxury of another crew, or money to spend on an outside consultant ask the most experienced person on the jobsite to review your methods and procedures. Your work will be looked at from a different perspective and detailed questions will be asked. A second opinion is always helpful.

CHECK CONTROL

If problems in layout are encountered, don't forget to check the project control. Are the project control points disturbed? Has heavy equipment been working next to the control points? Measure angles and distances at the control points to confirm their location. Recalculate the coordinates for the control points and compare them to the original calculations. Remeasure angles and distances from the primary control points to secondary control points. Any deviation from the original work should be thoroughly investigated and analyzed.

REVIEW TOLERANCES

One of the fundamental mistakes "green" field engineers make is they don't understand that tolerances vary for the different types of work on the jobsite. Rough staking of the site work might only need to be within a foot. Measuring those stakes to the same tolerance as anchor bolts is a waste of time and money. If tolerances aren't understood, ask several others on the site with experience. Use their judgement and yours to determine if the layout is acceptable. Reference *Chapter 4, Fieldwork Basics.*

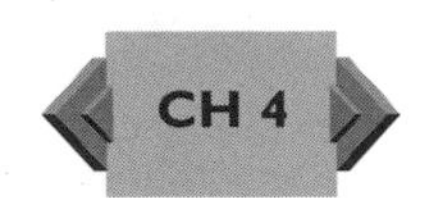

Don't try to achieve the impossible. Sometimes the tolerances found in the specifications are unreasonable. After all, architects don't build buildings. Review with others who have been field engineers and your local equipment dealer, to see if it is physically possible to achieve the stated tolerance. It may not be possible. If so, have the superintendent discuss the issue with the architect and owner to educate them about what is achievable.

CONSULT WITH SUBCONTRACTORS

Sometimes an excellent source for checking work is to talk to your subcontractors. See if their needs are being met by your layout. They have to make their work fit around the work of others and are sometimes the first to find out if things aren't fitting properly. They may not have said anything earlier since they are able to force things to fit. When in reality, their work would be proceeding better if the layout was exactly correct. Ask them. They may see something you haven't thought about.

SCHEDULE

Speak up when the two-week schedule is reviewed. A major souce of error in surveying layout doesn't come from the quality of measurements. Rather, it comes from a lack of planning the field engineer's time. When field engineers are supposed to be two or three places at one time, something will suffer. Ultimately, it may be the accuracy of the work. With a little better planning of the field engineer's time on the jobsite, adequate time should be allotted to each layout activity. If it isn't, justify your time requirements and speak up!

10 RULES OF THUMB

1. DOUBLE ALL ANGLES.

2. DO IT TWICE (REPEAT ALL PROCEDURES).

3. LONG BACKSIGHT AND SHORT FORESIGHT.

4. GET A HEIGHT OF INSTRUMENT (HI) FROM TWO BENCH MARKS ON JOBSITE WORK.

5. MEASURE DISTANCES BY CHAINING FORWARD AND REVERSE.

6. DOUBLE CENTER ALL POINTS LAID OUT.

7. CLOSE ALL LEVEL LOOPS ONTO A KNOWN BENCHMARK.

8. OBTAIN 1/2 OF WHAT IS FOLLOWING.

9. INSTRUMENTS MUST BE LEVEL WHEN THEY ARE BEING USED.

10. THE COMPENSATOR SHOULD BE WORKING IN AN AUTOMATIC LEVEL.

FIELD ENGINEERING MISTAKES TO AVOID

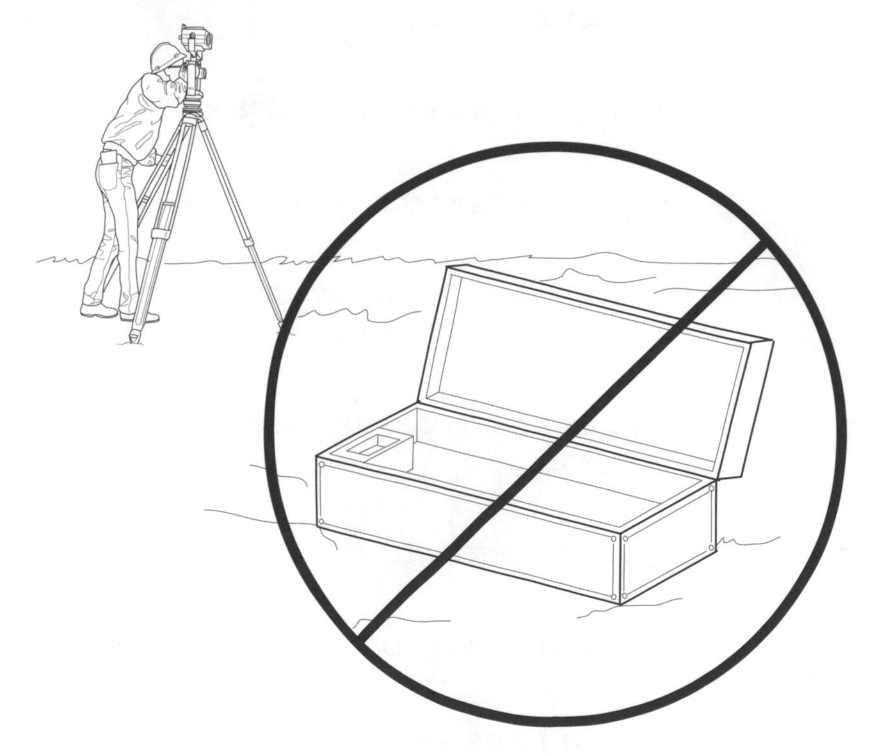

SCOPE

Over the last few years at seminars, superintendents and other field personnel have been asked a very simple question, "List the most common types of surveying layout mistakes you have seen field engineers make." The following is a list of their responses. These are direct quotations! A number of those listed were mentioned by several people. No effort has been made to identify the ones that were listed the most often, as all the mistakes should be avoided!

These lists are by no means inclusive. There are many, many more mistakes made in the layout of construction. As stated before, check, check, and recheck every measurement. Close back onto starting points. Double everything. Do it twice. Repeat the measurement. These are the only ways mistakes can be discovered and avoided.

GENERAL MISTAKES

General Mistakes
Transposing Numbers
Not recognizing handwriting
Reading an upside down 9 as a 6
Misinterpreting or giving incorrect hand signals

COMMON MISTAKES

TAPING

"Not taping in a straight line"

"Using poor measuring techniques"

"Reading tape improperly"

"Starting and measuring from the wrong point"

"Allowing too much sag in the tape"

"Not recognizing that every tape has a different zero end"

"Taping when it is windy"

"Not holding the tape the same way each time"

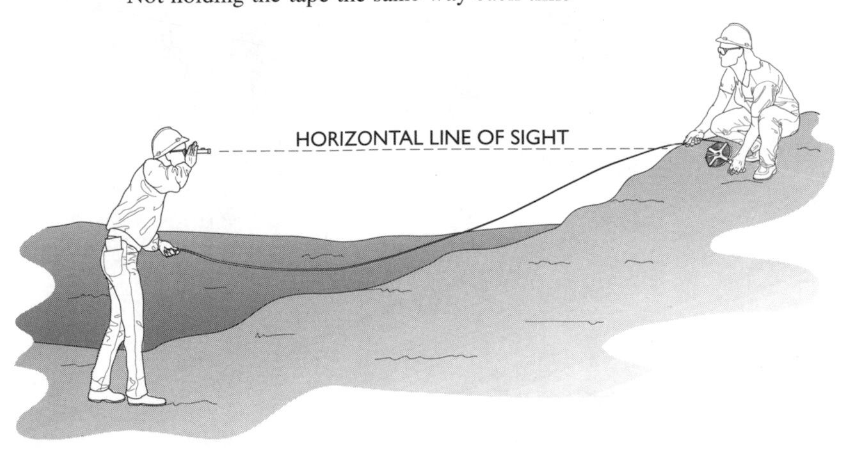

LEVELING

"Not holding the rod in a vertical position"

"Not fully extending the rod"

"Not ensuring that the instrument is level"

"Not setting up the instrument on a stable surface"

"Setting up on frozen ground and not checking level as the sun heats up"

"Setting the wrong datum elevation"

"Setting up high on loose dirt piles and not being able to keep it level"

"Not closing the loop"

"Incorrectly reading the rod"

INSTRUMENT

"Sighting on the wrong backsight"

"Throwing the instrument into the truck"

"Not calibrating the instrument"

"Not leveling the instrument"

LAYOUT

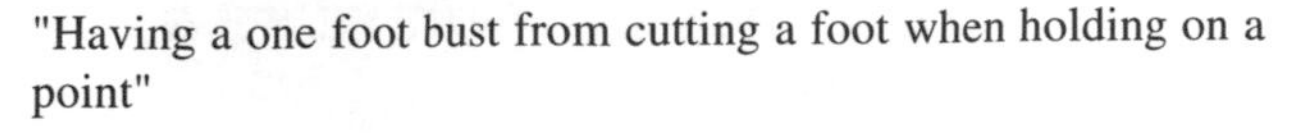

"Having a one foot bust from cutting a foot when holding on a point"

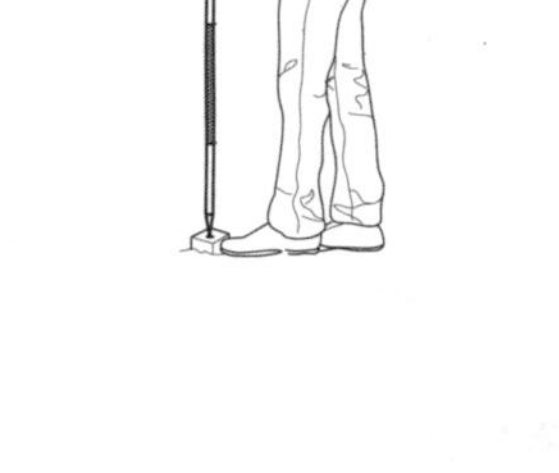

"Having a one tenth bust from cutting a tenth when holding on a point"

"Assuming surveyors' controls are correct"

"Not offsetting stakes far enough from the work"

"Setting points too close to where heavy equipment is working"

"Mistaking tenths for inches"

"Not marking the points the same way each time"

"Not measuring twice as a check"

Not going back to the starting point as a check"

"Not checking for square"

"Marking the wrong side of a level" (being off the width of the carpenter's level)

"Placing hubs in soft material"

"Placing hubs too short in length to be solid"

"Forgetting to offset points"

"Not checking, checking, checking"

"Communicating poorly"

"Improperly setting up the instrument "

"Misreading measurements"

"Allowing control points to move after post-tensioning and not rechecking"

"Allowing various people to take readings"

CALCULATION AND NOTEKEEPING

"Adding numbers incorrectly"

"Interpolating incorrectly"

"Calculating incorrectly"

"Incorrectly converting tenths to inches"

"Incorrectly converting inches to tenths"

"Erasing"

PLAN READING

"Not comparing total dimensions from the plans" (check 2 loops)

"Misreading the blueprints"

"Not checking the measurements on the print"

"Not ensuring latest plan revision matches early set"

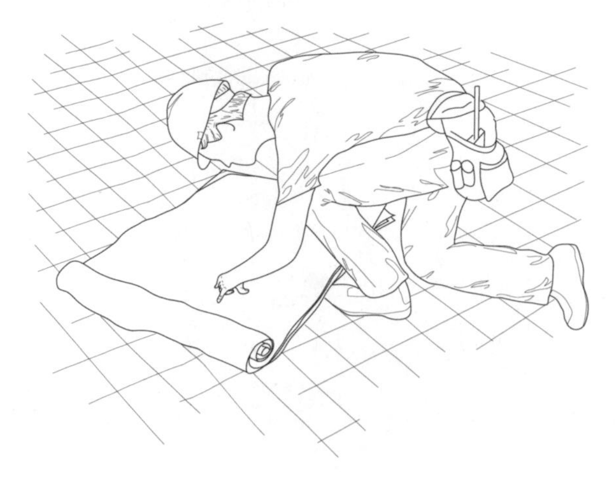

TEN COMMANDMENTS FOR FIELD ENGINEERING

I. Thou shalt plan all work, both field and office, lest thy oversights haunt thee forever.

II. Thou shalt make a thorough search of all records; for behold, they enlighten and reward the diligent in full measure.

III. Thou shalt analyze all research to provide fullness of knowledge from which floweth efficiency, productivity, and good works.

IV. Thou shalt do all work correctly once, lest thy superintendent's wrath be provoked and thy paycheck be revoked.

V. Thou shalt balance level rod on rounded TP's, and keepith backsights and foresights equal, wherefore to properly close the circuit.

VI. Thou shalt use standard note forms so all persons knowest what was completed by thee.

VII. Thou shalt double all measurments to assure purity of work, and measure distance with precision comparable to that for angles to close thy traverse within accuracies prescribed.

VIII. Thou shalt calculate closures in the field and immediately redo thy work which meeteth not prescribed accuracy, for it is there that thy work lies forever.

IX. Thou shalt prepare field books, drawings, and calculations to have but one interpretation; thy field books shall be complete, neatly executed, honest, and contain all necessary information, lest thy work is in vain and thy reputation be torn from thee.

X. Thou shalt be confident in thy competence, knowing what thou knowest and knowing what thou does not know; thou shalt posses courage to persist with judgment, and wisdom to support thy work, lest thy adversary smiteth thy reputation from thee, and smiteth thy pocket book with suit.

APPENDIX A

PRACTICE PROBLEMS

SCOPE

The following problems have been prepared to provide the user of this book with an opportunity to check their knowledge and understanding of the material presented in this book. Many problems are modeled after actual problems encountered in the field. Some problems can be solved very easily by reading the text. Others require that the knowledge gained be applied to actual construction situations. The problems follow the general organization of the book. Headings are provided to easily determine where particular types of problems are coming from in the book. Problems are numbered consecutively for convenience.

CHAPTER 1 - SCOPE AND RESPONSIBILITIES

1. Describe what a field engineer is and list the duties and responsibilities of a person in that position.
2. Discuss why field engineers should not perform property surveys.

CHAPTER 2 - COMMUNICATIONS AND FIELD ENGINEERING

3. The most important part of communicating is said to be the ability to listen. List some factors that improve the listening ability of field engineers.
4. What is the number for the following hand signals?

5. What are the meanings of the following hand signals?

6. What is the tolerance for the following types of construction stakes?
   a. rough grade stakes
   b. slope stakes
   c. curb and gutter
7. What is wrong with the following control information found on the wall of a construction site?
   a. off of "L" line
   b. feet above floor

CHAPTER 3 - GETTING STARTED AND ORGANIZED

8. Identify the most important activity of the field engineer.
9. List the common personal surveying tools used by most field engineers.

CHAPTER 4 - FIELDWORK BASICS

10. List the basic personal protective equipment of field engineers?
11. Identify the most important factor in the overall safety of the field engineer.
12. Develop a list of 10 mistakes or blunders that could occur in construction measurement..
13. Distinguish between errors and mistakes.
14. Discuss which errors can be eliminated and which errors can only be reduced in size by refined techniques.
15. What is the basic rule of measurement for a field engineer:
16. State the basic rule of thumb regarding tolerances in construction measurement.
17. List and discuss the cardinal rules of field book use.
18. State why erasures are absolutely not allowed in a fieldbook.
19. State why a combination of the methods of notekeeping is best.
20. State the simple rules regarding transporting an instrument in a vehicle.
21. What special storage precautions should be used when the climate is very cold or very hot?
22. How is the care of electronic instruments different from optical instruments?

23. Describe how to care for and maintain the following surveying tools: plumb bob, hand level, Gammon reel, chaining pins, range poles, prism poles, brush clearing equipment, steel chain, cloth chain, level rod , hand-held calculators, ni-cad batteries

24. State the left thumb rule for leveling an instrument.

25. Describe the process used to set up an instrument over a point with a plumb bob.

26. Describe the process used to "Quick Setup" an instrument over a point with an optical plummet.

CHAPTER 5- OFFICEWORK BASICS

27. List the basic information that should be shown on a site drawing.

28. State why a graphical scale should be used on all drawings.

29. Describe the common characteristics of contours.

30. State the purpose of lift drawings.

31. Using a set of plans for a building on a slab, prepare a lift drawing of the slab details by following the Lift Drawing Checklist in Chapter 5.

CHAPTER 6 - DISTANCE MEASUREMENT - CHAINING

32. List the tools a field engineer uses to measure a distance by chaining.

33. Outline the step-by-step procedure for chaining a distance.

34. State when it is necessary to use a hand level while chaining.

35. What is the process called when you use lengths shorter than 100 feet to measure between points?

36. List common mistakes that occur when chaining.

37. Distinguish between systematic errors and random errors in chaining.

CHAPTER 7 - DISTANCE MEASUREMENT - EDM

38. List potential uses of an EDM in construction.

39. What sighting procedure with the EDM can be used to eliminate complicated geometric calculations?

40. Outline the step-by-step procedure for measuring a distance with an EDM.

41. What is the greatest source of error with EDM measurement?

CHAPTER 8 - ANGLE MEASUREMENT

42. Illustrate the geometric relationships of any angle turning instrument.

43. Identify the three basic parts of the transit

44. State the basic order of the steps in turning a horizontal angle direct and reverse.

45. After turning two direct and reverse angles, you read an angle of 241° 21' 20".

What is the average angle turned?

46. What process allows for systematic errors in a transit's horizontal axis to be eliminated?

47. Distinguish between zenith angle and vertical angle and state why zenith angle is used on modern theodolites.

CHAPTER 9 - LEVELING

48. Illustrate the basic theory of differential leveling.

49. Define the following leveling terms.
    a. Benchmark
    b. Backsight
    c. Height of Instrument
    d. Foresight
    e. Turning Point

50. Describe the step-by-step differential leveling procedure.

51. State why the distance from the instrument to the backsight and the instrument to the foresight must be the same. (Balanced)

52. Complete the following set of differential leveling notes and perform the arithmetic check.

Closing a Loop from BM#6				
Station	**BS**	**HI**	**FS**	**ELEVATION**
BM#6				155.375
Station 1	1.255	156.630		
TP 1			3.11	153.52
Station 2	0.465	153.985		
TP 2			2.095	151.89
Station 3	0.13	152.02		
TP 3			0.245	151.775
Station 4	3.765			
BM#6			0.165	155.375

53. Complete the accompanying set of differential leveling notes and perform the arithmetic check for the following problem.

Station	BS	HI	FS	ELEVATION
BM 65	1.36			577.03
TP1	2.54		2.32	
TP2	2.55		2.23	
TP3	3.01		2.25	
TP4	3.02		2.35	
TP5	3.02		2.10	
TP6	2.98		2.01	
BM 65			5.24	

CHAPTER 10 - TOTAL STATIONS

54. Distinguish between a total station and an EDM.
55. List advantages of the total station as compared to a top-mount EDM.
56. Describe how the total station is used for measuring distances and angles.
57. List the features that are typically available on an Electronic Field Book.

CONSTRUCTION LASERS 11

58. What is LASER an acronym for?
59. List advantages of lasers as a construction tool.
60. Distinguish between fixed, rotating, and utility lasers.
61. Describe how it is possible to establish grade (elevations) with one person, using a lenker rod and Laser.

CHAPTER 12 - EQUIPMENT CALIBRATION

62. Illustrate with a sketch the following geometric relationships that exist in transits and theodolites.
    a. VA perpendicular to HA.
    b. VA perpendicular to PL.
    c. HA perpendicular to LOS.
    d. VCH perpendicular to HA.
    e. TL parallel to LOS.
63. The quick peg test is used to check which geometric relationship in a level?
64. Describe how to calibrate a bull's-eye bubble and a hand level.

CHAPTER 13 - MATH REVIEW AND CONVERSIONS

65. Convert the following feet and decimal parts of a foot to feet, inches, and eights.

Feet, tenths, hundredths	Feet, inches, eighths
222.20	
125.87	
42.42	
110.55	
15.92	

66. Convert the following feet, inches and fractions to feet and decimal parts of a foot.

Feet, inches, eighths	*Feet, tenths, hundredths*
25' - 4 3/8"	
44' - 11 5/8"	
123' - 6"	
88' - 3 1/4"	
222' - 2 1/2	

67. Convert the following degrees, minutes, and seconds to decimal degrees.

Degrees, minutes, seconds	*Degrees and decimal*
15° 25' 13"	
33° 53' 45"	
77° 23' 59"	
125° 48' 15"	
278° 42' 55"	

68. Convert the following decimal degrees to Degrees, minutes and seconds.

Degrees and decimal	*Degrees, minutes, seconds*
33.23456722°	
75.34698543°	
123.4500392°	
289.2222222°	
234.2700000°	

CHAPTER 14 - DISTANCE CORRECTIONS

69. A field engineer has measured the distance between column lines with a chain which is later discovered to be too long. In order to determine the actual distance between points, how should the correction be applied?

70. A field engineer used a chain which is too short to lay out a foundation of a large building. If she sets points at the prescribed distances from plans, how should the correction be applied?

71. What is the distance between two building control points measured (recorded) to be 600 feet, if the 100-foot chain was found to be too long by 0.015 feet?

72. A chain 100 feet long is laid on ground sloping 5%. What is the horizontal distance if the slope distance is 300 feet?

73. A 100-foot steel chain (known to be 99.93 feet) was used to measure between two building points. A distance of 147.44 feet was recorded at a temperature of 43°F. What is the distance after correcting for temperature and chain length error?

74. The slope distance between the two points is 24.776 feet and the slope angle is 1°17'. Compute the horizontal distance.

75. The slope distance between two points is 42.71 feet, and the difference in elevation between them is 3.56 feet. Compute the horizontal distance.

76. A distance of 328 feet was measured along a 2% slope. Compute the horizontal distance.

77. It is required to lay out a rectangular building, 75 feet wide by 100 feet long. If the 100' steel chain being used is 99.94 feet long, what distances should be laid out?

78. A concrete slab measuring 10 feet by 85 feet is to be laid out by a chain known to be 100.07 feet long under standard conditions. What distances should be laid out?

79. A 100' steel chain standardized at 99.98' was used to measure a distance between control points of 1275.36 feet when the field temperature was 87° F. The ground was sloping at 5%. What is this distance under standard conditions?

80. A steel chain known to be 99.94 feet under standard conditions is used to measure the distance between two control points. If a distance of 178.4 feet was recorded at 58° F, what distance should be measured at a temperature of 75° F?

81. A steel chain known to be 100.03 feet is used to measure the distance between two building corners. If the distance between the corners is supposed to be 268.33 feet and the field temperature is 97° F, then what distance should be laid out?

82. Two control points are known to be 487.63 feet apart. Using a 200' chain known to be 199.96 feet under standard conditions, what distance should be measured when the field temperature is 78° F?

CHAPTER 15 - TRAVERSE COMPUTATIONS

83. What is the angular closure for the following interior field angles of traverse ABCDEF, measured with equal precision?

    A. 87° 54' 12" B. 90° 32' 54" C. 102° 43' 31"

    D. 99° 24' 43" E. 156° 01' 55" F. 183° 23' 01"

84. Using the angles in the previous problem, how much adjustment is needed for each angle?

85. Using adjusted angles from above, and a direction for line AB of 247°, what is the direction for each line? Work clockwise.

86. Provide the back azimuth for the following azimuths:
    a. 132° 12' 07"
    b. 56° 52' 17"
    c. 311° 32' 42"
87. Convert the following azimuths into bearings:
    a. 351° 43' 52"
    b. 11° 32' 59"
    c. 326° 32' 52"
88. Given two successive azimuths of a traverse 69° 21' followed by 320° 11', what is the counterclockwise interior angle between them?
89. Given two successive bearings of a traverse N 41° 35' E followed by S 87° 36' E, what is the interior angle between them?
90. A 400' line bears N 88° 11' 07" E. What is the latitude and departure of the line?
91. A 656' line bears S 52° 53' 42" W. What is the latitude and departure of the line?
92. A 736.32' line has an azimuth of 78° 52' 13". What is the latitude and departure of the line?
93. A 1333.45' line has an azimuth of 203° 24'. What is the latitude and departure of the line
94. The closure in latitudes of a loop traverse is N 0.36 feet; the closure in departures is W 0.25 feet. What is the linear error of closure?
95. The unbalanced latitudes and departures of a closed traverse, and the lengths of the sides are given below. What is the precision?

Course	*Length*	*Latitude*	*Departure*
AB	357.65	N 326.41	W 146.18
BC	329.22	N 58.27	E 324.02
CA	423.6	S 384.52	W 177.70

96. For the following closed traverse WXYZ, what is the corrected departure of XY?

Course	*Direction*	*Length*	*Latitude*	*Departure*
WX	N 68° 12' E	340.2	126.3	315.9
XY	S 7° 54' E	261.7	259.2	36
YZ	S 51° 06'W	413.5	259.7	321.8
ZW	N 4° 25' W	392.5	376.6	30.3

97. A four-sided closed traverse ABCD has the following angles and distances.

A = 88° 30'	AB= 262.76'
B = 90° 22'	BC = 955.63'
C = 87° 00'	CD = 250.49'
D = 94° 08'	DA = 944.47'

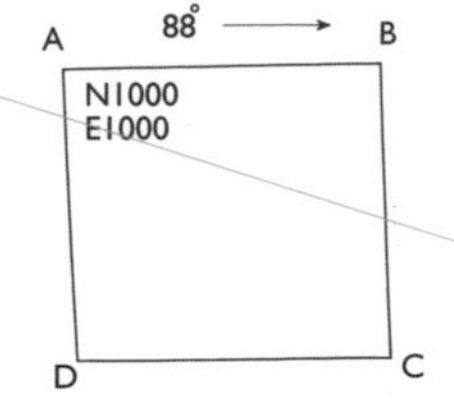

The direction of AB is an azimuth of 88°.

a. Perform a check for angular closure
b. Compute the direction for all sides. (Azimuths)
c. Compute the latitudes and departures.
d. Compute the linear error of closure and the precision ratio.
e. Using the compass rule, compute the adjustments for latitudes and departures and apply the adjustments.
f. Assuming coordinates of North 1000 and East 1000 for point A, compute the coordinates of B, C, and D.
g. Using the computed coordinates, compute the area in acres enclosed by the traverse.

98. A four-sided closed traverse LMNO has the following angles and distances.

L = 110° 38'	LM= 470.52'
M = 82° 56'	MN = 402.06'
N = 90° 37'	NO = 488.34'
O = 75° 50'	OL = 349.91'

M N 0° L N1000 E1000 O

The direction of LM is an azimuth of North 0°.

a. Perform a check for angular closure.
b. Compute the direction for all sides. (Azimuths)
c. Compute the latitudes and departures.
d. Compute the linear error of closure and the precision ratio.
e. Using the compass rule, compute the adjustments for latitudes and departures and apply the adjustments.
f. Assuming coordinates of North 1000 and East 1000 for point L, compute the coordinates of M, N, and O.
g. Using the computed coordinates, compute the area in acres enclosed by the traverse.

99. For the traverse shown, what is the distance & direction for each line? What is the area?

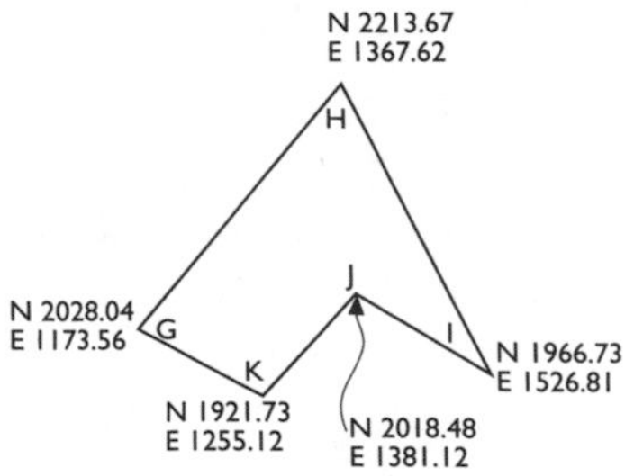

100. For the traverse shown, what is the distance & direction for each line? What is the area?

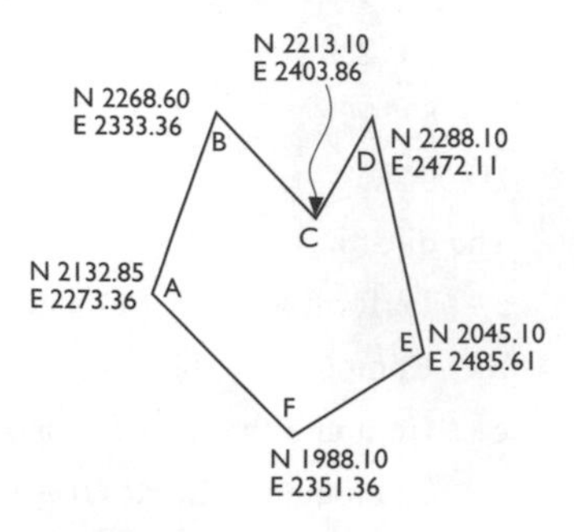

101. For the closed traverse ABCD, answer the following questions

Course	Bearing	Length	Latitude	Departure
AB	S 77° 48' E	76	16.06	74.3
BC	S 68° 14' W	135	50.1	125.4
CD	N 10° 26'W	42	41.3	7.6
DA	N 67° 03' E	63.71	24.8	58.67

a. What are the corrected departures?

b. What are the corrected latitudes?

c. What is the error of closure?

d. What is the approximate precision?

e. Assuming starting coordinates of North 1000 and E 1000 for point A, what are the coordinates of the other traverse points?

f. What is the area of Figure ABCD?

102. For the following coordinates, calculate the area using the coordinate method:

Point	Northing	Easting
A	1200	1400
B	1300	1100
C	1100	1200
D	1000	1000

103. For the adjusted latitudes and departures given below and with coordinates of Point A being North 1000 East 1000, calculate the coordinates of Point C. Calculate the area of traverse ABC using coordinates.

Course	Latitude	Departure	Point	N	E
			A		
AB	S 350	E 160			
			B		
BC	N 310	E 120			
			C		
CA	N 40	W 280			
			A		

CHAPTER 16 - COORDINATES IN CONSTRUCTION

104. Given the North and East coordinates of Point A (400, 100), and Point B (60, 470), what are the distance and bearing from A to B?

The next two questions are based on the information in the following table. The athletic department has decided to construct a tunnel from the basement of the new athletic building to the playing field in the football stadium. The distance and direction of the tunnel is needed for design purposes. Given the following traverse information, calculate the length and bearing for the tunnel from point basement to point field:

Side	Direction	Distance	N	S	E	W
Field - #1	N 65° 20' E	165.00	68.90		149.90	
#1 - #2	S 75° 15' E	110.90		28.00	106.40	
#2 - Basement	S 59° 15' W	270.00		138.00		232.00
Bsmt - Field	unknown	unknown				

105. What is the bearing from basement to field?

106. What will the length of the tunnel be?

For the illustration below, calculate the following:

107. The coordinates for the building corners and interior columns.
108. The coordinates of the control points.
109. The layout data (angle right and distance) to each of the building points if an instrument is set on CP4 and backsighted with 00°00'00" on CP1.

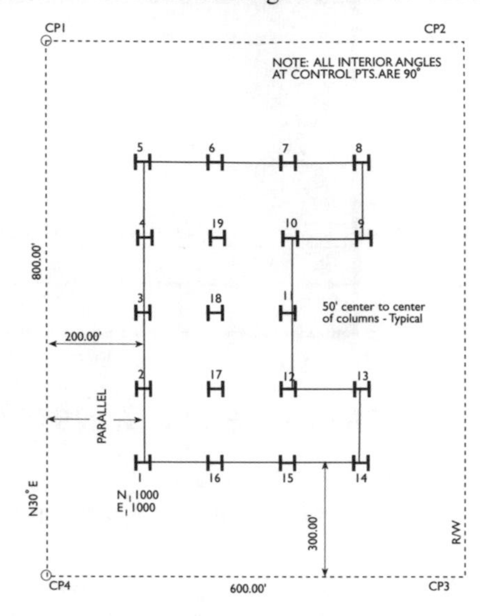

For the bridge shown below, calculate the following:

110. The coordinates for the corners on the abutments and the pier.
111. The coordinates of the control points on centerline and offset points A and B.
112. The layout data (angle right and distance) to each of the bridge points if an instrument is set on Point A and backsighted with 00° 00' 00" on Point B.

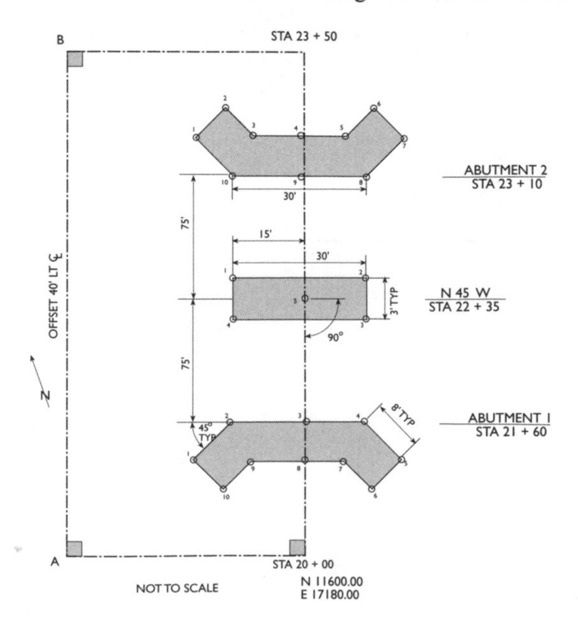

CHAPTER 17 - HORIZONTAL CURVE

113. Calculate the missing curve parts in the following table.

Curve #	"I"	Radius	º of Curve	Tangent	Length	External	Middle Ordinate	Long Chord
1	18° 54'	357.25						
2			2°30'		434.25'			
3		250.00		120.00				
4				200.00				350.00
5	2° 55'			225.50				

114. Given PI @ 9 + 87, "I" angle = 32°14', R = 800; compute tangent (T), the length of arc (L), and compute the stationing of the PC and PT

115. Given: PI @12 +73, "I" angle = 6° 19', Da = 7° , compute tangent (T). length of arc (L) and compute the stationing of the PC and PT.

116. Given: PI @ 15+55, "I" angle of 20° , R= 700, Compute the parts of the curve. T, L, LC, Da, E, MO, Stationing of the PC and PT.

117. Given: PI @55+ 24.776, "I" angle = 13°54' , and R = 400 feet, compute the deflection at every half-station.

118. Given: PI @ 11 +52.42, "I" angle = 25° 52'12", R = 2978 feet, compute the deflection at every 100' station.

119. Given: Da = 2.000 degrees, L = 432.62', and PC = 11 + 43.42, find the remaining components of the curve.

120. Two highway tangents intersect with a right deflection I angle of 31°15' 00" at PI sta. 22 + 22. A 03° 30' 00" horizontal curve (Da) is to be used to connect the tangents. Compute R, T, L, E, LC, and MO for the curve. Compute in tabular form the deflection angles to layout the curve at half stations. Compute the short and long chords.

121. If a simple circular curve has a length of curve 210', and the degree of curvature (Da) is known to be 21.000°, and the station of the PI is 125 + 79.62, what is the central angle of this curve?

122. Two parallel highways 3000' apart are to be joined by a reverse curve made up of two circular curves of equal radius. These curves are to have a (Da) of 3°. Determine all parts of these curves.

CHAPTER 18 - VERTICAL CURVES

123. Compute the gradient for each tangent of a pipeline profile and the elevation of each full station on the tangents.

Point	Station	Elevation
PVI	25+00	501.00
PVI	31+00	455.00
PVI	35+50	498.00
PVI	39+00	463.00
PVI	43+60	481.00

124. Determine the gradients between the points on the following highway profiles to three decimal places.

A. PI = 20+50 EL= 501.00
PI = 22+70 EL= 499.00
PI = 25+60 EL= 506.00

B. PI=51+00 EL=818.00
PI=52+30 EL=792.00
PI=55+10 EL=811.00
PI=57+00 EL=819.00

125. Given the following information, calculate the vertical curve and provide curve elevations.

Point	Station	Elevation	Length of Curve
PVI	10+00	611.32	
PVI	14+00	609.11	500'
PVI	21+00	611.58	500'
PVI	27+00	604.10	700'
PVI	32+52.57	617.302	

126. A +1.512% grade meets a -1.785% grade at station 31+50, elevation 562.00. Vertical curve = 700 feet. Calculate all parts of the vertical curve at full stations.

127. A -2.148% grade meets a +2.485% grade at station 21+50, elevation 435.00. Vertical curve = 800 feet. Calculate all parts of the vertical curve at half station intervals.

128. A grade of -3.35% intersects a grade of -1.412% at station 70+00 whose elevation is 512.85 feet and the distance back to the BVC is 500 feet and the distance to the EVC is 550 feet. Compute the elevations of the full stations on this unequal tangent vertical curve.

129. Given that L = 400 feet, g1 = -2.3%, g2 = +1.6%, PVI at 30+20, and elevation = 425.23, determine the location and elevation of the low point and elevations on the curve at even stations.

130. Given the following vertical curve data: PVI at 21+00; L = 700 feet g1 = +1.6%; g2 = -1.0%; elevation of PVI = 722.12 feet, compute the elevations of the curve summit, and of even full stations.

131. Sketch a vertical curve and label all of its parts.

132. A + 2.512% grade meets a -1.648% grade at station 18+50, elevation 445.00. Vertical curve = 600 feet. Calculate all parts of the vertical curve at full stations.

133. A -3.667% grade meets a +1.732% grade at station 45+00, elevation 312.00. Vertical curve = 700 feet. Calculate all parts of the vertical curve at half stations. Calculate the station and elevation of the low point of the curve.

134. A -1.153% grade meets a -2.432% grade at station 12+00, elevation 452.00. Vertical curve = 500 feet. Calculate all parts of the vertical curve at full stations. What are the station and elevation of the summit of the curve?

135. A +5.23% grade meets a -1.432% grade at station 63+00, elevation 345.00. Vertical curve = 900 feet. Calculate all parts of the vertical curve at full stations. Calculate the high point station and elevation.

CHAPTER 19 - EXCAVATIONS

136. For the road area shown, calculate the volume that was excavated between 32+0 and 35+0.

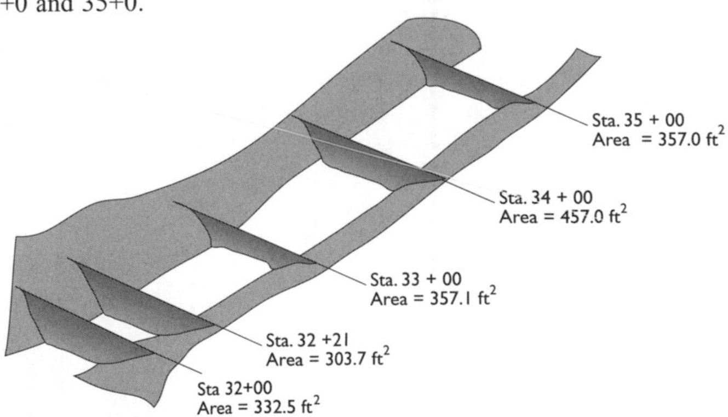

137. If station 22+00 has an end area of 245 square feet and station 23+00 has an end area of 334 square feet, what is the volume between the stations in cubic yards?

138. Calculate the volume of cut if the site in problem 144 is excavated to 610.00 feet.

CHAPTER 20 - DISTANCE MEASUREMENT APPLICATIONS

139. You have been assigned to pace a length across an open field. Along a given 600 feet distance, you pace 165, 164, 162, and 166 paces, respectively. Upon determining your pace length, you pace 205, 208, 206, 208, 206, and 207 paces, in walking an unknown distance AB. What is the length of AB?

140. Pick any survey point and reference it to four points. Prepare field notes for this activity.

CHAPTER 21 ANGLES AND DIRECTION APPLICATIONS

141. List two instances where bucking in on line would be needed by the field engineer.

142. List two construction layout activities that require "double centering."

CHAPTER 22 - LEVELING APPLICATIONS

143. Typically, rod readings taken during the process of profile leveling, or cross-sectioning, need to be read to the nearest _____.

144. Complete the following borrow pit leveling notes. The borrow pit grid is 50' by 50'. Plot the grid and the points and draw 1' contours.

Point	BS	HI	FS	IS	Elevation
BM 45	4.56				620.00
A-1				5.70	
A-2				7.30	
A-3				9.50	
A-4				8.20	
B-1				4.30	
B-2				8.00	
B-3				8.10	
B-4				7.50	
C-1				2.70	
C-2				7.50	
C-3				7.40	
C-4				5.70	
D-1				1.10	
D-2				7.30	
D-3				6.30	
D-4				4.00	
BM 45			4.56		

145. Complete the accompanying set of profile leveling notes and perform the arithmetic check for the following problem. Plot the profile.

Point	BS	HI	IS	FS	Elevation
BM 67	10.33				204.56
TP 1	9.55			3.89	
TP 2	7.63			3.1	
0+00			2.60		
0+50			2.50		
1+00			0.20		
TP 3	5.98			2.85	
1+50			1.50		
2+00			1.42		
2+50			0.04		
BM 68				2.47	

146. Complete the accompanying set of profile leveling notes and perform the arithmetic check for the following problem. Plot the profile using a horizontal scale of 1 inch = 30 feet and a vertical scale of 1 inch = 3 feet.

Point	BS	HI	IS	FS	Elevation
BM101	10.65				208.32
TP 1	9.47			4.32	
TP 2	7.36			3.85	
0+00			2.50		
0+50			2.1		
1+00			0.50		
1+50			1.60		
TP 3	5.40			2.85	
2+00			1.30		
2+14.23			1.72		
2+50			0.90		
BM102				2.54	

147. Complete the accompanying set of cross-section leveling notes and perform an arithmetic check for the following problem. Plot the cross-section at a horizontal scale of 1 inch = 20 feet and a vertical scale of 1 inch = 2 feet.

Point	BS	HI	IS	FS	Elevation
BM 21	4.25				571.68
TP 3	6.22			6.67	
6+00					
50 FT LEFT			3.10		
25 FT LEFT			3.50		
10 FT LEFT			3.20		
CL			2.90		
10 FT RIGHT			2.91		
25 FT RIGHT			2.70		
50 FT RIGHT			2.40		
TP 4	6.23			5.41	
BM 22				7.11	

CHAPTER 23 - CONSTRUCTION LAYOUT

148. Develop a layout plan for a "T" shaped house. Use random dimensions.
149. Develop a layout plan for a Strip Mall that is 1000' by 70'.
150. Develop a layout plan for the layout of 10 small (30' by 30' or smaller) buildings on the same site.
151. If the dimensions of a building are 57' 4 3/4" by 34' 7 1/2", what is the length of the diagonal?
152. A building has sides of 40' and 78'. What is the diagonal?
153. Determine the minimum amount of lumber needed to build "portable" batterboards.

CHAPTER 24 - BUILDING CONTROL AND LAYOUT

154. Distinguish between primary, secondary, and working control.
155. Outline the process involved in establishing primary control on a building site.
156. Using any available set of plans, develop the location of primary control and secondary control.
157. List ten construction activities in laying out a manufacturing plant that would need working control.

CHAPTER 25 - ONE PERSON SURVEYING TECHNIQUES

158. Develop a system for a field engineer working alone to measure the dimensions of a 40' by 50' building to lay out the corners.
159. Develop a procedure for a field engineer working alone to set and mark elevations on the inside of wall forms.

APPENDIX B

LEARNING ACTIVITIES

SCOPE

The exercises that are described here are intended to be completed within a time frame of 2 to 3 hours unless otherwise noted. Some exercises may, in fact, only take 30 minutes while others will take 3 hours or more. However, all exercises will take longer than indicated if the learner does not read the reference material and prepare a plan of attack for the exercise. This is especially true when the activity involves more than one person and even more so if three or more are involved. Planning is the key to successfully learning the activity and to developing the intended skills.

PACING

(Individual - 1 hour)

Objective: Determine pace length and use it to measure the dimensions of a building.

Procedure: Lay down a cloth tape and pace it's length. Divide to determine your pace length. Use your pace length and pace the sides of a building. Calculate the dimensions of the building. Obtain distances to within 1' in 50'. Obtain the distances within 1' in 50'.

Reference: Page 20-3

CHAINING TECHNIQUES

(2 or 3 persons - 2 hours)

Objective: Demonstrate the step-by-step procedure for chaining a distance. Use this process to determine distances between two points.

Procedure: Drive two hubs into the ground about 250 feet apart. Set range poles at each point. Establish chaining duties. Follow the step-by-step procedure outlined in the reference. Obtain a discrepancy ratio of 1/2000 or better.

Reference: 6-7

CHAINING A TRAVERSE

(2 or 3 persons - 2 hours)

Objective: Determine the lengths of the lines of a closed traverse.

Procedure: Use existing points or place at least 4 points in the ground around a field. Using proper chaining techniques, measure all distances forward and back. Obtain a discrepancy for each line of 1/3000 or better

Reference: 20-7

SETTING UP AND LEVELING OVER A POINT (PLUMB BOB)

(Individual - 1 hour)

Objective: Demonstrate the ability to set an instrument over a point with a plumb bob.

Procedure: Use an existing point or drive a hub in the ground. Follow the procedure in the reference. Level the instrument exactly and be within 0.01 of the point in less than 5 minutes.

Reference: 4-53

SETTING UP OVER A POINT (OPTICAL PLUMMET)

(Individual - 1 hour)

Objective: Demonstrate the ability to set an instrument over a point with an optical plummet.

Procedure: Using an existing point or set a point in the ground. Follow the procedure in the reference to set the instrument over the point and be within 0.01 of the point within 3 minutes.

Reference: 4-56

CLOSING THE HORIZON

(Individual - 2 hours)

Objective: Demonstrate the ability to turn horizontal angles with a transit or theodolite.

Procedure: Set the instrument over a point and select 3 or 4 adjacent points or well-defined objects. Turn independent direct and reverse angles between the points. Sum the values and compare to 360 . Be within 1 minute per angle of 360 .

Reference: 8-18

TRIGONOMETRIC LEVELING

(2 or 3 persons - 1 hour)

Objective: Demonstrate the ability to measure a vertical angle with a transit or theodolite and use it to calculate a horizontal distance

Procedure: Set a point at the bottom and top of a small hill (Between 100 and 200 feet apart). Set up a transit at the bottom of the hill and measure the vertical angle to the point at the top of the hill. Measure a slope distance to the top point from the instrument center. Calculate the horizontal distance. Compare to a distance obtained by breaking chain.

Reference: 8-23

MEASURING TRAVERSE ANGLES

(2 or 3 persons - 3 hours)

Objective: Demonstrate the ability to measure the interior angles of a traverse.

Procedure: Set up over each traverse point and turn 2 direct and 2 reverse interior angles between adjacent points. Record all angles. The sum of the interior angles should be within 1' minute per point of the theoretical geometric closure of (n-2)180 .

Reference: Page 8-27

CALIBRATION TESTS ON A TRANSIT OR THEODOLITE

(1 to 2 persons - 2 hours)

Objective: Perform calibration tests to determine if an instrument is in proper adjustment.

Procedure: Establish a location to perform the tests. Set up the instrument and check the plate bubble, the vertical cross-hair, the horizontal axis, the line of sight, and the telescope bubble. Maintain a record in the fieldbook of the results of the tests.

Reference: 12-11

DIFFERENTIAL LEVELING

(2 to 3 persons - 3 hours)

Objective: Transfer the elevation from a benchmark to a temporary benchmark.

Procedure: Locate a benchmark and determine where a temporary benchmark is needed. Utilize differential leveling procedures and run a level loop from the BM to the TBM. Close the loop. Be within the 0.05' when closing.

Reference: 9-15

SETTING BENCHMARKS

(Individual - 1 hour)

Objective: Determine the best location for benchmarks around a proposed construction site.

Procedure: Select any site as the location of a propose construction project. Walk around the site looking for permanent objects that could be used as permanent benchmarks during the duration of the project. Mark with a crayon the proposed benchmarks

Reference: 22-3

PROFILE LEVELING

(2 or 3 persons - 3 hours)

Objective: Determine the elevations at defined intervals along the ground.

Procedure: In the vicinity of a known benchmark, select two points 300 to 400 feet apart. Using a cloth tape, measure and mark every half station and other changes in grade along the line between the points. Set up an instrument and, using differential leveling techniques, level to the area of the profile line. Using profile leveling techniques determine the elevation of each point on the profile line. Close the loop.

Reference: 22-7

CROSS SECTIONING

(2 or 3 persons - 2 hours)

Objective: Determine the elevation of points perpendicular to an established profile line.

Procedure: Use an established profile line and profile leveling field notes. Use a hand level or an automatic level for determining the elevations. At each station on the profile, determine a perpendicular line and obtain the elevation of points 10, 20, 30, and 40 feet away from centerline and other definite changes in grade to the right and left of the station.

Reference: 22-23 and 25-9

QUICK PEG CALIBRATION TEST

(1 or 2 persons - 2 hours)

Objective: Determine if the line of sight of an instrument is horizontal.

Procedure: Set up and level between two utility poles. Make a mark on each pole. Set the instrument up close to one of the poles and read a rod held on both points. Read the rod to the nearest hundredth. Compare readings.

Reference: 12-22

MULTI-STORY LEVELING

(2 or 3 persons - 2 hours)

Objective: Transfer elevation from the ground up into a structure.

Procedure: Locate a local parking garage or a similar 3 or more story building. (Obtain permission.) Run a level loop from a nearby BM to the structure and close the loop. Using a steel chain, transfer the elevation up the structure. Use a leveling instrument and obtain an H.I. Determine an elevation of a point in the structure to the nearest hundredth.

Reference: 22-17

BORROW PIT LEVELING

(2 or 3 persons - 2 hours)

Objective: Establish a grid and determine the elevation of each grid point by use of differential leveling techniques.

Procedure: Locate an area at least 200 feet square. Using a right angle prism and a cloth tape, establish and mark a grid with 25' intervals. Locate a nearby benchmark and use differential leveling techniques to locate a temporary BM close to the site. Use grid leveling techniques to determine the elevation of the grid points to the nearest tenth.

Reference: 22-13

LASER LEVELING WITH A DIRECT ELEVATION ROD

(2 or 3 persons - 2 hours)

Objective: Utilize the laser level and a direct elevation rod to determine elevations of points.

Procedure: Establish a grid at 25 foot intervals. Locate a BM close to the site. Use a laser level and set the elevation of the BM on the direct elevation rod. Go around the grid and determine the elevation of the grid points to the nearest tenth.

Reference: 11-8

USING THE 3/4/5

(3 persons - 2 hours)

Objective: Layout a small building using 3/4/5 layout techniques to obtain right angles.

Procedure: Go to an open area 50' x 50' in size. Plan to layout a 30' x 40' building. Set a starting point and measure 30' along one side of the building. Use 3/4/5 layout techniques to measure the 40' and establish the 90 angle. Repeat for all corners of the building. Set points within 0.01'.

Reference: 23-20

SETTING A HUB

(2 or 3 persons - 1 hour)

Objective: Set line and distance on the center of a hub.

Procedure: Set up an instrument over a point. Sight and measure a distance of 50' to set a hub. Obtain a rough line and distance. Drive the hub into the ground. Check line and distance and adjust the location of the hub by hammering the ground to move it. Drive hub until flush. Check line and distance and adjust hub until line and distance intersect at the center of the hub. Measure to the nearest 0.01'.

Reference: Chapters 6 and 8.

BASELINE OFFSET

(2 or 3 persons - 2 hours field)

Objective: Establish a baseline and measure 90 off of the baseline to layout a small building.

Procedure: Establish a baseline parallel to a roadway. Set up on the baseline and turn angles to the corners of a 40' x 50' "L" shaped building. Set hubs at all corners. Check diagonals to ensure that the building is square. Measure to the nearest 0.01'.

Reference: 23-22

RADIAL BUILDING LAYOUT

(2 or 3 persons - 2 hours office, 2 hours field)

Objective: Locate the corners of a building using an EDM and theodolite by measuring angles and distances from one control point.

Procedure: Using a set of plans for any small building, calculate the distances and angles to be measured to locate the corners of the building from a control point. Set hubs and tacks at all corners. Check the layout by setting at another control point and measuring calculated angles and distances to the building corners.

Reference: 7-20

SETTING BATTERBOARDS

(2 or 3 persons - 3 hours)

Objective: Set batterboards at the corners of a small building.

Procedure: Obtain materials for building batterboards. Layout a building using the 3/4/5, baseline offset, intersecting lines, or radial layout techniques. Set an instrument at one corner and sight another corner. Build batterboards for the corners of the building. Check by measuring the diagonals between the string intersections.

Reference: 23-16

LAYOUT OF A HORIZONTAL CURVE

(2 or 3 persons - 1 hour office, 2 hours field)

Objective: Stake a curve using deflection angle and short chords.

Procedure: With an I = 35 ,an R = 600, and a PI station of 69+34.50 calculate the deflections and short and long chords to layout the curve at full and half stations. Select a spot in an open field to layout the curve. Locate the PI and PC and turn the I angle to locate the PT for a curve to the right. Set up on the PC and turn deflection angles and measure chords to locate the points on the curve. Check into the PT within 0.05'

Reference: 17-19

OFFSETTING A CURVE

(2 or 3 persons - 1 hour office, 2 hours field)

Objective: Stake a curve that is offset 50' to a centerline curve.

Procedure: Using the curve located in the exercise above, calculate new short and long chords. Set up on the PC, sight onto the PI or back tangent, turn a 90 angle and measure 50' to locate the offset PC. Move the instrument to that point and turn a 90 to obtain a tangent to the offset curve. Using the original deflection angles and the new chord distances, layout the curve. Check into an offset PT to within 0.05'.

Reference: 17-36

MOVING UP ON A CURVE

(2 or 3 persons - 2 hours)

Objective: Move up on a curve when an obstacle is encountered.

Procedure: Using the curve and curve data for the above exercise, assume that an obstacle was encountered when attempting to layout station 69+50. Move the instrument to station 69+00 and use "Moving up on a Curve" techniques to continue the curve layout while retaining the original notes.

Reference: 17-28

STAKEOUT OF A CURVE RADIALLY

(2 or 3 persons - 3 hours office, 2 hours field)

Objective: Layout a curve from any control point on a jobsite.

Procedure: Assume coordinates of North 1000, East 1000 for control point 5A, and North 966.59, East 1111.11 for the PC. Calculate the distance and direction between the two points. Using a direction S 45 W to the PI, calculate the coordinates for all points on the curve used in the previous exercise. Calculate the distance and direction from CP-5A to each point on the curve. Calculate data to layout the curve from CP-5A if the instrument is backsighted onto the PC. Locate all points to the nearest 0.01'.

Reference: 7-16, 15-24 and 17-19

SETTING SLOPE STAKES

(2 or 3 persons - 3 hours)

Objective: Set slope stakes for a highway.

Procedure: Go to the field and run a straight line for 200 feet, establishing stations at 50' intervals. Assume a nearby benchmark elevation of 696.00 and using profile leveling, determine the elevation of each station. Use the typical section shown at right for needed slope staking information. Use the slope staking procedure to set slope stakes to the right and left of the line. Set a lath at each point and write on it the cut and slope information. Set the stakes within a tenth.

Reference: 19-9

REFERENCING POINTS

(2 or 3 persons - 1 hours)

Objective: Reference control points so they could be found if covered with 1' of dirt or snow.

Procedure: Go to any established control point. Review the area for trees, utility poles, sidewalks, etc., as potential referencing objects. Follow the principles of good referencing practices and reference the control point. Record in a field book.

Reference: 20-10

ESTABLISHING POINTS BY INTERSECTION

(3 persons - 2 hours)

Objective: Locate a point by intersecting lines from two instrument setups.

Procedure: Layout a 100' by 100' building. Set hubs at the corners and at the midpoints of each line. Chose two adjacent lines and set an instrument up at the midpoint of each line. From the instrument setups, sight onto the midpoint of the opposite line of the building. Lower the line of sight of the instruments and set a hub at the intersection of the sight lines. Set a tack exactly on each line of sight.

Reference: 23-9

BUCKING IN ON LINE

(Individual - 2 hours)

Objective: Locate a point on line between two points.

Procedure: Place targets on two objects that are located in such a way that an instrument will be able to be set between them. Select two buildings, two utility poles, a building and a utility pole, or set two range poles in the ground. Take and instrument and estimate where line between the points is and set up the instrument. By trial and error, of sighting, plunging and sighting, determine the exact spot where the instrument is on line between the two points. Be exactly on line.

Reference: 21-3

DEVELOPMENT OF LIFT DRAWINGS

(Individual - 4 to 6 hours)

Objective: Review a set of drawings for a small commercial structure and develop a lift drawing for all poured walls or floor surfaces.

Procedure: Locate a set of drawings for a restaurant, service station, small factory, etc. Review the drawings and locate all embeds, pipe sleeves, equipment pads, blockouts for windows, doors, etc. Essentially locate any thing that will be penetrating the floor or walls. Prepare a drawing of the floor and walls and show the these penetrations. Provide dimensions that a field engineer could use to check their location.

Reference: 5-19

DEVELOPMENT OF A SITE DRAWING

(Individual - 8 hours of officework)

Objective: Plot data collected from the field to produce a site map.

Procedure: Use the data from the tables on the following pages to develop and plot a site drawing. Use a scale of 1"=10' and two sheets of 18" x 24" drafting paper (one for a rough working drawing and one for the finish drawing.) Calculate the coordinates of the traverse points and plot them on the working drawing. Plot the topographic and planimetric features. Plot the grid elevations. Interpolate and draw 1' contours. Use appropriate symbols to represent features. Trace points, features, and contours to the finish drawing. Include a border, title block, graphical scale, North arrow, legend, list of coordinates, distances and directions for each line, and point name. Use lettering guidelines and proper line quality .

Reference: 5-3 to 5-18

Traverse Data					
Course	**Direction**	**Distance**	**Point**	**North**	**East**
Line			A	5000.00	5000.00
AB	N 2°20' E	82.37'			
			B		
BC	N 53°30' E	61.71			
			C		
CD	S 67°30' E	100.21			
			D		
DE	South	37.29'			
			E		
EF	S 56°40' W	58.20			
			F		
FA	S 83°20' W	97.53			
			A		

Location of Topographic and Planimetric Data (Theodolite and EDM)			
Point	**Horizontal Angles Clockwise**	**Distance**	**Description**
	Instrument at,	CP A	Backsight B, 0° 00'00"
1	22° 15'	128.31	Centerline of 12' wide paved road
2	26° 22'	127.37'	Centerline of 12' wide paved road
3	31° 30'	127.12'	Centerline of 12' wide paved road
4	37° 13'	128.05'	Centerline of 12' wide paved road
5	44° 00'	130.88'	Centerline of 12' wide paved road
6	48° 46'	134.09'	Centerline of 12' wide paved road
7	53° 28'	138.37'	Centerline of 12' wide paved road
8	59° 10'	143.94'	Centerline of 12' wide paved road
9	62° 33'	150.49'	Centerline of 12' wide paved road

	Instrument at,	CP B	Backsight C, 0° 00'00"
10	53° 15'	140.34'	Treeline
11	57° 10'	136.75'	Treeline
12	59° 55'	130.00'	Treeline
13	63° 05'	124.32'	Treeline
14	68° 30'	118.25'	Treeline
15	78° 50'	113.05'	Treeline
16	80° 00'	103.10'	Treeline
17	85° 10'	95.00'	Treeline
18	92° 50'	89.15'	Treeline
19	100° 10'	85.00'	Treeline
20	105° 20'	83.50'	Treeline

	Instrument at,	CP D	Backsight E, 0° 00'00"
21	51° 10'	83.52'	Building Corner
22	59° 50'	74.78'	Building Corner
23	63° 20'	117.04'	Building Corner
24	66° 45'	92.25'	Building Corner
25	75° 30'	87.41'	Building Corner
26	78° 20'	106.78'	Building Corner
27	52° 30'	56.81'	Gravel Parking Lot Corner
28	67° 10'	86.09'	Gravel Parking Lot Corner
29	78° 20'	44.65'	Gravel Parking Lot Corner
30	83° 30'	80.00'	Gravel Parking Lot Corner

	Instrument at,	CP F	Backsight A, 0° 00'00"
31	19° 10'	56.62'	Pine Tree
32	48° 50'	73.00'	Pine Tree
33	55° 20'	15.61'	Pine Tree
34	60° 30'	60.00'	Pine Tree
35	75° 25'	62.50'	Centerline of 8' wide Gravel Drive
36	77° 10'	66.00'	Centerline of 8' wide Gravel Drive
37	79° 15'	73.10'	Centerline of 8' wide Gravel Drive
38	81° 20'	81.00'	Centerline of 8' wide Gravel Drive
39	82° 35'	86.44'	Centerline of 8' wide Gravel Drive
40	67° 10'	116.00'	Power Pole
41	94° 05'	72.1 '	Power Pole
42	153° 00'	65.00'	Power Pole
43	109° 10'	57.17'	Maple Tree
44	112° 20'	47.59'	Maple Tree
45	115° 30'	37.00'	Oak Tree

Grid Elevations - 0+00N, 0+00E is Located at Point A on the Traverse
Grid is oriented North

Point	0+00E	0+20E	0+40E	0+60E	0+80E	1+00E	1+20E	1+40E
1+20N	74.1	74.2	74.4	74.7	74.1	73.2	70.1	66.0
1+00N	76.4	77.63	79.4	78.8	76.8	76.8	72.5	68.8
0+80N	77.5	80.1	83.6	81.9	78.4	76.8	75.0	71.9
0+60N	76.4	79.0	82.0	80.2	77.6	76.5	75.5	75.4
0+40N	75.6	77.6	78.3	79.3	77.2	76.6	75.3	75.2
0+20N	74.0	74.2	75.2	76.0	75.8	76.6	75.2	75.1
0+00,N	73.7	73.8	74.3	74.4	74.9	74.9	75.0	75

APPENDIX C

GLOBAL POSITIONING SYSTEMS

SCOPE

Everyone is talking about Global Positioning Systems (GPS); automobile manufacturers, the trucking industry, shipping companies, the forest service, fishermen, hunters, hikers, professional surveyors, and of course, field engineers. GPS is revolutionizing the way we live and the way we work. Soon every automobile will have a GPS unit that will display to the driver the current location, distance to destination, location of service stations and restaurants, anticipated time of arrival, and much more. All one has to do is read any of the popular magazines related to automotive science to see how it will work. The uses of GPS are virtually limited only to imagination. The use of Global Positioning Systems is rapidly advancing to construction. Very large projects are using GPS for control work, and some applications in construction measurements and staking are beginning to occur. With on-site base stations working in conjunction with satellite signals, obtaining accuracy at the centimeter level is possible. GPS is an exciting development and will rapidly revolutionize the method of layout used by the field engineer.

BASICS

GPS as indicated earlier stands for Global Positioning System. It is a system that consists of approximately two dozen satellites orbiting the earth emitting distinct signals. These signals are received by the GPS units and calculations are performed to determine the location of the GPS unit in geodetic coordinates of longitude and latitude. The system was developed in the 1980's by the military as a method of rapidly determining the location of their forces. Fighter planes, ships, tanks, and infantry are equipped with GPS units that give their positions rapidly anywhere on earth. Obviously, this is a great advantage for commanders in the field as well as at headquarters in planning their strategy. Needless to say, as soon as the military started to use GPS, it was quickly recognized that this technology could be adapted to commercial use. Anyone who has a need to determine position or location is a potential candidate for GPS technology. Geodetic surveyors and professional surveyors were among the first to realize that control work could be accomplished effectively with GPS.

GPS ON THE CONSTRUCTION SITE

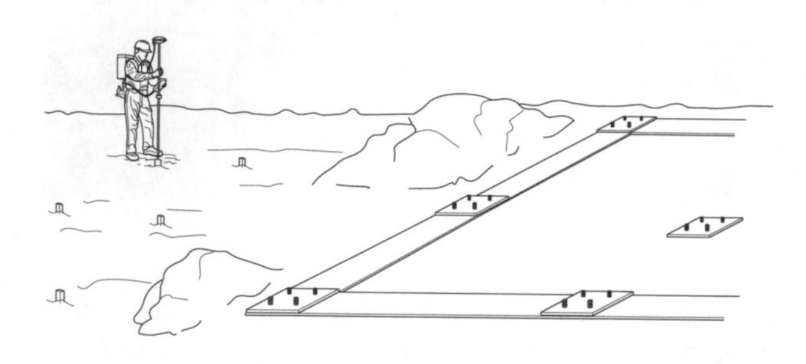

Effective use of GPS by surveyors wasn't readily available because of national security concerns that limited the precision attainable. The GPS signal information necessary to obtain the precision required for activities such as construction layout was not released. But, ingenuity has been working to overcome limited signal information. Through the use of base stations that are precisely located and are emitting their own signals, GPS receivers for precise locating of points can now achieve real-time coordinate location of points to the centimeter level. That is sufficient for most construction layout activities.

GPS AND THE FIELD ENGINEER

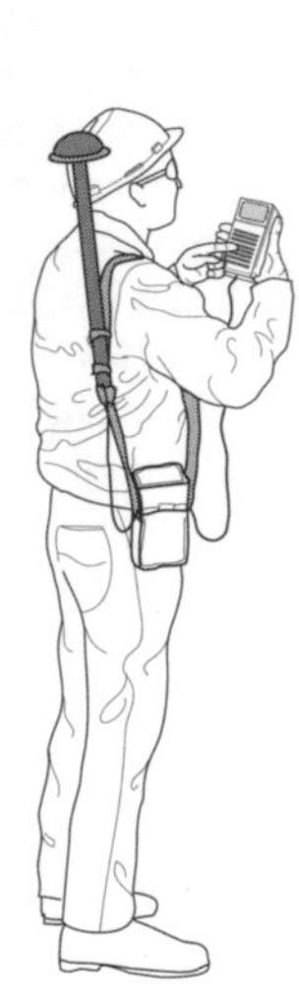

As always, it is the responsibility of the field engineer to be knowledgeable about the types of equipment that are available so the optimum choice of equipment can be made. At this time, almost all construction layout situations can best be performed using standard equipment such as the chain, EDM, automatic level, theodolite, or the total station. Currently there are few instances where GPS may be the instrument of choice for a construction application. But, that is changing.

Field engineers must keep up to date on the capabilities of GPS and should be constantly reading and searching for articles on its application to construction. Just like other electronic technology, GPS will get cheaper; it will get more precise; and it will be faster. GPS units for everyday construction use will become the norm.

Future field engineers will use GPS for the majority of their layout activities. Although it is difficult to imagine, there may come a time when conventional surveying instruments are no longer used. Every point and elevation will be set using GPS units.

When a corner is needed, the field engineer will input into the GPS computer; instantaneously the field engineer will be told how far and in what direction to go to set the point. When the point is reached, the GPS unit will indicate that a hub and tack can be set to mark the point. Wall alignment will be done by using hand held units preprogrammed with coordinates for the wall. Carpenters will be able to look at the display to determine which way to move the turnbuckles to align the wall. Ultimately just about any layout should be able to be performed using future GPS units. The future is GPS.

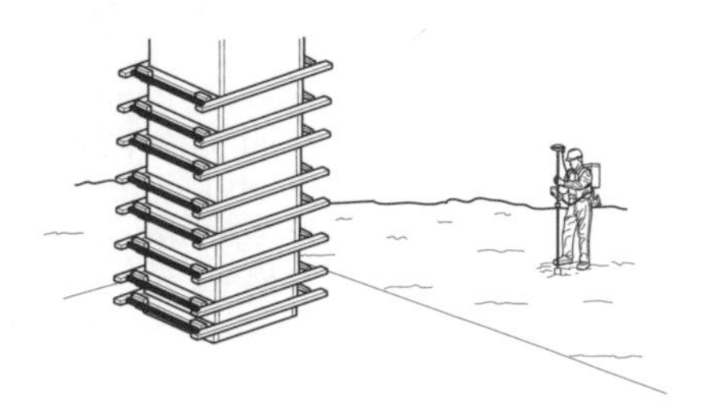

APPENDIX D

SURVEYING BOOK LIST

SCOPE

Undoubtedly many of the readers of this text will want to learn more about surveying practice and theory. Many books have been written on the subject of surveying that will provide the reader with valuable information. Some of the books that might be of interest are listed on the following pages.

American Society of Civil Engineers and American Congress on Surveying and Mapping. 1978. Definitions of Surveying and Mapping Terms, Manual No. 34.

American Society of Civil Engineers. 1985. Engineering Surveying Manual. Manual No. 64.

Anderson, James M., and Edward M. Mikhail. Introduction to Surveying. New York: McGraw-Hill.

Barry, B. A. 1988. Construction Measurement. New York: John Wiley and Sons.

Barry, B. A. 1978. Errors in Practical Measurement in Science, Engineering, and Technology. New York: John Wiley and Sons.

Benton, Arthur R., and Philip J. Taetz. 1991. Elements of Plane Surveying. New York: McGraw-Hill.

Brinker, R. C. 1978. 4567 Review Questions for Surveyors, 11th ed. R. C. Brinker, Box 1399, Sun City, AZ.

Brinker, R. C. and R. Minnick. 1987. The Surveying Handbook. New York: Van Nostrand Reinhold Co.

Brinker, R. C., B. A. Barry, and R. Minnick. 1981. Noteforms for Survey Measurements. 2nd ed. Rancho Cordova, CA: Landmark.

Buckner, R. B. 1984. Surveying Measurements and Their Analysis. Rancho Cordova, CA: Landmark.

Davis, R. E., et al. 1981. Surveying: Theory and Practice, 6th ed. New York: McGraw-Hill.

Evett, Jack B. 1991. Surveying. 2nd ed. Englewood Cliffs, NJ Prentice Hall.

Herubin, Charles A. 1991. Principles of Surveying. 4th ed. Englewood Cliffs, NJ: Prentice Hall.

Hickerson, T. F. 1967. Route Location and Design, 5th ed. New York: McGraw-Hill.

Jackson, W. P. 1979. Building Layout. Carlsbad, CA: Craftsman.

Kavanagh, Barry F. 1992. Surveying With Construction Applications. 2nd ed. Englewood Cliffs, NJ: Prentice Hall.

Kavanagh, Barry F. and S.J. Glenn Bird. 1992. Surveying: Principlesand Applications. 3rd ed. Englewood Cliffs, NJ: Prentice Hall.

Kissam, Philip. 1978. Surveying Practice. 3rd ed. New York: McGraw-Hill

Kissam, Philip. 1981. Surveying for Civil Engineers. 2nd ed. New York: McGraw-Hill.

Kratz, Kenneth E. 1991. Survey Crew Manual. Canton, MI. P.O.B

Landon, Robert P. 1994. Practical Surveying for Technicians. Albany, NY. Delmar

Leick, Alfred. 1990. GPS Satellite Surveying. New York. John Wiley and Sons

Lufkin Rule Co. 1972. Taping Techniques for Engineers and Surveyors. Apex, N.C.

Meyer, C. F., and D. W. Gibson. 1980. Route Surveying and Design, 5th ed. New York: Harper and Row.

McCormac, Jack C. 1991. Surveying Fundamentals. 2nd ed. Englewood Cliffs, NJ: Prentice Hall

Mikhail, E. M., and G. Gracie. 1981. Analysis and Adjustment of Survey Measurements. New York: Van Nostrand Reinhold Co.

Moffitt, F. H., and H. Bouchard. 1992. Surveying, 9th ed. New York: Harper & Row.

Pafford, F. W. 1962. Handbook of Survey Notekeeping. New York: John Wiley and Sons.

Parmley, Robert O. 1981. Field Engineer's Manual. New York: McGraw-Hill.

Stenstrom, Guy O. 1987. Surveying Ready Reference Manual. New York: McGraw-Hill.

Stull, Paul. 1987. Construction Surveying and Layout. Carlsbad, CA: Craftsman.

Van den Berg, John, and Ake Lindberg. 1983. Measuring Practice on the Building Site. Gavle, Sweden: International Federation of Surveyors.

APPENDIX E

GLOSSARY

SCOPE

Every text that is written has terms that are unfamiliar to the reader. This glossary is presented as an attempt to define terms that might not be understood. It is by no means all inclusive. There are so many terms in the construction and surveying area that it would take an entire book to define them all. See the book list in Appendix D for a listing of other books where needed definitions might be obtained. "Definitions of Surveying and Associated Terms," prepared by a joint committee of the American Congress on Surveying and Mapping and the American Society of Civil Engineers, has been used as the primary reference for assembling this list.

accuracy Refers to the degree of perfection obtained in measurements. Is a measure of the closeness to the true value.

accuracy (first-order) The highest accuracy required for engineering projects such as dams, tunnels, high-speed rail system, etc.

accuracy (second-order) The accuracy required for large engineering projects such as highways, interchanges, short tunnels, etc.

accuracy (third-order) The accuracy required for small engineering projects and topographic mapping control.

accuracy ratio The ratio of error of closure to the distance measured for one or a series of measurements.

acre 43,560 square feet

adjustment of data A process used to remove inconsistencies in measured or computed data. Provides a means of compensating for random errors which occur in all measurements.

adjustment of instruments The process of bringing the principle lines of an instrument into their designed geometric relationship.

allidade The part of a transit or theodolite that contains the telescope and plate levels.

as-builts Measurements made after a construction project is complete to provide the actual positions and features of the project.

axis A reference line about which a body rotates. A reference line for coordinates.

azimuth	The clockwise angle from the North reference meridian used to describe the direction of a line.
backsight (bs)	In leveling, a sight taken on the level rod held on a point of known elevation to determine the height of the instrument. When measuring angles, the sight taken on a point being used as the starting point of the angle.
baseline	A line established with great care that is used as a reference line for measurement of distances and angles during construction layout.
batterboards	Boards set at the corners of a building for stretching wire or string that marks the limit of construction. Typically the outside face of a wall. For trenching, a board set across the trench to carry the grade line.
bearing	The clockwise or counterclockwise angle measured off of North or South that is used to describe the direction of a line.
back bearing	The reverse direction of a line.
benchmark	A relatively permanent object with a known elevation used as a reference for leveling.
breaking chain	Measurements less than a full chain length that are used when chaining a slope.
calibration	The process of comparing an instrument or chain with a standard.
cardinal direction	The directions—North, South, East, and West.
chaining	The operation of measuring a distance on the ground with a chain or tape. Chaining and taping are used synonymously.
change of grade	The difference in elevation between existing ground line and previous ground line.
closed loop	A series of consecutive measurements that closes on the beginning point
closed traverse	A series of consecutively measured lines that start at a known point and end at the same point or another point that is known.
coefficient of expansion	A known amount a material will change in size because of a change in temperature.
constant error	An error that always occurs with the same sign and amount.
construction layout	A survey that is performed to locate designed structures on the ground.
contour	An imaginary line on a site plan that connects points of the same elevation.
contour interval	The spacing (elevation difference) between contours shown on a site plan.
control	A series of points coordinated and correlated together that serve as a common framework for all points on the site. Control can be horizontal and vertical.
coordinates (polar)	The distance and direction (Azimuth or Bearing) from one point to another.
coordinates (rectangular)	The linear values from the North and East axes which designate the location of a point.
correction	A value that is applied to a measurement to reduce the effect of errors.
cross-hairs	A set of wires or etched lines placed in a telescope used for sighting purposes.
cross sections	Elevations and distances measured along a line that is perpendicular to centerline and are plotted and used for design and volume computations.
cut	The removal of soil or rock to obtain a desired elevation or grade.
datum	A reference elevation such as mean sea level or in the case of some construction projects a benchmark with elevation 100.00.

degree — A unit of circular measurement equal to 1/360th of the circle. The unit of measurement for temperature.

deflection angle — The angle measured off of a prolongation of the preceding line.

departure — The East-West component of a line that is determined by multiplying the distance times the sine of the direction.

differential leveling — The leveling process of determining the difference in elevation between two points.

direction — The angle between a line and a chosen reference line. Can be an azimuth or a bearing.

double centering — A method of establishing a point by sighting direct and then plunging and marking the line of sight; reverse and then plunging and marking the line of sight; and averaging the marks to determine the true point.

earthwork — Construction operations connected with cutting or filling earth.

EDM — Electronic Distance Measurement is an instrument which times the transmission and reception of an electronic signal to and from a reflector and translates the signal wavelength and time to a distance.

elevation — The vertical distance of a point above a datum.

elevation angle — A positive or upward measured vertical angle off of the horizon.

error of closure — The difference between an actual measured location and its theoretical location determined mathematically.

field notes — The permanent detailed record made of field measurements and observations.

fill — The use of soil or rock to build up the ground to a desired elevation.

focus — The point in a telescope where the rays of light converge to form an image.

foresight — In leveling, a sight taken on the level rod to determine the elevation of any point. In angle turning, the sight taken on the ending direction of the angle.

frost line — The depth below the surface to which the ground becomes frozen.

geodetic surveying — Measurement in which the curvature of the earth is considered.

grade — The slope of the surface of the ground. Also used to designate that the desired elevation has been reached.

grade staking — The process of setting stakes that mark elevations to cut or fill.

gradient — The slope or rate of ascent or descent of a line.

grid — A set of evenly spaced lines drawn or laid out perpendicular to each other.

Gunter's chain — A measuring device composed of 100 metal links fastened together with rings. The length of the chain being 66 feet. The source of the term "chain" that is frequently used to describe measurement with a steel tape.

height of instrument (H.I.) — The elevation of the line of sight of the telescope above the datum plane in the differential leveling process.

height of instrument (h.i.) — The height of the instrument above the station point when using a top mount EDM or a total station. The height of the observer's eye above the station point when performing slope staking with a hand level.

horizon, closing the — Measurement of angles about a point such that the sum of the angles should be 360°.

horizontal axis — The axis about which the telescope rotates vertically.

hub — A 2 inch by 2 inch wooden stake from six inches to twenty four inches in length that is driven solidly into the ground. Is used to mark control points, building corners, centerline stations, centers of footers, etc.

index contour — On a topographic map, a contour line that is darkened or made bold.

indirect leveling — Measuring vertical angle and slope distances to determine the difference in elevation between the instrument and a point.

interior angle — An angle that is enclosed by two sides within a triangle or closed polygon.

intersection — The process of determining the location of a point by sighting from two or more points.

invar — An alloy of nickel and iron that has a very low coefficient of expansion.

invert — The low point on the inside circumference of a pipe.

land survey — A type of surveying that locates property lines, subdivides land, determines land areas. Describes land and provides information for the transfer of land.

latitude — In traverse computations, the North-South component of a line that is determined by multiplying the distance of the line by the cosine of the direction. Geodetically, it is the angular distance measured along a meridian from the equator to a position on the earth.

least count — The finest reading that can be made directly on the vernier of a transit or micrometer of a theodolite.

least squares — A mathematical method for the adjustment of observations based on the theory of probability.

leveling — Determining the difference in elevation between points.

level surface — A surface having all points at the same elevation and perpendicular to the direction of gravity.

line of collimination — A line which joins the optical center of the objective on a telescope and the intersection of the cross-hairs.

line of sight — The line extending from an instrument along which distinct objects can be seen. The straight line between two points.

longitude — The method of expressing the angular distance from the starting geodetic meridian at Greenwich, England. An angle measured at the poles between the starting meridian and any point on earth.

map — A paper representation at a reduced scale of the features on a part of the earth's surface.

mark — A call used to indicate that the listener should place a point on a hub, footing, monument, etc.

mean sea level — The average height of the surface of the sea measured over the complete cycle of high and low tides. (a period of 18.6 years)

meridian — A North/South reference line to which directions of all a survey are referenced.

mistake — A large difference from the true value of a measurement.

monument — A physical structure which marks the location of a survey point.

nadir — The point directly under the observer. The direction that a plumb bob points.

North — The primary reference direction for surveying. Represented as 000° as a North azimuth.

Offset	A measurement made to locate an auxiliary point or line for the purpose of preserving the location of the initial point if it is destroyed.
Offset Line	A line that is close to and roughly parallel to the main line.
optical plummet	A special device attached to an instrument that allows the operator to sight and locate the instrument directly over a point below the instrument.
parallax	The apparent movement of the cross-hairs caused by movement of the eye.
plane surveying	Surveying in which the curvature of the earth is not considered.
plotting	The transfer of survey data from field notes to paper.
plumb	The vertical direction. A line perpendicular to a horizontal plane.
plunge	To reverse the direction of the telescope of a transit around its horizontal axis.
point	A position which has not length, width or height. Also used synonymously with station.
position	The location of a point with respect to a reference system.
precision	The closeness of one measurement to another. The degree of refinement in the measuring process. The quality of the measuring operation.
profile	The graphical representation of the earth's surface performed by leveling and by plotting.
profile leveling	The process of determining the elevation of a series of points along a defined line.
prolongation	The lengthening or extension of a line in the same direction.
quadrant	A sector of a circle having 90 .
radian	The angle subtended at the center of a circle by an arc equal in length to a radius of the circle. It is approximately equal to 57° 17' 44.8".
random errors	Errors that are accidental in nature and always exist in all measurements. They follow the laws of probability and are equally high or low.
range pole	A slender wood or metal rod of varying length with a metal point, painted in alternating colors of red and white that is used as a sighting object at the ends of a line.
reference mark	A point or object which is measured to from a monument for the purpose of being able to relocate the monument if it is lost.
refraction	The bending of light rays as they pass through the atmosphere.
	The accumulation on the circle of an instrument of a series of measurement of the same angle.
resection	The process of determining the location of a unknown point by measuring distances and angles to at least three known points.
route survey	The establishment of control and construction stakes for the location of a line of transportation.
sideshot	A reading to a point that is not part of the main survey.
slope	The inclined surface of a hill. The inclined surface of an excavation or an embankment. A measurement of how much the ground varies fromhorizontal.
slope stake	A stake set at the point where the design slope meets the original ground.

stadia	A method of measurement using the optics of an instrument. An intercept between the stadia hairs on an instrument is calculated and used to determine distance.
station	A point whose location in route surveying is described by its total distance from the start of the project. Generally listed in terms of 100 foot intervals. A definite point on the earth whose location has been determined by surveying measurement. Also used synonymously with point.
surveying	The art and science of determining the relative position of points on, above, or beneath the surface of the earth by measurement of angles, distances, and elevations.
systematic error	Those errors that occur in the same magnitude and the same sign for each measurement of a distance, angle or elevation. Can be eliminated by mechanical operation of the instrument or by mathematical formula.
tangent	A straight line that touches a circle at one point.
tape	A ribbon of steel on which graduations are placed for measurement of distances. Used synonymously with chain.
taping	The process of using a tape to measure a distance on the ground. Also used synonymously with chaining.
traverse (closed)	A method of surveying measurement in which the directions and distances of the lines between a series of points is determined and used to calculate the positions of the points.
traverse (loop)	A traverse that starts and closes on the same point.
traverse (open)	A traverse that originates at a known position and ends at an unknown position.
traversing	The process of measuring distances and angles between traverse points.
topographic surveying	Measurements taken for locating objects and the elevation of points on the earth's surface.
triangulation	A method of surveying measurement where a baseline is established and the angles between stations are determined to calculate distances.
trigonometric leveling	The process of measuring vertical angles and slope distances to determine the difference in elevation between points.
turning point	A temporary point whose elevation is determined by differential leveling.
vertical	The direction in which gravity acts.
vertical angle	The angle measured up or down from the horizon.
zenith	The point directly above a given point on earth.
zenith angle	The angle measured downward from the zenith.

INDEX

A

B

C

D

E

F

G

H

I

L

M

T

U

V

W

Z